21st Offshore Symposium 2016

Emerging Offshore Technology and Deepwater Trends

Houston, Texas, USA
16 February 2016

ISBN: 978-1-5108-2643-4

Printed from e-media with permission by:

Curran Associates, Inc.
57 Morehouse Lane
Red Hook, NY 12571

Some format issues inherent in the e-media version may also appear in this print version.

Printed by Curran Associates, Inc. (2016)

For permission requests, please contact Society of Naval Architects and Marine Engineers (SNAME) Texas Section at the address below.

Society of Naval Architects and Marine Engineers (SNAME) Texas Section
c/o Arun Antony, Suite 350
15990 N. Barkers Landing Drive
Houston TX 77079
USA

Phone: (281) 877-6403
Fax: (281) 877-6797

aantony@houston-offshore.com

Additional copies of this publication are available from:

Curran Associates, Inc.
57 Morehouse Lane
Red Hook, NY 12571 USA
Phone: 845-758-0400
Fax: 845-758-2633
Email: curran@proceedings.com
Web: www.proceedings.com

TABLE OF CONTENTS

Closing Session II : Emerging Technology

Supplemental Proceedings

SNAME
THE INTERNATIONAL COMMUNITY
FOR MARITIME AND OCEAN PROFESSIONALS

Proceedings of the 21st Offshore Symposium, February 2016, Houston, Texas
Texas Section of the Society of Naval Architects and Marine Engineers

The Texas Section
The Society of Naval Architects and Marine Engineers

Proceedings of

The 21st Offshore Symposium

Emerging Offshore Technology and Deepwater Trends

February 16, 2016
Houston, Texas

SYMPOSIUM EXECUTIVE COMMITTEE

Symposium Chair	Manoj Jegannathan (Technip)
Symposium Vice-Chair	Samuel Linder (Granherne-KBR)
Technical Committee Chair	Mischa Dylewski (Granherne-KBR)
Technical Committee Vice-Chair	Daniel Dabrowski (Alan C. McClure Associates)
Publicity and Sponsorship Coordinator	Angela Zhu (ABS)
General Secretary	Abel Medellin (McDermott)
Committee Member	Ping Lu (ABS)

SYMPOSIUM ADVISORS

Sai Majhi (ExxonMobil)

Amir Izadparast (SOFEC)

Hisham Moideen (Houston Offshore Engineering)

Arun Antony (Houston Offshore Engineering)

SNAME TEXAS SECTION OFFICERS

Executive Committee	Hisham Moideen (Houston Offshore Engineering)
Auditor	Ray Fales (Granherne-KBR)
Chair	Ashish Bagaria (INTECSEA)
Vice-Chair	John Bandas (MARIN)
Secretary/Treasurer	Xiang Li (DMAR)
Web Admin	Zy Su (COTEC USA)
Membership Chair	Wan Wu (MODEC)
T&R Chair	James Brekke (ABS)
Arrangements	Daniel Dabrowski (Alan C. McClure Associates)

SILVER SPONSORS

BUREAU VERITAS

Williams®

EVENT SPONSORS

Luncheon: EXMAR OFFSHORE

Evening Reception: SOFEC

Digital Media: Lankhorst Ropes

Name Tag: WOOD GROUP

Morning Coffee Break: SABALCORE

Afternoon Coffee Break: Halton

Lanyards: SBM OFFSHORE

Delegate Bag: MARIN

MARIN

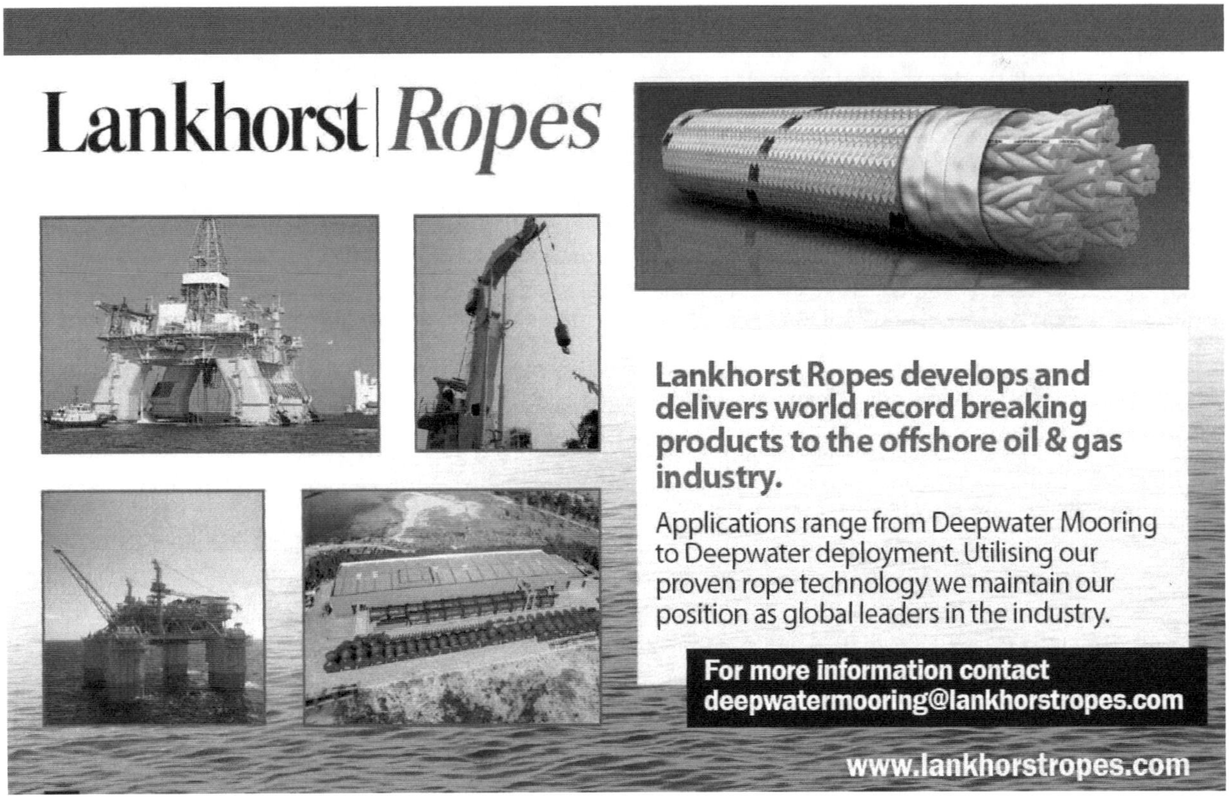
Lankhorst|Ropes
Lankhorst Ropes develops and delivers world record breaking products to the offshore oil & gas industry.
Applications range from Deepwater Mooring to Deepwater deployment. Utilising our proven rope technology we maintain our position as global leaders in the industry.
For more information contact deepwatermooring@lankhorstropes.com
www.lankhorstropes.com

Quality Assurance, Health, Safety & Environmental Services
Classification, Certification and Independent Verification
Consulting & Independent Analyses
Asset Integrity Management
Training & Outsourcing
Third Party Inspection & Services
Risk & Safety Analyses
Equipment Certification
ISM & SEMS Certification
Contact us at:
Bureau Veritas
390 Benmar Drive, Suite 100 - Houston, TX 77060
Tel.: 281.986.1300
houston.offshore@us.bureauveritas.com
www.us.bureauveritas.com
Move Forward with Confidence
BUREAU VERITAS

ONE INNOVATIVE SOLUTION
MANY APPLICATIONS
The current oil and gas industry environment demands an adaptable, efficient and proven solution. Through innovation and foresight, EXMAR Offshore delivers a flexible, cost effective solution that minimizes lead-times and accelerates first production.
EXMAR Offshore is perfectly placed to offer a re-deployable, semisubmersible platform designed to be a scalable solution for floating production systems.
Through the expertise of our people and strong partnerships with industry professionals we are able to engineer, procure and project manage viable solutions that maximize cost effectiveness in today's challenging environment.
EXMAR
OFFSHORE

WOOD GROUP
FRONTIER
by Wood Group
Experience. Independence. Continuity.
Frontier is a specialist engineering consultancy providing independent, innovative and responsive feasibility and conceptual field development studies executed by experienced engineers possessing early phase skills and mindsets.
Frontier leverages Wood Group competence in the total value chain from wellhead to processed fluid.
wgfrontier@woodgroup.com
www.woodgroupfrontier.com

SNAME
THE INTERNATIONAL COMMUNITY
FOR MARITIME AND OCEAN PROFESSIONALS

STORAGE AND OFFLOADING ASPECTS OF CONTAINMENT SYSTEM IN A CAP AND FLOW SCENARIO

THOMAS SELLERS
MARIN

TOM KOSTER
CONSULTANT ON BEHALF OF COBALT INTERNATIONAL ENERGY, INC.

CHANDRA DHIMAN
AET

Proceedings of the 21st Offshore Symposium, February 2016, Houston, Texas
Texas Section of the Society of Naval Architects and Marine Engineers

ABSTRACT

Events such as the Deepwater Horizon oil spill highlight the need for improved containment response systems in response to deepwater incidents. This paper outlines numerical analysis and bridge simulation work to determine the feasibility and weather operating limits for loading and offloading operations in the US Gulf of Mexico as part of an emergency response to a potential loss of well control requiring well flow back. This work was carried out by MARIN for HWCG, a consortium of sixteen deepwater operators committed to building a safe, comprehensive and rapid response system through industry collaboration and mutual aid. The response system studied includes four key surface vessels: a mobile offshore drilling unit (MODU), a production vessel, a storage vessel and an offloading vessel. This paper focuses on the storage and offloading vessels, which are held in place by tugs with dynamic positioning (DP), as these vessels are critical to the uptime of the response system. The numerical analysis consists of simulated hawser loads and holding tug tow line loads for two sizes of storage and offloading tankers and a range of wind, current, and wave conditions. The hydrodynamic models of storage tanker, offloading tanker, and holding tugs were then input into a bridge simulator to allow real time simulation under various conditions. An experienced tanker offloading mooring master, a DP process vessel Offshore Installation Manager (OIM), and positioning tug masters used the bridge simulator to confirm feasibility and develop procedures for position keeping of the storage tanker using tugs. Weather operating windows, hawser design and tug characteristics predicted by the numerical analyses were modified as a result of the bridge simulation work. The bridge simulator is now available for training tanker, process vessel, storage tanker, and offloading tanker personnel in advance of an actual emergency response event. This paper shows how integrating numerical analysis, a bridge simulator, and actual operator input can help to solve complex operational design challenges.

Keywords: SNAME, Offshore, Symposium, HWCG, Simulations, Real-time, Fast-time, Bridge simulations

INTRODUCTION

In the aftermath of the *Deepwater Horizon* event, offshore energy companies joined together to form HWCG with the mission of developing a comprehensive and rapid deepwater containment response system for the Gulf of Mexico. Today, HWCG is a consortium of sixteen deepwater operators committed to building a safe, comprehensive and rapid response system through industry collaboration and mutual aid.

The core components of HWCG's subsea incident response system is the capping stack, which can effectively shut in and cap a subsea release, see Figure 1.

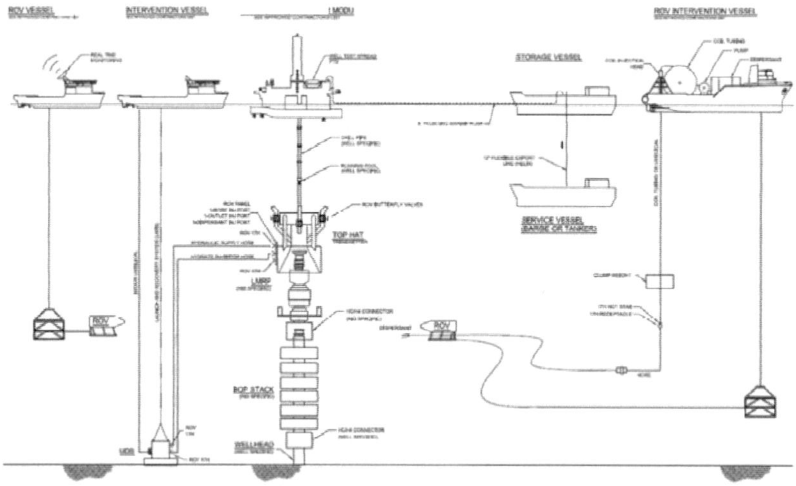

Figure 1: HWCG response system

In a situation where extreme well pressure may cause a well leak below the mud line, a flow and capture containment system will be deployed. In the flow back method, hydrocarbons are collected and safely transported via risers to process and storage vessels on the surface. HWCG has developed several "capture and flow" response plans that utilize a different combination of vessels depending upon the amount of flow back oil. The response method focused on for this paper makes use of four key surface vessels: a mobile offshore drilling unit (MODU), a production vessel, a storage vessel and an offloading vessel. The field layout of these vessels is shown in Figure 2.

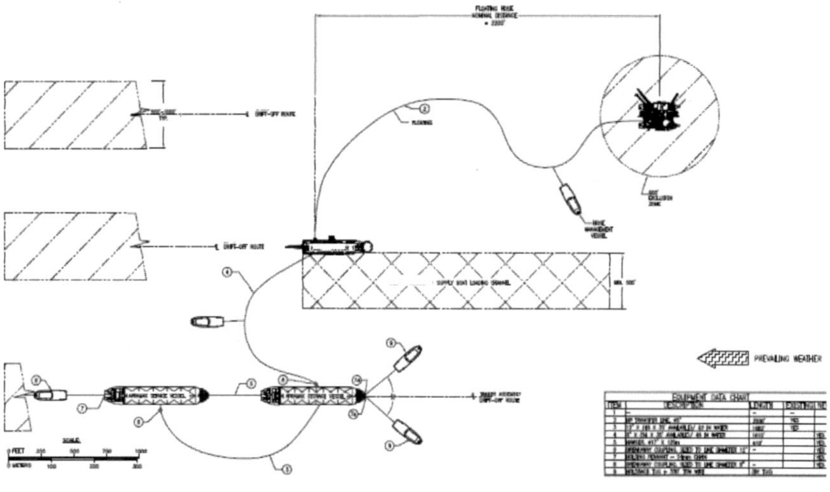

Figure 2: HWCG field layout

Both the production vessel, MODU and support tugs are station keeping with dynamic positioning (DP) systems. The storage vessel is held at position by two DP tugs at the bow and a hold-back tug at the stern. When the transport vessel arrives, the hold-back tug will disconnect from the storage vessel stern, and assist with the heading of the storage vessel. The offloading vessel will approach with two escort tugs, one push-pull tug at the bow and a hold-back tug connected to the stern.

To flow the captured oil between vessels, a Coflexip type connection exists between the MODU and production vessel. Floating marine hose connections are provided between the production vessel, storage vessel and when connected offloading vessel. When a change of weather requires the storage vessel to change heading, it will rotate with the production vessel and the MODU around the well. The station keeping and relative positioning of all connected vessels is therefore critical to ensuring a safe operation. Due to the lack of a DP system on the storage and offloading vessels, the weather limitations of the vessels held by two DP tugs will determine the expected uptime of the response effort.

A numerical analysis was combined with bridge simulations to study the feasibility and limits associated with the nautical aspects of the operation. Numerical simulations determined the weather limitations, hawser loads, and expected uptime and bridge simulations confirmed the feasibility and furthered the development of operational procedures for in-field maneuvering.

The study was carried out in four stages: development of the hydrodynamic database, frequency domain study, dynamic time-domain study, and bridge simulation study. Each stage of the study addressed different aspects of the operation. The frequency domain study gave insight into statistical weather limits of the offloading system, the dynamic time domain study further developed weather limits according to the DP capability of the assisting tugs and the dynamic hawser loads and the real-time simulations provided operational input into the procedures, procedural requirements and the overall feasibility of the maneuvers during approaches, departures and heading changes. This project combined different simulation methods with personnel from operational, engineering and management disciplines to provide an encompassing overview of a complex, multi faceted operation.

METHODOLOGY

The simulation method developed to study the nautical aspects of the HWCG response plan combined three stages of simulation to address the different aspects of the nautical response plan. Numerical simulations focused on the station keeping limits of the tug held storage and offloading vessels while real-time bridge simulations addressed the feasibility of the maneuvers required by the operation. Operational and engineering representatives from companies involved in the HWCG effort provided expert technical and operational input into the simulation models throughout the study.

The storage and offloading vessels for the HWCG response plan are Aframax size tankers, which were initially judged to be the largest size practical. There are no Jones act Aframax vessels. The Aframax storage vessel is not required to be Jones Act, however, if available, the offloading vessel should comply and will probably be a Handysize tanker or articulated tug barge (ATB). Numerical studies showed little difference in vessel motions and hawser loads between Handysize and Aframax offloading vessels. This paper will discuss the results for an Aframax offloading vessel.

A hydrodynamic database was developed for the vessels and carried between each phase of the analysis to provide consistent and comparable results at each phase of the study. An Aframax panel distribution was modeled at three drafts: ballast, intermediate and loaded, in order to account for each stage of the storage and offloading operation. The DP tugs used to hold the storage and offloading vessels at station were modeled after a loaded 120 ton bollard pull ASD anchor handler. Diffraction analysis was performed in order to obtain the wave forces and added mass and damping coefficients. The second order wave drift forces (in regular waves) and the low frequency wave drift forces (in regular wave groups) were obtained from the diffraction results through a direct pressure integration method. The wave drift forces were stored in the hydrodynamic databases in the form of quadratic transfer functions (QTF's). The hydrodynamic databases and all applied calculations are stored and solved, respectively, in the frequency domain. OCIMF wind and current coefficients were applied to the Aframax vessels and wind and current coefficients from MARIN's Dynamic Positioning Capability (DP CAP) database were used for the tugs.

The numerical simulations began with study of the mean loads in the frequency domain to give an initial estimate of the range of hawser loads from wind, waves and current on the storage and offloading vessels. Met ocean conditions for a location in the Gulf of Mexico were used to determine the environmental limits for the station keeping of the storage and offloading vessel, and to evaluate the hawser loads as a function of required uptime. Wave scatter diagrams with associated wind were combined with a directional probability for the Gulf of Mexico loop current to derive joint statistical data for the environment. The wind and waves were considered collinear and the wind and current were considered uncorrelated. A joint probability method was used to develop a hind cast environmental database consisting of 81,920 environments, each with different probabilities.

It was assumed that the maximum and mean hawser load during offloading occurs either at the start or at the end of the offloading operation. Therefore, the hawser loads and relative vessel angles were determined for the following two offloading configurations that consider the extremes of the offloading cycle.

1. Start of offload: Storage vessel loaded, offloading vessel ballasted
2. End of offload: Storage vessel ballasted, offloading vessel loaded

A tool was developed to calculate the mean heading of the storage and offloading vessel and then the respective environmental loads. At the equilibrium heading of each environment case, the environmental forces due to wind, wave and current were calculated to estimate the mean loads in the hawser and the relative angles between vessels. With the equilibrium heading of the offloading vessel, the loads in the hawser are transferred into an earth fixed force. This force is then added to the stern of the storage vessel and the equilibrium heading for the storage vessel is calculated. At this equilibrium heading, the environmental forces are added to the forces induced by the offloading vessel to calculate the total mean load on the storage vessel hawser.

To calculate the load sharing between the two tugs, it was assumed that the steady state relative angle between the tugs is 70° and the heading difference between each tug and the storage vessel is 35°. Therefore, the mean environmental loads on the combined vessels (FX and FY) are distributed over the tugs as follows:

$$F_{TUG1} = -\frac{FMX}{2\cos(35°)} - \frac{FMY}{2\sin(35°)} \quad (1)$$

$$F_{TUG2} = -\frac{FMX}{2\cos(35°)} + \frac{FMY}{2\sin(35°)} \quad (2)$$

This approach was applied to each loading condition to determine the required tug forces to keep the storage and offloading vessel at station. The DP capability plot for the tugs was compared to the environments of the mean load study to determine the station keeping environmental limits. It was assumed that half of the tugs available bollard pull could be dedicated to station keeping and therefore the DP capability is reduced by half.

The time domain analysis took the frequency domain mean load study a step further by considering the effect of the vessel motions on the hawser loads and the DP capability of the tugs. For the dynamic time-domain analysis, the intermediate draft of the storage vessel was considered without the offloading vessel. It was assumed that the offloading vessel will disconnect in severe weather conditions. A total of four configurations were simulated in the fast-time analysis. Configurations included single and dual station keeping tugs and the storage vessel with and without a holdback tug.

The configuration consisting of a two DP tugs set at 70° apart and hawser connected to the storage vessel with a holdback tug resulted in the most uptime. Due to the increased uptime and the complications of operating three tugs, this configuration was used during the bridge simulations. A full time domain model including DP systems, stern azimuthing thrusters (ASD), bow thrusters, and a catenary line to represent the hawsers was set up to determine the optimum layout and weather limits for this operation. To allow for better allocation in the DP model, the twin aft azimuthing thrusters were combined into one centerline mounted azimuthing thruster of equivalent power and the forward bow tunnel thruster and azimuthing thruster were simplified into one bow tunnel thruster of equivalent transverse power.

The environments studied in the dynamic analysis are derived from the critical environments determined in the mean load study. The two wind speeds in the dynamic analysis 11m/s (22kn) and 16m/s (32kn), have a 3.9% and 0.2% probability of exceedance, respectively. Wind and wave environments were applied head on to the storage tanker, while the 0.5m/s (1kn) and 1.0m/s (2.0kn) current was varied in 45° increments from northeast to south.

The frequency and time domain simulations addressed the station keeping ability of the storage vessel held by DP tugs. What the simulations did not consider was the maneuvering aspects of the operation. Discussions between mooring masters, tug masters and the OIM of the production vessel were held to plan the complete operation including the approach and departure of the storage and offloading vessels and how to implement a heading change of the hose connected vessels. The feasibility, procedures and requirements of the manoeuvres were developed through real time bridge simulations.

Hydrodynamic databases used during the numerical simulations were combined with a maneuvering model to represent the necessary vessels required to execute the simulations. The bridge simulations were carried out in two phases. The first round of simulations focused on developing and validating procedures for the offloading operations and for the initial approach and final departure of the storage vessel from the field. The second phase of simulations checked feasibility and developed operational procedures for a heading and position change manoeuvre involving all hose connected vessels in the field. The simulation field layout is shown in Figure 3.

Figure 3: View of real-time simulation field layout

A mooring master from AET used the bridge simulator to develop and validate operational procedures for the manoeuvres. It was decided the offloading vessel would not approach or depart from the storage vessel in environments that exceed the DP capability of the tugs. For this reason, the DP tugs were replaced by earth fixed bollards during the offloading vessel approach and departure simulations. The same hawser configuration model was used to connect the storage vessel to the bollards. In effect, this set up represented a hawser connection to a perfect DP tug model.

The second phase of real-time simulations focused on the heading change manoeuvre of the hose connected field vessels. To avoid overloading the DP systems, all vessels connected by the floating hose must update their respective headings and positions in response to weather changes. When a change of weather requires a significant heading change of a hose connected field vessel, all other connected vessels will need to respond with the appropriate heading and position change. This rotation around the well requires the production vessel and tug assisted storage vessel to move along a pre-defined track within the bounds of the floating hoses, Figure 4. Due to the lack of an onboard DP system, the storage vessel is the critical component during this manoeuvre and must rely on the two forward tugs and single aft tug during the manoeuvre. Moving the storage vessel to a new heading and position within the constraints of a floating hose requires constant interaction between the mooring master and tug masters. This kind of complex, empirical manoeuvre cannot be modelled using fast time simulation and must be input into a bridge simulator to allow real time interaction between the mooring master and the assisting tug masters. Full hydrodynamic tug models combined with the addition of tug masters from an offshore service company were added to the previous real-time simulation model to allow for the inclusion of the human factor and performance limitations of the assisting tugs. These additions to the simulation model provided a comprehensive and reliable understanding of the operation's feasibility.

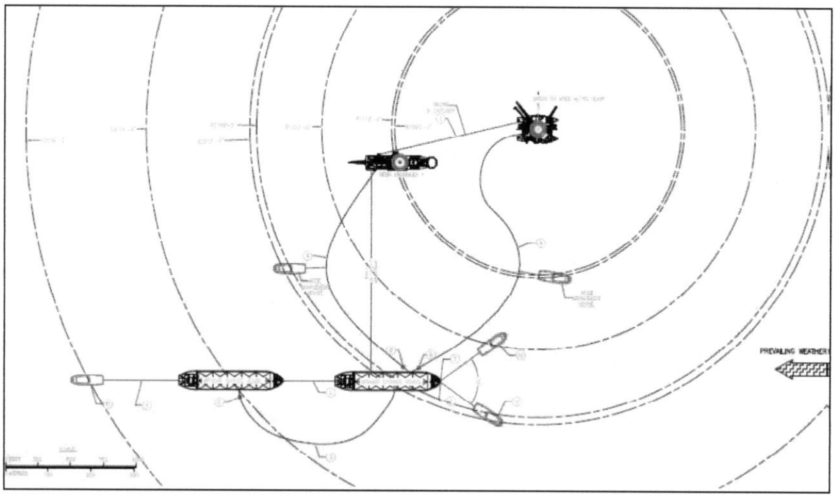

Figure 4: Maneuvering circles

Based upon discussions with the tug captains, the closed loop DP system will only be used when the weather is stable. During the heading change manoeuvre, the tugs will assist the storage vessel in a manual mode in order to remain in control of the tug and to reduce sudden high loads on the DP system caused by the hawsers. Due to the standard operational procedures of the tug masters, the heading change operation was performed without the use of DP systems.

The predefined maneuvering circles for the storage and production vessel were developed through coordination between HWCG, Cobalt, Xodus and MARIN. Maximum, minimum and mean maneuvering circles were produced for the storage and production vessels based upon the limitations of the floating hose. The maneuvering circles were then input into the nautical chart used by the mooring master and tug masters during the simulations.

The storage vessel began simulations at an equilibrium heading and position. During the simulations, the wind would begin to rotate up to 110 degrees at a rate of approximately 2°/minute. Squall scenarios were also simulated by increasing the wind speed and rotating the applied wind direction between 90°-110° over a period of 1 to 2 minutes. As the weather shifted and the storage vessel began to lose equilibrium the manoeuvre to change heading and position would begin. Depending upon the wind shift direction, the storage vessel would travel forward or aft along the maneuvering circle until reaching the correct position and heading relative to the production vessel and MODU.

The mooring master and two tug masters commanded their vessels from three individual simulator stations. Each station was equipped with the necessary controls and navigational equipment to perform the operation. The mooring master called bollard pull and pull direction commands to the tugs in response to the movement of the storage vessel. Depending upon the direction of travel, different tugs would lead the operation and the tug positioning would change accordingly. Once the wind direction became stable again, the tugs would move back into their original configuration and the mooring master would adjust the heading of the storage vessel and positioning of the tugs until the equilibrium heading was achieved.

RESULTS

Different simulation methods were applied to study the nautical aspects of the HWCG operation. The frequency domain mean load simulations provided a large scale statistical approach to expected uptime as a function of the allowable load on the storage vessel assisting tugs, time domain simulations took the mean load study a step further by including the effect of a dynamic system on the DP capability of the tugs and the real-time simulations assisted the operational personnel with development and verification of complex maneuvers.

The mean load study was the first step to determining the expected uptime of the offloading configuration. Figure 5 shows the resulting probability of non-exceedance for the loads in the hawser from the storage vessel to the DP unit holding the Aframax vessels at station. This probability indicates the possible uptime depending on the allowed load in the storage vessel single hawser, or the allowed load on a DP unit. In virtually all environments, the load is more than 49kN (5 tons), but in about 90% of the time the mean load is limited to 231kN (23.6 tons).

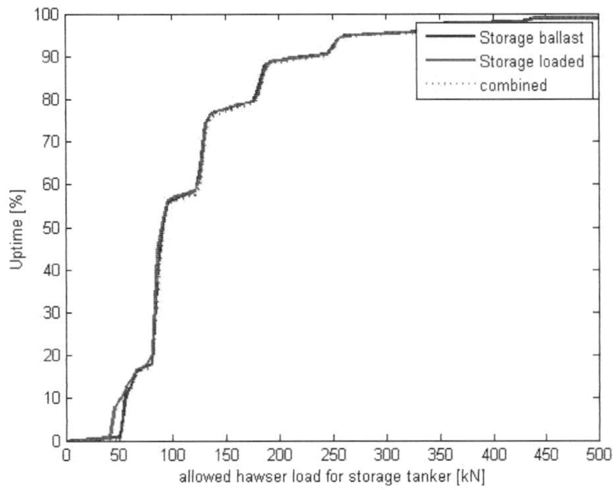

Figure 5: Uptime as a function of allowed load on DP unit

The largest hawser loads occur at the start of the offloading cycle with the storage vessel fully loaded and the offloading vessel ballasted. Figure 5 shows a staircase shape, induced by the bin sizes of the environmental data. This figure does not consider a hold back tug applied to the offloading vessel, thus allowing all vessels to weathervane freely. This approach results in relative angles between the vessels and angles between the hawser and centerline of the vessels. Figure 6, defines the four relative angles considered.

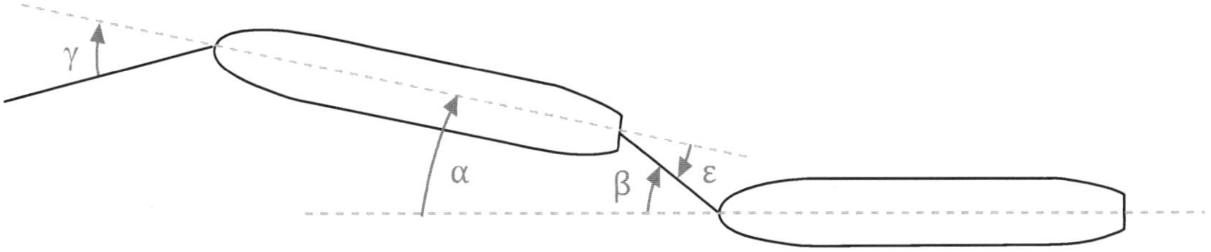

Figure 6: Four relative angles during offloading

Table 1 shows the non-exceedance probability for different criteria on all these angles. The last three columns labeled A100, A200 and A300 show the probabilities that all angles are within the criteria and the total load on the DP unit is below 100kN, 200kN and 300kN, respectively. The maximum mean angle and maximum hawser load for both loading conditions are used for every environmental condition.

Table 1: Probability of non-exceedance for offloading angle criteria

Relative Angle:	**α**	**β**	**ε**	**γ**	**A100**	**A200**	**A300**
< 10 deg	69.2%	40.4%	39.2%	60.5%	12.5%	25.0%	28.3%
< 20 deg	91.0%	80.1%	60.8%	80.2%	30.5%	50.6%	54.9%
< 30 deg	96.3%	88.6%	84.1%	89.2%	46.5%	73.3%	77.8%
< 45 deg	97.5%	90.6%	89.8%	94.4%	53.1%	80.9%	85.8%
< 180 deg	100.0%	100.0%	100.0%	100.0%	56.0%	88.6%	95.2%

The last row in Table 1 shows the probability with angles below 180°, which means no limitation on the angles and shows the same results as Figure 5. Comparison of this row with the rows above shows that if angles up to 45° are acceptable the uptime is hardly influenced by the angle criteria.

A tug at the stern can be used to better align the vessels and reduce the downtime due to large relative angles. Table 2 shows the increased uptime depending on tug pull capacity astern for the offloading vessel in all runs which exceed angles of 30°, results in additional uptime for case.

Table 2: Probability of non-exceedance with all relative angles below 30°

Mean DP load	**<100kN**	**<200kN**	**<300kN**
0 ton tug pull	46.5%	73.3%	77.8%
5 ton tug pull	46.7%	81.0%	85.9%
10 ton tug pull	46.7%	81.9%	87.3%
15 ton tug pull	46.7%	82.3%	88.7%
20 ton tug pull	46.7%	82.3%	89.7%

From Table 2 it is clear that increasing the tug pull beyond an arbitrary 200kN load on the DP system does not add any significant uptime. If mean environmental loads up to 200kN on the DP unit and mean angles up to 30° are acceptable, offloading from the Aframax storage vessel to an Aframax offloading vessel is possible up to 82.3% of the time.

The DP capability plot for the tugs was compared to the non-exceedance probabilities of the environments used in the mean load study to determine the station keeping environmental limits. It was assumed that half of the tugs available bollard pull is dedicated to station keeping and therefore the DP capability is reduced by half. Formulas 1 and 2 were used distribute the mean environmental loads on the combined vessels over tugs. The approach determined that the tug capacity is sufficient to maintain station in wind speeds up to 16m/s, or 99.5% of the time, however, the current causes large angles between the total force on the storage vessel and the longitudinal axis of the storage vessel. In 4% of cases, this angle is beyond 35° which indicates the specified tug configuration is not sufficient to keep the storage vessel at station.

The dynamic analysis provided further insight into the station keeping of the storage vessel. The environments studied in the dynamic analysis are derived from the critical environments determined in the mean load study. As observed during the mean load analysis, high currents resulted in large relative angles between vessels and between the applied force on the storage vessel and the longitudinal axis of the storage vessel. The effect of current speeds of

0.5m/s (1.0kn) and 1.0m/s (2.0kn) were therefore studied in increments of 45 degrees applied from northeast to south. The wind speeds and wave heights applied in the dynamic analysis were 11m/s (22kn) with an associated 2.5m wave and 16m/s (32kn) with an associated 4.5m wave. These wind speeds have a 3.9% and 0.2% probability of exceedance, respectively.

For the configuration with two forward tugs and an applied backward force on the storage vessel, fishtailing was observed in the high (16m/s (32kn)) wind and (4.5m) wave environments with opposing current. The coupled sway and yaw motions associated with fishtailing induced high hawser tensions. As a result of these observed hawser tensions the planned hawsers between the storage and service vessels were updated from a 981kN (100 ton) safe working load (SWL) hawser to a 1962kN (200 ton) SWL hawser. The sway and yaw of the fishtailing motion are significantly greater in the simulations without storage tanker stern thrust. The aft stern thrust reduced the fishtailing cases by 60%.

The DP capability of the tugs was also studied during the time domain simulations. As found in the mean load study the DP tugs have adequate mean thrust in all environments. However, thruster saturation (thrusters at maximum capacity) was observed for at least one time step in several high wind cases, especially those cases where fishtailing motions of the storage tanker occur. Thruster saturation occurred in 19% of studied environments. Two of the three occurrences were due to fishtailing of the storage vessel.

The dynamic time domain simulations proved the feasibility of tug provided station keeping for the storage vessel. The tug capacity is sufficient to maintain station in up to 16m/s (32kn) wind, 99.5% of the time. Two forward DP tugs combined with a 10 ton holdback tug could delay storage vessel fishtailing motions and subsequent overloading of the DP system to winds of 15m/s. Although, the configuration with two tugs produces favourable results and the redundancy to the operation is increased, the coordination between the two tugs becomes complicated. In addition to the station keeping role of the DP tugs, when a change in the weather requires the storage vessel to change heading, the tugs will be tasked with maneuvering the storage vessel to the new heading and position while staying within the limits of the floating hose. To simulate this type of complex, reactionary manoeuvre, real-time bridge simulators commanded by operational personnel were used. Meetings between mooring masters from AET, tug masters and engineering and management representatives from HWCG and MARIN were held to develop the operational procedures and discuss the required input for the real-time bridge simulations. Through the combination of engineering and operational personnel, the procedures and input for the manoeuvres were drafted and tested using a bridge simulator.

The real-time bridge simulations were divided into two phases, the first to study the approach and departure of the storage and offloading vessel and the second to study the feasibility of the heading change manoeuvre. The initial bridge simulations provided requirements for the operation, identified potential issues, and validated or in some cases, modified the initial operational and emergency procedures. Approach tracks, approach sectors, tug requirements and environmental limits were validated and the feasibility of the approach and departure operations in a crowded field was proven. The approach and departure of the storage and offloading vessel was possible in winds up to 10.3m/s (20kn), swells up to 2m and a crossed current of up to 1.0m/s (2kn).

During the second phase of bridge simulations, the tug capability was investigated further. ASD tugs and conventional tugs were used to assist the storage vessel during the heading change manoeuvres. Both tug types were sufficient for the operation; however the conventional tugs required approximately 25% more power than a similar ASD tug to keep the storage vessel at station. The tug masters preferred the ASD tugs for the heading change operation due to the increased manoeuvrability and reaction time. During squall scenarios, the tugs will require their rated bollard pull of 1177kN (120 tons) to manoeuvre the storage vessel. The tug captains successfully translated the storage vessel around the maneuvering circle while operating the tugs in a manual mode. The tug masters did not believe the operation could be performed in a DP mode due to the unpredictable movements of the storage vessel, the sudden hawser loads induced by the storage vessel motion and the time delay for the DP system to respond to changes.

Figure 7: Real-time bridge simulations

The simulations helped the operational personnel develop the necessary requirements to successfully complete the infield manoeuvres. Requirements included on board survey equipment to accurately show the position of all field vessels, the inclusion of the maneuvering circles on the field's electronic nautical chart (ENC), an equilibrium heading tool for the storage vessel mooring master and training for all tug and mooring masters involved with the response effort.

Environmental limits for the heading change operation were also determined through bridge simulations. In wind speeds less than 7.7m/s (14.9kn), the tugs have sufficient power to hold the storage vessel at a less than optimal heading. In higher wind speeds or once the port tug angle becomes limited due to the placement of the floating hose; the storage vessel will need to update heading and position. Depending on the direction of the environmental change, the storage vessel can either move forward or aft along the maneuvering circle. The production vessel and MODU will also need to update their headings and in some cases positions with the storage vessel. The vessels will need to maintain their relative positions and headings during the manoeuvre to ensure the floating hoses are not compromised. In current speeds greater than 1.0m/s (2.0kn) it is recommended to disconnect the hose from the storage vessel prior to commencing the heading and position change.

CONCLUSIONS

The Macondo incident demonstrated the complex nature of an offshore well blowout response. Multiple vessels are required to operate in close proximity and remain in the field for prolonged time periods. The DP drillship Discover Enterprise processed, stored and/or flared the recovered hydrocarbons during the Macondo effort. The HWCG response plan includes a MODU, a process plant on a DP vessel and conventional (non DP) Aframax storage tanker that relies on the DP capability of two DP anchor handler tugs and one hold back tug for station keeping. The introduction of a non DP vessel into the field of multiple hose connected vessels required a thorough study to check the feasibility of the nautical operation. Through a combined effort from engineers and operational personnel utilizing task specific simulation tools, a comprehensive study was performed to determine the feasibility, requirements and operational procedures for the nautical aspects of the HWCG response plan. Numerical simulations determined the station keeping limits for the DP tug held storage vessel in several configurations and bridge simulations utilized the expertise of operational and engineering personnel to confirm feasibility and develop procedures for offloading operations, positioning, and the translating the storage tanker using tugs.

The study confirmed the feasibility of the nautical operations embodied in the plan. This study demonstrates how the inclusion of engineering and operational input can be combined into numerical and bridge simulations to provide integrated solutions to complex operational design challenges.

ACKNOWLEDGEMENTS

The authors would like to acknowledge the work of the many people involved in the HWCG effort for their time spent providing input, feedback and solutions during the simulations. Particular acknowledgements are extended to the operational personnel without whom the real-time simulations would not have been possible.

INTERNATIONAL SHIPBUILDING AND REPAIR FACILITY IN NIGERIA WEST AFRICA

HERMAN J. SCHELLSTEDE
PRESIDENT, HERMAN J SCHELLSTEDE AND ASSOCIATES, INCORPORATED

Proceedings of the 21st Offshore Symposium, February 2016, Houston, Texas
Texas Section of the Society of Naval Architects and Marine Engineers

ABSTRACT

The Capricorn Maritime Limited group plans to build an extensive shipbuilding and repair center which will serve the region of West Africa. The Capricorn group has chosen to locate the shipbuilding and repair center at Ibaka, Mbo Lga., Akwa Ibom State, Nigeria near the city of James Town, Nigeria. Herman J. Schellstede & Associates, Inc. of New Iberia, Louisiana, U.S.A. has been awarded the technical contract to construct the shipbuilding and repair facility.

The west coast of Africa has proven oil reserves from Senegal in the north to Cameroon in the south. The reserves are located both onshore and offshore with a high degree of reserves lying near and in the Delta areas. Nigeria has long established their position as a leader in the oil and gas industry. Nigeria contributes 2.37 million barrels of oil per day to the marketplace. A large portion of Nigerian oil is expedited by marine vessels to various sales points throughout the world.

In recent years, deep water leases have been awarded in Nigeria. Many of the leases have been drilled and reserve reports have been produced. The three major markets which are offshore, near shore and onshore, require different drilling and completion equipment. Drill ships, jack-up rigs, semi-submersibles provide drilling in offshore waters. Drilling barges, both swamp and posted, are employed in near shore applications and land-based drilling rigs are employed onshore. The requirements to maintain and repair the drilling fleet are of great importance.

In order to support the offshore and near shore operations, various manner of crafts are required such as dredges, workboats, crew boats, tug boats and supply barges. These marine vessels are also joined by coastal and open sea cargo ships which transport the crude oil to the marketplace.

The government of Nigeria has changed and a new approach to the energy market is evident. The new government pledges that the "Nigerian content" laws will be upheld. (Nigerian oil and gas industry content development act NOGICD.) A detailed explanation of the development act is being enforced and being accepted throughout the country of Nigeria. The Capricorn facility considers the Nigerian development act and has directed their planning to accommodate the spirit of the laws.

The Capricorn facility, integrating proven management methods of the oil and gas industry, is designed to offer shipbuilding, repair and maintenance services for the shipping and oil and gas industries of the coastal region of West Africa and to also accommodate governmental contracts for the servicing of naval as well as commercial fleets.

An initial market survey and a production document for the shipyard facility review are also discussed herein in a preliminary manner. The facility, management and financial support are considered world-class and will result in a very profitable and acceptable return on investment.

The nine major shipyards, which have been studied, illustrate various building and expansion programs which were implemented to accommodate future contracts. Only a few shipyards reviewed exhibited controlled expansion events. The Capricorn project will provide preplanned expansion methods to control the required growth. The final design will also integrate the most advanced materials handling equipment and construction methods available today.

During the last 100 years, the United States has constructed major shipyards which were fundamental in the advancement of the United States as a prominent world leader. At the time of war, highly efficient shipyards were necessary and in great demand to provide vital servicing of vessels and equipment. The review of the U.S. shipyards will provide assistance in outfitting the Nigerian facility with the most up-to-date and efficient components including the expertly designed platforms equipped with modern advancements. As a result, the Capricorn facility will be one of the most innovative and efficient facilities built in the world today. The country of Nigeria will be the location of the first major shipbuilding and repair center constructed in recent times due to the urgent need for a comprehensive facility in this region.

In the near future, the region of West Africa will be very active in the transportation of crude oil and the drilling and production from offshore leases. The specialized marine vessels required for transportation of crude oil and drilling/production are well identified. The Capricorn shipyard will benefit greatly from the advancements of over 100 years of shipbuilding developments and innovations. A complete transformation of certain areas in the undeveloped properties and waterways of Ibaka, Mbo Lga., Akwa Ibom State, Nigeria is being undertaken to produce a world-class shipyard and repair facility which will provide support to the oil and gas industry of this region.

Keywords: SNAME, Offshore, symposium, shipbuilding, repair center, marine facility, West Africa, James Town, Nigeria, waste management, assembly, dry dock, graving dock, ship development, swamp drilling barge, oil and gas operations in Nigeria, repair services.

INTRODUCTION

Oil and gas discoveries occur near highly populated areas, while others are located in very remote and "harsh" environments. The common thread throughout all of the drilling sites, both onshore, near shore and offshore, is the construction, care and maintenance of the drilling rigs and associated equipment. Drilling rigs have been designed, constructed and supplied by U.S. manufacturers during the past 30 years. It is envisioned that new drilling rigs will be manufactured in Nigeria during the coming decades.

In many areas of the world, facilities have been developed to provide necessary services to the oil and gas industry as well as to the marine industry. The Capricorn facility will be constructed as a comprehensive and very efficient complex, which will ensure reliable expedited repair and maintenance of required equipment.

This document provides a summary of the Capricorn facility and the means and methods to employ Nigerian engineers, managers and skilled tradesmen in the development of this innovative complex.

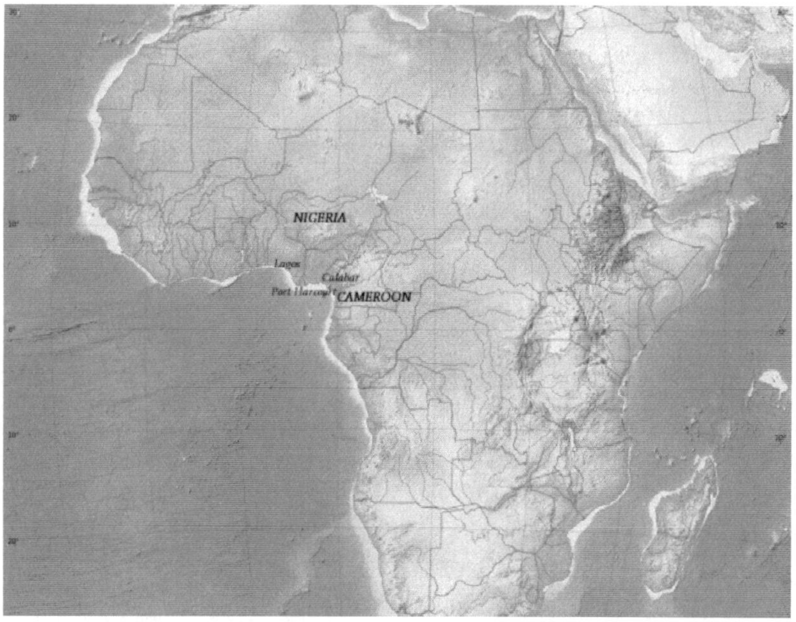

Figure 1 Location of Capricorn Shipbuilding and Repair Facility

METHODOLOGY

The construction of a major shipyard and repair center in an underdeveloped area must consider the necessary management and technical staff. The project requirements, culture and methods of providing services in the region of West Africa are quite different than in developed areas.

Figure 2 illustrates the proposed location of the future Capricorn facility in relation to the Bight of Bonny and Cross River area of Nigeria.

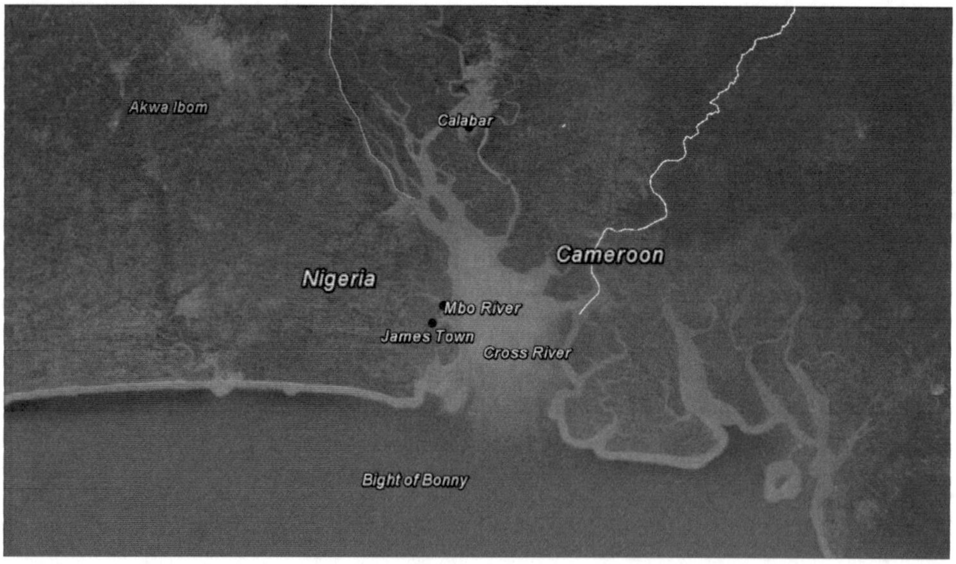

Figure 2 River Entrance into Proposed Capricorn Facility

Figure 3 illustrates the entrance area for the proposed location of the Capricorn facility. The area is underdeveloped and lacks any and all services which are typically available in populated, developed areas.

Figure 3 Cross River to Mbo River Entrance to Capricorn Shipyard Site

Figure 4 illustrates the proposed areas which will be transformed into the shipyard and repair center. The west facility will be the first area of development.

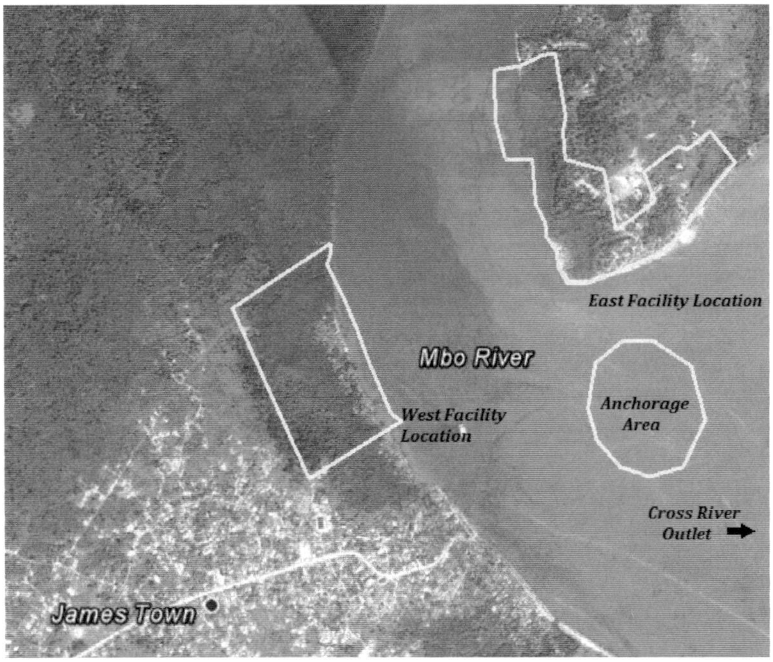

Figure 4 General Area which will be Developed to Accommodate the Capricorn Site at Akwa Ibom State, Nigeria

Figure 5 illustrates the first accommodations facility that will be constructed. This facility is comprised of a complete living facility for the first level workers and management team. This accommodations facility will be a part of the west facility location area.

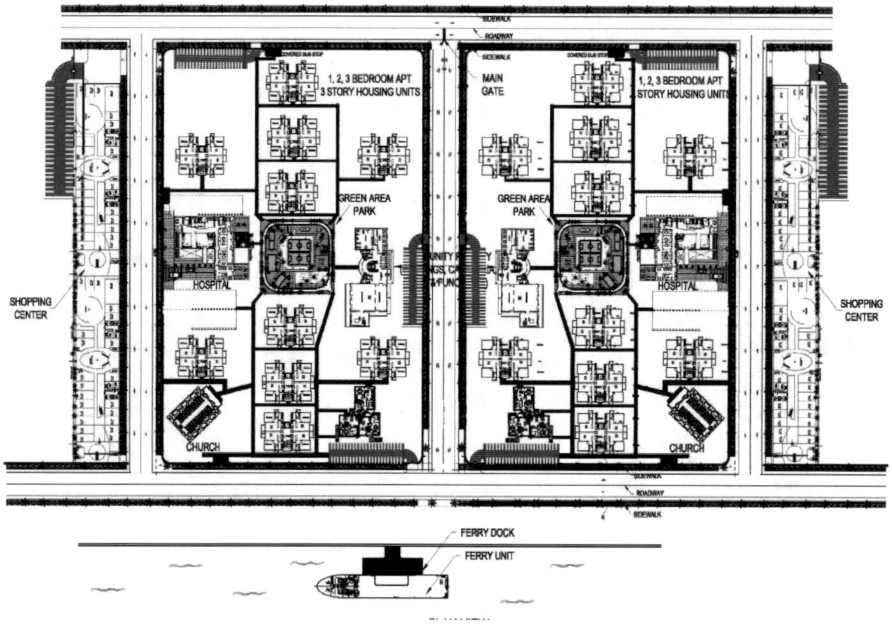

Figure 5 Capricorn Worker Accommodation Facility Layout

The construction of the accommodations area, designed to house 500 workers per complex, and the office complex, will be completed as an important initial task. The west facility, which is devoted to the onshore, near shore and offshore oil and gas industry, will be the first area to be constructed and will accommodate the following services:

- Crude oil storage
- Fuel storage
- Tubular repair, maintenance and storage
- Light tonnage vessel building, repair and maintenance
- Machinery and product repair and construction
- Electrical repair and maintenance area
- Vital equipment storage and marshalling area

The east and west Capricorn facilities are divided by a body of water which is considered a "safe harbor" providing a "sheltered" area to conduct oil and gas related marine repair operations. In addition, this body of water is dredged to accommodate vessel drafts and is an excellent location for the mooring of vessels, before or after service work is conducted. Other vital service features of this area include the following:

- Loading and offloading of heavy transit vessels
- Anchorage of major drilling vessels, i.e. jack-ups
- Large and small dry dock services
- Offshore terminal

- To provide security for vessels
- Harbor derrick barges

Capricorn will employ in-house services for the development, construction and operation of the facility which are presently nonexistent along the West African coastline. The west facility is based on a "stand alone" concept in that the complex can operate independently of the east facility, which is dedicated to servicing larger vessels. Both east and west facilities will be developed in the same manner with the preparation of the workers' accommodations as the initial task.

The east facility will accommodate the following services:

- Shipping
- Offshore platform deck construction and outfitting
- A.S.M.E pressure vessel manufacturing plant.
- Assembly of oil and gas offshore deck section (production)
- New construction graving dock and crane support system
- Vessel repair graving dock
- Plate cutting and metal forming plant
- Larger vessel component fabrication area
- Raw material storage areas
- Dock areas for imported goods
- Power plant for east facility service

Figure 6 (next page) illustrates the complete Capricorn facility including the east and west yards.

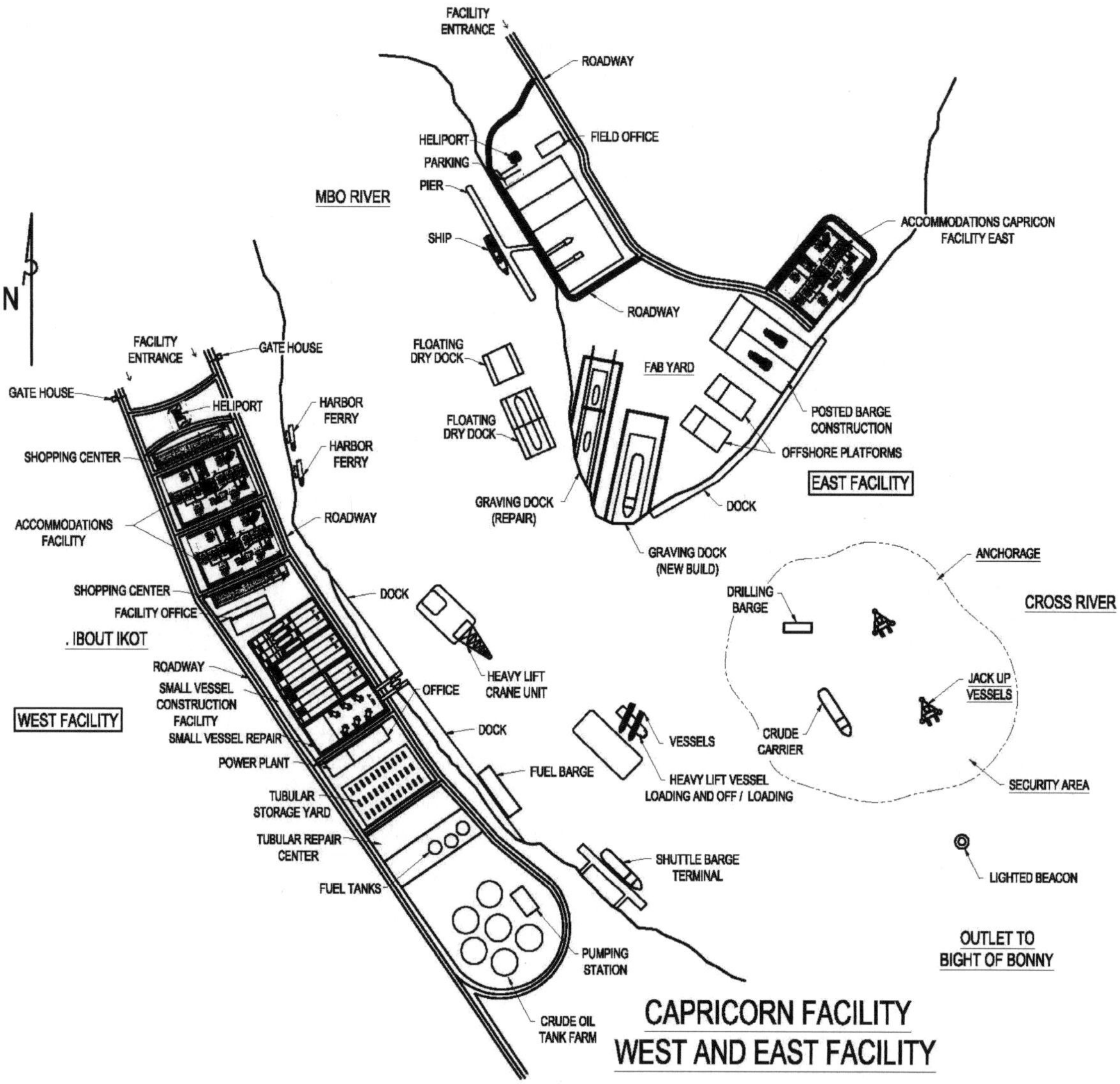

Figure 6 Capricorn West and East Facilities Layout

The Capricorn company, an indigenous Nigerian company, has studied the needs and requirements necessary to provide world-class services and to integrate the various talents of the local Nigerian work force in the implementation of an innovative shipbuilding and repair facility in the James Town region of Nigeria.

The Capricorn Maritime shipyard will provide two leasing campuses and one facility which will be located in New Iberia, Louisiana USA. The second company will be located at the shipyard site in Nigeria. The new frontier of eager, bright, young Nigerian workers is available to accommodate the needs of the Capricorn facility where they will be enrolled in programs in the two training areas. The shipyard companies will train the workers during a "hands on" apprenticeship in which they will obtain essential technical knowledge based on their respective disciplines. The training of the work force will begin during the construction phase. The workers will have the rare opportunity to observe and assist in the building of the Capricorn facility.

Nigeria has a large pool of engineers which, with customized training and direction, can accommodate the Capricorn work load. In addition, Nigeria also possesses a vast number of experienced managers who are capable of running the new facility.

Inherent in the overall plan, are services which are essential to the efficient operation of the facility. For example, one vital service of onshore, offshore oil and gas activities is the repair and maintenance of tubular goods. Drill pipe, drill collars, production tubing and casing are required in all stages of oil and gas production. The Capricorn facility will provide, as one of its many services, equipment to maintain the tubular goods. Another example is the machinery and product repair area. Onshore, near shore and offshore operations employ specialized drilling equipment. The facility has the capability of maintaining the necessary equipment for drilling which includes masts, traveling equipment, top drive, pumps, solids control system, electrical power and control systems. The facility has been designed to provide critical and vital services for the operation of drilling and production systems as well as to lessen costly downtime.

The above mentioned tubular repair, maintenance and product repair or "work cell" are some the many services which will prove to be highly advantageous to the operator or drilling contractor. The Nigerian work force will be trained and approved to conduct the above listed services with a world class level of quality control introduced to West Africa. As can be seen, the infrastructure, equipment and methods must operate in concert.

RESULTS

The mission of the Capricorn facility is to provide services for the shipping, repair and other offshore oil and gas operations along the coastline of West Africa and to employ Nigerian nationals in the planning, building, servicing and management of this facility.

The design of the Capricorn facility integrates over 60 years of historical, state-of-the-art technical, management, procurement and manufacturing methods employed in the development of the U.S. Gulf of Mexico oil and gas operations.

A description of the required workforce and the related financial information follows.

Table 1 and Figure 7 illustrate the required work force for the operation of the shipyard and repair center.

Table 1 Total Capricorn Facility Work Force

Location	Work Force
West Facility	650
Water Body Area	375
East Facility	970
Total	**1995**

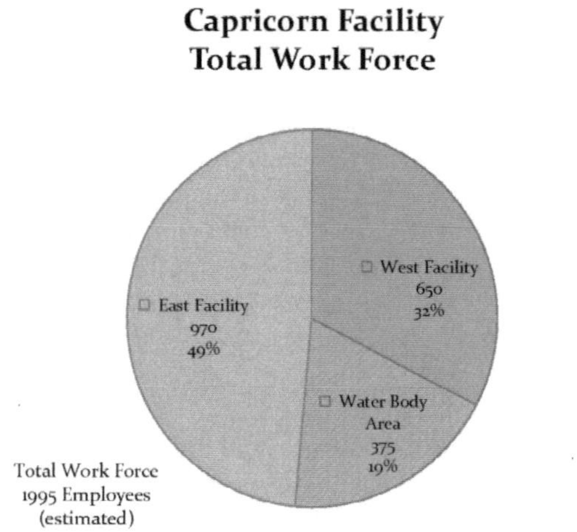

Figure 7 Capricorn Facility Total Work Force

Table 2 and Figure 8 illustrate the capital costs of the Capricorn facility.

Table 2 Total Capricorn Facility Capital Cost

Location	Capital Cost USD
West Facility	$ 117,120,000.00
Water Body Area	$ 69,540,000.00
East Facility	$ 179,340,000.00
Total	**$ 366,000,000.00**

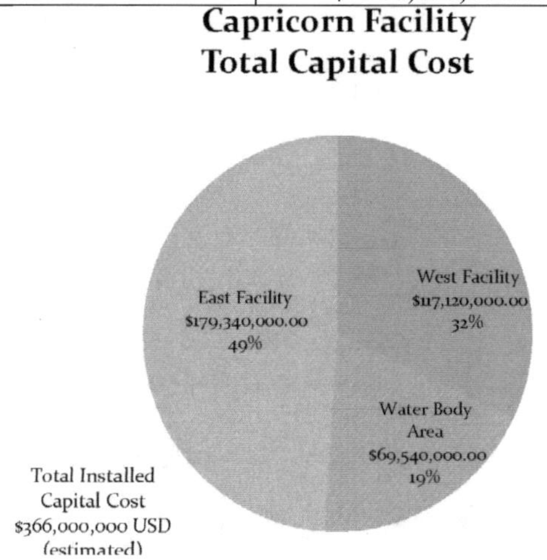

Figure 8 Capricorn Facility Total Capital Cost

Table 3 and Figure 9 illustrate the annual revenue that is produced from the Capricorn facility.

Table 3 Total Capricorn Facility Annual Revenue

Location	Annual Revenue USD
West Facility	$ 248,000,000.00
Water Body Area	$ 63,000,000.00
East Facility	$ 312,000,000.00
Total	**$ 623,000,000.00**

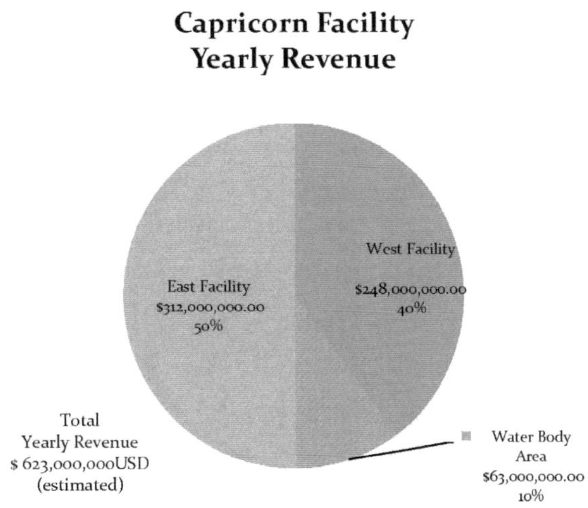

Figure 9 Capricorn Facility Total Annual Revenue

DISCUSSION

Studies and research efforts have been directed towards the design methods employed in the construction of shipyards for the last 100 years. An imbalance in the abilities of the shipyard to supply products in an efficient manner is due to ineffective materials marshalling, processing and fabrication methods. "Bottlenecks" occur in many areas of the studied shipyards as the result of inefficient planning and uncontrolled construction to accommodate specialized contracts.

The Capricorn facility has been designed to incorporate modern means and methods of fabrication and outfitting of platforms, drilling rigs and ships. The modern machining systems are comprised of computerized control systems employed for short and long manufacturing runs. The welding system and positioning systems are also state-of-the-art. These systems provide a great level of efficiency and incorporate a high degree of quality.

The Capricorn facility has also been specifically designed to accommodate equipment and to provide services that are presently not available in Nigeria. The infrastructure which is necessary to support oil and gas operations includes many specialized services which exist only in a few areas of the world. A great degree of these specialized companies are owned and operated in the U.S. Gulf of Mexico region (Louisiana and Texas, USA). For example, in order to produce a typical inland drilling vessel, over 360 separate items are required. The Capricorn group has developed direct relationships with specialized companies in the U.S. whereby rapid response, maintenance, and new builds can be accomplished.

The construction program will be implemented in an accelerated or "fast tracked" manner. The Capricorn planners have taken the following approach to expedite the construction of the facility:

a.) All critical work will be conducted at the shipyard site as per specifications.
b.) The Onyung City east and west yards will be constructed by local Nigerian contractors.
c.) Buildings, offices, materials handling and electrical load centers will be provided by the U.S.
d.) All specialized equipment for services to be offered will also be constructed and tested in the U.S.
e.) Special graving dock components will be constructed and tested in the U.S.
f.) Two floating dry docks will be furnished by the U.S.
g.) A heavy lift harbor crane and attending tug will be provided by the U.S.

h.) Metal preparation equipment will be provided by U.S. plate preparation systems including computer-guided cutting systems, metal preparation wheelabrator and coating systems.

i.) Automatic robotic welding systems will be provided from the U.S.

In order to achieve an on-time and on-budget program, work in Nigeria and the U.S. will occur in one event. The planning and upstream engineering has been progressing during the last two years. Final design and equipment procurement will be accomplished in early 2016.

Table 4 illustrates the Projected Construction Schedule of the Capricorn facility.

Table 4 Capricorn Facility Construction Schedule

2016	2017	2018	2019
• Planning • Engineering • Dredging • Worker Accommodations (Onyung City West Facility) • Start Training Area • Construction of Tubular Repair & Maintenance (West Facility) • Construction of Machine Shop and Special Product Repair (West Facility) • Construction of Light Tonnage Vessel Repair (West Facility)	• West Facility Component Area • West Facility Dock Area, Corporate Offices, Fuel and Crude Oil Storage • Heavy Lift Vessel Loading and Unloading Facility • Harbor Cranes Operations • Dry Dock Operations • Start Anchorage Operations • Start Construction of Phase I-East Facility • Start Worker Accommodations (East Facility)	• Start Repair Graving Docks • Start New Build Graving Docks • Start Drilling Rig Building and Fabrication Yard • Start ASME Pressure Vessel Plant • Start Ship Pier Assembly • Start Metal Preparation Plant	• Complete Both West and East Capricorn Facilities • Full Operation by the Fourth Quarter 2019

CONCLUSIONS

- A great need for a major shipyard facility exists in the nation of Nigeria.
- Market studies illustrate that Nigerian oil, located onshore, near shore and offshore, will be in great demand for several decades.
- The Capricorn facility provides a pathway for the residents of Nigeria to excel in providing a world-class shipyard and service center.
- The mission of Capricorn is to provide a building, repair and maintenance facility directed to the marine oil and gas industries.

View of Port Fourchon, Louisiana

ACKNOWLEDGEMENTS

Chief Dr. Mkgeorge Onyung, M.D., FAGP, 15 Udi Street, Osborne Estate, Ikoyi, Lagos, Nigeria, phone: +234 (1) 271 4434, website: www.jevkonoil.com

REFERENCES

Agbedo, Onyedika, "Why Nigeria Need to Establish a Mega Shipyard, by Industry Stakeholders", *The Guardian*, June 13, 2015.

Ekpo, Ekpo Imoh "Impact of Shipping on Nigerian Economy: Implications for Sustainable Development" *Journal of Educational and Social Research*, Volume 2 (7) October 2012.

Ijioma, E., "Dry Dock Facilities for Building and Repair of Merchant and Naval Ships" *Nigerian Society of Engineers Conference Paper*, 2011

SCREENING OF FIELD DEVELOPMENT OPTIONS USING THE ANALYTIC HIERARCHY PROCESS AND EVEN SWAPS ANALYSIS

YONGYAN WU, ROGER LU, TAO WANG, VISHNU VIJAYARAGHAVAN, ROLF EIDE
AKER SOLUTIONS, HOUSTON, TX

Proceedings of the 21st Offshore Symposium, February 2016, Houston, Texas
Texas Section of the Society of Naval Architects and Marine Engineers

ABSTRACT

Rational selection of a field development solution is a complex process, and it will significantly affect the overall project economics, schedule, and risk. Analytic Hierarchy Process (AHP) is a multi-criteria methodology which improves the qualitative and quantitative aspects of the decision-making process. However, AHP is based on pairwise comparison between different alternatives, which is a cumbersome process when the number of alternatives is large. Moreover, AHP views alternatives as monolithic, i.e. unchangeable ideas either to be eliminated or selected. This paper presents an improved screening process which combines AHP with an Even Swaps Analysis (ESA) method. The screening starts with developing potential alternatives based on different combinations of building blocks. Unfeasible or obviously inferior alternatives are eliminated and this yields a longlist of options. Instead of applying AHP to all alternatives directly, we apply AHP to the selection criteria to derive weighting factors for the criteria. These criteria are subsequently used to narrow the longlist down to a shortlist of options by consistent comparisons. Finally, the ESA method, which focuses on changes needed to improve alternatives instead of unilaterally eliminating them, is used to reach a final selection decision.

A generic case study is used to illustrate the application of the proposed screening process to select an offshore floating facility in a harsh environment. Hundreds of options are generated based on combinations of three building blocks. Elimination of unfeasible combinations results in a longlist of about twenty options. The AHP method is thereafter applied to generate criteria weights. A shortlist of five options emerges after initial screening. The ESA method is then used to balance pros and cons by changing the alternatives and studying the impact of the changes on various criteria. Eventually, by establishing a consequences table, the final concept is selected.

A robust concept screening methodology is critical for offshore development in deeper water and with complex reservoirs, which bring in increased technical, financial and execution risks. This paper provides one such methodology for concept selection in offshore field developments. The proposed method has also great potentials for complicated decision making problems in other areas.

Keywords: Analytic Hierarchy Process, Concept Screening, Even Swaps Analysis, Field Development, Offshore Floating Facility

INTRODUCTION

In recent times, the offshore industry has seen several large projects that require massive investments and long schedules to execute. The selection of a field development solution is a complex process comprising multiple disciplines including well data, metocean, bathymetry, soil, flow assurance, substructures, topside facilities, station-keeping (mooring, tendon, dynamic positioning), and transfer systems (riser, umbilicals, flowlines, shuttle tankers) etc. Concept selection typically occurs during the early phase of a project. Quality of the decision will significantly impact the overall project economics, schedule, and risk. Choosing an appropriate field development concept is critical to the success of a project.

The traditional scorecard method, which establishes a score matrix for each alternative in terms of different criteria and compares the summarized scores of each alternative, is an useful tool for simple decision making problems. However, for a complex decision problem such as field solution for offshore developments, one might have lots of options to choose from and many criteria or objectives to achieve. The traditional scorecard method, which compared all cases simultaneously, could be very subjective and it is difficult to maintain consistency in the comparisons(Saaty 1977; Linder 2014).

The Analytic Hierarchy Process (AHP) is a multi-criteria decision making (MCDM) method that helps the stakeholders to screen complex problems that have a multitude of conflicting and subjective criteria. This method was introduced by Saaty (1977), and is widely considered to be the most reliable MCDM method for screening and concept selection (Ishizaka and Labib, 2009). The AHP method has been applied in various diverse areas, including in offshore field developments (Linder 2014; Linder and Fales 2014). The AHP is based on constructing pairwise comparison matrices of alternatives. During decision making and concept selection, psychologists argue that it is easier and more accurate to express one's opinion on only two alternatives than simultaneously on all alternatives. This also allows consistency and cross checking between the different pairwise comparisons. Mathematically, it can be shown that the eigenvector of the comparisons matrices corresponds to the priority vector of the alternatives (Saaty 1977).

AHP has been widely adopted in many fields. However, the AHP method is a lengthy process when the number of the alternatives is large. This will make the whole screening process overwhelming. Moreover, the AHP method passively views alternatives as monolithic ideas that need to be either eliminated or selected, and does not provide for a dynamic mechanism to improve the alternative.

This paper presents an improved AHP method that is coupled with an Even Swaps Analysis (ESA). The screening starts with developing an initial list of potential alternatives for evaluation using the Field Option Generator (FOG) and building blocks. Infeasible or obviously inferior combinations are eliminated and this yields a reasonable number of options, the so-called "longlist". Instead of applying the AHP method to all alternatives directly, we apply it to the criteria to obtain the criteria weights, and to narrow down the longlist to a shortlist of options by consistent comparisons. Finally, the ESA, which focuses on improving alternatives instead of eliminating them, is applied to reach a final selection to be adopted for concept definition.

ANALYTIC HIERARCHY PROCESS

The AHP was first proposed by Thomas L. Saaty in 1970s, and has been extensively studied and developed. It has been applied in various fields, such as government, business, industry, healthcare, shipbuilding, offshore industry, and education (Saaty 1977; Ishizaka and Labib 2009; Linder 2014). The analysis process is outlined in this section.

Problem Modeling

The first step in the AHP method is to model the problem as a hierarchy, including three parts: goal, criteria and alternatives. An example for choosing a suitable offshore development concept screening hierarchy is shown in Figure 1, with three criteria and three alternatives.

The objective of building hierarchy model is to integrate large amounts of information into our understanding of the situation when we approach a complex decision problem.

As offshore concept section is a complex process, it is important to make sure that all possible alternatives are included.

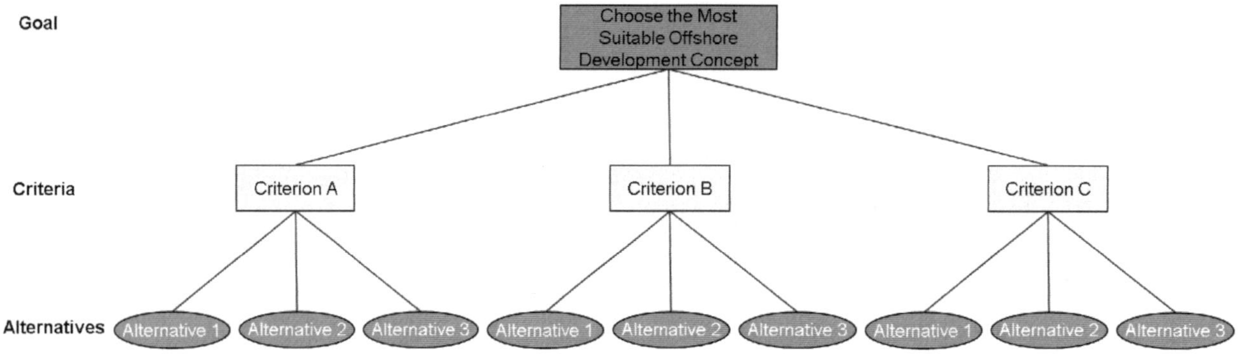

Figure 1 AHP Hierarchy Example

Pairwise Comparison

In contrast to the traditional scorecard method which uses a linear scale range, the AHP method uses a ratio scale, which requires no units in the comparison. The judgment is a relative value or a quotient *A*/*B* of two quantities *A* and *B*. Decision-makers provide a relative verbal appreciation in their more familiar experience instead of providing a numerical judgment (Saaty 1977).

The pairwise comparison scores will be documented in a pairwise comparison matrix. An example of a three criteria comparison matrix is shown Figure 2. The definition of the score ω_i in the pairwise comparison is shown in Table 1 (Saaty 1977). For example, to obtain the criteria *A* and *B* priorities, we evaluate their relative weights. If it is believed that criteria *A* is weakly more important than criteria *B*, then score of criteria *A* over *B* will be $\omega_1 = 3$, and the score of criteria *B* over *A* will be its reciprocal, i.e., 1/3, in the pairwise comparison matrix.

	Criterion A	Criterion B	Criterion C
Criterion A	1	ω_1	ω_2
Criterion B	$1/\omega_1$	1	ω_3
Criterion C	$1/\omega_2$	$1/\omega_3$	1

Figure 2 Pairwise Comparison Matrix Example

Table 1 Score Definition of Pairwise Comparison (Saaty 1977)

Intensity of Importance	Definition	Explanation
1	Equal Importance	Two activities contribute equally to the objective
3	Weak importance of one over another	Experience and judgment slightly favor one activity over another
5	Essential or strong importance	Experience and judgment strongly favor one activity over another
7	Demonstrated importance	An activity is strongly favored and its dominance demonstrated in practice
9	Absolute importance	The evidence favoring one activity over another is of the highest possible order of affirmation
2, 4, 6, 8	Intermediate values between the two adjacent judgments	When compromise is needed
Reciprocals of above nonzero	If activity i has one of the above nonzero numbers assigned to it when compared with activity j, the j has the reciprocal value when compared with i.	

Weight Derivation

After a pairwise comparison matrix is established, it can be proven mathematically that the eigenvector of the comparison matrix is its priority vector, which represents relative weights.

If the weight of each alternative or criterion is known as ω_i, where i is from 1 to n, and n is number of alternatives or criteria, the comparison matrix **A** can be defined as

$$\mathbf{A} = \begin{bmatrix} \omega_1/\omega_1 & \cdots & \omega_1/\omega_n \\ \vdots & \ddots & \vdots \\ \omega_n/\omega_1 & \cdots & \omega_n/\omega_n \end{bmatrix},$$

and the vector of the priorities as $\vec{\omega} = \begin{bmatrix} \omega_1 \\ \vdots \\ \omega_n \end{bmatrix}$, obviously we have:

$$\begin{bmatrix} \omega_1/\omega_1 & \cdots & \omega_1/\omega_n \\ \vdots & \ddots & \vdots \\ \omega_n/\omega_1 & \cdots & \omega_n/\omega_n \end{bmatrix} \begin{bmatrix} \omega_1 \\ \vdots \\ \omega_n \end{bmatrix} = n \begin{bmatrix} \omega_1 \\ \vdots \\ \omega_n \end{bmatrix}.$$

or

$$\mathbf{A}\,\vec{\omega} = n\,\vec{\omega}.$$

Therefore, $\vec{\omega}$ is eigenvector of comparison matrix **A**.

During a screening process, the priority vector $\vec{\omega}$ is unknown. We can obtain $\vec{\omega}$ by calculating eigenvector of the comparison matrix **A**. In other words, during the workshop, after experts agree on the pairwise matrix **A** based on their experience, the eigenvector will be the vector of priority, or weight.

Consistency and Sensitivity Test

In the AHP method, the priorities make sense only if derived from consistent or nearly consistent matrices. Saaty (1977) proposed a consistency index (CI), which is related to the eigenvalue method:

$$CI = \frac{\lambda_{max-n}}{n-1},$$

where λ_{max} is maximal eigenvalue.

The consistency ratio *CR* is given by:

$$CR = \frac{CI}{RI},$$

where *RI* is the random index (the average *CI* of 500 randomly filled matrices) which can be found in Table 2 (Saaty 1977; Ishizaka and Labib 2009; Jandova and Talasova 2013).

If *CR* is less than 10%, then the matrix can be considered as having an acceptable consistency.

Table 2 Random Indices

n	3	4	5	6	7	8	9	10	11
RI (n)	0.58	0.9	1.12	1.24	1.32	1.41	1.45	1.49	1.52

Final Ranking of Alternatives

In the traditional AHP method, a pairwise comparison shall be obtained for criteria to obtain the criteria priority vector (or weight vector), $\mathbf{V}_{\mathbf{critera}}$. Pairwise comparison matrix shall also be established for alternatives against each criterion. This will give a priority vector $\mathbf{V}_j$ for all alternatives in terms of each criteria, where $j = 1, \ldots, m$, and m is the number of criteria.

If we define local priority matrix

$\mathbf{A}_{\mathbf{option}} = [\mathbf{V}_1, \mathbf{V}_2, \ldots, \mathbf{V}_m]$,

the final ranking of the alternatives is obtained by synthesizing weight vector $\mathbf{V}_{\mathbf{critera}}$ with local priority matrix $\mathbf{A}_{\mathbf{option}}$:

Global Ranking = $\mathbf{A}_{\mathbf{option}} \cdot \mathbf{V}_{\mathbf{critera}}$

Note if we have m criteria, and n alternatives, during a workshop experts will be asked to make $m \times (m-1)/2$ times evaluation to fill out the pairwise comparison matrix to obtain $\mathbf{V}_{\mathbf{critera}}$, the eigenvector of criteria comparison matrix, and to evaluate $m \times n \times (n-1)/2$ times to obtain priority matrix $\mathbf{A}_{\mathbf{option}}$. This process could be overwhelming if the number of criteria and the number of alternatives are large. Therefore, in this paper, we use the AHP to obtain the weight vector $\mathbf{V}_{\mathbf{critera}}$, and then further screen it with Even Swaps Analysis.

EVEN SWAPS ANALYSIS

The Even Swaps Analysis (ESA) is a straightforward approach to multiple criteria decision making, proposed by John Hammond in 1998 (Hammond 1998). It requires a decision problem to be well structured and alternatives pool to be relatively small. Therefore, it is applied to further screen the shortlist after we have narrowed down all alternatives from a longlist to a shortlist. The algorithms are summarized in this section.

Creating Consequences Tables

With the screening problem structured, including goal, criteria, and alternatives, the first step for the ESA is to create a so-called "Consequences Table" (Hammond 1998). This is to have a clear picture of all the concepts and to evaluate how well they satisfy each criterion. To make rational swaps between concepts, it is important to use consistent terminology in describing all the consequences for a given criterion. The consequences in the table could be described in quantitative terms, e.g. cost or fatigue life. It could also be in qualitative terms such as "High", "Above Average", "Average", or "Low". In this paper, for simplicity, three ranks are chosen for each criterion: "High", "Medium" and "Low", with color codes as shown in Figure 3.

If an alternative does not meet the basic requirement of a criterion, it will be marked as "Constraint", and will be removed from further consideration.

Figure 3 Color Scale for the Consequences Table

"Dominated" Alternatives

After establishing the consequences tables, it is possible to eliminate one or more of the alternatives. If alternative *A* is better than alternative *B* on some criteria and no worse than *B* on all other criteria, *B* can be eliminated from further consideration. In such cases, *B* is said to be "dominated" by *A* – it has disadvantages without any advantages.

In certain situation, alternative *B* might be better than *A* in one or two criteria, but worse than *B* in most other criteria. If obviously the relatively small advantage in the one or two criteria doesn't out-weight the other disadvantages, we call alternative *B* is "practically dominated" by *A*.

Obviously, fewer the alternatives, fewer are the trade-offs that ultimately need to be made.

Even Swaps

After eliminating "dominated" and "practically dominated" alternatives, if we still have more than one alternative, and each will have some advantages and disadvantage relative to each other, the even-swap method offers a way to even out the advantages and disadvantages systematically until a clear choice is left.

As its name implies, an even swap increases the value of an alternative in terms of one criterion while decreasing its value in terms of another criterion. In other words, instead of passively viewing each concept as a monolithic idea to be either eliminated or selected, we strive to improve a concept in terms of certain criteria and check how this new concept ranks in other criteria. This provides a dynamic screening process and guarantees that the most reasonable concept will be selected during screening.

CASE STUDY

As a case study, this screening process is applied to select a Hull, Mooring, and Riser concept for an offshore floating production facility.

The field is located in a water depth of about 1,500 m. There is no requirement for drilling equipment or hydrocarbon storage on the hull. Some of the extraordinary requirements include longer than typical field service life (30+ years), and surviving storms greater than 1000-year return period.

All concept alternatives are defined in a systematic manner producing easily recognizable patterns. This allows the team to handle a large population of alternatives with a good oversight which are compiled in a Field Options Generator (FOG) matrix. Hundreds of options are generated based on three building blocks, viz., Import Transfer System, Substructure (Hull and Mooring), and Export Transfer System, as shown in Figure 4. Topsides is important for field development, but is not considered to be a differentiator for this hull, mooring, and riser screening study.

	Import Riser		Substructure (Hull&Mooring)		Export Riser
	Import Riser Tool Box		Substructure Tool Box		Export Riser Tool Box
a	SCR Large Diameter	A	Deep Draft Semi	a	SCR Large Diameter
b	Shaped SCR Large Diameter	B	Conventional Draft Semi	b	SCR Small Diameter
c	SLWR Large Diameter	C	TLP	c	SLWR Large Diameter
d	Flexible Riser (14" or 16")	D	Spar	d	Flexible Riser (14" or 16")
e	FSHR	E	Round Shaped FPU	e	FSHR
f	TTR	F	Ship Shaped FPU	f	TTR

Figure 4 Building Blocks of Floater and Riser

All options are grouped into "families" tied to the Substructure blocks based on a unique type of floater. This gives six families of concepts, consisting of Deep Draft Semi, Conventional Draft Semi, TLP, Spar, Round-Shaped FPU, and Ship-Shaped FPU. Invalid, inapplicable or evidently inferior options are dropped. This process yields a total of 20 longlisted options, as shown in Figure 5.

Case No.	Case ID	Descriptions of Import and Export Risers	Hull
1	aAa	SCR Import and Export	Deep Draft Semi
2	aAc	SCR Import, SLWR Export	
3	bAa	Shaped SCR Import, SCR Export	
4	Ac	Shaped SCR Import, SLWR Export	
5	cAa	SLWR Import, SCR Export	
6	cAc	SLWR Import and Export	
7	aAb	SCR Import, SCR small Export	
8	bAb	Shaped SCR Import, SCR small Export	
9	cAb	SLWR Import, SCR small Export	
10	cBc	SLWR Import and Export	Conventional Draft Semi
11	dBc	Flexible Import, SLWR Export	
12	dBd	Flexible Import and Export	
13	eBc	FSHR Import, SLWR Export	
14	eBe	FSHR Import and Export	
15	1C1	SCR Import and Export	TLP
16	aDa	SCR Import and Export	Spar
17	cEc	SLWR Import and Export	Round Shaped FPU
18	eEe	FSHR Import and Export	
19	cFc	SLWR Import and Export	Ship Shaped FPU
20	eFe	FSHR Import and Export	

Figure 5 Longlisted Alternatives

The 20 longlisted options is a reasonable number to be screened with aid of the AHP. Eight criteria are used for the initial screening, viz., Feasibility, Technology Readiness, Flexibility, Fabrication, Transportation and Installation, Cost and Schedule, Operability and HSE Risk. Each concept is ranked from "Weak" to "Excellent" in meeting the criteria. To further differentiate the difference between these two limits, scores between 1 and 10 are also used in the criteria definition, with 1 as "Weak in meeting criteria", 10 as "Excellent in meeting the criteria", and 6-9 as above medium, and 2-5 as below medium.

The pairwise comparison matrix for criteria is shown in Table 3. From the matrix we obtain the criteria weights from the eigenvector of the matrix, as shown in Table 4. Note that the sum of the weight is and should be 1 (Saaty 1977; Ishizaka and Labib 2009).

The pairwise comparison matrix in Table 3 has a maximum eigenvalue of 8.54. The consistency analysis has showed that the consistency ratio (*CR*) of the proposed pairwise comparison matrix is 5.5%, indicating an acceptable level of consistency (Saaty 1977; Ishizaka and Labib 2009).

Table 3 Pairwise Comparison Matrix for Criteria

	Feasibility	Technology Readiness	Flexibility	Fabrication	T&I	Cost & Schedule	Operability	HSE Risk
Feasibility	1	3	4	7	7	2	3	1
Technology Readiness	1/3	1	0.5	2	2	3	2	1/3
Flexibility	1/4	2	1	2	2	2	3	1/2
Fabrication	1/7	1/2	1/2	1	1/2	1/2	1/3	1/5
T&I	1/7	1/2	1/2	2	1	1/2	1/3	1/5
Cost & Schedule	1/2	1/3	1/2	2	2	1	2	1/2
Operability	1/3	1/2	0.33	3	3	1/2	1	1/3
HSE Risk	1	3	2	5	5	2	3	1

Table 4 Criterion Weights

Criteria	Weight
Feasibility	0.28
Technology Readiness	0.11
Flexibility	0.13
Fabrication	0.04
T&I	0.05
Cost & Schedule	0.09
Operability	0.08
HSE Risk	0.23

Based on the weights of the criteria, the ranking for the 20 "longlisted" concepts are conducted in a workshop format. The deep draft Semi with shaped SCR and SLWR are selected as the reference cases, and all the other 19 cases are compared against it. This is to improve the screening efficiency and to maximize screening consistency. After obtained all the scores of the 20 cases, the top two of Deep Draft Semi, the top two of Conventional Draft Semi, and the Spar are shortlisted.

To compare the shortlisted options, a consequences table is established during workshop, as shown in Table 5. Based on the consequences table, conventional draft Semi with flexible riser concept is eliminated from further consideration as it is obviously dominated by the deep draft Semi with shaped SCR and SLWR, as shown in Table 6. Similarly, the deep or conventional draft Semi with SLWR concepts are dropped as they are dominated by the deep draft Semi with shaped SCR and SLWR.

Table 5 Consequences Table

Sketch	Discription	Feasibility	Technology Readiness	Flexibility	Fabrication	T&I	Cost & Schedule	Operability	HSE Risk
	Hull: Deep Draft Semi Import Riser: Shaped SCR Large Dia. Export Riser: SLWR Large Dia.								
	Hull: Deep Draft Semi Import Riser: SLWR Large Dia. Export Riser: SLWR Large Dia.								
	Hull: Conventional Draft Semi Import Riser: SLWR Large Dia. Export Riser: SLWR Large Dia.								
	Hull: Conventional Draft Semi Import Riser: Flexible Riser (14" or 16") Export Riser: Flexible Riser (14" or 16")								
	Hull: Spar Import Riser: SCR Large Dia. Export Riser: SCR Large Dia.								

Merit Scale: High | Medium | Low

Table 6 Eliminating Dominated Options

Sketch	Discription	Feasibility	Technology Readiness	Flexibility	Fabrication	T&I	Cost & Schedule	Operability	HSE Risk	
	Hull: Deep Draft Semi Import Riser: Shaped SCR Large Dia. Export Riser: SLWR Large Dia.	✓	✓	✓	✓	✓	✓		✓	← Dominant
	Hull: Deep Draft Semi Import Riser: SLWR Large Dia. Export Riser: SLWR Large Dia.									
	Hull: Conventional Draft Semi Import Riser: SLWR Large Dia. Export Riser: SLWR Large Dia.									
	Hull: Conventional Draft Semi Import Riser: Flexible Riser (14" or 16") Export Riser: Flexible Riser (14" or 16")	✗	✗	✗	✗	✗	✗		✗	← Dominated
	Hull: Spar Import Riser: SCR Large Dia. Export Riser: SCR Large Dia.									

When comparing a Spar to the Semi, the Spar is worse than or even to the Semi in all criteria except for Technology Readiness. As both hull concepts have pros and cons, instead of passively viewing each concept to be eliminated or selected, we improve the concepts by trading off those advantages and disadvantages using Even Swaps Analysis. The ESA provides an opportunity to focus on balancing pros and cons by changing the alternatives to meet certain criteria. During the workshop, it was agreed that HSE is a priority, and the experts are asked the question "What can we do to change the Spar concept, such that the Spar will have similar HSE Risk as the Semi?" This is documented in the Consequences Tables as shown in Table 7. To improve the HSE Risk criterion performance, it is agreed that Spar footprint needs to be increased, and this gives an alternative case. However, such an increased footprint leads to difficulties in the layout. Furthermore, no yard has proven experience to build a Spar of such size, and there is no transportation and installation solution for such a Spar. Therefore, Feasibility, Flexibility, Fabrication, T&I all become "constraints".

From the ESA, it is concluded that the Semi concepts are more favorable compared with Spar. Therefore, Spar concepts are removed from further consideration.

Eventually, the concept of Deep Draft Semi with SLWR is recommended as the solution for field development.

Table 7 Even Swaps Analysis between Spar and Semi

Case ID	Sketch	Discription	Feasibility	Technology Readiness	Flexibility	Fabrication	T&I	Cost & Schedule	Operability	HSE Risk
bAc	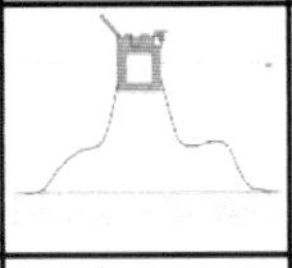	Hull: Deep Draft Semi Import Riser: Shaped SCR Large Dia. Export Riser: SLWR Large Dia.								
cAc	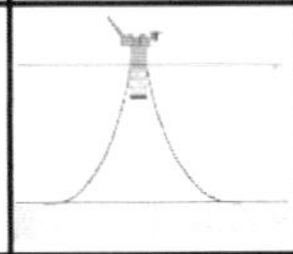	Hull: Spar Import Riser: SCR Large Dia. Export Riser: SCR Large Dia.								
aAa_ Alternative	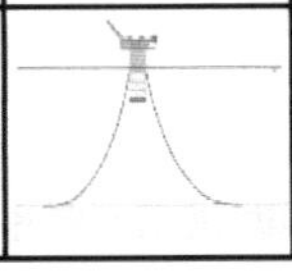	Hull: Spar (Larger footprint) Import Riser: SCR Large Dia. Export Riser: SCR Large Dia.	Constaint		Constaint	Constaint	Constaint			

CONCLUSIONS

An improved screening process is presented in this paper and its efficacy is demonstrated by the means of a case study. All possible alternatives are defined in a systematic manner by combining individual building blocks. Dropping infeasible or impractical alternatives yields a reasonable number of longlist alternatives. This longlist is screened through ranking against a reference case using criteria weights derived by the AHP method to yield a shortlist of alternatives. This shortlist is further evaluated using ESA, which provides a dynamic screening process to further explore the possibility of improving alternatives by making tradeoffs among criteria, and to arrive at a final concept selection for the intended field development.

For offshore applications in deep water with increased technical, financial and execution risks, rigorous concept screening is critical. This paper presents a robust and valuable screening methodology that has broad application in offshore and other industries.

ACKNOWLEDGEMENTS

The authors want to express their appreciation to Aker Solutions for supporting the study work and allowing the publication of this paper.

REFERENCES

Hammond, JS, Keeney, RL, Raiffa, H., "Even Swaps: A Rational Method for Making Trade-offs", Harvard Business Review, 76(2), pp. 137-150, 1998.

Ishizaka, A., Labib, A., "Analytic Hierarchy Process and Expert Choice: Benefits and Limitations", OR Insight, 22(4), pp. 201-220, 2009.

Jandova, V., Talasova, J., "Week Consistency: A New Approach to Consistency in the Saaty's Analytic

Hierarchy Process", Acta Universitatis Palackianae Olomucensis. Mathematica, 52(2): pp 71-83, 2013.

Linder, S. F., "Field Option Development Applications of the Analytic Hierarchy Process", Proceedings of the 19th Offshore Symposium, Texas Section of the SNAME, pp. 1-12, Houston, TX February 2014.

Linder, S.F., Fales, R, "Application of a Field Development Decision-Making Methodology", Deep Offshore Technology International Conference, Aberdeen, Scotland. 2014.

Saaty, T., "A Scaling Method for Priorities in Hierarchical Structures", Journal of Mathematical Psychology 15(3), pp. 234-281, 1977.

ECONOMICALLY VIABLE APPLICATIONS OF CFD IN SEAKEEPING

NICHOLAS A. BARCZAK
NAVAL ARCHITECT, ALAN C. MCCLURE ASSOCIATES

JEFFEREY D. REIFSNYDER
NAVAL ARCHITECT, ALAN C. MCCLURE ASSOCIATES

ROGER C. HATFIELD
PRESIDENT, GOLD COAST YACHTS

Proceedings of the 21st Offshore Symposium, February 2016, Houston, Texas
Texas Section of the Society of Naval Architects and Marine Engineers

ABSTRACT

Introduction

Alan C McClure Associates (ACMA) performed a seakeeping analysis on a Tandem Catamaran using computational fluid dynamics (CFD). The challenge was how to perform a seakeeping analysis in a commercially viable fashion. ACMA utilized an unsteady RANS method to model regular waves of various wave encounter periods interacting with the vessel. ACMA then post processed results to create response amplitude operators (RAO) from the simulations.

This paper reviews the strategy for optimization of the mesh refinements within the wave zone, the timestep, and simulation run time. The results of this optimization yielded a careful balance between computational cost and accuracy. The CFD analysis was supported by validation and error analysis, which the paper summarizes. Results also include a computational budget for seakeeping analysis, visualizations of volume fraction dispersion, and plots of response amplitude operators. All results include thorough discussion on successes and challenges with maintaining simulation quality within the project budget.

This work revealed the current economic limitations of CFD. At this point, CFD is not yet ready for unrestricted application to seakeeping. Limited scope applications are possible, but generalized seakeeping applications require extensive project budgets.

Keywords: SNAME, Offshore, Symposium, CFD, Seakeeping, Cost, Economic, RANS, Mesh

INTRODUCTION

The advancement of new technology endures two stages of acceptance before full application in the commercial world. First the concept is proven and established, usually in an academic setting. Equally important is the second trial, when commercial industry determines if the technology is economically viable. This is the current cusp for seakeeping analysis using Reynolds Averaged Navier Stokes (RANS) Computational Fluid Dynamics (CFD). Alan C. McClure Associates (ACMA) had the opportunity to perform a RANS CFD analysis on a unique Tandem Catamaran. (Figure 1) ACMA gained extensive insight into limitations of CFD for seakeeping analysis.

Figure 1 Concept Photo of Tandem Catamaran (Gold Coast Yachts, March, 2015)

The authors present this CFD analysis within the context and concerns of a commercial business. It is already demonstrated that seakeeping analysis within CFD is *technically possible.* (Guo, Steen, & Deng, March 2012) But for the next step in widespread acceptance it must become *economically feasible*. This paper reviews the CFD seakeeping project: the challenges, results, and modeling compromises necessary to complete the analysis within a commercial budget. This addresses how to keep a CFD seakeeping analysis economically feasible.

MOTIVATION FOR CFD SEAKEEPING

To prepare for a seakeeping RANS CFD analysis, ACMA engineers first critically examined the alternatives. Was model testing more appropriate? Or perhaps traditional boundary element methods? Engineers debated if the complexity of RANS CFD was truly necessary.

ACMA selected RANS CFD methods by process of elimination. The project did not allow sufficient budget or schedule for a battery of towing tank tests to determine seakeeping response. The unusual hull structure posed significant challenges for accurate modeling with traditional boundary element methods. The close proximity of the hulls raised concerns of strong wave interactions between hulls. RANS CFD exhibited its own challenges, but CFD presented the least technical and logistical hurdles at project inception. All other methods were excluded due to schedule and budget, or due to inescapable modeling restrictions. Only CFD offered the flexibility for this unique situation.

After deciding on RANS CFD, ACMA also debated modeling an irregular seastate or a series of regular waves. The ultimate goal was to apply the CFD results to multiple wave spectra, requiring response amplitude operators (RAOs). ACMA conducted a test case to determine the requirements for frequency decomposition of an irregular seastate. The decomposition required a Fourier transform from a CFD time history to capture both amplitudes and phase angles for numerous wave periods. ACMA's test quickly revealed little probability of success with this method. An accurate Fourier transform required a simulated time history of several minutes duration. Mere seconds of simulation time already required large computational costs. Instead, ACMA chose to simulate the vessel response with a series of regular waves at various wave periods. The regular wave approach allowed direct evaluation of each individual wave frequency. Isolating the wave frequencies ensured accuracy of results for calculation of body response amplitude operators (RAO) and phase angles.

METHODOLOGY

The seakeeping analysis required a multi-step approach. ACMA engineers completed the following steps to analyze the seakeeping response.

1. Simulate regular waves for each of the wave periods and record body response motions
   a. Setup simulations
   b. Verify simulation methodology
   c. Establish mesh independence
   d. Establish time independence
   e. Perform error analysis
2. Post process time history of body responses
3. Determine amplitude of steady state body responses
4. Calculate response amplitude operator (RAO) based on input wave amplitude
5. Apply specified wave spectrum to RAO and determine body response for various probabilities of exceedance

Only the first step is addressed in this paper. Steps two through five were all post processing and very similar to post processing experimental results from a towing tank. The fundamental challenge behind this project was establishing feasible CFD simulations that produced accurate results for post processing. Table 1 summarizes general simulation parameters. Ocean waves presented onerous requirements for refined simulation meshes and extensive time domains. This quickly resulted in exorbitant simulation costs. To combat the problem of large simulation costs, ACMA engineers incorporated extensive simulation optimizations, which shall be discussed in the following sections.

Table 1 Simulation Summary

Item	Model	Details
Vessel Dimensions	Length Beam Depth	30.1 m 5.1 m 4.0 m
Time Variation	Implicit-Unsteady	1st order interpolation
Turbulence Model	None	Laminar flow assumed
Momentum Equations	RANS Momentum	2nd order interpolation
Free Surface Method	Volume of Fluid (VOF)	Two phases: water and air
Total Simulation Time	Varies	Typically set as 2 s of initialization time, plus 3.5 – 4.5 wave periods
Average Body Courant Number	Less than 0.20	Typically 0.05 – 0.10. Required for stable simulation convergence.
Wave Directions	3	Beam, bow-quarter, bow
Wave Periods	11 – 12 per direction	34 production runs total

CFD BOUNDARY CONDITIONS

Figure 2 shows the domain boundaries and applied boundary conditions. The shape of the domain boundaries changed radically with wave direction. This was a key optimization. The simulation cell count depended on the domain size at the waterline. ACMA engineers utilized elliptical shapes to minimize the domain volume away from the waterline. The domain size also shrank as wavelength decreased. Each simulation modeled a series of regular waves at a single period, independent of other wave periods. At the higher wave periods, wave length decreased and the larger domain was not necessary.

Reducing the domain size controlled computational costs, but presented a danger to accuracy. Smaller domain volumes may over-constrain the simulation. But ideal large domain volumes drastically increased the mesh size. Shrinking the boundaries also reduced the time before waves reflected off the domain boundary and disturbed the response results. To mitigate this danger, ACMA engineers utilized a damping boundary condition downstream of the body. Selection of domain size required a careful balance between simulation quality and simulation costs.

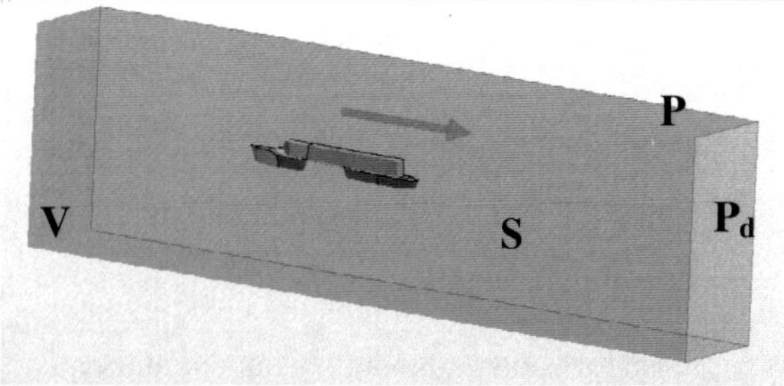

Bow Waves
Wave Period: 7.24 s

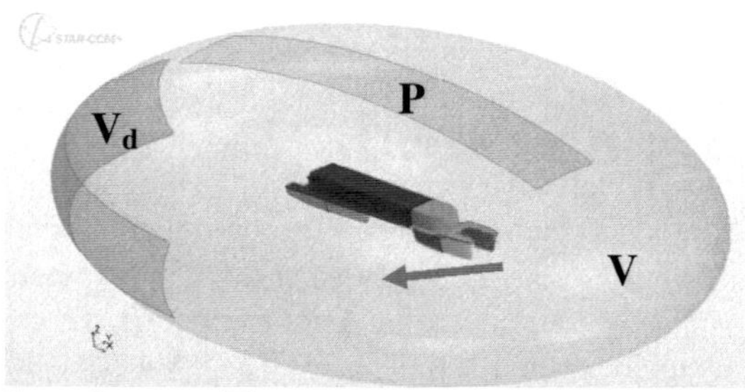

Bow Quarter Waves
Wave Period: 6.52 s

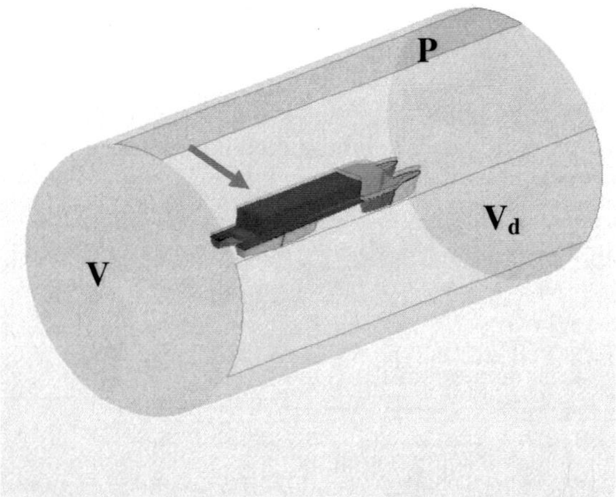

Beam Waves
Wave Period: 3.11 s

Legend:
V: Velocity inlet, wave specified
P: Pressure outlet
d: Wave damping applied to boundary
S: Symmetry plane

Arrow shows direction of wave propagation.

Half-model geometry used in bow wave case

Figure 2 Domain Boundaries

CFD MESH REFINEMENTS

The mesh refinements were the second major optimization for CFD simulations. The CFD simulation required three major mesh refinements.

1. Surface refinement on the body, depending on body curvature and proximity of surfaces
2. Anisotropic refinement of volume mesh on the waterline. The extent of refinement depended on the wave height.
3. Isotropic refinement of a volume around the body with a vertical extent of twice the wave height, minimum. This refinement captures the interactions between the wave and body.

The wave refinements were the critical element for simulation feasibility. Simulation quality demanded a very small cell size in the region of the wave. Cell size was typically 5% - 10% of wave amplitude. This refinement region must extend vertically for the full wave height and horizontally for the entire extent of the simulation domain. The majority of the simulation mesh will be concentrated within this refinement. Accurate wave propagation required this refinement of small cells to extend across the entire domain. Judicious optimization of this region was critical.

The wave refinement could be optimized with anisotropic refinements. Figure 3 shows an example of the simulation mesh, highlighting the wave refinement. Horizontal and vertical cell sizes were nearly uniform in the direction of wave propagation, but the cells were less refined along the width of the wave crest. This anisotropic refinement reduced the cell count when applied over the entire simulation domain.

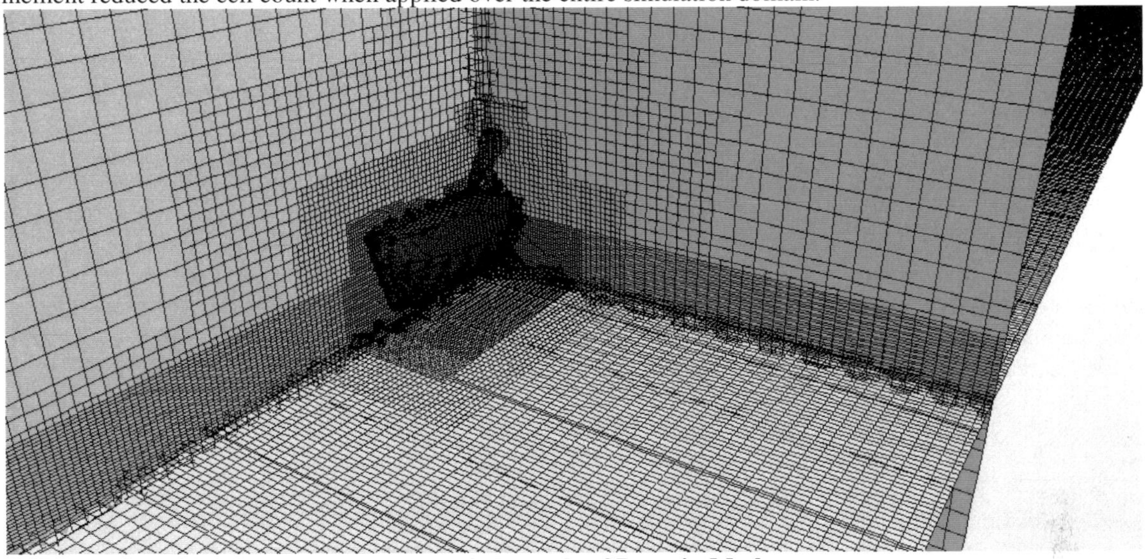

Figure 3 Example of Domain Mesh
Bow Waves, Showing Cut Planes Through Domain Volume

The anisotropic refinement was not suitable for capturing interactions between the wave and the body. This required isotropic mesh refinement in the region near the body. The isotropic refinement must be even smaller than the wave refinement to fully capture all detailed wave interactions.

The final element of mesh optimization was adjustment with the wave amplitude. Regular ocean waves can only sustain a maximum wave slope before collapsing. The authors recommend the steepest wave slope should be no more than 1:10. To avoid collapsing waves, the input wave amplitude must be reduced as shorter wave periods were modeled. The reduced wave amplitude required a smaller cell size in the wave refinement. This can be offset by a smaller vertical extent of wave refinement region. The height of the wave mesh refinement should remain proportional to the wave height. The cell size and refinement extents must be adjusted for every wave period to control simulation costs.

DFBI MESH DISPLACEMENT

The seakeeping simulations required the body to move in response to the fluid forces. This in turn required mesh displacement through dynamic fluid body interaction. (DFBI) There were multiple ways to implement DFBI mesh deformation in the simulation. ACMA considered local mesh morphing and single unit mesh displacement.

In local mesh morphing, the mesh nodes within a small region of the domain shift. The cells shift and change relative to each other. All nodes outside that region remain fixed. This has an advantage of improved simulation quality. Simulation quality remained uncompromised outside the region of local mesh morphing. There were definite advantages given the large domain size. It was critical that the waves propagate from the domain boundaries to the body without excess distortion. Morphing a local section of the mesh could achieve this.

Unfortunately, local mesh morphing was not the feasible solution. Mesh morphing required solving an additional differential equation to update positions of the mesh nodes. Due to the tight coupling between vessel motion and fluid forces, this morphing must be completed at every iteration. The end result was that simulation time increased drastically. Computing time for a single timestep escalated from seconds to minutes. Mesh morphing was simply infeasible for the project budget. The only feasible option was single unit motion of the domain.

With single unit displacement, all mesh nodes in the domain moved in a rigid body manner according to translation and rotation of the vessel. Each cell maintains the same relative configuration and shape. This method provided the fastest solver performance and greatly reduced simulation costs. The downside was that rotating the mesh resulted in large nodal displacement when far away from the center of rotation. This large displacement adversely affected simulation quality at these large distances. ACMA engineers validated this method and discovered that under specific circumstances the method could produce accurate results. Part of this validation included a detailed error analysis of the simulation.

ERROR ANALYSIS

As part of the quality control for this project, ACMA performed a formal error analysis on the simulation. Table 2 summarizes all error sources that ACMA identified and quantified. In this paper, error is a technical term referring to deviation of the simulated solution from a true correct solution. The true solution is unknown, but ACMA proceeded by assuming some level of error inherent in the simulation. ACMA tested several different simulation settings in a controlled manner and examined relative change in simulation results. By comparing these different settings, ACMA estimated the total simulation error, which originated from various sources. Different techniques were employed to estimate each source of simulation error.

Table 2 Error Analysis

Error Source	Value	Remarks
Mathematical Model Error	0.0%	Zero, based on validation comparison. See next section of paper.
Convergence Error	0.00055%	Based on converged simulation residuals
Discretization Error	4.1%	
Timestep Error	8.8%	
Total Error	**13.3%**	**Product of all above numbers, taking each error as an added safety factor (e.g. 1.041 x 1.088 . . .)**

The Convergence Error was derived from monitored simulation residuals. During simulation execution, ACMA engineers monitored the heave force on the vessel for each iteration. This monitored value was converted into a residual. Engineers recorded the average residual value from a representative simulation, which yielded a conservative estimate for the Convergence Error.

To determine the discretization error, ACMA performed a mesh independence study. ACMA ran a simulation test case for several different mesh sizes and compared the change in results. ACMA checked variation based on four mesh qualities.

1. Number of elements in mesh
2. Average element size in mesh
3. Order of convergence of mesh
4. Grid convergence index (GCI)

The order of convergence and grid convergence index were based on an enhanced method of the Richardson extrapolation. (Slater, 2008) (Celik, 2008) (Roache, 1998)

GCI was the principal method of error estimate, and ACMA calculated this for four different mesh sizes. Figure 4 shows the GCI plotted against element size. Element size is shown in non-dimensional format relative to principal simulation length scales. The GCI plot showed a clear and strong trend of decreasing error. ACMA engineers conclusively estimated a discretization error of only 4.1% for this simulation.

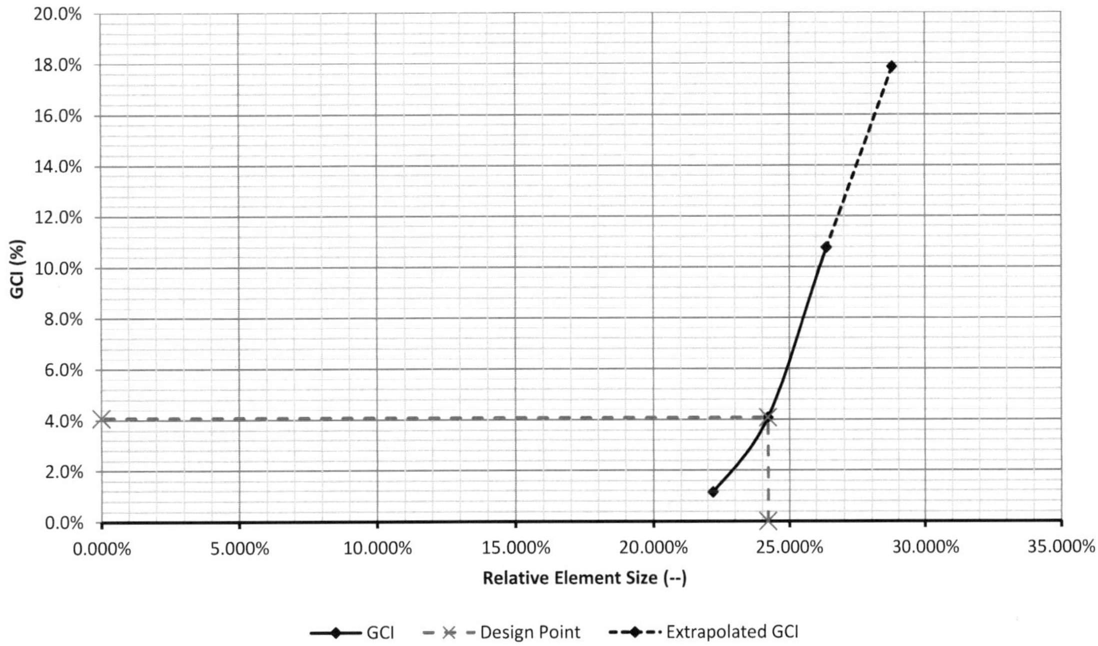

Figure 4 Mesh Independence Results – Grid Convergence Index

ACMA further checked simulation accuracy within the time domain. The seakeeping problem required discretization of the time continuum and discretely stepping through time. This introduced another element for simulation error. ACMA performed a second mesh independence study on the simulation timestep, using methods similar to the previous mesh independence study. This resulted in a time discretization error of 8.8%, based on GCI. ACMA intentionally accepted a larger timestep error because it permitted a larger simulation timestep. This was critical to maintaining reasonable computational budgets for all the production runs considered.

Quantification of all major error sources allowed ACMA high confidence in the accuracy of the simulation. This defined the bounds of simulation accuracy and ensured quality of the final results. However, even the most thorough error analysis remains an estimate based on simulation comparisons. ACMA coupled this with a validation effort to confirm the accuracy of the error analysis and resulting simulations.

VALIDATION

ACMA engineers validated the CFD simulation by comparing simulation results to a simple 2D seakeeping experiment. The validation experiment was a body of triangular cross section exposed to regular waves. (Fonseca, Guedes Soares, & Incecik, 1997) The body was ballasted to float in the inverted position. (Figure 5) The floating body responded with motions in heave, roll, and sway. Figure 5 diagrams the experimental setup.

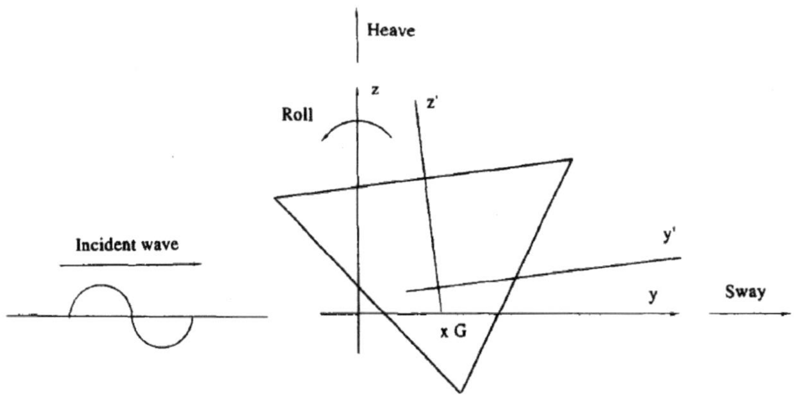

Figure 5 Diagram of Experimental Setup (Fonseca, Guedes Soares, & Incecik, 1997)

This test case provided ideal experimental comparison for several reasons.

1. The experiment allowed a 2D CFD simulation for comparison. The 2D simulation was computationally inexpensive and allowed rapid evaluation to identify optimal simulation settings.
2. The experiment specifically selected a triangular cross section because it generated nonlinear responses to waves.
3. The experiment tested responses in regular waves, which was the same scenario simulated for the Tandem Catamaran.

ACMA worked to reproduce the experimental results with CFD simulations, and engineers encountered several challenges. Accurate simulation required extremely small cell sizes and small time steps. If these same modeling guidelines were applied to the 3D case of the Tandem Catamaran, it would result in extremely large computing costs. However, ACMA engineers discovered that accurate simulation results were possible if the body motions were limited to just heave motion. More importantly the cell size and timestep requirements were far less demanding after imposing the limit to heave motion. With this constraint, ACMA engineers achieved reasonably accurate comparison to experimental results. (Figure 6)

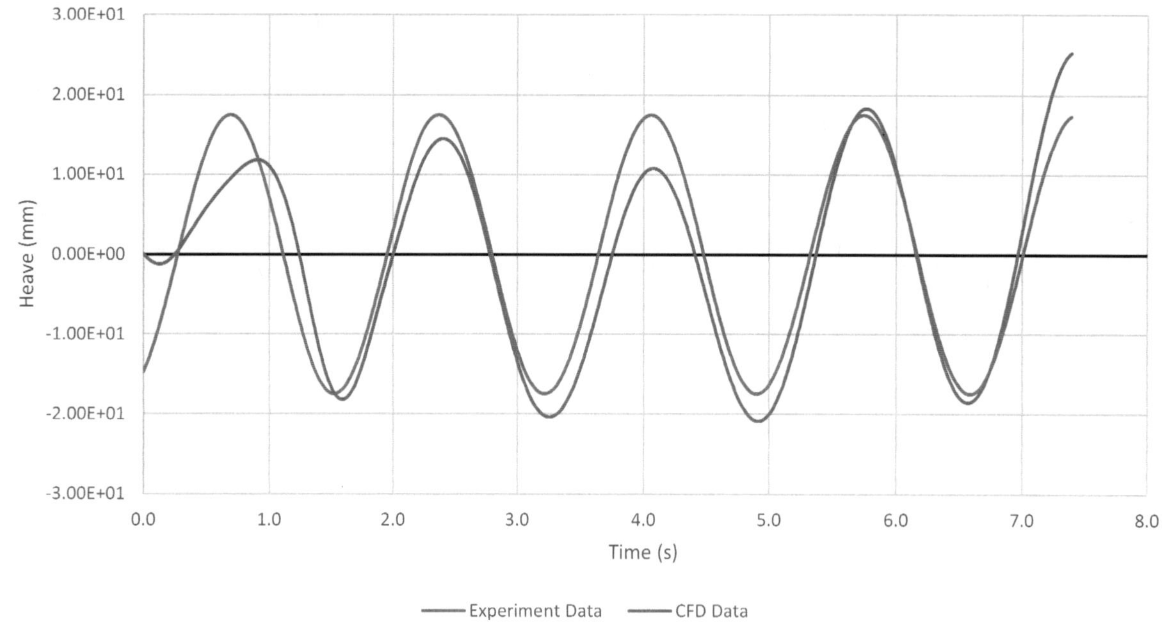

Figure 6 Results of Experimental Comparison

For better presentation in this paper, the authors adjusted the phase angle of the experimental data to synchronize with the CFD data. In any case, ACMA's main interest consisted of steady state response in seakeeping, and the only important items were the average response amplitude and average response period. Table 3 demonstrates that these average values compare exceptionally well.

Table 3 Validation Data Comparison

Item	Experiment Value	CFD Value	Error
Average Response Amplitude	17.47 mm	17.04 mm	2.42%
Average Response Period	1.68 s	1.62 s	3.90%

The estimated discretization errors exceeded the actual errors from experimental data. The discretization errors in Table 2 also applied to this validation study. This showed that discretization errors were a conservative estimate and the accuracy of the CFD simulation was reliably bounded within the limits of the error analysis. Hence, ACMA recorded zero mathematical model error, due to the conservative estimate from other error sources.

The constraint of heave-only motion greatly limited the wave cases that were comparable within the experiment. But the special case of the Tandem Catamaran still allowed a full range of simulation possibilities. The Tandem Catamaran was very stable and stiff in the water. It had low mass and low moment of inertia. The simulated wave environment was not large compared to the vessel size. All elements suggested that the Tandem Catamaran would exhibit low rotational motion, effecting a heave dominated response. Given these indications, ACMA engineers pressed forward with simulation of the Tandem Catamaran. The vessel was allowed three degrees of freedom: heave, pitch, and roll. However, ACMA engineers anticipated that pitch and roll should be very small.

The validation effort provided clear modeling guidelines for a feasible simulation and confirmed simulation accuracy within the limits of estimated errors. Next ACMA focused on review and analysis of simulation results.

RESULTS: COMPUTATIONAL BUDGET

The analysis delivered a unique result in the form of a computational budget. This tool had significant implications for planning future CFD seakeeping analyses. The first pragmatic question for a CFD seakeeping project is an accurate estimate of the computing costs. The computational budget provided a tool to predict simulation run costs based on input wave parameters.

ACMA discovered a definite correlation between wave period and CFD computing costs. The computing effort can be expressed as core-hours: hours of computation multiplied by the number of computing cores used. ACMA determined a response amplitude operator (RAO) for core-hours based on the simulations. (Figure 7) This RAO was specific to the given vessel size, but reasonable parametric extrapolations can adapt the RAO curve to any vessel size. The RAO curve does include all optimizations for mesh refinements and domain sizing, both of which changed with wave period. Utilizing Figure 7 ACMA engineers can estimate the computational cost to simulate a given wave environment, and the curve clearly illustrates which wave periods should be avoided to maintain feasible simulation costs.

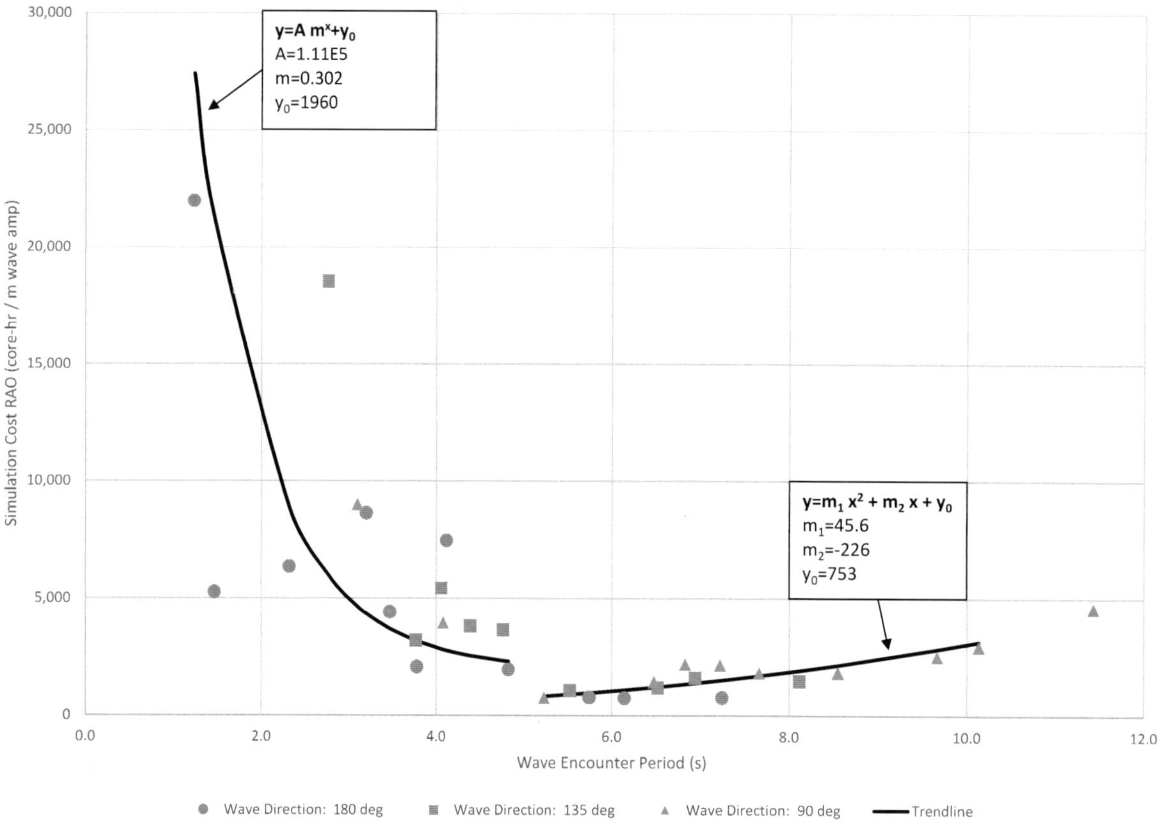

Figure 7 Cost of Individual CFD Simulations

RESULTS: FREE SURFACE DIFFUSION

Free surface diffusion was a major problem for simulation quality. Figure 8 shows the effects of free surface diffusion. At the initial timestep, the free surface was crisply defined by the initial conditions. As the simulation stepped through time, the motion of the mesh resulted in vertical diffusion of the volume fraction across the waterline cells. The only solution to this problem was to decrease cell size and decrease timestep to maintain small Courant numbers, typically below a Courant Number of 0.1.

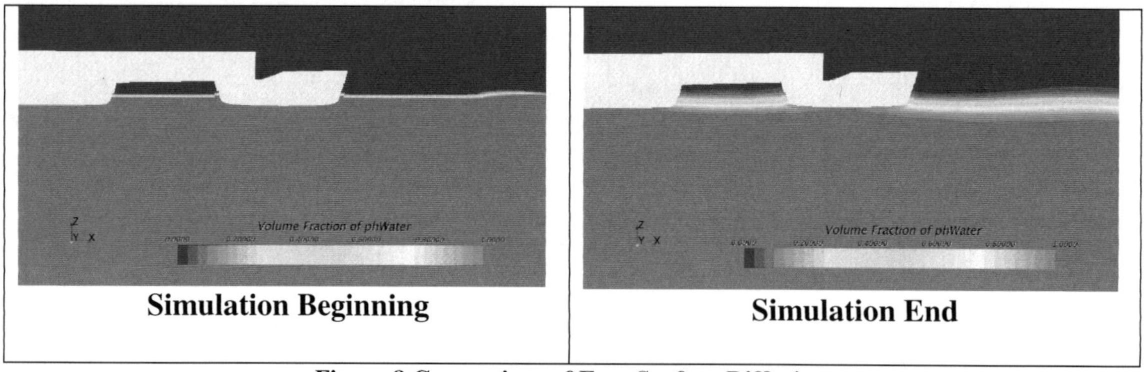

Figure 8 Comparison of Free Surface Diffusion
Bow Waves

The problem arose with computational costs at short wave periods. As wave periods became shorter, mesh density and computational costs increased drastically. A short wave period required a small wave amplitude, leading to small cell sizes and smaller timesteps. Too small for a feasible simulation. ACMA engineers decided to select a coarser mesh size at these short wave periods.

At these short wave periods, coarser cell sizes were necessary to achieve any simulation results within a normal commercial budget and schedule. This greatly reduced computing time. But coarsening the mesh adversely affected simulation quality. ACMA engineers carefully considered the consequences of a coarser mesh. The cell size only increased for the waterline refinement region. All other regions maintained the same cell sizing criteria. Remember that cell size was relative to wave amplitude. For the short wave periods considered, waterline cell size was already an order of magnitude smaller than long wave periods.

Engineers could not estimate how much mesh coarsening would increase free surface diffusion and the resulting effect on vessel forces. The mesh independence study established that a coarser mesh had a predictable change on the simulation answer. Highly exaggerated answers were not likely for small increases in waterline cell size. Thankfully, the intended application appeared to mitigate the risks from any potential inaccuracies. These short wave periods carried small risk to the total vessel response. These wave periods were not expected to dominate the vessel response, given the minimal wave energy available at these short periods. Engineers accepted limited mesh coarsening as a reasonable modeling compromise.

Unfortunately, the side effects from a coarser mesh were far larger than initially anticipated. The larger cell size lead to excessive free surface diffusion. The body forces still matched the body displacement, but the pressures were diffused over a much larger surface area. This effect was most apparent in cases leading to large angular motion. The angular motion generated large vertical displacements in the mesh far away from the body, leading to extreme diffusion. (Figure 9)

The simulation results demonstrated that mesh coarsening should be assiduously avoided. It is preferable to exclude a wave period before considering mesh coarsening. Also, numerical diffusion was far more sensitive to mesh displacements resulting from angular body motions. Strictly linear body motions showed minimal numerical diffusion. To maintain a crisp free surface, avoid short wave periods and angular body motion.

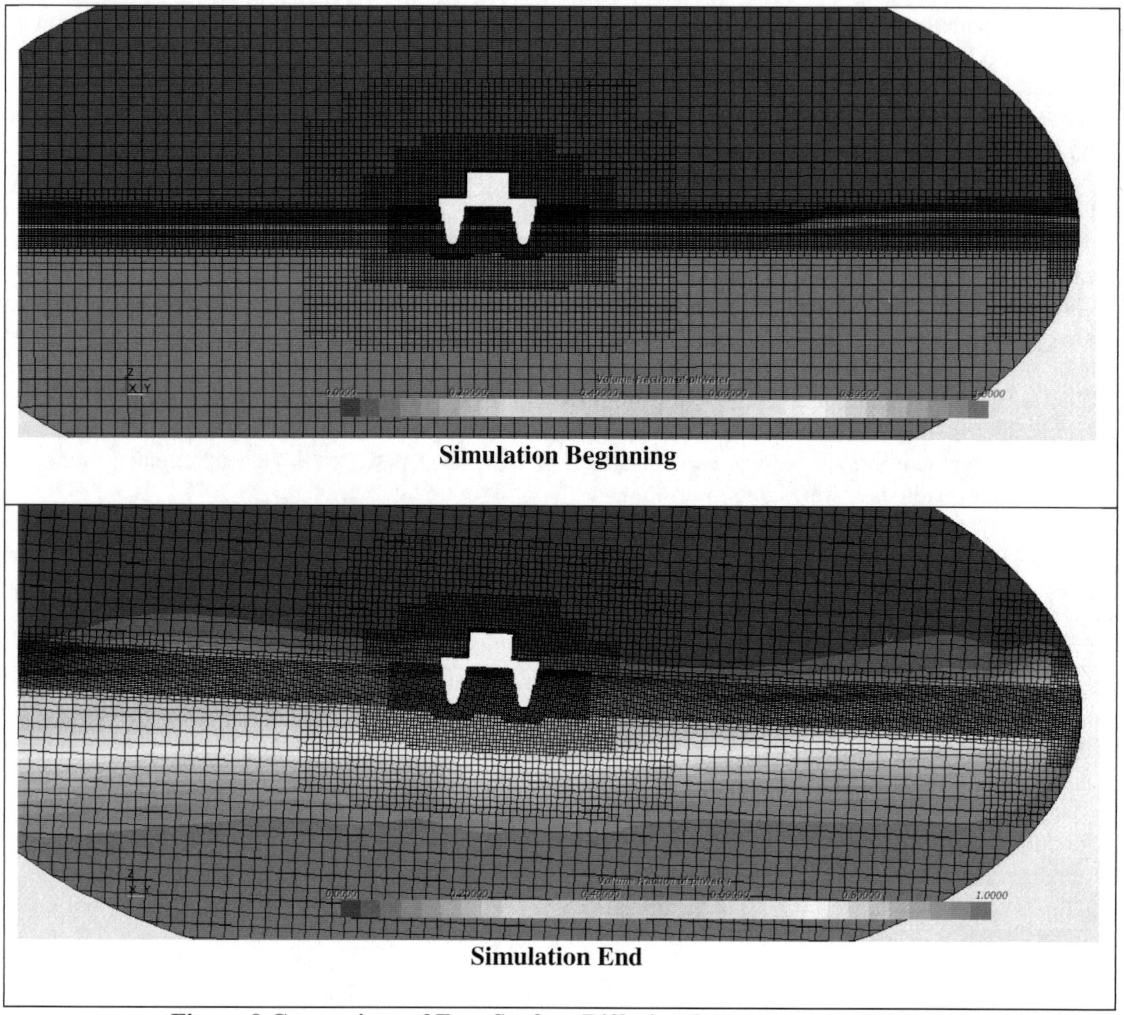

**Figure 9 Comparison of Free Surface Diffusion Due to Angular Motion
Beam Waves**

RESULTS: RESPONSE AMPLITUDE OPERATORS

All evidence indicated that the selected CFD seakeeping method required avoidance of short wave periods. Within this work, the short wave periods introduced numerous problems. The excessive free surface diffusion presented as excessive RAO values near the short wave periods.

Analysis of the response amplitude operators (RAO's) for the Tandem Catamaran illuminated a corollary: avoid vessels with short natural periods. This vessel showed natural periods of less than 3.0 s. ACMA did not expect any concerns regarding resonant conditions, due to the lack of available wave energy at these periods. But the RAO results from CFD exceeded this pragmatic interpretation.

Figure 10, Figure 11 and Figure 12 show the heave, pitch, and roll RAO curves for the vessel, respectively. The wave periods were nondimensionalized as multiples of the vessel natural period in heave. The figures clearly show a resonant response near the heave natural period. The results were highly suspect at these resonant responses.

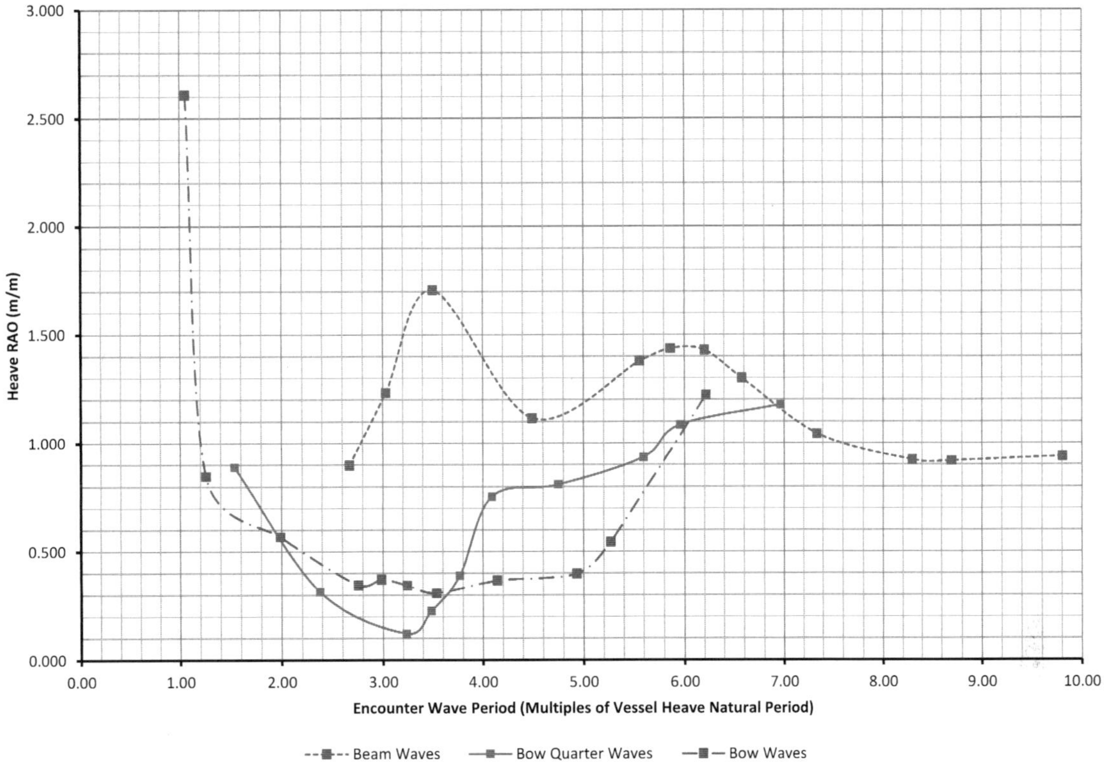

Figure 10 Heave Response Amplitude Operator

The CFD predictions for heave RAO mostly showed typical behaviors. At very long wave periods, the vessel trended to a heave RAO of 1.0, matching the wave amplitude. This was most obvious with the beam waves. The effect was less obvious on the bow and bow-quarter waves because the encountered wave period did not reach as low for the range considered.

The vessel response also varied with wave direction as expected. The vessel showed RAO peaks for beam waves because the hulls presented a large perpendicular surface area, leading to larger responses. Comparatively, the bow and bow-quarter cases showed a reduction in RAO for the shorter wave periods. The skinny hulls of the Tandem Catamaran were largely unaffected by waves at these short periods. The shorter wave periods produced a steady average force over the length of the hull.

The final notable feature was a drastic peak in the RAO curves near the heave natural period. One expects some peak here. The CFD accurately captured the resonant condition. However, the excessive peak of the RAO indicated that some non-linear physics were not captured. ACMA engineers concluded that at resonant conditions, laminar flow assumptions were not valid. Turbulent flow physics should be utilized at resonance, to capture increased fluid viscosity.

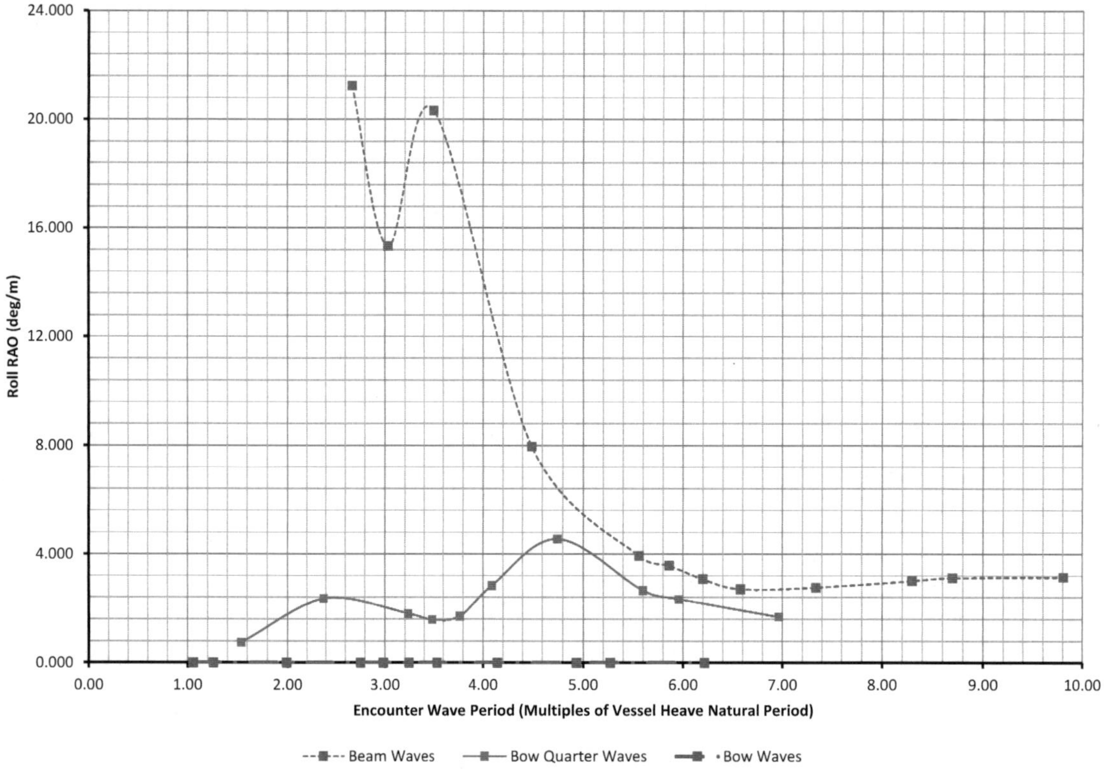

Figure 11 Roll Response Amplitude Operator

The roll RAO demonstrated that the vessel had a very stiff roll response. At shorter wave periods, roll was exaggerated. This large roll RAO raised concerns. Previous validation efforts assumed negligible roll angles, which was necessary to minimize free surface diffusion. The large RAO violated this assumption. Based on the peak RAO for roll in beam waves, ACMA engineers concluded that the location of the peak was correct. However, the magnitude of the RAO peak was likely inaccurate. This error rose from two primary sources: free surface diffusion, and insufficient roll damping due to laminar flow assumptions.

The bow-quarter waves also showed an RAO peak in roll, though greatly reduced from beam waves. The coupling with pitching motion alleviated the vessel motions. The reduced roll motions also reduced inaccuracies from CFD errors. The large difference between roll RAO for beam and bow-quarter waves underscores the importance of minimal angular motion for accurate CFD seakeeping within a commercial budget and general application. Allowing accurate angular motion requires a substantial increase in budget or careful modeling specific to each individual scenario.

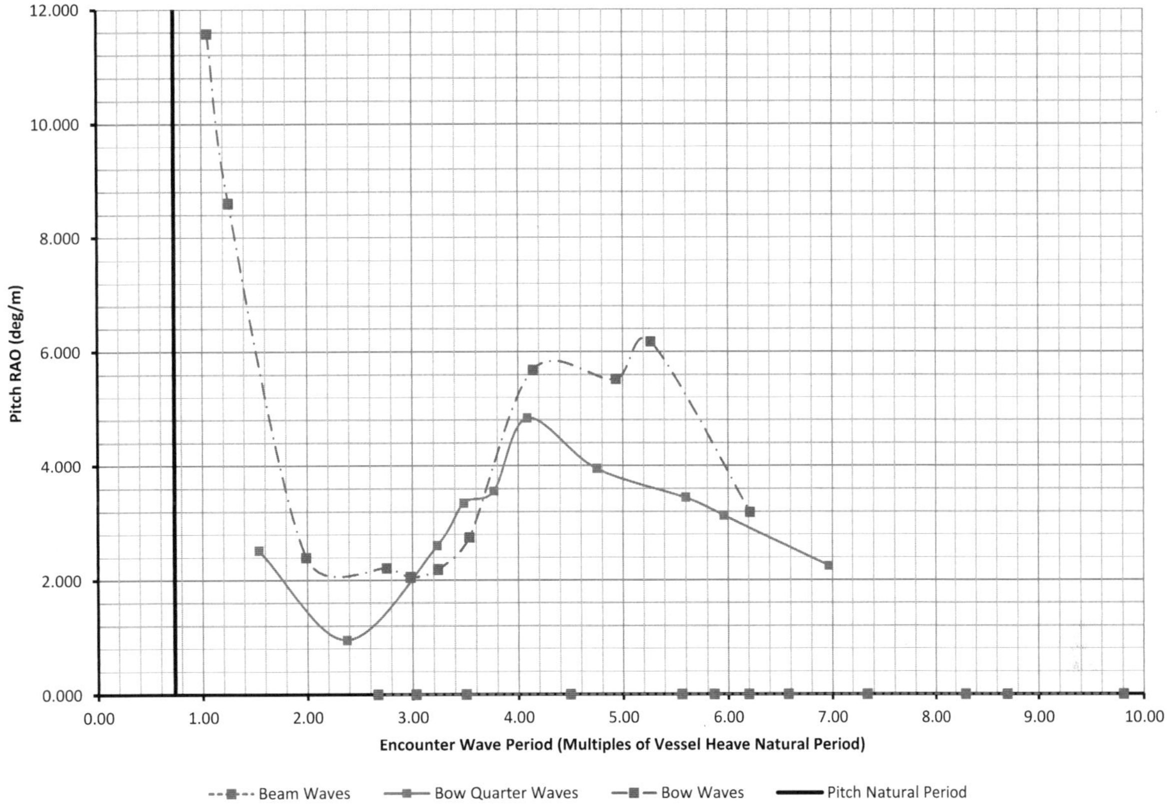

Figure 12 Pitch Response Amplitude Operator

The CFD analysis of the pitch RAO showed the same critical features already shown in previous RAO plots. At long wave frequencies, the pitch RAO dropped. It showed a typical RAO peak at medium wave periods. And near the heave natural period, we saw resonant peaks in both bow and bow-quarter RAO curves. ACMA engineers expected a resonant peak near the heave natural period. By coincidence, the natural period for pitch response closely matched the heave natural period.

The pitch resonant peak reintroduced all the previous problems of excessive resonant response: free surface diffusion and failure to include additional damping from turbulent viscosity. Considering all RAO curves, ACMA identified several sources of error for the resonant response condition.

- Resonance occurred at a short wave period, requiring small wave amplitudes which resulted in small vessel motions. Calculation of the RAO involved division of one small number by another small number. Numerical errors were more significant with such small quantities.
- At these short wave periods, free surface diffusion exaggerated the vessel motions.
- Large vessel motions should consider fully turbulent physics to capture additional damping from increased turbulent kinetic energy.

The vessel motions at natural periods are highly suspect. ACMA engineers decided to keep these as part of the RAO curve so that any inaccuracies result in conservative overestimation of vessel motions. Experimental model testing would be required to verify the simulation predictions in this region. Project budget and schedule did not allow for such testing. Beyond the actual results, analysis of the RAO curves produced valuable modeling guidance: avoid simulating vessel resonance conditions in seakeeping.

MODELING GUIDELINES

Experience with this seakeeping project revealed that RANS based seakeeping analysis is only cost effective within a selected set of applications. An ideal RANS seakeeping analysis will contain the following simulation properties.

1. A large body that only requires heave motions or remains fixed
2. Minimal body motions that result in minimal mesh displacement
3. Long period waves with an encounter period longer than approximately 3.0 s
4. Little to no body rotation (1 deg. or less) in response to wave interaction
5. Low amplitude waves without the danger of potential nonlinear interactions

Possible ideal scenarios may include: wave run-up on a fixed offshore platform; wave diffraction around a breakwater; or even wave spreading within a shallow water harbor. Shallow water effects and changing bottom topography are easily modeled with RANS CFD methods. The main criteria is the avoidance of large mesh displacements.

Some scenarios require mesh displacements. There are numerous strategies to achieve mesh displacement, and each strategy requires a careful modeling plan. Exercise caution when assessing these CFD projects. Depending on the environmental conditions, these scenarios pose a danger of becoming economically infeasible. The key is to avoid mesh morphing, which changes the position and volume of individual cells. Ideally, one should utilize mesh displacement, which moves all cells as a single rigid body and maintains constant volume of each cell. This minimizes simulation computation time.

Economic execution of a CFD seakeeping analysis requires careful selection of applicable scenarios and attentive setup of each individual CFD analysis. The most important step is to establish an appropriate size for the region of waterline refinements on the mesh. This refinement region must shrink in size proportional to the wave amplitude. The authors can recommend the following settings for waterline mesh refinements.

1. Cell size should be approximately 5%-10% of the wave amplitude. Confirm this with a mesh independence study that specifically examines the waterline refinement.
2. Waterline cells should be isotopically sized near the body.
3. Away from the body, mild anisotropic sizing is possible. Cell length in wave propagation direction should be no more than 1.5x cell vertical depth. Cell width along width of wave crest should be no more than 3x the cell vertical depth.
4. Cell size requirements should remain proportional to the wave amplitude. If wave amplitude shrinks, the cell sizes must also reduce.
5. The depth of the wave refinement region must reduce with wave amplitude. This adjustment is critical to controlling simulation cell count. Without this reduction, the smaller cell sizes on the waterline quickly lead to massive cell counts.

Beyond the wave refinements, the authors can also provide several general recommendations for completion of a seakeeping analysis.

- If the end goal is RAO results, this permits arbitrary selection of the CFD wave amplitude. Select a small wave amplitude that will not introduce non-linear effects. But avoid too small of an amplitude, which may get dominated by viscous effects. Consider several possible wave amplitudes and carefully select the final design amplitude based on estimated performance for the entire range of wave periods anticipated.
- Wave amplitude is governed by a maximum wave slope of 1:10. Anything beyond this slope and the simulation begins to predict collapse of the wave train. Consequently, input wave amplitude may reduce with shorter wave periods.
- Estimate the required computational cost for all planned simulations and optimize simulation settings before commencing production runs. Adjust mesh and timestep settings for each individual simulation.
- Anticipate a large post processing effort beyond the initial CFD simulations. Post processing requires conversion of several thousand time histories into frequency domain RAO's and eventual application of design wave spectrums.

- Avoid vessel resonant conditions. They lead to large non-linear effects with questionable results quality.
- Laminar flow assumptions are not applicable if any resonant conditions are encountered. Fully turbulent RANS become necessary for these conditions.

FURTHER WORK

The current work revealed several opportunities for future improvements in CFD seakeeping analysis. One of the primary limitations is linear wave forces. The post processing efforts only considered linear wave forces. Many commercial seakeeping programs calculate up to second order wave forces with improved accuracy of results. Additional post processing efforts for CFD could include second order wave forces. This would broaden the range of applicable scenarios greatly.

The second effort for future work is reconsideration of the CFD solver methodology. Full RANS simulations with volume of fluid (VOF) present extensive detail for including applicable physics. But this is only beneficial if the physics can be matched with equally refined mesh settings. There is no benefit to sacrificing discretization quality for the sake of including additional physics equations. In seakeeping, the tradeoff of mesh resolution with computational cost is paramount. A less advanced CFD solver may allow better mesh resolution for the same computational cost. Future work should consider single phase flow with boundary conditions to model the free surface. The physics are less advanced, but a faster solver permits better mesh refinements which may ultimately improve the total simulation accuracy within time and budgetary constraints.

CONCLUSION

CFD requires a careful balance between simulation accuracy and computational cost. This CFD based seakeeping analysis truly pushed the limits of this balance. The work presented in this paper attempted a seakeeping analysis based on simulation of regular waves. This paper identified several modeling guidelines that mitigate computational costs and permit economically feasible simulations. The optimizations were extensive and limited the range of applicable scenarios. CFD seakeeping simulation is feasible for specialized applications, but general applications still require larger computational resources or further development in modeling techniques.

ACKNOWLEDGEMENTS

The authors thank Gold Coast Yachts for the opportunity to work on this project.

REFERENCES

Celik, I. (2008). Procedure for Estimation and Reporting of Uncertainty to Discretization in CFD Applications. *Journal of Fluids Engineering, 130.*

Fonseca, N., Guedes Soares, C., & Incecik, A. (1997). Numerical and Experimental Study of Large Amplitude Motions of Two-Dimensional Bodies in Waves. *Applied Ocean Research*, 35-47.

Gold Coast Yachts. (March, 2015). *Tandem Catamaran: Concept Design.* St. Croix, U.S. Virgin Islands: Hull #111.

Guo, B. J., Steen, S., & Deng, G. B. (March 2012). Seakeeping Prediction of KVLCC2 in Head Waves. *Applied Ocean Research, 35*, 56-67.

Roache, P. J. (1998). *Verification and Validation in Computational Science and Engineering.* Albuquerque, NM: Hermosa Publishers.

Slater, J. W. (2008, July 17). *Tutorial on CFD Verification and Validation.* Retrieved August 06, 2014, from NPARC Alliance CFD Verificationand and Validation Web Site: http://www.grc.nasa.gov/WWW/wind/valid/tutorial/tutorial.html

SNAME
THE INTERNATIONAL COMMUNITY
FOR MARITIME AND OCEAN PROFESSIONALS

CFD TECHNIQUES FOR PREDICTING VIV OF A CIRCULAR CYLINDER

WORAKANOK THANYAMANTA
NUMERICAL MODELLING SPECIALIST, OCEANIC CONSULTING

MUHAMMED ISLAM
DIRECTOR OF TECHNICAL SOLUTIONS, OCEANIC CONSULTING

Proceedings of the 21st Offshore Symposium, February 2016, Houston, Texas
Texas Section of the Society of Naval Architects and Marine Engineers

ABSTRACT

Vortex-Induced vibrations (VIV) of a circular pipe or a riser are caused by flow separation and vortex shedding from the riser when exposed to ocean currents. This phenomenon is frequently observed in the field for drilling risers as well as production risers. Such phenomenon is not desirable as bending stresses due to the vibrations can cause significant fatigue damage to the system. Riser fairings and/or strakes are generally used in the field to streamline the flow and eliminate VIV. Accurately predicting riser VIV responses with or without VIV suppression devices is one of the key challenges for designers and researchers in the relevant offshore industry.

In this study, the VIV benchmark study organized by ITTC Ocean Engineering Committee in 2013 was revisited. A commercial RANS-based CFD code was used to predict drag, lift, and Strouhal number of flow past a fixed long circular cylinder at Reynolds numbers close to the drag crisis. Both 2D and 3D cases were studied. Two turbulence models, namely LES and DES were used. Grid and time-step sensitivity analyses were also conducted. Results were compared with the ITTC benchmark data. This study shows that it is often insufficient to consider VIV of a long symmetrical cylinder as two-dimensional system due to three-dimensional nature of vortex structures. In addition, accurate VIV predictions also require flow in the viscous sub-layer to be fully resolved. These requirements lead to excessive number of cells to carry out CFD simulations. In order to determine effective domain size and practical mesh discretization for VIV simulations, this study included an investigation into the effects of length to diameter ratio of the computational domain. It was found that increasing length to diameter ratio to 2 improved drag and lift predictions significantly and the results agreed well with the benchmark data.

Preliminary simulations were also conducted with a circular cylinder fitted with a generic strake to evaluate the VIV characteristics. Results were compared with experimental data where a close match was obtained for both drag and lift. This also revealed the effects of the strake on the VIV of the pipe. For the present cases, major flow features including pressure, velocity, and vorticity fields are also presented. Three-dimensional effects and unsteadiness

were well captured in both turbulence models. LES-based turbulence models seem to be the key to better solve and predict the flow problem numerically. However, their considerable computational demand still does not allow applications for engineering design purposes.
Keywords: SNAME, Offshore, Symposium, VIV, CFD

INTRODUCTION

In offshore operations, subsea circular pipes or risers may experience undesirable bending stresses caused by vortex-induced-vibration (VIV). Such vibration caused by flow separation and vortex shedding off the circular structures in ocean currents may lead to significant fatigue damage to the system. Attempts have been made to alleviate the problem by means of VIV suppression devices including riser fairings and strakes. Various numerical models have been developed in order to provide accurate VIV predictions. Computational Fluid Dynamics (CFD) is one of the numerical tools that have been used to simulate the VIV phenomena. CFD codes that solve transient, three-dimensional, full Navier-Stokes and continuity equations along with appropriate turbulence model were expected to be suitable for capturing the three-dimensional and unsteady flow fields that are the basis to the VIV problem. However, applications of such codes for engineering design purposes are often hindered by significant computational demand.

In the past decade, efforts have been made to understand such complex phenomenon of the VIV. Numerous experimental studies were conducted to study VIV of fixed and elastic bluff bodies as summarized in a review by Gabbai and Benaroya (2005). An important characteristic of the VIV found from experimental studies is the three-dimensional nature of the vortices generated from uniform flow past two dimensional bluff body. This is caused by instability of the shear layers that leading to vortex shedding that is not constant along the length of the body (Bearman 1984). In model experiments, two-dimensionality of flow is also influenced by the length to diameter ratio of the cylinder model. In CFD, an infinitely long cylinder can be modelled using two- or three-dimensional grids. However, two-dimensional grid may not be sufficient due to the three-dimensional nature of the unsteady flow field. For three-dimensional grid, demand for computational resources increases with the size of the fluid domain. It is therefore of interest to determine a minimum length of the domain that is sufficient to capture such three-dimensional effects without incurring excessive computational expense.

In this study, a commercial CFD code, STAR-CCM+, was used to predict drag and lift induced by flow past stationary long circular cylinder. The main focus of the study was on determining practical techniques for setting up CFD models for such computationally demanding problems as VIV. In order to do so, 2D and 3D simulations were investigated as well as grid and time-step sensitivity analyses. Results were compared with the experimental data provided by ITTC Ocean Engineering Committee for the VIV benchmark study organized in 2013. Based on the VIV benchmark study, Large Eddy Simulation (LES) and Detached Eddy simulation (DES) were found to provide better results than Unsteady Reynolds-Averaged Navier-Stokes models (URANS) (ITTC, 2014). Therefore, LES and DES models available in STAR-CCM+ were chosen to be used in this study. In order to validate ability of the code to predict VIV of stationary smooth cylinder, Reynolds numbers near the critical range needed to be considered. For a Reynolds number within a range of $3x10^5$ to $3.5x10^5$, flow separation is delayed due to instability of flow in the shear layer. This delay of separation can be observed from the abrupt reduction of the drag coefficient (ITTC, 2014). It is crucial for a CFD code to be able to capture this phenomenon. As such, the simulations were conducted in this study at Reynolds numbers close to the drag crisis. Preliminary simulations were also conducted for a circular cylinder fitted with a generic strake in order to evaluate behaviours of flows around the cylinder with and without the strake. In addition to drag force, lift force, and Strouhal number, key flow features including pressure, velocity, and vorticity fields resulting from the CFD simulations are also presented in this paper.

MODEL GEOMETRY

Two cases were considered in this study: VIV of a smooth circular cylinder and VIV of a cylinder equipped with strake. For the smooth circular cylinder, geometry used for the ITTC's VIV benchmark study was used here.

The benchmark data were obtained from model experiments as described in ITTC (2014). The cylinder model was 3.52 m long with a diameter of 0.2 m as shown in Figure 1. The model was towed at approximately 1.7 m depth below the calm water surface. As such, the model was fully submerged and the effect of the free surface was minimal.

For the strake case, a cylinder with outside diameter of 0.111 m was equipped with generic strake geometry. The strake had a pitch of approximately 12D and a height of 20% D as shown in Figure 2.

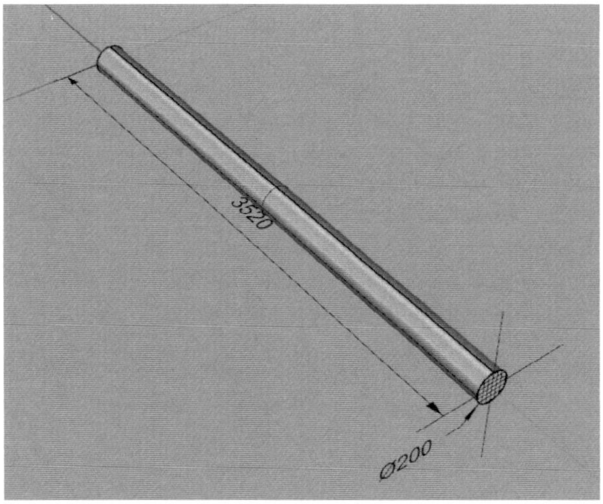

Figure 1 3D geometry of smooth circular cylinder (in mm)

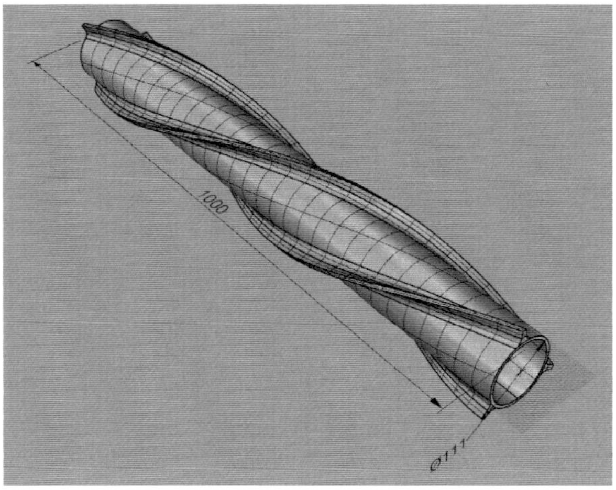

Figure 2 3D geometry of cylinder with strake (in mm)

NUMERICAL MODELS AND MESH SETUP

STAR-CCM+ is a commercially available finite-volume-based CFD code that applies the full Navier-Stokes and continuity equations (CD-Adapco, 2014). The governing equations were solved using a segregated iterative method based on SIMPLE algorithm. The model was constructed with single phase incompressible fluid with freshwater properties. The free surface was not considered in all of the simulations. Implicit Unsteady time-solving scheme with second-order time accuracy was applied. Choice of turbulence model plays critical role in VIV simulations. As mentioned above, two turbulence models were used in this study. These are LES with WALE

Subgrid Scale model and a DES model. DES model is a combined method of the LES and RANS. It was developed to overcome limitations of RANS models that are not sufficient for modeling flow with large separation and the computational cost imposed by the LES (Spalart, 2009). The choice of the DES model used in this study was Spalart-Allmaras Detached Eddy model with improved delayed detached eddy simulation (IDDES) formulation option.

Smooth cylinder

The mesh blocks used in the simulations were set up using unstructured trimmer mesh. As shown in Figure 3, the fluid flowed from the left boundary (negative x-direction) to the right boundary (positive x-direction). The left boundary was assigned a velocity inlet whereas the right boundary was assigned a pressure outlet. Fluid velocity was also specified at the top and the bottom boundaries. Boundaries in the z-direction were assigned symmetry condition representing the length of the cylinder that extended beyond the z-boundaries of the fluid domain. Mesh refinements were applied to the volume mesh in vicinity of the cylinder models as well as on the cylinder surface. For surface refinement, the number of grid points on the cylinder's perimeter depended on the base size specified for mesh generation. For a base size of 0.04 m leading to a cell size of 1.25 mm on the cylinder surface, there were approximately 500 cells on the perimeter. Mesh refinement volume was created around the cylinder and extended downstream in order to capture the wake behind the cylinder. Prism Layer Mesher was used to create thin layers of cells parallel to cylinder surface. In order to ensure that the flow in the viscous sub-layer was fully resolved, the prism layer grids were refined such that wall y+ values were maintained below one. This was achieved by using 30 grid points in the prism layer.

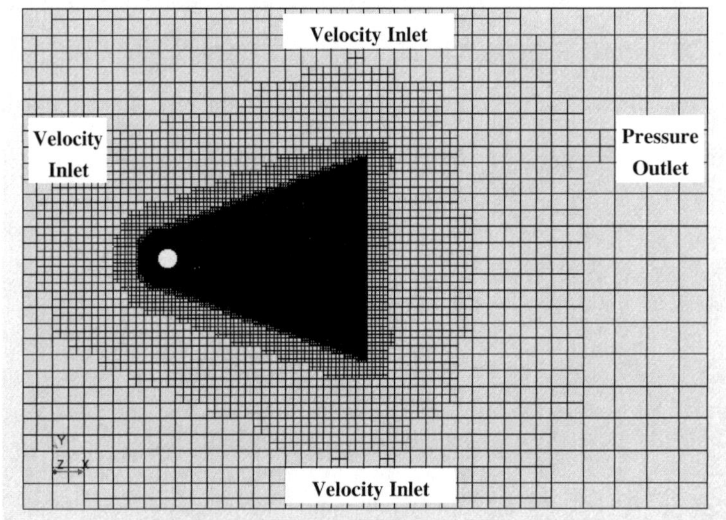

Figure 3 Mesh block and boundary conditions

For the smooth circular cylinder, both 2D and 3D mesh blocks were applied. The 2D mesh block was generated in such a way that it contained one cell in the z-direction creating a quasi-3D condition. As shown in Figure 4, a small domain length in the direction of the cylinder's axis (z-direction in this case) of 1 mm was used to provide a small grid size applicable for the LES model near the cylinder as well as to avoid inverse grid ratio close to the cylinder. A high resolution of at least 20 cells per thickness of the boundary layer in all directions is required for the LES to be accurate (Spalart, 2009). Such resolution could be achieved as the base size was reduced to a sufficiently small value. In order to investigate three-dimensional effects, the length of the domain in the z-direction was varied for the 3D mesh block. Domain lengths of 10%D, 20%D, 100%D, and 200%D were used. The total number of cells for the mesh block with 200%D length was approximately 6.6 million cells. Provided the cell size of 1.25 mm on the cylinder surface for a base size of 0.04 m, there were approximately 320 cells across the length of the domain for the 200%D length.

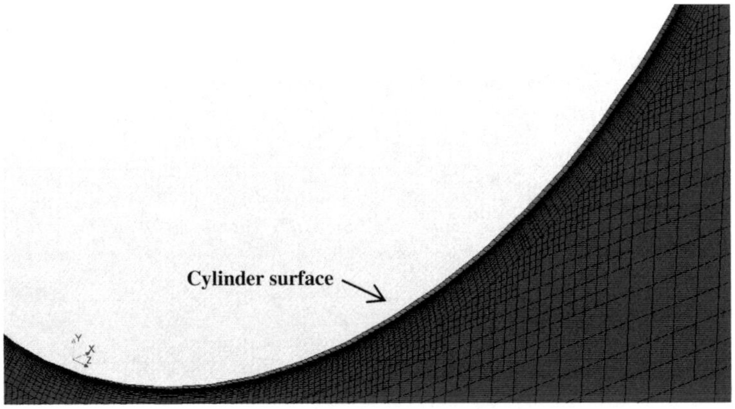

Figure 4 2D mesh block near the cylinder surface for smooth cylinder case

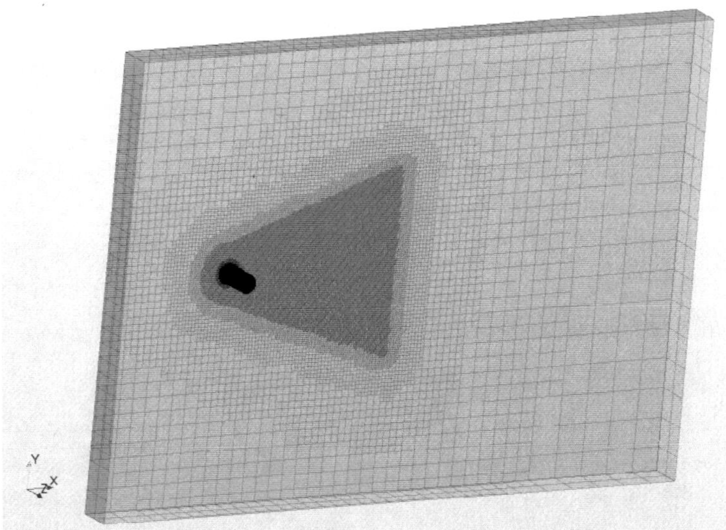

Figure 5 3D mesh block for smooth cylinder case

Cylinder with strake

Strake is a common device used to break up the current eddies and hence suppress VIV. A smooth cylinder of the same size as the cylinder equipped with the strake as described in the model geometry was therefore considered as well in order to evaluate performance of the strake. The strake was removed from the original pipe leaving a bare cylinder with an outer diameter of 0.111 m. For each of these configurations, a 3D mesh block was set up similarly to those described above for the ITTC's smooth cylinder case. A domain length of 50% of the pitch was used. Grid refinements were applied to the volume mesh near the cylinder as well as on the model surface. Figure 6 and Figure 7 show mesh discretization near the cylinder models for the case with the strake in place and the case with the strake removed, respectively.

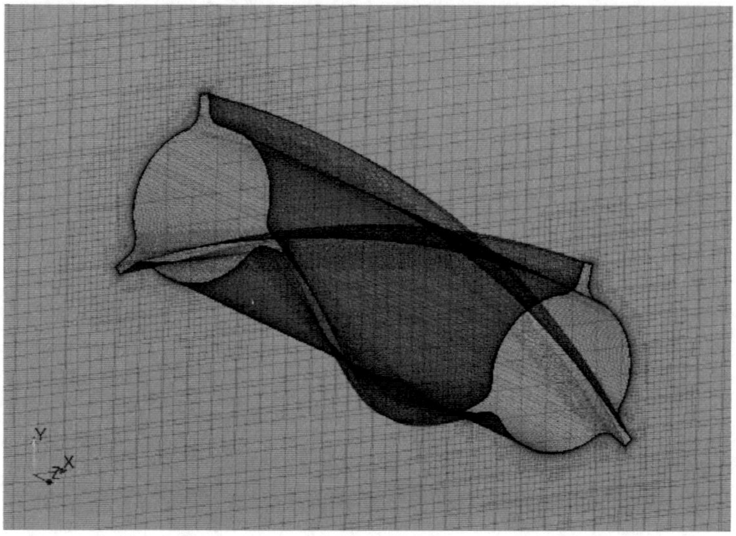

Figure 6 Mesh block for cylinder with strake installed

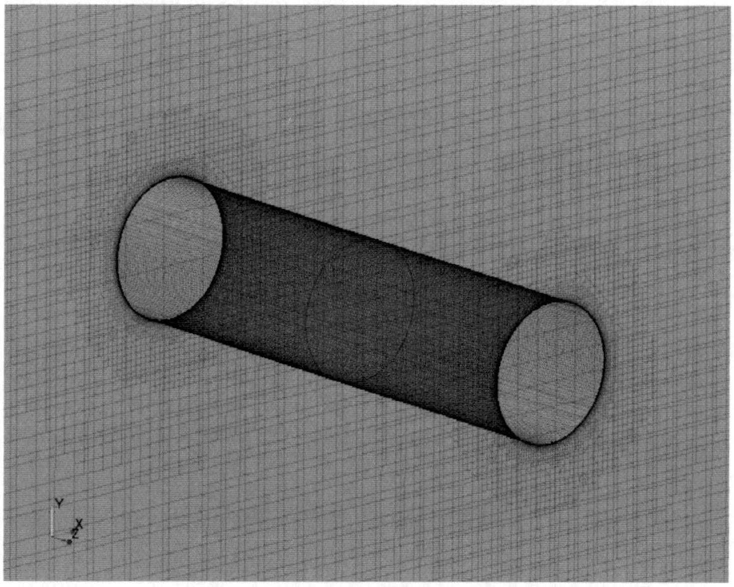

Figure 7 Mesh block for bare cylinder

SENSITIVITY STUDIES

For the sensitivity studies, two Reynolds numbers were selected close to the drag crisis. The values selected were 1.25E5 and 7.57E5 which were slightly lower and higher than the critical Reynolds number, respectively. Simulations were then performed for combinations of various parameters as summarized in Table 1. As seen in the table, numerical model parameters that were evaluated include domain length, grid size, time step size, and turbulent model. Results were compared against the ITTC benchmark data including drag coefficient, lift coefficient and Strouhal number calculated as below:

$$C_D = \frac{F_D}{0.5\rho U^2 D} \tag{1}$$

$$C_L = \frac{F_L}{0.5\rho U^2 D} \tag{2}$$

$$S_t = \frac{f_s d}{U} \tag{3}$$

where F_D is the mean of the drag force, F_L is the RMS of the lift force, ρ is the density of the fluid, U is the velocity of the fluid, and D is the diameter of the cylinder. Strouhal number, S_t, was calculated based on frequency of the lift force signal which represents shedding frequency, f_s, of the vortex. The Figure 8 to Figure 10 show results from the sensitivity studies including drag, lift, and Strouhal number, respectively.

A grid sensitivity analysis was initially performed on 3D models. The base size was reduced until convergence in solution was obtained. It was found from the study that a base size of 0.04 m provided acceptable level of accuracy. Reducing the base size to 0.028 m did now show significant improvement in the drag and lift predictions as shown in Figure 8 and Figure 9. Therefore, the base size of 0.04 m was used in further sensitivity analyses.

Also from the figures, it is clear that the 2D model was not sufficient to simulate VIV. The drag and lift results were highly overestimated. LES was found to provide results that agreed better with the experimental data compared with the DES.

For sensitivity to temporal discretization, time step size values were initially selected based on Strouhal numbers obtained from the experiment. Strouhal number is a function of predominant frequency of vortex shedding, flow velocity, and diameter of the cylinder (Gabbai and Benaroya, 2005). Initial time step size was selected for each of the Reynolds numbers to provide 50 to 100 time steps per one shedding period. These were 0.02 sec for the Reynolds number of 1.26E5 and 0.002 sec for the Reynolds number of 7.57E5. These initial time step sizes were then reduced in order to investigate sensitivity of the solution to time step size. As seen in Figure 8 to Figure 10, reducing the time step sizes by half for both flow speeds did not affect the solutions much, particularly the drag force. A slight difference was observed in the lift force for the low speed case (Re = 1.26E5).

Finally, using the optimal base size and time step size obtained from the sensitivity analyses, the length of the domain was varied in order to investigate the three-dimensional effects. It was found that prediction of the drag force was less sensitive to the length of the domain compared with prediction of the lift force. Moreover, the domain length had less influence on the results at high Reynolds number. In all of the cases, a considerable improvement was obtained when the length of the domain was increased to 200%D. Shedding frequency was also predicted quite accurately by the code as represented by the Strouhal numbers. As seen in Figure 10, Strouhal number predicted for the high Reynolds number case seemed to be slightly underpredicted compared with the experimental data. However, it was within an acceptable range of Strouhal numbers for rigid smooth circular cylinders as published in the literature such as Lienhard (1966).

Table 1 Summary of simulations conducted for sensitivity studies

Reynolds number	Domain length	Grid base size (m)	Time step size (sec)	Turbulent model
7.57E5	10%D	0.04	0.002	LES
			0.001	LES
		0.028	0.001	LES
	100%D	0.04	0.002	LES, DES
	200%D	0.04	0.002	LES
1.25E5	10%D	0.04	0.02	LES
			0.01	LES
		0.028	0.02	LES
	100%D	0.04	0.02	LES
			0.01	LES, DES
	200%D	0.04	0.01	LES

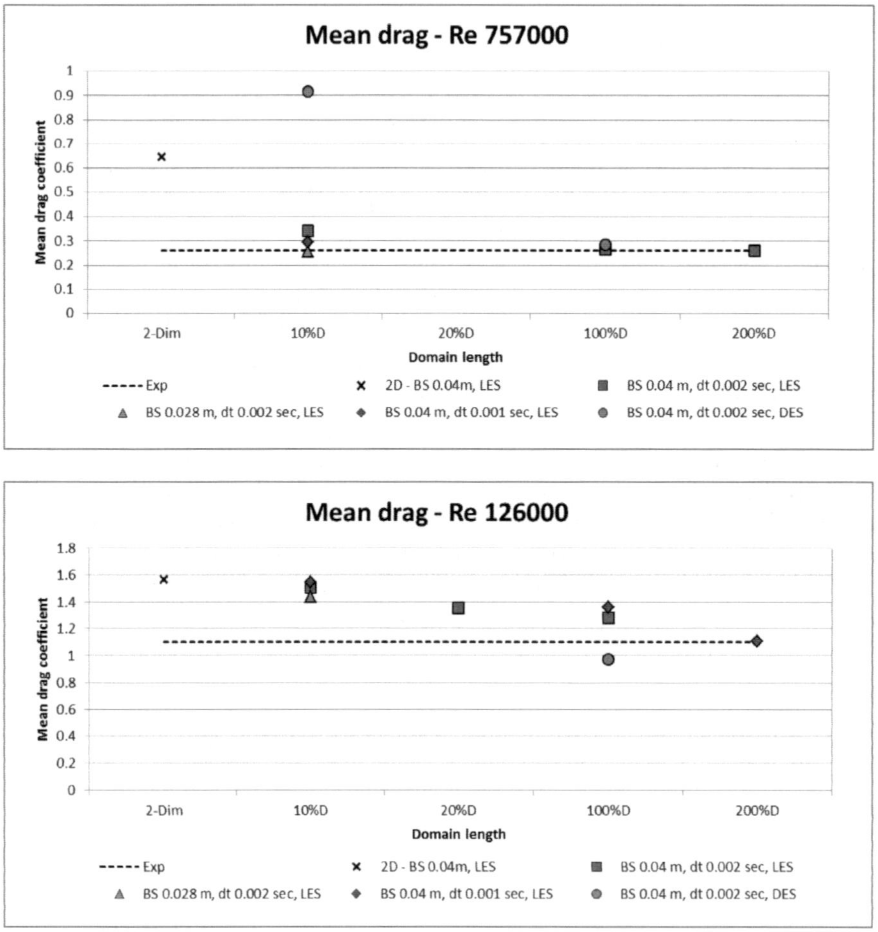

Figure 8 Sensitivity of drag force prediction to various parameters

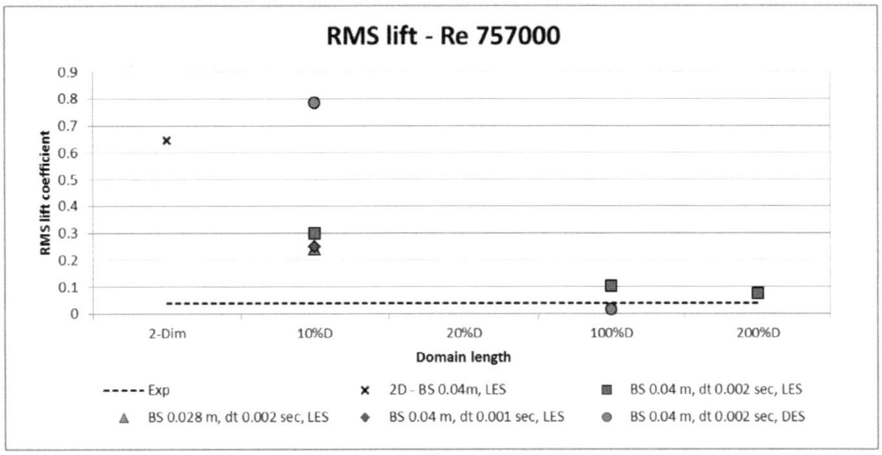

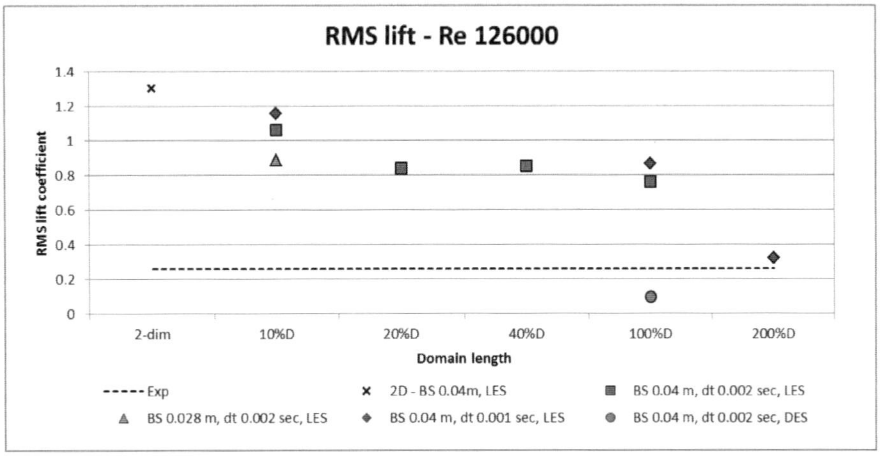

Figure 9 Sensitivity of lift force prediction to various parameters

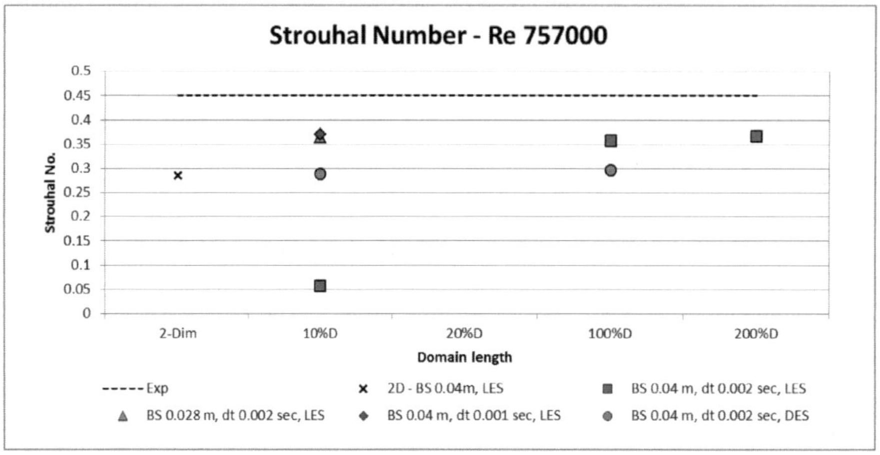

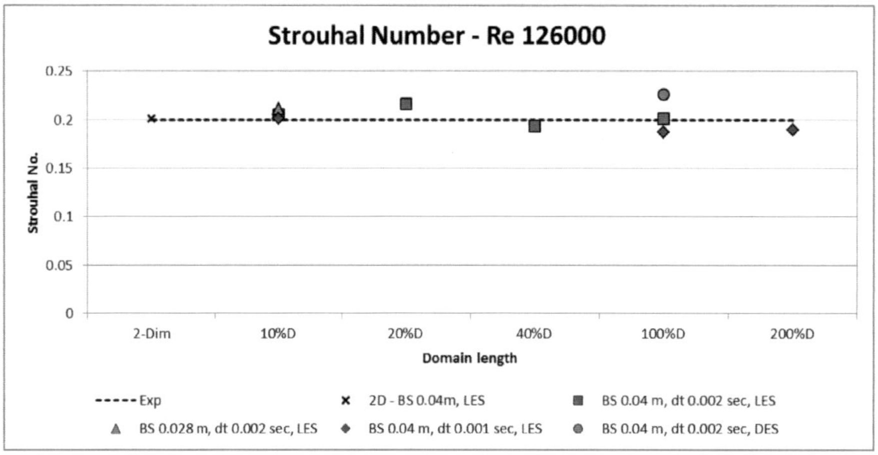

Figure 10 Sensitivity of Strouhal number prediction to various parameters

RESULTS AND DISCUSSIONS

Based on outcomes of the sensitivity studies, the CFD code was validated at various Reynolds numbers. Four additional Reynolds numbers of 0.63E5, 3.15E5, 2E6, and 5E6 were investigated on the ITTC's smooth cylinder. The 200%D mesh block generated using the base size of 0.04 m was used for these simulations. Only the LES model was used for these cases as it was found to provide better results than the DES. Results were compared against the ITTC benchmark data as shown in Figure 11 for the drag predictions and Figure 12 for the lift predictions. Other numerical results submitted as part of the ITTC benchmark study in 2013 are also shown in the Figure (ITTC, 2014).

Considering the cases near the critical Reynolds number, Figure 13 to Figure 15 compare vorticity isosurface, velocity, and pressure of the flow around the cylinder for the cases of Re = 7.57E5 and 1.26E5. The plots show higher oscillation and unsymmetrical distribution of the flow parameters in the flow for the Re = 1.26E5 case. This is reflected well in the larger magnitude of the lift force.

The ITTC benchmark data did not include the high Reynolds number cases of 2E6 and 5E6. The results from these two CFD simulations were compared against data published in Yeon et al. (2013) only in terms of drag coefficient. The drag coefficients calculated by the CFD were 0.29 and 0.40 for the Reynolds numbers of 2E6 and 5E6, respectively. It was found that the drag prediction for the case of Reynolds number of 2E6 agreed quite well with experimental data. Drag forces measured in different experimental programmes seemed to vary drastically for high Reynolds numbers. For the Reynolds number of 5E6, drag coefficients of approximately 0.55 and 0.8 were obtained from the experiments referenced by Yeon et al. (2013). Regardless, the drag coefficient predicted by the CFD at such Reynolds number was still underestimated.

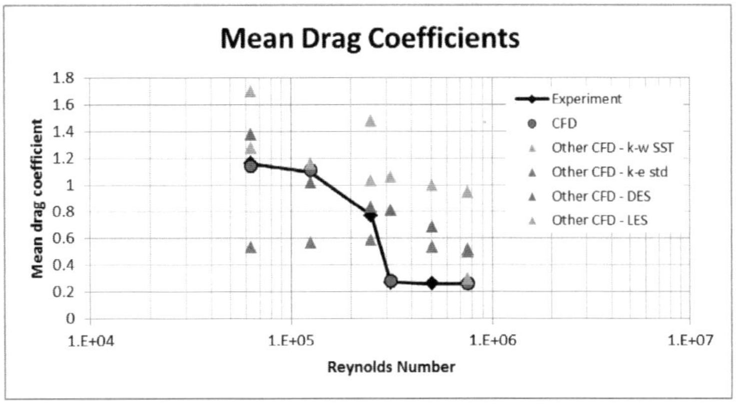

Figure 11 Comparisons of drag coefficients between CFD and experimental data

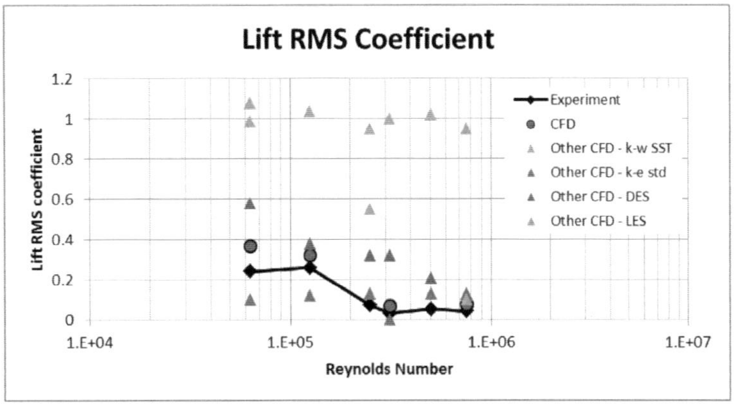

Figure 12 Comparisons of lift coefficients between CFD and experimental data

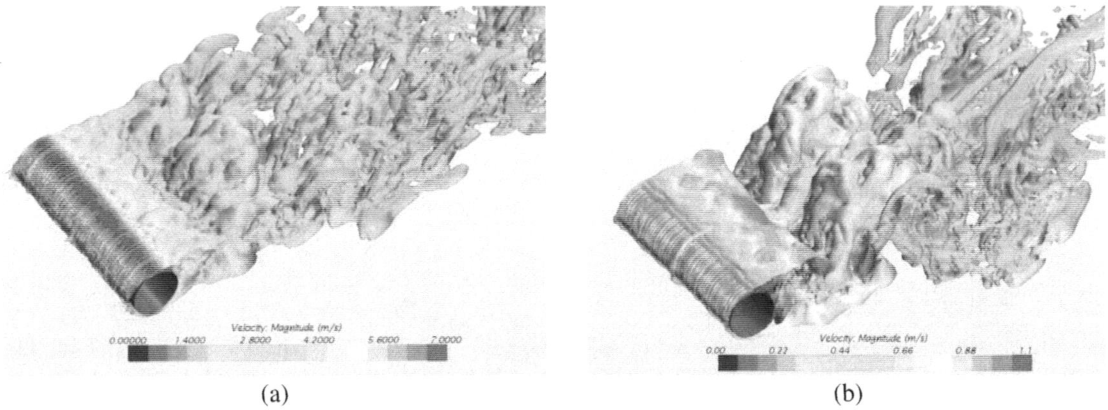

(a) (b)

Figure 13 Vorticity isosurface (a) Re = 7.57E5 (b) Re = 1.26E5

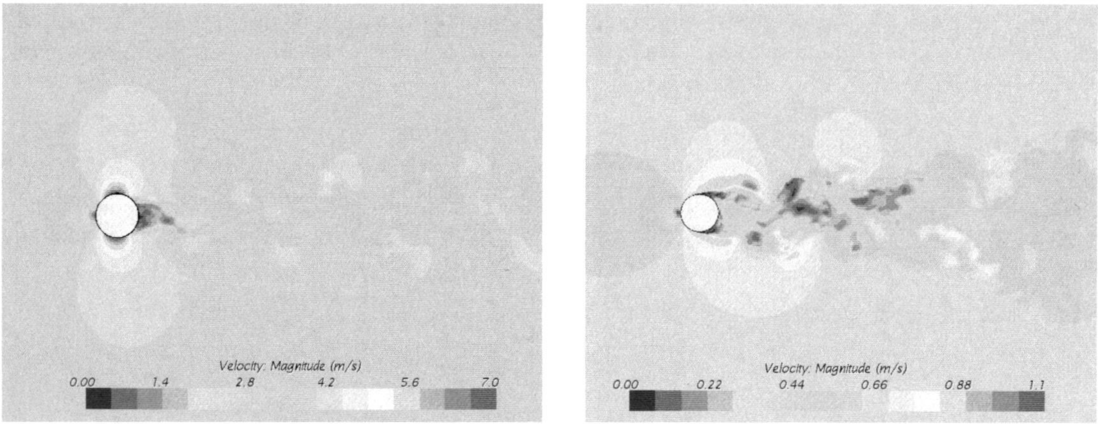

(a) (b)

Figure 14 Velocity profile (a) Re = 7.57E5 (b) Re = 1.26E5

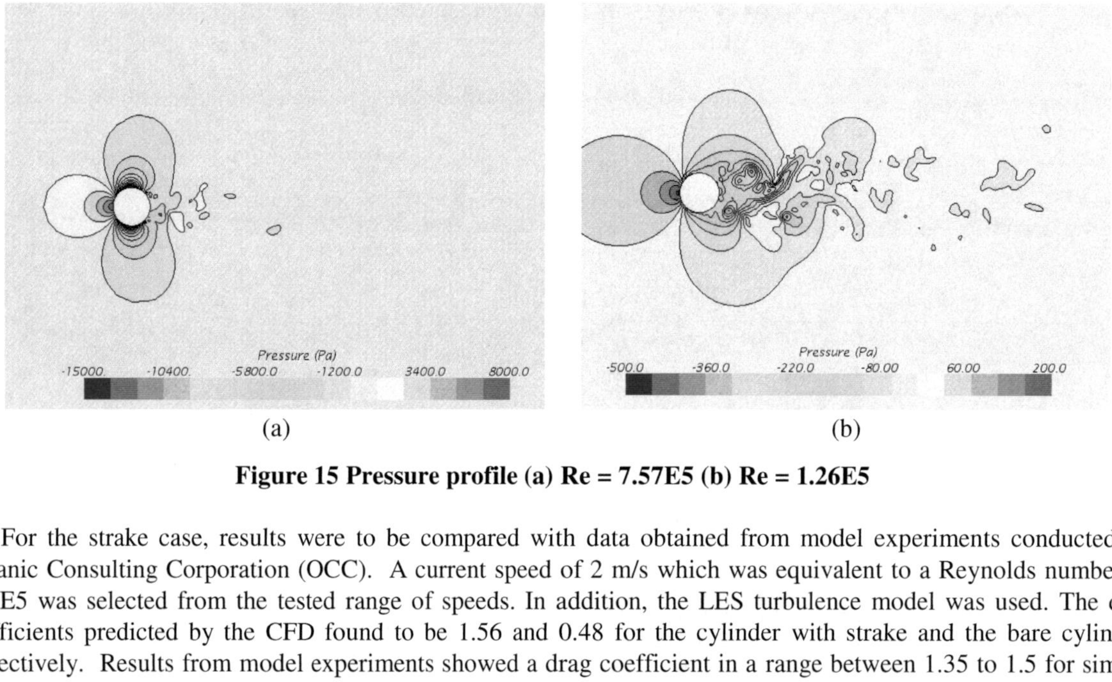

(a) (b)

Figure 15 Pressure profile (a) Re = 7.57E5 (b) Re = 1.26E5

For the strake case, results were to be compared with data obtained from model experiments conducted by Oceanic Consulting Corporation (OCC). A current speed of 2 m/s which was equivalent to a Reynolds number of 2.22E5 was selected from the tested range of speeds. In addition, the LES turbulence model was used. The drag coefficients predicted by the CFD found to be 1.56 and 0.48 for the cylinder with strake and the bare cylinder, respectively. Results from model experiments showed a drag coefficient in a range between 1.35 to 1.5 for similar strake geometry at the same current speed of 2 m/s. The predicted drag for the bare cylinder was found slightly low. This might be due to insufficient number of grid points on such small diameter of the cylinder. The lift coefficient obtained from the CFD was reduced from 0.15 to 0.09 when the strake was equipped. Figure 16 to Figure 18 show plots of vorticity isosurface, velocity, and pressure around the cylinders.

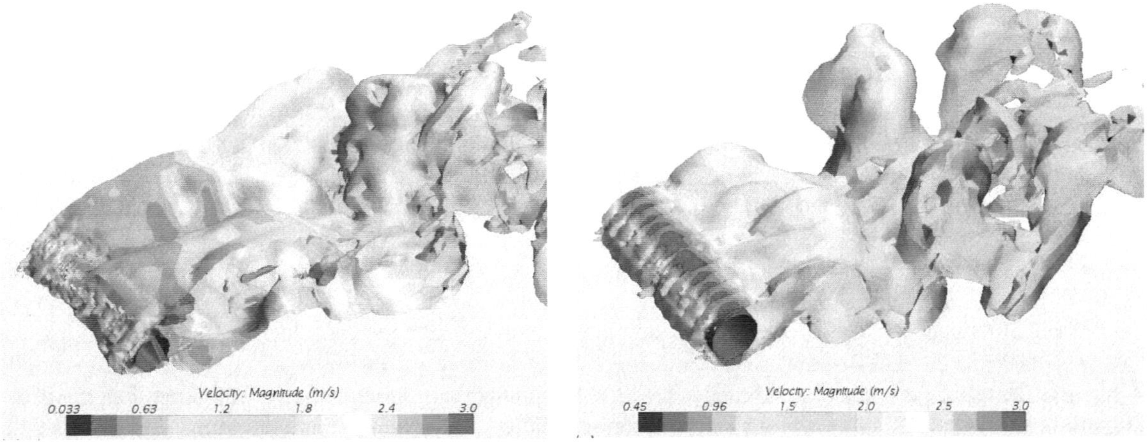

(a) (b)

Figure 16 Vorticity isosurface (a) cylinder with strake (b) bare cylinder

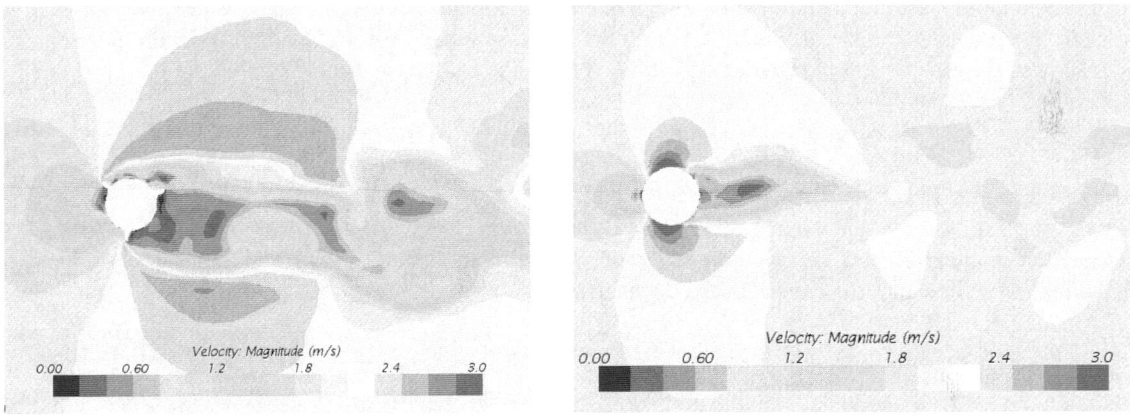

(a) (b)

Figure 17 Velocity profile (a) cylinder with strake (b) bare cylinder

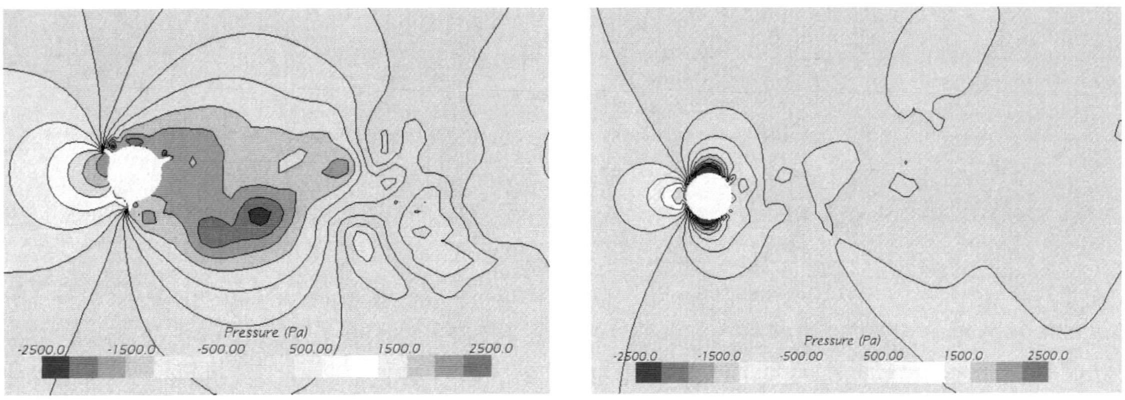

(a) (b)

Figure 18 Pressure profile (a) cylinder with strake (b) bare cylinder

CONCLUSIONS AND RECOMMENDATIONS

This paper presents techniques for setting CFD simulations for such complex flow problems as VIV. Numerical model provides an advantage of modelling a long symmetrical as an infinitely long cylinder without having to include the entire length of the cylinder. Extensive sensitivity analyses were conducted of various numerical model parameters including domain length, grid size, time step size, and turbulent model. This study shows that it is not sufficient to model VIV of a long symmetrical cylinder as two-dimensional system due to three-dimensional nature of vortex structures. The minimum domain length required in order to capture the three-dimensional effects was found to be 200%D. The number of time steps between 50 to 100 steps per vortex shedding period was found sufficient. Moreover, LES turbulence model was found to provide accurate VIV prediction. Drag, lift, and Strouhal numbers predicted by the CFD simulations performed at various Reynolds number ranging from below the critical range to considerably large value were found to agree well with the ITTC benchmark data and/or published experimental data except for the highest Reynolds number considered of 5E6 where the drag coefficient was highly underpredicted. Further investigation is recommended in order to determine optimal grid and time step size required for such high speed flow. In addition, efforts should be made to further extend the domain length as well as to refine the grid size in order to ensure convergence in the solutions for all of the cases.

CFD simulations were also conducted on a cylinder equipped with generic strake. Results matched well with OCC's model test data of similar geometry. CFD results of the cylinder with strake compared with those of a bare cylinder showed the effects of the strake on suppressing VIV of the cylinder.

ACKNOWLEDGEMENTS

The authors would like to thank Fleetway Inc. for providing support on this work and publication of this paper. Many thanks to Trevor Harris for assistance on data analysis.

REFERENCES

Bearman, P. W., "Vortex Shedding from Oscillating Bluff Bodies", Annual Review of Fluid Mechanics 16, pp. 195-222, 1984.

CD-Adapco, STAR-CCM+ Version 9.06 User Guide, 2014.

Gabbai, R. D. and Benaroya, H., "An Overview of Modeling and Experiments of Vortex-Induced Vibration of Circular Cylinders", Journal of Sound and Vibration 282, pp. 575-616, 2005.

ITTC Ocean Engineering Committee, "Final Report and Recommendations to the 27th ITTC", Proceedings of the 27th International Towing Tank Conference – Volume I, Copenhagen, Denmark, 2014.

Lienhard, J. H., Synopsis of Lift, Drag, and Vortex Frequency Data for Rigid Circular Cylinders, Bulletin 300, Technical Extension Service, College of Engineering Research Division, Washington State University, 1966.

Spalart, P. R., "Detached-Eddy Simulation", Annual Review of Fluid Mechanics 41, pp. 181-202, 2009.

Yeon, S. M., Yang, J. and Stern, F., "Large Eddy Simulation of Drag Crisis in Turbulent Flow Past a Circular Cylinder", Proceedings of ITTC Nantes Workshop on Wave Run-up and Vortex Shedding, Nantes, France, 2013.

(The Manuscript for Abstract ***SNAME_OS16_06*** has been retracted by the Authors)

SNAME
THE INTERNATIONAL COMMUNITY
FOR MARITIME AND OCEAN PROFESSIONALS

Uni-directional and Alternating Flow around 2-D Rigid Cylinder - Experimental and Computational Study

SPYROS A. KINNAS
PROFESSOR, THE UNIVERSITY OF TEXAS AT AUSTIN

GUANGYAO WANG
GRADUATE STUDENT, THE UNIVERSITY OF TEXAS AT AUSTIN

YE TIAN
POST-DOCTORAL ASSOCIATE, THE UNIVERSITY OF TEXAS AT AUSTIN

Proceedings of the 21st Offshore Symposium, February 2016, Houston, Texas
Texas Section of the Society of Naval Architects and Marine Engineers

ABSTRACT

Turbulent flows past a cylinder is one of the most intensively investigated problem in fluid mechanics. The staggered shed vortices in the wake region exert oscillatory forces to the cylinder, and consequently excite the so called vortex-induced vibrations (VIV). Prediction and suppression of VIV are of special interest to the offshore industry because the cylinder represents a vast number of structural components in the offshore hydrocarbon exploitation.

Our previous work, reported in the 20th Offshore Symposium of Texas SNAME by Wang et al, focused on the study of flow around rigid cylinders in uniform flow, with both Particle Image Velocimetry (PIV) experiment and Computational Fluid Dynamics (CFD) simulations. PIV measurements of the flow field at the downstream of the cylinder were first presented. The boundary conditions for CFD simulations were measured in the PIV experiment. Then the PIV flow was compared with both RANS (2D) and LES (3D) simulations performed with ANSYS Fluent. The velocity fields from PIV and CFD were compared in the time domain.

In this work, the time-averaged velocity profiles and Reynolds stresses are analyzed. It is found that, in general, LES (3D) gives a better prediction of flow characteristics than RANS (2D).

Subsequently, RANS simulations are performed in the case of alternating flow around a rigid cylinder and the results, for various combinations of Reynolds and Keulegan-Carpenter numbers, and the predicted drag force, as a function of time, is compared with that based on the classical Morison equation, as well as with experimental values reported in the API Manual.

Keywords: Particle image velocimetry (PIV), Computational Fluid Dynamics (CFD), vortex-induced vibrations (VIV), Alternating Flow around Cylinder, Morison's equation

INTRODUCTION

Turbulent flow around a cylinder is a complex phenomenon within fluid mechanics, to which a lot of researchers have devoted efforts. When the natural frequency of the cylinder approaches that of the vortex shedding at the downstream of the cylinder, vortex-induced vibrations (VIV) occur and can lead to fatigue-driven failure (Stetson, 2013). It is meaningful to model VIV, and there is a large amount of literature addressing the problem of VIV prediction.

In the field of ocean engineering, VIV is a very common phenomenon. When ocean currents pass over large aspect ratio cylinders, such as risers, pipelines, mooring lines, cables, unexpected failure of such cylinders, due to VIV, can have catastrophic impact. Although few examples of cylinder failure due to VIV exist in the literature, VIV is still regularly considered when designing such cylinders. A lot of experiments have been conducted on the large aspect ratio cylinders in the past several years. While these experiments contributed a lot to the research of VIV, experiment has limitations as well, such as devices availability, capacity limits, model scale limit, difficulty of current profile generation, cost concerns, etc. (Huang, 2011). Under such conditions, Computational Fluid Dynamics (CFD) can be an effective alternative to experiments in studying large aspect ratio cylinders.

This work is to study the flow around a cylinder with both PIV experiment and CFD simulations. In the PIV experiment, measurements have been performed around the cylinder and at the boundaries of the flow field. In CFD simulations, using ANSYS-Fluent, both RANS (2D) and LES (3D), will be adopted and compared with the PIV measurements. In addition, the assessment of Morison's equation, applicable in the case of alternating flow (due to the action of waves) around cylinders will be discussed. The unsteady force in the direction of inflow from CFD simulations will be compared with that from Morison's equation (Morison et al., 1950), and the resulting values of the drag and the inertia coefficients will be compared with those measured in previous experiments, as presented at the API manual.

METHODOLOGY

1. PIV Experiment and CFD Simulations

1.1 PIV experiment setup

The experiment was performed at the flume in the Fluids and Hydraulics Lab of the Department of Civil, Architectural and Environmental Engineering at the University of Texas at Austin (Figure 1). The flow field downstream of the cylinder was captured with a two-dimensional PIV system which was manufactured by Dantec Dynamics. The PIV system is comprised of a Nd:YAG laser and a NanoSense Mk III CCD camera taking images at 1000Hz in single frame mode. The water was seeded with 10 μm diameter, silver-coated, neutrally buoyant, hollow glass spheres. More details of this experiment are given in Wang (2015).

A uniform inflow (0.35m/s) was used around a cylinder of 0.018m (Re=6300). The cylinder is set across the span of the flume, perpendicular to the flow and above the boundary layer (Figure 2 and Figure 3). Based on the experience of previous experiments, the adaptive methods were applied. The adaptive PIV method iteratively optimizes the size and shape of each interrogation area to better adapt to local flow gradients and seeding densities (Zhang and Lee, 2005). According to the manual of Dantec Dynamics, the following settings were found to provide accurate velocity fields: 16×16 pixel grid step size, a low pass Gaussian filter (k=3), peak height validation (minimum=0.15), peak height ratio validation (minimum=1.15), and S/N ratio (4.0).

Due to the limitation of the mobility of the laser beam, we can only measure a part of the region above the cylinder and the whole region under the cylinder. The measured region of flow field was shown in Figure 4. In the simulations of CFD, the boundary conditions of the measured region were required. The mean horizontal velocity was 0.35m/s and the standard deviation was 0.007m/s. So the turbulence intensity was 0.007/0.35=2%.

The region which we were really interested in was the one around the cylinder, especially downstream the cylinder. Finally, the flow field around the cylinder was measured. For the purpose of quality of data, two measurements were performed at the same region around the cylinder.

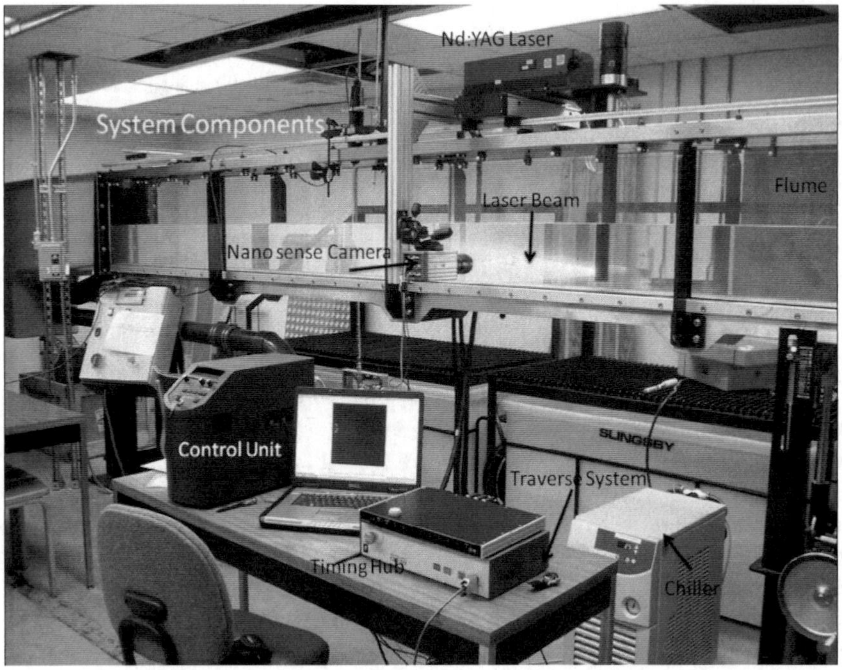

Figure 1 UT CAEE's flume and PIV setup

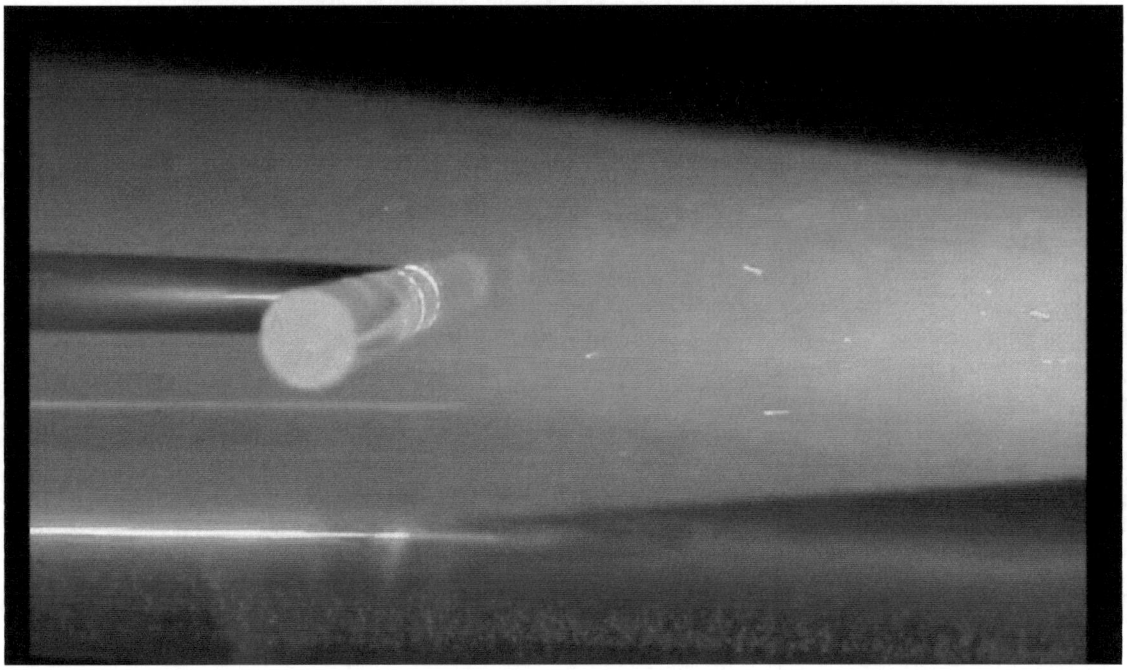

Figure 2 Cylinder during measurements, flow is from the left, the laser beam originates to the right

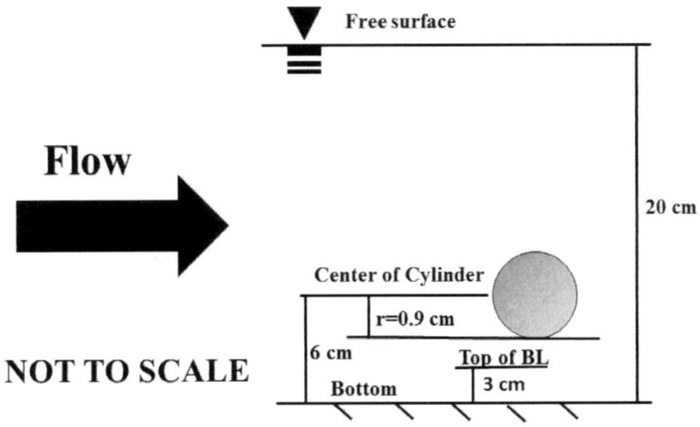

Figure 3 Schematic of cylinder in flume

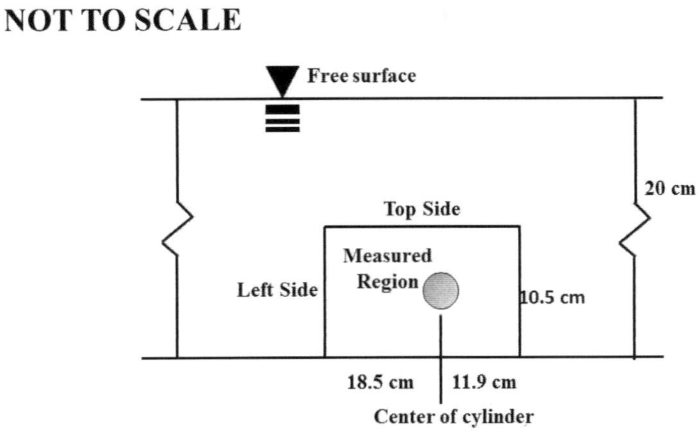

Figure 4 Measured region of the flow field

1.2 Mesh and boundary conditions

The domain of CFD simulations was the same as the measured region in the PIV experiment. Meshes (2D and 3D) replicating the dimensions of the measured region were created in ANSYS ICEM CFD meshing software. The 2D mesh consisted of 45,000 cells (Figure 5). The boundary condition for left hand side was velocity-inlet with the velocity profile from the PIV experiment. The upper boundary was also velocity-inlet applying the velocity profile from the PIV experiment. The turbulence intensity for the left and upper boundaries was 2%. The bottom and cylinder were walls. The right hand boundary was pressure-outlet. The 3D mesh (Figure 6) consisted of 30 layers in the spanwise direction and the spacing was constant (0.01m). The velocity profiles of the left hand side and upper boundary kept the same at all layers in the spanwise direction. So the two boundaries which corresponded to the vertical walls of the flume were symmetry, instead of wall. All other boundary conditions were the same as those in the 2D mesh.

Figure 5 2D mesh of the domain

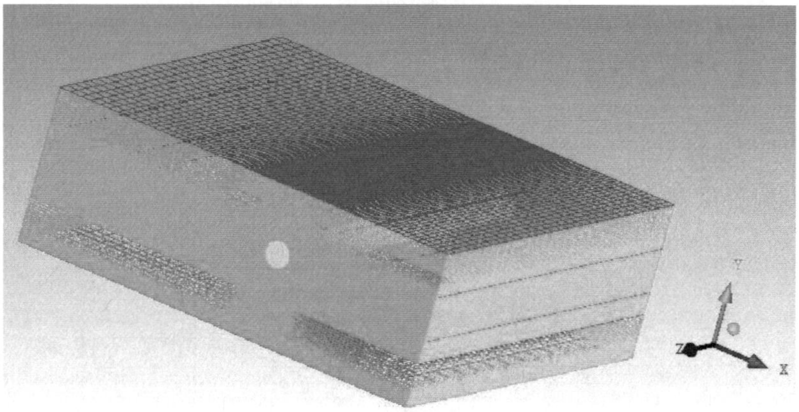

Figure 6 3D mesh of the domain

1.3 Turbulence models

In CFD simulations, two turbulence models were applied: Reynolds-averaged Navier-Stokes (RANS) and Large Eddy Simulation (LES). In RANS, the two-equation $k-\omega$ shear-stress transport (SST) eddy viscosity model was be applied. In LES the subgrid-scale model was Smagorinsky-Lilly (Al-Jamal and Dalton, 2004). In RANS, the 2D mesh was applied. In LES only the 3D mesh can be used. In RANS, Y+ values around the cylinder were less than 1.25. In LES, Y+ values around the cylinder were less than 1.50. The flow around cylinder was periodic and unsteady, so the influence of the initial condition and the iterative error of the solution of each time step must be reduced to negligible levels (Eca et al., 2014). Thus, the total simulation time was 12 seconds and the time step size was 0.002 second. In addition, there were 50 iterations in each time step.

2. Assessment of Morison's Equation

2.1 CFD simulations

As shown in Figure 7, this problem is in two dimensions. The cylinder is subject to an alternating flow with $U = U_m \cdot \cos(\omega t)$. The main task of this research is to study and compare cases with different combinations of Re and KC. The evaluations of C_d and C_M will be conducted based on the results from CFD. Basically RANS is applied in this research. In RANS, the two-equation k-ω shear-stress transport (SST) eddy viscosity model will be applied. In order to study the effects of turbulence model on the result, the Reynold's Stress Model (RSM) is applied, which is a higher level, elaborate turbulence model. In RSM, the eddy viscosity approach has been discarded and the Reynolds stresses are directly computed. The exact Reynolds stress transport equation accounts for the directional effects of the Reynolds stress fields. There are totally 21 different combinations of Re and KC. There are three different Reynolds numbers: 1,070, 20,000 and 300,000. There are seven different Keulegan-Carpenter numbers (Keulegan and Carpenter, 1958) : 2, 6, 12, 20, 30, 40 and 50.

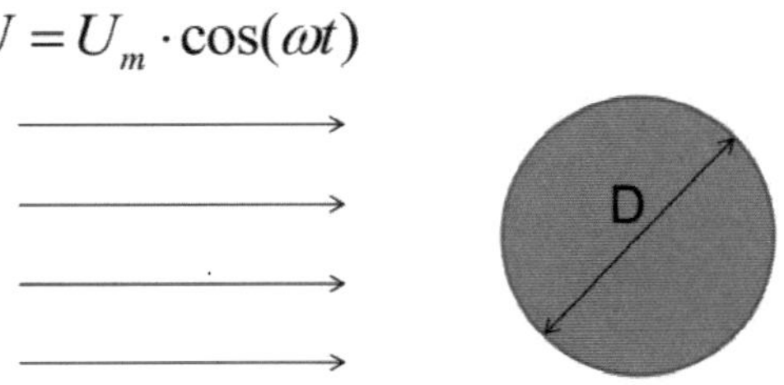

Figure 7 Two dimensional cylinder in alternating flow

2.2 Mesh and boundary conditions

One two dimensional mesh is created in the ANSYS CFD ICEM meshing software. The center of the cylinder is at the origin of the coordinate system. The geometry is symmetric about both x and y axes. W=100D and L=100D. There are 108,000 cells in this mesh (Figure 8). The boundary condition for the left side of the geometry is velocity-inlet. The realization of the periodic velocity is through the application of user defined function (UDF). The top and bottom sides of the geometry are set as symmetry. The right side is set as pressure outlet. Finally, the boundary condition for the cylinder is no-slip wall.

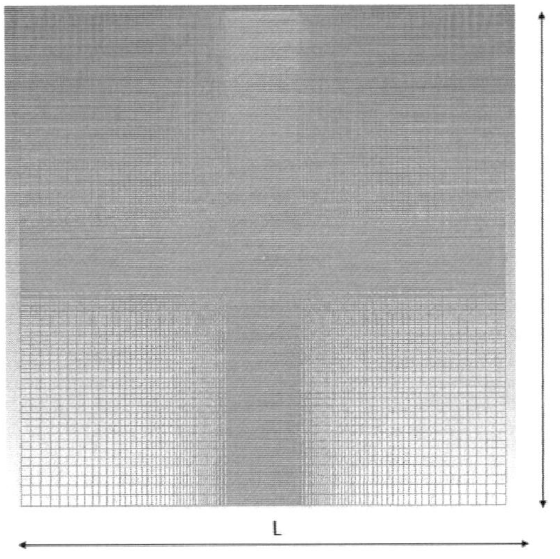

Figure 8 Two dimensional mesh

2.3 Evaluation of C_d and C_M

We first define that $C_x = \frac{F}{\frac{\rho}{2}U_m^2 D}$. The time history of C_x can be obtained from the application of monitor in ANSYS Fluent. From Morison equation, C_x can be given as

$$C_x = C_D \cos(\omega t)\,|\cos(\omega t)\,| - \frac{\omega \pi D C_m \sin(\omega t)}{2U_m} \quad (1)$$

Then the integral of C_x is performed as Equation (2).

$$\int_0^{\frac{T}{2}} C_x dt = \int_0^{\frac{T}{2}} C_D \cos(\omega t)\,|\cos(\omega t)\,| - \frac{\omega \pi D C_m \sin(\omega t)}{2U_m} dt \quad (2)$$

From the above integral, C_M can be calculated as Equation (3)

$$C_M = -\frac{U_m}{\pi D}\int_0^{\frac{T}{2}} C_x dt \qquad (3)$$

The second integral is performed as Equation (4)

$$\int_0^{\frac{T}{2}} C_x \cos(\omega t)dt = \int_0^{\frac{T}{2}} C_D \cos^2(\omega t)\,|\cos(\omega t)\,| - \frac{\omega\pi D C_m \sin(\omega t)\cos(\omega t)}{2U_m}\,dt \qquad (4)$$

From the second integral, C_D can be calculated as Equation (5)

$$C_D = \frac{\int_0^{\frac{T}{2}} C_x \cos(\omega t)dt}{\int_0^{\frac{T}{2}} C_D \cos^2(\omega t)|\cos(\omega t)\,|dt} \qquad (5)$$

RESULTS

1. Time-averaged velocities and Reynolds stresses

The time-averaged velocities and Reynolds stresses are analyzed and compared at three different sections (Figure 9). The first section is at the downstream of the cylinder, 2 diameters (0.036m) away from the center of cylinder. The second section is above the cylinder, 1 diameter (0.018m) away from the center of cylinder. The third section is at the downstream of cylinder, 2.5 diameters (0.045m) away from the center of cylinder. Since the PIV experiment is two dimensional, only three Reynolds stresses are analyzed, $\langle u'^2\rangle$ $\langle v'^2\rangle$ and $\langle u'v'\rangle$. Representative comparisons between PIV and CFD at station 1 are shown in Figures 10 to 12. More comparisons at stations 1, 2, and 3 are provided in Wang (2015). Additionally, comparisons of the measured and predicted velocities in time are provided in Wang et al (2015) and Wang (2015).

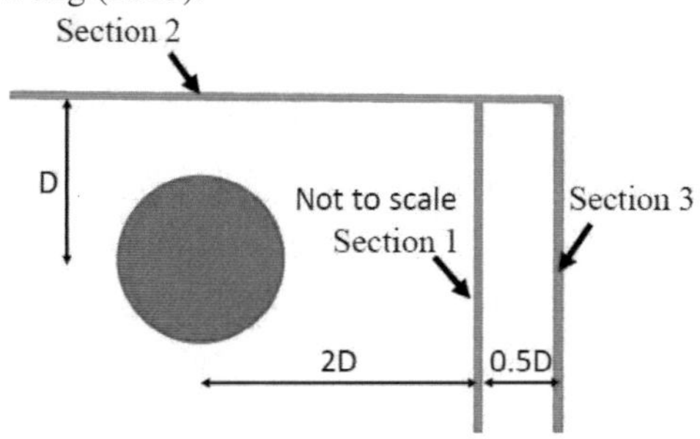

Figure 9 Stations where the time-averaged velocities and Reynolds stresses were measured

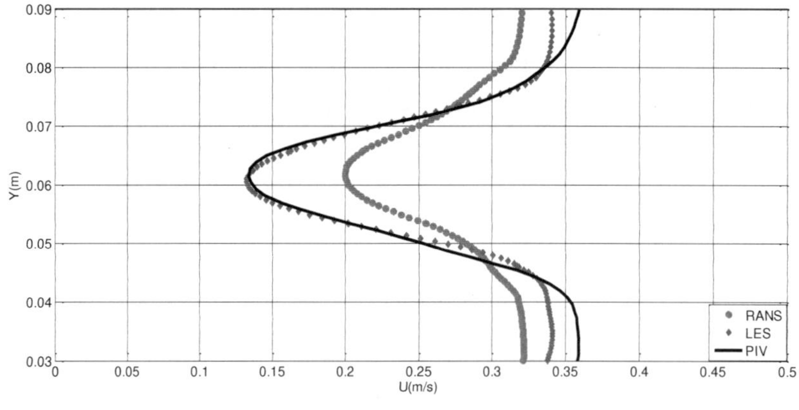

Figure 10 Time-averaged horizontal velocity at section 1

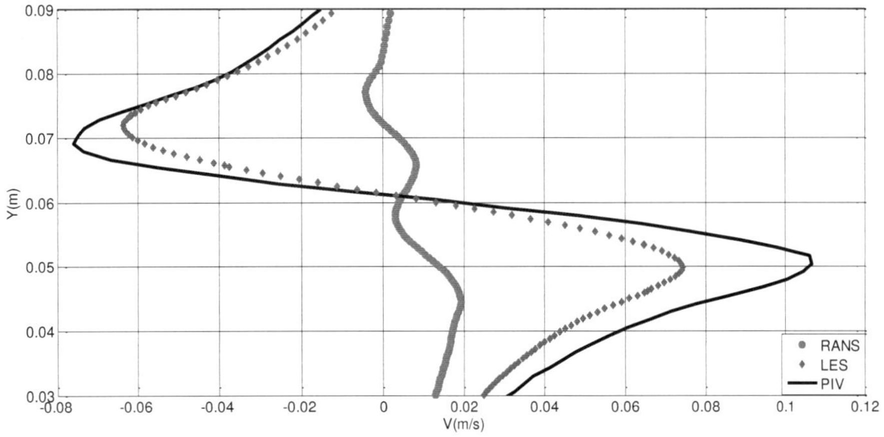

Figure 11 Time-averaged vertical velocity at section 1

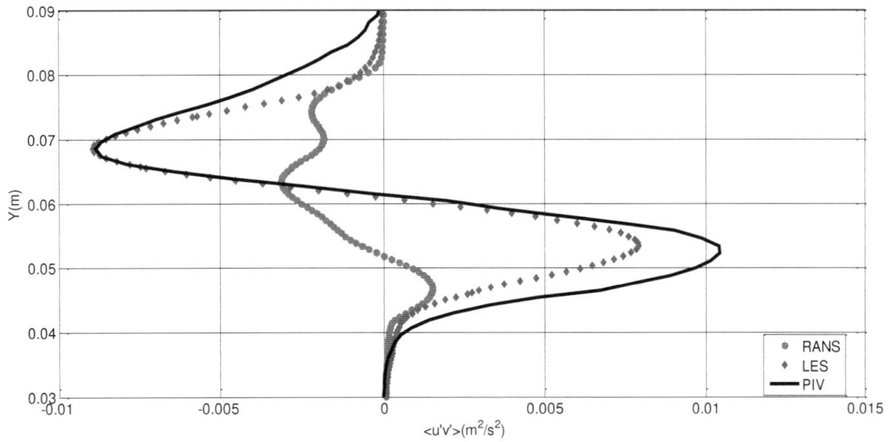

Figure 12 Profile of $\langle u'v' \rangle$ at section 1

From the comparisons shown here, as well as those in Wang (2015), it can be concluded that the predicted velocities from LES are much closer to those measured, than those from RANS.

2. Correlation between CFD and Morison equation

2.1 Comparison between RANS and Morison equation

For low KC numbers (≤ 12), the profiles of C_x from CFD are smoother than those of high KC numbers (≥ 20). So the Morison equation has a better correlation with CFD for low KC numbers. For high KC numbers, the correlation between CFD and Morison equation is better for high Re. Only plots with Re=300,000 are shown in Figures 13 to 19.

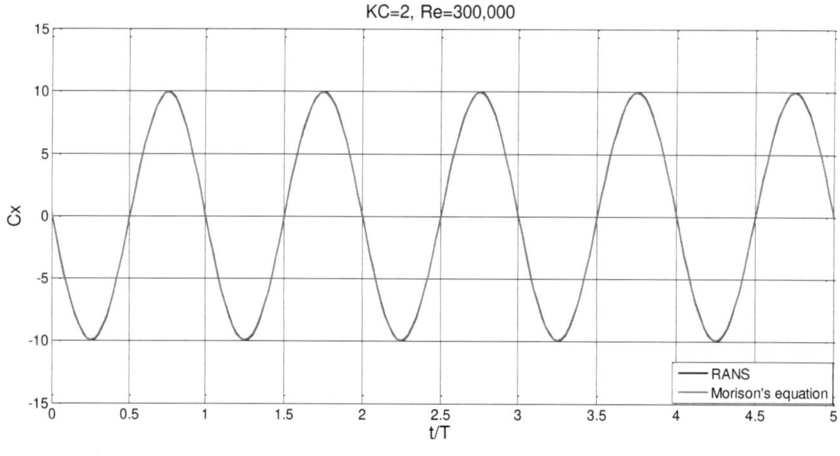

Figure 13 C_x for $KC = 2$ and $Re = 300{,}000$

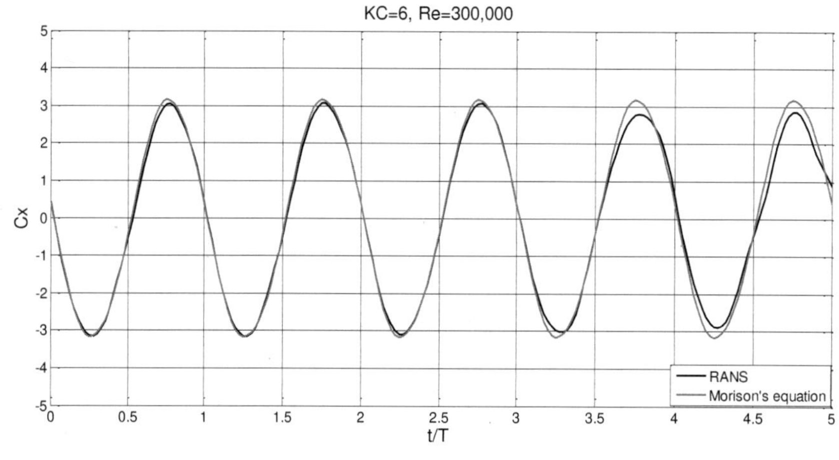

Figure 14 C_x for $KC = 6$ and $Re = 300{,}000$

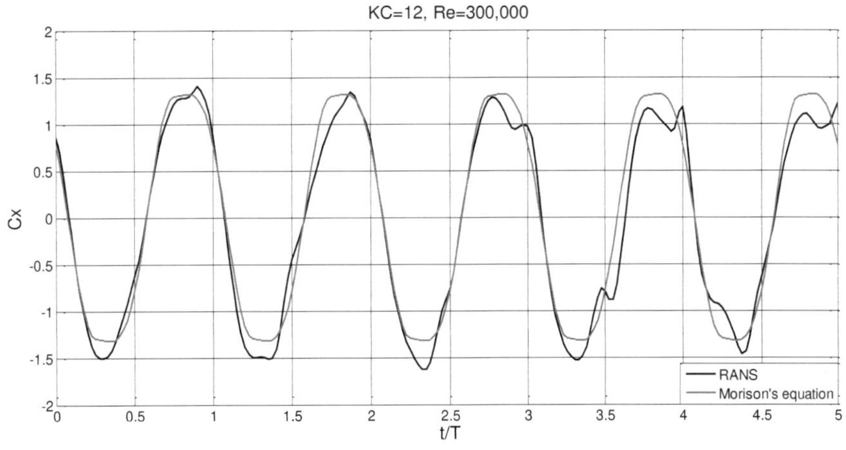

Figure 15 C_x for $KC = 12$ and $Re = 300{,}000$

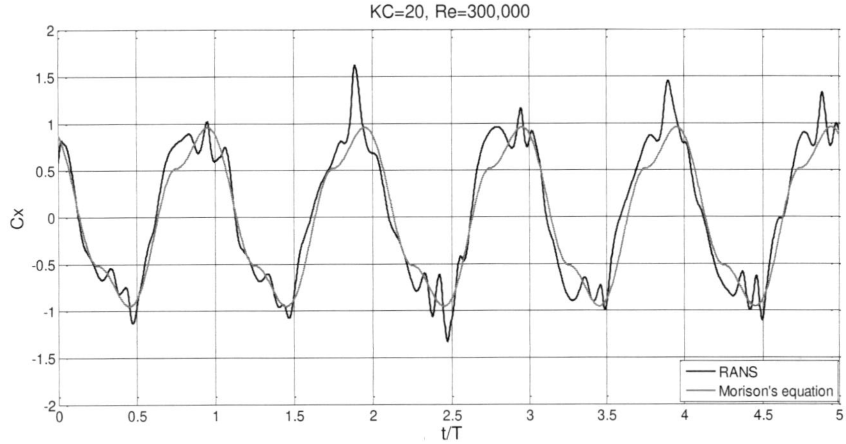

Figure 16 C_x for $KC = 20$ and $Re = 300{,}000$

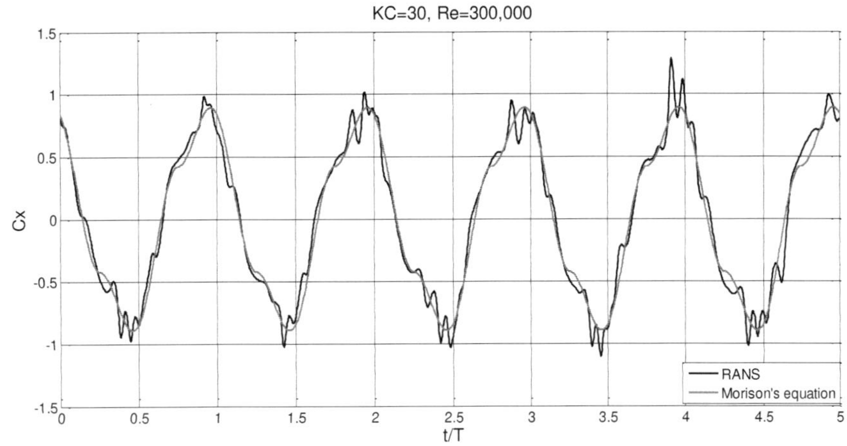

Figure 17 C_x for $KC = 30$ and $Re = 300{,}000$

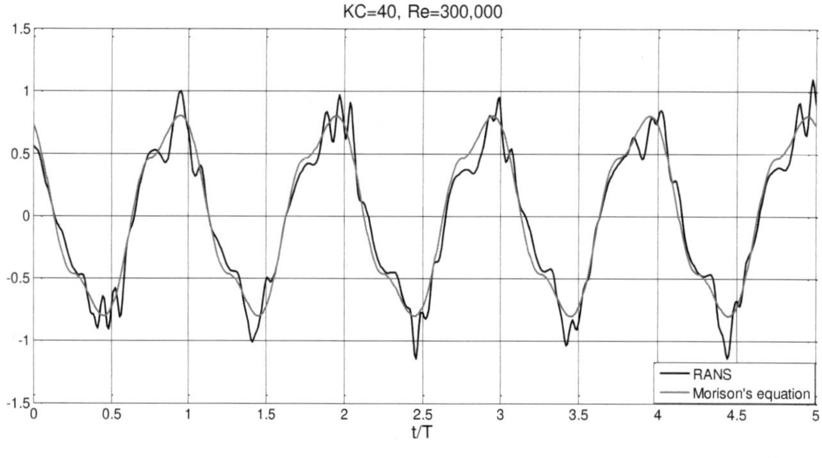

Figure 18 C_x for $KC = 40$ and $Re = 300{,}000$

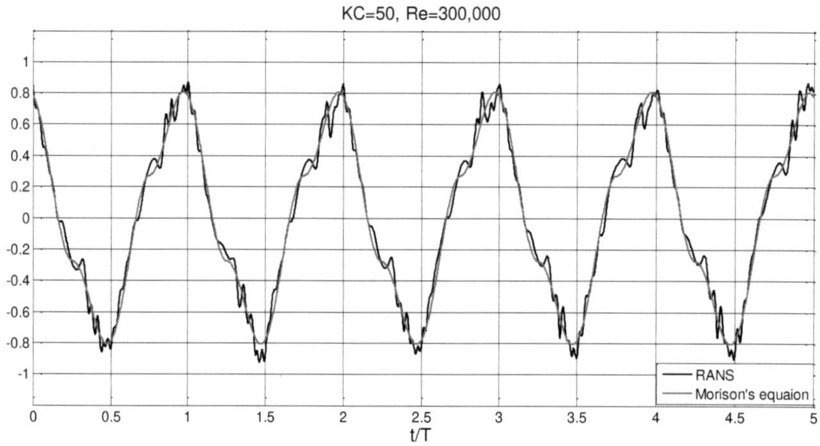

Figure 19 C_x for $KC = 50$ and $Re = 300{,}000$

2.2 Assessment of C_d and C_M

Based on the data from different combinations of Re and KC, we can get the corresponding values of C_d and C_M. The relation between KC and C_D is shown in Figure 20 for three different Reynolds numbers (1,070, 20,000 and 300,000). The relation between KC and C_M is shown in Figure 21 for three different Reynolds numbers (1,070, 20,000 and 300,000).

American Petroleum Institute (API) also provides the values of C_d and C_M corresponding to different KC numbers. For most offshore structures in the extreme design environment, Reynolds number are well into the post-critical flow regime, where C_d, C_M and C_{ds} for cylinders are independent of Reynolds number (RP2A-WSD 2000). Thus, the results of C_d and C_M for Re=300,000 from this study and those from API are compared. Figure 22 shows the comparison of C_d for KC<12. C_d is normalized by C_{ds}, which is 0.6 for smooth cylinders and K denotes KC. Figure 23 shows the comparison of C_d for larger KC numbers, where KC is normalized by C_{ds}. Figure 24 shows the comparison of C_M for KC<12. Figure 25 shows the comparison of C_M for larger KC numbers, where KC is normalized by C_{ds}.

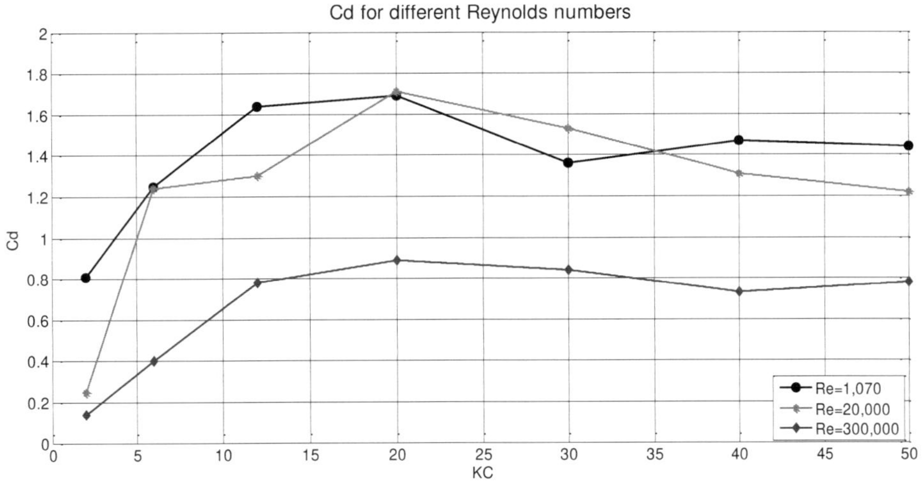

Figure 20 C_d vs. KC

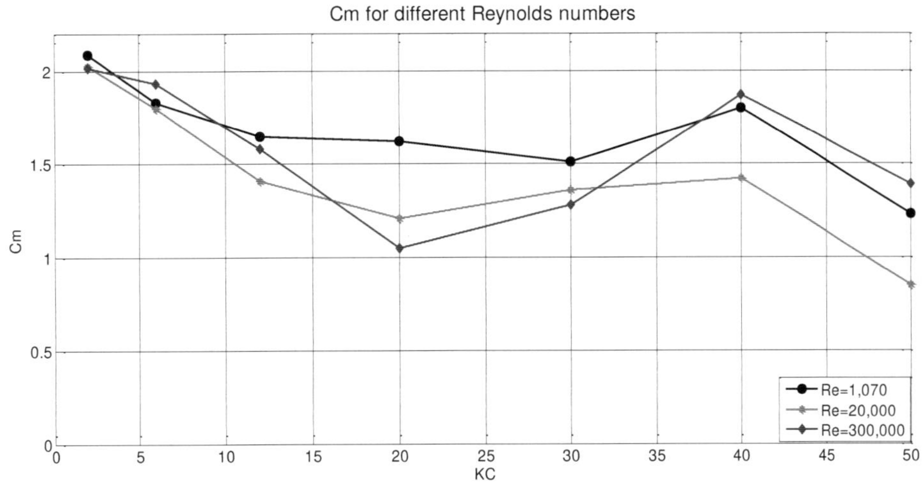

Figure 21 C_M vs. KC

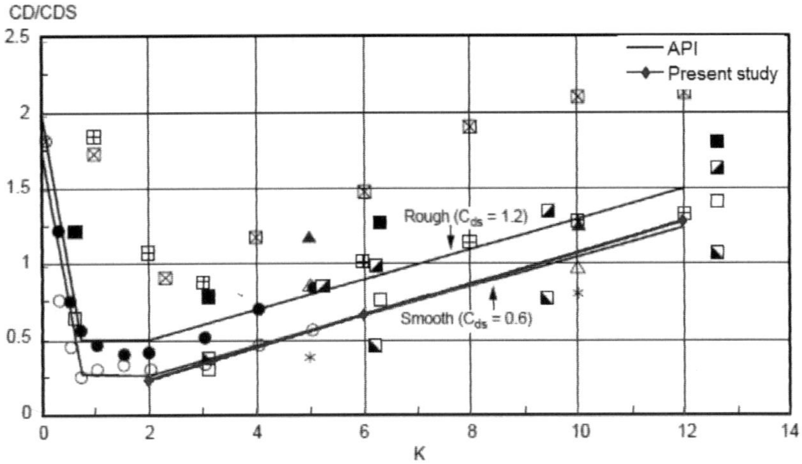

Figure 22 Comparison of C_d for KC<12

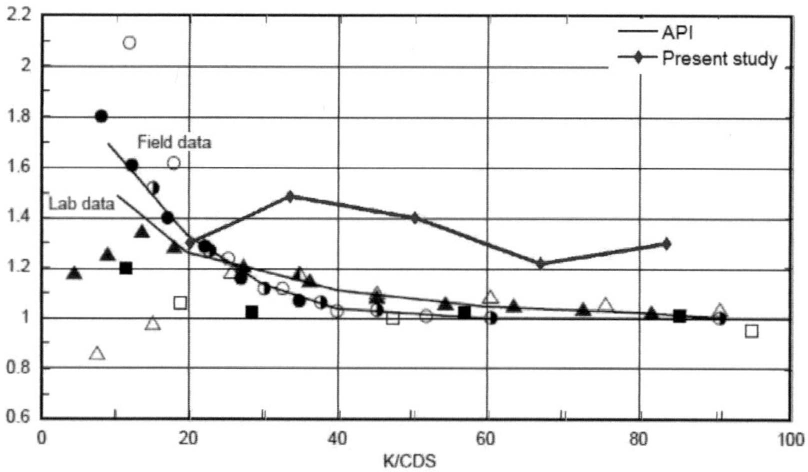

Figure 23 Comparison of C_d for larger KCs

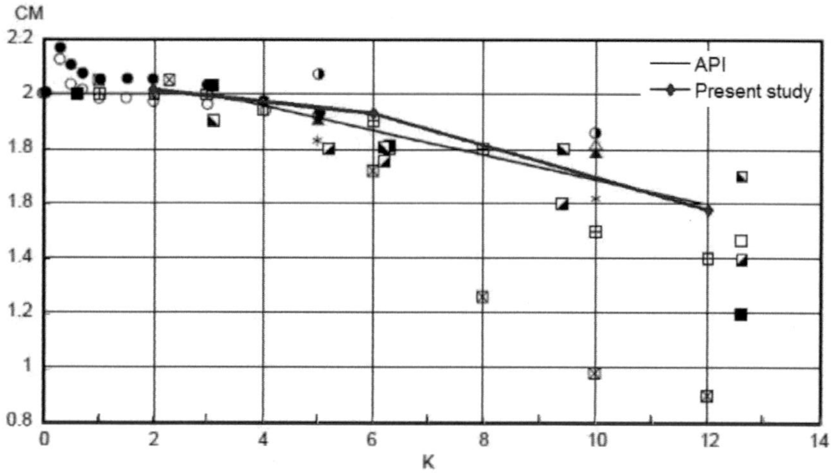

Figure 24 Comparison of C_M for KC<12

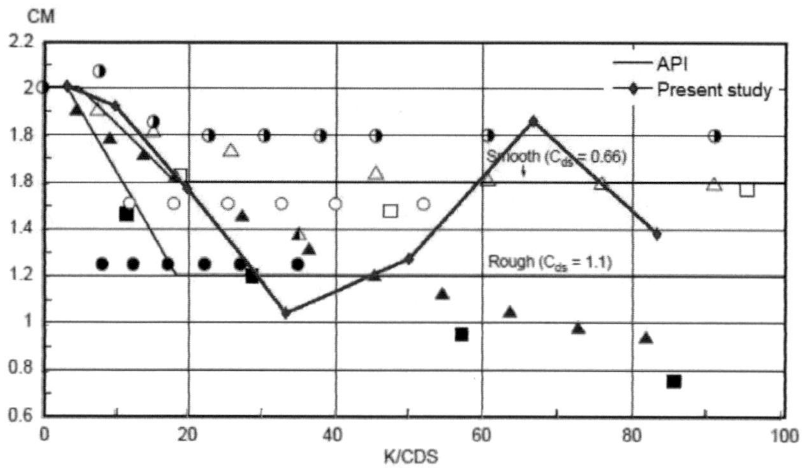

Figure 25 Comparison of C_M for larger KCs

2.3 Effect of turbulence models

In order to study the effect of turbulence models on the result, several cases are run with the Reynolds Stresses model and compared with the results from RANS, using the k-ω shear-stress transport (SST) model, as shown in Figures 26 to 29. When KC=2, the difference between the results from the two models is negligible. With the increase of KC, the effect of turbulence models on the results becomes more significant. Also at low Reynolds number, the effect is more significant than at high Reynolds number.

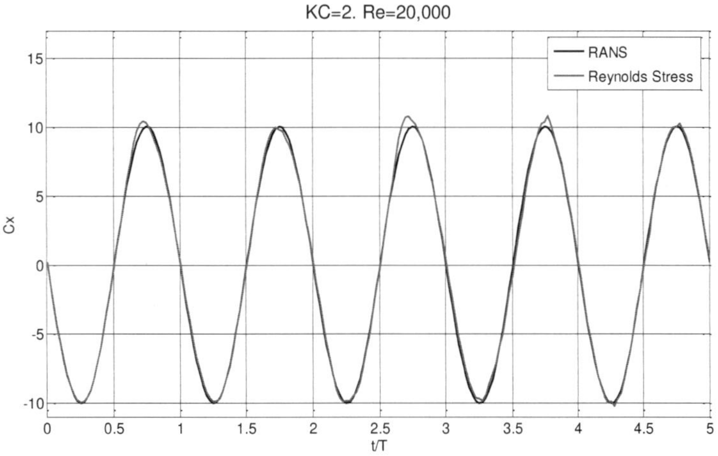

Figure 26 Comparison of C_x for KC=2, Re=20,000 from two turbulence models

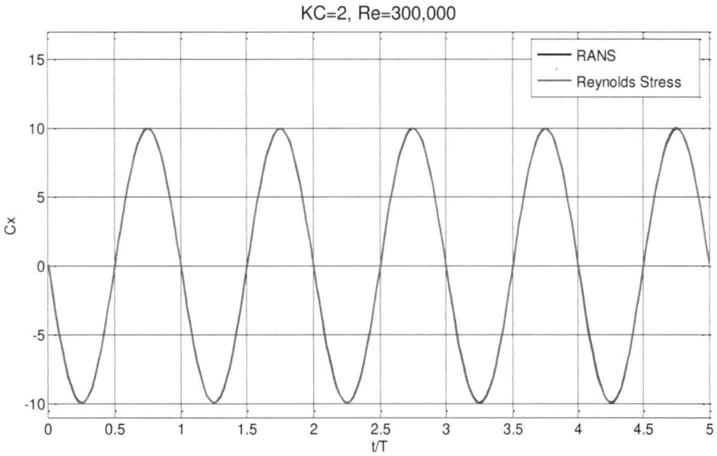

Figure 27 Comparison of C_x for KC=2, Re=300,000 from two turbulence models

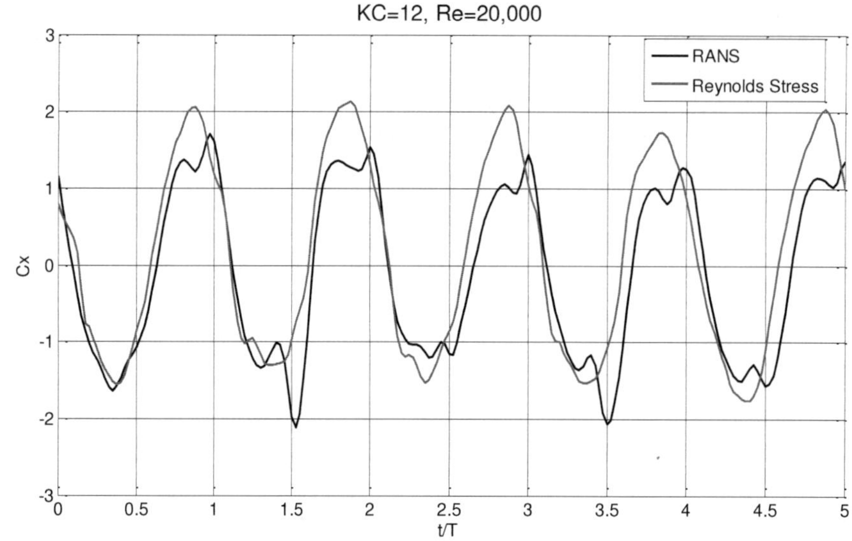

Figure 28 Comparison of C_x for KC=12, Re=20,000 from two turbulence models

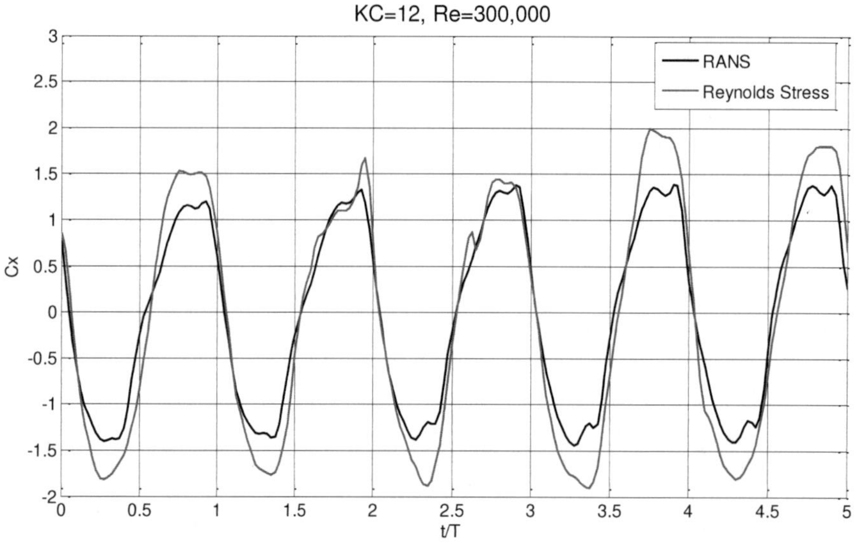

Figure 29 Comparison of C_x for KC=12, Re=300,000 from two turbulence models

2.4 Least Squares method to evaluate C_d and C_M

The method of least squares is an alternative way to evaluate C_d and C_M. It can be done by the application of Curve Fitting Tool in Matlab. The comparisons between the two methods to evaluate C_d and C_M are made for three different combinations of KC and Re, as shown in Figures 30-32. In addition, the corresponding values of C_d and C_M from two methods are shown in Table 1.

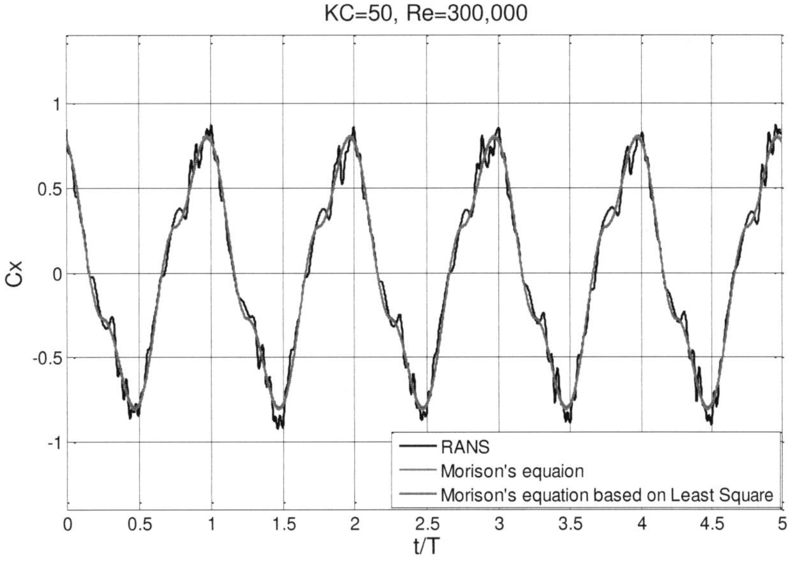

Figure 30 Comparison of C_x for KC=50, Re=300,000 from two methods

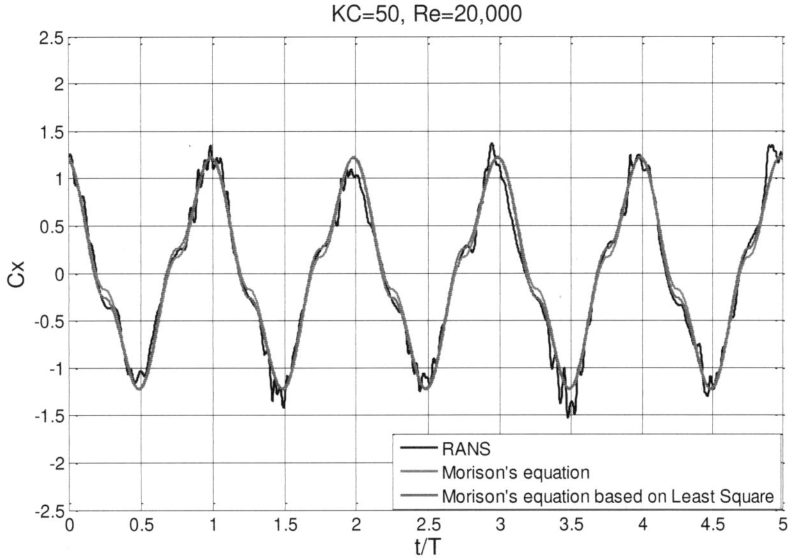

Figure 31 Comparison of C_x for KC=50, Re=20,000 from two methods

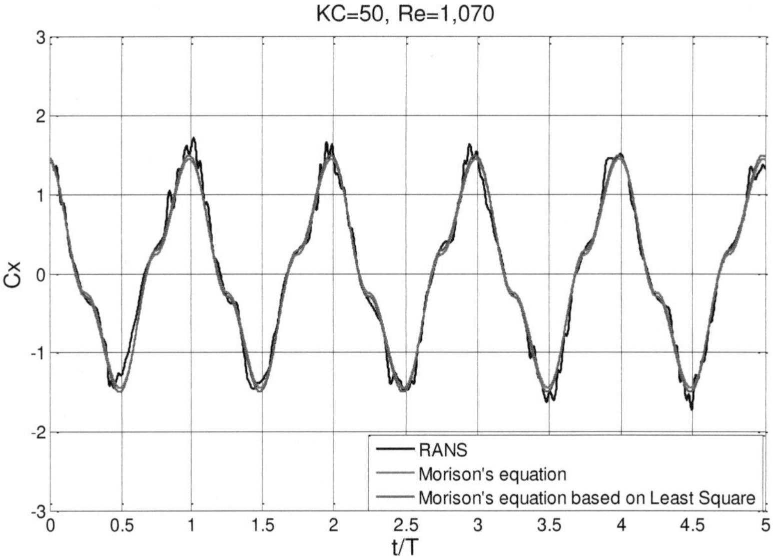

Figure 32 Comparison of C_x for KC=50, Re=1,070 from two methods

Table 1 Comparison of C_D and C_M from two methods

		Least Squares Method	Our method
KC=50	C_D	0.77	0.78
Re=300,000	C_M	1.36	1.39
KC=50	C_D	1.21	1.22
Re=20,000	C_M	1.29	0.85
KC=50	C_D	1.48	1.44
Re=1,070	C_M	1.47	1.23

CONCLUSIONS

At first, the flow around a cylinder inside a flume was measured by PIV and computed via RANS and LES, and comparisons of the predictions were made with each other, as well as with the measurements.

The mean values of the measured velocities and the Reynolds stresses were compared with those predicted by RANS and LES, with the latter being applied on the 3-D geometry of the flume and cylinder. The predictions from LES were found to be in much closer agreement to those measured, than those from RANS.

Then, a RANS simulation of a cylinder subject to alternating flow was used to assess the validity of Morison's equation. Despite the well-known oversimplification of Morison's equation, this method is found to be useful, based on the comparison between results of the Morison equation and those from CFD simulations for different combinations of Reynolds and Keulegan-Carpenter numbers. Nevertheless, for several combinations of Re and KC numbers, the detailed time history of the predicted inline force from ANSYS Fluent is quite different from that provided from Morison's equation.

The behavior of the drag and inertia coefficients, predicted from RANS, for various Reynold and Keulegan-Carpenter numbers is studied, and the results are compared with those from previous experiments as presented in the API manual. It is found that for KC<12, the results from this study can match those provided by API very well for both C_d and C_M. However, for KC>12, the results from this study tend to provide larger values of C_d and C_M than those of API. A possible reason for this discrepancy might be that the flow domain needs to be enlarged for higher KC numbers.

In this study, the range of Keulegan-Carpenter number is 0~50. However, in offshore engineering, the KC number can reach 100. So further analyses of cases for larger Keulegan-Carpenter numbers are needed, which could better meet the requirements of designing offshore structures. Finally, it was found that for some combinations of Re and KC numbers, the turbulence model can have a significant effect on the predictions, and thus this topic, along with the application of LES in the case of alternating flow, deserves further study.

ACKNOWLEDGEMENTS

We wish to thank the people who helped make this work possible: the members of Ocean Engineering Group (OEG), especially, graduate students Mr. Yiran Su and Mr. Weikang (Allen) Du. We also wish to thank Mr. Zhihao Li, graduate student in OEG, for his help with the writing of this paper. This work was partly supported by the office of the Offshore Technology Research Center at UT Austin.

REFERENCES

Al-Jamal, H., and C. Dalton. "Vortex induced vibrations using large eddy simulation at a moderate Reynolds number", Journal of Fluids and Structures 19.1, 73-92, 2004

Eca, L., G. Vaz, G. Rosetti and F. Pereira. "On the numerical prediction of the flow around smooth circular cylinders", Proceedings of the ASME 2014 33rd International Conference on Ocean, Offshore and Arctic Engineering (OMAE2014), San Francisco, California, 2014

Huang, Z. "CFD simulation of riser VIV", Texas A&M University, 2011

Keulegan, G. H., and L. H. Carpenter. "Forces on Cylinders and Plates in an Oscillating Fluid", Journal of Research of the National Bureau of Standards 60(5): 423-440, 1958

Morison, J. R.; O'Brien, M. P.; Johnson, J. W.; Schaaf, S. A. (1950), "The force exerted by surface waves on piles", Petroleum Transactions (American Institute of Mining Engineers) 189: 149–154]

API. "Recommended practice for planning, designing and constructing fixed offshore platforms-working stress design", RP2A-WSD, 2007

Stetson, P., "PIV Measurements of Flow-field Downstream of a Cylinder with and without Fairing and Comparison with CFD", MS thesis, Ocean Engineering Group, CAEE Department, UT Austin, 2013

Wang, G. "PIV Experiment and CFD Simulation of Flow around Cylinder", MS thesis, Ocean Engineering Group, CAEE Department, UT Austin, May 2015 (also UT-OE Report 15-01)

Wang, G., Tian, Y., and Kinnas, S.A. (2015) "PIV vs. CFD for flow around cylinder", 20th Offshore Symposium, Society of Naval Architects and Marine Engineers (SNAME), Houston, Texas, February.

Zhang, Wei, and Sang Joon Lee. "PIV measurements of the near-wake behind a sinusoidal cylinder", Experiments in Fluids 38.6, 824-832, 2005

SNAME
THE INTERNATIONAL COMMUNITY
FOR MARITIME AND OCEAN PROFESSIONALS

CRITICAL ISSUES IN THE DESIGN OF FOWT MOORING SYSTEMS WHEN COMPARING TO OIL&GAS INDUSTRY STANDARDS

MARTIN DUMONT
MOORING ENGINEER, BUREAU VERITAS

OLIVIER CARTIER
HEAD OF OFFSHORE FLOATING UNITS SECTION, BUREAU VERITAS

AUDE LEBLANC
RENEWABLE ENERGY - OFFSHORE SENIOR SURVEYOR, BUREAU VERITAS

CRISTINA BOUILLON
SUBSEA ENGINEER, BUREAU VERITAS

Proceedings of the 21st Offshore Symposium, February 2016, Houston, Texas
Texas Section of the Society of Naval Architects and Marine Engineers

ABSTRACT

The development of renewable energy through wind turbine systems at sea induces new engineering challenges. Mooring systems for Floating Offshore Wind Turbines (FOWT) are one of the most critical issues for the design process of such units. Over the past decades, the offshore industry has created and developed innovative and cost-effective solutions for the mooring of Oil and Gas (O&G) units such as FPSOs, semi-submersible platforms or TLPs. However the specificities of FOWT need to be carefully considered when trying to adapt the actual state-of-the-art to these new types of units.

This paper aims to highlight some of the main differences between mooring systems of FOWTs and O&G units. Comparison studies have been performed between several standards and rules in order to emphasize key points in the design process. Specific constraints related to FOWT implantation and in-service operations have also been analyzed. Relevant key points, such as redundancy of mooring lines and extreme analysis assumptions, have been selected and specifically studied, in order to give recommendations to the designer.

The redundancy of the mooring components is a recognized requirement in the O&G offshore industry. However, for small units, such as FOWT, with a limited number of mooring lines, this requirement may become too restrictive. This paper proposes alternative solutions.

Some of the assumptions usually made to perform fatigue and extreme mooring analysis for large unit such as FPSOs may not be valid for smaller units as FOWTs. This paper studies the compliance with different usual assumptions, such as the line dynamic influence. It also proposes an alternative method for extreme tension calculations.

The main objective of this paper is to propose guidelines to the industry for the design of mooring systems of FOWTs according to the results of these analyses, through the Bureau Veritas NI572 "Classification and Certification of Floating Offshore Wind Turbines". Even though the knowledge coming from the O&G industry is an important background, this paper tends to show that many points of the design may require specific attention.

Keywords: FOWT, Mooring systems, standards

INTRODUCTION

Nowadays, the need to develop renewable energy leads engineer to develop and design new systems. Considering the advantages of offshore fields (large amount of winds, weak visual impact), the development of Offshore Wind Turbines offers consistent solution for electricity production. To reach offshore location with larger water depth, fixed structures may not always be relevant. For these reasons, Floating Offshore Wind Turbine (FOWT) projects emerged started in the recent past years.

The development of such innovative systems raises many challenges for the industry. In particular, the design of the FOWT mooring system is one of the biggest issues faced by the engineers. The experience coming from O&G is of course to be considered, for instance through standards and classification societies rules and guidance notes. However, the specificity of FOWTs has to be taken into account when designing the mooring system. All these potential key points have to be identified and analyzed with great care.

The methodologies used for the extreme analysis of the mooring system are often based on several assumptions. Because of their dimensions, FOWT mooring systems may not always comply with all these assumptions. For instance, the line dynamic effect can be neglected or studied only through sensitivity studies for O&G. However, it may become much more crucial for the design of these small floaters. This paper proposes a method to account for these effects under certain assumptions.

The redundancy of a mooring system is a recognized requirement of the O&G industry. However, the risk associated to the failure of the mooring system for a small unmanned unit such as FOWT is lower. Moreover, this system may be composed by a small number of lines, which can lead to unrealistic requirements when trying to reach redundancy.

STANDARDS AND RULES FOR MOORING SYSTEMS OF O&G UNITS

Over the past decades, many standards and rules have been developed for the mooring systems of O&G units. Thanks to experience, analyses and calculations tools, classification societies are now able to give consistent recommendations for the design of mooring systems. A lot of these requirements are largely recognized in the O&G industry and can be found in international standards such as ISO, API or NORSOK documents. When thinking about mooring systems of innovative floating bodies such as FOWTs, it is important to first consider the state-of-the-art for the design of offshore mooring systems. The Table 1 presents some of the main standards and rules from classification societies regarding mooring systems for permanent and mobile offshore units.

Table 1 : List of standards and rules

American Petroleum Institute (API)	API RP 2SK
International Organization for Standardization (ISO)	ISO 19901-7
International Electrotechnical Commission (IEC)	IEC 62600-10 (in development)
NORSOK	N-001
Bureau Veritas	NR493
DNV	DNV-OS-E301
ABS	Guide for MOU

A huge quantity of information regarding mooring systems design is contained into these standards and rules. Obviously, a large part is still relevant for FOWTs and it is highly recommended to use these documents as guidance for the design. As examples, the definition of environment (wave, wind and current), the requirements for mooring line components in terms of fatigue life and design tensions, and the corrosion consideration are some of the issues where the experience coming from large O&G units is often relevant for small floating bodies such as FOWTs.

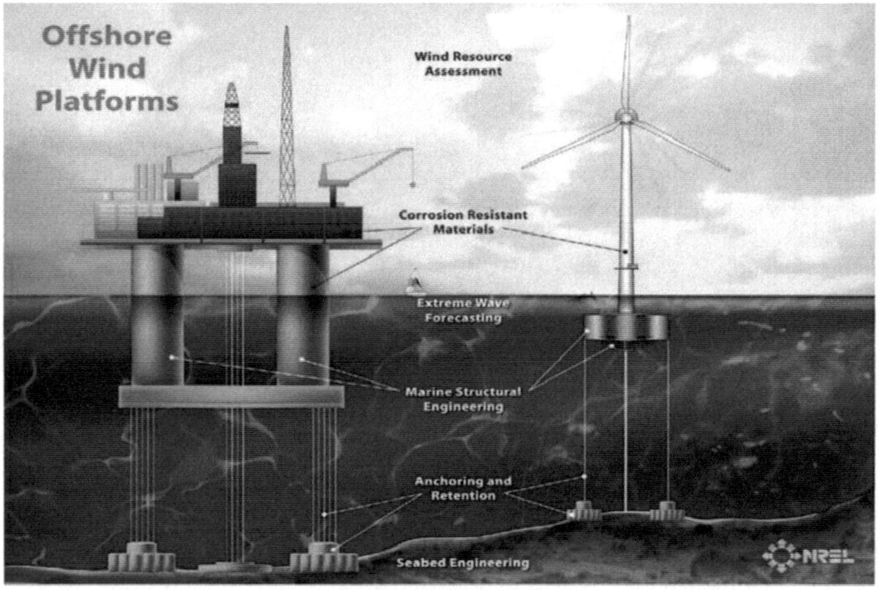

Figure 1: From O&G units to FOWTs (with the courtesy of NREL)

However, taking the O&G rules and standards without thinking about the differences with FOWTs would lead to large inconstancies in the design. Indeed, a large range of discrepancies between these mooring systems can be highlighted. Obviously, the dimensions of the floaters are the first difference to be taken into account. The responses of the floating bodies to the environment are clearly linked to its size and it becomes difficult to apply systematically the same calculation methods. As an example, for an FPSO, the line weight can be considered as negligible when studying the motions of the unit. This is not always a relevant assumption for a single FOWT considering the small floater displacement.

Considering these differences, the designer of FOWT mooring system has to list all the relevant key factors that have to be taken into account when trying to adapt the rules and standards from O&G. This paper aims to highlight two of these potential specificities and to give recommendations through the BV rules NI572.

APPROXIMATION OF LINE DYNAMICS

For a large O&G unit with catenary mooring system, the line dynamic may often be neglected. Indeed, the floater motions are usually not sensitive to line dynamic for traditional O&G systems because the mass/cable weight ratio is very large. For small floating bodies such as offloading buoys or FOWTs, this assumption does not seem relevant anymore. Indeed the contribution of line dynamic to extreme tensions can become larger and crucial for the design. In these cases, a sensitivity study is not enough to fully consider this effect.

However, a fully coupled analysis (where dynamic anchor leg response affects the first and second order motions of the floaters) is a time-consuming process. So it may not be relevant regarding time and cost constraints for the designer. Moreover, the large number of seastates that are usually to be analyzed is not consistent with the required computing time for dynamic calculations. The BV mooring rules propose a methodology to study several seastates using approached calculations such as quasi-dynamic calculations (based on dynamic responses of the floaters to static behavior of the line). In order to account for dynamic effect but keeping this fast type of calculations, the Bureau Veritas has developed a methodology presented here as an option for the designer.

1. Method

This method does not pretend to replace a fully coupled analysis. As explained below, it is based on several assumptions that have to be verified. The main purpose of this method is to give the possibility to perform a scanning analysis, in order to identify the dimensioning seastates that will be studied through a fully coupled analysis.

The idea is to divide the tension into two components, as presented in the equation (1), assuming that low and wave frequency motions and tensions do not interfere.

$$T_{dyn} = T_{dyn_{LF}} + T_{dyn_{WF}} \qquad (1)$$

Where:

- T_{dyn} is the total tension computed by a fully coupled analysis
- $T_{dyn_{LF}}$ is the low frequency part of the tension computed by a fully coupled analysis
- $T_{dyn_{WF}}$ is the wave frequency part of the tension computed by a fully coupled analysis

The second assumption is to consider that the low frequency part of the tension is not influenced by the dynamic of the line, hence we can write:

$$T_{dyn_{LF}} \approx T_{quasidyn_{LF}} \quad (2)$$

Where:

- $T_{quasidyn_{LF}}$ is the low frequency part of the tension computed by a quasi-dynamic analysis

One's assumes that $T_{dyn_{WF}}$ can be derived from $T_{quasidyn_{WF}}$ by a constant coefficient (DAF) which represents the increase of tension due to dynamic effect. Then $T_{dyn_{WF}}$ can be written as:

$$T_{dyn_{WF}} = DAF * T_{quasidyn_{WF}} \quad (3)$$

Where:

- $T_{quasidyn_{WF}}$ is the wave frequency part of the tension computed by a quasi-dynamic analysis
- DAF is a Dynamic Amplification Factor defined as per BV NR493

$$DAF = \frac{\sigma_{dyn_{WF}}}{\sigma_{quasidyn_{WF}}} \qquad (4)$$

Where:

- $\sigma_{dyn_{WF}}$ is standard deviation of the wave frequency part of the tension computed by a fully coupled analysis
- $\sigma_{quasidyn_{WF}}$is the standard deviation of the wave frequency part of the tension computed by a quasi-dynamic analysis

Finally the equation (1) becomes:

$$T_{dyn} \approx T_{quasidyn_{LF}} + \frac{\sigma_{dyn_{WF}}}{\sigma_{quasidyn_{WF}}} * T_{quasidyn_{WF}} \quad (5)$$

2. Calculations

$T_{quasidyn_{LF}}$, $T_{quasidyn_{WF}}$ and $\sigma_{quasidyn_{WF}}$ can be calculated quickly by neglecting the line dynamic influence. The in-house Bureau Veritas Software Ariane allows the user to compute this type of values through a quasi-dynamic analysis. Wave frequency motions are calculated in frequency domain and added at each time step to the calculation of low frequency motions in time domain. Thanks to these fast calculations (few seconds for 3 hours simulated), a lot of seastates can be studied. The only issue comes from $\sigma_{dyn_{WF}}$, that cannot be directly computed with quasi-dynamic calculations. Two steps are then required for the estimation of $\sigma_{dyn_{WF}}$:

- For a couple of representative offsets identified from many sea-states by quasi-dynamic analysis, one's compute RAOs tensions in a line dynamic software through a fully coupled dynamic computation. The RAOs are computed for each line of the mooring system around the mean position (several offsets have to be checked) on regular wave (in time or frequency domain).

- A spectral calculation is to be performed afterwards for each mean offset and the associated environments and RAOs. In this manner $\sigma_{dyn_{WF}}$can be assessed.

In conclusion, for each seastate the total tension as per equation (5) can be issued in order to classify all the environments and identify the most critical ones. These last may afterwards be tested by fully coupled dynamic simulations for validation.

REDUNDANCY

All the standards and Class Rules presented in the Table 1 impose redundancy for permanent mooring systems. In other words, the system is to be able to survive the worst combinations of environments with one line broken. This requirement makes particular sense for O&G units where a total loss of the mooring system could lead to catastrophic consequences (loss of human lives, oil spill, etc.). However, it may not be always relevant for small structure with low consequence class. Indeed, FOWTs are usually unmanned units, and the destruction of electric cables does not have the same environmental impact than risers and oil pipes. Moreover, for such small units, the mooring systems may be composed of a small number of lines. As an example, for a three lines system, the requirement of redundancy becomes a huge engineering constraint leading to non-acceptable costs increases.

Figure 2: Example of FOWT mooring systems with a small number of lines

From a classification society point of view, it seems difficult to impose the redundancy in any case considering the specificities of FOWTs. The challenge is then to propose a reliable alternative to the designer, but keeping acceptable level of reliability. This alternative has also to be easy to use for the designer. Indeed, large changes in the methodology could lead to a lot of additional computations, increase of computing time, potential errors and higher complexity of the studies.

An idea could be to increase the requirements in terms of extreme environment. For O&G units, the mooring system is to be able to resist to an environment corresponding to a 100 years return period (RP) for both intact and damage (one line broken) conditions. For non-redundant systems, it is necessary to reduce the probability of failure. In this study, it has been considered the following requirement: non redundant systems have to resist to seastates corresponding to a 1000 years RP with the usual safety factors (=1.67 in most standards and rules).

However, due to the difficulty to get reliable metocean data, this requirement seems complicated to apply for the designer. Usually, metocean engineers provide the environmental conditions corresponding to a return period of 100 years. From this consideration, an increase of safety factors to be applied with a 100 RP seems easier to use. In order to reach the same level of reliability than the "1000 RP requirement", this increase is to correspond to the difference between the 100 years RP and the 1000 years RP. The following study proposes the calibration of safety factors corresponding to this recommendation.

1. Methods

The calculation of the safety factor increase is performed according to the equation (6).

$$Td = \alpha * T(RP100) = SFactual * T(RP1000) \qquad (6)$$

Where:

- Td is the design tension calculated using the BV NR493 methods
- α is the safety factor to be calibrated
- $T(RP100)$ is the maximum tension occurring in the mooring lines when facing a 100 years return period environment
- $SFactual$ is the actual minimum required safety factor for intact study and is equal to 1.67 according to API, ISO and BV rules
- $T(RP1000)$ is the maximum tension occurring in the mooring lines when facing a 1000 years return period environment.

For the calibration of α, four mooring projects and three analytical distributions have been used. $T(RP100)$ and $T(RP1000)$ are computed for each project and distribution. Then α can be calculated from the equation (7).

$$\alpha = \frac{SFactual * T(RP1000)}{T(RP100)} \qquad (7)$$

2. Mooring projects and distributions

Four projects have been studied. They covered a large range of properties such as mooring configuration, line materials, water depths, shapes and dimensions:

Table 2: Mooring projects

	Floater Shape	Mooring configurations	Materials	Water depth (m)
Project 1	Ship shape	Spread - catenary	Chain	58.4
Project 2	Round shape	Spread - taut	Chain - Fiber ropes	400
Project 3	Tripod	Spread - catenary	Chain - Fiber ropes	36.5
Project 4	Barge	Turret - catenary	Chain - wire ropes	147

For these projects, the following environments have been applied for the calculations (see table 3). They are directly extracted from Metocean data reports.

Table 3: Metocean data

	Wave 1		Wave 2		Wind	Current
	Hs(m)	Tp(s)	Hs(m)	Tp(s)	V (m/s)	V(m/s)
Project 1 RP100	2.5	16	1	6.5	10	0.7
Project 1 RP1000	3	16.48	1.75	8	12	0.84
Project 2 RP100	15.6	17.2	x	x	23	0.59
Project 2 RP1000	18.72	20.64	x	x	27.6	0.708
Project 3 RP100	9.49	15.2	x	x	x	x
Project 3 RP1000	11.388	16	x	x	x	x
Project 4 RP100	14.6	15.7	x	x	26.6	0.61
Project 4 RP1000	17.52	17	x	x	31.92	0.732

Three Weibull distributions with shape factor = 1, 1.2 & 1.5 are also used to simulate the design tension (Td) response to the environment. In these analytical distributions, Td can be decomposed into two components as shown in the equation 2.

$$Td = Tmean + Tdy \qquad (8)$$

Where:

- $Tmean$ is the static part of the tension without environment, also called pre-tension
- Tdy is the dynamic part of the tension including the environment, assumed to be a function of the wave height Tdy =f(Hs²). This dynamic part of the tension is not to be mixed up with T_{dyn} as defined above in the Approximation of Line Dynamic. Tdy is the design tension when subtracting the pretension.

To evaluate the influence of the pretension on the results, calculations have been computed with $Tmean$ =0, 1 and 10 (related to Tdy for a return period of 100 years).

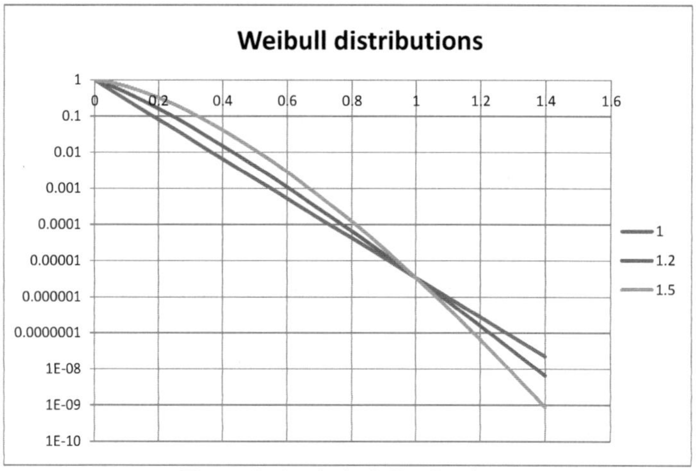

Figure 3: Analytical distributions

3. Results

The calculations for mooring projects have been performed using the BV in-house mooring software Ariane. $T(RP100)$ and $T(RP1000)$ have been computed for the four mooring project and for the analytical distributions.

Then, the safety factors α are calculated according to equation (7). The results are presented in the Table 4 and Table 5.

Table 4: Safety factor for mooring projects

	α
Project 1	2.03
Project 2	1.71
Project 3	2.01
Project 4	2.02

Table 5: Safety factors for analytical distributions

	α		
	Tmean=0	Tmean=1	Tmean=10
Weibull (shape 1)	2.34	2.00	1.73
Weibull (shape 1.2)	2.21	1.94	1.72
Weibull (shape 1.5)	2.09	1.88	1.71

4. Conclusion

From the mooring projects, it can be noted that the project 2 presents a lower value of α. One explanation could be the larger water depth (400 meters) giving a higher pretension, hence reducing the impact of the environment increase. This trend is confirmed by the analyses of the analytical Weibull distributions. The largest Tmean (=10) gives the smallest values of α (around 1.72), really close to α value of project 2. Tmean=0 gives the highest value of α, which is consistent with the idea that the pretension (or mean tension) reduces the influence of the environment.

It can also be highlighted that the shape of the Weibull distributions has an impact on the results. Comparing with the mooring projects, it seems that the shape 1 is closer to the real projects used for this study.

From all these considerations, the value of α has been fixed at α=2 as being the more representative value. It corresponds to an increase of 20% of the safety factor coming from O&G background for mooring systems with redundancy. Even if for FOWTs the required RP is equal to 50 years, the same increase can be applied. This value is then recommended in the BV NI572 for the design of non-redundant mooring systems.

CONCLUSIONS

Experience from O&G project can bring many things to the designers of FOWT mooring systems. Standards and Rules such as BV NR493 contain many information and recommendations that are relevant for these new innovative units. However, the specificities of FOWT have to be considered with care when trying to adapt these rules and methodologies.

Through the BV NI572, the Bureau Veritas proposes specific methods and criteria taking into account some of the challenging aspects of FOWTs. In particular, the mooring system is usually one the most critical issues.

A method of calculation for extreme tension has been developed in this paper to account for the line dynamic effect and to keep the advantage of fast quasi-dynamic calculations. Under some assumptions, the method allows to select the worst seastates from a large number of environment combinations, in order to limit the amount of fully coupled calculations.

The compliance with redundancy requirements has also been studied and new criteria have been proposed for non-redundant systems. From this study, an increase of 20% of the safety factors is proposed for non-redundant systems.

A lot of different key points non-presented in this paper are also to be studied with the greatest care for FOWTs mooring design, such as offsets calculations or the use of new materials. More information for the design of FOWT can be found in the BV NI572.

ACKNOWLEDGEMENTS

Special thanks to Anne-Laure DEFOY, for her large contribution to the development of the Approximation of Line Dynamic method.

REFERENCES

American Petroleum Institute, "Design and Analysis of Stationkeeping Systems for Floating Structures", API RP 2SK, 2015.

International Organization for Standardization, "Petroleum and natural gas industries - Specific requirements for offshore structures - Part 7: Stationkeeping systems for floating offshore structures and mobile offshore units", ISO 19901-7, 2013.

Bureau Veritas, "Classification of mooring systems for permanent and mobile offshore units", BV NR493, 2015.

Bureau Veritas, "Classification and Certification of Floating Offshore Wind Turbines", BV NI572, 2015.

SNAME
THE INTERNATIONAL COMMUNITY
FOR MARITIME AND OCEAN PROFESSIONALS

ON THE MONITORING OF MOORING SYSTEM PERFORMANCE

SUE WANG
SENIOR MANAGING PRINCIPAL ENGINEER, ABS

PING LU
SENIOR ENGINEER, ABS

Proceedings of the 21st Offshore Symposium, February 2016, Houston, Texas
Texas Section of the Society of Naval Architects and Marine Engineers

ABSTRACT

Detailed inspection of mooring systems in deep water is not easy, to say the least. As a safety critical element of an offshore floating unit, it is critical to understand the performance and degradation, if any, of the system. There is a consensus within the offshore industry that monitoring of the mooring system is important. Monitoring methodologies and devices have been developed and applied to some mooring systems. A certain level of requirements for a monitoring of mooring systems appears in standards and class societies' rules. However, gaps exist between the applications of the monitoring systems. This paper is to provide a review of the state-of-the-art in monitoring of mooring systems with particular focus on the effectiveness of monitoring methods that may depend on the characteristics of a mooring system, objectives, and many other conditions. Specifically, the influential parameters for the performance of mooring systems at a given environmental site and for different types of mooring systems are studied.
Keywords: Monitoring, Mooring System, FPSO, Global Performance, Failure Detection

INTRODUCTION

The number of floating production systems (FPS) has been continuously growing over the years. Currently, 349 floating production units are in service, on order or off-field/being remarketed (MaCaul, 2014, see Table 1). Stationkeeping for the floating production systems is generally achieved through the use of a mooring system, which will stay at a fixed location for the design life without regular dry docking. Mooring systems are exposed to a wide variety of environmental conditions and can deteriorate over time. The main degradation of mooring systems can be the result of: corrosion, wear, installation damage, flaws in manufacturing, material incompatibility, design errors, etc. The offshore industry has come to rely on regular in-service inspection of mooring systems as part of the integrity assurance program, but the available processes and tools are not always suitable, reliable, or sensitive enough. Mobile Offshore Drilling Unit (MODU) mooring systems can be inspected in air as they are retrieved or deployed, but that option is not normally available on an FPS.

A majority of mooring systems have performed to their design specifications; however, a number of failure incidents have occurred in recent years. Several references (Brown, et al, 2005, Ma et al, 2013, D'Souza and Maihi, 2013) have reported mooring system failures involving multiple line failure, loss of stationkeeping, emergency production shutdown, damage to risers with an associated small amount of hydrocarbon leak and others. Although none of the failures resulted in complete failure of the mooring system or in significant loss of hydrocarbon to the environment, the financial consequences were quite high.

Table 1 Floating Production Systems

Total oil/Gas FP in service, on order or available	**320**
Ship-shaped	*211*
Production barge	*10*
Semis	*48*
Spars	*23*
TLP	*28*
Total floating LNG processing system in service or on order	**29**
Liquefaction floaters (all on order)	*4*
Regasification floaters	*25*
Total Floating Storage units total in service on order or available	**102**

Among those reported failure incidents, a high number of failures occurred at an interface or discontinuity (e.g., connector, spelter socket, splice). The reports also indicate that corrosion, installation control, and cooperation management have been major contributors to several failure incidents.

These incidents suggest that there is a strong need to improve reliability of mooring systems. It has become even more important considering an increased number of FPS and a number of existing facilities are aging and several of them are approaching their design life or in some cases exceeding the original design duration. Although improvement in each related aspect (e.g., design, installation, maintenance, inspection, monitoring, operation) could positively contribute to the reliability of the mooring systems, experience indicates that there is room for improvements in the areas of mooring integrity management, inspection, and monitoring programs. A rigorous mooring integrity management plan, improved inspection, and monitoring program by owners and operators could greatly help improve the reliability.

It is imperative, however, to understand the limitations of an in-service inspection system. Improved quality of mooring components, installed as per the design, is a more effective way to improve reliability than are current underwater inspection techniques. These techniques will likely improve over time, and the value of current visual techniques at finding excessive corrosion and mechanical damage should not be dismissed, but there are significant limitations as to what can currently be achieved. Effective mooring monitoring can supplement inspection to provide information on mooring system performance, anomalies, mooring line loading for fatigue assessment, etc. and can have many benefits to the integrity of the mooring system. This paper mainly focuses on the monitoring of mooring systems.

REVIEW OF MOORING SYSTEMS MONITORING

Monitoring is a continual examination of a part or component of a facility to detect changes in performance. This can be to identify the onset and/or progression of specific degradation or simply to identify when component failure has occurred. As with inspection, it is essential that the degradation and failure modes and rates are understood as this forms the basis for selecting the appropriate technology and the monitoring frequency if it is not a continuous process. In-service monitoring for mooring systems is implemented for various reasons, in particular for mooring leg failures.

A number of mooring system monitoring methodologies have been developed to measure the mooring line load directly or indirectly, or to detect the failure of a mooring line. There are also more comprehensive monitoring and advisory systems that have been reported as being in use in existing mooring systems. The following descriptions of monitoring technologies have been included to help create and improve understanding of what is being done at present and proposed for the future. Table 2 provides the summary of discussed monitoring methods.

1. Load Cell

In theory, mooring line load monitoring is the most straightforward way to detect mooring line failures. There are systems which use fixed chainstoppers that have been outfitted with load cells underneath their base. There are also load cells installed in mooring lines. Load cells are an existing and mature technology; however, experience has indicated that accuracy, reliability and robustness are major issues in using the load monitoring system, especially underwater where access to the mooring line and instrument is very difficult if not possible. The power and signal transmission cables are areas of particular weakness for systems exposed to long-term offshore loading conditions. Signal drift can be detected up to a point; however, it is not always possible to determine what is due to instrument drift versus what is a slow change in the real mooring line load. Another problem can be the system losses between the chainstopper where the load cell is installed and the mooring line tension away from the facility. Not only is it difficult to ascertain the frictional losses, but they may not be fixed over time. A new development in recent years is the use of in-line load cells with a load cell housed in a protective casing, making it better suited for offshore installation. Data transfer is conducted via an acoustic transmitter.

2. Inclinometers

A simple inclinometer could measure mooring line angles. In calm weather, if any of the mooring line angles have changed to a significant extent, there is a possibility of a line failure. It can also be used for catenary systems to derive an approximation of the mooring line (quasi-static) tension by measuring the angle of the mooring line at or near the top connection to the platform (Ukani, et al, 2012). However, the dynamic and nonlinear effects are not included in the calculated tension from angle.

3. Global Positioning System (GPS) and Gyro

Theoretically, the abnormal changes of the offset of an FPS should be able to indicate the failure of a mooring line. Some units have installed GPS / Differential Global Positioning Systems (DGPS) for position monitoring. However, the effectiveness of using offset monitoring to detect mooring line failure depends on many factors such as the characteristics of the mooring system, water depth, monitoring of environmental conditions, reliability of GPS and Gyro. But in general, overall offset monitoring and recording using GPS and Gyro is cheap. The offset information combined with knowledge of environment and mooring system behavior could, at least, provide triggers for further inspection.

4. Integrated Monitoring System

An integrated monitoring system may include systems that monitor many different parameters and onboard forecast systems. The monitoring system monitors environmental conditions, such as wind, wave and current, FPS motions, mooring line angle and load. The forecast system may include the prediction of FPS motions and mooring line load. The data acquisition system uses a central server onboard to collect, process, and store the data from the various instruments. The system is relatively expensive and requires a lot of data processing.

A relatively simple integrated monitoring system including the measurements of vessel's position and mooring line angle has been installed on a FPSO. The mean mooring line tension is derived using the measured mooring line angles and vessel's position data. The calculated mooring line tension is displayed and compared with preset thresholds.

<u>Code and Standard on Monitoring Requirement</u>

- API RP 2SK: "*Moored floating units should be equipped with a calibrated system for measuring mooring line tensions if the operation requires mooring line adjustment, and line tensions should be continuously displayed at each winch. For units that do not require a tension measurement device, a device for detecting mooring failure should be considered.*"

- API RP 2SM: "*During operations, mooring line tensions or geometry should be monitored regularly, and the mooring lines should be adjusted, if needed to maintain stationkeeping performance.*"

Table 2 Summary of Monitoring Systems

	Load cell	**Inclinometer**	**GPS**	**Integrated system**
Maturity	Has been used in a number of applications	Has been used in a limited number of applications	Has been used in a number of applications	Has been used in a limited number of applications
Intent	Monitor mooring line load for the detection of line failure, over loading, or fatigue assessment	To measure mooring lines angles for the use of detecting line failure or as inputs to the mooring lines load assessment	To monitor and measure FPSO locations and hence to derive the mooring line load	Monitor and detect mooring line overloading, failure, and provide operation advisory
Application	All type mooring systems; difficulty for submerged turret mooring system	All types of mooring systems, most suitable for catenary system for derived line tension	All types of mooring systems	All types of mooring systems
Deployment	New and existing systems. for inline load cell; retrofit is relatively difficult	New and existing units	New and existing systems	New and existing systems.
Advantage	Long time, very well used method for load measurement. Relatively simple system, low cost.	Direct measurement of line angle and not affected by other parameters. Relatively simple system with low cost. Signal could be acoustic transmitted, low maintenance.	Easy to install and relatively low cost. Equipment on board the vessel and easy to maintain	Comprehensive system for real time monitoring and provide advisory for operations
Disadvantage	Measurement is not reliable especially when the mooring line load is not very high. Many other factors affect the output of the measurement, such as friction, temperature, signal drafting, recalibration, etc. durability of transmission cables	Can only derive quasi static mooring line tension. Dynamic and nonlinear effects are not included. Need careful calibration for the model of mooring line angles and mooring line load.	The relationship between vessel's position and mooring line load needs to be carefully studied to have a clear understanding between the measured positions and mooring line load. Environment measurement (wind, wave, and current) may be necessary.	Relatively expensive; A lot of data processing; The system is relatively new and the effectiveness of the advisory system is not known yet.

- UK HSE: "*The dutyholder should measure and record mooring line tensions. Where this is not practicable then suitable arrangements should be in place to verify the integrity of the mooring system and detect a line breakage. Various surveillance systems are discussed in the Oil and Gas UK guidance. Where none of these is practicable, evaluation of mooring tension may be inferred from the positional information. In this case the installation's position must be continuously monitored.*"

- IACS UR D3: "*Means are to be provided at the individual windlass control positions to monitor anchor line tension, windlass power load and to indicate amount of anchor line payed out.*"
- ISO 19901-7:"*Moored floating structures shall be equipped with a calibrated system for measuring mooring line tensions if the operation requires mooring line adjustment, and line tensions shall be continuously displayed at each winch. For permanent floating structures that do not require a tension measurement device, a means of detecting mooring failure shall be installed.*"
  "*If operation of the floating structure requires mooring line adjustment, mooring lines shall be equipped with a system for measuring line payout.*"
  "*If serviceability requirements impose constraints on the floating structure offset, the structure shall be equipped with a system to monitor its position.*"

EFFECTIVE USE OF MONITORING SYSTEMS

There has been considerable discussion in industry meetings about the effectiveness of mooring system monitoring, and particularly in determining mooring line tensions. It is important to realize that not every system will work on every type of mooring system. As an example, a monitored GPS-based system will be able to quickly detect windward line failures on a conventional catenary moored facility. The same GPS system may not be able to quickly detect a windward line failure on a taut moored grouped system, and could completely fail to identify a leeward mooring line failure. Among many factors, the following three are quite important and need to be established during the design stage.

- Clearly defined monitoring objectives, such as analysis tool validation, detection of abnormal behavior, line failure detection, or others.
- Evaluation of mooring system performance under various environmental conditions
- Understand the monitoring system, such as functionalities, limitations, reliability, ease of use, etc.

Using a monitoring system to detect abnormal behaviors of a mooring system is a complex issue since there could be many factors such as normal or abnormal phenomena, especially when a mooring line has a problem at the sea bottom or below the sea bed. For such a case, most available monitoring methods will not be able to identify the problem. Therefore, the limitations of monitoring systems should be clearly identified. For other influencing factors, a systematic study of the mooring system performance could positively contribute to the effectiveness of monitoring systems.

1. Floating unit hull type and motion characteristics

Mooring system response characteristics not only depend on the mooring type and pattern, but are also related to the hull type of the floating units. Ship-shape, semi-submersible and spar are three major hull types of floating production units, and their motions characteristics can be summarized in Table 3.

Table 3 Motion and Load Characteristics of Floating Unit

Vessel type	Heave motion (natural period (s))	Directional sensitivity	Wind load sensitivity	Wave load sensitivity	Current load sensitivity
Semi-submersible	Small (>20)	Low	High	Low	Medium
Spar	Small (>20)	Low	High	Low	High
Ship-shape	Large (5-12)	High	Medium	High	Medium

A series of studies on global analysis conducted in three phases of the DeepStar JIP from 1999 to 2003 (API RP 2SK, 2008) has found that the dynamic offset and tension in terms of total offset and tension decreases with increasing water depth for all types of floaters. The dynamic mooring line tension in terms of the total tension for ship-shaped FPSO appears to be less sensitive to change in water depth than other floating hull forms.

2. Mooring system response sensitivity study

In this study, responses of a turret moored ship-shaped FPSO and a moored spar were analyzed (see Figure 1). The study is to find the changes of the mooring line load, vessel offsets, and mooring line angles, which are the parameters that can be monitored, when the mooring systems experience one-line damage from intact conditions.

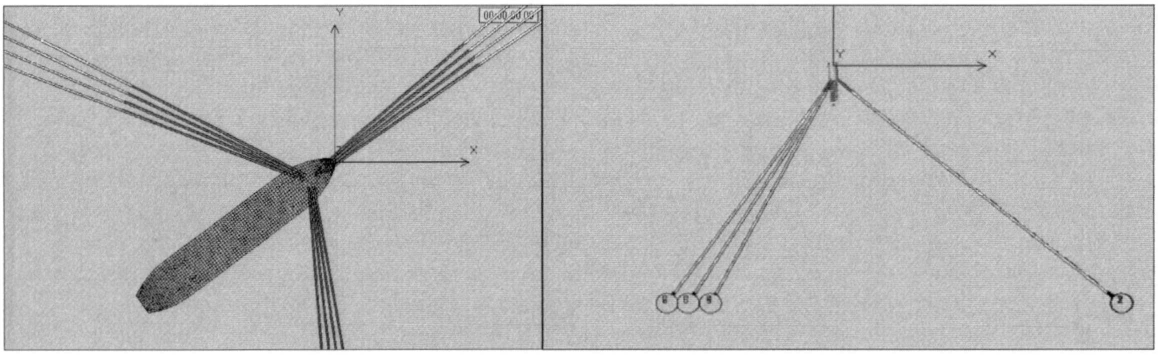

Figure 1 Numerical model of mooring systems

100-year hurricane wind of the Gulf of Mexico was used for the study (see Table 4). Wind, wave and current are assumed in co-linear condition and in the direction in line with one of the mooring line groups (see Figure 2).

Table 4 100-year Hurricane Wind

Water depth (m)	1000
Significant wave height (m)	11.98
Peak wave frequency (rad/s)	0.42
Mean wind speed (m/s)	49.65
Surface current speed (m/s)	1.34

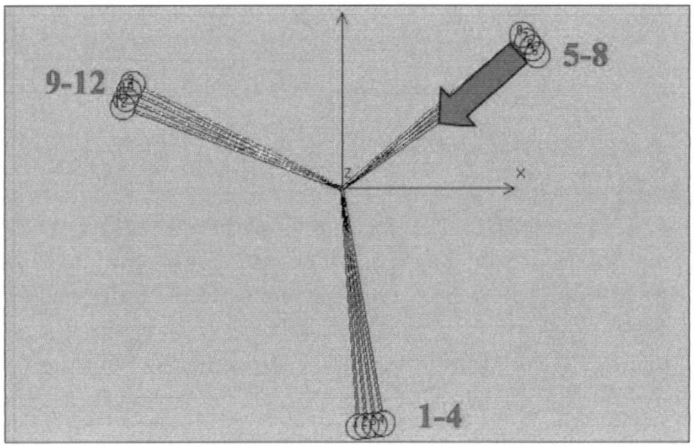

Figure 2 Environmental direction

Figures 3, 4 and 5 depict the mooring line tension, mooring line top angle and turret offset, respectively, for the turret moored system for intact and one line damage conditions. The mooring system consists of chain and fiber rope sections, and is a taut or semi-taut mooring system. The mooring line top angle is not very sensitive to the change of mooring line tensions, and may not be a good indicator for the evaluation of mooring line performance. For the mooring line tension, it can be seen that the broken mooring line load is mostly taken by the mooring lines within the cluster; changes of mooring line tension in other clusters are relatively small.

Vessel's offset is a global response that reflects the overall characteristics of the mooring systems; therefore, the damage of different lines does not make much difference on the total offset (see Figure 5).

Another analysis case includes the same environmental conditions, but comes from a different direction (see Figure 6). Figures 7, 8 and 9 are the corresponding mooring line tension, mooring line top angle and turret offset, respectively.

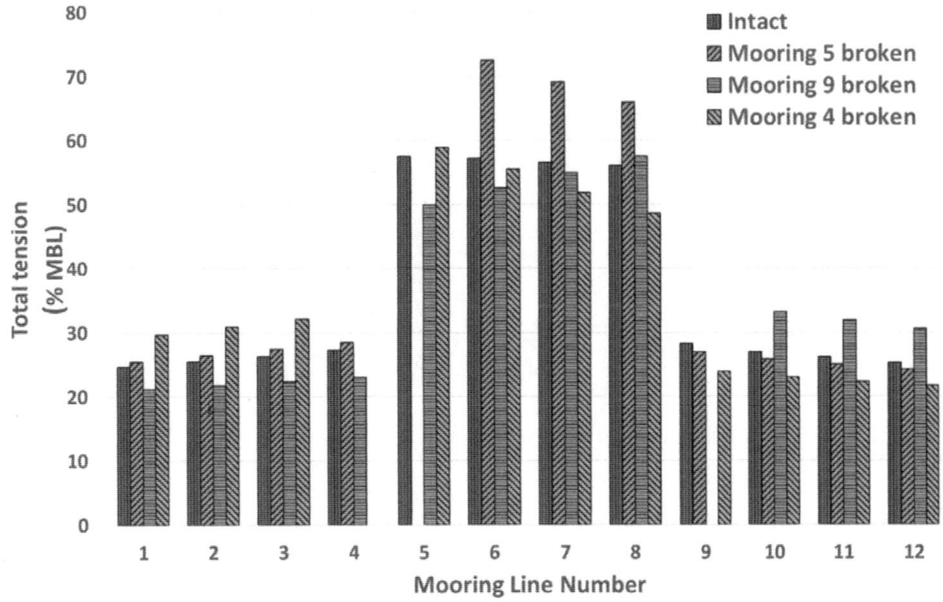

Figure 3 mooring line tension of turret mooring system

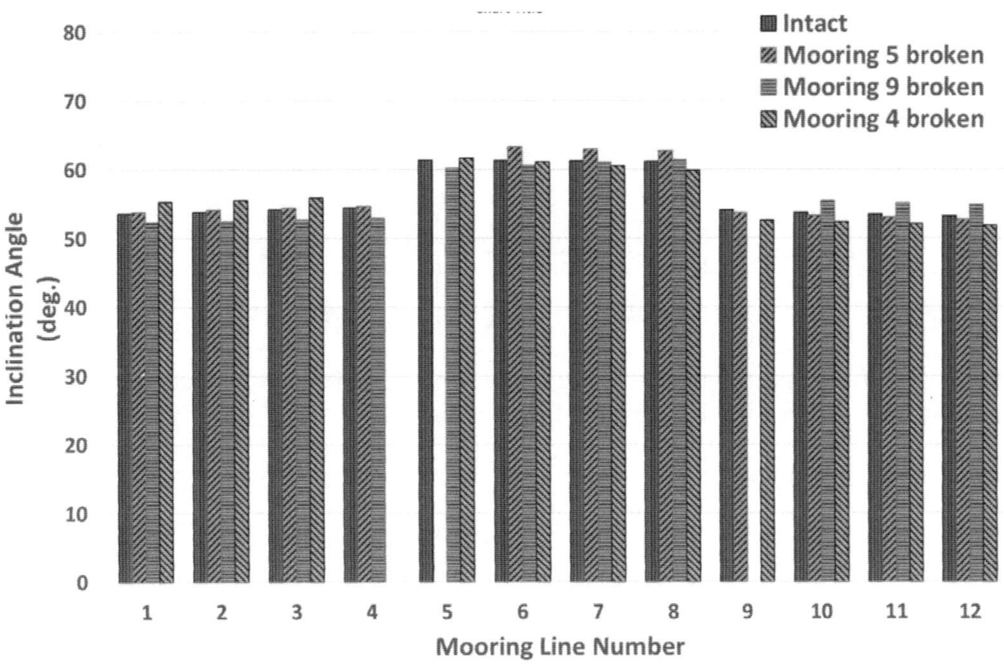

Figure 4 Mooring line top angle of turret mooring system

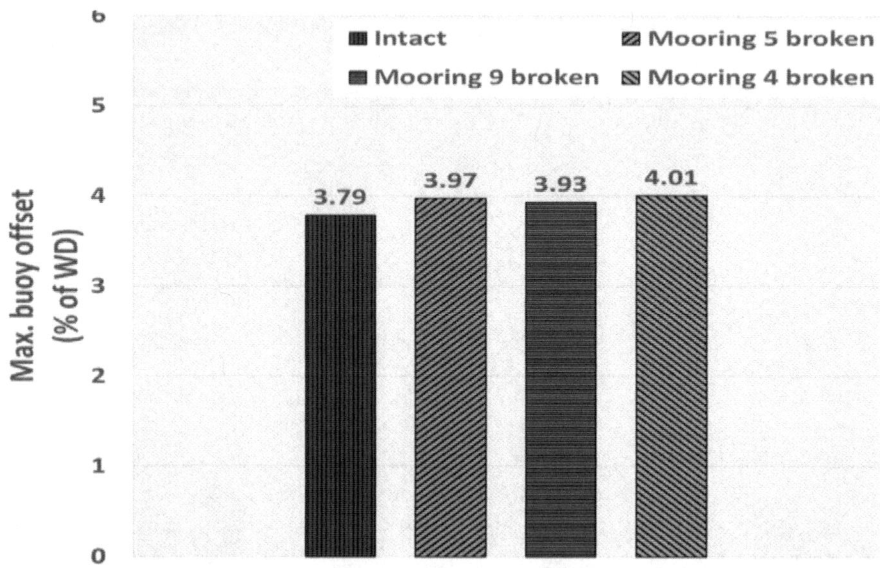

Figure 5 Turret offset of turret mooring system

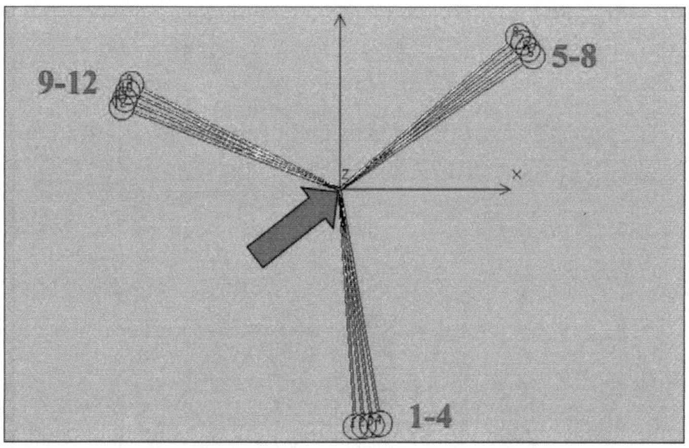

Figure 6 Turret mooring system with load toward a mooring line cluster

It can be seen that the performance of the mooring line cluster including lines 5, 6, 7 and 8 is not sensitive to one-line damage condition, since those mooring lines in the cluster are not taking much of the environmental loads.

Analysis results for the floating spar show similar trends. Since the mooring system for spar only has three mooring lines in each cluster, the change of mooring line load and spar's offset is more visible than that of a turret mooring system.

For a continued monitoring system, shift of mooring line tension and offset may be captured (see Figure 10). Please note that the results are for 100-year hurricane condition in the Gulf of Mexico and the change of the mooring system responses due to one-line damage may be noticeable. This may not be true for a benign environmental condition. Understanding the performance of the mooring system globally and in individual components are important for the planning of the monitoring system and the alarming system that is perhaps one of the most important goals for the monitoring.

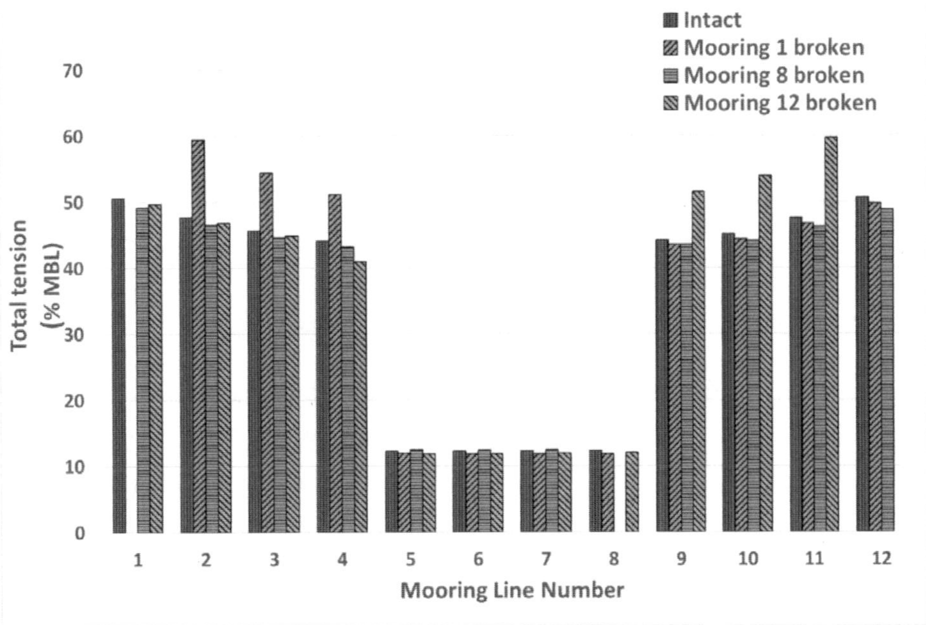

Figure 7 Mooring line tension for environment condition 2

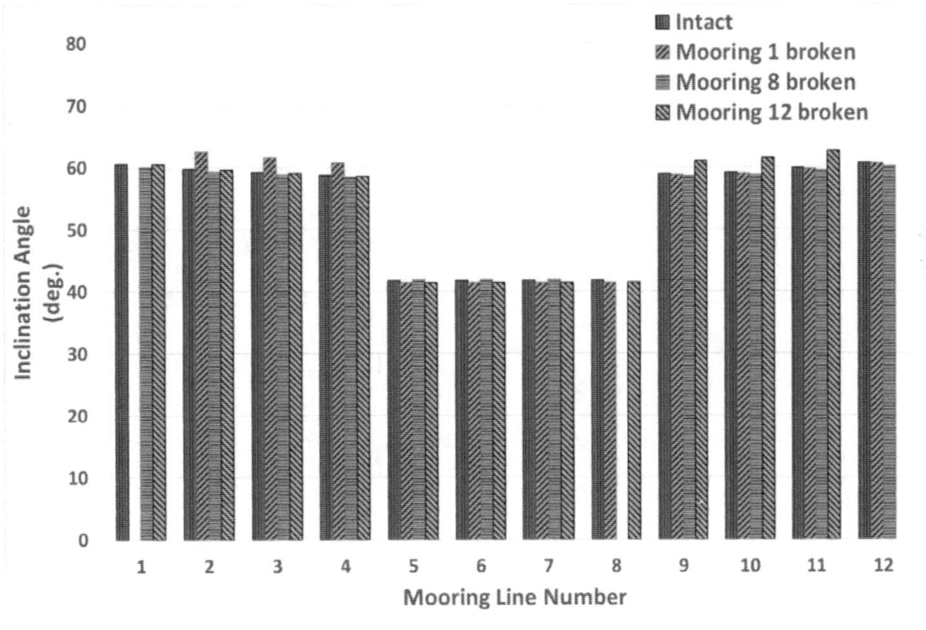

Figure 8 Mooring line top angle for environment condition 2

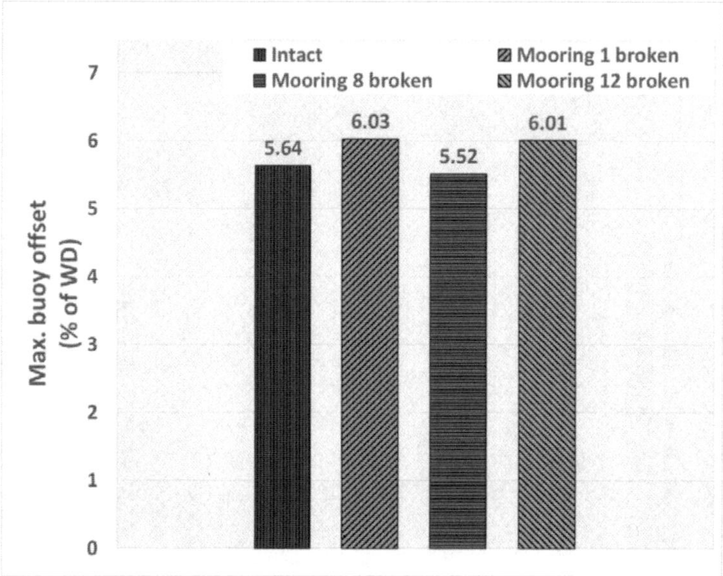

Figure 9 Turret offset for environment condition 2

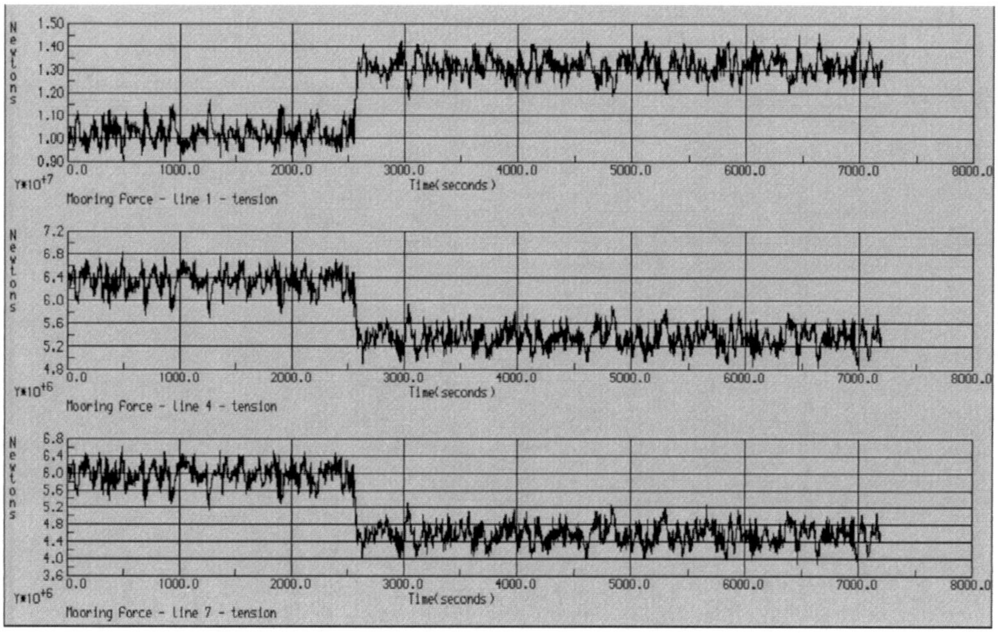

Figure 10 Mooring line tension time history

CHALLENGES OF MONITORING SYSTEMS

Real time mooring monitoring systems can have many benefits to the integrity of the mooring system, including:

- Reveal anomalies of mooring system for further investigation
- Provide warnings on mooring line failures
- Provide information for RBI, modification of inspection and repair plans
- Verification of numerical model for mooring strength analysis and understanding mooring system

performance

- Fatigue damage assessment using measured mooring line load, verification against fatigue analysis and condition assessment for possible life extension

However, it should be understood that, so far, the reliability and the effective use of monitoring systems are still limited. Much progress has been made in terms of the number of installed monitoring systems and technical advancements of monitoring systems. However, there are still many challenges, especially in the following areas.

1. Planning and Cost of Monitoring

The cost of monitoring is not insignificant and, therefore, methodologies have to be considered very carefully to determine how to achieve the optimum useful information at the lowest practical cost. For monitoring to be effective, the following aspects should be considered at the beginning:

i) Information needed to understand performance (e.g., extent of corrosion, fatigue tensions, mechanical damage, extreme storm tensions, loss of line)
ii) What to monitor to provide the understanding needed
iii) Reliability and accuracy of the related equipment and systems
iv) Establishment of baselines
v) Effective method of data processing, reporting and alarming for the ability to indicate degradation or failure of monitored system
vi) Interface design for ease of understanding of the monitoring parameters by operational personnel

A monitoring system is most easily incorporated into a facility at the design and construction phase. It appears that more operators are realizing the benefits of having such a monitoring system for an FPS, hence new builds now tend to include this in the design. The most challenges are for existing units to be retrofitted. The cost of retrofitting not only includes the cost of monitoring systems, but also the cost of installations which is often very expensive, especially for underwater installation. The cost and the effects of the monitoring system will continue to be a challenging topic.

The measurement equipment for a mooring monitoring system is subject to harsh environments, especially for underwater installations. They are often out of reach for inspections and maintenance. It could be difficult to determine if an abnormal signal is due to failure of the instrument or the behavior of the measured system without redundancy or a lot of systematic analysis and comparisons. A possible solution to this issue, at least in part, may be to use multiple monitoring methods – almost the case of one system monitoring the other.

2. Data Processing and Data Management

Over many years and for the multiple parts of an offshore facility, the amount of data gathered inevitably becomes substantial. It is important therefore that computerized data management systems be used to store, analyze, interrogate, and trend the captured data. In many cases, it is the comparison of different data that allows a good assessment to be made, and the data management system must allow this to be done.

Data processing and data management may have received less attention than those basic requirements during the planning and design process, which could result in large volumes of data being generated but then not used, or taking a long time to sort out needed results since the amount of data is overwhelming. A mooring line failed on one instrumented North Sea FPSO, but it took two weeks of data processing from the other lines to reveal the tension spike that confirmed it was a real failure rather than an instrumentation fault.

No sensors can be 100% accurate, and thus modeling tools are necessary to discard useless information and extract meaning out of the useful data. It is equally important to interpret the results in a way that is easy to understand. Visualization through visual imagery has been regarded as an effective way to communicate both abstract and detailed analysis.

3. Competence and Training

Mooring is a passive stationkeeping system, and the operations personnel, in general, do not seem to have an active role in mooring system management. However, providing knowledge and training to operations personnel regarding the mooring system is important even for a passive stationkeeping system, especially with a monitoring system on board. Operations procedures and failure response procedures should be clear and easily understood by the operations personnel, and they should be kept onboard, preferably within the maintenance and emergency response manuals.

4. Gaps in Mooring Monitoring Technology
There are a limited number of properly monitored mooring systems currently in existence. There are different views as to why this is the case. Among others, the lack of the direct link between the monitored parameters and specific onboard operations guidelines is, perhaps, the reason for insufficient impetus for industry to deploy suitably robust systems to work effectively. Notwithstanding this, there is a much better chance that suitable systems will be developed and deployed in the near future. The gap most likely to be filled first is the real time detection of failed mooring line(s).

CONCLUSIONS

The consequence of a mooring system failure can be very high. With the increased number of FPS installed and a growing number of aging FPS, monitoring of mooring systems becomes more important to the safe operation of the FPSs. However, monitoring of mooring system performance is complicated due to the many factors that could affect the performance of the mooring systems as illustrated in the paper. Inspection of a mooring system is difficult to say the least. Therefore, clearly defined monitoring objectives and the establishment of a design philosophy for the monitoring system are essential. In comparison with mooring system performance monitoring, mooring line failure monitoring/detection is relatively easier, but still not simple. The system needs to be tailored to the facility, and needs to be carefully monitored so that it is working properly and giving the results expected.

Advanced monitoring technologies, such as acoustic transmission, etc., could improve the reliability of monitoring systems. In addition, the following aspects are also important in an effective use of monitoring systems.

1. Mooring System Operations Guidelines
Specifications on mooring system operations manuals should be included in industry guidelines and standards. A mooring system is considered a passive stationkeeping system in comparison with a dynamic positioning system. There is often no detailed operations manual developed for the mooring system. However, experience suggests that it is important to have a well-developed operations manual for a mooring system, especially for a complex system. The following items may be included in the operations manual:

- Capability of mooring system under different conditions (intact, one-line failure, or other degraded cases)
- Expected response of the system under the same conditions (vessel offset, motions, etc.)
- Mooring risk studies should be kept on board for reference when unexpected events occur
- Response to monitoring status or alarms
- Emergency response
- Data records
- Incident report

2. Document of Mooring System Management Philosophy
It is recommended that a mooring system management philosophy document be developed for a mooring system. It should be the foundation upon which the design is undertaken, fabrication specifications are developed, installation is carried out, and the facility operated. Many of the other documents needed to ensure a reliable mooring system will inherently grow from this base document (e.g., the MIM, emergency procedures, mooring line failure detection method). It should include at least the following aspects:

- Design philosophy
  - Recognition that manufacturing is critical and QA/QC is fundamental to reliability
  - A holistic view of the purpose and limitations of the mooring system
  - Expected performance of the mooring system

- Installation verification, both the methods to be used and the results of the surveys
- Monitoring of mooring system to ensure that the system response is as expected from the analyses undertaken during the design stage
- Data gathering, codifying, use, and identification of anomalies
- Emergency response
- Methodology for determining when the MIM document should be updated

Over the years, mooring systems have been evolving in terms of numbers, size, water depth, new materials, operation conditions, etc., and a number of systems have become older. The offshore industry faces increased uncertainties of the performance of the mooring systems. Effective mooring monitoring can supplement inspections to provide information on mooring system performance, anomalies, mooring line loading for fatigue assessment, etc. and can have many benefits to the integrity of the mooring system. The drivers for inspect and monitor mooring systems range from regulatory compliance to dutyholder driven activities that are integral to their integrity management plan. The prescribed standards with the supplement of performance-based guidelines would be beneficial to the improvement of overall safety of the mooring systems.

REFERENCES

American Petroleum Institute, "Design and Analysis of Stationkeeping Systems for Floating Structures", API RP 2SK, 2005.

American Petroleum Institute, "Design, Manufacture, Installation, and Maintenance of Synthetic Fiber Ropes for Offshore Mooring", API RP 2SM, 2014.

D'Souza, R., Majhi, S., "Application of Lessons Learned From Field Experience to Design, Installation and Maintenance of FPS Moorings" OTC 24181, 2013

HSE, "Offshore installation moorings", Offshore Information Sheet No 4/2013

IACS, "General design parameters", IACS UR D3 Rev 5, 2012

ISO, "Stationkeeping systems for floating offshore structures and mobile offshore units", ISO 19901-7, 2013

Ma, K.T., Shu, R,. Smedley, P., L'Hostis, D., Duggal, A.,"A Historical Review on Integrity Issues of Permanent Mooring Systems", OTC 24025, 2013

Mooring Integrity JIP Reports, Phase 1, 2005; Phase 2, 2012

Ukani, S., Maurel, W., Daran, R., 'Mooring Lines – Integrity Management', Offshore Technology Conference, OTC-23369, Houston, TX, USA, 2012

SNAME
THE INTERNATIONAL COMMUNITY
FOR MARITIME AND OCEAN PROFESSIONALS

MOORING SYSTEM DESIGN AND ANALYSIS FOR THE DELTA HOUSE SEMISUBMERSIBLE

ZHIGANG TIAN
LEAD NAVAL ARCHITECT, EXMAR OFFSHORE COMPANY

BRANKA RADANOVIC
NAVAL ARCHITECTURE MANAGER, EXMAR OFFSHORE COMPANY

OTTO DASILVA
VICE PRESIDENT OF ENGINEERING, EXMAR OFFSHORE COMPANY

GLEN AUTHEMENT
PROJECT MANAGER, EXMAR OFFSHORE COMPANY

Proceedings of the 21st Offshore Symposium, February 2016, Houston, Texas

Texas Section of the Society of Naval Architects and Marine Engineers

ABSTRACT

The paper presents the design and analysis of a spread mooring system for the Delta House Floating Offshore Installation (FOI), which is a production semi-submersible installed in Mississippi Canyon 254 in the Gulf of Mexico in 1350 m water depth. The unit was designed by Exmar Offshore Company (EOC) and owned and operated by LLOG Exploration LLC.

In the design and analysis of the mooring system, one major challenge is modeling polyester rope change-in-length characteristics. Conventional method to envelope maximum vessel offset and maximum line tension using lower and upper bound stiffness is compared to results based on nonlinear change-in-length characteristics from full scale sub-rope tests. Mooring chain fatigue analysis considering vessel Vortex Induced Motion (VIM) and tension Root Mean Square (RMS) amplification due to Out-of-Plane bending (OPB) are performed. Significant effects of VIM and OPB on chain fatigue are found.

Keywords: Semi-submersible, Mooring Design and Analysis

1 INTRODUCTION

The mooring system for the LLOG Delta House FOI is designed according to API RP 2SK [Ref 1] and classed by DNV GL. This spread mooring system consists of 12 chain-polyester-chain lines and 12 suction piles. One hundred year hurricane condition obtained from Delta House Metocean Criteria Study [Ref 2] is used as the base design environment. One thousand year hurricane is used as a robustness check. Loop current conditions are also considered in the mooring strength and fatigue analysis.

Mooring strength analysis is performed using fully coupled dynamic analysis in time domain. Lower ("Post-installation") and upper bound ("Storm") stiffness of the polyester ropes is used to envelop the maximum offset and line tension, respectively. Additionally, to achieve more accurate predictions, nonlinear polyester stiffness obtained through model tests is used in the analysis. In addition, the analysis examines the effect of permanent elongation on line tension and vessel offset.

For the mooring chain fatigue analysis, environmental bins, including loop currents, are created from wave scatter diagrams provided in the metocean study. Time series of the line tensions in these environmental bins are obtained and processed to obtain the low and high frequency tension RMS. Fatigue damage due to the low and high frequency tension is calculated and combined according to API RP 2SK method. Additionally, mooring chain fatigue damage due to vortex induced motion (VIM) and bending at fairleads is estimated with VIM model test results [Ref 3] and API RP 2SK recommendations.

Overall, the performance of the mooring system meets the offset criterion required by the riser system and line tension criterion set by API RP 2SK. The mooring system in a 100 year hurricane environment is the governing case while the loop current environment produces less demanding line tension and vessel offset. The mooring chain fatigue life is much longer than the required 10 times of the 25 year service life.

2 DELTA HOUSE SEMI-SUBMERSIBLE

The Delta House semi-submersible is a column stabilized FOI. The hull features the EOC patented pontoon "lip" in front-of-columns configuration, which improves vessel heave motion performance. The vessel also has unique five-sided columns, which have been shown to reduce VIM of semi-submersibles and increase hull stability efficiency.

The semi-submersible has an operating draft of 30 m and a displacement of 38,075 m3. The column width is 14.4 m and the pontoon is 88.8 m long with a cross section of 10.2 x 7.2 m.

3 MOORING SYSTEM AND DESIGN PRINCIPLE

The taut mooring system is designed according to API RP 2SK. The system consists of 12 mooring lines with a group of three lines per corner of the ring pontoon. Each mooring line is composed of top and bottom chain and polyester rope, along with connecting chain links, subsea connector and a suction pile.

The payout of the R4 studless top chain is determined to be 235 m such that the top of the polyester rope is deep enough to avoid direct exposure to sunlight. It also has to be long enough to haul into the chain locker to compensate for permanent elongation of the polyester rope during the 25 year service life. Similarly, the R4 studless bottom chain is chosen to be 230 m. Such a bottom chain length ensures that the polyester rope of the taut mooring system maintains enough clearance to the seabed, even in 1000 year hurricane environments.

The chain diameter is 137 mm, including consideration of a corrosion factor of 0.4 mm per year during the 25 years of service life. The chain has a minimum breaking strength (MBS) of 14,955 KN. The polyester rope is manufactured by Bexco and it has a diameter of 241 mm with a minimum breaking load of 16,677 KN. The effective polyester rope length is 2000 m per line.

As discussed later in Section 6.3, polyester rope has noticeable permanent elongation/construction stretch under first loading. This permanent elongation impacts the polyester rope length specification for procurement and manufacture and may also impact the mooring line installation process. Polyester rope vendors typically need to back calculate rope manufacture length based on specification and estimated construction stretch under mooring installation load. Overestimation of the construction stretch will result in a shorter than expected rope length, which may cause difficulties in mooring line installation (excessive messenger chain and larger handling vessel needed for hook up); underestimation of the construction stretch can cause excessive chain-on-board and thus reduce payload.

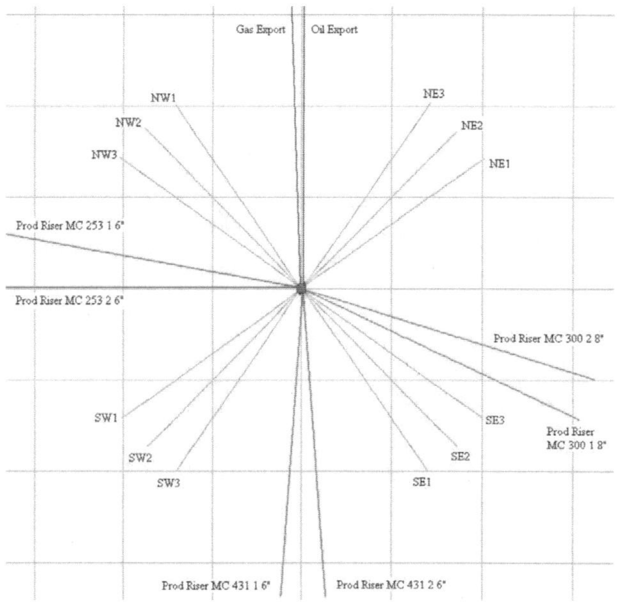

Figure 1 Mooring and Riser System Layout

Figure 1 presents an illustration of the layout of the designed mooring system.

4 ENVIRONMENTS

Site specific environmental data [Ref 2] is used in the design of the mooring system. Wind dominated hurricane conditions, rather than wave dominated or current dominated hurricanes, govern the design, as they produce the greatest mooring line tensions at the fairleads and maximum vessel offsets. Table 1 presents the 100 year and 1000 year hurricane parameters used in the design.

In the Gulf of Mexico, loop currents may play an important role in a mooring system design. Loop currents can sustain up to one month and this sustained duration will cause significant creep elongation of the polyester rope and, hence, greater vessel offset. Creep elongation is estimated based on polyester rope test results and considered in the current design and analysis of the mooring system, though it is found that loop current does not govern for this project. Table 2 shows the site-specific loop current data used in the analysis.

For fatigue analysis, the scatter diagram obtained in the metocean study is used to generate environment bins for time domain simulations. The tension time history is then used to generate tension RMS for fatigue damage calculation according to API RP 2SK.

Table 1 Environments

Hurricane Conditions dominated by WIND				
Return Period (yrs)	Wind	Wave		Current
	(m/s)	Hs (m)	Tp (sec)	(m/s)
100	43.68	15.22	15.69	1.87
1000	57.82	19.35	17.34	2.62

Table 2 Loop Current

Return Period (yr)	Current (m/s)	
	1-day Average	1-hour Average
5	1.04	1.20
10	1.25	1.37
50	1.53	1.62
100	1.60	1.68

Table 3 API RP 2SK Mooring System Integrity Criteria

Analysis Type	Quasi-Static		Dynamic	
	%MBS	FoS	%MBS	FoS
Intact	50%	2.00	60%	1.67
Damage	70%	1.43	80%	1.25

Table 4 Vessel Offset Criteria (Percentage of Water Depth)

Offset limits	
Intact	Damage
6%	8%

5 DESIGN CRITERIA

Mooring system integrity is established as per API RP 2SK. Maximum mooring line tension criteria are summarized below in Table 3. The criteria allow for different maximum line tensions based on the analysis methodology and the mooring system status considered.

- Intact: All mooring lines are intact and operational;
- Damaged: One mooring line is broken and/or non-operational.

The maximum allowable vessel offset is established based on riser design requirements and shown in Table 4. In addition, polyester rope lower end is required to have no contact with seabed.

Mooring line fatigue life is required to have a safety factor of 10, equivalent to 250 years, as Delta House has a service life of 25 years.

6 MOORING SYSTEM MODELING AND ANALYSIS

Fully coupled dynamic analysis of the mooring system is conducted in OrcaFlex [Ref 4]. This includes static analysis to determine the vessel equilibrium positions and time-domain simulations (three-hours) to calculate vessel motions and mooring line tensions under the action of wind, wave and current. The dynamic analysis considers:

- mooring line dynamic behaviors;
- current and wave loads on the mooring lines;
- coupling effect between vessel motion and mooring line dynamics;
- motion history effect, or the so-called "memory" effect.

Static analysis is first carried out to provide a starting position for the dynamic simulation. In the dynamic analysis, a build-up stage is used to smoothly ramp up the environments and vessel motions. This helps reduce the transients that are generated by the change from the static position to fully dynamic motion.

The equation of motion of the vessel and mooring system is solved for given initial values and boundary conditions in the time domain analysis. As an example, a single Degree of Freedom (DOF) equation of motion is shown below:

$$[M + A(\infty)]\ddot{\eta} + \int_0^\infty K(\tau)\dot{\eta}(t-\tau)\, d\tau + C\eta = F_{env}(t) + F_{moor}(t)$$

Here, M is the vessel displacement and η is the motion to be solved for; $A(\infty)$ is the added mass at infinite wave frequency and $K(\tau)$ is the Impulse Response Function (IRF); $A(\omega)$ and $B(\omega)$ are the added mass and the radiation damping coefficients at frequency ω that can be obtained in radiation and diffraction analysis using WAMIT [Ref 5]; C is the hydrostatic coefficient, representing the hydrostatic restoring effect; $F_{env}(t)$ and $F_{moor}(t)$ are the environmental force and the mooring forces, respectively.

$A(\infty)$ and $K(\tau)$ are estimated with the following equations using the added mass and radiation damping obtained in radiation analysis.

$$A(\infty) = A(\omega) + \frac{1}{\omega}\int_0^\infty K(\tau)\sin(\omega\tau)\, d\tau$$

$$K(\tau) = \frac{2}{\pi}\int_0^\infty B(\omega)\cos(\omega\tau)\, d\omega$$

In theory, $A(\infty)$ is constant if $A(\omega)$ and $B(\omega)$ can be estimated at all frequencies and full integration is used. However, the analysis estimates $A(\infty)$ and $K(\tau)$ by using $A(\omega)$ and $B(\omega)$ calculated in WAMIT at a finite number of discretized frequencies and a limited duration for integration (i.e. the cutoff time). Then $A(\infty)$ is taken as the mean of each estimated value at the discretized frequencies to reduce the approximation effect on the final simulation results. In addition, the cutoff time is set sufficiently long so that the error in the added mass and damping load estimation is minimal. More details regarding evaluating the $A(\infty)$ and $K(\tau)$ are referred to Cao [Ref 12].

6.1 Environmental loads

To calculate wave loads, radiation and diffraction analysis is first performed to obtain the wave load RAOs and QTFs using WAMIT. Then the time history of first order wave loads can be calculated with specified wave time history, which can be generated with site specific wave parameters and random amplitude/phase methods. Similarly the wave drift forces can be determined with the calculated QTFs and Newman's approximation.

For wind load estimation, wind and current drag coefficients are determined with semi-empirical formulations and verified by the wind model tests performed at Force Technology in Denmark [Ref 6]. The one-hour mean wind speed along with an API wind spectrum is used to produce the time history of wind speed.

The current load calculation is straightforward with available current drag coefficients obtained from model tests [Ref 3, 6]. The analysis considers the current speed profile to more accurately model the current load on the mooring lines.

6.2 Polyester change in length characteristics

One of the most challenging problems in the mooring analysis is modeling the change-in-length characteristics of the polyester rope, which depends on material properties, construction method, load history, frequency and amplitude. The polyester rope stiffness can be determined through change-in-length full scale sub-rope tests, typically conducted in a later phase of the procurement, manufacturing and delivery process of the fiber ropes.

In this project, the change-in-length model tests are conducted according to API RP 2SM [Ref 7]. Figure 2 below illustrates the tension-elongation curves measured at different steps of the sub-rope tests.

To better utilize the model test results, a methodology based on API RP 2SM is utilized and is shown in Figure 3. As shown in Figure 3, the tension-elongation relationship is highly nonlinear and it depends strongly on the loading history. The rope elongation behavior under first loading varies significantly from subsequent loading conditions. Permanent elongation shows up when the rope is subject to the first loading. The majority of the permanent elongation can be "removed" through initial tensioning during mooring installation. Subsequent permanent elongation can be compensated by hauling in top chain and re-tensioning the mooring system after a storm.

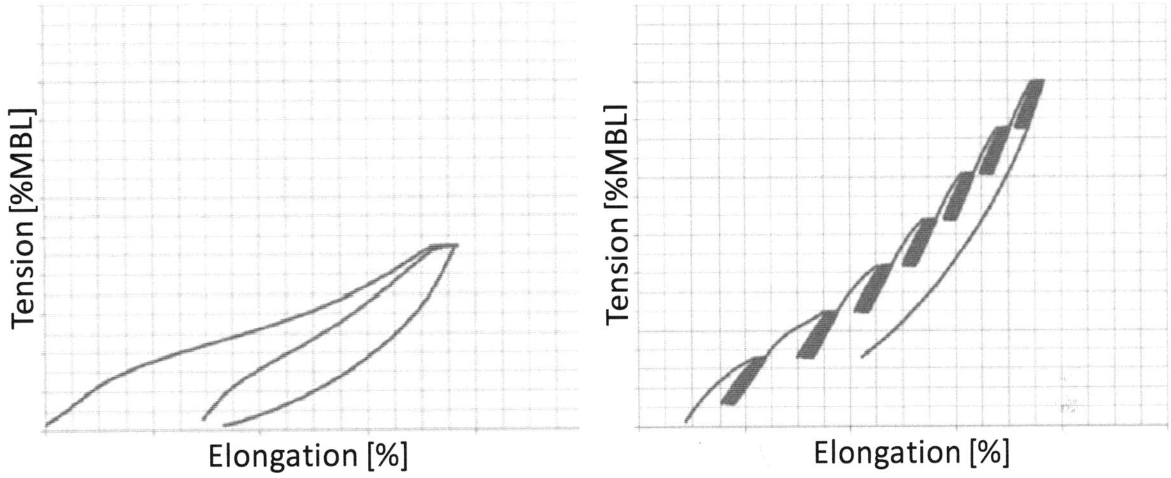

Figure 2 Sample results of Tension-Elongation curve recorded in different steps of the test [Ref 8].

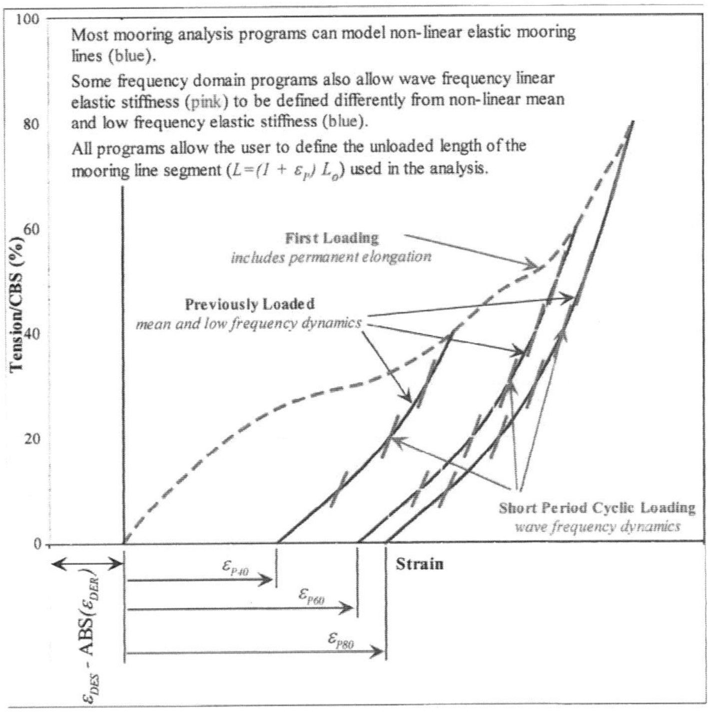

Figure 3 Idealized change-in-length characteristics of a polyester rope [Ref 7, Figure B-1].

The nonlinear tension-elongation curve after the first loading can be used to determine polyester rope static and dynamic stiffness needed for mooring analysis using commercial software. Figure 4 shows the calculated static and dynamic stiffness based on model tests. In the figures, T_{inst} represents the maximum expected line tension of initial tensioning during the installation. T_{int} and T_{dmg} are the maximum expected tension of the intact and one-line damaged mooring system in design extremes, respectively. LF and WF indicate the dynamic stiffness due to low frequency and wave frequency exciting forces, respectively.

It is a conventional practice to use a Post installation and Storm stiffness to envelope the actual stiffness for conservative estimation of the maximum vessel offset and the maximum line tension, respectively. According to the full scale sub-rope test results, post installation stiffness is chosen as 10xMBL and the storm stiffness is chosen as 45xMBL.

However, for more accurate prediction of the vessel offset and line tension, more realistic models for the polyester rope change-in-length behavior shall be used. As an example a tension-elongation curve for OrcaFlex input is constructed based on determined static and dynamics stiffness and expected tension levels through an iterative process (see Figure 5).

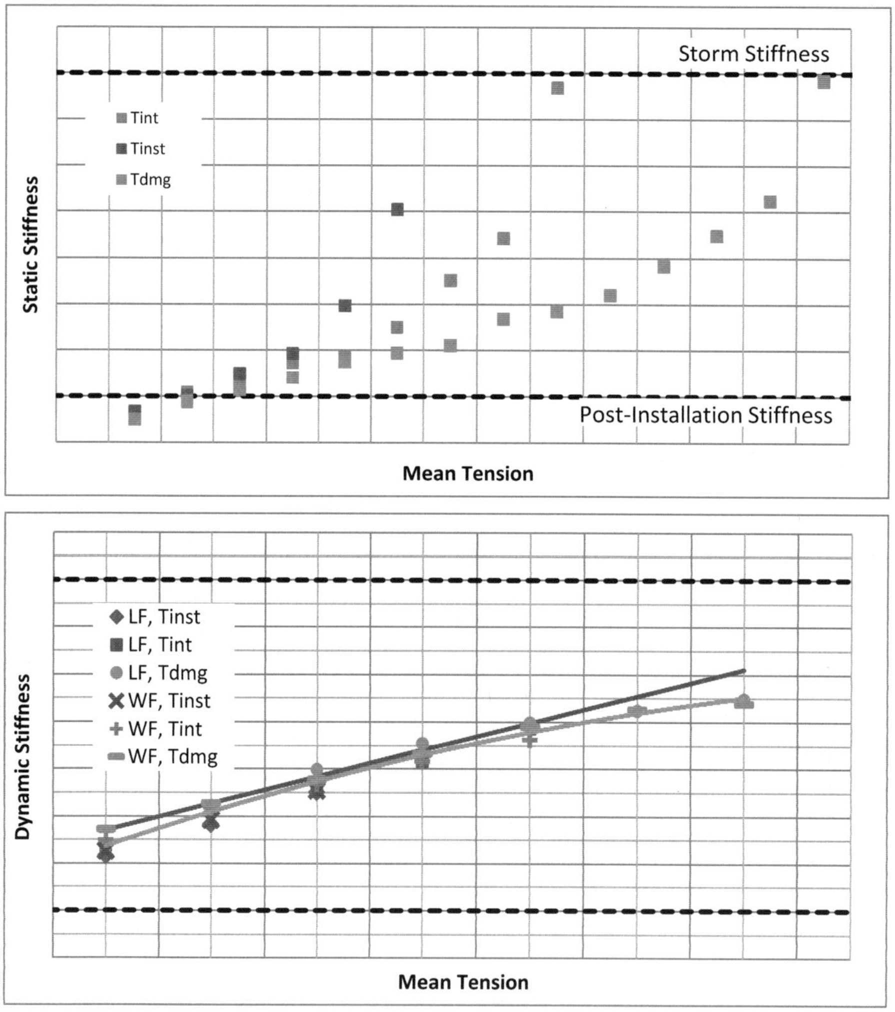

Figure 4 Polyester stiffness obtained from model tests.

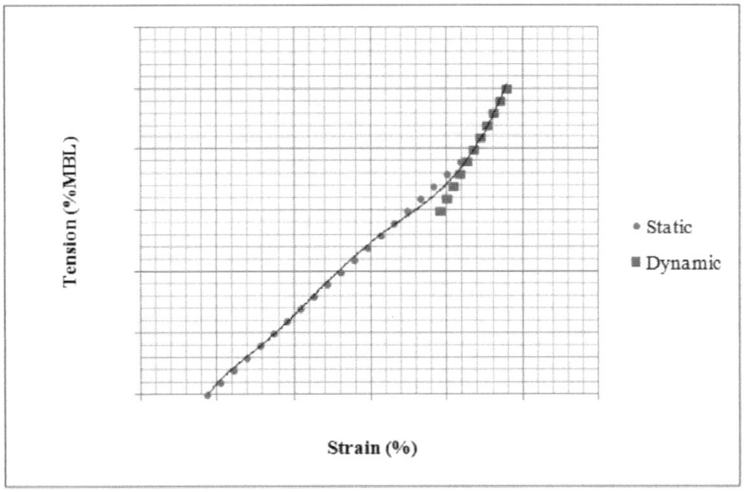

Figure 5 Constructed polyester rope tension-elongation curve for dynamic analysis.

6.3 Fatigue analysis

Fatigue analysis of each component, i.e. chain, polyester rope and connectors, is performed individually. In this paper, focus is given to the fatigue analysis of the top chain at the fairleads, where out of plane bending could happen. The fatigue damage is calculated using API RP 2SK chain fatigue curve for studless chains:

$$NR^M = K$$

Here, N is number of cycles; R is ratio of tension range to reference breaking strength; M = 3 and K = 316 for common studless chain links according to API RP 2SK.

- **Vortex-induced motion (VIM)**

VIM results are obtained from model tests [Ref 3]. Figure 6 below provides the non-dimensional VIM as a function of current speed. As can be seen, Delta House has minimal VIM due to the five-sided columns, especially for the 45 degree heading. Note that D is the column width, 14.4 m.

In given environment bins for fatigue analysis, VIM introduces additional tension RMS, which is combined with low frequency tension RMS in the assessment of fatigue damage. The VIM introduced tension RMS is determined through a static analysis by offsetting the vessel from its equilibriums.

- **Tension RMS amplification due to bending**

Bending, particularly Out of Plane Bending (OPB), is known to reduce mooring chain fatigue life [Ref 9, 10]. However, limited official guidance exists on how to calculate the bending-introduced fatigue damage to chain links at fairleads. Recently, BV [Ref 9] published a guide for estimation of the OPB effect on chain fatigue life; the guide provides procedures that involve extensive FEA and a complex parametric model based on JIP data.

API RP 2SK suggests applying an amplification factor of 1.2 to tension RMS to account for bending effect for chain links at seven (7) pocket right fit fairleads. In this project, this simplified method is used in the assessment of chain link fatigue life.

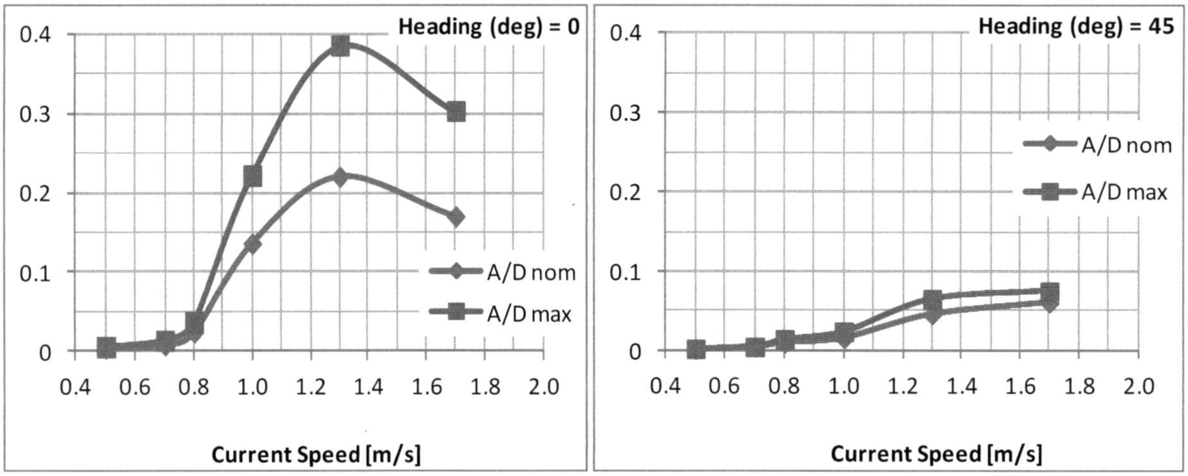

Figure 6 VIM of the Delta House Semisubmersible.

7 RESULTS

7.1 Maximum vessel offset and line tension

Figure 7 (a) presents the maximum mooring line tension at fairleads in various environments. Clearly, the designed mooring system satisfies the API tension safety requirements, less than 60% for intact and less than 80% for one-line damaged mooring in 100 year hurricanes. One thousand year hurricanes are used in the analysis for a robustness check and the mooring system passes this check. For this project, loop currents produce smaller line tensions and are not the governing environmental condition for the mooring system design.

Comparing the maximum tension estimated with storm stiffness to that with the nonlinear tension-elongation curve, the former estimation is overly conservative, which could potentially lead to an overly designed mooring system.

For the analysis of a damaged mooring system in 100 year hurricanes and intact mooring in 1000 year hurricanes, the maximum tension exceeds the installation load, 40%MBS; therefore, permanent elongation due to this first loading may occur (see Figure 3). Through an iterative process, it is estimated that the permanent elongation is 0.25% and 0.6% of the polyester rope length for the two load cases, respectively. With this additional permanent elongation added to the mooring payout to the windward mooring lines, the maximum line tension shows noticeable decrease (up to 3%MBS) and the maximum vessel offset increases by 0.3% to 0.6%WD (see Figure 7 (b)).

Figure 7 (b) shows the maximum vessel offset in various environments. As shown, the vessel offset also satisfies the API vessel offset criteria, less than 6% for intact mooring and less than 8% for damaged mooring system in 100 year hurricanes. Again, the loop current does not control the mooring system design.

The post installation stiffness predicts a reasonable estimation of the maximum vessel offset, as such estimation is only slightly greater than the prediction using nonlinear tension-elongation relationship based on full scale sub-rope tests. Particularly, when the permanent elongation due to first loading (see Figure 3) is taken into consideration, the estimated offsets using post-installation stiffness and model test stiffness are very close to each other.

The lower end of the polyester rope is examined to determine its clearance to seabed. It is found that the rope lower end of the intact mooring system has a clearance of at least 20 m and 1 m in 100 year and 1000 year hurricanes, respectively. For the one-line damaged case, the rope lower end has a seabed clearance of at least 10 m in 100 year hurricanes.

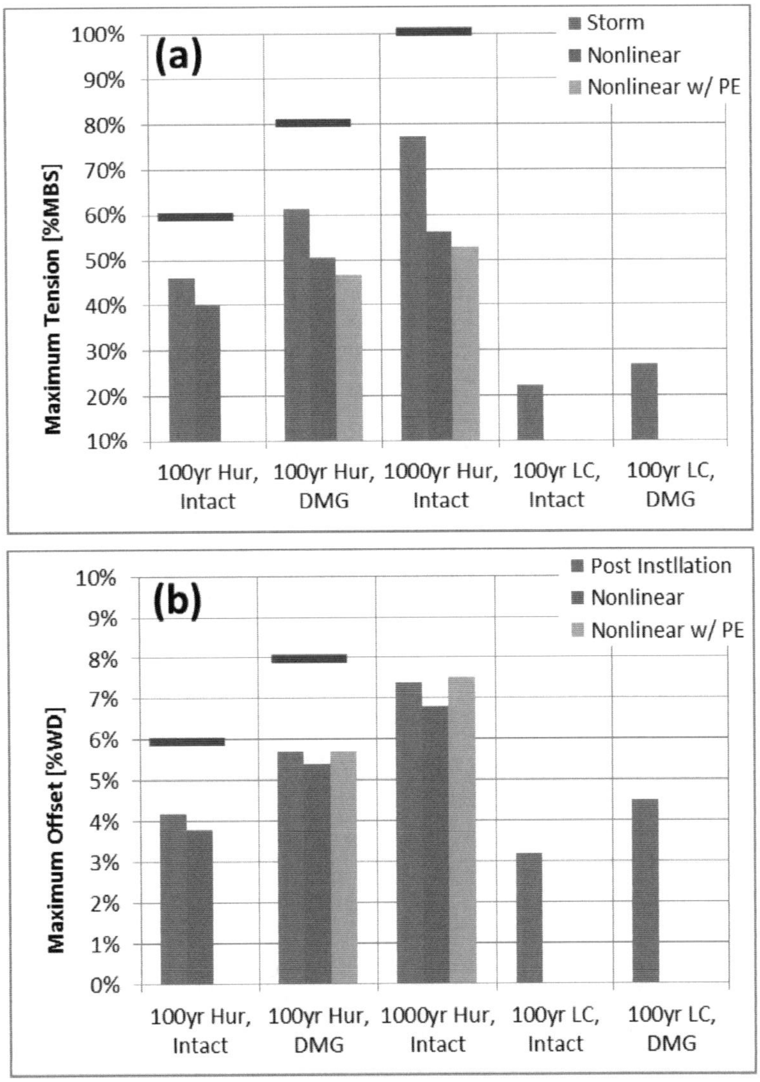

Figure 7 Maximum line tension (a) and maximum vessel offset (b) in various environments

7.2 Fatigue life

The service life of the unit is 25 years and a safety factor of 10 is required for the fatigue life of the mooring lines, i.e. a minimum of 250 years of fatigue life is required. Figure 8 shows the calculated fatigue life safety factor for the mooring chain at fairleads of all mooring lines. As shown, the safety factor is far greater than 10.

The results show that VIM motion can reduce fatigue life by up to 55% and it is necessary to include the VIM effect in mooring fatigue analysis. For reference, Kramer et al. [Ref 11] reported a fatigue life increase by up to three times when a reduced VIM design curve was used in their assessment of VIM effects on mooring line fatigue damage. Details can be found in the reference [11].

Also obvious in the figure is the safety factor deduction due to bending at fairleads. The bending effect is taken into account by using a tension RMS range amplification factor in fatigue damage assessment. It is observed that this bending effect can reduce the safety factor up to 40%. For a sensitivity study, the tension RMS amplification factor due to bending is systematically increased to estimate its impact on fatigue life. As shown in Figure 9, the fatigue life safety factor appears to exponentially decay as the amplification factor linearly increases.

Current methodologies and guidelines [Ref 9] for estimation of OPB effects on mooring line fatigue damage involves extensive FEA analysis of the tension history of the local chain at fairleads and complex parametric models based on JIP model test results. It may be desirable to simplify the estimation of OPB effects by using a single tension RMS amplification factor for this purpose. However, this is beyond the scope of the current project.

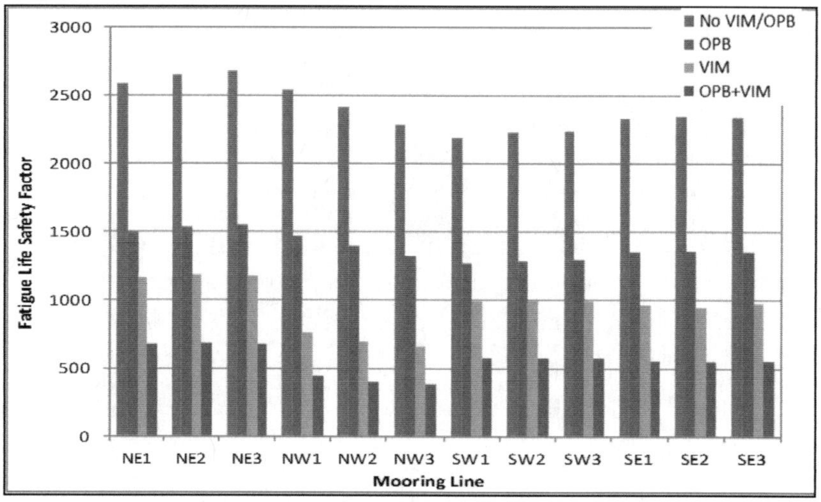

Figure 8 Mooring chain fatigue life estimation

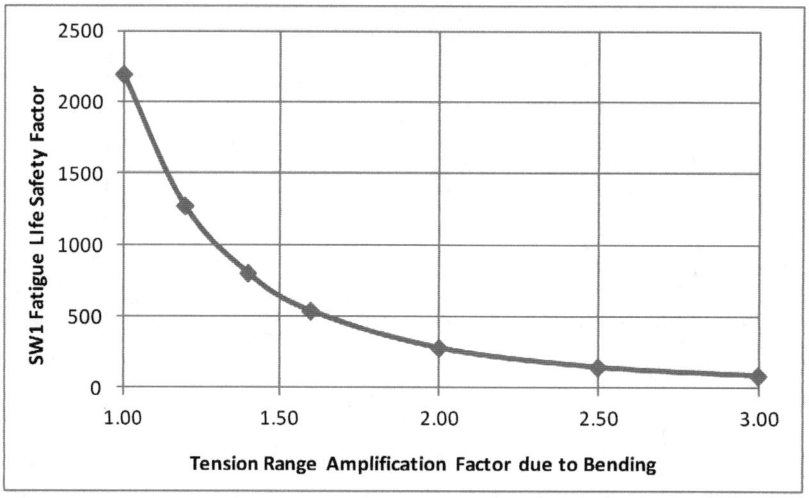

Figure 9 Dependence of chain fatigue life on tension RMS amplification due to bending

8 CONCLUSIONS

The paper presents the design and analysis for the Delta House semisubmersible mooring system. It is found that the system satisfies the vessel offset criteria set forth by steel catenary risers and line tension criteria according to API RP 2SK. The fatigue life safety factor of the mooring system is well beyond ten (10) required by the standard.

The polyester rope change-in-length characteristics post a main challenge in the design and analysis of the mooring system. The construction stretch/permanent elongation has to be taken into consideration in the rope specification, manufacturing and installation process.

Conventional upper bound ("storm") stiffness, 45xMBL, used in this study yields an overly conservative estimation of mooring line tension. This storm stiffness may be reduced if allowed by polyester rope model test results. On the other hand, the lower bound ("post installation") stiffness used in the design and analysis, 10xMBL, produces a reasonable estimation of maximum vessel offsets.

Mooring chain fatigue analysis results demonstrate that VIM and OPB effects on fatigue damage is significant: VIM reduces fatigue life by up to 55% and an amplification factor of 1.2 in tension RMS due to bending can decrease fatigue life up to 45%. Therefore, these effects have to be included in the fatigue analysis.

ACKNOWLEDGEMENT

The authors thank for the permission from LLOG Exploration LLC to publish the mooring system design and analysis results. The authors also appreciate the internal support by EOC in preparation and presentation of the manuscript.

REFERENCE

1. Design and Analysis of Station Keeping Systems for Floating Structures, API Recommended Practice 2SK (Third Edition), American Petroleum Institute, October 2005
2. Delta House Metocean Criteria Study, Forristall Ocean Engineering, Inc., March, 2012
3. Hydrodynamic Model Tests of the Delta House semisubmersible platform, project No LLOG 11-7700, Part I – Model Production, Resistance and Drag Tests and VIM Tests, Revision 2, April 2013
4. OrcaFlex (version 9.6d) User Manual, Orcina Ltd., Cumbria, UK, 2012
5. WAMIT (version 7) User Manual, Massachusetts Institute of Technology, Cambridge, MA, USA, 2012
6. Delta House Semi-Submersible Wind Tunnel Tests, Force 112-26703 Rev 1.0, Force Technology, Dec, 2012
7. Recommended Practice for Design, Manufacture, Installation, and Maintenance of Synthetic Fiber Ropes for Offshore Mooring, API Recommended Practice 2SM (Draft Version), American Petroleum Institute, 2013
8. Change-in-Length testing of subrope for Qualification testing of 1780 ton polyester mooring line for the Opti Delta House project. DNV Doc. No.: 168JQ8B-20/2013-5361, Bexco N.V. Nov 2013
9. Fatigue of Top Chain of Mooring Lines due to In-plane and Out-of-Plane Bendings, BV Guidance Note NI604 DT R00 E, Bureau Veritas, Oct 2014
10. OPB Chain Fatigue: Mechanism and JIP, The 19th Offshore Symposium Keynote Presentation, P. M. Vargas, Houston, TX, Feb 2014
11. M. Kramer, W. Ma, N. Xu, S. Gupta, J. Abadin, VIM induced Mooring Fatigue Prediction for Floating Production Facilities: Line Dynamics and Comparison with Field Measurements. Deep Offshore Technology International, DOT-2015.
12. Y. Cao, A procedure for evaluation, assessment and improvement of added mass and radiation damping of floating structures, OMAE2008-57275.

MODU MOORING DESIGN CONSIDERATIONS IN THE ARCTIC ENVIRONMENT

GAURAV SINGHAL, PHD, PE
FLOATING SYSTEMS LEAD, WOOD GROUP KENNY

JUN CHENG
SENIOR CONSULTANT, WOOD GROUP KENNY

ANTONY CROSTON
DRILLING & WELLS MANAGER, WOOD GROUP KENNY

ALAN WHOOLEY
VP RISERS & FLOATING SYSTEMS, WOOD GROUP KENNY

Proceedings of the 21st Offshore Symposium, February 2016, Houston, Texas
Texas Section of the Society of Naval Architects and Marine Engineers

ABSTRACT

The Arctic region contains one of the largest oil and gas reserves in the world. However, the harsh and unpredictable environment poses extreme challenges to drilling and exploration operations in the region. Cold temperatures coupled with sea icing result in limited uptime for various offshore activities. In addition, constant threat from icebergs may require moored mobile offshore drilling units (MODUs) to be equipped with quick-release mooring systems in order to disconnect and move out of harm's way (Newfoundland Offshore Petroleum Installations Regulations, SOR/95-104).

While a lot of attention has focused on MODU mooring design criteria in the Gulf of Mexico in light of recent hurricanes (Petruska et al. 2007), such criteria are not properly defined for Arctic regions. This paper addresses critical aspects for mooring design in the Arctic, particularly extreme temperatures (typically below 0°C) and icing that are not emphasized in current mooring design codes (API-RP-2SK, DNVGL-OS-E301). Typically, mooring designers assume air density of 1.225kg/m^3 for computing wind loads, which corresponds to 15°C air temperature. The Arctic environment, on the other hand, characteristically has cold, relatively dry air coupled with strong winds. In such conditions, air density is much higher, thereby increasing the wind load on the MODU. Icing on the superstructure may further increase the wind load due to increase in projected wind area. These two effects (air density and icing) may increase the net wind load on the MODU by around 10-15% depending upon the ambient conditions. It is therefore important to apply correct physics for MODU mooring design in the Arctic. The paper also provides commentary on reasonable ways to combine the design storm condition (typically 100-year) with extreme temperatures in order to reduce conservatism in design, while allowing safe and robust operation.

Keywords: Arctic, MODU, Mooring Design, Extreme Temperature

INTRODUCTION

The Arctic is estimated to hold the world's largest remaining untapped gas reserves and some of its largest undeveloped oil reserves. According to United States Geological Survey (USGS), an estimated 90 billion barrels of oil, 1669 trillion cubic feet of natural gas, and 44 billion barrels of natural gas liquids remain unexplored in the Arctic region. About 84% of these deposits are expected to occur in the offshore areas (USGS 2008). The USGS study included only those resources that are believed to be recoverable using existing technology. Although the current offshore technology may be viable to harvest these reserves, there would be a need to drill a large number of wells over a relatively short period of time, thereby accentuating the need for year-round operations (Kokkinis et al. 2010).

The Arctic environment is typically characterized by extreme cold temperatures and excessive icing. The extreme environment therefore poses unique challenges to year-round exploration and drilling operations. Most of the exploration and drilling activities in the Arctic have traditionally been in shallow water (<100m water depth). In the early 1970s, drilling in the Arctic region was performed using ice-reinforced drill ships. The early drill ships were limited to summer and early fall seasons (classified as open water season), but the open water season was, at times, extended through use of ice breakers. Soon after, in the mid-1970s, Kulluk – a second generation conical drilling unit which is arguably regarded as the pioneer in Arctic drilling operations – was purpose-built to extend the open water season (Gaida et al. 1983). Due to its innovative conical design, Kulluk offered the capability to operate under pack ice conditions, something that was not viable using earlier drill ships. Kulluk's design also incorporated unique features in its mooring system. Particularly, the mooring system was equipped with rig anchor release by Interocean for quick disconnection and the mooring lines were routed through hull to reduce ice impact.

The success of Kulluk provides vast experience and a knowledge base for the next generation of floaters for year-round operations in the Arctic environment. Recent advancements in offshore technology have propelled the industry interest towards even more challenging deep-water Arctic areas. Some recent studies on purpose-built floaters for the deep-water Arctic include Technip's disconnectable Arctic Spar (Sablok et al. 2011), SBM Atlantia's Arctic FPSO (Li 2012), and Circular FPSO (Srinivasan and Sreedhar 2011).

While Arctic floater design has garnered significant industry attention, current industry practice on mooring system design is still inadequate to address the design of Arctic mooring systems (Kwan et al. 2014). To address mooring-specific issues in the Arctic, ABS sponsored an industry workshop on mooring systems for Arctic drilling and production operations in 2012. The workshop concluded with a strong need for a joint industry project (JIP) to address mooring-specific challenges in the Arctic. In June 2013, the Arctic mooring JIP was launched by Kwan Engineering Services and Global Maritime with an expected completion in early 2016 (Kwan et al. 2014).

While the Arctic mooring JIP will cover important aspects pertaining to mooring design and hardware including ice management, the main purpose of this paper is to provide technical information to complement ongoing industry efforts for Arctic mooring design that have not yet been addressed. This paper addresses critical aspects for mooring design in the Arctic, particularly the impact of extreme temperature (typically below 0°C) and icing on wind loads that are not systematically addressed in relevant codes (e.g. API-RP-2SK, DNVGL-OS-E301). Some guidance on selection of suitable air temperature is given, for instance, in DNVGL-OS-C101 which defines the design air temperature as "the lowest mean daily temperature to which the structure may be exposed to during installation and operation". Note, however, that the DNV standard pertains to design of offshore structures, particularly during material selection and temperature load determination, and not for wind load determination for mooring design.

The paper is organized as follows. 'Wind Load Consideration' describes the relevant parameters important for wind load calculation. 'Icing Consideration' highlights the impact of superstructure icing from mooring design standpoint. 'Arctic Mooring Design Case Study' summarizes the comparison of mooring analysis results for a generic MODU under Arctic and tropical conditions. 'Summary and Discussion' provides the summary and concluding remarks.

WIND LOAD CONSIDERATION

For mooring design, the mean wind loads on MODU superstructure are typically computed based on Equation (1).

$$F_{wind} = \frac{1}{2} C_D A \rho_{air} v^2 \quad (1)$$

where F_{wind} is the wind force in Newtons, C_D is the wind drag coefficient, A is the projected wind area in m^2, ρ_{air} is the density of air in kg/m^3, and v is the wind velocity in m/s (mooring designers typically use 1-hour average wind velocity at 10 m elevation with suitable gust spectrum). For most mooring software, wind load is inputted in units of force/velocity^2 as indicated in Equation (2).

$$\frac{F_{wind}}{v^2} = \frac{1}{2} C_D A \rho_{air} \quad (2)$$

In subsections below, the parameters on the right hand side of the above equation are discussed in more detail as it pertains to Arctic application.

Drag Coefficient and Projected Area

Drag coefficient and projected area are a function of rig size, shape, and geometry. Drag coefficients are usually presented in non-dimensional form and are typically obtained from wind tunnel test, but can also be computed through commercial software or hand calculations. Projected wind area is also normally reported in the wind tunnel test report, but can easily be computed based on vessel general arrangement drawings in case model test data are not available.

For Arctic application, the MODU superstructure is typically winterized to sustain extreme cold temperatures and to prevent superstructure icing. Derrick winterization for a typical MODU is illustrated in Figure 1. Due to winterization, there is not only an increase in the projected area but also in the drag coefficients due to change in geometry which affects the surrounding flow regime.

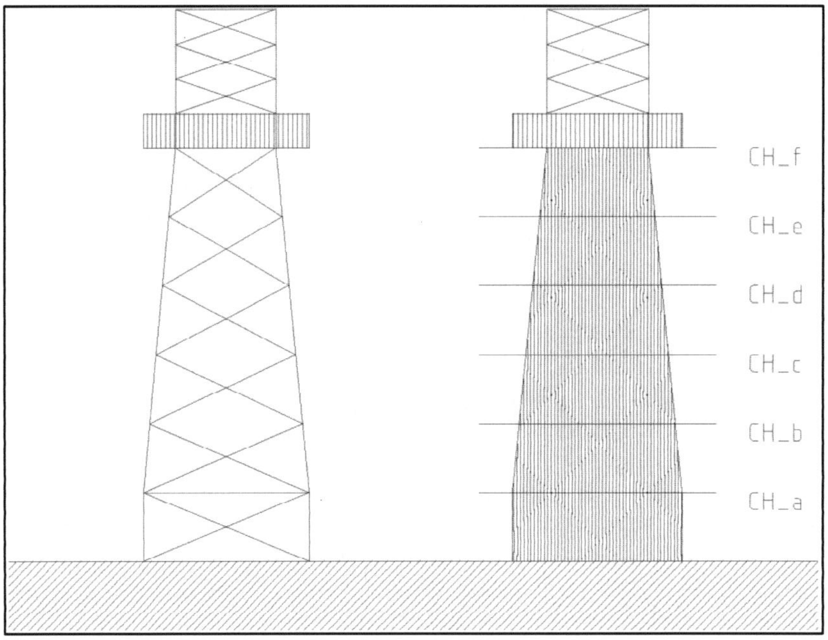

Figure 1 Example of MODU Derrick without and with Winterization

If the MODU is purpose-built for Arctic operation, the wind tunnel test is performed on the rig with and without winterization. As a result, all relevant physical phenomena are accurately captured. On the contrary, if the MODU is originally designed for service in tropical regions and later proposed for Arctic operations, the wind tunnel test data with winterization may not be available. In that situation, the mooring designer must rely on approximate methods to compute the revised drag coefficients including the effect of winterization. As stated earlier, the mooring designer should consider two aspects:

- The overall increase in projected area, and
- The revised drag coefficient.

Computing the revised projected area is rather straightforward, however revising the drag coefficient can be cumbersome. The designer should consider appropriate wind profile along with suitable height coefficient as recommended in API RP2SK (2005) and DNVGL-RP-C205 (2014).

Air Density

Air density is not dependent on rig shape or size but rather on the MODU's operational location. In some cases, wind tunnel test data may specify air density based on ambient conditions during the model test (typically 15°C to 20°C); however, the ambient conditions may or may not reflect the actual conditions at the operational location. Mooring designers typically assume air density at ambient conditions, such as 15°C, for computation of the right hand side of Equation (2). This assumption is valid and conservative for MODUs operating in tropical and sub-tropical environments where air temperatures are usually in the ambient range. However, this assumption is not valid for MODU operations in the Arctic as temperatures are substantially lower. As the Arctic mooring designer, therefore, one must consider the prevalent conditions that the MODU is expected to experience during operation.

Air density is a function of pressure (or equivalently, elevation at or above sea level), temperature, and relative humidity (RH). Table 1 provides the variation of air density with temperature and humidity for dry (0% RH) and saturated (100% RH) air at 10 m elevation above sea level. The chosen elevation corresponds to wind measurements that are typically relative to 10 m elevation. Two trends can be noticed from data presented in Table 1:

- Air density increases with decreasing temperature, and
- The effect of humidity on air density decreases with decreasing temperature.

It can be seen that air density at 0°C is about 5.5% higher than that at 15°C. The impact of humidity on air density is insignificant at lower temperatures.

Table 1 Variation of Air Density[1] with Temperature and Humidity

Temperature (°C)	Dry Air Density[2] (kg/m^3)	Saturated Air Density[3] (kg/m^3)
-20	1.393	1.392
-15	1.366	1.365
-10	1.340	1.338
-5	1.315	1.313
0	1.291	1.288
5	1.268	1.263
10	1.245	1.239
15	1.224	1.216
20	1.203	1.192

Note: 1) Air density is computed at 10m elevation above mean sea level
2) Dry air corresponds to 0% relative humidity
3) Saturated air corresponds to 100% relative humidity

Some locations in the Arctic can be well below 0°C for much of the winter, with temperatures as low as -40°C recorded in Beaufort Sea (Brower et al. 1977). By way of example, statistics of monthly air temperature data measured from Canadian Buoy 44251, located offshore Newfoundland (Canada) in water depth of about 70m, are shown (data are obtained from Environment Canada website; https://weather.gc.ca/canada_e.html). Data from this location are selected as they provide reasonably long measured continuous data (about 17 years from 1998 to 2015) within the Arctic region. This dataset is of sufficient duration to obtain long-term statistical inferences on variation in air temperature and thereby air density for Arctic mooring applications. Table 2 shows the long-term monthly statistics of air temperature recorded at Buoy 44251. The coldest temperatures occur during winter months (December – March), with lowest temperature of -12.4°C recorded in January. Monthly variation of air density at buoy 44251 is calculated based on the monthly mean temperature and shown in Figure 2. Base case air density at 15°C is also shown for comparison (blue line) and is only found to occur during month of July at buoy 44251. For other months, the air density can be up to 6% higher than the base case. It is thus important to consider location-specific air density when performing mooring design in such environments.

Table 2 Long-Term Monthly Statistics of Air Temperature at Buoy 44251

Month	Mean (°C)	Minimum (°C)	Maximum (°C)	Standard Deviation (°C)
January	0.0	-12.4	9.4	3.6
February	-0.7	-10.5	9.3	3.5
March	0.3	-11.2	9.4	3.3
April	1.9	-6.2	10.0	2.6
May	4.2	-2.3	16.1	2.0
June	7.6	0.4	22.7	2.1
July	12.1	5.8	25.0	2.1
August	14.7	8.3	19.5	1.9
September	13.3	5.4	19.0	2.3
October	9.7	-0.1	18.8	2.4
November	5.8	-3.4	14.2	2.7
December	2.5	-9.1	11.3	3.2

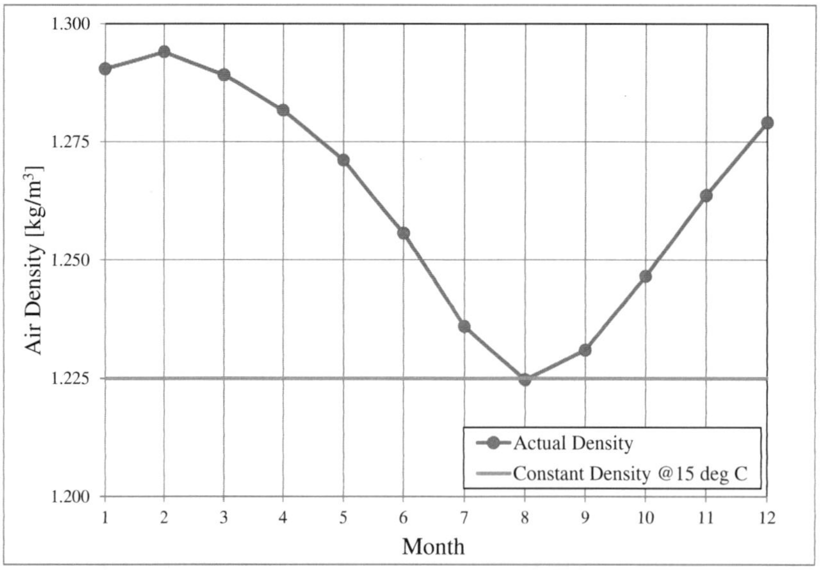

Figure 2 Monthly Density Variation at Buoy 44251, Offshore Newfoundland

ICING CONSIDERATION

The impact of superstructure icing should also be taken into account for Arctic operations. According to ISO 19901-7:2013, Section 7.2.7 on atmospheric icing: "Where applicable, an assessment shall be made of atmospheric icing, including sea spray icing. Increased wind area due to superstructure icing shall be considered in the calculations of wind actions on a floating structure."

ISO 19901-7: 2013 does not provide specific guidance on how to quantify the impact of icing on wind area. However, DNVGL-OS-A201 provides some guidance on icing weight estimate as well as the projected area as shown in Table 2. Depending on the extent of icing, considerations to MODU stability, in addition to the impact on mooring design due to increase in projected area from icing, may also be required. Note that if rig is winterized, some superstructure may not be directly exposed to icing.

Table 2 Guidance on Icing Load and Projected Area for Semisubmersibles (DNVGL-OS-A201 2010)

Icing load (kg/m²) to be applied to horizontal surfaces and projected lateral (vertical) area[1]		
	Horizontal Surfaces	**Projected Lateral (Vertical) Area**
> 24 m from WL	10	10
> 18 to 24 m from WL	30	10
> 12 to 18 m from WL[2]	40	10
> 6 to 12 m from WL[2]	80	20
0 to 6 m from WL[2]	120	30

[1] For the purpose of this Rule, the waterline (WL) shall be taken as the load line most relevant to the operational condition for which the stability case is being calculated
[2] For horizontal surfaces with active anti-icing systems, the icing weight load in that area may be set to 30 kg/m^2.

ARCTIC MOORING DESIGN CASE STUDY

In order to investigate the potential impact of increased air density on mooring design, a case study for a generic semisubmersible MODU is presented. For this investigation, symmetric mooring layout with eight mooring lines is considered. The angle between each line is 45°. An all-chain system with 3.5 inch stud diameter and R4 grade is considered. The MODU is also assumed to be equipped with thrusters to provide assistance to the mooring system. Note that the selected mooring configuration is representative of typical MODU mooring systems implemented worldwide.

Mooring analysis is performed in frequency-domain using commercially available software. Only survival mooring analysis corresponding to 100-year return period storm condition is presented. Particularly, two analysis cases are considered:

- Mooring design using air density at 15°C temperature (air density of 1.225 kg/m^3)
- Mooring design using air density at 0°C temperature (air density of 1.294 kg/m^3)

The results of chain utilization for the above cases are presented in Figure 3 for headings between 0°-360°. It can be seen that if the MODU mooring system is designed based on 15°C temperature, the utilization criterion is satisfied (i.e. utilization less than unity after appropriate safety factors are applied). However, if air density at 0°C is considered, then the utilization criterion is not satisfied. It is also worth pointing the significant discrepancy in utilization for key headings of 67.5° and 292.5°, where the utilization is under predicted by around 25% using established air density. As discussed earlier, at lower temperature the air density becomes higher, thereby inducing an increase in overall wind load on the MODU. Note that rig winterization impact is not considered for this case study. But it is noted, in passing, that wind loads will further increase due to winterization. In order to satisfy the mooring requirements, some considerations include use of higher chain size (or grade) and additional mooring lines.

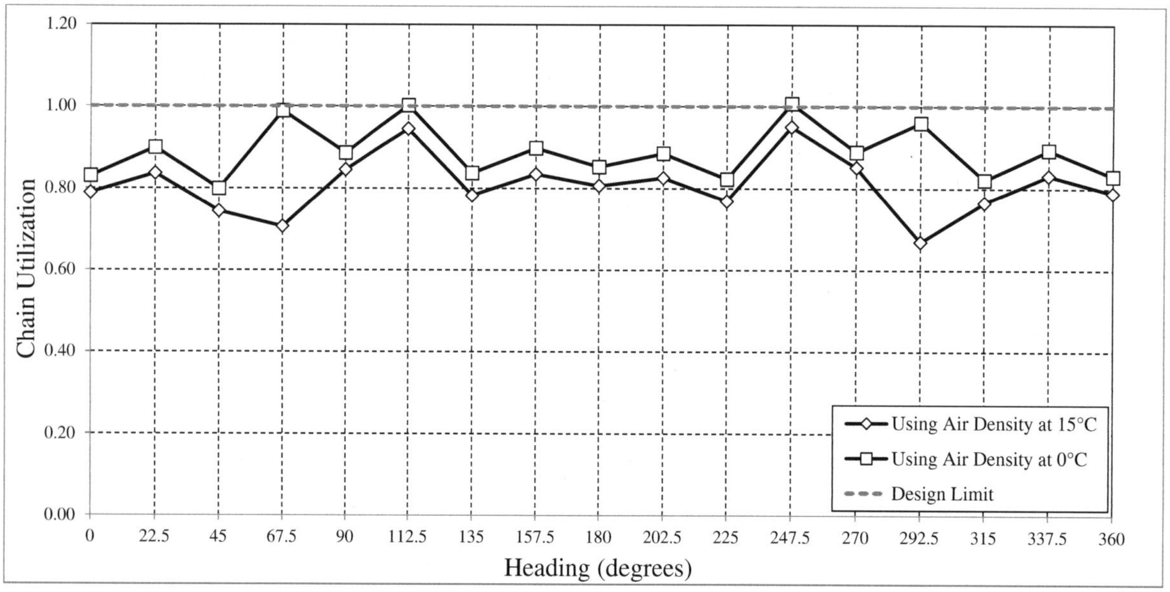

Figure 3 Mooring Chain Utilization vs. Heading for a Generic MODU with 8-point Mooring

SUMMARY AND DISCUSSION

This paper discusses critical aspects of Arctic mooring design that have not been addressed in current mooring design codes. While the recently initiated Arctic mooring JIP will cover topics geared more towards mooring hardware including disconnection and reconnection in the event of icebergs, the current study provides technical considerations relevant to mooring analysis in Arctic extreme environment.

Particularly, the impact of air density – as a result of extreme cold temperatures – on wind load is addressed. It is shown that as temperatures decrease, the air becomes denser thereby increasing the net wind load on the MODU. Other aspects affecting the wind load calculation are drag coefficients and projected area. If the MODU is winterized, then changes in geometry will impact not only the projected wind area but also the flow regime thereby changing the drag coefficients. The case study showing the comparison of mooring chain utilization with and without proper air density is also presented. It is concluded that use of proper air density in Arctic mooring design may not satisfy the mooring chain utilization requirements and therefore other considerations regarding proper chain size, number of lines, etc. may need to be invoked for ensuring safe MODU operation.

In concluding remarks, a brief discussion regarding selection of appropriate air temperature to compute correct air density for Arctic mooring analysis is provided. Typically, MODU mooring analysis requires survival condition assessment for 100-year storm condition which is composed of worst combination of wind, wave, and current (API RP 2SK, DNVGL-OS-E301). Extending this requirement to include air temperature for Arctic mooring application – i.e. use of 100-year extreme low temperature along with 100-year wind, wave, and current – may be very conservative as it is highly unlikely (from statistical viewpoint) to have the coldest temperature occur at the same time as the 100-year storm event. For the case study presented in this paper, the 100-year event was used in conjunction with monthly mean temperature; this selection may not appear conservative enough. A more realistic combination could be combining the 100-year storm event with 10-year or 1-year extreme temperature, while still permitting safe and robust MODU operation. These requirements will need to be addressed in mooring design codes to become standard practice for Arctic applications.

ACKNOWLEDGEMENTS

The authors thank Wood Group Kenny for allowing the publication of material presented in this paper.

REFERENCES

American Petroleum Institute, "Design and Analysis of Stationkeeping Systems for Floating Structures", API RP 2SK, 2005.

Brower, W. A., Diaz, H.F., Prechtel, A.S., Searby, H.W., and Wise, J.L., "Climatic Atlas of the Outer Continental Shelf Waters and Coastal Regions of Alaska", Volume I Gulf of Alaska, 1977.

Det Norske Veritas GL AS, "Design of Offshore Steel Structures, General (LRFD method)", DNVGL-OS-C101, 2015.

Det Norske Veritas AS, "Environmental Conditions and Environmental Loads", DNV-RP-C205, 2014.

Det Norske Veritas GL AS, "Position Mooring", DNVGL-OS-E301, 2015.

Det Norske Veritas GL AS, "Winterization for Cold Climate Operations, DNVGL-OS-A201, 2015.

Gaida, K.P., Barnes, J.R., and Wright, B.D., "Kulluk-An Arctic Exploratory Drilling Unit", Offshore Technology Conference, OTC-4481-MS, 1983.

International Organization for Standardization, "Stationkeeping Systems for Floating Offshore Structures and Mobile Offshore Units", ISO 19901-7:2013.

Li, G., "Arctic FPSO: Technical Feasibility and Challenges", OMAE2012-83028, 2012.

Kokkinis, T., Brinkmann, C. R., Ding, J., and Fenz, D.M., "Arctic Deepwater Development Drilling Design Considerations", ICETECH10-115-RF, 2010.

Kwan, C.-T., Bond, J., Yu, H., and Morandi, A., "Arctic Mooring Systems – The Past, Present, and Future", Arctic Technology Conference, OTC 24581, 2014.

Petruska, D., Stone, B., Kwan, T., Zimmerman, E., Devlin, P., Wishahy, M., and O'Connor, P., "Improved Moored MODU Design Codes for Hurricane Season", Offshore Technology Conference, OTC 18900, 2007.

Sablok, A., Ramachandran, M., and Kim, J. W., "Disconnectable Arctic Spar", Arctic Technology Conference, OTC 22139, 2011.

Srinivasan, N., and Sreedhar, M. S., "Circular FPSO for Arctic Deepwater", Arctic Technology Conference, OTC 22136, 2011.

United States Geological Survey, "Circum-Arctic Resource Appraisal: Estimates of Undiscovered Oil and Gas North of the Arctic Circle", USGS Fact Sheet 2008-0349, 2008.

SNAME
THE INTERNATIONAL COMMUNITY
FOR MARITIME AND OCEAN PROFESSIONALS

MULTIPHASE FLOW-INDUCED VIBRATION ANALYSIS FOR SUBSEA RIGID TIE-IN SPOOL

ZHILIANG LI, MENG-LUNG LIU, HAO SONG, GWO-ANG CHANG
OFFSHORE TECHNOLOGY, AMERICAN BUREAU OF SHIPPING

Proceedings of the 21st Offshore Symposium, February 2016, Houston, Texas
Texas Section of the Society of Naval Architects and Marine Engineers

ABSTRACT

Multiphase flow-induced dynamic loads in rigid tie-in spool are mainly generated by fluid density variation, slugging, and direction change at bends. When the frequencies of the induced dynamic loads are close to the natural frequencies of the rigid tie-in spool, resonance will occur. This may cause significant strength and fatigue issues, and should therefore be included in the rigid tie-in spool system design.

In this paper, a procedure for assessing structural response of a tie-in spool using fluid-structure interaction (FSI) method is introduced to investigate the internal multiphase flow-induced vibration and fatigue damage due to fluid density change, slugging, and direction change at bends. A numerical example to demonstrate the procedure are also included.

In the procedure, the rigid tie-in spool structural mode shapes and natural frequencies are first determined by FEA. Then, the inlet slug frequencies are selected based on the natural frequencies of the tie-in spool. Finally, coupled CFD and structure analysis are carried out to calculate the stress and stress range at the critical locations of the rigid tie-in spool for strength and fatigue evaluation. In the coupled CFD and structure analysis, fluid-induced loads due to fluid density variation, slugging, and flow direction change at bends are all included.

For the numerical example, a typical tie-in spool is selected. Star CCM is used as CFD software for multiphase flow-induced load analysis. ABAQUS is used as structure software for mode shape and FSI analysis for strength and fatigue evaluation.

The proposed procedure together with the numerical example provides an in-depth understanding of the resonance effect on strength and fatigue for a typical rigid tie-in spool. It also demonstrates the importance of applying the FSI method to the rigid tie-in spool system design when slugging and resonance are expected.

Keywords: Multiphase flow, tie-in spool, slug, FSI, resonance

INTRODUCTION

Multiphase flow in a pipe is different from single phase flow due to its unsteady flow behavior and induced dynamic loads to the pipe. To predict multiphase flow unsteady behavior and induced dynamic loads, we need to understand fluid properties of each phase in the multiphase flow and the interaction between different phases. For multiphase flow inside a rigid tie-in spool, the induced dynamic loads are mainly from fluid unsteady density variation, slugging, and direction change at bends.

It has been revealed that slug flows tend to generate fatigue damage due to the transient pressures caused by motion of liquid slugs. Taitel and Dukler (1976) came up with a method to determine the slug frequency of two-phase flow in a horizontal pipeline, in which they defined the slug frequency based on the characteristics at the inlet. Lu et al. (2014) came up with a method to assume the downstream pressure fluctuation as the outlet boundary condition to excite a tie-in spool structure. Other studies have set a resonance case where the slug frequency approaches to pipe structural natural frequency, which usually drives to a high structure response and potential risk of failure. Similarly, bends are key regions in forming slugs and experience large transient pressures which can induce high levels of vibrations (Nair et al., 2011).

In this paper, a procedure for assessing the structural response of a tie-in spool, with both pipe and bend sections, using fluid-structure interaction (FSI) method is introduced to investigate the internal multiphase flow-induced vibration and fatigue damage due to fluid density change, slugging, and direction change at bends. A numerical example to demonstrate the procedure is also included.

In the procedure, the rigid tie-in spool structural mode shapes and natural frequencies are first determined by FEA. Then, the inlet slug frequencies are selected based on the natural frequencies of the tie-in spool. Finally, coupled CFD and structure analysis are carried out to calculate the stress and stress range at the critical locations of the rigid tie-in spool for strength and fatigue evaluation. In the coupled approach, fluid-induced loads due to fluid density variation, slugging, and flow direction change at bends are all included.

In the numerical example, a typical tie-in spool is selected. Star CCM is used as CFD software for multiphase fluid-induced load analysis. ABAQUS is used as structure software for mode shape and FSI analysis to calculate stress and stress range for strength and fatigue evaluation.

The proposed procedure together with the numerical examples provides an in-depth understanding of the resonance effect on strength and fatigue for a typical rigid tie-in spool. It also demonstrates the importance of applying the FSI method to the rigid tie-in spool system design when slugging and resonance are expected.

ANALYSIS PROCEDURE

The procedure for assessing structural response of a tie-in spool using the fluid-structure interaction (FSI) method is introduced as shown in Figure 1.

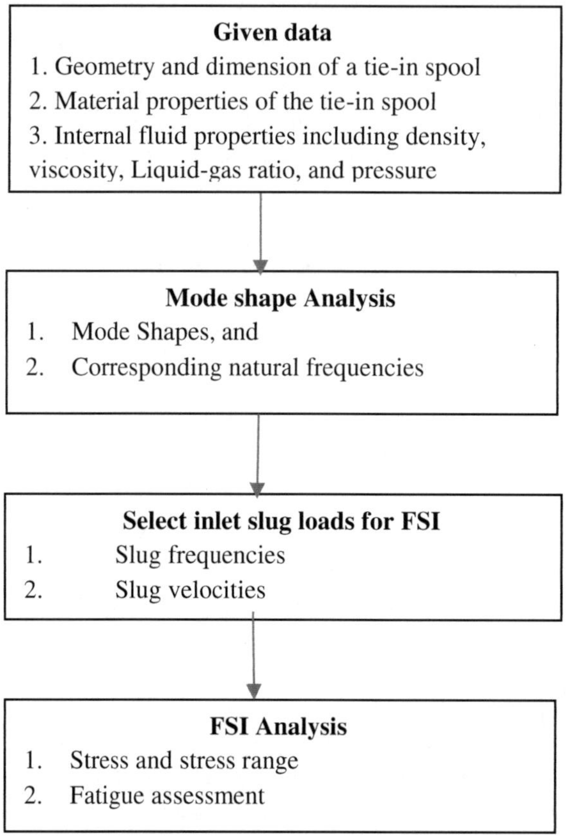

Figure 1: procedure for assessing structural response of a tie-in spool using FSI method

NUMERICAL EXAMPLE

In the numerical example, a typical tie-in spool is selected. Star CCM is used as CFD software for multiphase fluid-induced load analysis. ABAQUS is used as structure software for mode shape and FSI analysis to calculate stress and stress range for strength and fatigue evaluation. For simplification, added mass and damping effect to the tie-in spool structural response due to sea water is not included.

Given Data

Given data in the numerical example includes tie-in spool geometry and dimension, spool structural properties and internal fluid properties.

Geometry and dimension of Tie-in spool

A typical rigid M-shaped tie-in spool geometry as shown in Figure 2 is selected in the numerical analysis. Pipe inner diameter is 0.14224 m (5.6 in), pipe wall thickness is 0.03048 m (1.2 in), and bend radius is 0.42672 m (16.8 in) for all six bends.

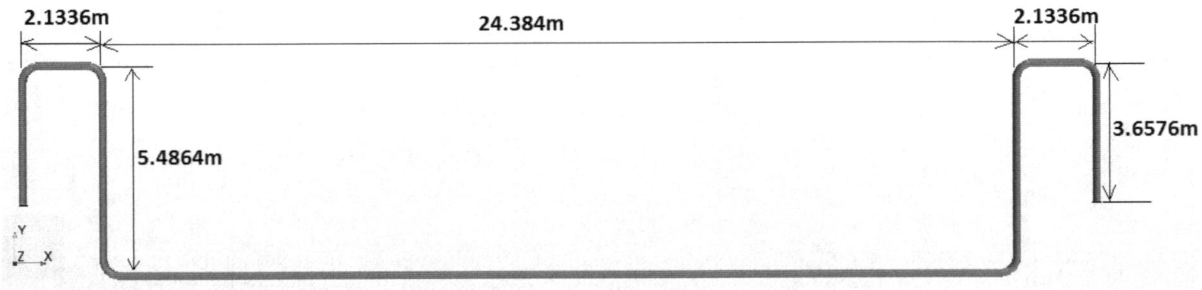

Fig 2. M-shaped tie-in spool configuration.

Material properties of tie-in spool

Spool structural material properties are listed in Table 1.

Table 1. Properties of pipe material

Parameters	Value
Density	7850 kg/m^3
Young's Modulus	210 GPa
Poisson's Ratio	0.3

Internal fluid properties

Internal fluid properties at the inlet of the tie-in spool are listed in Table 2.

Table 2. Internal fluid properties

Parameters	Value
Liquid density	978.14 kg/m^3
Liquid dynamic viscosity	8.8871E-4 Pa-s
Methane density	Ideal gas law
Methane viscosity	1.11906E-5 Pa-s
Liquid gas ratio	1:1
Internal outlet boundary pressure	10 MPa

Mode Shape Analysis

Mode shape analysis is carried out to determine mode shapes and corresponding natural frequencies of the tie-in spool. In the analysis, zero effective tension and fixed boundary conditions at both ends are assumed. For different end boundary conditions, mode shapes and their corresponding natural frequencies will vary. The results of the first four modes in air and water are shown in Figure 3 and their corresponding natural frequencies are listed in Table 3. The first vertical mode with frequency of 1.4015 Hz (period = 0.7 second) is considered the dominant natural frequency for FSI analysis.

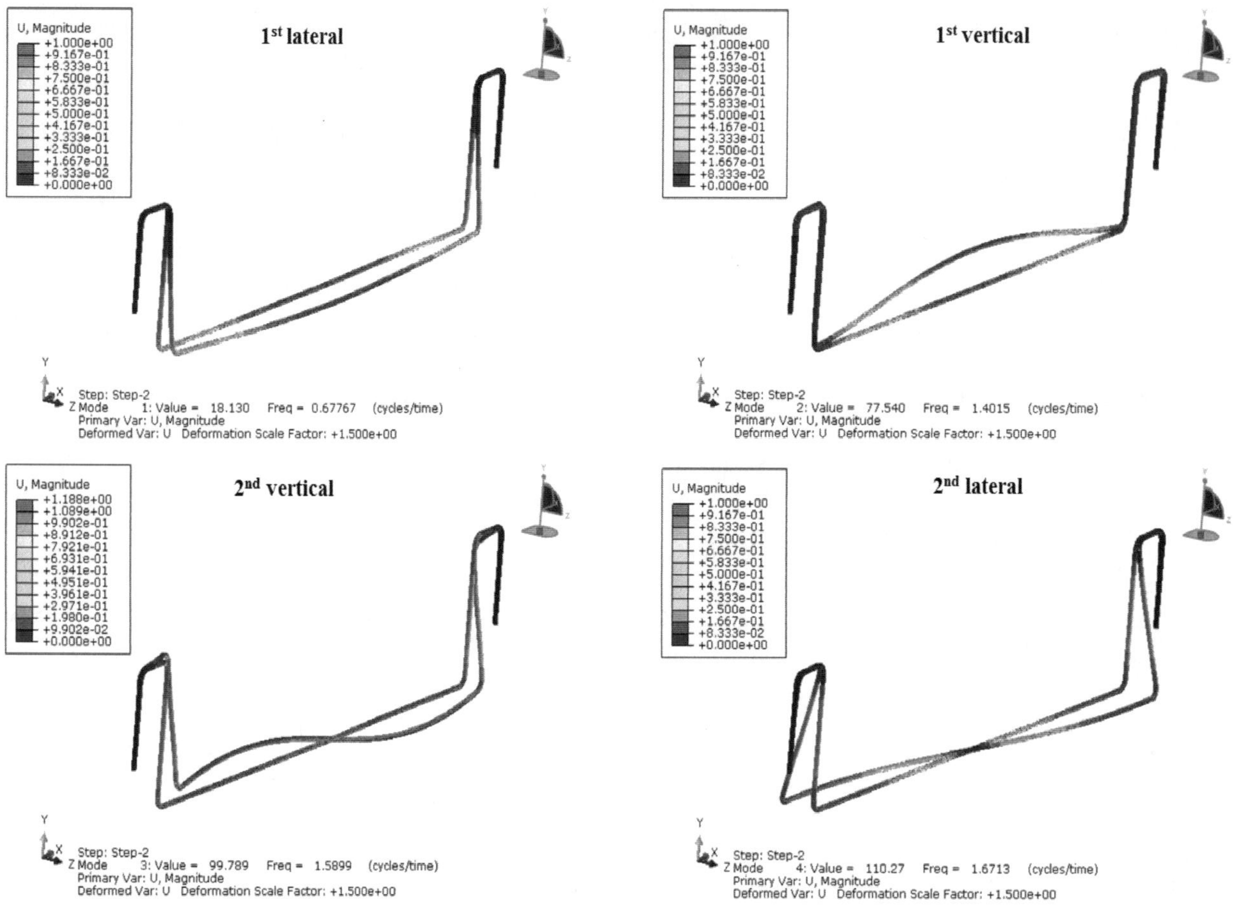

Figure 3. Mode shapes of the first 4 modes

Table 3. Mode shapes and natural frequencies (Hz)

Mode No.	Mode shape	Natural frequencies (Hz)	
		Flooded in Air	Flooded in Water
1	Lateral 1st Mode	0.67767	0.60861
2	Vertical 1st Mode	1.4015	1.2586
3	Vertical 2nd Mode	1.5899	1.4278
4	Lateral 2nd Mode	1.6713	1.5010

Select Inlet Slug loads for FSI

The inlet slug is assumed as pure liquid flow and pure gas flow entering the tie-in spool inlet alternatively for each half of the slug period (T_0) with a constant flow inlet velocity (V_0). Figure 4 exhibits the volume of fluid (VOF) profile of liquid for inlet slug at $T_0 = 0.7$ sec. Gas (methane) enters the inlet for the first half of each slug period, then liquid enters the inlet for the second half of each slug period. The overall liquid-gas volumetric ratio of 1:1 is selected for easy to trigger slug resonance in this selected tie-in spool geometry.

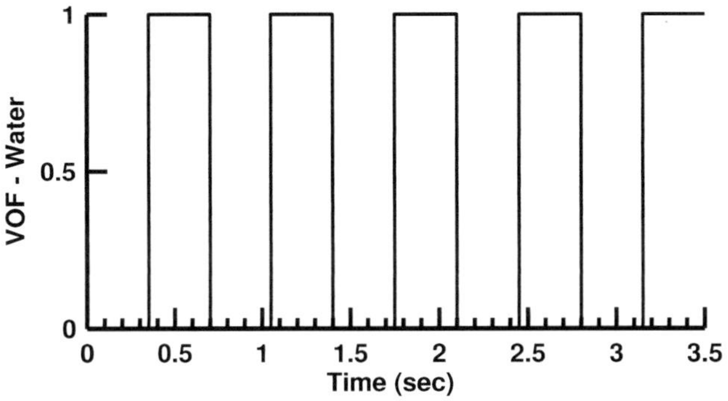

Fig 4. Slug inlet boundary conditions for $T_0 = 0.7$ sec.

To study the effects of the inlet slug frequencies and velocities separately, two sets of inlet slug load cases listed in Table 4 and Table 5 are selected for FSI study. For the first set of load cases in Table 4, inlet flow velocity $V_0 = 10$ m/s is chosen and inlet slug periods are varied from 0 sec to 2.8 sec. For the second set of load cases in Table 5, resonance case of $T_0 = 0.7$ sec ($f = 1.4015$ Hz) is chosen and inlet flow velocities are varied from 5 m/s to 20 m/s.

Table 4. Study cases with constant inlet flow velocity

Study Case	**Inlet flow velocity** (V_0) **(m/s)**	**Inlet slug period** (T_0) **(sec)**
T 2.8	10	2.8
T 1.4	10	1.4
T 1.0	10	1.0
T 0.7	10	0.7
T 0.0	10	0.0 (no inlet slug)

Table 5. Study cases with constant inlet slug period

Study Case	**Inlet flow velocity** (V_0) **(m/s)**	**Inlet slug period** (T_0) **(sec)**
5 m/s	5	0.7
10 m/s	10	0.7
20 m/s	20	0.7

FSI ANALYSIS AND RESULTS DISCUSSION

For no inlet slug case with slug period = 0 sec (case T 0.0), the inlet fluid is mixed with 50 % gas and 50 % liquid with velocity of 10 m/sec. The analysis results are shown in Figure 5 - Figure 7. In Figure 5, the volume fraction of liquid contour indicates that liquid fraction (in red color) is accumulated in the first horizontal pipe section and followed by a gas fraction (in blue color), which indicates a hydrodynamic slug is formed there. After the flow falls down through the 2nd vertical pipe and enters the middle horizontal pipe section, both the hydrodynamic slug and stratified wavy patterns (gas and liquid phases are separated) exist in the downstream. This observation indicates that slug may occur in a tie-in spool for multiphase flow even without inlet slug.

The slug and wavy flow formed in the tie-in spool induces pressure fluctuation and structure vibration. Time history of the fluctuation pressure as shown in Figure 6 are tracked at the bottom of the spool pipe for three locations as monitoring points; they are at the left, right and center in the middle horizontal pipe. In Figure 6, the pressure range drops from left to right due to shear loss along the pipe and constant pressure boundary condition at outlet. At the center point, the average pressure fluctuation period is approximately 0.7 sec (the natural period of the first vertical mode), and the peak pressure variation range is 30 kPa. The induced vertical vibration in Y direction (Y displacement) at the center monitor point is plotted in Figure 7, which indicates the peak Y displacement variation range is of 6 mm, and average Y displacement variation range is of 2 mm.

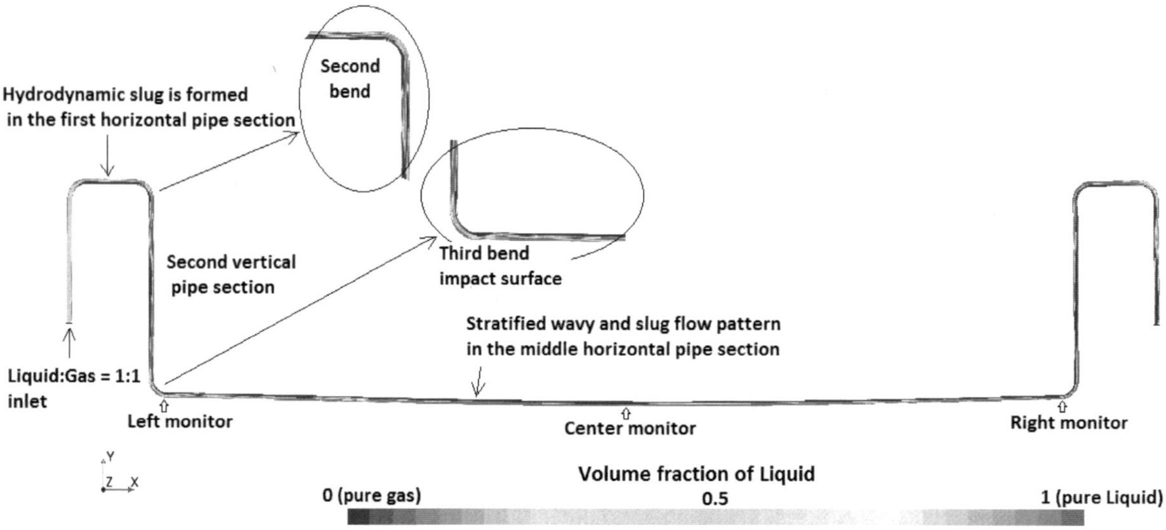

Fig 5. VOF-liquid contour for case T 0.0

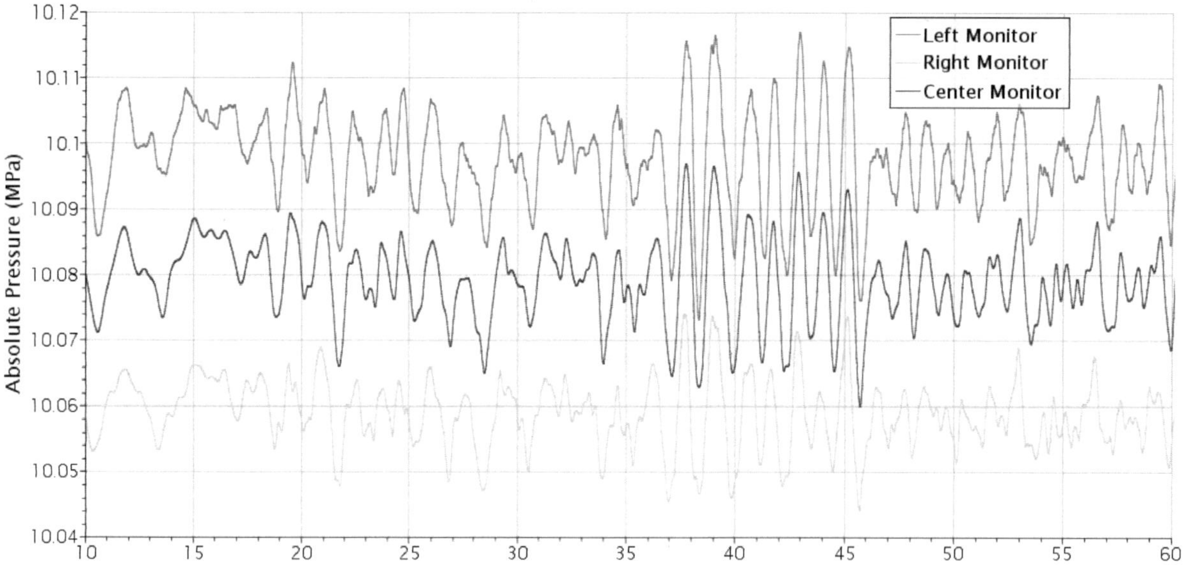

Fig 6. Pressure time history at three monitor locations for case T 0.0

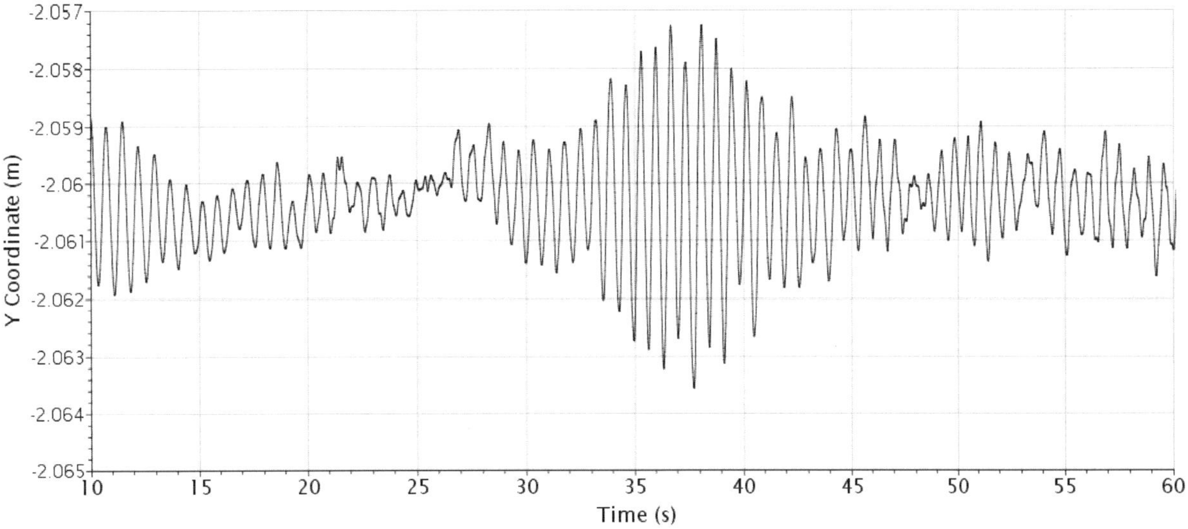

Fig 7. Y displacement time history at center point for case T 0.0.

The results of the study case of 20 m/s (listed in Table 5) with inlet slug period of 0.7 sec (natural period of the first vertical mode) are shown in Figure 8 ~ Figure 11, which represent a typical resonant response. In Figure 8, time history of Y displacement at the center monitor point is plotted. In the plot, instant time point A, B and C in a half cycle are selected when Y displacement are the minimum, medium and maximum, respectively. The corresponding VOF-liquid contours to these three instant time points are shown in Figs 9A-9C and pressure contour are shown in Figs 10A-10C.

Figures 9A-9C indicate the slug are propagated from inlet to the 2nd vertical downfall section, and to the mid horizontal section during this half cycle. At the instant time A, the slug is in the 2nd vertical downfall section. In the instant time B, the slug is half way passing through the 3rd bend section. In the instant time C, the slug has entered the mid horizontal section completely. The pressure contour in Figures 10A-10C indicate pressure variation along the tie-in spool during this half cycle. The time history of the pressure fluctuation at three monitor points for case of 20 m/s is shown in Figure 11, and the maximum pressure vibration range at the center monitor point is 80 kPa.

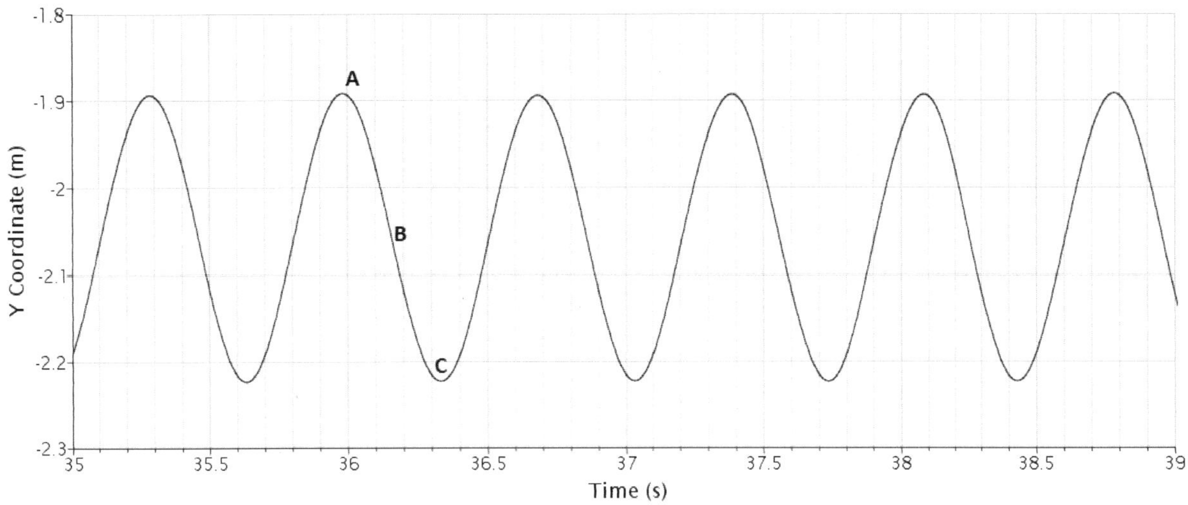

Fig 8. Y displacement time history at center monitor point for case 20 m/s

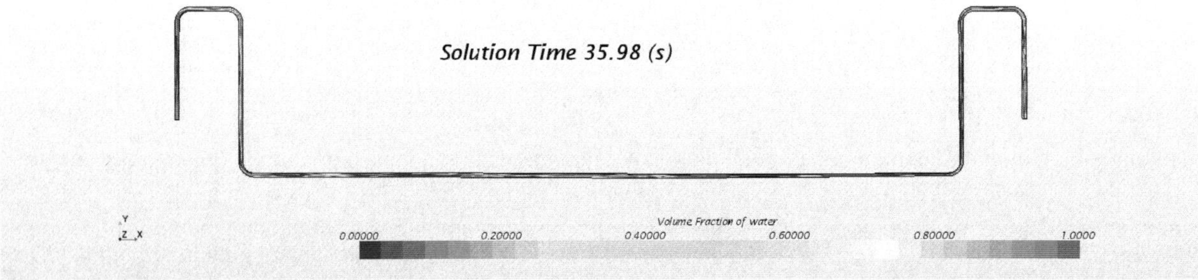

Fig 9A. VOF-liquid contour at instant time A

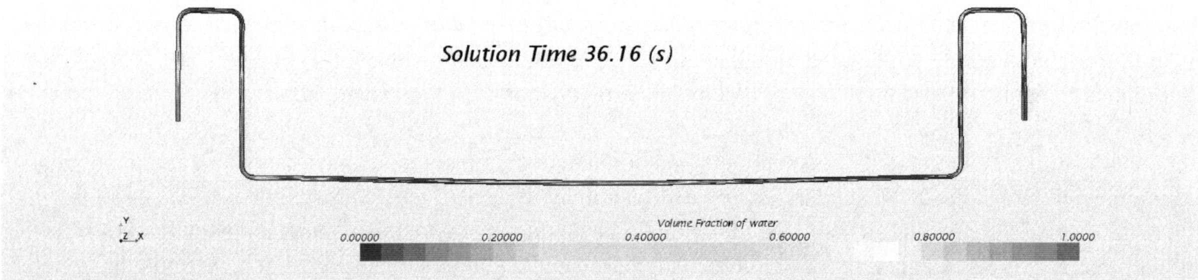

Fig 9B. VOF-liquid contour at instant time B

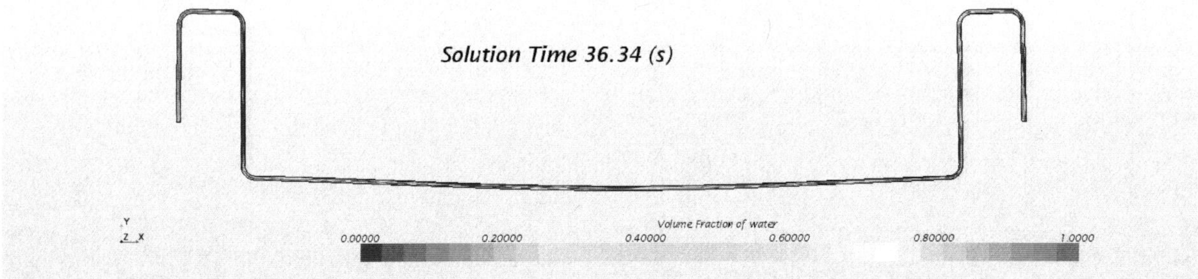

Fig 9C. VOF-liquid contour at instant time C

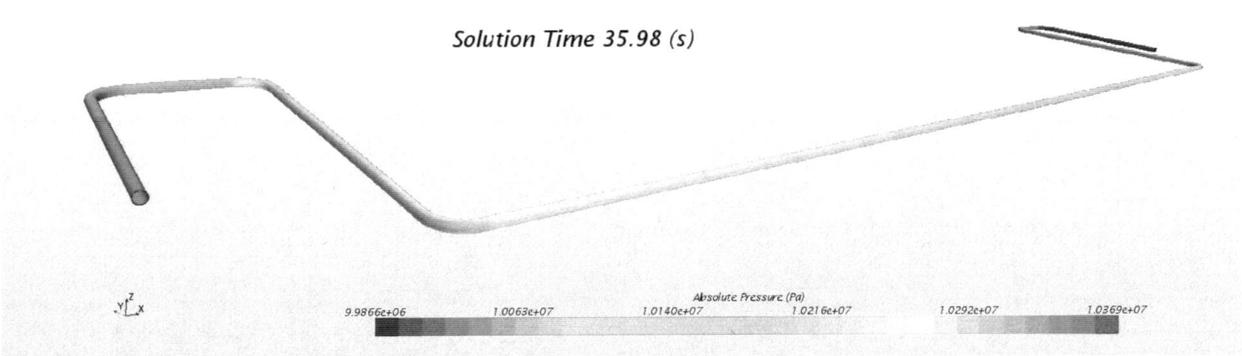

Fig 10A. Pressure contour at instant time A

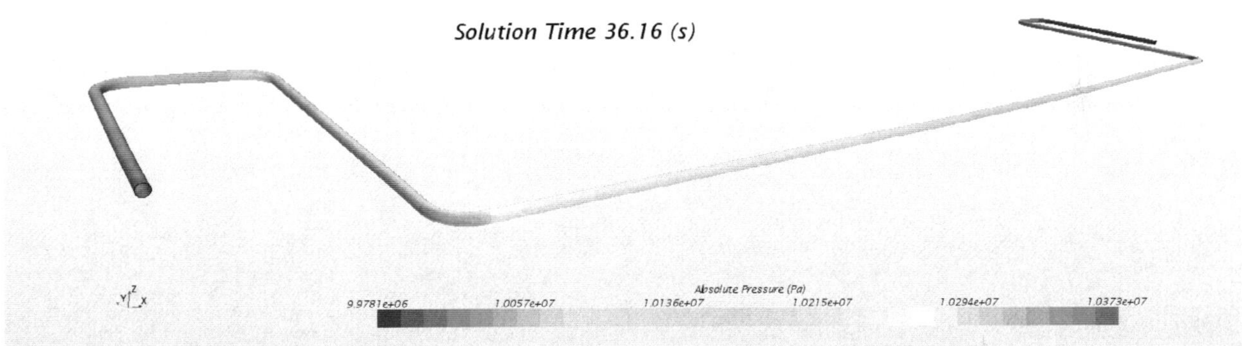

Fig 10B. Pressure contour at instant time B

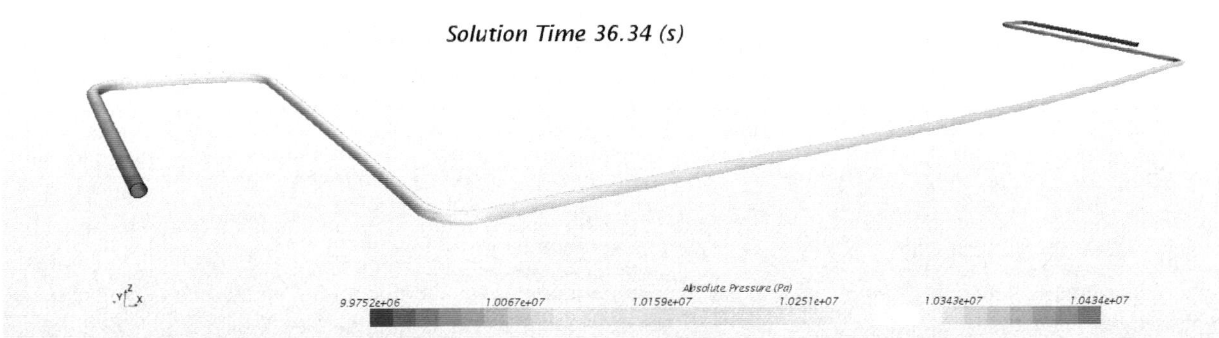

Fig 10C. Pressure contour at instant time C

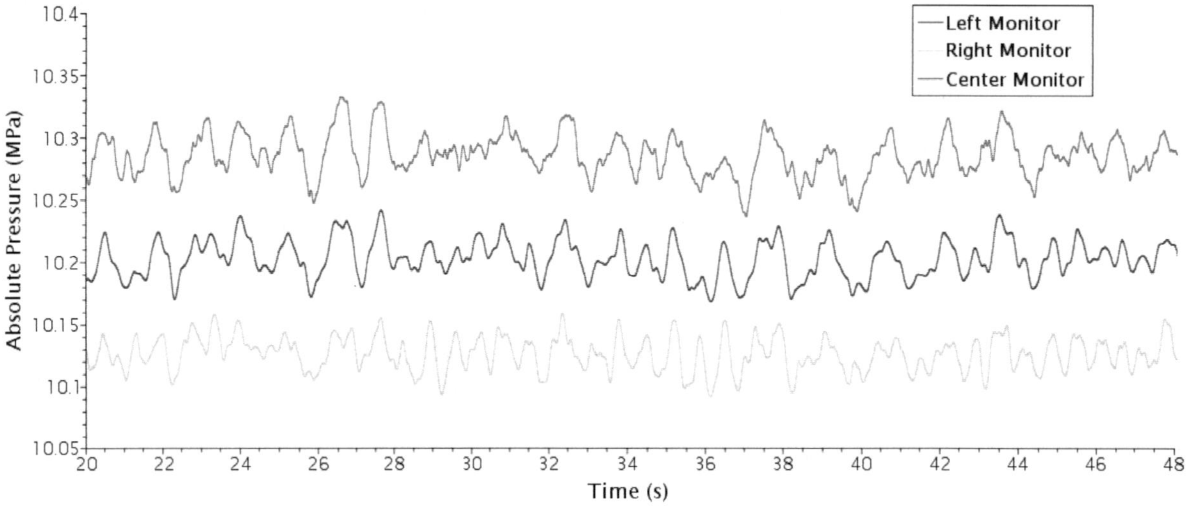

Fig 11. Pressure fluctuation time history at three monitor points for case 20 m/s

The pressure time history at three monitoring points for case 10 m/s (V_0 = 10 m/s and T_0 = 0.7 sec) as listed in Table 5 is shown in Figure 12, and the maximum pressure variation at the center monitor point is of 30 kPa.

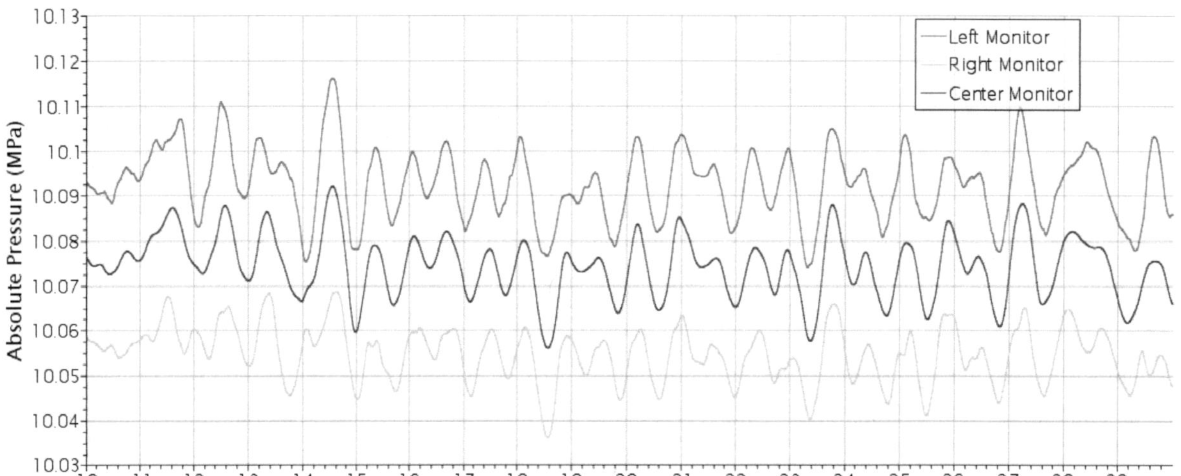

Fig 12. Pressure fluctuation time history at three monitor points for case 10 m/s

For all study cases with constant inlet slug period of 0.7 sec as listed in Table 5, the maximum Y displacement and pressure variation ranges at the center monitor point are summarized in Table 6. From the table, we can conclude that maximum Y displacement and pressure variation ranges increase significantly as slug velocity increases from 5 m/s to 20 m/s.

Table 6. Maximum Y displacement and pressure variation ranges for cases listed in Table 5

Study Cases	Inlet flow velocity(m/s)	Slug inlet period(sec)	Max Y variation range at center point (mm)	Max Pressure variation range at center point (kPa)
5 m/s	5	0.7	10	15
10 m/s	10	0.7	68	30
20 m/s	20	0.7	300	80

For all study cases with constant flow velocity but varying slug period as listed in Table 4, the maximum Y displacement and pressure variation ranges at the center monitor point are summarized in Table 7. In the table, it is observed that the highest Y displacement variation range reaches to 68 mm as inlet slug period equals to 0.7 sec (the resonant period) as expected. For no inlet slug case, the maximum Y displacement variation range can reach to 6 mm due to slug formation occurred in the tie-in spool.

Table 7. Maximum Y displacement and pressure variation ranges for cases listed in Table 4.

Study Cases	Inlet flow velocity(m/s)	Inlet slug period (sec)	Max Y variation range at center point (mm)	Max pressure variation range at center point (kPa)
T 2.8	10	2.8	4	40
T 1.4	10	1.4	12.6	62
T 1.0	10	1.0	8.2	36
T 0.7	10	0.7	68	30
T 0.0	10	No inlet slug	6	30

In Table 7, it is note that the highest pressure variation range occurs at case of T 1.4 when the imposed inlet slug period (1.4 sec) is twice of the natural vibration period (0.7 sec). Because, in this case, liquid slug impacts the third bend and oppresses the middle horizontal section of the tie-in spool every other natural vibration period, and the lower vibration circle and high vibration circle occurs alternatively as shown in Figure 13 B "Y displacement range time history". The maximum pressure occurs at the lower trough of every high vibration circle when the water slug impact the third bend during that circle, and the minimum pressure occurs at the higher trough of every lower vibration circle. This accumulative and alternative impacting effect maximizes the pressure variation range in this case.

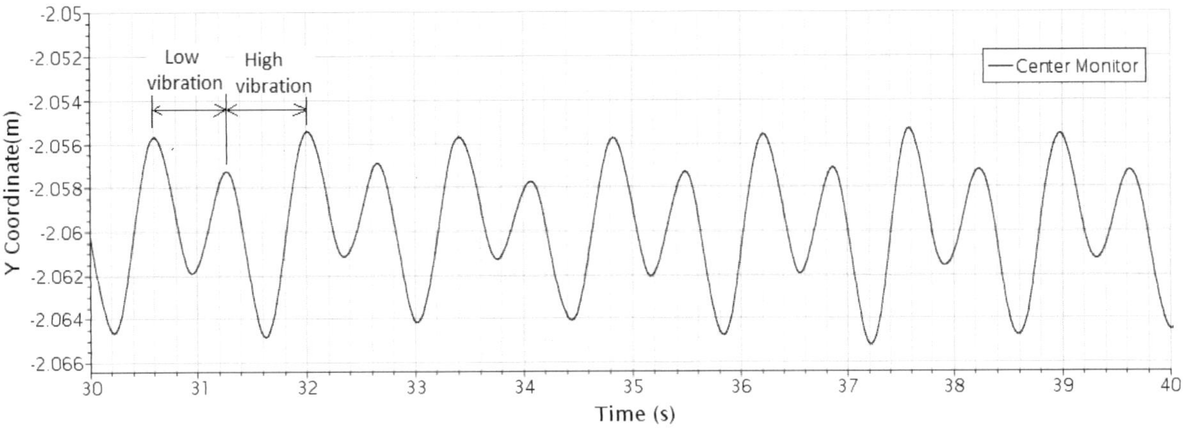

Fig 13 A. Y displacement range time history at center monitor point for case of T 1.4.

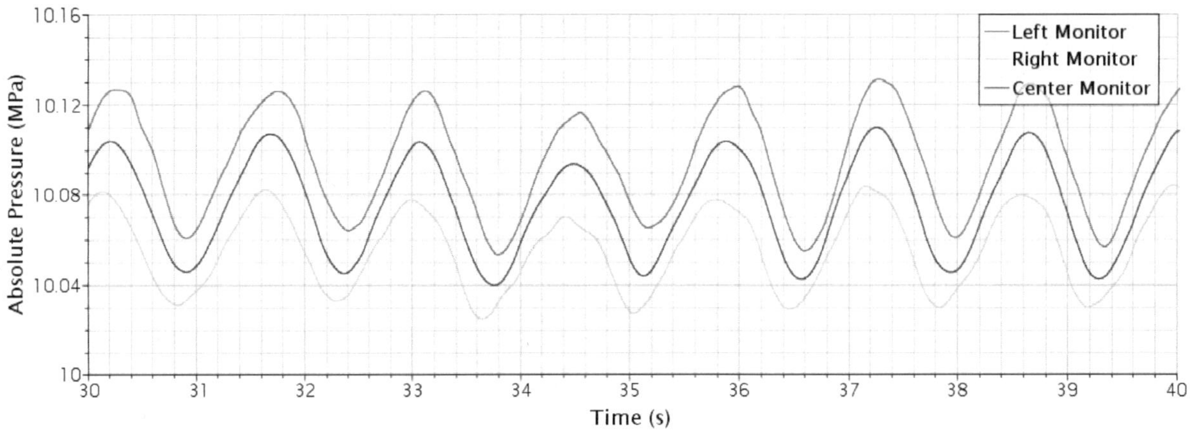

Fig 13 B. Pressure fluctuation time history at three monitor points for case of T 1.4

Structural Vibration

Forced vibrations are observed to the tie-in spool due to different inlet slug periods and different inlet slug velocities. When the inlet slugging period is a multiple of the fundamental natural period, both the first and second vertical modes may be triggered, vibration shapes of combined vertical modes are observed as shown in Fig 14. Similarly, a lower inlet flow velocity is likely to trigger the first vertical mode while a higher inlet flow velocity would possibly trigger the second vertical mode.

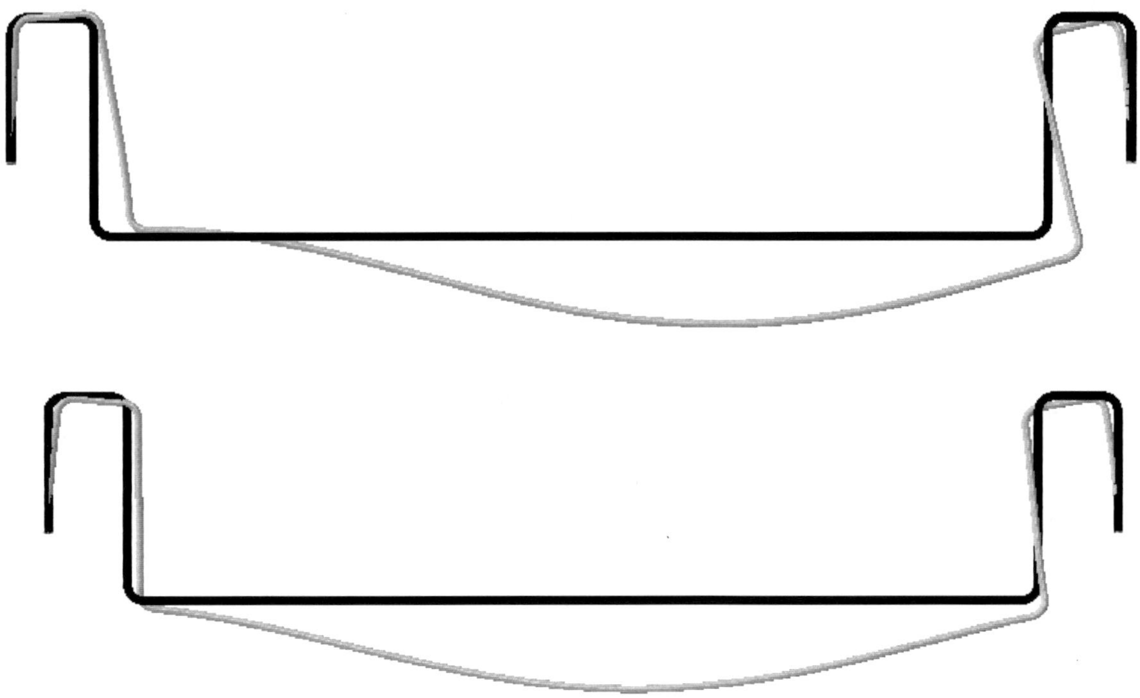

Fig 14. Forced vibration curves for combined first and second vertical modes

Stress and Fatigue Assessment

Two primary hot spot locations with high stress are found at the left and right corners' intradoses as illustrated in Fig 15. These two corners can experience severe repeating bending moments. Therefore, their fatigue performance should be assessed in accordance with the corresponding stress ranges. Fig 16 shows a typical stress time history at these two locations.

Fig 15. Stress hot spot locations

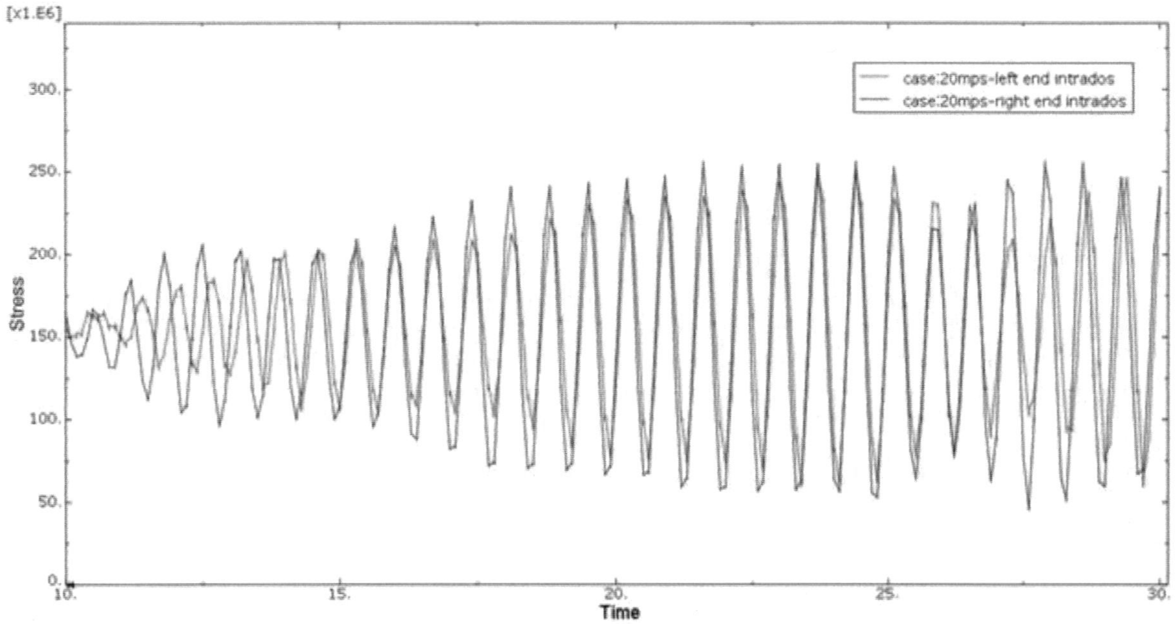

Figure 16. Stress time history for case of 20 m/s

Fatigue S-N curve of T (CP) for a tubular joint is selected from the ABS *Guide for Fatigue Assessment of Offshore Structure* as shown in Figure 17. Fatigue damage per year is calculated to assess the tie in spool fatigue performance. In the calculation, fatigue design factor (FDF) and stress concentration factor (SCF) are not applied.

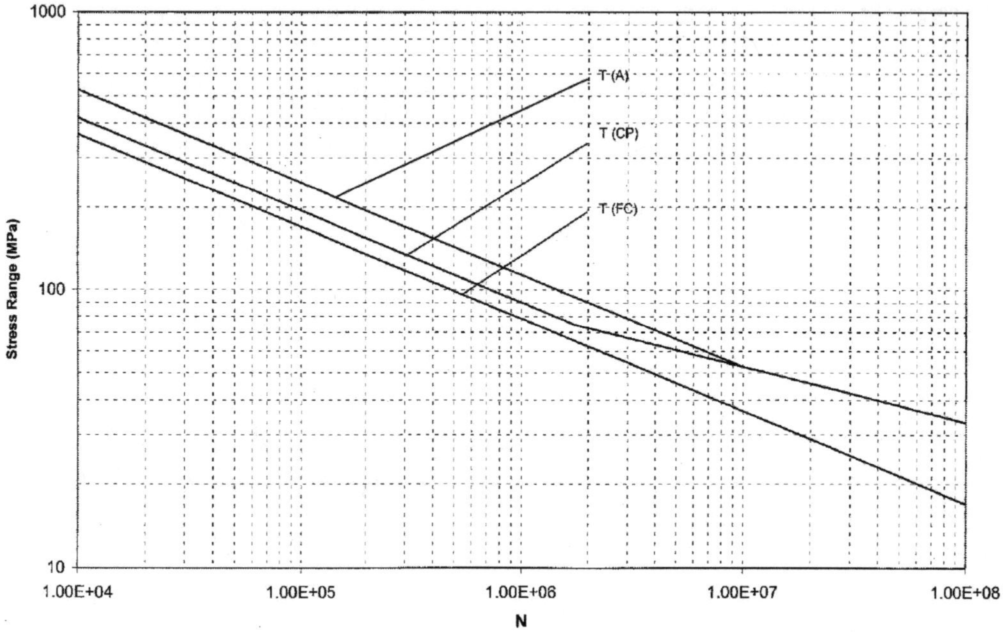

Figure 17. S-N curve from ABS *Guide for Fatigue Assessment of Offshore Structures*

Structural responses in terms of displacement range at the spool center location as well as stress ranges and fatigue damage per year at the worst location are summarized in Table 8 for sensitivity study only. Please note the structural response of the tie-in spool in this study does not include the added mass and damping effect from the sea water. Since the added mass and damping effect will induce additional inertia and damping resistance to the structural response, less displacement and stress ranges results than those in Table 8 are expected.

Table 8. Tie in spool displacement, stress and fatigue assessment

Study Case	Displacement Rang (mm)	Stress Range (MPa)	Fatigue Damage per year	Corresponding Location
T 0.0	6	9.37	8.03E-04	inlet
T 1.0	8.2	15.44	6.83E-03	inlet
T 1.4	12.6	14.11	3.11E-03	inlet
T 2.8	4	6.72	3.81E-05	inlet
5 m/s	10	10.30	1.29E-03	inlet
10 m/s	68	57.14	6.78E+00	inlet
20 m/s	300	139.90	1.69E+02	outlet

From the Table 8, it is observed that the higher displacement range, the higher stress range, and the higher the fatigue damage. Very high fatigue damage results are observed for high slug velocities cases of 10 m/s, and 20 m/s due to structural resonance (inlet slug period coincides with the natural period of the first vertical mode), which demonstrates the importance applying the proposed procedure using FSI method for rigid tie-in spool design when slug induced resonance is expected. The relationship between the slug velocities and stress range is shown in Figure 18. From the figure, we can also conclude that the hot spot stress range increases as the inlet slugging velocity increases, Therefore, limiting the slugging velocity is helpful to reduce fatigue damage.

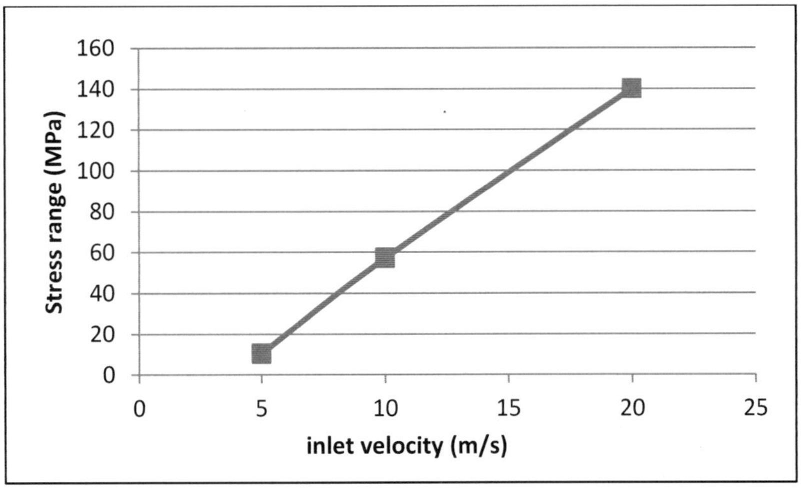

Fig 18. Relationship between Inlet slugging velocity and Induced Stress Range

For constant inlet flow velocity of 10 m/s, the relationship between the slugging period and the induced stress ranges is shown in Figure 19. For case T 0.0 with no inlet slugging, because slug is formed in the spool, the stress range is only slightly less than the case of 5 m/s with inlet slug of 0.7 sec.

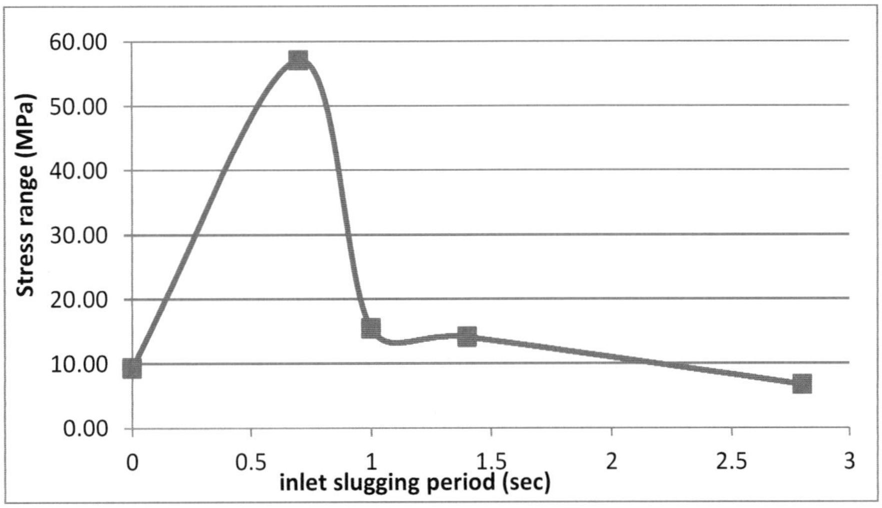

Fig 19. Relationship between Inlet Slugging Period and Induced Stress Range

CONCLUSIONS

In this paper, a procedure for assessing structural response of a subsea tie-in spool using a fluid-structure interaction (FSI) method is proposed. A numerical study is also carried out to implement the procedure and investigate the effects of slugging loads on the strength and fatigue of the spool structure. Based on the studies, the following conclusions can be made:

1. The proposed analysis procedure using FSI method has demonstrated the importance to a typical rigid tie-in spool system design when resonance due to slugging is expected.
2. Hydrodynamic slug may occur in a tie-in spool for multiphase flow and incur resonance to the tie-in spool even without inlet slug.

3. Modal analysis is necessary to check if resonance due to slugging is susceptible. The slugging load period should not be close to natural period of the first few fundamental modes or the multiples of periods of those fundamental modes.
4. The fatigue damage increases when the inlet slugging velocity increases. Therefore, limiting the slugging velocity is helpful to reduce fatigue damage.
5. Added mass and damping effect from the surrounding water of the tie-in spool will be included in our future study for better structural response assessment in term of displacement and stress ranges. In addition, VIV from sea bottom current will also be included in the future study using FSI method to examine its effect on tie-in spool structural response.

REFERENCES

ABS, "Guide for Fatigue Assessment of Offshore Structures", April, 2003.

Nair, A., Chauvet, C., Whooley, A., Eltaher, A., and Jukes, P., "Flow Induced Forces on Multi-Planar Rigid Jumper Systems", OMAE 2011-50225, Rotterdam, Netherlands, 2011

Lu, Y., Liang, C., Manzano-Ruiz, J., Janardhanan, K., Perug, Y., "FSI of Flow-induced Vibration in Subsea Jumper Subject to Downstream Slug and Ocean Current", OMAE 2014-24120, San Francisco, California, 2014

Taitel, Y, Dukler, A. E., "A model for predicting flow regime transitions in horizontal and near horizontal gas-liquid flow", AIChE Journal, V 22, Issue 1, Jan 1976, 47–55.

THE INTERNATIONAL COMMUNITY
FOR MARITIME AND OCEAN PROFESSIONALS

RISER SIZING FOR HPHT APPLICATIONS

ASSEM MAHMOUD
PROJECT SPECIALIST, WOOD GROUP KENNY
MIKE MAO
SENIOR CONSULTANT, WOOD GROUP KENNY
STEVEN WANG, PHD
STAFF CONSULTANT, WOOD GROUP KENNY
CHENGYE FAN, PHD
STAFF CONSULTANT, WOOD GROUP KENNY

Proceedings of the 21st Offshore Symposium, February 2016, Houston, Texas
Texas Section of the Society of Naval Architects and Marine Engineers

ABSTRACT

For the past decade, offshore oil and gas observed more challenging realms of the industry which include ultra-deep waters, harsher environments, and higher pressures and temperatures. The most notable of such is the challenges of high pressure high temperature (HPHT) conditions which are still regarded as uncharted territories and have prompted the industry to deliver safe and reliable solutions. Subsea design for HPHT conditions presents numerous engineering challenges which can limit conventional technology designs. For riser systems, HPHT conditions can require higher than usual pipe wall thickness, which can lead to pipe manufacturing and fabrication issues as well as riser payload and hang-off system challenges to the floating facility.

Conventional design codes will be reviewed for riser sizing, hence suitable assumptions will be derived for riser sizing methodologies in HPHT applications. The purpose of this paper is to discuss the conventional thin-wall and thick-wall riser wall sizing designs and provide recommendations the appropriate sizing of the riser in HPHT applications.

Keywords: SNAME, Offshore, Risers, HPHT, Sizing

INTRODUCTION

Over the past decade, high pressure high temperature (HPHT) hydrocarbon bearing wells have been recognized as the new frontier in offshore oil and gas recovery. Coupled with ultra-deep waters and harsh environments, delivering safe and reliable pressure containment systems in HPHT conditions, along with the steadfast risks of offshore hydrocarbon drilling and production, have become the next set of challenges the industry must overcome. Riser systems are particularly susceptible to HPHT environments due to the substantial increase in wall thickness which presents knock-on effects as discussed in this paper.

In accordance with 30 CFR 250.807(b), HPHT is defined as the anticipated surface pressure or shut-in tubing pressure (SITP) greater than 15,000 psig or the flowing temperature is equal to or greater than 350 degrees Fahrenheit (177 degrees Celsius) on the seafloor for a well with a subsea wellhead or at the surface for a well with a surface wellhead (§250.807, 2015).

Riser design for HPHT applications is critical and onerous where each design scenario may present a different set of challenges. Riser design considerations, such as impact of reservoir fluid pressure and temperature, fatigue performance and material selection, riser strength degradation and payload impact on the vessel, are some of the key design criteria which must be studied and optimized in order to deliver a robust riser design for HPHT environments.

A robust riser design and delivery hinges on an appropriate riser sizing methodology. Determination of an accurate wall thickness is a critical first step in the design. For HPHT applications, the use of conventional riser sizing formulas may prove detrimental. Conventional riser sizing codes may not be suitable due to uniform hoop stress across the wall thickness; thus, resulting in an overly conservative wall thickness.

This paper describes the challenges and limitations with employing certain riser sizing formulas used for riser sizing in conventional and non-HPHT environments by comparing the methodologies and applications of the stress calculations of certain key formulas proposed by riser design industry standards and codes. By investigating the methodology and application of said formulas, the paper aims to provide recommendations for a suitable methodology for riser sizing in HPHT environments based on the comparison of the formulas and the discussion provided herein.

Risers, whether rigid, catenary or flexible, experience fatigue loads due to vessel motions and wave loading as well as vibrations caused by current loading during their design life. Risers also experience stresses due to static and dynamic loading during their design lives which may result in high utilizations that can exceed a riser's structural capacity. Thus the fatigue and strength performance of the riser are critical to its design. The impact on the riser strength capacity and fatigue performance will be investigated and discussed in the example case study provided.

RISER SIZING METHODOLOGY

In order to achieve a thorough and robust design, the riser pipe wall thickness should be checked for all possible failure modes, which include burst, collapse, bending, buckling, and the failure under combined loads. For a riser in HPHT conditions, burst due to the internal overpressure is the dominant scenario for the riser sizing design.

To prevent riser burst failure under high internal pressure, the pipe hoop stress should be governed based on pipe burst capacity. This paper investigated the three main issues in HPHT sizing: the method of hoop stress calculation, the selection of design factor and the determination of pipe burst capacity.

STRESS CALCULATION FORMULAS

For the internal overpressure case, the hoop stress is calculated for the burst pressure check. There are two categories of methods for calculating hoop stress: the thin wall formula, also called Barlow's equation, is based on the assumption of uniform hoop stress across pipe wall and the thick wall formula (Lame's equation) gives an exact solution for nonlinear hoop stress variation. The different formulas for hoop stress calculation in current riser design codes are summarized. Based on ASME B31.4 (ASME B31.4, 2012) and ASME B31.8 (ASME B31.8, 2014), the hoop stress is calculated using Barlow's formula for thin wall pipe.

$$\sigma_h = \frac{PD}{2t} \tag{1}$$

Where:
P is the internal pressure;
D is the pipe OD;
t is the pipe wall thickness.

For the thick wall pipe, a revised formulation is applied as per the following.

$$\sigma_h = \frac{P(D-t)}{2t}. \tag{2}$$

Based on API 2RD (API STD2RD, 2013) and API 1111 (API RP1111, 2015) the average stress across the pipe wall thickness is calculated.

$$\sigma_h = \frac{(P_i - P_o)D}{2t} - P_i \qquad (3)$$

Where:
P_i is the internal pressure;
P_o is the external pressure.

Lame's equation for pipe sizing presents the hoop stress and radial stress across the pipe wall. The pipe hoop stress is calculated using the following formula.

$$\sigma_h = \frac{P_i r_i^2 - P_o r_o^2}{r_o^2 - r_i^2} + \frac{(P_i - P_o) r_o^2 r_i^2}{(r_o^2 - r_i^2) r^2} \qquad (4)$$

The pipe radial stress is calculated using the following formula.

$$\sigma_r = \frac{P_i r_i^2 - P_o r_o^2}{r_o^2 - r_i^2} - \frac{(P_i - P_o) r_o^2 r_i^2}{(r_o^2 - r_i^2) r^2} \qquad (5)$$

Where:
r_i is the pipe inner surface radius;
r_o is the pipe outer surface radius;

The hoop stress is calculated for a 10 ¾" OD x 0.4" WT riser for 5,000 psi internal pressure. The hoop stress calculated based on ASME B31.8 (ASME B31.8, 2014) thin wall theory is overly estimated. API 2RD (API STD2RD, 2013) has a good representation using the average stress across the pipe wall. For the 5,000 psi pressure case, the maximum hoop stress is found at the inner surface, which has less than 2.6 ksi difference from the average value. However, for the 20,000 psi (20 ksi) internal pressure case, the maximum hoop stress across the wall is more than 11 ksi than the average hoop stress calculated based on API 2RD (API STD2RD, 2013), which presents a 12.6% difference. For the same 20 ksi internal pressure case, the radial stress is also doubled at the inner surface compared to average radial stress across the wall. It is also noted in the current riser sizing methods the radial stress is excluded from the pipe sizing consideration. Figure 1 presents a comparison of the pipe hoop stress of the three formulas used.

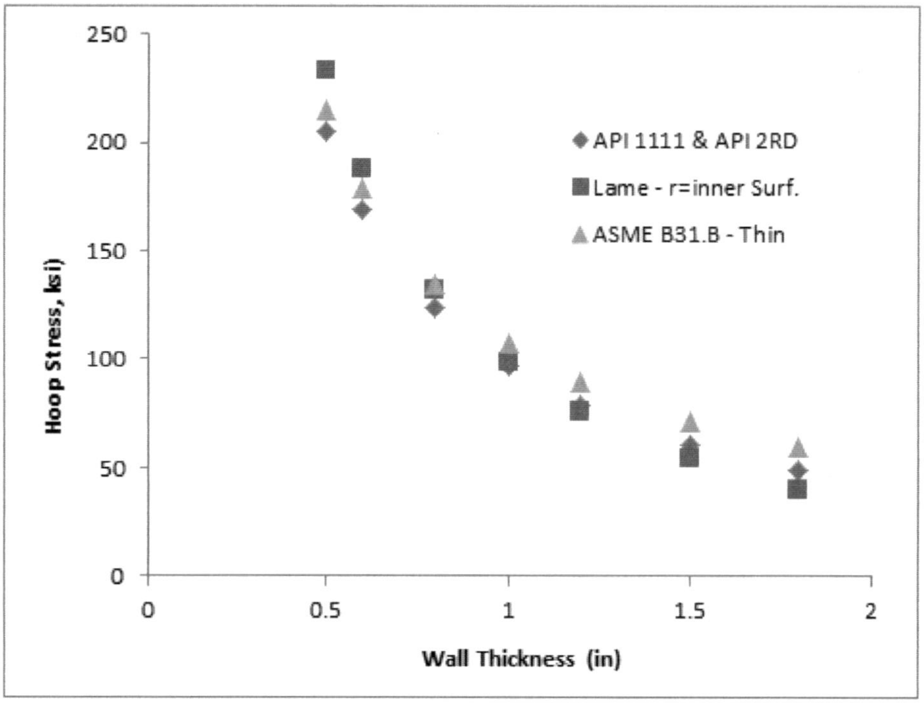

Figure 1 Pipe Wall Thickness versus Hoop Stress

The hoop stress calculation using ASME B31.8 (ASME B31.8, 2014) thick wall formula is very close to the maximum hoop stress across the pipe wall, which is found at the pipe inner surface. Figure 2 and Figure 3 compare the hoop stress across the wall of the pipe for a 10 ¾" x 0.4" WT riser with 5,000 psi internal pressure and a 10 ¾" x 1" WT riser with 20,000 psi internal pressure. Figure 4 shows the radial stress calculated using API 2RD (API STD2RD, 2013) for a 10 ¾" x 1" WT riser with 20,000 psi internal pressure. ASME B31.8 (ASME B31.8, 2014) thin wall or thick wall sizing formulations do not consider the pipe radial stress; hence, they are not shown in the aforementioned figure.

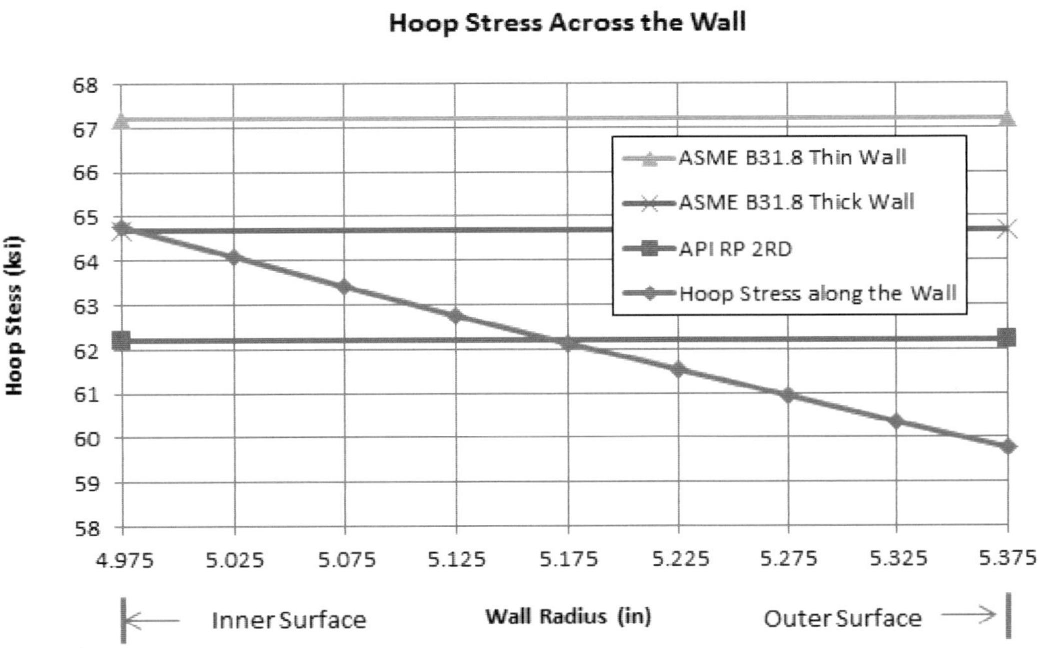

Figure 2 10 3/4" x 0.4" Riser with 5,000 psi Internal Pressure Hoop Stress Calculation

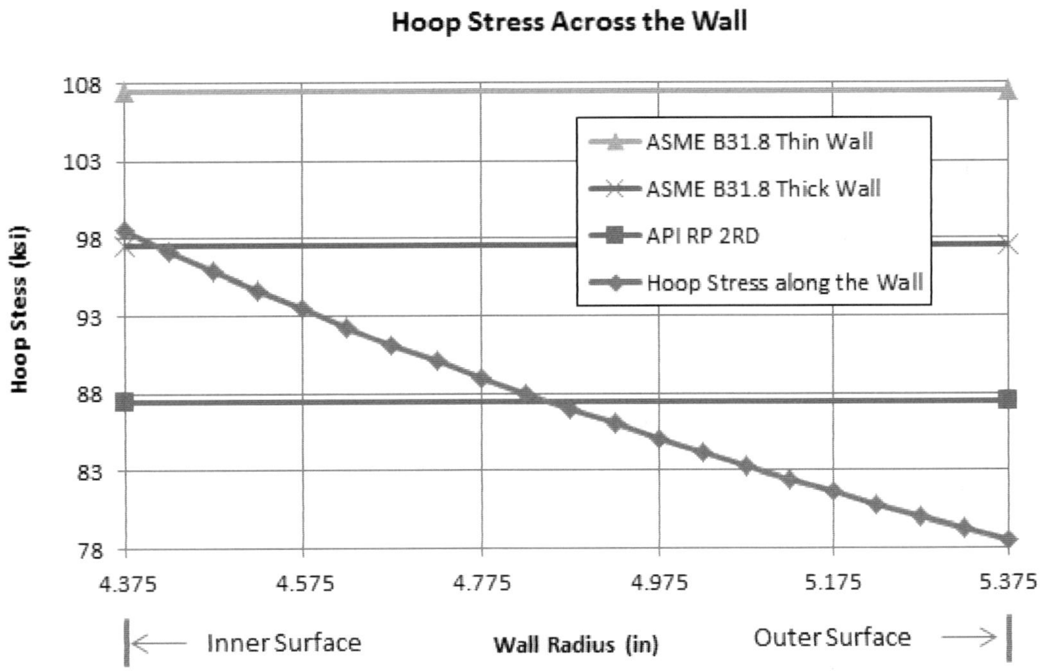

Figure 3 10 3/4" x 1" Riser with 20,000 psi Internal Pressure Hoop Stress Calculation

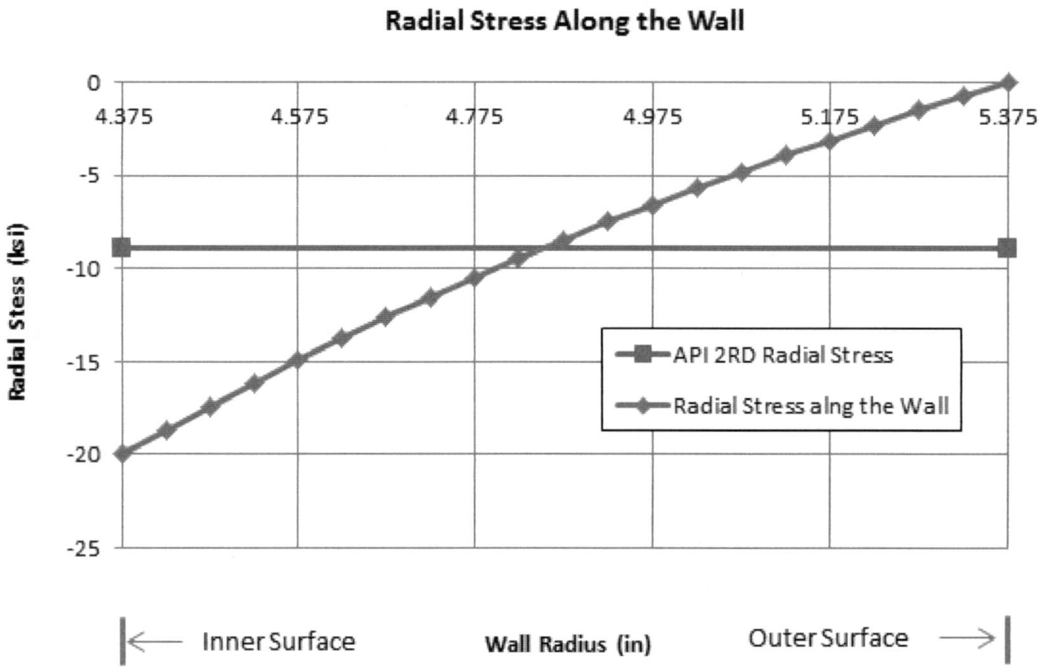

Figure 4 10 3/4" x 1" Riser with 20,000 psi Internal Pressure Radial Stress Calculation

RISER BURST CAPACITY

Burst pressure is a characteristic of riser pipe. API 1111 (API RP1111, 2015) specifies the minimum burst pressure of the pipe, which is determined by the equation $p_b = k(S + U)\ln(D/(D - 2t))$, where k is a parameter to account for variability in mechanical properties and wall thickness, and is equal to 0.45 for API 5L (API Spec5L, 2012) or API 5CT (API Spec5CT, 2011) pipe. The burst pressure formula proposed by API 1111 is based on theoretical considerations confirmed by many burst tests. However, the burst tests are mainly for the thin walled pipe. For both thin wall pipe and thick wall pipe, the pipe burst capacity is theoretically calculated using finite element software ABAQUS® and compared to the burst pressure using API 1111 formulation.

Figure 5 shows the pipe finite element model used. A partial section of the full cylinder is modeled. Symmetry boundary conditions are applied at the ends to ensure that the behavior of the full cylinder is represented. Plane strain element CPS4R is used. Only the internal pressure is applied in the inner surface. NLGOEM option is used to allow large geometry deformation. The pipe section is modeled using the suitable element size. The pipe radial stress and hoop stress are read at the riser inside and outside surfaces, which are close (hoop stress difference less than 0.1%, radial stress less than 1.5%) to the theoretical value using Lame's equation for the elastic material. Two pipe sizes are modeled: 10 ¾" OD x 0.5" WT representing thin wall pipe and 10 ¾" OD x 1" WT representing the thick wall pipe.

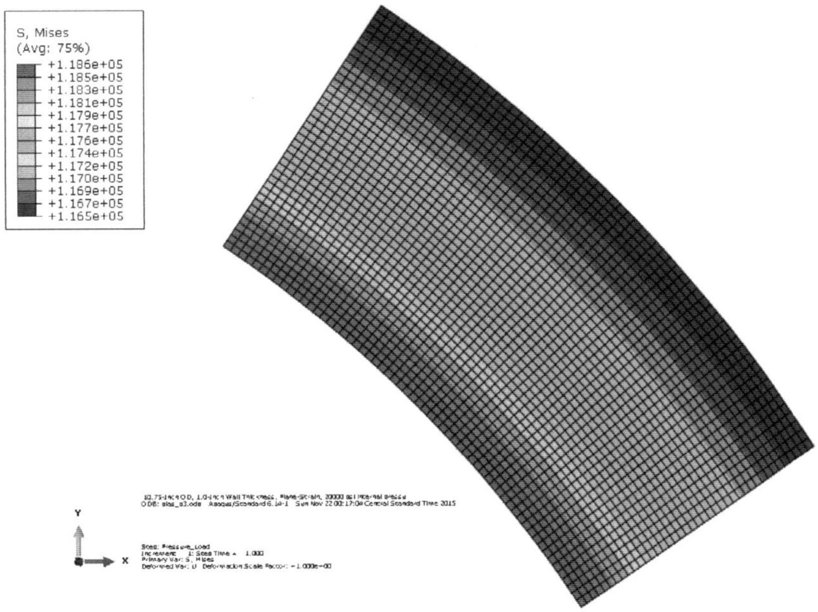

Figure 5 Pipe Finite Element Model

Both elastic and elastic-plastic materials are used. The elastic material is applied to test the riser stress, which is comparable to the theoretical value. For the pipe burst capacity calculation, the stress-strain curve is assumed using Ramberg-Osgood Parameters. For HPHT conditions, the yield strength derating factors should be considered based on ASME B31.8 (ASME B31.8, 2014). For high temperatures up to 449.6 °F (232 °C), the derating factor is 0.867. The stress-strain curve for 110 ksi material (C110) during the normal temperature and extreme high temperature is plotted in Figure 6. The pipe burst pressures under different temperatures are obtained using the different stress-strain curves.

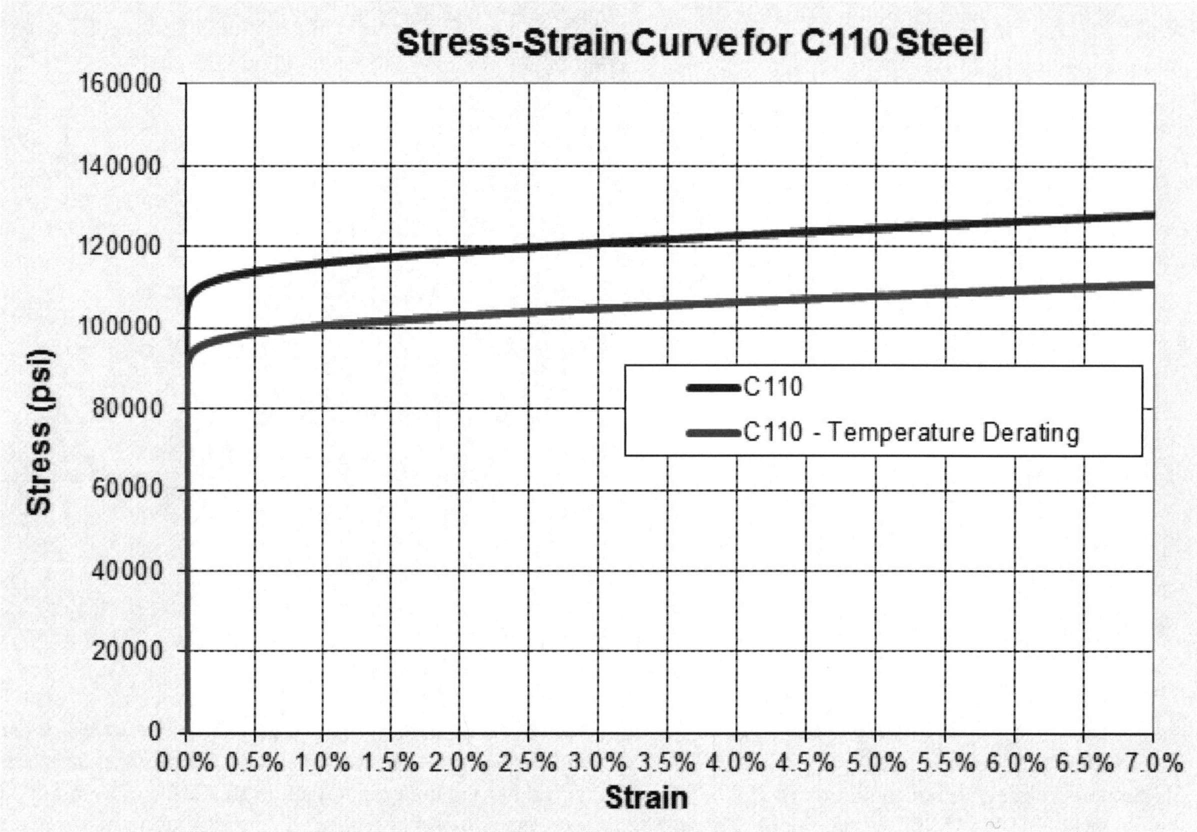

Figure 6 110 ksi Material Stress-Strain Curve

When the von Mises stress at the outside surface reaches the pipe tensile strength capacity, the failure occurs by the plastic collapse based on the method suggested by Fu and Kirkwood (Fu et al., 1995).

The pipe burst pressure from the finite element analyses are reported, which are compared to the burst pressure calculated using API 1111 formulation.

Using the elastic material, the pipe is applied with 20 ksi internal pressure and 27 ksi internal pressures. The radial stress at the inner surface is close to the applied pressure. The radial stress at the external surface is close to zero in this case. The maximum hoop stress is found at the inner surface, which is proportionally increased when the internal pressure increases. Figure 7 compares the elastic material hoop stress and the radial stress in 20 ksi and 27 ksi internal pressures.

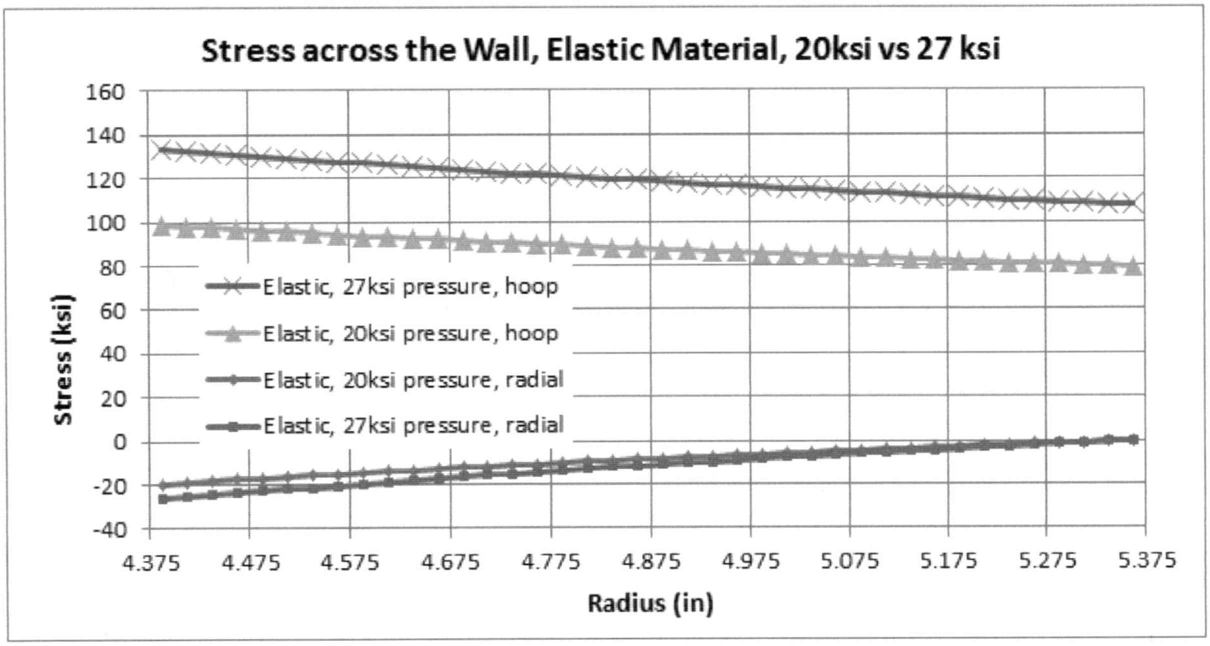

Figure 7 Hoop Stress versus Radial Stress for Elastic Material in 20 ksi and 27 ksi Internal Pressure

Comparing the elastic to the elastic-plastic material, the pipe is applied with 27 ksi internal pressure. The radial stress at the inner surface is close to the applied pressure. The radial stress for both the elastic and elastic-plastic materials is almost identical, which converges to close to zero at the external surface. On the other hand, the maximum hoop stress for the elastic material is found at the inner surface of the pipe, while the maximum hoop stress for the elastic-plastic material is found at the outer surface of the pipe. This suggests that using the stress strain curve in Figure 6, the hoop stress distribution is different for elastic and elastic-plastic materials when 27 ksi internal pressure is applied. The maximum hoop stress is found at the external surface for the elastic-plastic material. On the contrary, the maximum hoop stress is found at the internal surface for the elastic material.

Figure 8 compares the elastic material to the elastic-plastic material in terms of hoop stress and radial stress in 27 ksi internal pressures.

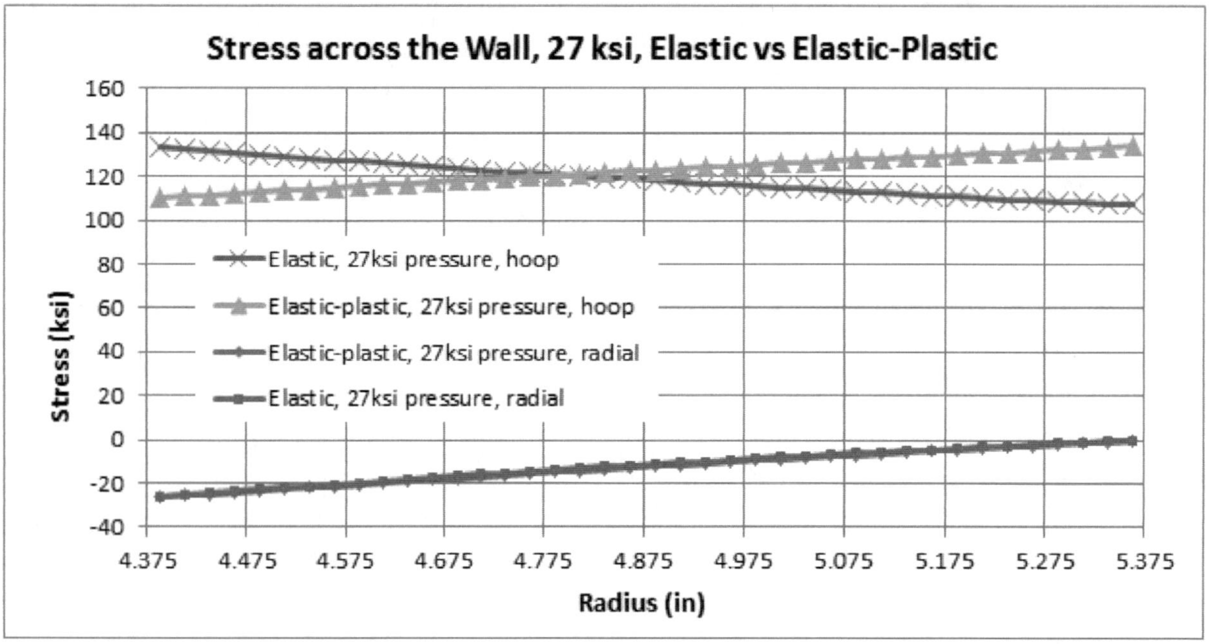

Figure 8 Hoop Stress versus Radial Stress for Elastic and Elastic-Plastic Materials in 27 ksi Internal Pressure

Thin wall pipe burst pressure is estimated and compared to the thick wall pipe burst pressure in Figure 9. A 10 ¾" OD x 0.5" WT pipe (thin wall) is selected and the 12.8 ksi burst pressure is calculated. For the 1" wall thickness pipe (thick wall), the burst pressure is 27 ksi. Although two pipes have different burst pressures, the hoop stress and von Mises stress across the wall thickness are very close when the pipes reach critical burst. This confirms the failure occurs when hoop stress or von Mises stress reaches the critical value at the pipe external surface, regardless of pipe wall thickness in elastic-plastic material.

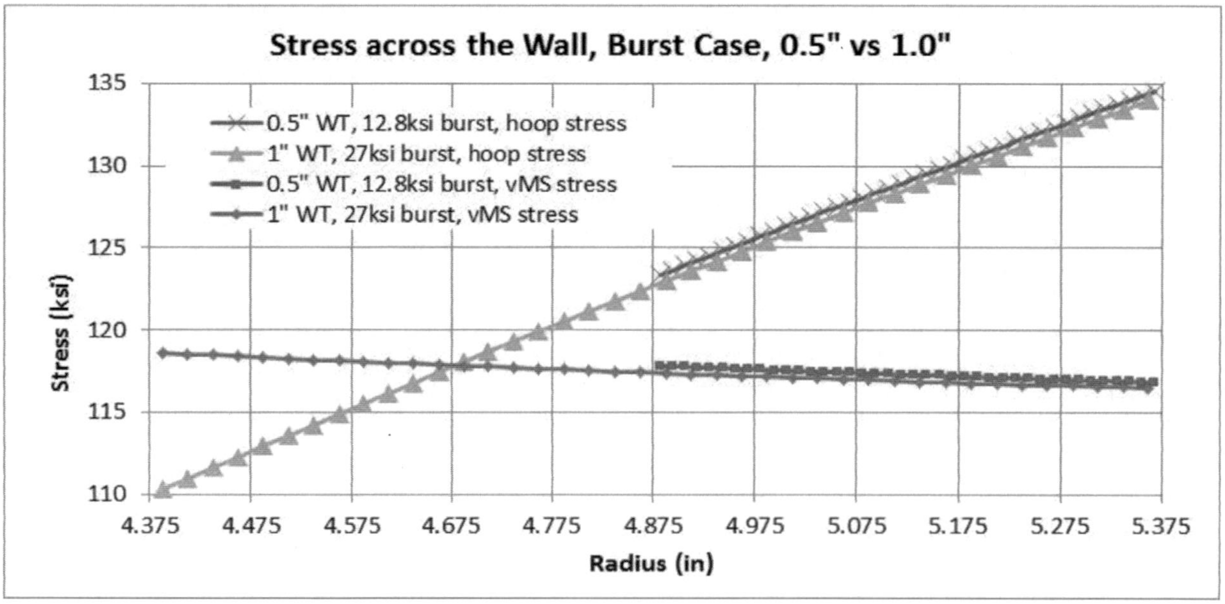

Figure 9 Hoop Stress versus von Mises Stress for Thin Wall and Thick Wall Pipe in Burst Conditions

High temperature has an impact on riser burst capacity; hence, C110 material hoop stress and von Mises stress are compared by applying a temperature derating factor. The same size thick wall pipe used (10 ¾" OD x 1" WT) under 232°C high temperature has a yield stress derating factor of 0.867. The riser pipe burst capacity is estimated using the revised stress-strain curve considering the temperature effects. The burst pressure is reduced from 27 ksi to 23.5 ksi. Figure 10 presents said comparison in the pipe burst conditions.

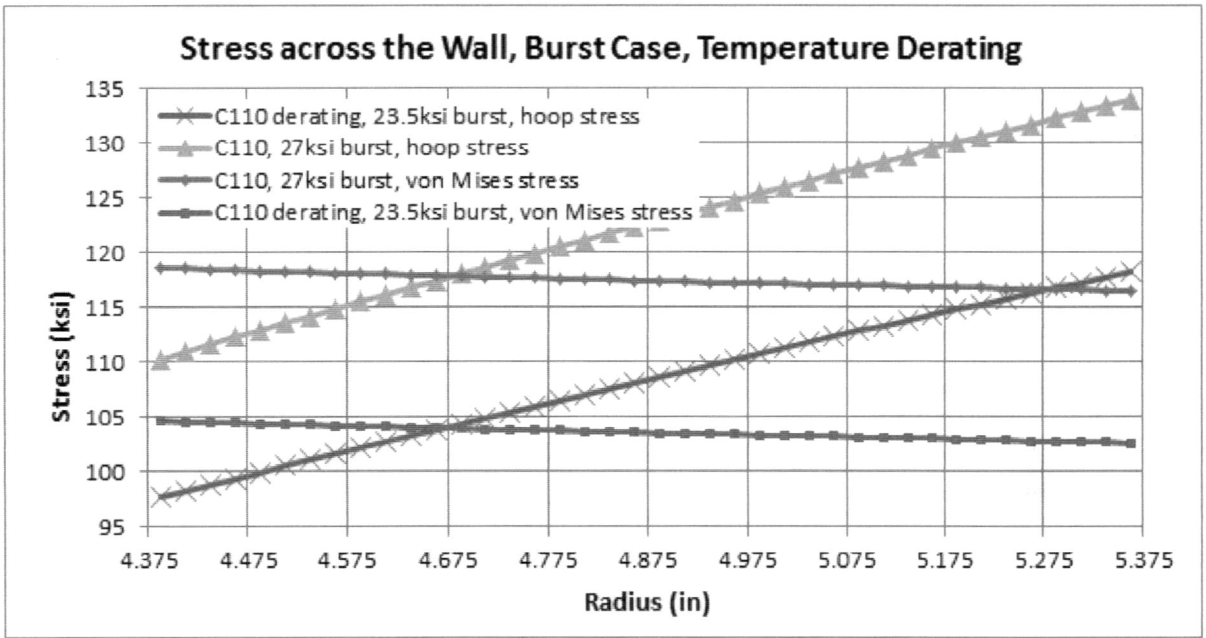

Figure 10 Hoop Stress versus von Mises Stress for Thick Wall Pipe with Temperature Derating in Burst Conditions

RISER SIZING DESIGN FACTORS

Certain riser wall thickness sizing design factors are based on extensive testing and field application history which are recommended to ensure a safe and robust riser wall thickness sizing formulation. Such design factors however might be intrinsic and might be only applicable to pipe geometry and manufacturability; such as misalignment and tolerance factors. However, other extrinsic design factors can apply but are limited to thin wall riser pipe in the current riser design codes.

In order to determine an appropriate design factor for wall thickness sizing of thick wall riser pipe, the pipe burst capacity from the finite element analysis is compared to the API formulations (API RP1111, 2015). The results indicate that API 1111 conservatively estimates the pipe burst capacity by recommending a 0.45 design factor (k). Table 1 compares the burst pressure calculated from the finite element analysis and the formulas provided in API 1111 (API RP1111, 2015).

Table 1 Burst Pressure Comparison

Riser Pipe	API 1111			Analysis Results
	k=0.45 0% tolerance	k=0.45 -12.5% tolerance	k=0.5 -10% tolerance	
	Burst Pressure, ksi			
10 ¾" OD x 1" WT (Thick Wall)	21.3	24.7	26.6	27
10 ¾" OD x 0.5" WT (Thin Wall)	10.1	11.6	12.5	12.8

DESIGN BY RULE VERSUS DESIGN BY ANALYSIS

Based on the above comparisons the question of whether using API 2RD average hoop and radial stress is suitable for high pressure cases can be provoked. Since the riser sizing is generally achieved using design by rule methods, the riser strength and fatigue will be verified in the detailed finite element analyses. For ductile material like metal, the yielding of materials begins when its von Mises stress reaches a critical value of material yield stress.

The von Mises stress curves for 110 ksi yield strength material are plotted in Figure 11. For the combined longitudinal stress and hoop stress located inside the curve, the von Mises stress will be less than 110 ksi yield stress. According to API 2RD (API STD2RD, 2013), the average radial stress across the wall is around 10 ksi for a 20,000 psi internal pressure case, which produces a von Mises stress curve with high capacity for hoop stress and longitudinal stress combinations. This will be less conservative when the von Mises stress at the pipe inner surface (20 ksi radial stress) is checked but will be more conservative than the von Mises stress at the pipe outer surface (0 ksi radial stress).

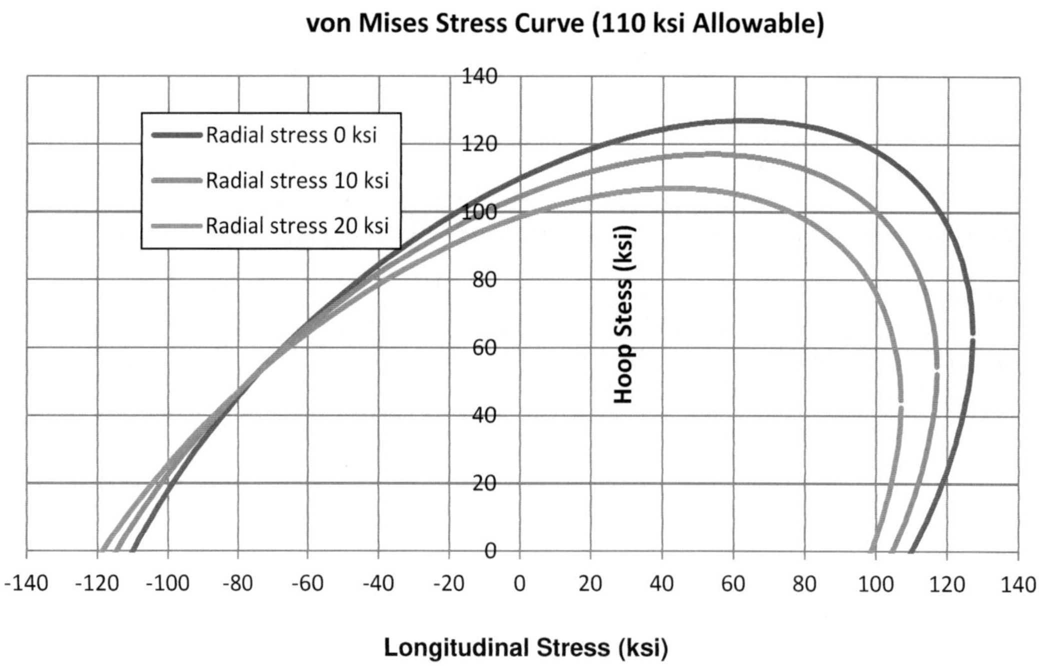

Figure 11 von Mises Stress Envelope for 110 ksi Yield Strength Material

The von Mises stress curves for 110 ksi yield material are plotted in Figure 12. For the combined radial stress and hoop stress located within the envelope, the von Mises stress will be less than 110 ksi yield stress. According to API 2RD (API STD2RD, 2013), the average radial stress across the wall is around 10 ksi for the 20 ksi internal pressure case, which is the average stress across the pipe. The API 2RD stress is located in the middle of the stress plot across the wall. When the longitudinal stress is positive, the API 2RD stress is below the von Mises stress curve, which means the von Mises stress is less than the allowable yield stress. However, for some radial stress and hoop stress across the wall near the inner surface, the stress combination is outside of the von Mises stress curve, which indicates the von Mises stress will be more than the allowable stress.

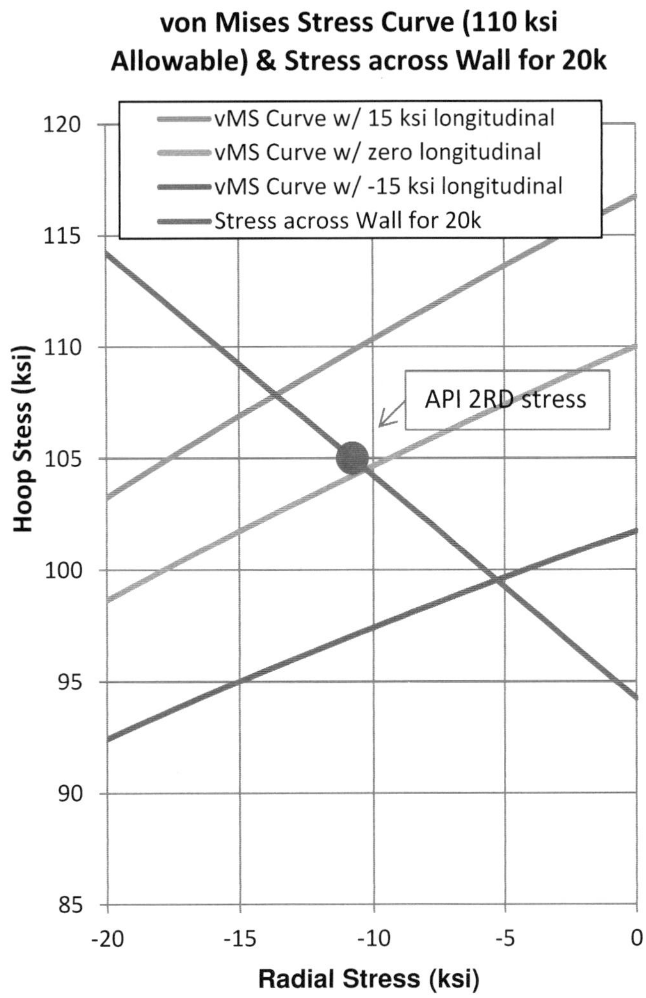

Figure 12 von Mises Stress Curve for 110 ksi Yield Strength Material

RISER SYSTEM OPTIMIZATIONS FROM CASE STUDY PERSPECTIVE

Using the conventional riser wall thickness sizing methodologies, a conservative wall thickness will be estimated based on the pipe burst capacity. The riser will be designed with heavier wall thickness, which can increase the payload to the host platform, but leaves more margins on the riser strength and fatigue performance. The riser sizing can be optimized for the riser global performance. However, for the high temperature case, the riser material yield strength is reduced considering the derating effects. For the high temperature condition, the temperature derating factor on the riser burst capacity should be taken into account.

RISER MANUFACTURING AND MATERIAL LIMITATIONS IN HPHT

Riser fatigue performance is directly correlated to the riser pipe base metal and the quality of the welds. However, riser thick wall design (D/t < 15) may pose feasibility issues in terms of procurement and fabrication as well as pipe mechanical properties and fatigue performance. This section describes the challenges imposed by thick wall riser designs on riser pipe procurement, fabrication and fatigue performance in HPHT environments.

Although welding a thick-wall riser pipe might be achievable, it can pose knock-on challenges in other aspects of the procurement and fabrication processes. It is not commonplace to find and procure API standard riser pipe (API Spec5L, 2012 / API Spec5CT, 2011) with a heavy wall thickness. Thus a special ordered riser pipe must be provided as a bespoke option if a thick-wall riser is deemed necessary. This does not imply that riser pipe with a thick wall has not been designed, manufactured and employed to date.

The thick-wall riser pipe can present challenges in the welding process as it provides a greater heat sink from which variations in material properties may occur across the wall of the heat affected zone if heat input requirements cannot be maintained across the pipe ruling section. In order to ensure a high quality full penetration weld, thick-wall riser joints may require more weld passes when compared to thin wall joints. This may increase the riser joint's susceptibility for flaws in the weld which can also increase the time to detect, characterize and remediate said flaws. The probability of this scenario occurring is high for the thick-wall welded riser joints due to the increase of probability in critical flaw sizes with the greater number of weld passes. Risers are fatigue sensitive structures subject to several components of cyclic loading such as those induced by vessel motions and current induced VIV. Thus, the riser requires high quality and high fatigue performance pipe-to-pipe or pipe-to-connector welds, typically of a DNV C Classification, with minimal stress concentration. Such expectations are generally common for all risers regardless of function, size, water depth or environment. In order to achieve the high quality weld requirement, full scale fatigue resonance testing for welded riser joints is necessary for qualification of the riser welding procedure. Even though fatigue testing might be a rigorous and costly process, it is necessary to ensure the riser fatigue life meets or exceeds the project requirement and a safe and robust design can be achieved. The riser weld fatigue performance depends on several factors in which to minimize the stress concentration. Such factors include out-of-roundness, wall thickness tolerances and joint misalignment. In order to reduce misalignment, stricter pipe end tolerances must be implemented. Thick-wall riser pipe inevitably possesses greater wall thickness variations than the standard pipe sizes and may pose a challenge in the management of pipe end tolerances.

The challenges presented above can significantly impact the vendor selection process, supply chain management and the project costs which in turn will most likely cause delays in the overall project schedule. However, in order to mitigate for said impacts, alternative riser fabrication and construction methods can be adopted. Mechanical connections in the form of non-welded threaded and coupled (T&C) riser connectors can provide a solution for thick-wall riser connections. Although limited to upset pipe, there are numerous benefits for using T&C riser connections which include fast make-up and improved fatigue performance; however, the most notable and advantageous benefit for HPHT applications is the use of high strength steel. The high strength steel has a higher burst capacity which in turn can result in smaller wall thickness requirements compared to lower strength steel grades. Numerous premium connection suppliers provide thick wall designs in high strength steel grades such as T95, P110 or Q125. The mechanical strength of such steel grades is high due to the effects of the chemical composition and hardenability of the alloy. The chemical composition and stress relief treatment impact the material fracture toughness which is critical for the riser fracture and fatigue performance. Therefore, tighter control on mechanical and material properties should be implemented for steel grades used for dynamic riser applications in HPHT. Hydrogen induced embrittlement should also be minimized in the high strength steels by active control of the material hardness.

CONCLUSIONS

The use of Barlow's equation may not be appropriate and can lead to an overestimated wall thickness for thick wall scenarios (Ghosh, et al., 2012). On the other hand, using Lame's equation requires an iterative solution to determine the required pipe wall thickness. It is recommended to use ASME B31.8 thick wall formula for hoop stress calculation for HPHT conditions, in order to prevent the pipe inner surface from entering yield first. However, according to the finite element analysis performed herein, when considering the pipe burst capacity, the von Mises stress should be checked at the inner surface of the pipe as well as the outer surface since the stress is relatively constant across the wall thickness. On the other hand, the hoop stress is found to be governing at the external surface of the pipe for both thin wall and thick wall pipe.

The burst pressure formula proposed by API 1111 is based on theoretical considerations confirmed by many burst tests. The FEA results indicate the pipe burst capacity is mainly governed by the hoop stress or von Mises stress at the external surface, and API 1111 underestimates the pipe burst capacity.

The results indicate that for the thin wall pipe, the burst pressure from FEA matches well with API 1111 results. However, for the thick wall pipe, API 1111 underestimates the pipe burst capacity.

ACKNOWLEDGEMENTS

We would like to personally thank Wood Group Kenny for allowing the publication and the presentation of the material in this paper. We would also like to thank Dr. Chii-Ren Lin for his assistance and support in writing this paper.

REFERENCES

American Petroleum Institute, "Dynamic Risers for Floating Production Systems", API STANDARD 2RD, 2013.

American Petroleum Institute, "Design, Construction, Operation, and Maintenance of Offshore Hydrocarbon Pipelines (Limit State Design)", API RP 1111, 2015.

American Society of Mechanical Engineers, "Gas Transmission and Distribution Piping Systems", ASME B31.8, 2014.

American Society of Mechanical Engineers, "Pipeline Transportation Systems for Liquids and Slurries", ASME B31.4, 2012.

American Petroleum Institute, "Specification for Line Pipe", API SPEC 5L, 2012.

American Petroleum Institute, "Specification for Casing and Tubing", API SPEC 5CT, 2011.

Fu, B and Kirkwood, M.G. 1995b, "Determination of failure pressure of corroded linepipes using the non-linear finite element method". In: R. Dennis ed. Pipeline Technology. Proc. 2 nd International pipeline technology conference, Ostend, Belgium, 11-14 Sept. 1995. Amsterdam: Elsevier. Vol 2. 1-10. ISBN 0444 82197X.

Ghosh, Rupak, Gioielli, Paulo and Kan, Wan, "Design considerations for HP/HT steel catenary risers". In Offshore Magazine, May 01, 2012.

§250.807, "Additional requirements for subsurface safety valves and related equipment installed in high pressure high temperature (HPHT) environments", <http://www.ecfr.gov/cgi-bin/textidx?SID=b0c9e5574a3c05174354b4b7f9665258&node=se30.2.250_1807&rgn=div8>

INVESTIGATION OF STEEL LAZY WAVE RISER FLEXIBLE JOINT ANGLE FOR A SEMISUBMERSIBLE FPU IN VERY HARSH ENVIRONMENT

ALAA M. MANSOUR[1]
MARINE ENGINEERING MANAGER, INTECSEA INC.

SHANKAR U. BHAT
TENDON AND RISER MANAGER, INTECSEA INC.

CHENG PENG
NAVAL ARCHITECT, INTECSEA INC.

DHARMA THEJA R. PASALA
RISER ENGINEER, INTECSEA INC.

Proceedings of the 21st Offshore Symposium, February 2016, Houston, Texas
Texas Section of the Society of Naval Architects and Marine Engineers

ABSTRACT

One of the key design parameters of Flexible Joint as the Top Termination Unit (TTU) of Steel Catenary Riser (SCR) or Steel Lazy Wave Riser (SLWR) is the maximum predicted dynamic angular motion of the flexible joint. It is important for the Flexible Joint rotation angle not to exceed the current qualification limits to reduce project risks and avoid additional cost and potentially schedule impact.

In this paper, the design rotational angle requirements for a Flexible Joint TTU of a steel lazy wave riser hosted on a Semisubmersible Floating Production Unit (FPU) in Western Australia harsh environment is investigated. Fully coupled time domain analysis is used to develop the motion time history of the FPU that is used in detailed Finite Element Analysis (FEA) of the riser to predict the riser performance and the Flexible Joint rotational angle. 10,000 year survival condition is investigated. Far, Near and Cross conditions of the riser are analyzed. Contributions of the motion components (participation factors) to the Flexible Joint rotational angle are estimated. Offset, heave and roll/pitch modes of vessel motions are considered. Percentage contributions to the Flexible Joint rotation angle from mean, low frequency and wave frequency for each component are extracted.

Numerical results are presented to illustrate the significance of each of the motion components to the Flexible Joint rotation angle. The paper concludes with recommendations on the key design considerations and optimizations of the semisubmersible hull design that should be implemented to reduce the flexible joint angle so that it remains within the qualified limits.

Keywords: Flexible Joint, Steel Lazy Wave Riser, Semisubmersible

INTRODUCTION

Semisubmersible FPUs can offer large and ventilated deck area which has its operational and safety advantages. Topsides-hull integration is performed at quayside and the integrated unit is wet towed in an upright position to the installation site, offering safer and significantly cheaper installation operations. The semisubmersible FPU is less restrictive in terms of payloads. Moreover, its motion response in harsh environment is suitable to support steel risers at deepwater locations. Steel risers are more durable and reliable than flexible risers. Therefore they are often preferred over flexible risers.

Currently, there are no semisubmersible FPUs installed in offshore Western Australia. With the huge gas discoveries in this area, many operators are considering semisubmersible as a viable option for their field developments. There are several unique characteristics to the offshore Western Australian region that should be considered in designing a semisubmersible. These characteristics include severe cyclonic events, persistent swell, remoteness of the area, severe locally formed cyclones and the calcareous soil type.

The remoteness of the area and the potential for severe locally formed cyclones makes it difficult to evacuate the people onboard for cyclones. Therefore, robust design in this area is a key factor for the success of such developments. The consequences of any incidents can be serious and costly, resulting in delayed production, damage to environment, loss of the facilities, and even worse, the potential loss of many lives. These unique features and the design robustness requirements make the semisubmersibles with steel risers an ideal field development option in offshore Western Australia.

Building robustness in the semisubmersible design especially when personnel stay onboard during cyclonic events requires the system to be designed to survive events with return periods larger than the typical 1,000 year return period adopted for the survival events in the GoM. Events with 10,000 year return periods need be adopted as the survival events in Western Australia. Events with 10,000 year return periods have higher significant wave height and longer peak period, both of which increase the semisubmersible motion response especially in the heave direction. This coupled with a design location of moderately deep water depth (<1,500 m), make the conventional SCR strength design very challenging. In addition, the persistent swell in the Western Australia area makes the SCR fatigue design more challenging compared to the GoM SCR designs.

There are potentially two different solutions to overcome the above mentioned SCR strength and fatigue challenges. One option is to improve the semisubmersible motion by using a Very Deep Draft Semi (VDDS) >50m and maintain a conventional SCR design and the second option is to use conventional Semi design (draft~30m) and adopt a Steel Lazy Wave Riser (SLWR).

As part of this study, the two solutions were developed and compared. It has been found that the VDDS with SCR is a costly option and imposes challenges on the fabrication due to the deep draft and high column freeboard driven by the airgap requirements. Therefore, conventional semisubmersible design with SLWR is adopted. An all-steel mooring system is adopted to simplify the foundation pile design.

In the SLWR design, the steel riser is configured as a lazy wave to moderate the stresses in the touchdown zone. The SLWR systems are relatively new and the industry has limited experience with these systems. SLWRs have been installed in two projects to date, and buoyancy has been installed in the upper catenary of SCR in another project. A review of recent advances in SCR and SLWR technologies is presented in Mekha and Bhat, 2013.

Two types of Top Termination Units (TTUs) are currently qualified and available to connect the top of the SCR or SLWR risers to porches fitted to the hosting semisubmersible hull. These are Flexible Joint (FJ) or Titanium Stress Joint (TSJ).

A typical Flexible Joint assembly sketch is provided in Figure 1. The bellow and thermal barrier shown in Figure 1 are optional, and the need for them is usually assessed on a case by case basis during detailed design depending on maximum and minimum temperatures and CO_2 content of the bore fluid. The bellows unit in a

Flexible Joint serves several purposes. It works in conjunction with a thermal barrier to insulate the elastomer layers from the high or low temperature of the bore fluid. It also isolates the elastomer from contact with the bore fluid. In the present study, the thermal barrier and bellows are needed for the production risers, and bellows may also be needed for the export riser depending on the CO_2 content in the export system.

Flexible Joints have been designed, manufactured, and delivered by Oil States Industries (OSI), up to 24 inch size. Maximum free rotation angle range in a Flexible Joint delivered to date is +/-25 degrees, and this was done for 8-inch, 10-in and 20-inch nominal riser sizes in the Independence Hub Project (Mekha, 2007).

SCR stress joints are mainly made of titanium. The titanium stress joint (TSJ) has typically been designed for rotation angles of about 15 degrees or less. In the present study, the rotation angle requirements are significantly larger, and a significant amount of dynamic fatigue loading is expected due to persistent swells in offshore Western Australia. Preliminary sizing of TSJs was carried out, but the sizes required were outside the envelope of what has been done in prior projects. Therefore, Flexible Joint was selected as the more feasible option for TTU.

Flexible Joints which are designed for a certain value of free rotation angle (such as 25 degree) can withstand a few degrees of over-rotation without failure and leakage of contents to the environment. The over-rotation capacity is typically 3 to 4 degrees, and when required, this is typically investigated and confirmed by project specific FE analysis. If the dynamic motion is such that the maximum dynamic rotation angle exceeds the 25 degree qualified limit, the performance of the flexible joint including the incidents of over-rotations is investigated and the risk associated with the plastic deformation is evaluated. If the risk is not acceptable, the current design needs to be modified to allow larger angle and Flexible Joint prototype qualification program for the new design will be required.

In this paper a case study is presented for a conventional Semisubmersible FPU located in Western Australia harsh environment and hosting SLWR. Fully coupled time domain analysis is used to develop the motion time history of the FPU that is used in detailed Finite Element Analysis (FEA) of the riser to predict the riser performance and the Flexible Joint rotational angle. 10,000 year survival condition is investigated. Far, Near and Cross conditions of the riser are analyzed. In 10,000 yr survival condition, the SLWR Flexible Joint is found to undergo a dynamic rotation exceeding the currently qualified limits. Contributions of the motion components (participation factors) to the Flexible Joint rotational angle are estimated. Offset, heave and roll/pitch modes of vessel motions are considered. Percentage contributions to the Flexible Joint rotation angle from mean, low frequency and wave frequency for each component are extracted. Numerical results are presented to illustrate the significance of each of the motion components to the Flexible Joint rotation angle. The paper concludes with recommendations on the key design considerations and optimizations of the semisubmersible hull design that should be implemented to reduce the flexible joint angle so that it remains within the qualified limits.

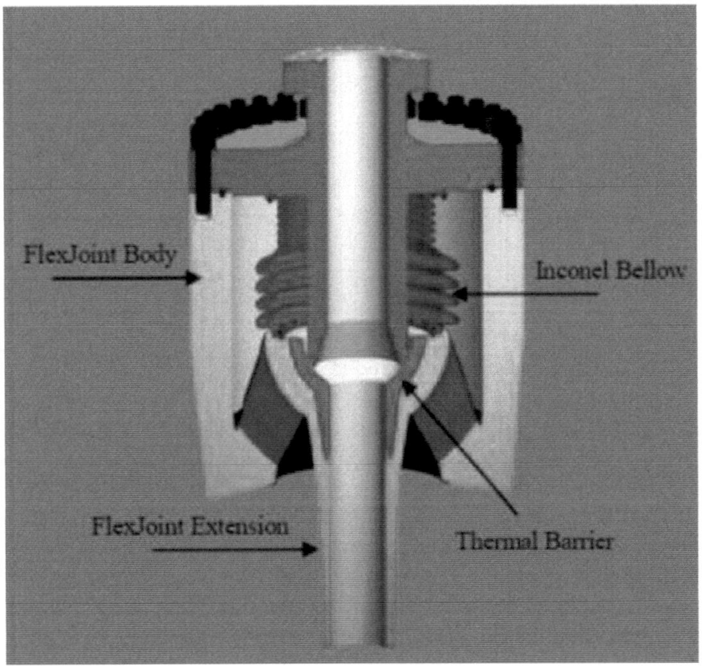

Figure 1 Flexible Joint TTU (Courtesy of Oil States Industries)

RISER PROPERTIES AND CONFIGURATION

The riser system that is presented in this paper is a 16-inch export riser connecting to a 20-inch export pipeline tie-in to shore or to a FLNG facility. X65 material is assumed for the riser pipe. This is found to be the most critical riser. The design pressure is 3,000 psig and the minimum wall thickness requirement including the required corrosion allowances (1.6 mm) is 0.75-inch. Based on strength, fatigue and interference considerations, the selected wall thickness for the export riser is increased to 1.1-inch in the top 1,000 m of the SLWR. The riser has 3 mm thick 3LPP coating and no requirement for insulation coatings. The selected TTU for this riser is flexible joint due to the relatively large diameter and high dynamic motion. The riser has full strake coverage except in the buoyancy zone. The buoyancy zone is 450m long. 18-degree hang-off angle was selected for this riser to enhance fatigue feasibility,. The configuration of the riser is shown in Figure 2.

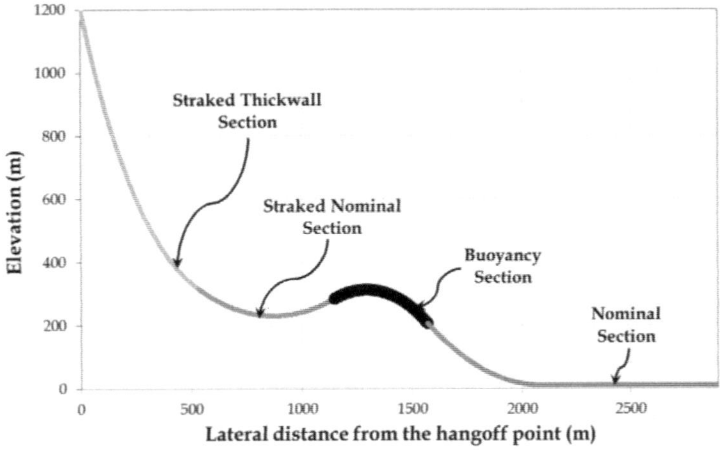

Figure 2: Steel Lazy Wave Riser Configuration

RISER PERFORMEANCE

Strength analyses have been carried out for various load cases in Far, Near, and Cross directions, using the vessel motion time histories. Strength utilizations of the riser are found to be well within the allowable limits. However, the Weibull extreme of the flexible joint (FJ) rotation angle in a 10,000 year survival event is predicted to be 29 deg (including 1.5 deg installation tolerances). The extrema of the FJ rotation angle are +29/-20 deg, considering Far, Near and Cross cases.. As noted earlier in this paper, the maximum angle range in a FJ delivered to date is +/-25 degrees. Since the maximum predicted Flexible Joint rotation angle is 4 degrees greater than the free rotation angle qualification limit, a detailed study is performed in this paper to determine the factors that resulted in a large FJ rotation angle.

FJ rotation is defined in two orthogonal directions, the two principal directions of the semisubmersible, X and Y as shown in Figure 3. The far, cross and near directions are defined with reference to the riser heading. The correlation between the vessel motion components and the FJ rotation is discussed in this section. Only the critical case results (Far case) are discussed to keep the discussion short.

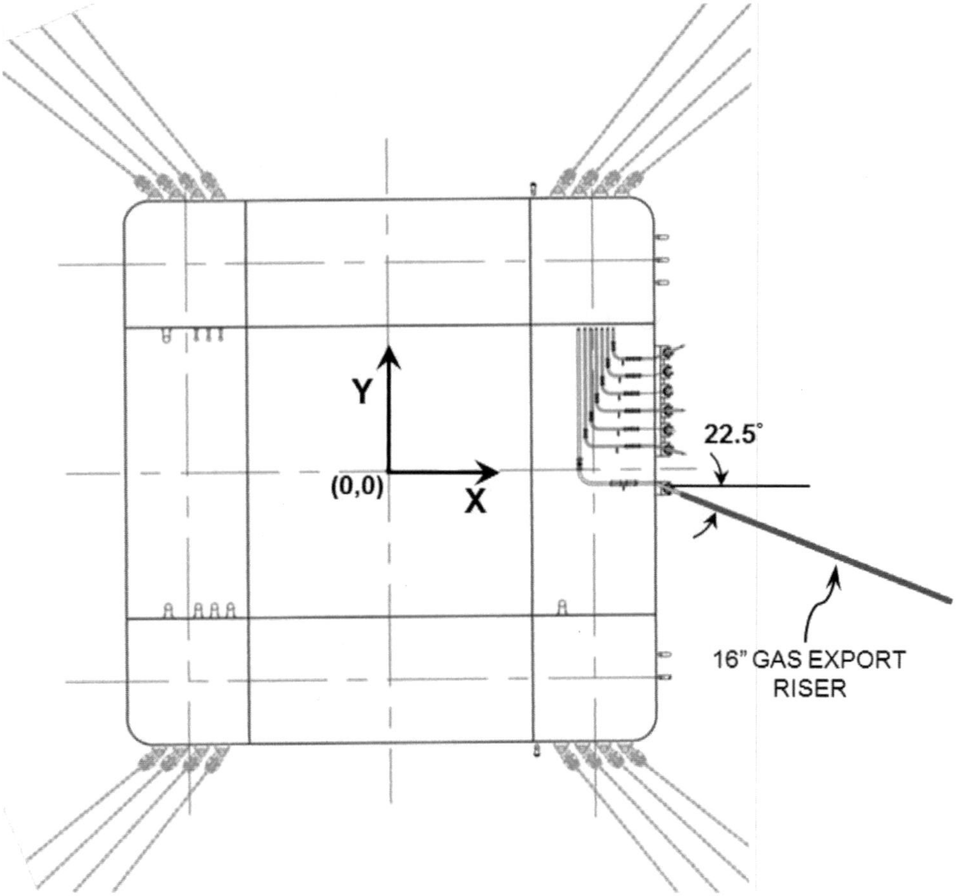

Figure 3: Heading Definition

The FJ rotation angle component (Rx : rotation in the X-Z plane) time history versus the roll motion time history during the interval where the maximum FJ rotation angle occurs is presented in Figure 4. The maximum rotation angle Rx is found to be 27 deg with a mean value of 8.5 degrees. Weibull extreme for the Rx component is found to be 25.8 deg. The Rx component is the dominant contributor to the total FJ angle, although the maximum

FJ rotation is not in the X-Z plane. Detailed analysis of the contributions of various motion components (participation factors) to the Flexible Joint rotational angle are estimated by analyzing the Rx rotation time histories.

Comparison of power spectral densities (PSD) of the FJ angle (Rx) versus the roll motion, shown in Figure 5, indicates that the wave frequency roll motion has a major contribution to the FJ wave frequency Rx rotation angle. Snapshots of the wave frequency and low frequency FJ angle Rx versus roll motion time histories for the interval where the maximum dynamic FJ Rx rotation angle occurs are extracted and presented in Figure 6. It can be seen that about 6.2° out of the wave frequency FJ Rx angle of 12.5° are the contribution of the wave frequency roll motion. Low frequency roll motion contribution is found to be 1.8° out of the 6.8° total low frequency FJ Rx angle (after subtracting the mean angle of 8.5 degree).

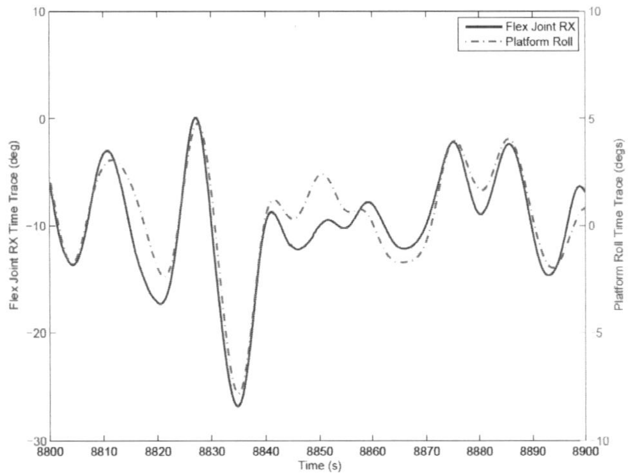

Figure 4: Comparison of Flexible Joint Angle and Semisubmersible Roll Motion Time History

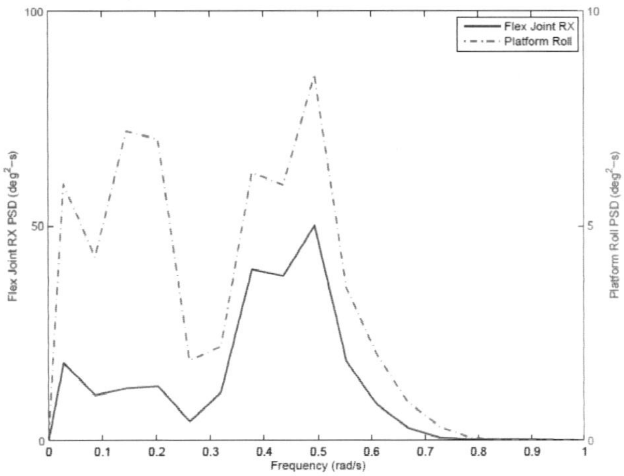

Figure 5: Comparison of Flexible Joint Angle PSD and Semi-Roll PSD

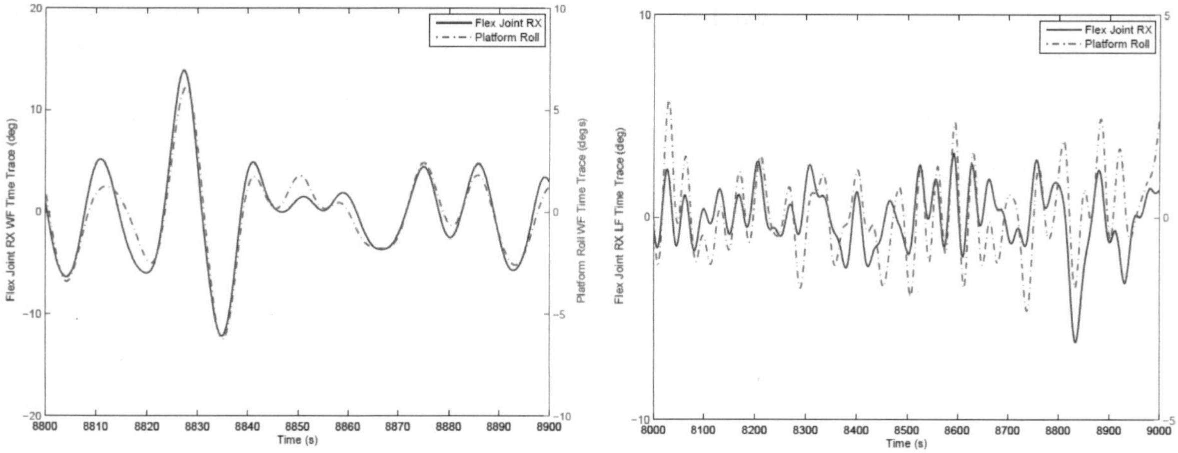

(a) WF FJ Angle and WF Semi-Roll (b) LF FJ Angle and LF Semi-Roll

Figure 6: Comparison of Flexible Joint Angle and Semi-Roll in Time Domain

To better understand the contribution of the sway motion to the FJ angle, a static analysis is performed and the correlation between FJ rotation angle and static sway is developed (shown in Figure 7). At the 10,000 yr platform mean offset of 15.1% of water depth, the corresponding FJ angle is found to be 7.25°. This indicates that the majority of the mean FJ angle of 8.5 degrees is dominated by the sway mean offset.

The FJ rotation Rx angle PSD, excluding the platform roll motion contribution, versus sway motion PSD is shown in Figure 8. Wave frequency and low frequency FJ Rx angle versus sway motion during the interval where the maximum FJ angle occurs are presented in Figure 9. It can be seen that Sway has a major contribution to the flex joint rotation angle and its low frequency contribution is more pronounced over the wave frequency.

To get the contribution from dynamic sway, the analysis is performed without the sway motion dynamic components (wave frequency and low frequency sway motions are eliminated). The Weibull extreme for the FJ rotation Rx angle from this analysis is processed and found to be 20.2 deg. This indicates that the sway dynamic response is contributing about 5.6 deg. Next, the analysis is repeated including the sway low frequency motion. The FJ rotation Rx Weibull extreme is found to be 23.8 deg indicating that the contribution from the sway low frequency motion to the FJ Rx rotation is 3.6 deg and sway wave frequency motion is contributing about 2.0 deg to the FJ rotation angle.

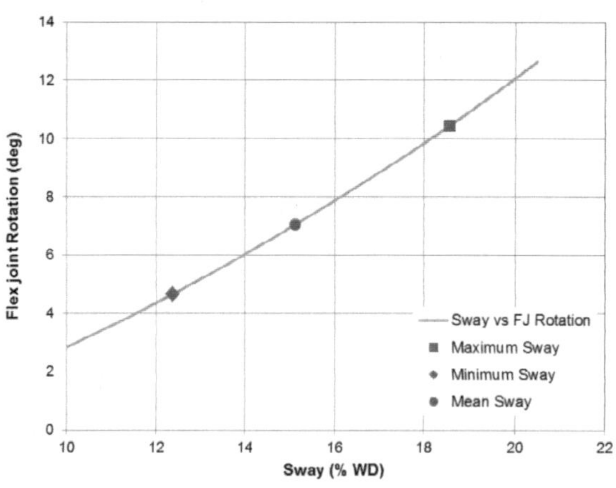

Figure 7: Sway Contribution to the Flexible Joint Angle [Static Case]

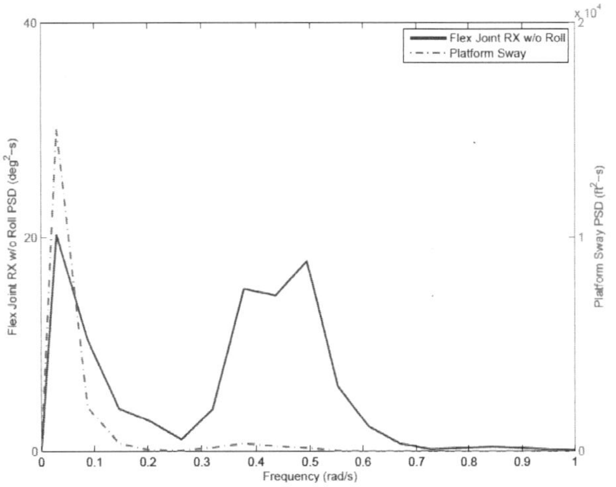

Figure 8: Comparison of Flexible Joint Angle PSD and Semi-Sway (without Roll) PSD

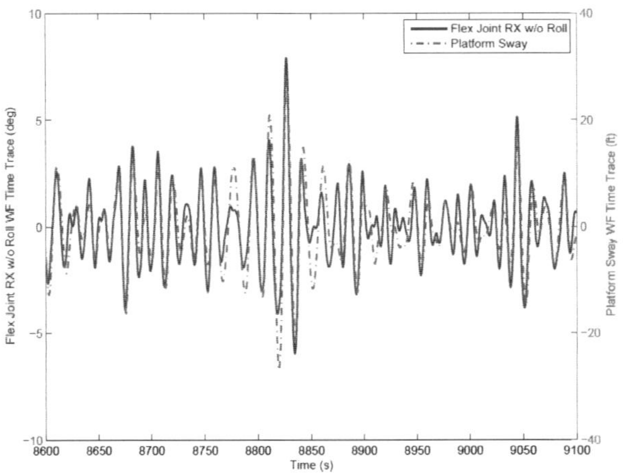

(a) WF FJ Angle and WF Semi-Sway

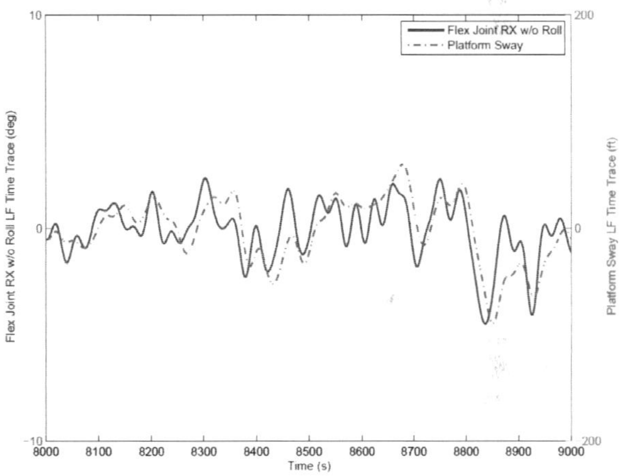

(b) LF FJ Angle and LF Semi-Sway

Figure 9: Comparison of Flexible Joint Angle and Semi-Sway (without Roll) in Time Domain

The relationship between the FJ angle and the static heave motion at the mean offset position is developed to understand the heave contribution to the FJ rotation. This relationship is presented in Figure 10. For a static heave motion corresponding to the predicted LF heave response of 2.3 m, the FJ rotation is found to be negligible (<0.15 deg) while the FJ angle corresponding to the wave frequency heave is found to be around 1.5 deg.

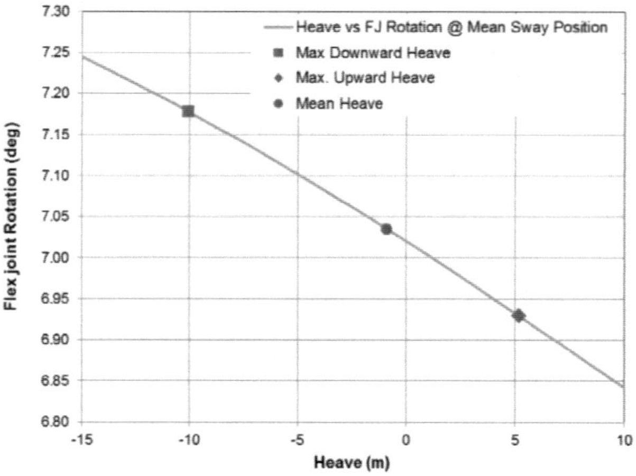

Figure 10: Heave Contribution to the Flexible Joint Angle [Static Case]

The PSD of the Flex Joint (FJ) rotation (excluding the roll motion) versus heave motion is presented in Figure 11. The wave frequency and low frequency time history comparison of FJ rotation angle, in the time window in which maximum FJ rotation angle occur are shown in Figure 12. To further confirm the heave contribution, the riser analysis is repeated without the vessel heave for one of the critical realization. The total contribution from heave (including LF and WF components) is found to be only 1.77 deg as shown in Figure 13.

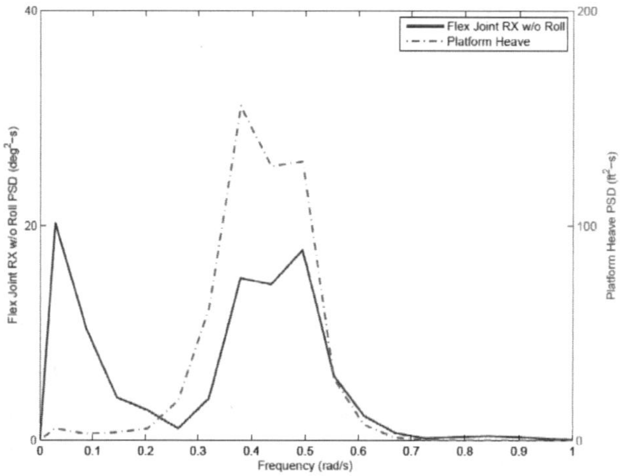

Figure 11: Comparison of Flexible Joint Angle PSD and Semisubmersible Heave (without Roll) PSD

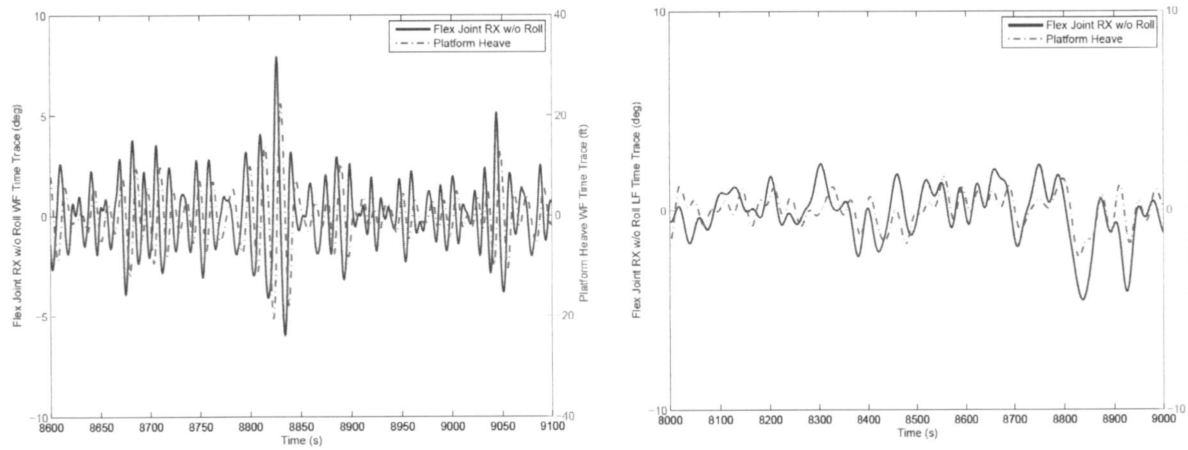

(a) WF FJ Angle and WF Semi-Heave (b) LF FJ Angle and LF Semi-Heave

Figure 12: Comparison of Flexible Joint Angle and Semi-Heave (without Roll) in Time Domain

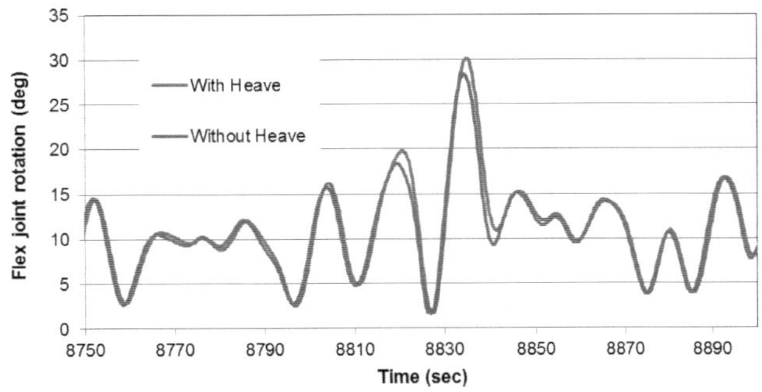

Figure 13: Heave Contribution to the Flexible Joint Angle [Dynamic Case]

Based on the above discussion, the results from the detailed investigation of the effects of various components of platform motions on the maximum FJ Rx Rotation angle in the Far load case are summarized in Table 1.

Table 1: Contribution of the Response Components to the Flexible Joint Rotation

Response Component	**Contribution to FJ Rx Rotation Angle**
Mean offset	35% (~8.5°)
Roll wave frequency	26% (~6.2°)
Roll Low frequency	7% (~1.8°)
Sway Wave Frequency	8% (~2.0°)
Sway Low Frequency	16% (~3.8°)
Heave Wave Frequency	6% (~1.8°)
Heave Low Frequency	negligible

OPTIMIZATION OF THE FLEXIBLE JOINT ANGLE

The mean FJ rotation angle can be reduced by reducing the mean offset. This can be achieved with a stiffer mooring system. One way to increase the mooring lateral stiffness in an all-steel mooring system is by adding mooring lines. A second way could be by adding in-line buoys to the mooring system. Both options are relatively expensive.

Mean angle can also be reduced by balancing the seating angle between Far and the Near conditions. This option has been discussed with the FJ supplier. Although this is likely to be a feasible solution, FJ supplier is not in favor of this solution since it has not been done to date in a new design, despite such initial tilts occurs unintentionally during actual riser installations.

Next largest contribution to the FJ angle is from wave frequency roll. This contribution can be reduced by generally improving the platform motion through adopting deeper semisubmersible draft. Reducing wave frequency roll in this way is not an ideal solution due to its knock on effect on the already critical air gap of semisubmersibles in harsh environment. Another significant contributor is the low frequency roll motion which is recommended to be minimized through optimizing the system roll restoring stiffness and the selection of the mooring fairlead elevation.

The wave frequency sway motion for a semisubmersible with a given displacement is generally a function of the column spacing and changing this parameter has significant technical and commercial impacts to the project. Therefore, controlling the FJ rotation through this parameter is not a practical solution.

Low frequency sway can be reduced with stiffer mooring system. As noted earlier, increasing the stiffness of an all-steel mooring system will be expensive. Optimizing the wave frequency heave response also helps the FJ rotation but the knock-on effect on the airgap design should be carefully addressed.

Adopting polyester mooring system is an alternative solution to reduce the offset and help the FJ rotation angle, however, it may exert additional loads on the foundation, making the design of the foundation pile in the typical Western Australian calcareous soil challenging.

CONCLUSIONS

Detailed study on the effect of the motion components on flexible joint rotation is performed to better understand the contributing factors to the rotation angle and explore ways to reduce the maximum flexible joint rotation.

Detailed time domain global motion and riser analysis is performed for a 16" OD steel lazy wave riser hosted in a conventional semisubmersible offshore Western Australia. An all-steel mooring system is adopted to simplify the foundation pile design. Flexible Join angle performance in 10,000 year survival condition is thoroughly investigated.

For this case study, it has been found that mean offset is the main contributor to the mean flexible joint angle with about 35% contribution. Wave frequency roll and low frequency sway motions are the second and third contributors of 26% and 16%, respectively. Wave frequency heave, wave frequency sway and low frequency roll contributes about 7% each indicating that the platform design is optimized for these three components.

It can be concluded that controlling the platform mean and low frequency offset is the key to reduce the flexible joint angle. This can be achieved by adopting stiffer steel mooring system or adopting polyester mooring system. Optimizing the wave frequency and minimizing the low frequency roll responses help when optimizing the flexible joint angle performance. Among the different contributing components, heave is found to be the least contributor to the flexible joint rotation angle.

ACKNOWLEDGEMENTS

The authors would like to express their gratitude to INTECSEA Company for sponsoring the preparation of manuscript.

REFERENCES

American Petroleum Institute, "Recommended Practice for Planning, Designing and Constructing Fixed Offshore Platforms", API RP 2A, 2005.

Bhattacharya, B., Basu, R., and Ma, K., "Developing target reliability for novel structures: the case of the Mobile Offshore Base", Marine Structures 14, pp. 37-58, 2001.

Hamilton, J., Holub, C. and Blunt, J., "Simulation of Ice Management Fleet Operations Using Two Decades of Beaufort Sea Ice Drift and Thickness Time Histories", Proceedings of International Society of Offshore and Polar Engineers (ISOPE), Maui, Hawaii. 2011

Mekha, B., and Bhat, S., "Newer Frontiers in the Design of Steel Catenary Risers for Floating Production Systems", Paper No. 11562, Proceedings of 32nd International Conference on Ocean, Offshore and Arctic Engineering (OMAE), Nantes, France, 2013.

Mekha, B., "Independence Hub Flowline SCRs: Design, Fabrication and Installation Challenges", OTC Paper No. 18584, 2007.

THE STUDY OF DRILLSHIP DRILLING IN SHALLOW WATER (500FT-2000FT) AND DRIFT OFF ANALYSIS

SHIYU CHEN
SR.ENGINEER, P.E., ENSCO PLC, HOUSTON, TX

CHRIS STEWART
DIRECTOR, SUBSEA ENGINEERING, ENSCO PLC, HOUSTON, TX

XIAOCHUN SHEN
MANAGER, ENGINEERING ANALYSIS, ENSCO PLC, HOUSTON, TX

Proceedings of the 21st Offshore Symposium, February 2016, Houston, Texas
Texas Section of the Society of Naval Architects and Marine Engineers

ABSTRACT

Dynamically positioned offshore drill-ships and semisubmersibles are commonly used for drilling operations in deep water (more than 3000ft water depth). For shallow water depth (500ft (152.4m) to 2000ft (609.6m)), moored semisubmersibles were often used to perform the drilling operations. Because of the shallow water depth, in case of an emergency, dynamically positioned drill-ships and semisubmersibles may have limited watch circles to enable safe disconnection of the riser system from the wellhead.

With superior mobility and loading capacity, dynamically positioned drilling vessels are used more frequently for offshore drilling. In recent years, there are growing interests in using them for shallow water drilling. In this paper, the feasibility and limitation of using dynamically positioned drilling vessels for shallow water operation are studied. A drill ship was selected to perform the riser and drift off analysis. The water depth was varied from 500ft to 2000ft, with mud weight at 8.56ppg and 16ppg, respectively. Typical Gulf of Mexico metocean conditions were applied to a global riser model using DeepRiser Software. Based on the results of the study, a minimum water depth was found with sufficient watch circles for drilling operation. Below the minimum water depth, the watch circle may be insufficient, or not available.

Keywords: Drillship, Shallow Water, Drilling Riser, Riser Analysis, Drift Off Analysis, Watch Circles

INTRODUCTION

More and more dynamically positioned drilling vessels are used for offshore drilling. Dynamically positioned drilling vessels are quick to move from well to well, and can escape when a hurricane condition is developing at the areas of operation (comparing to moored semisubmersibles). Recently, there are growing interests in using them in shallow water drilling applications, due to the time-saving and cost advantage associated with rig move and preparation for drilling. One disadvantage is that the watch circles in the case of lost power or DP system failure may become very small in shallow water such that there may not be enough time to disconnect the riser from the wellhead.

In this paper, the possibility of drill ship drilling in shallow water was studied. A drill ship was used to do the riser and drift-off analysis in shallow water. Water depth was varied from 500ft to 2000ft. Mud weight changed from 8.56ppg to 16ppg, and typical GOM metocean data and DeepRiser software were used for the analysis.

METHODOLOGY

This paper presents the results of the studies for a drill ship drilling in shallow water (500ft-2000ft). A drill ship was used to perform riser and drift-off analysis in shallow water and under a set of typical GOM ocean environments. Water depth was varied from 500ft to 2000ft, and mud weight was varied from 8.56ppg to 16ppg. Through riser and drift off analysis, a minimum water depth was found that drill ship may perform the drilling operation with a sufficient watch circle.

RISER AND DRIFT-OFF ANALYSIS RESULTS

An Ensco Drill ship was used (as shown in Figure 1) to perform the riser and drift off analysis in shallow water to demonstrate the methodology and show the minimum water depth that drill ship may operate under the assumptions stated.

Figure 1 Drillship

The following GOM metocean data was used for the riser analysis (as shown in Table 1 and Table 2).

Table 1 Metocean Data for 1yr and 10yr Winter Storm

Return Period	1 Winter	10 Winter
Hs (ft)	10.8	19.4
Tp (s)	8.0	10.3
Shape	Jonswap	
Gamma	1.6	2
WS (ft/s) 1hr, 10m	48.0	68.4
Current		
Depth (ft)	Speed (ft/s)	
0	0.50	0.60
200	0.50	0.60
300	0.30	0.30
2000	0.30	0.30

Table 2 shows the metocean data for Loop Current (95% Non-Exceedance)

Table 2 Metocean Data for Loop Current (95% Non-Exceedance)

Loop Current and Associated Wind and Wave			
Hs (ft)	7.78		
Tp (s)	6.93		
Shape	Jonswap		
Gamma	N/A		
WS (ft/s) - 1hr @ 10m	37.5		
	Current (ft/s)		
Depth (ft)	1yr return	10yr return	50yr return
0	3.72	5.39	6.45
82	3.72	5.38	6.44
164	3.69	5.34	6.39
246	3.29	4.77	5.70
328	2.76	4.00	4.78
410	2.40	3.47	4.16
492	2.14	3.10	3.71
574	1.91	2.77	3.31
656	1.71	2.48	2.96
738	1.54	2.22	2.66
820	1.42	2.06	2.46
902	1.31	1.89	2.26
984	1.19	1.73	2.07
1066	1.11	1.61	1.93
1148	1.03	1.50	1.79
1230	0.99	1.43	1.71
1312	0.92	1.34	1.60
1394	0.86	1.25	1.49
1476	0.82	1.18	1.41
1558	0.79	1.14	1.36
1640	0.74	1.07	1.28
1722	0.70	1.01	1.21
1804	0.67	0.96	1.15
1886	0.66	0.92	1.10
1968	0.66	0.88	1.05
2000	0.66	0.86	1.02

Table 3 shows the metocean data for Drift-Off study. DeepRiser software (Version 3.1.1) was used to build the drilling riser model and run the riser and drift-off analysis.

Table 3 Metocean Data for Drift-Off Study

Tp	s	8
Hs	ft	5, 10, 15
Wind Speed	ft/s	5, 15, 30, 50
Surface Current	knots	0.5, 1, 2, 3

Riser operability study was performed following the design matrix listed in table 4

Table 4 Riser Operability Analysis Matrix

WD (ft)	Heading (deg)	Mud Weight (ppg)	Environment Condition	Tension Setting (Trec) (kips)	Offset (% of WD)
500	0	8.5	1yr Winter Storm	832	-10% to +10%, 1% increment
	0	8.5	10yr Winter Storm	832	-10% to +10%, 1% increment
500	0	16	1yr Winter Storm	846	-10% to +10%, 1% increment
	0	16	10yr Winter Storm	846	-10% to +10%, 1% increment
500	0	8.5	1yr Loop Current	832	-10% to +10%, 1% increment
	0	16	1yr Loop Current	846	-10% to +10%, 1% increment
1000	0	8.5	1yr Winter Storm	1033	-10% to +10%, 1% increment
	0	8.5	10yr Winter Storm	1033	-10% to +10%, 1% increment
1000	0	16	1yr Winter Storm	1090	-10% to +10%, 1% increment
	0	16	10yr Winter Storm	1090	-10% to +10%, 1% increment
1000	0	8.5	1yr Loop Current	1033	-10% to +10%, 1% increment
	0	16	1yr Loop Current	1090	-10% to +10%, 1% increment
1500	0	8.5	1yr Winter Storm	1234	-10% to +10%, 1% increment
	0	8.5	10yr Winter Storm	1234	-10% to +10%, 1% increment
1500	0	16	1yr Winter Storm	1334	-10% to +10%, 1% increment
	0	16	10yr Winter Storm	1334	-10% to +10%, 1% increment
1500	0	8.5	1yr Loop Current	1234	-10% to +10%, 1% increment
	0	16	1yr Loop Current	1334	-10% to +10%, 1% increment
2000	0	8.5	1yr Winter Storm	1436	-10% to +10%, 1% increment
	0	8.5	10yr Winter Storm	1436	-10% to +10%, 1% increment
2000	0	16	1yr Winter Storm	1578	-10% to +10%, 1% increment
	0	16	10yr Winter Storm	1578	-10% to +10%, 1% increment
2000	0	8.5	1yr Loop Current	1436	-10% to +10%, 1% increment
	0	16	1yr Loop Current	1578	-10% to +10%, 1% increment

Riser Drift Off study was performed following the design matrix listed in table 5

Table 5 Riser Drift Off Analysis Matrix

WD (ft)	Mud Weight (ppg)	Heading (deg)	Tension-Trec (kips)	Wind Speed (ft/s)	Significant Wave Height (ft)	Surface Current (knots)
500	8.5	5	832	5, 30, 50	5, 10, 15	0.5, 1, 2, 3
500	16	5	846	5, 30, 50	5, 10, 15	0.5, 1, 2, 3
1000	8.5	5	1033	5, 30, 50	5, 10, 15	0.5, 1, 2, 3
1000	16	5	1090	5, 30, 50	5, 10, 15	0.5, 1, 2, 3
1500	8.5	5	1234	5, 30, 50	5, 10, 15	0.5, 1, 2, 3
1500	16	5	1334	5, 30, 50	5, 10, 15	0.5, 1, 2, 3
2000	8.5	5	1436	5, 30, 50	5, 10, 15	0.5, 1, 2, 3
2000	16	5	1578	5, 30, 50	5, 10, 15	0.5, 1, 2, 3

Static and dynamic analysis were conducted on the drilling riser under connected drilling, and connected non-drilling conditions to verify its suitability at the specified water depth and environmental criteria in accordance with API RP 16Q (see Table 6).

For the static aspect of the analysis, the DeepRiser Software was used and API RP 16Q recommended calculations were adopted to determine the following:

- Develop Riser Spaceout incorporating wall thickness from collapse check.
- Estimate the Expected Slip Joint Stroke-out
- Calculate Minimum or Recommended Tension Required - min pull required to maintain positive tension on riser.

Tension settings for Risers from 500ft to 2000ft WD are shown in table 4.

For the dynamic aspect of the analysis, the DeepRiser software was used to verify riser stability, and analyze the loads and effect on the riser and casing due to the region's offshore environment, including the following:

- Verify Minimum Effective Tension for LMRP Disconnect with various mud densities
- Maximum Von-Mises Stresses on the drilling riser
- UFJ Mean & Maximum Angles
- LFJ Mean & Maximum Angles
- Stroke-out Dynamic Extents
- Effects of Offsets from Well Center (up to 10% of Water Depth were analyzed)

Table 6 API RP 16Q Riser Limits

Design Parameter	Riser Connected		Riser Disconnected
	Drilling	Non-drilling	
Mean Flex Joint Angle (deg)	2	N/A	N/A
Max. Flex Joint Angle (deg)	4	90% of limit	90% of limit
Riser von Mises Stress (ksi)	0.67 * σy	0.67 * σy	0.67 * σy
Min. Top Tension Setting	Tmin	Tmin	N/A
Max. Tension Setting (kips)	90% of DTL	90% of DTL	N/A

Tmin: Minimum Top Tension Setting to Prevent Buckling of the Riser.

DTL: Dynamic Tension Limit

Table 7 summarizes the results of the riser operability study.

Table 7 Shallow Water Riser Operability Summary

Riser Operability								
Water Depth	Heading	Mud Weight	Operability in % WD			Operability in ft.		
(ft)	(deg)	(ppg)	1yr Winter Storm	10yr Winter Storm	1yr Loop Current	1yr Winter Storm	10yr Winter Storm	1yr Loop Current
500	0	8.5	-4% to 4%	-3% to 3%	-5% to 3%	-20 to 20	-15 to 15	-25 to 15
		16	-4% to 4%	-4% to 3%	-5% to 3%	-20 to 20	-20 to 15	-25 to 15
1000	0	8.5	-3% to 3%	-4% to 3%	-5% to 2%	-30 to 30	-40 to 30	-50 to 20
		16	-4% to 3%	-4% to 4%	-5% to 2%	-40 to 30	-40 to 40	-50 to 20
1500	0	8.5	-3% to 3%	-3% to 3%	-5% to 1%	-45 to 45	-45 to 45	-75 to 15
		16	-4% to 3%	-4% to 3%	-5% to 2%	-60 to 45	-60 to 45	-75 to 30
2000	0	8.5	-3% to 3%	-3% to 3%	-5% to 1%	-60 to 60	-60 to 60	-100 to 20
		16	-4% to 3%	-4% to 3%	-5% to 2%	-80 to 60	-80 to 60	-100 to 40

For drift-off study, the limits are established to meet the criteria summarized in Table 8. These limits were derived from API guidance, and the rig specifications.

Table 8 Drift-Off Criteria

Item	Available	Limit	Reference
TJ Stroke (ft)	65.0	65	Vendor spec
Tensioner wire travel (ft)	50.0	50	-
Upper Flex (degree)	15.0	15.0	-
Lower Flex (degree)	10.0	10	-
Casing Bending Stress	56 ksi Yield	80%*(Yield) = 44.8 ksi	API RP 17G
Other Component Stress	80 ksi Yield	2/3*(Yield) = 53.3 ksi	API Spec 16A, 16F
Wellhead Bending Moment	4000 kips-ft	4000 kips-ft	Representative
Time from red alert to EDS (sec)	N/A	63	Representative
Time from yellow to red alert (sec)	N/A	60	Representative
Heave (ft)	N/A	Per RAO and Hs	Vessel RAO
Tide (ft)	N/A	2	Standard GOM
Safety Margin (ft)	N/A	2	Representative

A representative 20ft yellow watch circle and the timing of 60s between yellow and red alerts were selected to gauge the results on a relative basis. The drift-off analyses were performed in accordance with these requirements.

- The yellow watch circle size is summarized in Table 9. Cells with green background indicate the yellow watch is acceptable for the environments.
- In general, the yellow watch circle is influenced jointly by wind speed, current speed and wave magnitude. It also decreases with water depth and worsened environments.

Table 9 Shallow Water Drift-Off Analysis Yellow Watch Circle Summary

Drift-Off Study Matrix-Yellow Watch Circle Size													
Water Depth (ft)		500											
Significant Wave Height (ft)		5				10				15			
Surface Current Speed (knots)		0.50	1.00	2.00	3.00	0.50	1.00	2.00	3.00	0.50	1.00	2.00	3.00
Wind Speed (ft/s)	5	N/L	N/L	13.37	N/A	N/L	N/L	13.37	N/A	N/L	N/L	13.37	N/A
	30	18.69	11.58	0.39	N/A	18.69	11.58	0.39	N/A	18.69	11.58	0.39	N/A
	50	N/A	N/A	N/A	N/A	N/A	N/A	N/A	N/A	N/A	N/A	N/A	N/A
Water Depth (ft)		1000											
Significant Wave Height (ft)		5				10				15			
Surface Current Speed (knots)		0.50	1.00	2.00	3.00	0.50	1.00	2.00	3.00	0.50	1.00	2.00	3.00
Wind Speed (ft/s)	5	N/L	N/L	46.61	1.63	N/L	N/L	46.82	1.63	N/L	N/L	46.61	1.73
	30	40.36	34.15	9.56	0.00	40.36	34.15	9.56	0.00	40.36	34.36	9.38	0.00
	50	2.61	1.83	0.01	N/A	2.61	1.83	0.01	N/A	2.61	1.83	0.01	N/A
Water Depth (ft)		1500											
Significant Wave Height (ft)		5				10				15			
Surface Current Speed (knots)		0.50	1.00	2.00	3.00	0.50	1.00	2.00	3.00	0.50	1.00	2.00	3.00
Wind Speed (ft/s)	5	N/L	N/L	N/L	11.56	N/L	N/L	N/L	11.81	N/L	N/L	N/L	11.56
	30	58.43	56.26	26.73	3.35	58.43	56.26	26.73	3.35	58.43	56.26	26.73	3.35
	50	9.92	8.59	2.81	N/A	9.92	8.59	2.81	N/A	9.92	8.59	2.81	N/A
Water Depth (ft)		2000											
Significant Wave Height (ft)		5				10				15			
Surface Current Speed (knots)		0.50	1.00	2.00	3.00	0.50	1.00	2.00	3.00	0.50	1.00	2.00	3.00
Wind Speed (ft/s)	5	N/L	N/L	N/L	29.86	N/L	N/L	N/L	29.86	N/L	N/L	N/L	29.86
	30	73.80	74.06	48.20	13.71	73.80	74.06	48.20	13.72	73.80	74.06	48.20	13.72
	50	19.02	17.91	10.14	1.38	19.02	17.91	10.14	1.38	19.02	17.91	10.40	1.49

Note: 1. N/L - No Limit
2. N/A - Not Available
3. Cells with green background indicate acceptable yellow watch circles.

Table 10 shows the watch circles in 1yr winter storm conditions, and Table 11 shows the watch circles in 1yr loop current conditions.

Table 10 Drift-Off Analysis Watch Circle Summary for 1Yr Winter Storm in Shallow Water

1yr W Storm First Limits:

Water Depth	Mud (ppg)	Heading (deg)	P.O.D.		Red		Yellow		Limiting Criteria
			Time (sec)	Dist. (ft)	Time (s)	Dist. (ft)	Time (s)	Dist. (ft)	
2000	8.56	5	210.0	150.9	147.0	75.8	87.0	27.1	Max. Casing Bending Stress
	16	5	207.0	139.8	144.0	68.2	84.0	22.3	Max. Casing Bending Stress
1500	8.56	5	189.0	119.5	126.0	55.7	66.0	15.2	Max. Casing Bending Stress
	16	5	188.0	112.7	125.0	50.9	65.0	12.6	Max. Casing Bending Stress
1000	8.56	5	163.0	87.9	100.0	35.1	40.0	5.1	Max. Casing Bending Stress
	16	5	164.0	84.4	101.0	32.6	41.0	4.2	Max. Casing Bending Stress
500	8.56	5	127.5	52.9	64.5	14.2	4.5	0.0	Max. Casing Bending Stress
	16	5	131.0	52.2	68.0	13.7	8.0	0.0	Max. Casing Bending Stress

Table 11 Drift-Off Analysis Watch Circle Summary for 1Yr Loop Current in Shallow Water

1yr Loop Current First Limits:

Water Depth	Mud (ppg)	Heading (deg)	P.O.D.		Red		Yellow		Limiting Criteria
			Time (sec)	Dist. (ft)	Time (s)	Dist. (ft)	Time (s)	Dist. (ft)	
2000	8.56	5	197.5	152.4	134.5	82.5	74.5	27.9	Max. Casing Bending Stress
	16	5	191.5	141.4	128.5	72.6	68.5	21.3	Max. Casing Bending Stress
1500	8.56	5	170.5	118.7	107.5	54.7	47.5	11.1	Max. Casing Bending Stress
	16	5	167.0	111.2	104.0	48.4	44.0	8.0	Max. Casing Bending Stress
1000	8.56	5	141.5	83.5	78.5	29.0	18.5	1.3	Max. Casing Bending Stress
	16	5	141.5	80.0	78.5	26.4	18.5	0.9	Max. Casing Bending Stress
500	8.56	5	109.5	47.4	46.5	9.0	N/A	N/A	Max. Casing Bending Stress
	16	5	112.0	46.2	49.0	8.3	N/A	N/A	Max. Casing Bending Stress

CONCLUSIONS

1. Drilling operation may be performed under GOM 1yr winter storm environment conditions. However, physical operating envelope decreases with water depth, the rig's DP capability should be verified to be able to maintain the small footprint required by the riser system.
2. The drill ship is capable of operating at 2,000ft (610m) water depth under normal GOM environment conditions.
3. At 1,500ft (457m) water depth, operation is limited by the watch circle requirements. In particular, the yellow watch circle is sufficient only under benign environment conditions (for example, reduced wind speed).
4. Rig footprint and the watch circle size may restrict the practicality of operating a drillship at 1,000ft (305m) and shallower water depths.

These conclusions are based on the assumptions, limitations and results presented in this report. Additional analysis should be performed if any condition has been changed.

ACKNOWLEDGEMENTS

Thanks for Ensco plc for supporting this research and study.

REFERENCES

American Petroleum Institute, "Recommended Practice for Design, Selection, Operation and Maintenance of Marine Drilling Riser Systems", API RP 16Q, First Edition, Nov. 1, 1993.

American Petroleum Institute, "Specification for Marine Drilling Riser Equipment", API Specification 16F, First Edition, Aug. 2004.

SNAME
THE INTERNATIONAL COMMUNITY
FOR MARITIME AND OCEAN PROFESSIONALS

EVALUATION OF GREEN WATER LOADS ON OFFSHORE STRUCTURES USING A NUMERICAL WAVE BASIN

BARCAROLO DANIEL, COUTY NICOLAS, BERRY LUKE, JACQUIN ERWAN
HYDROCEAN

GUILCHER PIERRE-MICHEL
NEXTFLOW SOFTWARE

LEDOUX ALAIN
TOTAL

LEFEBVRE THIMOTHEE
TECHNIP

LEGREGEOIS NICOLAS, BOUTROT JONATHAN, DERBANNE QUENTIN
BUREAU VERITAS

L. BEGUIN, GUILLAUME DUCROZET, DAVID LE TOUZE
ECOLE CENTRALE NANTES

Proceedings of the 21st Offshore Symposium, February 2016, Houston, Texas
Texas Section of the Society of Naval Architects and Marine Engineers

ABSTRACT

Green Water and Wave impacts are amongst the most severe and dangerous loads that effect offshore structures. Even if many design procedures are well established, they can over or underestimate the structural loading in complex conditions as they are outside of their application scope. This therefore leads to over designed and expensive structures, or to under designed structures leading to dangerous situations. The calculation of complex structure loadings is therefore a key issue for engineering companies.

HydrOcean, in cooperation with Ecole Centrale de Nantes and NextFlow, has developed the SPH-flow software, based on the SPH method.

This paper details the development and validation of a methodology dedicated to the evaluation of green water loads on offshore structures. This methodology was developed in cooperation with Total and Technip. Comparisons with experimental results are provided. The following aspects will be addressed in this article:

- *Validation of wave propagation by the SPH method and development of a forcing with a Higher-Order Spectral method to simulate open sea.*
- *Validation on slamming and green water cases. The experiments were performed in the wave tank of the Ecole Centrale de Nantes. SPH calculations are performed under similar conditions, to compare the results obtained with the experiments.*
- *Application of the developed and validated methodology for the assessment of green-water event on offshore structures.*

INTRODUCTION

During storms and severe sea states, damages suffered by ships or offshore structures can be particularly important, possibly putting human lifes in danger, destroying equipment on board, or leading to environmental pollution. Examples of such events regularly occur and the offshore industry, as well as engineering teams and classification societies are fully aware of the need in improving the evaluation of extreme loads. These extreme loads cover a wide range of events, like the violent impact of a wave onto the ship hull –slamming impact-, the water flooding of ship decks, blowing away all the equipment –greenwater phenomenon-, or the sloshing inside the oil tanks.

Such events on ships result from an interaction between a wave train and a structure. From a numerical point of view, one has to solve two different kinds of problem with very different typical spatial lengths and time durations. It involves wave generation and propagation onto the structure, with highly nonlinear waves, irregular sea states, whose typical length and duration are respectively 300m and 10s. Eventually, the interaction between the liquid and the structure leads to impacts, with typical length and duration 5 orders smaller (1mm and 1ms). Moreover, some compressibility effects could occur in the gas, and even in the liquid during impact. Hydroelasticity (deformation of the structure under the fluid loads) should also be addressed. The full problem involves a lot of physics, and the present paper is a first step towards the simulation of all phenomena occurring during impact.

Regarding the impact simulation, special attention has been paid to the Smoothed Particle Hydrodynamics (SPH) method. Indeed, the SPH method offers advantages over classical numerical methods when simulating impact type problems. No connectivity is required for the free surface, enabling the simulation of violent flows with possible fragmentation and interface reconnection. The Lagrangian formulation cancels the interface diffusion, resulting in a sharp definition of interfaces between gas, liquid and structure. Moreover, SPH method can be applied to any continuum description, resulting in an ability to easily approach multi-physics. Therefore, SPH method can theoretically solve in a fully coupled way, the compressible multi-phase structural interaction problem occurring during impact phenomenon.

As the SPH method is best suited for fast dynamics, a careful attention need to be paid for slow dynamics like wave propagation. It would be completely inadequate to use the SPH method or any CFD software in order to simulate the propagation of a wave generated in a wave flume and then the impact phenomenon (complete problem). The propagation phase can be solved accurately by software based on potential theory (BEM for example), much faster than by CFD. However, such potential-theory-based software can obviously not simulate the local interactions between liquid, gas and structure that happen during impacts. Therefore, the strategy developed is based on a coupling between two codes sharing the different tasks. A potential-theory based code is in charge of the incident flow (non-perturbed wave train) and the SPH method is in charge of the local interactions simulations during the impact in the vicinity of the impacted structure. A specific forcing algorithm between potential solvers and SPH-flow was developed in order to reduce as much as possible the numerical diffusion of the wave trains between the wave generator and the structure. As a consequence, this method allows to simulate open sea problems, providing the incident wave field from a potential solution, and providing the fact that the diffracted wave field from the structure does not reflect on the inlet/outlet wave boundaries.

The paper is organized as follow: the different solvers used throughout the paper are presented. The coupling algorithm is explained and applied to a simple focused wave propagation case. The methodology is then applied to the simulation of greenwater events for a fixed simplified FLNG ship. Comparisons with experimental values are provided

I. SOLVERS

A. *SPH-flow*

SPH-Flow is a multi-purpose, multi-physics CFD Software. It has been developed by ECN and HydrOcean through several PhD works (see [1], [2], [4], [5], [6], [8]). The solver was first developed for fluid flow simulations dedicated to complex non-linear free surface problems. SPH-Flow has now been extended to multi-fluid, multi-structure, fluid/structure interaction and viscous flows. It relies on state-of-art meshless algorithms, such as Riemann solver upwinding, MUSCL reconstruction and gradient renormalization for increasing the order of the formulation.

SPH-Flow solver is parallelized to reduce the elapsed time of a computation by use of cpu resources available on clusters. This distributed memory parallelization relies on a domain decomposition: the computational domain is divided in as many sub-domains as allocated processes. The communications between processes are achieved through the MPI library.

Through this paper, only the liquid and rigid structures are considered. The gas phase is not taken into account, and the liquid phase is considered as incompressible. Thus no compressibility effects (gas or liquid) can be addressed with the current numerical modeling.

The system of conservation laws corresponds to the isentropic Euler equations and is written in conservative form:

$$L_{v_0}(\phi) + div\left(F_E(\phi) - v_0\phi\right) = S \qquad (1)$$

where Φ is the vector of conservative variables, FE the Eulerian flux matrix and S the source term per unit of volume. The system of equations is written in an arbitrary moving reference frame (Arbitrary-Lagrange-Euler (ALE) description) where denotes the transport velocity field.

To close the system, a Tait Equation of State relating pressure to density is used:

$$p = \frac{\rho_0 a_0^2}{\gamma}\left[\left(\frac{\rho}{\rho_0}\right)^{\gamma} - 1\right] \qquad (2)$$

where ρ0, a0 and γ are respectively the nominal density, the nominal speed of sound and the isentropic constant for the considered fluid.

Fluid domain is discretized by a set P of particles. Each particle i has a location xi, a volume of control wi and carries its own properties (velocity, pressure, etc.). A weak formulation associated to (1) by use of SPH spatial derivative operators leads to the following scheme for Euler equations:

$$\frac{d}{dt}(w_i\phi_i) + w_i\sum_{j\in P} w_j 2G(\phi_i,\phi_j,n_{ij})B_{ij}.\nabla W_{ij} = w_i S_i \qquad (3)$$

Upwind numerical flux G is given either by exact or approximate Riemann solvers in the moving framework. Extension to second order with MUSCL algorithm [10] is performed with linear reconstruction on pressure and velocity. The term Bij stands for the symmetrised renormalization matrix [11], [12].

Ordinary Differential Equations are marched in time with classical 4th order Runge-Kutta scheme, or 3rd order Strong Stability Preserving Runge-Kutta scheme. Time step is restricted by a CFL-like condition on acoustic waves.

The complex shapes of ships are solved with a general accurate boundary formulation [7].

B. Potential solutions

Several potential solutions can be used, depending on the complexity of the generated wave train. Two possibilities are mainly used for the coupling algorithm with SPH-flow, depending if the wave train is regular or irregular, and 2D or fully 3D. The two solutions rely on a fully nonlinear theory.

1) Fully nonlinear regular wave train solution (Rienecker and Fenton)

Rienecker and Fenton derived in 1981 a numerical solution to the problem of fully nonlinear wave trains, restrained to the regular case [9]. This solution relies on the stream function and a Fourier series decomposition. A simple solver based on this formulation was developed at Ecole Centrale Nantes, and provides a fast solution for regular waves, up to the wave breaking limit. As an output, this solver provides velocity and pressure on any point of the fluid domain.

2) Fully nonlinear irregular wave train solution (HOS solver)

The HOS solver developed by the LHEEA Labs of Ecole Centrale Nantes is a spectral potential method. HOS stands for Higher Order Spectral. It relies on a decomposition of the solution onto a set of basis functions. The spectral formulation provides quasi analytic solutions for the wave propagation problem with high accuracy and small cpu times. No simplification is involved in the system of equations, allowing solving highly nonlinear transient wave trains [3].

Typical applications are the generation and propagation of regular waves, irregular waves, 2D and 3D waves, the numerical reproduction of experimental waves from wave tank, the simulation of oceanic waves based on statistical approaches, and the propagation of focused and freak waves.

The main limitation of the HOS solver relates to the introduction of a body into the fluid domain. Thus, the solver is restricted to the problem of generation and propagation, without any possibility to compute directly the interaction with a body. A second limitation relates to the univocal definition of the free surface, avoiding the simulation of wave trains beyond wave breaking (neither fragmentation nor reconnection of the free surface).

As an output, this solver provides also velocity and pressure maps on a regular grid, on the whole fluid domain.

II. SPH-potential solver weak coupling

Once the HOS or Rienecker&Fenton solution is known for the wave propagation, this solution is used as an input for the SPH-flow computation.

A view of the coupling principle is presented in Figure 1. The wave solution is solved with the potential solver for the complete physical domain, without body. This solution corresponds to the unperturbed wave train, and is denoted as the incident wave train. The physical domain can be either a 2D/3D numerical wave tank, or an open sea domain. The wave profile can be generated through a wavemaker motion history to reproduce an experiment, or through an ocean wave spectrum with periodic conditions to reproduce an irregular open sea state.

The local interactions between the wave train and the structure are solved with SPH-flow, only for the truncated domain. The coupling between the two solvers is set up by imposing the potential incident wave solution onto the fluid domain boundaries of the SPH calculation (dotted lines). Then the waves propagate numerically from the outward boundaries to the impact area.

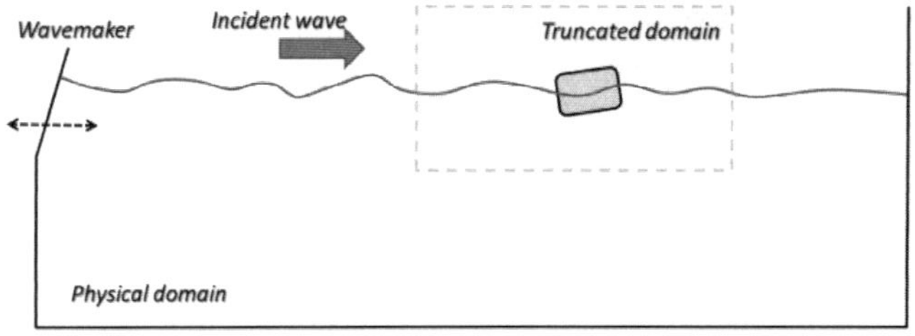

Figure 1. Coupling between potential flow solvers and SPH-flow.

These boundaries must be far enough of the area of interest, or the temporal window must be selected to ensure that no diffracted field will perturb the upstream and downstream boundaries. Despite these constraints, it is possible to simulate only a small part of the initial wave tank and reduce the overall cpu time. During the SPH computation, the inlet and outlet boundaries are continuously updated from the potential solution.

A. Inlet/Outlet boundaries for the SPH computation

As mentioned, the SPH fluid domain can be a smaller domain around the structure of interest (ship, oil rig, etc), in order to reduce the total number of particles, and then the cpu time. The main problematic is the treatment of these boundaries, which are no more wall boundaries, but inlet or outlet boundaries. Actually, in a Lagrangian framework, particles near the inlet/outlet boundaries will move with the wave velocity field, continuously going inside and outside following an elliptic trajectory (see Figure 2). The Lagrangian framework makes the treatment of particles more complex, because the position x_b of the inlet/outlet boundary is kept fixed during the whole computation (Eulerian framework). At any time, a particle can be considered as "free", in the computational domain, or "forced", in the inlet/outlet area. The (green) forced particles are updated at each time step with the incident field, and advected in time with this velocity field. The (yellow) particles inside the computational domain are considered as "free" and evolve according the classical SPH scheme, interacting with free and forced particles. As the velocity and pressure fields are fully known over the inlet/outlet area, it is possible to build MUSCL gradients and renormalization matrices also on the forced particles, to keep the same numerical scheme as in the interior domain. Particles move following the velocity field, and can be alternatively marked as free or forced, even inside a time step. The inlet/outlet area is meshed once at the beginning of the computation: the width of the inlet/outlet area must be sufficient to ensure at least a 2-Radius completion and to avoid the emptying of the area because of the Stokes drift.

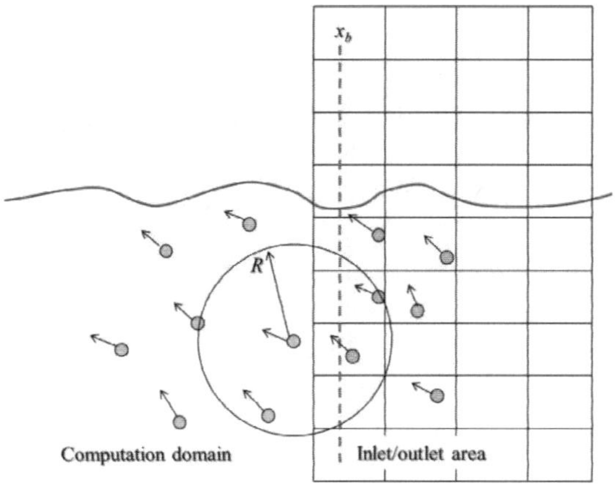

Figure 2. Interpolation of potential flow solution on a regular grid. Free particles in the computational domain are solved with classical SPH scheme. Particles in Inlet/Outlet area follow prescribed incident wave field.

A consistent formulation for the treatment of compressible inlet/outlet boundaries should rely on subsonic inlet/outlet conditions. In general, a constant velocity is imposed at inlet, and a constant pressure at outlet. Other properties of the fluid at boundary are determined with compatibility relations between the interior solution and the boundary state. The simplest solution adopted here is to impose the potential flow solution in terms of velocity and pressure. The underlying hypothesis is the fact that the perturbed area located around the impact location, does not have the time to diffract and propagate on the inlet/outlet boundaries. One must keep in mind that this solution is not fully consistent, with too much variables imposed at boundary, but is sufficient from an industrial point of view.

In practical terms, the incident fields are computed in two different ways. For the Rienecker&Fenton solution, the coefficients of the Fourier decomposition are stored then used by SPH-flow so as to reconstruct the velocity and pressure on each particle. In the case of the HOS solution, the reconstruction step is cpu time consuming, and a faster solution is adopted. The velocity and pressure fields are computed on a Cartesian grid during the potential calculation. This grid covers the area of the inlet/outlet boundary (Figure 2). This grid is saved on a file, for several time steps.

During the SPH calculation, the computation of variables on the particles is performed in two steps. A first time linear interpolation gives the incident grid at the current time. Then, a space bilinear interpolation gives the velocity and pressure at any required position. The horizontal space step can be different from the vertical one, allowing mesh refinement for the free surface area.

B. Validation of the coupling algorithm

The coupling procedure presented hereinbefore should enable accurate simulations of focused wave trains with SPH-flow, providing the potential solution. As a validation, it appears interesting to compare:

- on one hand the complete calculation performed through the potential solver,
- on the other hand the SPH solution coupled with the potential inlet/outlet boundaries, after restarting from a sooner time.

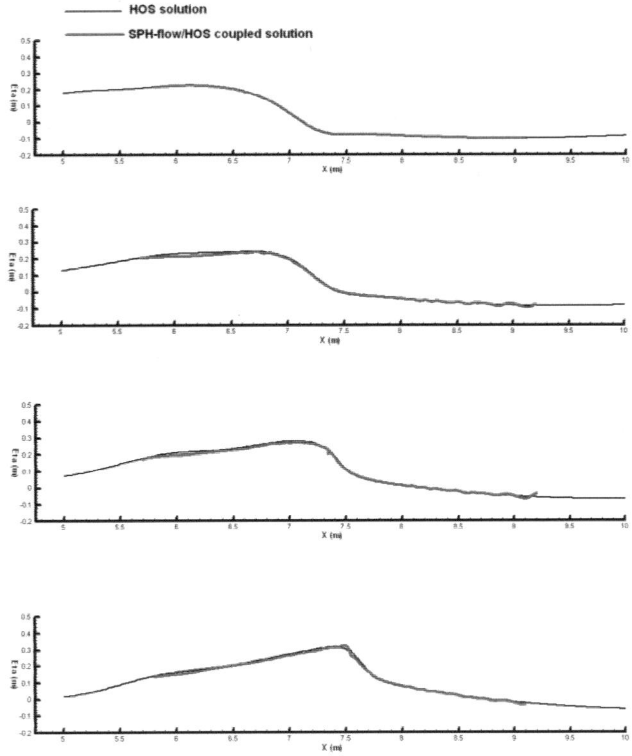

Figure 3. Comparison of free surface profiles from HOS solution (black line) and SPH-flow/HOS coupled solution (red line), at various instants (0.125s between two snapshots).

In order to check the validity of the coupling algorithm, a focused wave is studied. The free surface deformation is used as an indicator to validate the coupling methodology. The wave is generated following a Bretschneider spectrum with characteristic wave height Hs=0.14m and average period Tp=3s. Figure 3 shows a comparison of the free surface profiles computed by HOS solver and SPH-flow/HOS solvers at different instants.

For this computation, the typical distance between SPH particles is around Δx=2.10-3 m, resulting in 2.106 particles for the whole domain. The agreement between the SPH/potential coupled solution and the potential solution is fairly good, despite some discrepancies at the inlet and outlet boundaries. Nevertheless, this kind of approach is sufficient for wave/structure interactions, as far as the perturbed field is not reflected onto these open water boundaries.

III. Simulation of extreme waves impacts on a FLNG

The objective of this part is to demonstrate that the SPH method coupled with HOS forcing is an efficient method for simulation of structures under extreme sea. The simulation presented hereafter is compared to the experiment conducted at Ecole Centrale Nantes.

A. Experimental set-up

Experiments were performed on a stiffly fixed FLNG model (see Figure 4) in the oceanic wave tank of Ecole Centrale Nantes, which is 50 meters long, 30 meters wide, and equipped with a wavemaker able to generate multidirectional waves. Green water effect was studied under a diversified set of regular front waves. Regular wave profiles were chosen to reduce numerical and experimental uncertainties, in order to validate the numerical tool.

The model is instrumented using a series of wave probes located in the fore deck in order to measure elevation of shipping water, as well as a set of pressure transducers, installed through the wavebreaker to observe slamming events (see Figure 5). The model deck is 1.1m wide.

Figure 4. Set-up in the Ecole Centrale Nantes oceanic wave tank.

Incident water wave is chosen in order to be numerically reproducible and to not surge, its wave length of about 7.3m and amplitude 0.44m at model scale.

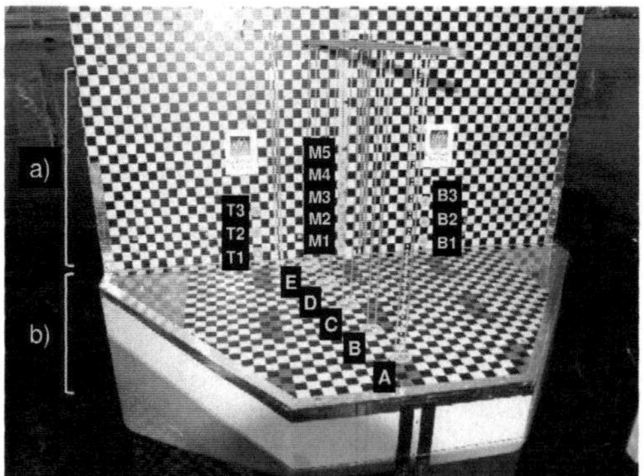

Figure 5. a) Position of pressure probes on the breakwater b) Position of wave probes on the fore deck.

B. Experimental behavior of greenwater event

Those environmental conditions lead to greenwater events (see Figure 6). Time separating two frames is T/8 where T is the period of incoming waves (T = 2.2s model scale).

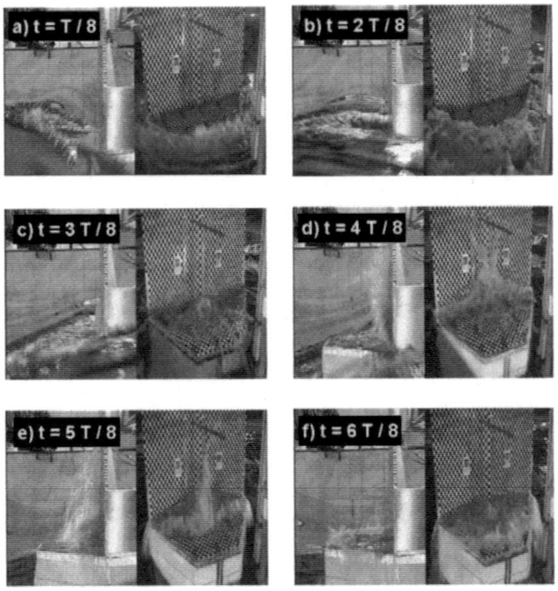

Figure 6. Greenwater event on the fore deck.

The greenwater can be described into 6 steps:

(a) Wave elevation at the bow exceeds effective freeboard.

(b) A large amount of water is plunging onto the fore-part of the deck and some water invades the deck sideways.

(c) A large amount of water at the centre of the deck results from convergence of flows from the sides and from the bow. Over 20cm (model scale) of shipped water is measured.

(d) Flows from the sides are meeting at the centre of the wall, creating a vertical water-jet. Just after, water from the front impacts the wall. Some water escapes from the deck at the edge of the wall.

(e) The centred water column on the wall starts to collapse, the shipped water is mostly close to the wall.

(f) Water first escapes through deck's sides, before the water evacuates the deck until next incident wave. Two maxima of water height are thus measured on the deck.

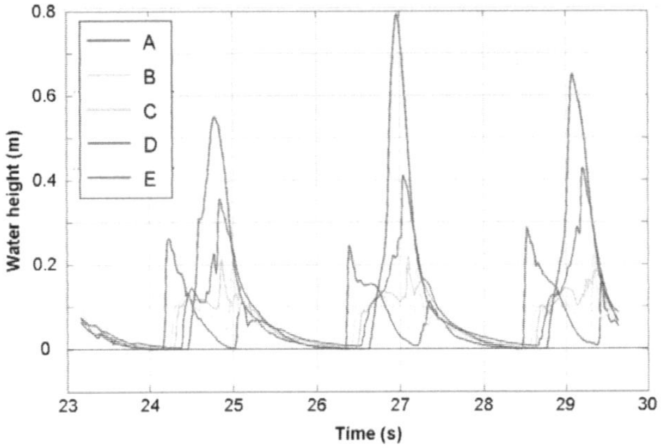

Figure 7. Time history of water height on deck (model scale).

C. Experimental measures

1) Water on deck elevation

Measures of water elevation on deck (see Figure 7) show very good repeatability at the frequency of incident wave. It gives also valuable qualitative and quantitative information about incident and returning flows (two maximum are observed for the probes located near the bow). Time of presence and spatial distribution of the amount of greenwater are also measured.

2) Pressure impacts on the wavebreaker

Pressure measures are presented in Figure 8. These results stress out the repeatable qualitative behaviour of pressure pattern during an impact. The pressure experiences a rapid (in the range of a few milliseconds) rise towards a maximum value, followed by a slower drop, related to dynamic pressure caused by the running-up of water along the wall. Finally, dynamic pressure due to the collapse of water along the wall is recorded.

The symmetric response of the sensors located at port (B1, B2, B3) and those located at starboard (T1, T2, T3) argues in favour of a precise experimental set-up. As observed, values of pressure peaks are not fully repeatable, due to incident conditions, but timescales as well as local maximum of dynamic pressure following the impact are reproducibly determined. These abilities ensure experimental results to be leveraged for comparisons with SPH. In this respect, average pressure peaks are computed (see Fig. 9), corresponding to the first incident waves (before any reflexion from the basin).

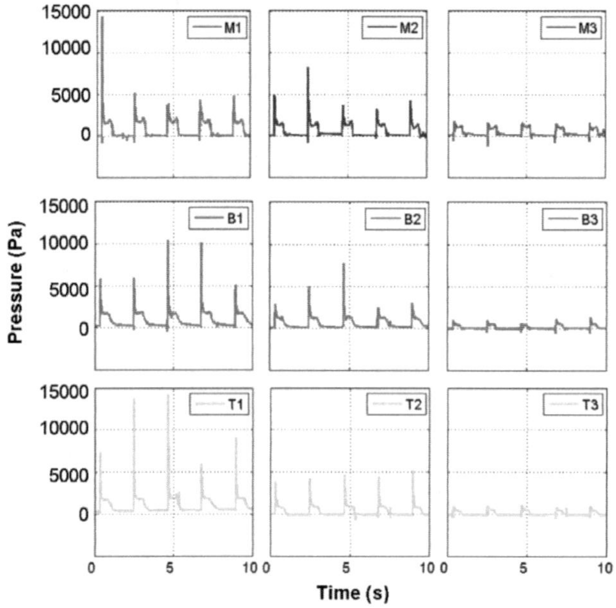

Figure 8. Time history of pressure impacts on the wavebreaker (model scale).

These experimental measures confirm that altitude from the corner formed by the wavebreaker and the deck is a key factor for pressure impact assessment in greenwater events. The pressure is maximum at bottom of the wavebreaker, where the incoming deck wave hits the wall at first. This pressure peak is related to the rapid change in momentum, from horizontal to vertical direction, and is comparable to the incompressible pressure peak occurring during a Wagner-type impact (wedge drop test).

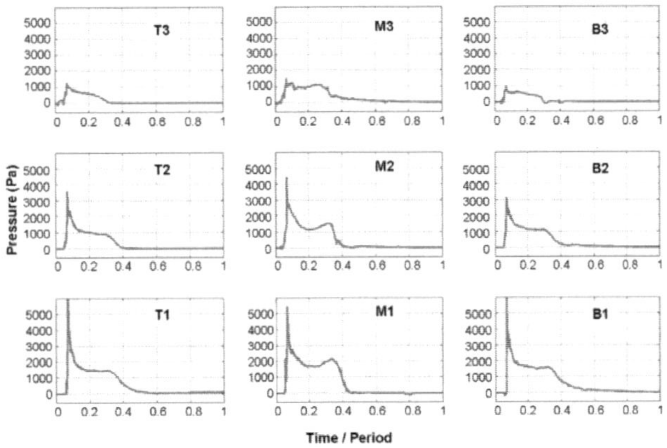

Figure 9. Average pressure peaks for pressure sensors during a period (model scale).

Table 1 sums up quantitative pressure statistics on impacts that occurred during the experiment. <Ppeak> stands for the average value of pressure peaks, σpeak its standard deviation, Ppeak+ and Ppeak- respectively the maximum and minimum measured value of pressure peaks. The maximum of pressure, at probe M1, is fairly good measured, with a ratio σpeak/<Ppeak> equals to 15%.

Table 1. Peak pressure statistics at model scale

	$\langle P_{peak} \rangle$ [Pa]	σP_{peak} [Pa]	P_{peak}^{+} [Pa]	P_{peak}^{-} [Pa]
M1	5 440	810	6 170	4 630
M2	4 380	1 830	6 100	2 440
M3	1 510	330	1 920	1 120
B1	6 630	2 280	9 510	4 150
T1	9 780	3 080	12 230	5 350
B2	3 140	650	3 940	2 460
T2	3 590	1 540	5 490	2 240
B3	990	270	1 340	680
T3	1 250	250	1 470	900

D. Simulation

1) Set-up

SPH calculation is performed at model size in order to compare the results obtained with the experiments, and avoid any possible scaling effect. The configuration of the simulation domain is sketched in Figure 10. As the problem is symmetric, only the port side of the simplified FLNG is simulated. The particles discretisation is space variable with a typical interparticle size ranking from 5mm to 5cm with 3% of growth rate, leading to approximately 25 million particles. The fine area is located just in front of the fore deck, in order to capture adequately the greenwater event. The adopted discretisation ensures a resolution of 150 points per wave length and 100 points within the deck length.

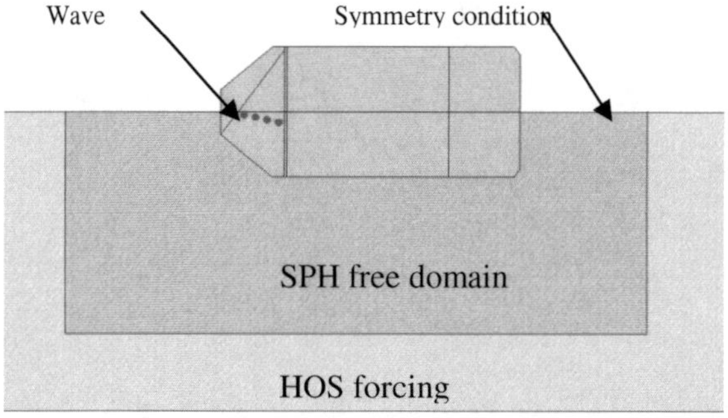

Figure 10. Domain configuration sketch.

The speed of sound is chosen in order to fulfil both a maximum of 1% density variation at tank bottom, and a low Mach number in jets on the free surface. The hydrostatic condition appears to be the most prohibitive, leading to a speed of sound equal to 45m/s for 2m of water depth. The kernel chosen is 4th order Wendland with ratio Radius/Δx equal to 4. The time advance is solved with Strong-Stability Preserving Runge-Kutta (8,3) scheme with CFL number equal to 1.5.

2) Relaxation step

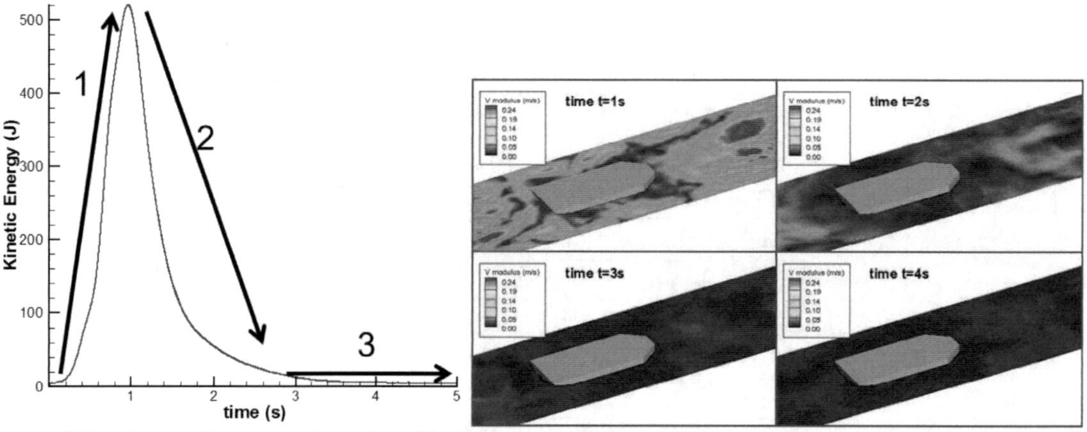

Figure 11. Relaxation step. Top: temporal evolution of liquid kinetic energy. Bottom: Velocity modulus on solid boundary in place of free surface, at various instants.

A first particle mesh is created inside the whole truncated domain, a parallelepipoid in the present case. The FLNG hull is then added and particles inside the ship are removed with a specific algorithm. The resulting set of particles turns out not to be in discrete linear hydrostatic equilibrium and would lead to acoustic noise if simply used as an input for the SPH computation.

An intermediate step of numerical relaxation is then introduced through the packing algorithm [13] to find at least a constant discrete equilibrium. A rigid boundary is added on top of the free surface to avoid the escape of particles. The total liquid kinetic energy is monitored to determine the end of the relaxation step.

The relaxation computation can be characterised by three steps (Figure 11): 1/ the kinetic energy reaches a maximum value, due to the initial positions of particle which does not fulfil the condition $\nabla 1=0$ at the discrete level ; 2/ the intrinsic numerical viscosity of the scheme helps to find a discrete equilibrium with a rapid decrease of kinetic energy ; 3/ the particles arrangement is almost stabilized as indicated by the slow decrease of kinetic energy.

The resulting particle mesh, with the stabilized positions, is used as an input with consistent initial conditions of hydrostatic equilibrium and zero velocity profile for the physical computation of wave-structure interaction.

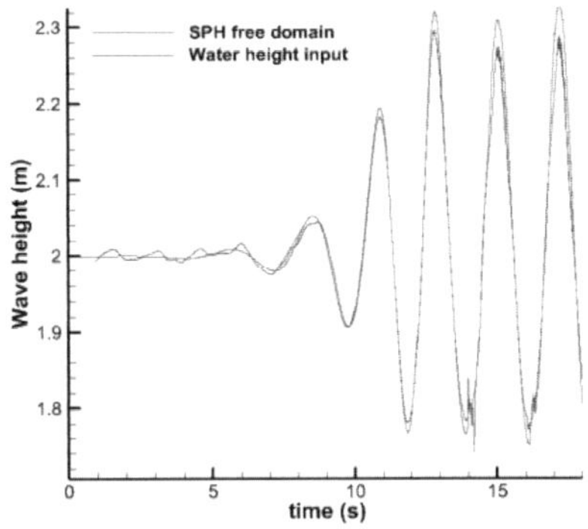

Figure 12. Comparison of HOS reference wave probe and SPH wave probe in the non-perturbed area.

3) Results

The numerical incident wave in the free domain is compared with the theoretical wave forcing computed by HOS at the same probe position on Figure 12. The agreement is fairly good in terms of period and synchronisation, without important phase shifting. The amplitude is also well captured, with a damping of the amplitude inferior to 10%. The SPH simulation, as well as the HOS incident field, starts from rest, and the incoming wave must propagate from the inlet boundary to the structure. In consequence, half part of the computation is dedicated to the transient propagation of the incoming wave train. However, the use of a truncated domain helps to reduce the global elapsed time by two ways: the overall number of points is reduced according to the reduction of volume simulated, and the diminution of distance between the wave generation point and the structure allows to decrease the number of points per wavelength because the numerical damping is less important.

Three greenwater events are simulated. Comparisons are done for wave probes A, C and E respectively located on the deck from the bow to the breakwater, see Figure 5. The two maxima shape corresponding to the incident deck wave and returning wave, as well as the temporal evolution are very well captured by the SPH simulation. However the thinness of the converged jet and the impact induced spray are not obtained on wave probe E (see Figure 13). This could be explained by two reasons: the quantity of water flowing on the deck is underestimated, as indicated by the 10% difference on the crest wave (Fig. 12), or the resolution is still not sufficient and the numerical upwind dissipation introduced by Riemann solvers disperses the kinetic energy of the wave on deck, resulting in a too viscous behaviour.

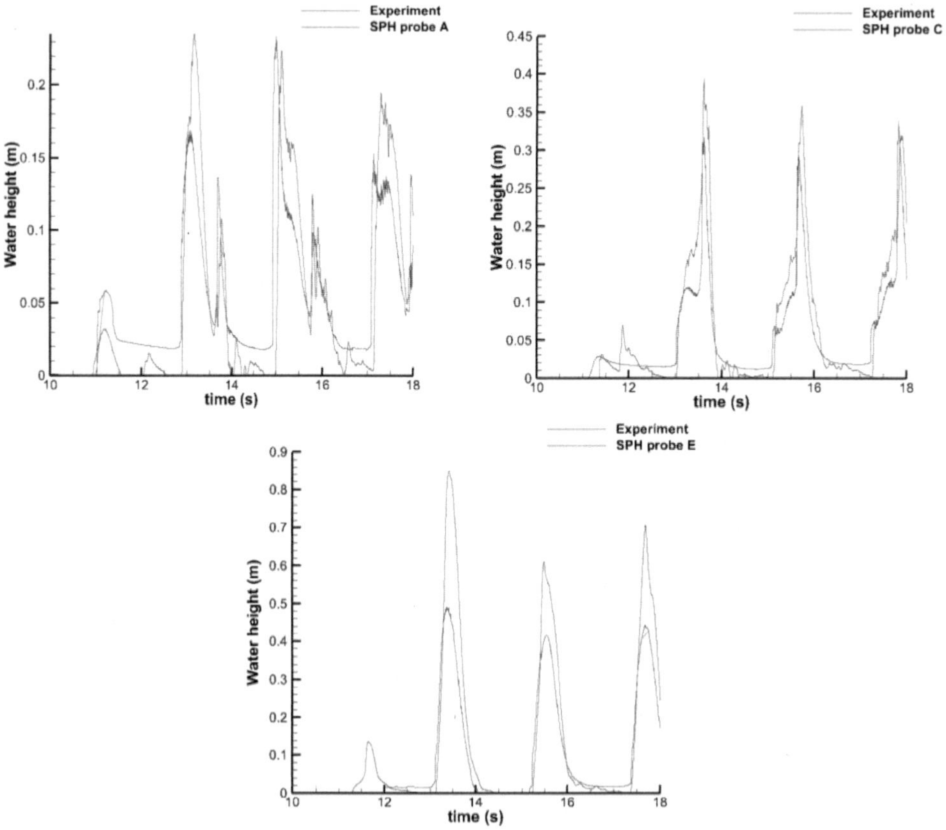

Figure 13. Water height on the front desk for probe a) A, b) C and c) E.

The global flow is qualitatively captured, as figured on Fig. 15, with the 6 stages described in the experimental section. High number of neighbours, 256 for ratio Radius/Δx equal to 4, is very useful to propagate the swell but it leads fluid to be numerically hard to scatter. As a result, some water remains on the deck between two greenwater events. The use of smaller particles (smaller Δx) should avoid this issue.

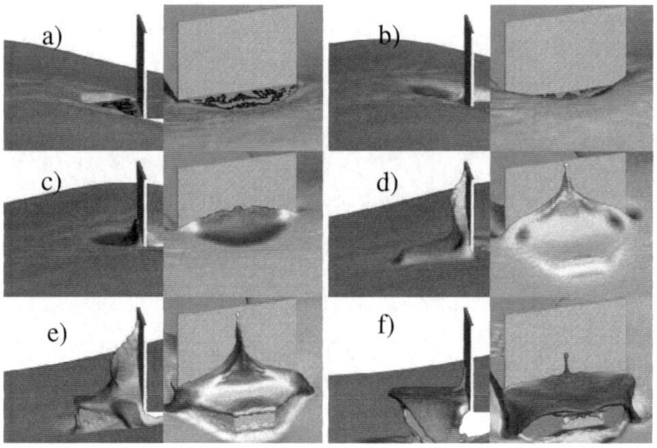

Figure 14. Free surface colored with velocity magnitude at approximatively same times as Figure 6

CONCLUSIONS

An SPH monofluid version of the SPH-flow software is being developed by HydrOcean and ECN, for the special use of liquid impact simulations in the case of wave-structure interactions. An important step forward was achieved recently with the development of a coupling between potential solvers for the wave generation problem, and SPH-flow for the impact simulation. The potential solution relies on state-of-the-art solvers, to be able to take into account irregular nonlinear open sea states representative of real conditions.

The coupling between solvers is simply enforced by imposing the incident unperturbed wave solution on the boundaries of the SPH computational domain. Special care is required to manage the Lagrangian motion of particles at the inlet/outlet boundaries. This coupling allows simulating only a small part of the fluid domain around the impacted structure, as far as the diffracted flow does not reach the inlet/outlet boundaries. Further developments are under consideration, to combine general subsonic inlet and outlet conditions with the incident wave solution. This would enable the diffracted wave field to evacuate through the open boundaries, allowing larger time windows and smaller computational domains.

A unidirectional regular wave train impacting a FLNG-type structure was simulated in 3D. Same general characteristics of the greenwater events as those observed during the experiments conducted at oceanic wave basin of Ecole Centrale Nantes were observed:

- The flooding of dry deck by the incoming wave, from the bow as well as from the sideways
- The convergence of the bow flow and sides flows towards an horizontal impacting jet located at the centre of the deck
- The impact at the bottom of the breakwater, resulting in a string impact and the formation of a vertical upward jet
- The collapse of the vertical jet, with evacuation of water through the deck sides.

These good comparisons with experimental results bring credibility to the numerical simulations, at least with their ability to capture the relevant phenomena. The water height on deck is well captured in terms of temporal synchronization as well as maximum amplitude. Since these validations, HydrOcean has been successfully applying this methodology on projects for the offshore industry.

Nonetheless, some path of improvement remain. HydrOcean is working on integrating a dynamic particle refinement procedure which will allow maintaining the same level of precision of the simulations presented below but with less particle. Also, HydrOcean is working on developing and validating an equivalent dam-break approach that could be applied also as a first approach to assess the forces acting on some structures of the deck.

ACKNOWLEDGEMENTS

HydrOcean and ECN would like to acknowledge the support of TOTAL and TECHNIP, within the sponsorship of CITEPH R&D Program, dedicated to research and innovative projects in Exploration and Production. The results presented are part of CITEPH-66-2010 Project. In addition, the contributions of the following individuals and organizations are recognized and greatly appreciated:

- F. Bonnefoy for his invaluable contribution for the experimental overview and his wise recommendations.
- J.-L. Toularastel and S. Lambert who helped for experimental preparation as well as conception of the structure.
- This work was granted access to the HPC resources of TGCC under the allocation 2015- t20152a7491 made by GENCI
- Use of IFREMER Caparmor cluster for set-up of numerical model and initial validations.
- This work was supported by a grant from the Swiss National Supercomputing Centre (CSCS) under project ID s398.
- We acknowledge PRACE for awarding us access to resources on Cray XE6 "HERMIT" based in Stuttgart, Germany at GCS in HLRS.

REFERENCES

[1] Cra J.-B. Deuff, "*Extrapolation au réel des mesures de pressions obtenues sur des cuves modèle réduit*", PhD Thesis, Ecole Centrale de Nantes/GTT, France, 2007.

[2] M. Doring, , "*Développement d'une méthode SPH pour les applications à surface libre en hydrodynamique*", PhD Thesis, Ecole Centrale de Nantes, France, 2005.

[3] G. Ducrozet, F. Bonnefoy, D. Le Touzé, P. Ferrant, "*A modified High-Order Spectral method for wavemaker modeling in a numerical wave tank*", European Journal of Mechanics – B/Fluids, Vol. 34,pp. 19-34, 2012.

[4] N. Grenier, "*Modélisation numérique par la méthode SPH de la séparation eau-huile dans les séparateurs gravitaires*", PhD Thesis, Ecole Centrale de Nantes, France, 2009.

[5] P.-M. Guilcher, "*Contribution au développement d'une méhode SPH pour la simulation numérique des interactions houle-structure*", PhD Thesis, Ecole Centrale de Nantes, France, 2008.

[6] M. de Leffe, "*Modélisation d'écoulements visqueux par méthode SPH en vue d'application à l'hydrodynamique navale*", PhD Thesis, Ecole Centrale de Nantes, France, 2011.

[7] M. de Leffe, D. Le Touzé, B. Alessandrini, "*Normal Flux Method at the boundary for SPH*", Proc. Of the 4th Spheric Workshop, pp. 149-156, 2009, Nantes, France.

[8] G. Oger, "*Aspects théoriques de la méthode SPH et applications à l'hydrodynamique à surface libre*", PhD Thesis, Ecole Centrale de Nantes, France, 2006.

[9] M. M. Rienecker, J. D. Fenton, "*A Fourier approximation method for steady water waves*", J. Fluid Mech., 1981, vol. 104, pp. 119-137.

[10] B. Van Leer, , "*Towards the Ultimate Conservative DifferenceScheme. V. A Second-order Sequel to Godunov's Method*", Journal of Computational Physics, vol. 32, pp. 101-136, 1979.

[11] J. P. Vila, "*SPH renormalized Hybrid Methods for Conservation Laws : Applications to Free Surface Flows*", Meshfree Methods for Partial Differential Equations II, 43, 2005.

[12] J. P. Vila, "*On Particle Weighted Methods and Smooth Particle Hydrodynamics*", Mathematical Models and Methods in Applied Sciences, 1999.

[13] A. Colagrossi, B. Bouscasse, M. Antuono, S. Marrone, *"Particle packing algorithm for SPH schemes",* Computer Physics Communications, Volume 183, Issue 8, August 2012, Pages 1641-1653

SNAME
THE INTERNATIONAL COMMUNITY
FOR MARITIME AND OCEAN PROFESSIONALS

COLUMN CONFIGURATION AND WAVE ENHANCEMENT IN THE WELLBAY OF OFFSHORE FLOATING PRODUCTION UNITS

ALAA M. MANSOUR[1]
MARINE ENGINEERING MANAGER, INTECSEA INC.

DHIRAJ KUMAR
NAVAL ARCHITECTURE SPECIALIST, INTECSEA INC.

Proceedings of the 21st Offshore Symposium, February 2016, Houston, Texas
Texas Section of the Society of Naval Architects and Marine Engineers

ABSTRACT

Maintaining sufficient airgap between the wave crest and the bottom of deck steel in the wellbay area of column stabilized floating production units is one of the key design challenges for such structures and especially so in the harsh environments survival conditions.

In this paper, the effect of column configuration on the wave enhancement has been investigated. A detailed parametric study is performed to examine the influence of the column cross-sectional shape, column spacing relative to wavelength and column orientation relative to wave propagation direction on the wave enhancement. A four-column structure is considered in this parametric study. Platforms with circular and rectangular columns are investigated and compared. The results are calibrated with available model test results. For the selected configuration, time domain coupled analysis including first and second order wave elevation transfer functions are used.

A new column design with hybrid cross-sectional configuration is developed which significantly reduces the wave enhancement in the wellbay area. Numerical results are presented to illustrate the benefit from this new design.

The paper concludes with discussion on the constructability of the new configuration compared to the conventional design.

Keywords: Airgap, Wave Enhancement, Column Cross Section, Column Stabilized floaters

INTRODUCTION

Maintaining a positive airgap between the bottom of the superstructure and the enhanced wave crest due to the presence of the structure is very important for the survivability of the column stabilized offshore floating structures. Alternatively, the superstructure should be designed to withstand the wave slamming and its effect on the different elements of the structure. Increasing the deck clearance distance above the Mean Sea Level (MSL) to maintain positive airgap could be a costly solution due to the added structural weight, increased wind loads, increased vertical center of gravity and the knock-on effect on the platform stability and mooring loads.

The recommended minimum deck clearance criteria requires a 1.5 m (5 ft) clearance to main steel in extreme conditions (100-year return period), combined with zero or greater clearance in survival (1000-year minimum return period) conditions (API RP 2T). Local wave effects in these conditions can be dealt with by local strength design.

The estimation of the relative wave elevation between the wave crest and the bottom of deck steel in the wellbay area of a column stabilized floating platform is very complex. This is mainly driven by the very complicated wave field formed as a result of the wave structure-interaction in the wellbay area and the increased wave steepness due to wave-current interaction. This is in addition to the already complicated motion response due to wave, wind and current.

Large loads are to be expected if waves hit the platform deck. The magnitude of the loads is proportional to the ''negative airgap'' and the relative slamming velocities between waves and structure, see e.g. (Baarholm et. al. 2001).

Linear diffraction theory has been commonly used as a simplistic approach for predicting the diffracted wave fields in the wellbay area of offshore structures. However, this approach is found to under predict the wave enhancement and run-up. Niedzwecki and Duggal, 1992 performed small-scale model tests for run-up on a bottom mounted circular column and concluded that except for a very low wave steepness the linear diffraction theory under predicts the wave run-up. Second order method has been adopted by other authors such as Lee and Newman, 1994.

The most reliable approach to simulate this problem to date is through scale model testing with special attention given to the current simulation and the wave current interaction. There are a few challenges associated with this approach. The current spatial and temporal variation generated in the basin, the scaling effect, limitation in the number of cases to be performed and the effect of white water splash and run up on the wave probe readings. Typically a visual correction is required to estimate the airgap based on the green water. The alternative approach is the Computational Fluid Dynamics (CFD). The latter approach also has its own shortcomings due to the computational burden and the reliability of the calibrated model.

This paper focuses on understanding the key design parameters that affects the wave enhancement in the well-bay area of a column stabilized floater to provide guidance on how to optimize the design to reduce the wave enhancement and improve the airgap. Parameters including column cross-sectional shape, column spacing relative to wavelength and column orientation relative to wave propagation direction for a four-column structure are considered. Columns with circular versus rectangular cross sectional shape are analyzed. Also, columns in diagonal versus orthogonal orientations are investigated and compared. The results are calibrated with the available model test results.

New column design with hybrid cross-sectional configuration is developed and may significantly reduce the wave enhancement in the wellbay area. The new design is compared to a selected configuration based on the parametric study. Time domain coupled analysis including first and second order sum and difference frequency wave elevation transfer functions are used in this analysis. Numerical results are presented to illustrate the benefit from this new design.

The constructability and cost associated with the new configuration compared to the conventional design are discussed.

THE PARAMETRIC STUDY

In this parametric study, a four column semisubmersible platform with conventional hull form (Figure 1) has been considered as the column stabilized structure. The supported topside and riser payloads and environmental conditions considered in this study are presented in Table 1.

Column Spacing Effect

Hull with columns of square and circular cross section has been initially considered to investigate the effect of column spacing on the produced wave enhancement. First order three-dimensional diffraction analysis is performed using the industry recognized WAMIT software (WAMIT, Inc.).

The column spacing is varied incrementally and the remaining hull design parameters are optimized to provide the required displacement and produce similar motion responses with maximized hull efficiency accordingly. Excel based with Visual Basic GUI tool utilizing simplified Morrison equation with iterative solver is developed and used in the initial optimization process. Then the motion is verified using the full diffraction analysis.

The transparency ratio, T_r, defined as the ratio of the column face-to-face spacing relative to the equivalent column diameter has been estimated and used as the parameter to be correlated to the wave enhancement factor. Wave enhancement factor defined as the ratio of the predicted wave crest in the wellbay area to the design crest elevation is derived at several locations. The points of interest and environmental headings are defined in Figure 2.

Table 1: Design Payload and Environmental Condition

Area	Central Gulf of Mexico	-
Topside Payloads	28,000	Mt
Riser payloads	2,000	Mt
Survival Condition	1,000	Year
Significant Wave Height (Hs)	65	ft
Significant Peak Period	16	sec
Max Crest Elevation	62	ft

Figure 1: Conventional Semisubmersible Hull Form

The estimated wave enhancement factors versus the transparency ratio, T_r, relationship is derived and presented in Figure 3 for circular columns and square columns at critical locations in different headings. It can be seen that the wave enhancement factor is inversely proportional to the transparency ratio (Tr) for both circular and square columns. The enhancement factors are significantly higher for T_r values less than 2. Therefore, in future projects, optimization study including enhancement factor should be performed with a target value of T_r greater than 2.0.

The wave enhancement in 135° heading is more than that in 90° heading with circular column configurations while the wave enhancement is almost similar in 135° and 90° headings for square columns. The circular columns shows more pronounced enhancement compared to square columns configurations in 135°. The wave enhancement at the center of the wellbay area is small and generally similar between square and circular column configurations. The wave enhancement versus T_r has been calibrated based on available model testing in database. The calibrated relationships are presented in Figure 4. Similar conclusions can still be derived.

Reviewing Figure 3 and Figure 4, it can be seen that a square-circular hybrid (Tear Drop) column configuration could lead to the lowest wave enhancement in all environmental headings.

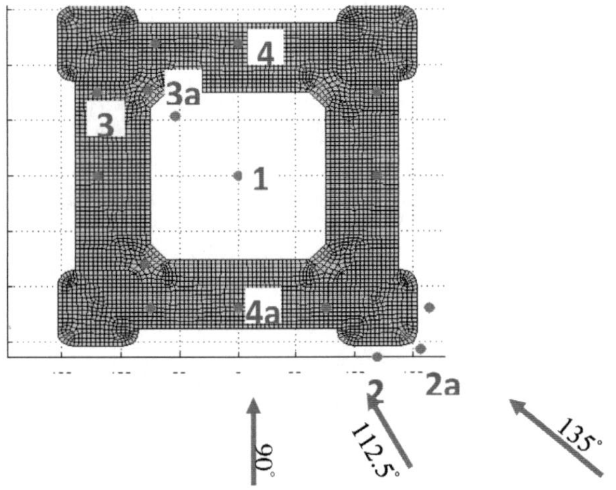

Points	Description
1	Point at Center of Platform
2a	Upwave Points in front of Column
2b	
3a	Downwave Points in front of Column
3b	
4a	Upwave Point at Center of Pontoon
4b	Downwave Point at Center of Pontoon

Figure 2: Infield Points for Wave Enhancement Calculations and Heading Definition

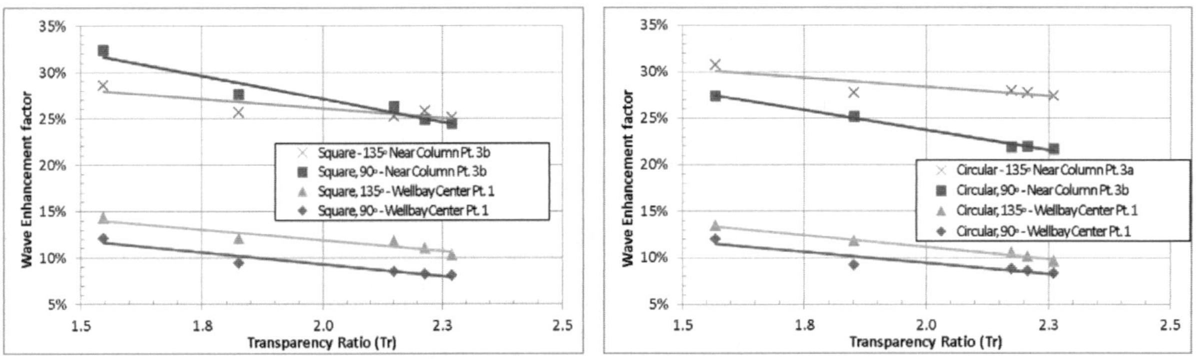

Figure 3: Wave Enhancement versus Transparency Ratio (Tr)

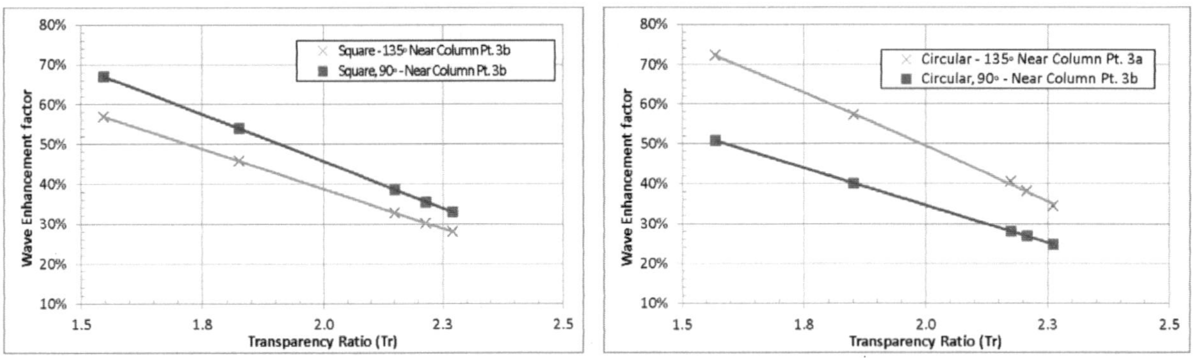

Figure 4: Wave Enhancement versus Transparency Ration (Tr) Calibrated with Model Tests

Column Cross-Section and Orientation Effect

One of the optimized design for the column spacing based on the results presented in Figure 3 and Figure 4 with T_r of 2.25 has been further considered to investigate the effect of the column cross sectional shape and orientation on the wave enhancement in the well-bay area. The following designs have been considered:

a) Hull with orthogonal square columns
b) Hull with quartering square columns
c) Hull with Circular columns
d) Hull with hybrid square-circular columns (tear-drop with vertex angle of 90°)
e) Hull with tear-drop cross sectional shape with vertex angle of 45°

All the designs supports same payload and provides equivalent cross sectional area and column spacing. The hull draft has also been maintained for all configurations. The different hull forms considered in this study are presented in Figure 5.

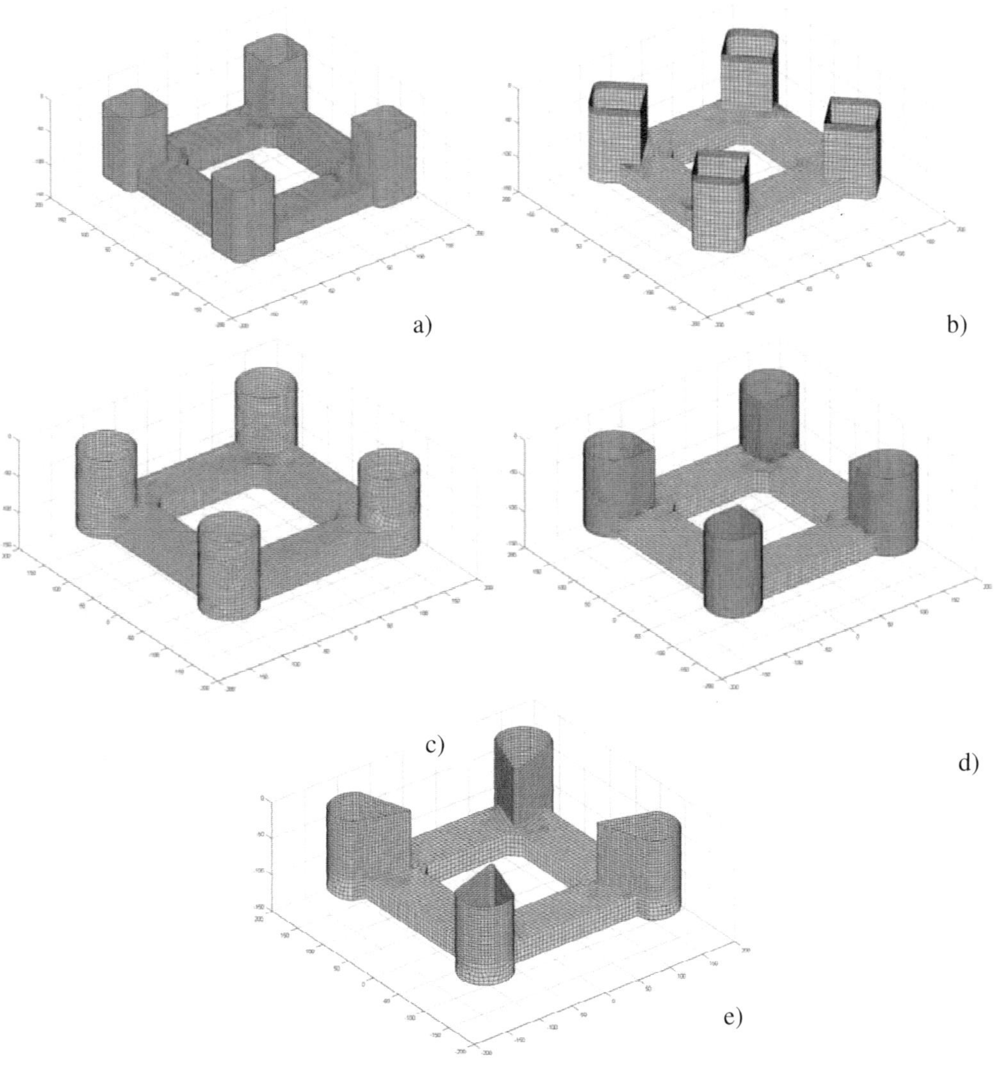

Figure 5: Hull Forms used in Column Cross-Section and Orientation Parametric Study

Again the different hull forms have been considered in WAMIT linear analysis and the wave enhancement has been estimated for the 1,000 year environment presented in Table 1 and at the different locations shown in Figure 2. The results are compared in Figure 6 for the different hull forms in both 90° and 135° headings. The results have been calibrated using the available model testing data. While no model testing data is available for the Tear-Drop column configurations (d) and e) Figure 5), the calibration coefficients from the circular and square column configurations has be applied based on the shape, the wave heading relative to the surface and the location of the point of interest.

The calibrated wave enhancement relationships at the critical locations are presented in Figure 7. It can be seen that at the critical location near the face of the down-wave column, the hull form of Tear-Drop column with 45° vertex provides the lowest wave enhancement among all the different configurations under consideration in the 135° critical headings with 56% and 40% reduced calibrated wave enhancements compared to those from the square-diagonal and circular column configurations, respectively. It also provides 18% and 21% improvement in the wave enhancement factor compared to the trear-drop-90°-vertex ad square-orthogonal column configurations, respectively. In the 90° heading, at the critical location close to the column surface, the circular column configuration is found to be the best in terms of lowest wave enhancement factor however by only 11% compared to the tear-drop-45°-vertex configuration. The tear-drop-45°-vertex configuration shows 12% better enhancement factor compared to the square-orthogonal configuration. The heading perpendicular to the flat surface of the tear-drop-45°-vertex column has been further analyzed. The wave enhancement factor in this case is found to be similar in magnitude to that from 90° heading. While it is expected that this heading should provide more critical wave enhancement compared to the 90° due to the flat surface orientation relative to the wave propagation direction, however, the standing waves in 90° between the up and down wave columns made the situation slightly worse relative to this perpendicular direction. At the center of the wellbay area the enhancement is found to be much smaller and the tear-drop-45°-vertex configurations are predicted to provide the lowest enhancement among the different configurations in both 90° and 135° headings (Figure 6).

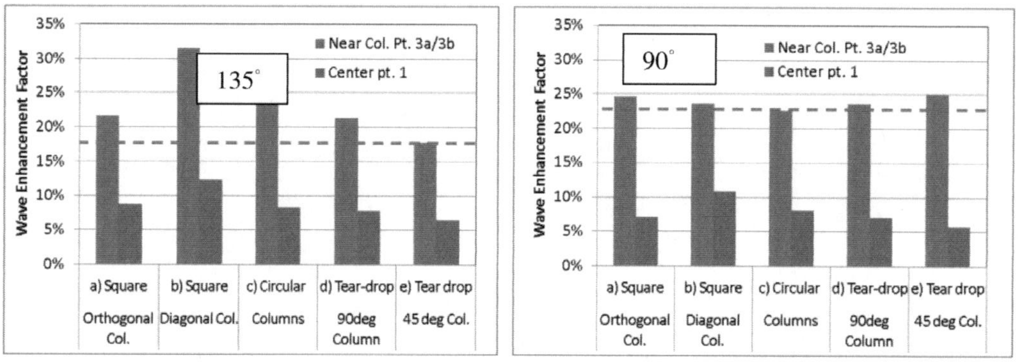

Figure 6: Column Cross-Section Shape and Orientation versus Wave Enhancement Factor

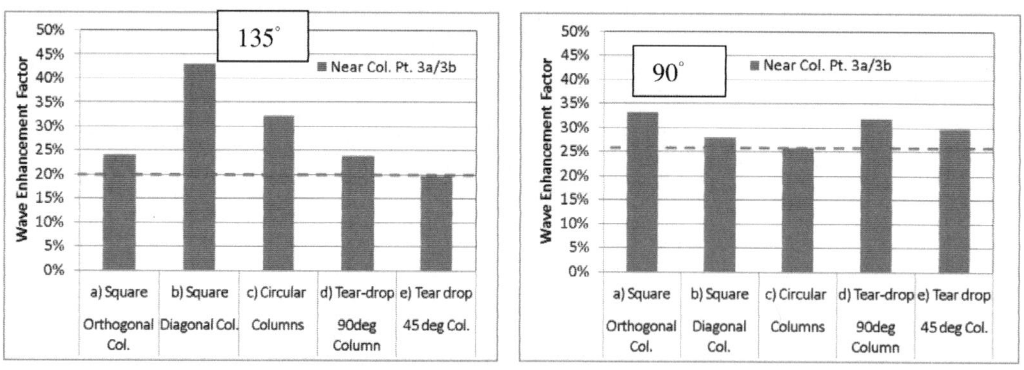

Figure 7: Column Cross-Section Shape and Orientation versus Wave Enhancement Factor Calibrated with Model Test

The maximum wave enhancement factors irrespective of the heading from all configurations are summarized and compared in Table 2. Considering design crest elevation per Table 1 and assuming that all the platforms will be designed to clear the waves in 1,000 year condition presented in Table 1, the deck structure for the Tear-Drop-Vertex 45° can sit about 3 ft, 9 ft, 3 ft and 2ft lower than that for square-orthogonal, square-diagonal, circular and tear-drop-90°-vertex configurations, respectively (Table 2). It has to be noted that the data presented in this section is based on the optimized transparency ratio, T_r of 2.25. Should a smaller T_r is adopted in the design for other reasons, the relative benefit in the wave enhancement between the different configurations will be more pronounced. The enhancement factors are also function of the wave length and in different areas of the world the enhancement factors will be different, however, the inter-relationship between them is expected to be similar for the different environments. The tear-drop shape and vertex angle can be further optimized on a project-by-project basis and according to the topside footprint to further minimize the wave enhancement in the wellbay area and further reduce the deck elevation. Reducing the deck elevation has significant benefits such as increased platform stability, reduced wind loading, lower mooring loads, lower steel weight and reduced cost.

Table 2: Maximum Wave Enhancement Factors

Configuration	Max. Wave Enhancement Factor in the Wellbay	Estimated Deck Elevation relative to Case e)
a) Square-Orthogonal	33.5%	+3 ft
b) Square-Diagonal	43.0%	+9 ft
c) Circular	33.5%	+3 ft
d) Tear-Drop-90°-Vertex	32.5%	+2 ft
e) Tear-Drop-45°-Vertex	29.0%	Reference Case

TIME DOMAIN SIMULATION WITH WAVE ELEVATION SUM AND DIFFERENCE FREQUENCY

To further validate the inter-relationship of the wave enhancement from the different column cross sections and orientations, configurations a), c), d) and e) have again been considered in time domain fully coupled analysis to predict the wave enhancement including sum and difference frequency wave elevation. WINPOST software with hydrodynamic coefficients and full Quadratic Transfer Function predicted using WAMIT first and second order solutions are used for this purpose. Risers and mooring systems have been considered in the stiffness matrix in WAMIT analysis while they are simulated using finite element method in WINPOST time-domain simulation. The wave enhancement has again been calculated at the points of interest presented in Figure 1. The WINPOST model for configuration a) and d) are shown in Figure 8.

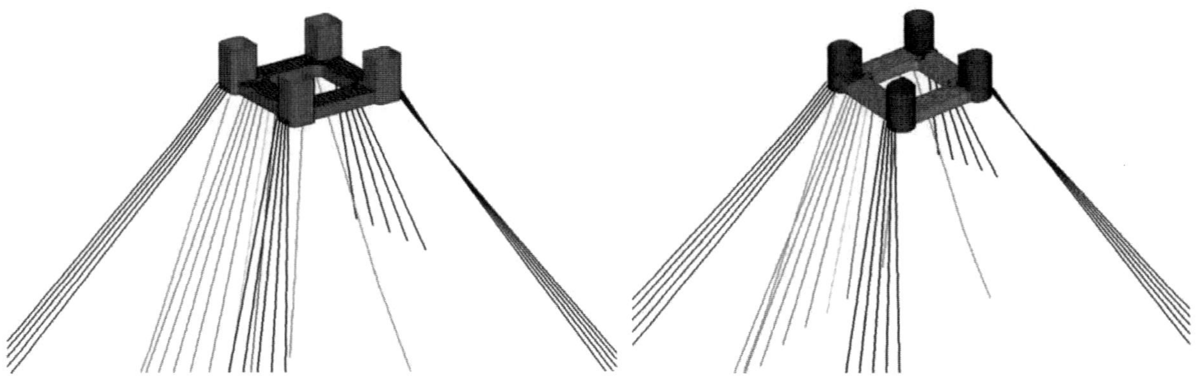

Figure 8: WINPOST Time Domain Coupled Model – Hull with Square-Orthogonal Columns (left) and Tear-drop 90° Vertex (Right)

The wave enhancement for configuration a), c) and d) relative to the wave enhancement for configuration e) in the critical wave heading of 135° at down-wave column are compared to the frequency domain wave enhancement results calibrated with model test and presented earlier in Figure 7. The comparison is shown in Figure 9. The figure shows while the magnitude of the relative wave enhancement is not exactly matching between the two approaches, the results do follow the same trend. Relative wave enhancement in 135° heading at other locations including the center of the well-bay (Point 1) and the up-wave points (2a and 2b) are derived and presented in Figure 10. The Tear-Drop-45°-Vertex configuration proves to be the best among the tested configurations, especially at down-wave column and up-wave column locations with reduction as high as 29% at the up-wave column over the configuration with square-orthogonal columns.

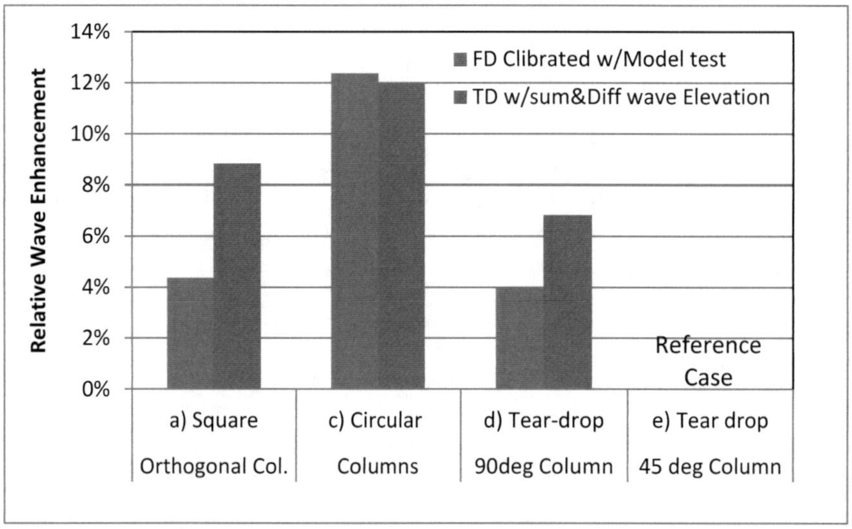

Figure 9: Relative Wave Enhancement at Down-wave Column Calibrated Frequency Domain versus Time domain with Sum and Difference Frequency Wave Elevation in 135° Wave Heading

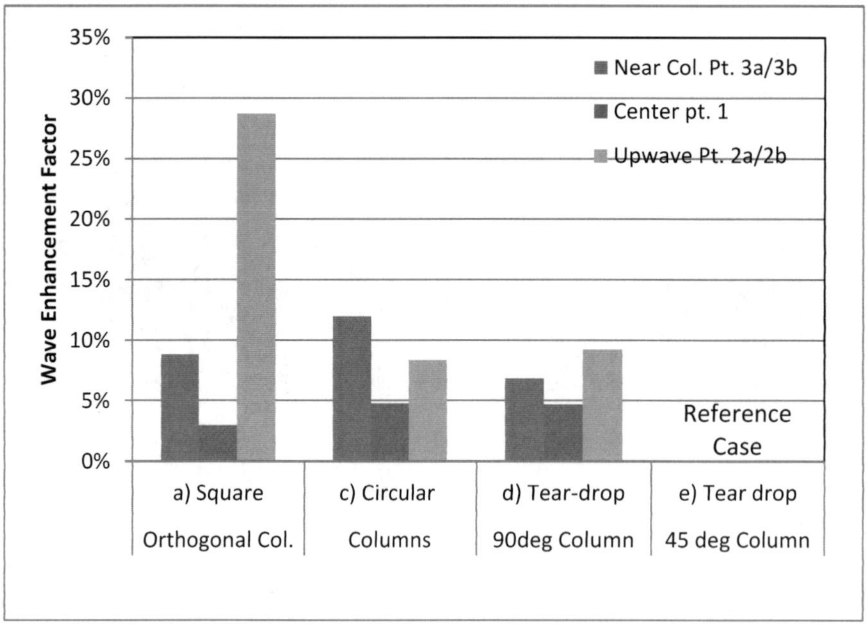

Figure 10: Relative Wave Enhancement at Critical Locations for 135° Wave Heading

TEAR-DROP COLUMN CONSTRUCTABILITY

The hybrid column design is a combination of flat and cylindrical surfaces. Both have been extensively used and fabricated in many past column stabilized floaters. The key fabrication concerns will be related to the access for welders and inspectors. For these reasons, it is recommended that the Tear-Drop column vertex angle to be determined to facilitate the welding and inspection process during fabrication. Providing a rounded corner at the vertex with a minimum radius of 2.0 m is recommended for that reason. The intersection of the column with the pontoon can be facilitated by adopting offset pontoons as presented in this paper. No significant fabrication cost delta compared to circular or square columns are anticipated for adopting the tear-drop column shape.

CONCLUSIONS

The paper investigated the effect of the column shape, orientation and spacing on the wave enhancement inside the well-bay area of typical column stabilized floaters with four columns and ring pontoons. A relationship between wave enhancement and transparency ratio T_r (column spacing relative to equivalent column diameter) has been derived using frequency domain diffraction analysis and the results are calibrated with available model testing data. It has been found that the wave enhancement factors are inversely proportional to the transparency ratio (T_r) for both circular and square column configurations. The enhancement factors are significantly higher for T_r values less than 2. Therefore, in future projects, optimization study on wave enhancement factor should be performed with a target value of T_r greater than 2.0.

The wave enhancement in 135° heading is found to be more than that in 90° heading with circular column configurations while the wave enhancement is similar in 135° and 90° headings for square columns. The circular columns shows more pronounced enhancement compared to square columns configurations in 135° heading. The wave enhancement at the center of the well-bay area is found to be small and generally similar between square and circular column configurations.

A new column design with a Tear-Drop cross section is introduced and the wave enhancement from this new configuration is compared to that from the orthogonal square columns, the diagonal square columns and the circular columns. The effect of the Tear-Drop column vertex angle on the wave enhancement is also investigated. The new design with 45 deg. vertex angle is shown to significantly reduce the wave enhancement in the well-bay area compared to other configurations. Reducing the deck elevation has significant benefits such as increased platform stability, reduced wind loading, lower mooring loads, lower steel weight and reduced cost.

Fully coupled time-domain analysis with sum and difference frequency wave elevation QTFs has been considered and further confirmed the advantages of the Tear-Drop configuration over the alternative designs. The Tear-Drop configuration also significantly improved wave enhancement at the up-wave columns which will reduce wave column overtopping.

Fabrication considerations are discussed and recommendations are made to facilitate the fabrication of the new column design.

ACKNOWLEDGEMENTS

The authors would like to express their gratitude to INTECSEA WorleyParsons for sponsoring the preparation of this manuscript. Thanks are also due to Chunfa Wu and Bill Greiner with INTECSEA for their careful review and valuable comments on this manuscript.

REFERENCES

American Petroleum Institute, "Recommended Practice for Planning, Designing and Constructing Tension Leg Platforms", API RP 2T, 2010.

Baarholm RJ, Faltinsen OM, Herfjord K, "Water impact on decks of floating platforms", Proceedings of the Eighth International Symposium on Practical Design of Ships and other Floating Structures, Shanghai, China, 2001.

Lee C-H, Newman JN, "Second-order wave effects on offshore structures", Proceedings of the Seventh International Conference on Behavior of Offshore structures, 1994.

Manuel, L. "Air gap assessment for floating structures in random seas", Department of Civil Engineering University of Texas at Austin, Austin, TX 78712

Niedzwecki JN, Duggal AS, "Wave runup and forces on cylinders in regular and random waves", Journal of Waterways, Port, Coastal, Coastal Engineering, ASCE 1992;118:615–34.

WAMIT "User Manual Version 6.3", WAMIT Inc.

WINPOST "User Manual"

COMPARATIVE STUDY OF A PAIRED-COLUMN SEMI-SUBMERSIBLE AND A TRUSS SPAR WITH DIRECT VERTICAL ACCESS FEATURE IN ULTRA-DEEP WATER OF WEST GULF OF MEXICO

BRAJESH KUMAR
SENIOR NAVAL ARCHITECT, HOUSTON OFFSHORE ENGINEERING, AN ATKINS COMPANY

JUN ZOU
DIRECTOR OF NAVAL ARCHITECTURE, HOUSTON OFFSHORE ENGINEERING, AN ATKINS COMPANY

Proceedings of the 21st Offshore Symposium, February 2016, Houston, Texas
Texas Section of the Society of Naval Architects and Marine Engineers

ABSTRACT

A truss Spar with direct vertical access merit has been designed and installed in ultra-deep water of Gulf of Mexico for a few years. This paper intends to carry out a comparative study of a Paired-Column Semi-submersible (PC-Semi) platform with the same topsides payloads, TTRs tensions, Ram-style tensioners with 28-ft stroke range and same SCRs tensions in the same metocean criteria with same global performance and mooring design criteria.

PC-Semi hull form development and evolution for ultra-deep water dry tree applications in central Gulf of Mexico have been undertaken for a few years through RPSEA ultra-deep water program. Extensive and comprehensive model testing including wind tunnel tests, wave basin tests and VIM tests has been completed successfully. Feasibility and advantages of PC-Semi for dry tree application with large topsides payloads in central GoM have been published and established for years. PC-Semi platform Approval in Principal has been awarded by DNV.

Application of PC-Semi concept as an alternative to Spar for West GoM with fewer top tension riser counts and less topsides payloads than those in central GoM was carried out in this study to identify advantages of hull form and scalability of PC-Semi concept. PC-Semi hull and mooring configurations are sized and validated through comprehensive global performance analysis consisting of offsets, rotations, vertical motions, minimum airgaps, maximum accelerations and TTR strokes and mooring analysis. After validation analysis, key figures of hull and mooring configurations along with the detailed global performance results of PC-Semi are summarized and discussed. Features of pre-service conditions, such as feasibility of vertical wet tow and quayside integration, of both hull forms are also addressed and highlighted.

Keywords: Dry Tree Semi-submersible, West GoM, Alternative Solution, Direct Vertical Access, Ram-Style Tensioner, Perdido Truss Spar

INTRODUCTION

As of today, the proven hull form with dry tree or direct vertical access feature have been a Spar for water depth more than 6,000 ft. The semi-submersible hull form is known to offer larger deck space and quayside integration features. For more than a decade, development of dry tree semi-submersibles as an alternative hull form is very active. In general, two distinct approaches are summarized as follows:

1) To employ a traditional hull form to adopt an innovative riser design, such as, CVAL (Compliant Vertical Assess Risers) by Mungall et al (2004) and non-conventional production riser stroke range of 11 m (35 ft) or longer. The later approach was considered by William and Leverette (2009), William et al (2010) and Wang (2012).
2) To utilize the proven Ram style tensioner with a regular stroke range and invent a more appropriate vertical motion hull form, such as Technip EDS (Sablok et al, 2011), Floatec ESemi (Murray et al, 2008) and HOE PC-Semi (Zou et al, 2008).

The Paired-column semi-submersible (PC-Semi) hull form was first introduced by Zou (2008) and it was selected by RPSEA for dry tree development in ultra-deepwater in 2009. Since then, PC-Semi developments, maturations and validations, such as wind tunnel tests, wave basin model tests and VIM model tests, have been continued for a few years. In 2014, PC-Semi approval in principle was awarded by DNV GL.

For the Perdido development, a Spar was selected by Shell during concept selection in 2005 and the installation was completed in 2008 after Technip USA was awarded the EPC contract in 2006. Herema Marine Contractors (HMC) was selected for the transportation and installation (except for the dry tow from Finland). Perdido, scheduled for first oil in 2010, is Shell's first Spar production platform.

Shell's Perdido truss Spar is the world's deepest offshore oil drilling and production platform which is located in 8,000 ft of water in Alaminos Canyon block 857, West of GoM. Perdido is the world's deepest direct vertical access spar with peak production of more than 100,000 boe/d. Direct vertical access (DVA) reduces drilling costs, simplifies workovers and facilitates access to subsea equipment.

In the past few years, dry tree semi-submersible developments focused on large topsides payload (42,500 st), the worst metocean criteria, such as central GoM and large TTRs count; details are given in RPSEA study report (2009). In this paper, a case with medium topsides payload, the less challenging metocean criteria, such as West of GoM and less TTRs count, same as those of Perdido, was selected and studied to explore feasibility and scalability of PC-Semi.

As illustrated in Figure 1, Perdido spar has a draft of 505 ft and three-level deck while the proposed PC-Semi has a draft of 140 ft and two-level deck. In order to have a better understanding of the PC-Semi hull form features, a brief introduction on development philosophies is presented as below:

- Maintaining good features of a typical deep draft Semi, such as quayside integration and operator friendly large deck area
- Conventional hull/deck structure and well bays
- Conventional Spar Ram-Style tensioner (off-the-shelf) as in Perdido Spar
- No jack-up & no moving parts
- No solid ballast, no horizontal wet tow and no upending for installation

PC-Semi has the unique feature to maintain quayside integration as well as in-place motion requirements by decoupling dependencies of a) deeper draft (for better heave motion for DVA); b) wider column spacing (for meeting stability requirements for pre-service and in-place conditions); c) best supporting position for deck structure (for optimizing deck steel); d) design cycles of topsides and hull (once the inner column spacing is determined, topsides layout can proceed independently). The methodology and tools used for the analysis are detailed in Das and Zou (2015).

A PC-Semi with the same topsides payload, top tension risers count and Ram-style tensioner as in Perdido Spar was sized for this comparative study. PC-Semi hull and mooring configurations are analysed and validated through comprehensive global performance analysis consisting of offsets, rotations, vertical motions, minimum airgaps, maximum accelerations and TTR strokes and mooring analysis. Key figures of hull and mooring configurations along with the detailed global performance results of PC-Semi are summarized, compared and discussed. Features of pre-service conditions, such as feasibility of vertical wet tow and quayside integration, are also addressed and highlighted.

The outline of the paper is as follows. First, descriptions of the study case including hull, TTRs, SCRs and moorings are highlighted. Second, the numerical results of key design parameters are summarized. Third, discussions on both hull features are presented. Finally, main conclusions are drawn.

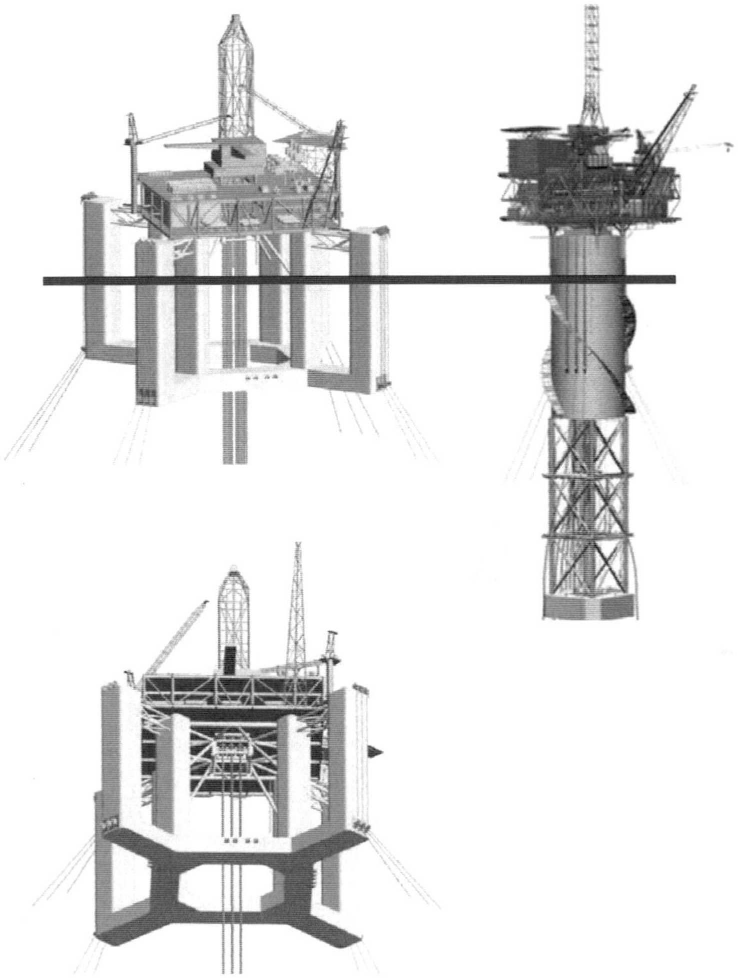

Figure 1 A truss Spar and a PC Semi Hull Representation

STUDY CASE

Study case has been formed for PC-Semi based on Perdido Spar data collected from public domain on topsides payloads, metocean criteria, riser tensioners, load cases/design criteria and moorings/foundations. A brief summary is given in section for conveniences.

Topsides Payloads

Based on 2012 worldwide survey of Spar, DDCV and Madoc Vessels by Offshore Magazine, [October, 2012], topsides dry and operating weights of Perdido are summarized in Table 1. Topsides dry and operating weight in Table 1 include deck structure steel.

Topsides Payload	Perdido
Facility Dry Weight (st)	12401
Operating Weight (Facility + Rig) (st)	20117

Table 1 PC-Semi Topsides Payload

Metocean Criteria

Liapis et al (2010) summarized 10-yrp winter storm, 100-yrp and 1,000-yrp hurricanes significant wave height (Hs), peak period (Tp) and surface current velocity (Vc) shown in Table 2. Associated wind speeds and current profiles were not given in the paper. The missing data were substituted by API-RP-2MT (2014).

Environment Conditions	Design Category	Wave			Wind	Current
		Hs (ft)	Tp (s)	γ	Ws (ft/s)	Vc (ft/s)
10-yrp Winter Storm	A	20.0	10.9	2.1	63.5	1.18
100-yrp Hurricane	B	40.0	14.2	2.3	124.4	5.74
1,000-yrp Hurricane	S	53.0	16.3	2.3	155.5	7.94

Table 2 Metocean Parameters for West Gulf of Mexico

Tensioner Stroke and Load Curve

The RAM style top tensioned riser system on Spars have been described and presented by Perryman et al (2005) and Smith et al (2005). Perdido TTR tensioners in prototype have been illustrated in Figure 2 (Larsen, H.A., 2010).

Figure 2 Perdido Spar Riser Tensioner

Both Perdido Spar and PC-Semi have the same tensioner properties modelled over the entire 28 ft stroke range. The tensioner max up stroke (11.9 ft) and min down stroke (16.1 ft) from the nominal position and the resultant tensioner load / stiffness plotted in Figure 3 is referenced to Liapis et al (2010). The tensioner stroke range and nominal stiffness are summarized in Table 3.

Tensioner Characteristic Parameters		
Tensioner Stroke Range	(ft)	28
Max Up Stroke from Nominal Position	(ft)	11.9
Min Down Stroke from Nominal Position	(ft)	16.1
PR Tensioner Nominal Vertical Stiffness, each	(kips/ft)	25
DR Tensioner Nominal Vertical Stiffness	(kips/ft)	38
Total Tensioner Nominal Vertical Stiffness	(kips/ft)	163
PR Tensioner Nominal Load, each	(kips)	1,098

Table 3 Tensioner Stroke and Nominal Stiffness

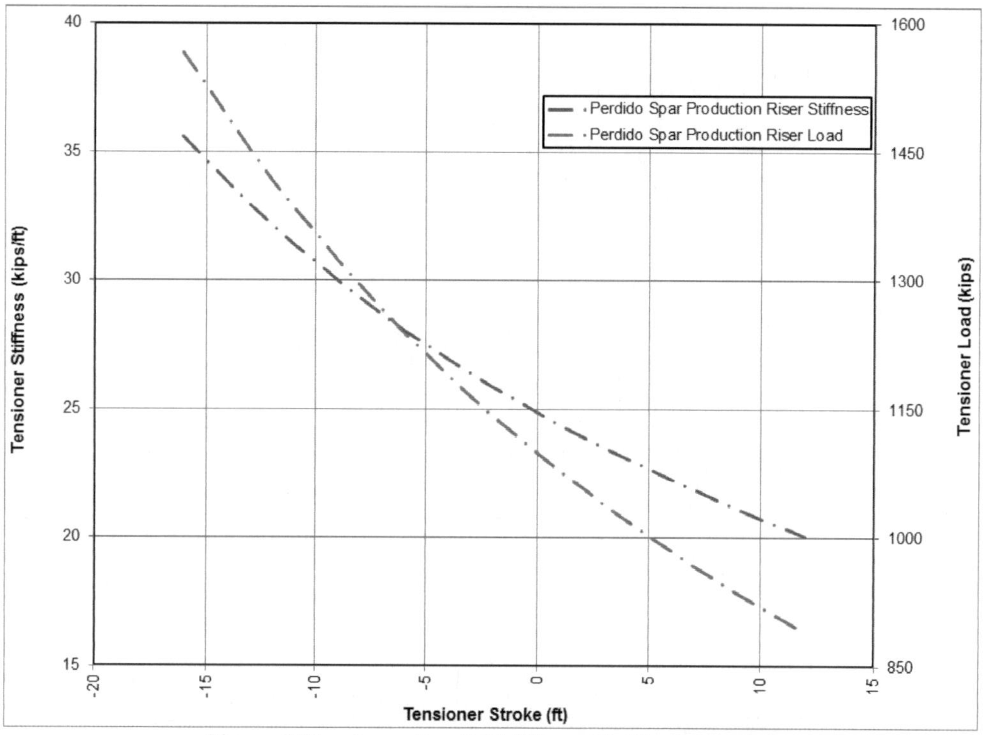

Figure 3 Tensioner Load / Stiffness Vs Stroke Range

Load Cases and Design Criteria

In this paper, load cases, as presented in Table 4, covered categories A, B and S and intact and damage conditions in various environmental conditions. Global performance (GP) criteria and TTR stroke is highlighted in Table 5. In addition, mooring line minimum factor of safety criteria based on API and ABS guidelines are summarized in Table 6.

Design Condition	Design Category	Hull/Mooring Condition	Sea State
Operating	A	Intact	10-yrp Winter Storm
Design Extreme	B	Intact	100-yrp Hurricane Wave
		One Mooring Line Damage	10-yrp Winter Storm
		Two Adjacent Tanks Damage	10-yrp Winter Storm
Survival	S	Intact	1,000-yrp Hurricane Wave
		One Mooring Line Damage	100-yrp Hurricane Wave

Table 4 Design Environmental Conditions for PC-Semi

Characteristic GP Response Description	GP Design Criteria
Maximum Horizontal Offset	< 7% WD
100-yrp Min Airgap at Lowest Deck Bottom	5 ft
1,000-yrp Min Airgap at Lowest Deck Bottom	0 ft
Max Tensioner Stroke in Survival Condition	≤ 28 ft

Table 5 Global Performance (GP) Design Criteria

Design Cases	Intact		One-line Damage	
	Chain	Poly Rope	Chain	Poly Rope
100-yrp Extreme Events	1.67 (60% of MBL)	1.82 (55% of MBL)	1.25 (80% of MBL)	1.43 (70% of MBL)
1,000-yrp Survival Events	1.00 (100% of MBL)	1.11 (90% of MBL)	NA	NA

Table 6 Mooring Tension Design Criteria

PC-SEMI CONFIGURATION

Based on specified topsides payloads, metocean criteria, global performance design criteria, TTR stroke requirement and mooring strength and fatigue requirements, PC-Semi hull has been sized and iterated until all design targets having been met and a robust design is reached. The hull configuration presented in this paper is the final after all the validation work.

Hull Configuration

Hull configuration principal dimensions as well as weights and loads are summarized in Table 7 and hull configuration was illustrated in Figure 4.

Design Case		5PRs+1DR+SCRs	0PR+0DR+0SCR
Draft	(ft)	140.0	
Displacement	(st)	74,745	
Water Depth	(ft)	8,000	
Number of Inner Columns		4	
Inner Column Height	(ft)	175.0	
Inner Column Freeboard Height	(ft)	59.0	
Inner Column Side Length	(ft)	40.0 x 32.0	
Inner Column c/c Span	(ft)	150.0	
Number of Main Pontoons		4	
Main Pontoon Width	(ft)	44.0	
Main Pontoon Height	(ft)	24.0	
Number of Outer Columns		4	
Outer Column Height	(ft)	181.0	
Outer Column Side Length	(ft)	40.0 x 42.0	
Number of Connecting Pontoons		4	
Connecting Pontoon Width	(ft)	40.0	
Connecting Pontoon Height	(ft)	24.0	
Distance between Inner/Outer Columns (Length of Connecting Pontoons)	(ft)	55.0	
Deck Dimensions (2 Level Deck)	(ft)	212.0 (N-S) x 205.0 (E-W)	
Deck Height (BOS to TOS)	(ft)	40.0	
Deck Post above Inner Column	(ft)	3.0	
Deck Still Water Clearance from MWL to Bottom of Lower Deck	(ft)	62.0	
Total weight	(st)	67,758	72,708
Vertical C.G. from Base (Mass Only)	(ft)	97.9	93.3
Total TTRs+DR Tensions	(st)	3,150	0
TTR Tensioner Vertical Stiffness at Nominal Position, each	(kips/ft)	25	0
DR Tensioner Vertical Stiffness at Nominal Position	(kips/ft)	38	0
Total Tensioner Vertical Stiffness at Nominal Position	(kips/ft)	163	0
SCRs + Umbilicals Vertical Tension	(st)	1,800	0
GMtc (Free Surface Correction, Mass VCG)	(ft)	20.47	25.12
Mooring Vertical Load	(st)	2,037	

Table 7 PC-Semi Hull Key Particulars with/without TTRs+SCRs

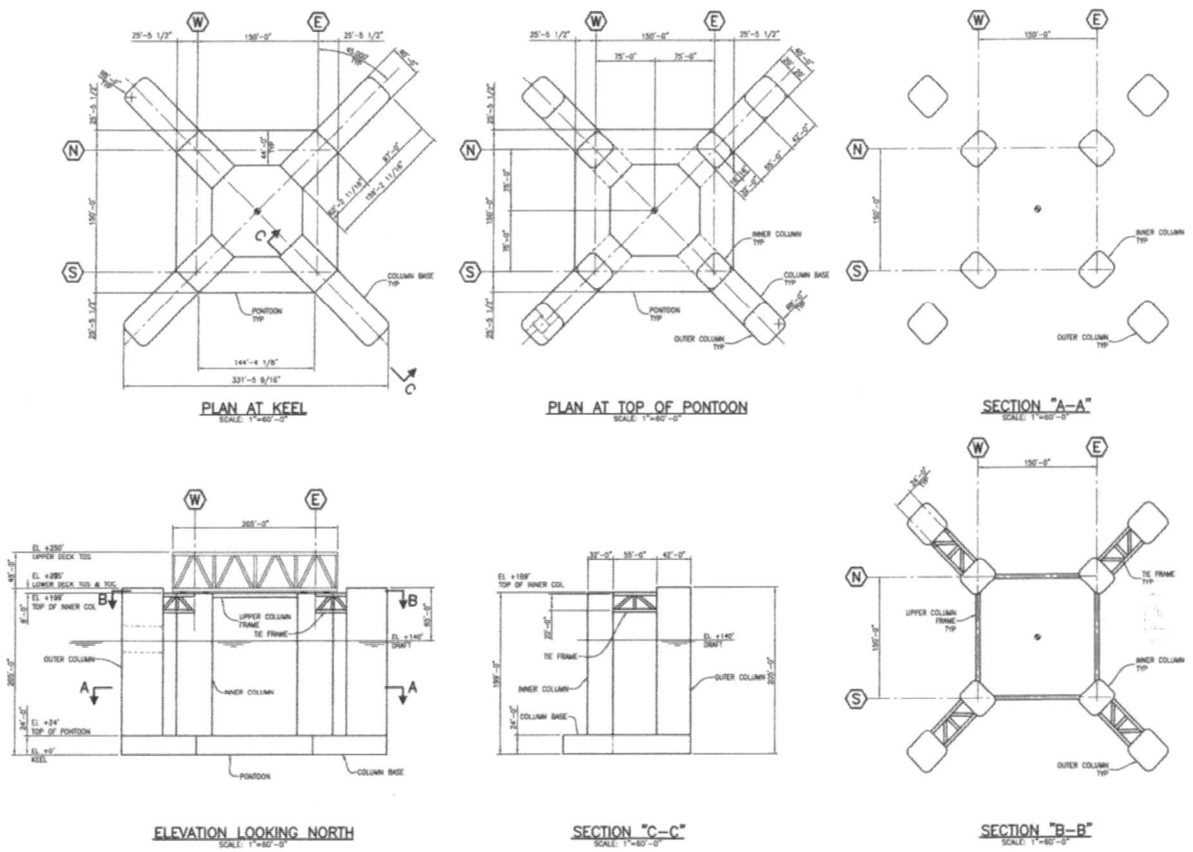

Figure 4 PC-Semi Hull Configuration

Mooring Configuration

Lohr and Smith (2010) highlighted Perdido Spar mooring system as well as foundation piles. PC-Semi adopted taut chain-polyester-chain mooring system. Except PC-Semi has 8 mooring lines in a 4x2 pattern as shown in Table 8 whereas Perdido Spar comprises 9 mooring lines in a 3x3 pattern.

Table **8** summarizes PC-Semi mooring azimuths. Table 9 highlights the mooring size comparison for the PC-Semi and Perdido Spar.

In order to ensure a robust mooring design, steel unbalanced catenary risers (SCRs) horizontal load is estimated to be 400 kips for PC-Semi and mooring pretensions have been accordingly adjusted.
3/8" corrosion allowance for top/bottom chain has been accounted for the calculation. Flat sea bed at 8,000 ft water depth is assumed at the mooring installation site. Polyester rope static and dynamic mooring stiffness are assumed to be 15xMBL and 30xMBL respectively.

Group No.	Fairlead Location	Line No.	Azimuth (deg)
1	NE	1	47.5
		2	42.5
2	SE	3	-42.5
		4	-47.5
3	SW	5	-132.5
		6	-137.5
4	NW	7	-222.5
		8	-227.5

Table 8 PC-Semi Mooring Azimuths

Mooring Components	Description	Uncorroded Diameter (in)		Length (ft)	Wet Wt. (lbs/ft)	MBL (kips)	EA (kips)
		Perdido Spar	PC-Semi	PC-Semi	PC-Semi	PC-Semi	PC-Semi
Platform Chain	Studless R5	*5.280*	5.625	450	239.25	4094	330750
Polyester Rope	ScanRope	*9.680*	10.000	11,800	7.30	4200	Static: 63000 Dynamic: 126000
Ground Chain	Studless R5	*5.280*	5.625	300	239.25	4094	330750

Table 9 Perdido Spar Vs PC-Semi Mooring Size and PC-Semi Mooring Properties

NUMERICAL RESULTS

In this section, key numerical results including motion RAOs, wave upwelling RAOs; wind force and moment coefficients; extreme offsets, extreme downward motions, extreme rotations, extreme accelerations, min airgaps, TTRs strokes and mooring extreme tensions and min factor of safety are presented and highlighted.

RAOs (Surge, Heave, Pitch and Wave Upwelling)

PC-Semi surge, heave and pitch motion RAOs with and without risers are shown in Figure 5 to Figure 7. PC-Semi (with TTRs+SCRs) vertical relative motion RAO at the most critical airgap location is shown in Figure 8. As the heave natural period is within the wave frequency range, both heave and vertical relative motion RAOs are wave frequency and hence (wave) sea state dependent and correspond to 100-yrp hurricane wave in the figures below.

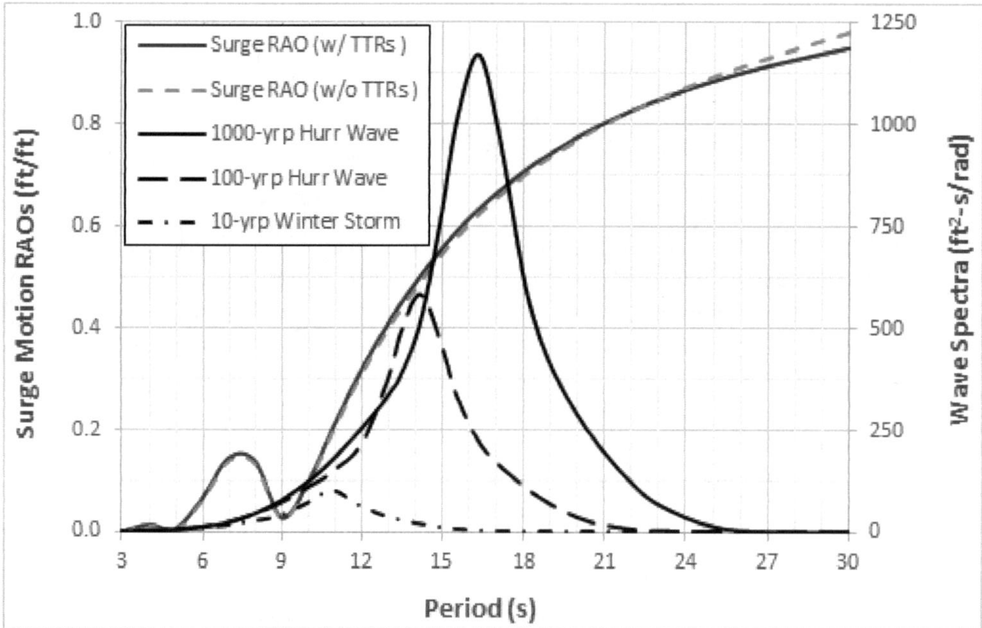

Figure 5 PC-Semi: CG Surge Motion RAOs - 180deg

WAMIT (potential theory) does not consider viscous effect resulting in deep cancellation which is negated in model test results of Semi heave motion RAO in extreme sea states. In Figure 6, it has been observed that no deep cancellation was found since viscous effects were considered and employed to represent physical phenomena. The methodology of how to model viscous effects in a consistent manner has been given in Das and Zou (2015). In addition, it can be seen that TTRs riser counts (TTR tensioner stiffness) play a crucial role on heave motion response characteristics.

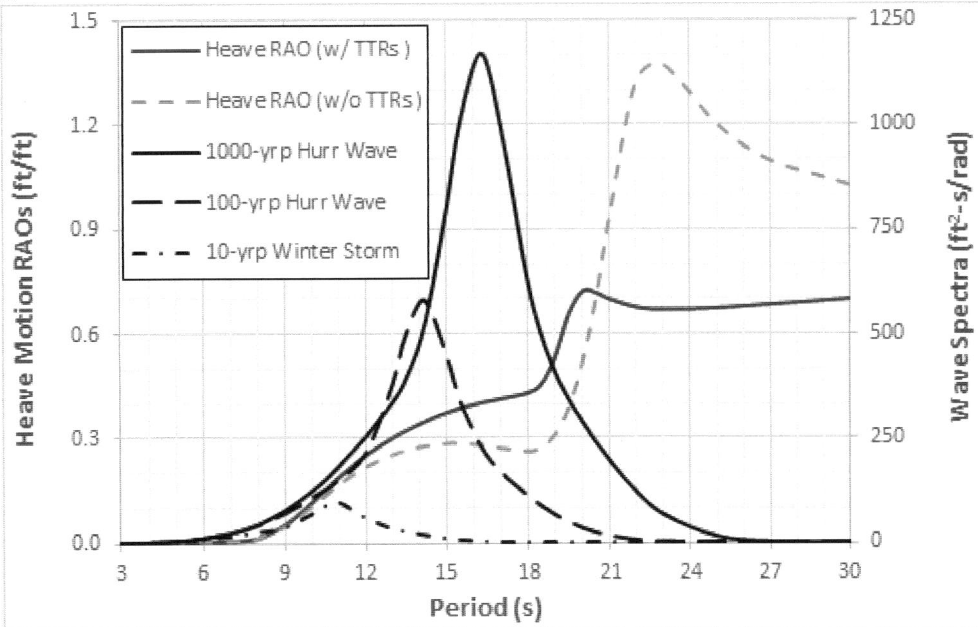

Figure 6 PC-Semi: CG Heave Motion RAOs - 180deg

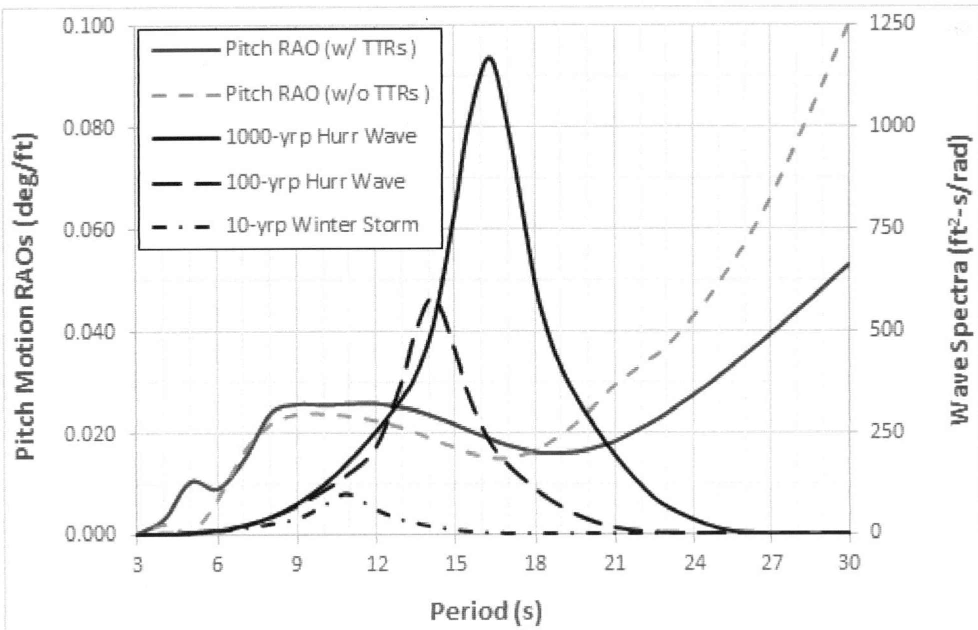

Figure 7 PC-Semi: CG Pitch Motion RAOs - 180deg

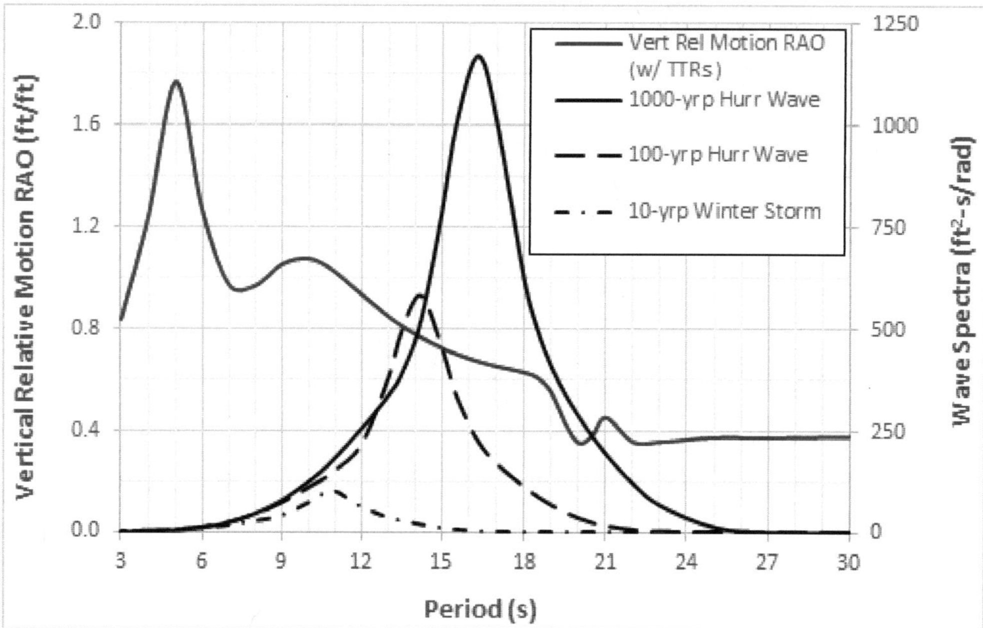

Figure 8 PC-Semi (with TTRs): Wave Upwelling RAO near Inner Down Wave Column - 180deg

Wind Loads

Perdido Spar 3-level deck illustrated in the supplement to OIL&GAS Journal (May, 2010) was the basis for PC-SEMI 2-level deck projected area for wind force and moment coefficient calculation plotted in

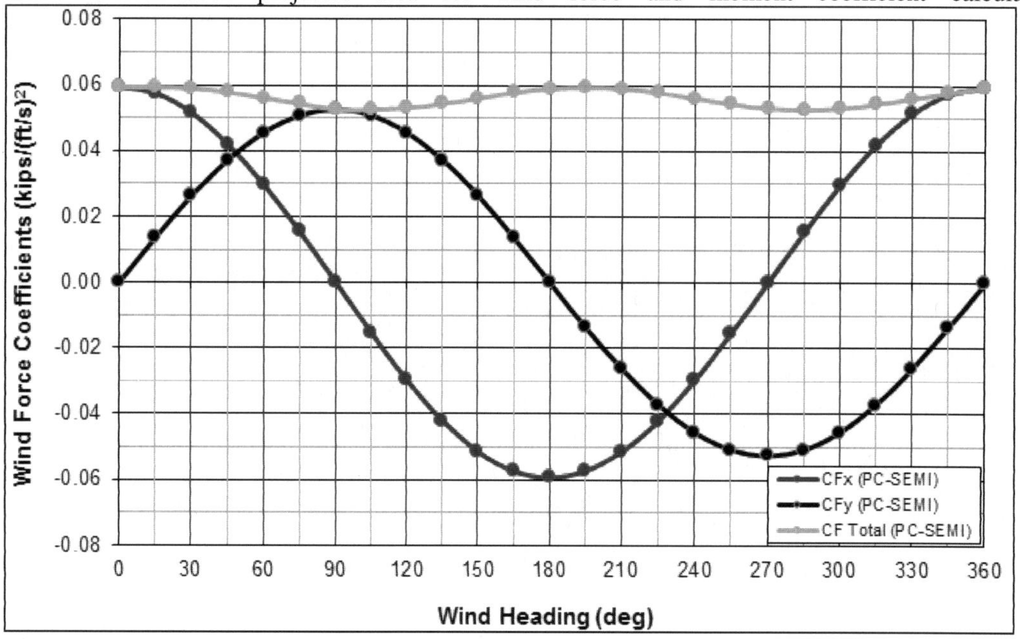

Figure 9 and Figure 10 respectively.

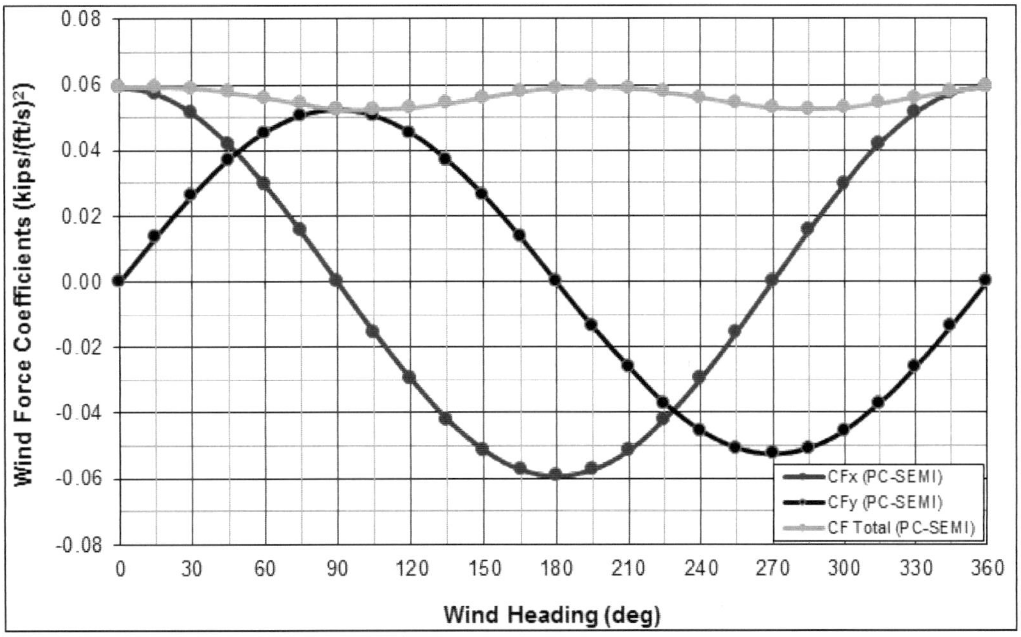

Figure 9 PC-Semi: Wind Force Coefficients

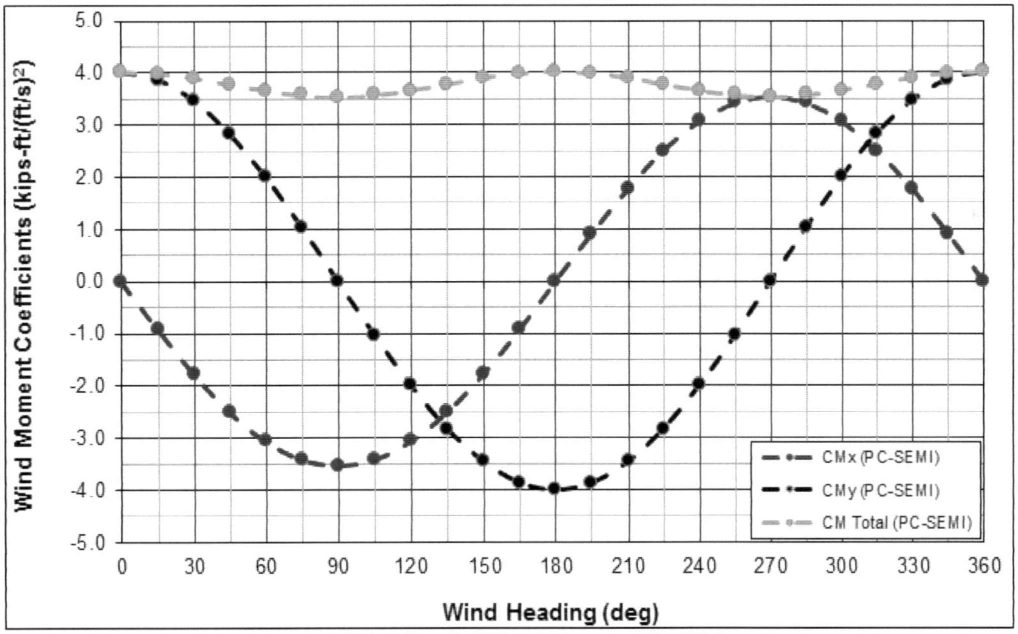

Figure 10 PC-Semi: Wind Moment Coefficients

Extreme Offsets

Table 10 summarizes PC-Semi extreme offsets for 10-yrp winter storm and 100-yrp hurricane wave intact and damage conditions and 1,000-yrp hurricane wave intact condition. The summary table illustrates that there is more than 1%WD (percentage of water depth) margin from the design criteria for all design categories in west GoM. The design category details are summarized in Table 4.

Design Category	Environment	Hull/Mooring	Offset (ft)	%WD	Criteria (%WD)	Satisfy (Y/N)
A	10-yrp Winter	Intact	81.3	1.02%	< 3%	(Y)
B	100-yrp Hurr	Intact	262.5	3.28%	< 5%	(Y)
B	10-yrp Winter	Up-wave Mooring Damaged	188.4	2.36%		(Y)
S	100-yrp Hurr	Up-wave Mooring Damaged	475.2	5.94%	< 7%	(Y)
S	1000-yrp Hurr	Intact	455.6	5.70%		(Y)

Table 10 PC-Semi (with TTRs+SCRs) Extreme Offsets

Extreme Vertical Motions

Table 11 summarizes PC-Semi extreme heaves for 10-yrp winter storm and 100-yrp hurricane wave intact and damage conditions and 1,000-yrp hurricane wave intact condition. The two damaged tanks locations and elevation in the collision zone on the outer column are shown with dotted lines in Figure 4. The minimum airgap and TTR tensioner strokes are proportional to the extreme heave motions and the GP analysis shows these heave motions are appropriate for minimum airgap and 28 ft stroke range.

Design Category	Environment	Hull/Mooring	Heave (ft)
A	10-yrp Winter	Intact	-3.14
B	100-yrp Hurr	Intact	-11.97
B	10-yrp Winter	2 Down Wave Tank Damage	-4.48
S	100-yrp Hurr	Up-wave Mooring Damaged	-12.57
S	1000-yrp Hurr	Intact	-17.36

Table 11 PC-Semi (with TTRs+SCRs) Extreme Heave

Extreme Rotations

Table 12 summarizes PC-Semi extreme rotations for 10-yrp winter storm and 100-yrp hurricane wave intact and damage conditions and 1,000-yrp hurricane wave intact condition. Vertical motions are induced by PC-Semi extreme rotations which are much less than those of a typical (~8 and ~12 deg for category B and S respectively) west GoM Spar.

Design Category	Environment	Hull/Mooring	Rotation (deg)
A	10-yrp Winter	Intact	2.07
B	100-yrp Hurr	Intact	5.44
B	10-yrp Winter	2 Down Wave Tank Damage	4.82
S	100-yrp Hurr	Up-wave Mooring Damaged	6.21
S	1000-yrp Hurr	Intact	6.93

Table 12 PC-Semi (with TTRs+SCRs) Extreme Rotations

Extreme Accelerations

Table 13 and Table 14 summarizes PC-Semi extreme horizontal and vertical accelerations respectively at the top of the deck for 10-yrp winter storm and 100-yrp hurricane wave intact and damage conditions and 1,000-yrp hurricane wave intact condition. Extreme deck accelerations are crucial for machineries as well as for operators. PC-Semi extreme horizontal accelerations are much less than those of a typical (~0.3g and ~0.4g for category B and S respectively) West GoM Spar.

Design Category	Environment	Hull/Mooring	Horz. Accel. at Decktop CG (g)
A	10-yrp Winter	Intact	0.071
B	100-yrp Hurr	Intact	0.184
B	10-yrp Winter	2 Down Wave Tank Damage	0.114
S	100-yrp Hurr	Up-wave Mooring Damaged	0.170
S	1000-yrp Hurr	Intact	0.206

Table 13 PC-Semi Extreme (with TTRs+SCRs) Horizontal Accelerations

Design Category	Environment	Hull/Mooring	Vert. Accel. at Decktop CG (g)
A	10-yrp Winter	Intact	0.029
B	100-yrp Hurr	Intact	0.070
B	10-yrp Winter	2 Down Wave Tank Damage	0.029
S	100-yrp Hurr	Up-wave Mooring Damaged	0.070
S	1000-yrp Hurr	Intact	0.083

Table 14 PC-Semi Extreme (with TTRs+SCRs) Vertical Accelerations

Extreme Strokes

Table 15 summarizes PC-Semi extreme tensioner strokes for 10-yrp winter storm and 100-yrp hurricane wave intact and damage conditions and 1,000-yrp hurricane wave intact condition. The maximum up stroke, minimum down stroke and total stroke range is within the Perdido Spar tensioner design limit summarized in Table 3.

Design Category	Environment	Hull/Mooring	Stroke Description	Tensioner Stroke (ft)
A	10-yrp Winter	Intact	Max. Up Stroke	2.71
			Min. Dn Stroke	-3.08
			Total Stroke	*5.79*
B	100-yrp Hurr	Intact	Max. Up Stroke	6.55
			Min. Dn Stroke	-13.50
			Total Stroke	*20.05*
B	10-yrp Winter	2 Down Wave Tank Damage	Max. Up Stroke	5.33
			Min. Dn Stroke	-2.92
			Total Stroke	*8.25*
S	100-yrp Hurr	Up-wave Mooring Damaged	Max. Up Stroke	4.04
			Min. Dn Stroke	-17.17
			Total Stroke	*21.21*
S	1000-yrp Hurr	Intact	Max. Up Stroke	8.39
			Min. Dn Stroke	-18.24
			Total Stroke	*26.63*

Table 15 PC-Semi (with TTRs+SCRs) Extreme Tensioner Strokes

Min Airgap

Table 16 summarizes PC-Semi minimum airgap for 100-yrp and 1,000-yrp hurricane wave intact conditions. The minimum airgap is calculated at the most critical location at the bottom of lower deck steel which shall have no water impact in survival condition.

Design Category	Environment	Hull/Mooring	Min. Airgap (ft)	Criteria (ft)	Satisfy (Y/N)
B	100-yrp Hurr	Intact	18.4	5.0	Y
S	1000-yrp Hurr	Intact	2.1	0.0	Y

Table 16 PC-Semi (with TTRs+SCRs) Minimum Airgap

Extreme Mooring Tensions and Min Factor of Safety

Table 17 summarizes PC-Semi maximum tensions and factor of safety for all design categories. The mooring line tensions also include line dynamics. Line tension at the top of the platform chain, polyester rope and bottom chain satisfy the mooring tension design criteria summarized in Table 6 and there is no grounding of the polyester rope in any load condition.

Design Category	Environment	Mooring	Top Chain Tension (1) (kips)	Platform Chain (1)	Polyester Rope (1)	Bottom Chain (1)
A	10-yrp Winter	Intact	Max Tens (kips)	1107.8	1027.5	969.0
			Safety Factor	3.70	4.09	4.23
			≥ Criteria (Y/N)	Y	Y	Y
B	100-yrp Hurr	Intact	Max Tens (kips)	2082.2	2006.2	1948.3
			Safety Factor	1.97	2.09	2.10
			≥ Criteria (Y/N)	Y	Y	Y
B	10-yrp Winter	Up-wave Mooring Damaged	Max Tens (kips)	1754.6	1677.5	1619.4
			Safety Factor	2.33	2.50	2.53
			≥ Criteria (Y/N)	Y	Y	Y
S	100-yrp Hurr	Up-wave Mooring Damaged	Max Tens (kips)	3187.7	3114.0	3055.1
			Safety Factor	1.28	1.35	1.34
			≥ Criteria (Y/N)	Y	Y	Y
S	1000-yrp Hurr	Intact	Max Tens (kips)	3014.6	2940.6	2883.5
			Safety Factor	1.36	1.43	1.42
			≥ Criteria (Y/N)	Y	Y	Y

(1) Mooring line tension includes tensions due to line dynamics.

Table 17 PC-Semi (with TTRs+SCRs) Extreme Mooring Tensions and Factor of Safety

DISCUSSIONS

After validation analysis, the feasible hull and mooring configuration have been developed. As an alternative solution to Perdido Spar, a few key observations are summarized and discussed first. In the past 7 years, the major developments and maturations of PC-Semi are highly focused on central GoM which requires configuration with large topsides payload and large TTR riser counts. However, the objective of this study is to design a platform with medium topsides payload, less TTR counts and less harsh metocean criteria. Scalability of PC-Semi is also discussed and highlighted.

Scalability of PC-Semi

As documented in RPSEA study report (2010) and Das and Zou (2015), key figures of PC-Semi at central GoM and west GoM are summarized and presented in Table 18.

	PC semi	
Items	Central GoM	West GoM
Water depth (ft)	8,000	8,000
TTR riser counts	12	6
Topsides payload (st)	42,500	24,000
Platform draft (ft)	175	140
Column freeboard (ft)	75	62
Total column height (ft)	250	202
Number of moorings	16	8
Displacement (st)	108,000	74,744

Table 18 Key Figures of PC Semi at Central GoM and West GoM

In Table 18, Topsides payload, which includes TTRs tensions and SCRs/Umbilicals vertical tensions, is different from definition of topsides payload given in Table 1 for comparison of topsides payload defined in central GoM. From Table 18, 175 ft PC-Semi draft ensures suitable heave motions for 28 ft TTR tensioner stroke at central GoM while 140 ft PC-Semi draft provides adequate heave motions for 28 ft TTR tensioner stroke at west GoM. In addition, 75 ft PC-Semi column freeboard satisfies min airgap criteria at central GoM while 62 ft PC-Semi column freeboard is needed at West GoM. PC-Semi column height at central GoM is 48 ft more which poses more challenges on construction, integration, transportation and installation than that of PC Semi at West of GoM.

In addition, topsides payload of this study is about 56% of the RPSEA study payload and this hull requires 8 moorings for station-keeping. For all key figures shown in Table 18, it clearly indicates that PC-Semi hull has a good scalability feature which is suitable for a wide range of topsides payloads in distinct metocean conditions while maintaining high structural and mooring efficiencies.

Perdido Spar and Its Alternative PC Semi

Topsides payload, hull steel weight, platform draft, displacement, number of moorings, line sizes, solid ballast and transportation/installation are compared and summarized in Table 19.

Items	Perdido Spar	PC Semi
Topsides payload (st)	20,117	20,117
Hull steel (st)	22,678	19,400
Platform draft (ft)	505	140
Solid ballast	10,300	NA
Displacement (st)	93,000	74,744
Number of moorings	9	8
Chain size (in)	5.280	5.625
Polyester size (in)	9.680	10.000
Quayside integration	No	Yes
Wet tow	Horizontal	Vertical
Offshore deck/hull mating	Yes	No

Table 19 Key Features of Perdido Spar and PC Semi

Table 19 clearly illustrates that a Spar has very deep draft (505 ft) with solid ballast and requiring horizontal tow and hull upending and offshore deck and hull mating while PC-Semi only carries water ballast with relatively shallow draft (140 ft) with quayside integration capability, vertical tow and one less mooring line. With same topsides payloads, TTRs tensions and TTR tensioners, hull steel is about 3,000 st less. Thus, it is believed that PC Semi is an economic alternative to a truss spar in ultra-deep water GoM.

CONCLUSIONS

A comparative study of PC-Semi and Perdido Spar with the same topsides payload, TTR tensions and tensioners as well as similar SCRs and Umbilicals horizontal and vertical loads has been undertaken and detailed study results have been presented and discussed. From validation analysis, a few key conclusions about the proposed PC-Semi hull are drawn as follows:

- Suitable for supporting Perdido Spar TTR tensioner
- Adequate airgap in 100-yrp and 1,000-yrp hurricanes
- Needs only 8 chain-polyester-chain mooring system to have adequate station-keeping capability in 100-yr and 1,000-yr hurricanes
- Has good structural efficiency and with quayside integration feature.

Overall, the proposed alternative hull is a feasible and economical solution for future application in west GoM.

ACKNOWLEDGEMENTS

The author would like to thank the management of Houston Offshore Engineering, an Atkins Company for their support of the research on alternative dry tree solutions and for allowing the publication of the study results.

REFERENCES

[1] API RP 2MET, Derivation of Metocean Design and Operating Conditions, First Edition, , American Petroleum Institute, November, 2014.
[2] Das, S and Zou, J., "Characteristic Responses of a Dry Tree Paired-Column and Deep Draft Semi-submersible in Central Gulf of Mexico", SNAME2015, Feb. 2015.
[3] Larsen, H. A., "Intervention System for the Perdido SPAR Wet Tree Wells", SPE/ICoTA Coiled Tubing and Well Intervention Conference and Exhibition, The Woodlands, USA, March 2010.
[4] Liapis S., Bhat S., Caracostis C., Webb C., Lohr C., "Global Performance of the Perdido Spar in Waves, Wind and Current: Numerical Predictions and Comparison With Experiments", Proceedings of the ASME 2010 29th International Conference on Ocean, Offshore and Arctic Engineering OMAE2010, Shanghai, China, June 2010.
[5] Lohr C. and Smith K., "Perdido Development Project: Spar & Moorings", Offshore Technology Conference (OTC), Houston, USA, May 2010.
[6] Mungall, C., Haverity, K., Bhat, S., Andersen, D., Sarkar, I., Wu, J. and Martensson, N., "Semi-submersible Based on Dry Tree Platform with Compliant Vertical Assess Risers", Offshore Technology Conference 2004, Houston, Texas, USA. OTC 16199.
[7] Murray, J.J., Yang, C.K., Chen C.Y., and Nah, E. "Two Dry Tree semi-submersible Designs for Ultra-Deepwater Post-Katrina Gulf of Mexico," Proceedings of the 27th International Conference on Ocean, Offshore and Arctic Engineering, OMAE2008-57462.
[8] Offshore Magazine, "2012 worldwide survey of Spar, DDCV and MinDoc Vessels", October, 2012.
[9] Perryman, S., Gebara, J., and Botros, F., and Yu, A., "Holstein Truss Spar Design Challenges and Innovations", Offshore Technology Conference 2005, Houston, Texas, USA. OTC 17292.
[10] Research Partnership to Secure Energy for America; RPSEA (2009-2010). "Ultra Deepwater Dry Tree System for Drilling and Production", RPSEA UDW 1402, Stage 1 Final Report & Stage 2 Summary Report, H08130-G-RPT-GN-15002 & H08130-G-RPT-GN-15003.
[11] Sablok, A., Edelson, D., Kim, J., and Luo, M., "Extended Draft Semi-submersible", Society of Petroleum Engineers, Brasil Offshore Conference and Exhibition, Macae, Brazil, June 14-17, 2011. SPE 143202.

[12] Smith, T., Allen, T., and Yu, A., "Holstein Spar Supported Vertical Riser and Dry Tree System," Offshore Technology Conference 2005, Houston, Texas, USA. OTC 17254.

[13] Wang, T., "Dry Tree Semi-submersible for Deepwater Gulf of Mexico," INTSOK, Norwegian Oil and Gas Partners, March, 2012.

[14] Williams, N., Leverette, S., Bian, S., Large, S. and Cao, P., "Fourstar Dry-Tree Semi-submersible Development", Proceedings of the 29th International Conference on Ocean, Offshore and Arctic Engineering, OMAE2010-20465.

[15] Zou, J., "Dynamic Responses of a Dry Tree Semi-submersible Platform with Ram Style Tensioners in the Post-Katrina Irregular Seas." International Society of Offshore and Polar Engineering Conference 2008, Vancouver, Canada.

[16] WAMIT® v6.4, "The State of the Art in Wave Interaction Analysis", WAMIT Inc., Chestnut Hill, Massachusetts. 2006.

DETERMINISTIC BREAKING WAVE SIMULATION FOR OFFSHORE APPLICATIONS

JOOP A. HELDER
SENIOR PROJECT MANAGER, MARIN

TIM BUNNIK
SENIOR RESEARCHER, MARIN

Proceedings of the 21st Offshore Symposium, February 2016, Houston, Texas
Texas Section of the Society of Naval Architects and Marine Engineers

ABSTRACT

Extreme wave events are generally taken into account in the design of offshore structures, to assure integrity and safety of the structure. To accurately predict extreme wave impact loading, a realistic modeling of all aspects involved is required. CFD simulations have recently shown more and more promising results in this field, however at present most of the industry still relies on detailed model testing for wave impact predictions. Especially when impacts due to steep (nearly) breaking waves are considered, accurate numerical modeling is challenging. The difficulty starts with a realistic numerical representation of the incoming breaking wave, which is essential for a realistic simulation but by no means easy to obtain.

This paper presents an approach towards obtaining deterministic breaking waves numerically, and applying these in offshore applications. The presented method allows to numerically replicate a critical extreme wave (impact) event, and therewith opens doors for CFD to a whole new field of applications. An extreme event that was measured during a wave basin test can be replicated numerically to run design variations, or a design wave defined from Met Ocean data can be reproduced to simulate its impact on a structure. The possibilities of the new approach can be optimized by implementing it in a CFD simulation tool that allows for dynamic mooring line simulation, such that mooring failures under the impact of an extreme wave event can be studied numerically.

Examples of how the proposed method can assist the industry in future extreme environment modeling are presented, and results are compared to model basin results for the impact of a (numerically replicated) extreme breaking wave on a flexible mono pile structure.

Keywords: SNAME, Offshore, Symposium, CFD, breaking waves, wave impact, green water, extreme loading, numerical simulation

INTRODUCTION AND MOTIVATION

Accurate predictions of phenomena like air gap, green water velocities and impact loads are vital for a proper design of any offshore structure. Up to a few years ago, such predictions hardly could be made numerically. At that moment, CFD simulations were still to computationally costly, and lower order numerical methods could not accurately enough predict the strong nonlinearities inherent to these phenomena. Over the past few years however, CFD capabilities have leaped forward and recent work has shown promising results in the field of violent free surface flow simulation for offshore applications. Not only for (bottom) fixed structures accurate correspondence is found between model test and CFD (Iwanowski et al., 2014) and (Pakozdi et al., 2015), but also for floating structures interactively moving in waves impressive results for global performance analysis were reported, again validated by model tests (Wu et al., 2014). It should be clear that the latter is much more challenging numerically, as a realistic representation of the mooring system has to be included in the modeling, and simulation has to account for phenomena such as incoming-diffracted wave interaction and low frequent motion of the structure.

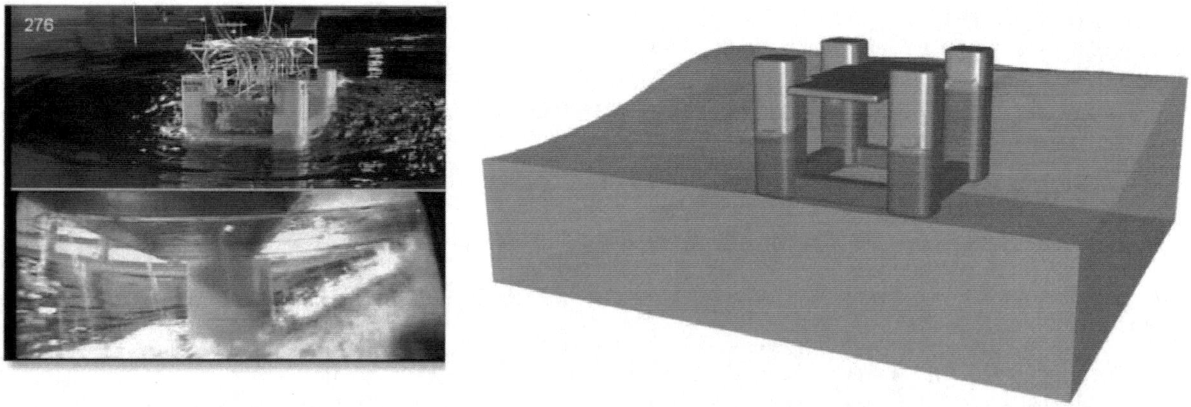

Figure 1 Illustration example of the status of CFD capabilities in the field of violent free surface flow for offshore applications: Simulation of interactively moving TLP with realistic mooring modeling in steep waves (right), mimicking wave basin tests (left).

One of the critical issues related to applying CFD simulation for the mentioned phenomena commercially, is that one needs to have confidence in the reliability of the results. Therefore, a thorough validation of the numerical results needs to be in place. Indeed, for most of the reported studies validation was performed for regular waves, or using spectral formulations. In a way, this sound counterintuitive, as phenomena such as green water and wave impacts do not typically occur in regular seas, but mainly during a steep or even (nearly) breaking wave (group) passing. These isolated events however are very difficult to validate numerically, as replicating the exact wave condition is not straightforward. Even copying the exact wave flap motions of a wave basin test in CFD is not guaranteed to result in an identical copy of the basin wave at the location of the structure, due to numerical propagation error and the unstable character of steep irregular waves. An accurate matching however is crucial for a proper validation, as velocities in extreme breaking waves show a significant increase compared to those in regular waves, and the sensitivity of the results to small deviations in these velocities and steepness of the passing wave crest can be significant. In this paper, exactly for the reason discussed above, a methodology was devised to accurately replicate a steep breaking wave (group) deterministically in CFD. The application area of the method however goes far beyond validation purposes. As the method can replicate an extreme wave event from a basin test, it allows for a more detailed flow analysis or even design variations to be conducted of the selected wave basin event, after the wave basin tests are completed.

DETERMINISTIC BREAKING WAVE RECONSTRUCTION METHODOLOGY

When wanting to reconstruct an existing (or *target*) wave profile numerically, typically two approaches can be followed. The first approach assumes that the to-be-reconstructed wave originated from a wave basin. The motions of the wave maker can then be copied in the CFD model to generate the waves in an identical way as in the basin. The second, more general approach uses an additional wave profile from an upstream location as inflow boundary condition to generate the target wave profile. Unfortunately, using either of these approaches does not naturally guarantee a satisfactory reconstruction of the existing profile. Typically, in wave basin testing the to-be-reconstructed wave profile is located at a large distance from the wave maker, making the first approach susceptible to discretization errors which can result in dispersion errors and loss of amplitude. Additionally, propagating waves in CFD over such large distances can be very computationally expensive. The disadvantage of the second approach is that the wave orbital velocities at the inflow boundary have to be estimated based on surface elevation and no exact formulation exists for random irregular waves. It should be noted here that for a (breaking) wave impact it is essential that the profile of the incoming wave is reconstructed with a very high accuracy in the CFD model, as a small difference in wave steepness may lead to large differences in impact load. The above considerations might be too strict when performing e.g. global performance analysis with CFD, where criteria on the error in reconstruction are less severe.

The proposed methodology presented in this paper uses the second approach (the most general of the two), for which an iterative wave calibration procedure was developed to have a better match between the target wave profile and the numerically reconstructed wave. To estimate the wave orbital velocities at the inflow boundary, the method proposed by (Stansberg et al., 2008) is used:

$$\vec{u}(z) = \vec{u}^{(1)}(z) + \frac{\partial \vec{u}^{(2)}}{\partial z} \quad z \leq 0 \tag{1}$$

$$\vec{u}(z) = \vec{u}^{(1)}{}_{z=0} + z\frac{\partial \vec{u}^{(1)}}{\partial z}|_{z=0} + \frac{\partial \vec{u}^{(2)}}{\partial z}|_{z=0} \quad z > 0$$

Where $\vec{u}^{(1)}$, $\vec{u}^{(2)}$ stand for the first and second-order wave orbital velocity as originally formulated by (Sharma et al., 1981), and z=0 corresponds to the calm water surface. The inflow wave profile is split into linear and second-order components using 2nd order wave theory. Any inaccuracies in this second-order velocity estimate and remaining dispersion and dissipation errors of the CFD model are to be eliminated by the iterative wave calibration procedure. The methodology is as follows:

1. Wave profiles at inflow boundary and target location are available.
2. Kinematics of the wave at the CFD inflow boundary are estimated from the inflow wave profile using 2nd order wave theory.
3. The wave is propagated through the (typically 2D) CFD domain.
4. The target wave profile is compared to the wave obtained by the CFD.
5. Based on the difference between target profile and simulation the inflow is corrected.
6. This iteration continues until a satisfactory agreement between the target and simulated wave profile is obtained.

The correction of the inflow is based on a Fourier decomposition of the target and simulated wave profile at the target location:

$$F_{corr} = \frac{\hat{\eta}_{required}}{\hat{\eta}_{simulated}} \tag{2}$$

Where $\hat{}$ stands for the Fourier operator. The inflow wave is then corrected as follows:

$$\eta_{inflow} = \left(\hat{\eta}_{inflow}\, F_{corr}\right)^{\vee} \tag{3}$$

Where $(\ )^{\vee}$ stands for the inverse Fourier operator. This complex correction factor will result in an amplitude correction and a phase shift of the wave elevation at the inflow. In order to avoid end effects in the FFTs a ramp function was applied to the beginning and the end of the wave profile time trace. As indicated in step 3., typically a 2D computational domain is used for the iteration process. This assumption is valid as long as the to-be-reconstructed wave is long crested, and reduces computational time significantly.

Figure 2 shows a typical result of this iteration process, as obtained for the example discussed later on in this paper. After 10 iterations a satisfactory agreement between the target and simulated wave was found. Special attention was paid to the (temporal) steepness of the wave around the time of impact ($\frac{\partial \eta}{\partial t}$) since this is the driving factor behind the impact.

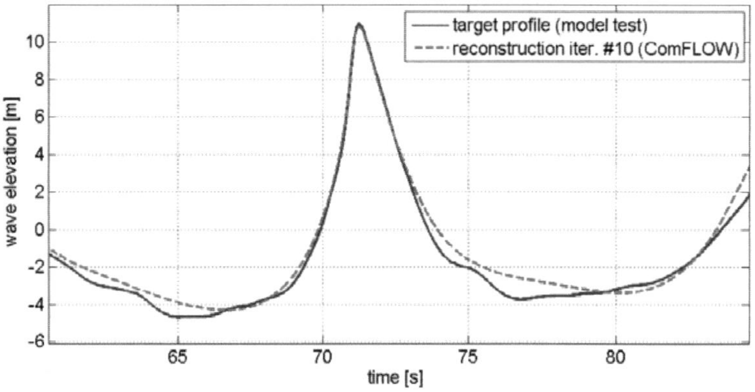

Figure 2 To-be-reconstructed wave elevation profile (blue) and result of the wave reconstruction methodology (red). In total, 10 iterations were needed to obtain the depicted result. Focus of the reconstruction was on the steepness (rise time) and height of the extreme crest.

After satisfactory agreement is achieved between the reconstructed and target wave profile, the resulting wave kinematics can be used as initialization and inflow boundary condition for a 3D simulation, including structure. Hereto, the 2D wave kinematics should be extended in the 3rd direction, and for the 3D simulation typically a shorter time window and a smaller domain in the propagating wave direction should be chosen, to reduce computational costs.

This section is ended by noting that the presented methodology is independent of the CFD simulation tool used, and can in principle be implemented in any CFD method (that accommodates the necessary functionalities). At MARIN, the methodology was implemented in ComFLOW, a CFD simulation tool specifically designed for efficient simulation of violent free-surface flow problems. ComFLOW has been continuously and actively developed at the university of Groningen together with the offshore industry and MARIN within a series of joint industry projects. A description of the algorithms, functionalities and recent developments can be found in (Luppes et al., 2013) and (Veldman et al., 2014, 2015). The results that are presented in this paper are obtained using the methodology as implemented in ComFLOW.

FIELDS OF APPLICATION

As the output of the presented wave reconstruction methodology simply provides a calibrated inflow and initial condition for the numerical simulation at hand, the method's field of application can be considered only limited by the functionality of the used CFD tool (as mentioned in the previous section, at MARIN the methodology was implemented in the CFD simulation tool ComFLOW). In this section, two application fields in the area of green water and wave impacts are discussed, for which it is believed a large gain is to be made by applying the wave reconstruction methodology. Of course, in addition to the two discussed application areas, many other fields of application can gain from the capability of using deterministic waves in CFD simulations.

Reconstruction based on extreme wave basin event

As a first field of application, the reconstruction of extreme events as observed during wave basin tests is discussed. At present, most of the industry relies on detailed model testing for wave impact predictions. During such model tests, it is not unusual to make some discoveries that were not foreseen in the design process beforehand. During testing however, it is not always possible to react upon these discoveries immediately. Some of the observations are caused by nonlinearities that are difficult to tune in common-practice design models, even when model test data is available. As an example, one could think of unforeseen green water impacting structures on deck, or unexpected slamming events on an external turret. By applying the presented wave reconstruction methodology, these events can be accurately recreated within the CFD model. Typically, a short time window from a long-duration model test is selected and reconstructed. This makes it possible to:

- acquire additional information (e.g. loads, pressures, green water velocities) in non-instrumented areas on the structure,
- study the effect of design variations (e.g. impact mitigation devices),
- obtain a better understanding of the flow details during the extreme wave event.

At MARIN, this approach has recently been successfully applied for both (bottom) fixed and floating structures. An example of the former will be discussed in the next section. For the latter, in addition to reconstructing the wave profile from the basin data, also the 6DOF measured motions of the structure need to be prescribed in the CFD tool and synchronized with the reconstructed wave profile. This functionality is available within the CFD tool used in this paper (ComFLOW), however at the moment of writing this is unfortunately not the case for all commercially available CFD tools. It is stressed that for floating structures, in order to capture the event accurately it is of crucial importance that the synchronisation is very accurately performed. Depending on the character of the event, diffraction and radiation effects might be present and need to build-up. In that case it is recommended to reconstruct (and simulate) a slightly larger time window around the extreme crest, to allow for incoming-diffracted wave interactions. As practical examples in this application area, the optimization of breakwaters and green water deflectors, analysis of green water velocities building up alongside the vessel, global forces on free-fall lifeboats and impacts on non-instrumented on-deck structures are mentioned.

To conclude this section it is remarked that in the near future it is likely that CFD-based numerical wave basins will also be used as screening tool for detecting extreme wave impact events. These tools typically apply large grid spacing and time stepping in order to simulate large time windows, and as such do not directly predict impact and green water events in detail. Extreme wave events can however be identified during such simulations, and can be studied in detail afterwards by applying a similar methodology as described above. One of the challenges of such an approach will be reducing the computational costs of the screening by reducing the order of complexity of the simulation model, and finding proper screening criteria for the relevant limit states.

Reconstruction based on undisturbed wave profile (from database or simulation)

The presented wave reconstruction methodology can also be used when no wave basin data is available. For instance, when wave conditions at the foreseen location of the structure are known, 3-hour realisations of the undisturbed waves (obtained from e.g. a wave data base or by non-linear potential flow methods) can be screened based on case-specific selection criteria. As a result, the selected time window and to-be-reconstructed wave can be reconstructed in CFD.

One of the most challenging parts of this approach is finding appropriate selection criteria for the to-be-reconstructed wave, based on only the undisturbed wave elevation time trace. For (bottom) fixed structures with little incoming-diffracted wave interaction (such as jackets or mono piles), a selection criteria can relatively easily be found in e.g. the undisturbed wave height or wave steepness. However for structures moving interactively with the waves, finding a selection criteria based on only the undisturbed wave elevation is often not realistic, as the occurrence of wave impact or green water events is influenced by the wave- and low frequency motions of the floater. In that case, as discussed in the previous subsection, one could consider using lower order methods to predict the motions of the interactively moving floater, during the 3-hour wave realisations. From the results, possible extreme wave impact or green water events can then be more easily identified and reconstructed within a detailed CFD simulation. Such an approach is being developed at MARIN at the moment, and will also make it possible to perform wave impact analysis on a statistical level. Directly using undisturbed wave elevation profiles to select critical wave conditions is at the moment applied at MARIN for wave-in-deck analysis on jackets and impact analysis on mono pile structures.

APPLICATION EXAMPLE: BREAKING WAVE IMPACT ON A FIXED MONOPILE

As a practical example on how the proposed methodology can assist the industry in effective design and assessment of wave impact related events, the reconstruction methodology is used to simulate a deterministic extreme breaking wave impacting a mono pile structure. The relevance of the application example is illustrated in Figure 3, and by realizing that most offshore mono pile structures are located in water depths were breaking waves occur. Results for both a rigid and flexible mono pile structure are presented and validated against model tests, in terms of horizontal wave impact loading and flexible response, respectively. As the example is presented to illustrate the way the methodology can be applied in an industrial application and not as a validation case in itself, certain details of the simulation and wave basin setup are omitted intentionally for clarity of reading.

Figure 3 Extreme breaking wave impacting a mono pile structure offshore (left, www.flyingfocus.nl), and in MARIN's shallow water wave basin (right).

Validation of breaking wave impact load

From a 3-hour irregular wave basin test performed with a rigid mono pile, a short time window was selected around the maximum horizontal impact load that was recorded. Figure 4 depicts the 3-hour time trace, and the selected time window around the maximum impact.

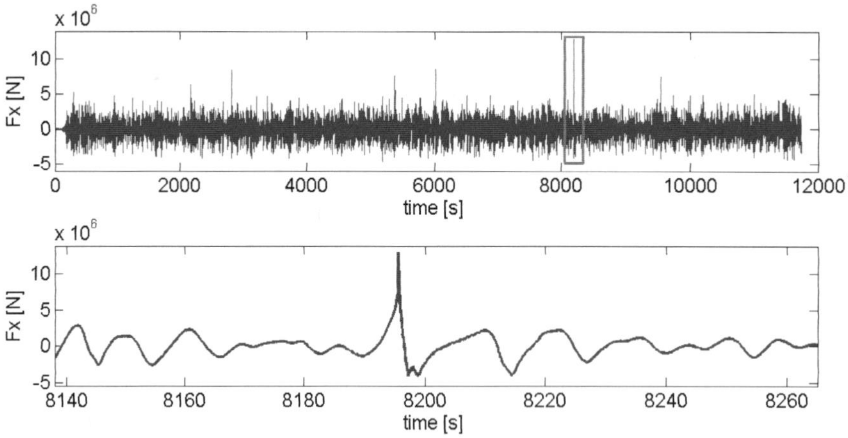

Figure 4 Time evolution of horizontal load on mono pile during 3-hour wave basin tests (above), and the time window selected for reconstruction (below and above in red).

The corresponding incoming wave profile at the location of the mono pile was reconstructed using the methodology described in this paper. For this particular case each iteration took about 13 minutes on a single core, where a uniform grid of 800mx45m was used in resp. the propagating wave (X) and vertical (Z) direction, using cells of 1mx1m. The to-be-reconstructed time window was set to 130 seconds. Satisfactory agreement between target and reconstructed wave profile was achieved after 10 iterations. Figure 2 depicts the result of the reconstruction; a near perfect match was obtained. Note that the horizontal (i.e. time) axis in Figure 2 is shifted compared to Figure 4, as the start time of the target wave profile was chosen approximately 8125 seconds after the start of the wave basin measurement.

After the iteration process was completed, the wave kinematics were used as a boundary condition in a 3D domain including the mono pile structure. Figure 5 shows the setup of the 3D computational domain. Compared to the 2D domain used for the wave reconstruction, the 3D domain was reduced in the propagating wave direction from 800m to 250m. In the transverse (Y) direction, the 3D domain was set to be 30m wide, and a symmetry plane was used at Y=0. Grid verification was performed on uniform grid sizes of 1.0mx1.0mx1.0m, 0.67mx0.67mx0.67m, 0.5mx0.5mx0.5m and 0.25mx0.25mx0.25m. Only the results on the coarsest grid showed slight deviations from the results obtained on the other 3 meshes. The depicted mesh in Figure 5 has a uniform grid size of 0.5mx0.5mx0.5m. Absorbing boundary conditions were set at the outflow. To reduce computational costs further, not the entire time window (130s) was simulated, but only a duration of 30 seconds around the time of impact. It is further remarked that for this particular case, a schematic sand wave was included in the modeling (see Figure 5). This sand wave was also present in the model tests and during the 2D iteration procedure for reconstructing the incoming wave.

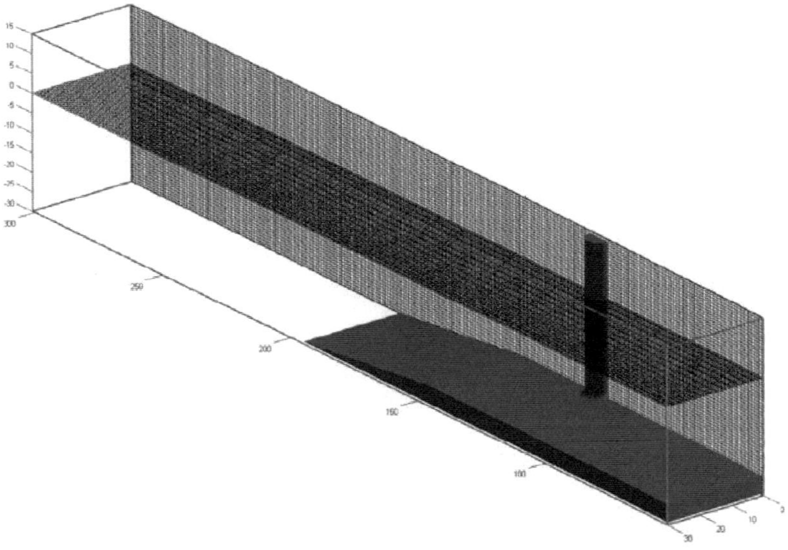

Figure 5 Computational domain for extreme wave impact simulation on mono pile.

The resulting wave impact load on the mono pile structure using the deterministically reconstructed breaking wave on the depicted mesh is shown in Figure 6. The figure confirms the additional value of the breaking wave reconstruction; a very accurate match is found between the measurement and simulation. Without the accurate match in incoming wave condition, a similar agreement between wave basin result and simulation would have been practically unachievable for such an extreme breaking wave event. With the obtained confidence in the accuracy of the results, design variations of the mono pile could be performed within the CFD in order to e.g. reduce the horizontal loading on the mono pile.

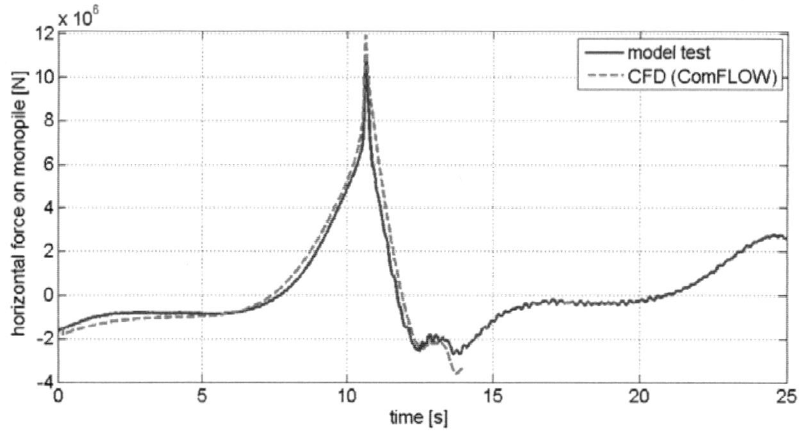

Figure 6 Horizontal impact on mono pile structure during extreme breaking wave event. Comparison between wave basin result (blue) and CFD simulation using deterministic breaking wave reconstruction (red).

Validation of flexible response of mono pile

To take the example one step further, also the response of a flexible mono pile subject to an extreme breaking wave is validated. Hereto, not the wave impact load but the dynamic acceleration at the top of the mono pile (118.8 m above the seabed) is compared with experiments results. To start, again a short time window was selected around the moment of impact, now from a wave basin tests with flexible mono pile. The right image in Figure 3 illustrates the severity of the impact - a large amount of wave run-up is generated due to high pressures in the spray root. For the selected case, 6 iterations were needed to reconstruct the incoming wave profile with sufficient accuracy. Again a 3D simulation was started, using the wave kinematics of the final iteration as initial and inflow boundary condition. Figure 7 shows a snapshot of the 3D simulation. Again a grid refinement study was performed, and again a uniform grid size of 0.5mx0.5mx0.5m was found to be more than adequate to assure convergence of the results.

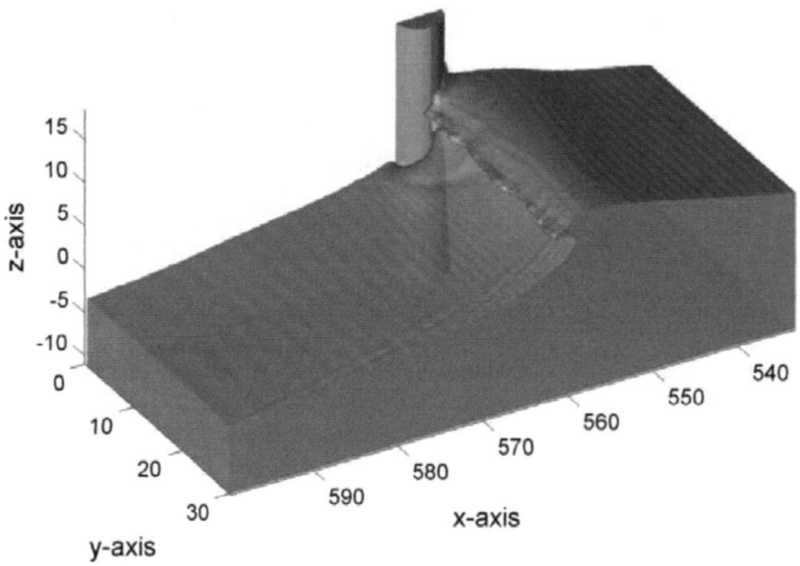

Figure 7 Simulation of extreme breaking wave impacting a mono pile. The breaking wave was deterministically reconstructed to match a target wave elevation profile from the wave basin test.

Assuming the influence of the response on the loading process to be negligible, the mono pile was kept rigid in the 3D CFD simulations. The dynamic response was then calculated by a one-way coupling between the CFD and a structural model of the mono pile. Hereto, a simple structural response model was constructed of the mono pile, based on a representation with mode shapes, see for example (Sireta et al., 2012). First, the modal excitation was computed by integrating the wave impact pressures on the rigid mono pile multiplied by the mode shape over the wetted surface of the mono pile. Subsequently, the model responses could be calculated and compared to the model test measurements. It is remarked that a 2-way coupling using a modal approach is being developed within the CFD tool ComFLOW at the moment. Figure 8 shows the measured and computed horizontal acceleration at the top of the mono pile (118.8 m above seabed). The figure shows both the time trace and spectrum of the acceleration, and shows accelerations reaching a maximum of more than ½ G. Overall a very good agreement is found between measured and computed accelerations, again confirming the additional value of the breaking wave reconstruction. Also here the accuracy of the results should provide enough confidence to apply design changes within the CFD method, to reduce for instance the accelerations at the top of the monopile by adjusting the mode shapes of the structure.

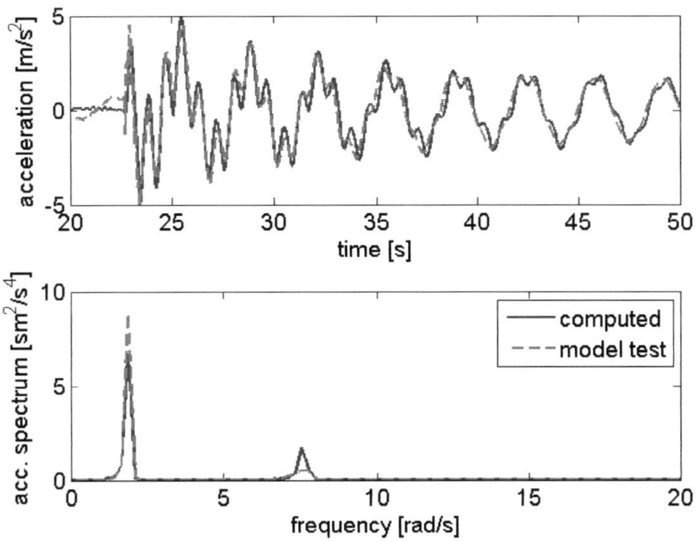

Figure 8 Comparison of experimental results and computed values for the horizontal acceleration at the top of the mono pile (118.8m above seabed). Time traces (above) and spectrum (below).

CONCLUSIONS AND FUTURE WORK

In this paper, a methodology for deterministically reconstructing extreme (breaking) waves within CFD is presented. This methodology allows for a validation of CFD for extreme wave impact and green water related applications. As such, the method can provide the industry with a greater confidence in applying CFD for the afore mentioned application areas. Besides being a powerful tool for the validation of extreme breaking wave impact simulations, the presented methodology can also be used to assist at several stages in the design of an offshore structure. For instance, wave basin results can be complemented by reconstructing the most extreme events within CFD simulations and applying design variations. Alternatively, at an earlier stage in the design, available wave data at the foreseen location of the structure can be used to select extreme wave events that can be reconstructed numerically.

The potential of the reconstruction methodology is illustrated by simulating an extreme breaking wave impact on both a rigid and flexible offshore structure. The simulations show that, using the presented methodology, a very accurate correspondence between CFD and model tests can be achieved in terms of impact loading and dynamic response, for an extreme breaking wave event.

The presented methodology is independent of the CFD simulation tool used, and can in principle be integrated in any CFD method (that accommodates the necessary functionalities). The results presented in this paper were obtained by implementing the methodology in the ComFLOW CFD tool, specifically developed for efficient simulation of violent free-surface flow problems.

As future work, the methodology will be used to validate wave kinematics in extreme wave crests and during impacts, as predicted by CFD. Hereto, PIV measurements on breaking wave crests and impacts on a deck-type structure will be carried out by MARIN.

ACKNOWLEDGEMENTS

This work was partially performed within the ComFLOW-3 JIP, ComMotion JIP and WiFi JIP.

REFERENCES

Iwanowski, B., Vestbostad, T., Lefranc, M., "Wave-in-deck load on a jacket platform, CFD calculations compared with experiments", Proceedings of the ASME 2014 33rd International Conference on Ocean, Offshore and Arctic Engineering (OMAE). San Francisco, CA, USA, 2014.

Luppes, R. van der Heiden, H.J.L., van der Plas, P., Veldman, A.E.P. and Duz, B., "Simulations of Wave Impact and Two-Phase Flow with ComFLOW: Past and Recent Developments", Proceedings of 23rd International Society of Offshore and Polar Engineers (ISOPE). Anchorage, Alaska, USA, 2013.

Pakozdi, C. et al., "Estimation of Wave in Deck Load Using CFD Validated Against Model Test Data", Proceedings of 25th International Society of Offshore and Polar Engineers (ISOPE). Hawaii, USA, 2015.

Sharma, J. and Dean, R.G., "Second-Order Directional Seas and Associated Wave Forces", J. Soc. of Petr. Eng., SPE, , pp. 129-14. 1981.

Sireta, F-X. et al., "Hydroelastic response of a ship structural detail to seakeeping loads using a top down scheme". Proceedings of the ASME 2012 31st International Conference on Ocean, Offshore and Arctic Engineering (OMAE). Rio de Janeiro, Brazil, 2012.

Stansberg, C. T., Gudmestad, O. T., and Haver, S. K., "Kinematics Under Extreme Waves", Journal of Offshore Mechanics and Arctic Engineering, 130. 2008.

Veldman et al., "Turbulence modeling, local grid refinement and absorbing boundary conditions for free-surface flow simulations in offshore applications", Proceedings of the ASME 2014 33rd International Conference on Ocean, Offshore and Arctic Engineering (OMAE). San Fransisco, CA, USA, 2014.

Veldman, et al., "Turbulence Modeling for Locally-Refined Free-Surface Flow Simulations in Offshore Applications", Proceedings of 25th International Society of Offshore and Polar Engineers (ISOPE). Hawaii, USA, 2015.

Wu, G. et al., "Benchmark of CFD modeling of TLP Free Motion in Extreme Wave Event", Proceedings of the ASME 2014 33rd International Conference on Ocean, Offshore and Arctic Engineering (OMAE). San Fransisco, CA, USA, 2014.

THE INTERNATIONAL COMMUNITY
FOR MARITIME AND OCEAN PROFESSIONALS

PREDICTION OF DESIGN LOADS FOR STRUCTURE DESIGN OF FPSO TURRET

YANBIN BAI
MOORING & RISER ENGINEER, COTEC INC.

ZHIYONG SU
NAVAL ARCHITECT, COTEC INC.

XIAOCHUAN YU
ASSISTANT PROFESSOR, UNIVERSITY OF NEW ORLEANS

HUI SHEN
NAVAL ARCHITECT, COTEC INC.

YAXIN SONG
PRINCIPAL STRUCTURE ENGINEER, COTEC INC.

YONG LUO
PRESIDENT, COTEC INC.

Proceedings of the 21st Offshore Symposium, February 2016, Houston, Texas
Texas Section of the Society of Naval Architects and Marine Engineers

ABSTRACT

Internal turret system is one of the most popular single point mooring systems that is uniquely integrated with the FPSO hull and connects with both mooring lines and risers/umbilicals. Hydrodynamic aspects of various single point mooring systems have been intensively researched and studied. However, industry standards of deriving the various load components and their correct combination for the purpose of turret structure design are not well documented. This paper presents the procedures of deriving the turret design loads for the structure design of an internal turret using a roller upper bearing system. Based on a sample case of a deepwater FPSO operating in a harsh environment, detailed descriptions of the various load components and their impacts on structural design and mechanical system selection are examined. A summary of the design cases and loads combinations are presented. In addition, various design parameters and their criticalities are studied. The paper concludes that the correct prediction of turret design loads is an important step towards the safe design of Deepwater FPSOs.

Keywords: Internal turret system, FPSO, Turret design loads

INTRODUCTION

Up to 2014, more than 150 Floating Production, Storage and Offloading (FPSO) system has been deployed worldwide in the water depth of 13m to over 2600m. Luo (2005) also pointed out that water depth is not a stopper to the development of FPSO in deepwater Gulf of Mexico. So far FPSO has become the most widely adopted and the most versatile floating production system in offshore oil & gas industry. Due to the seasonally alternating meta-ocean environmental condition at most offshore oil fields, single point mooring (SPM) system introducing a weather vane advantage to FPSO is the most popular stationkeeping technology. Regarding deepwater SPM solution, external and internal turret mooring systems have been developed for deepwater FPSO. To accommodate eleven risers/umbilicals, the internal turret mooring is chosen for the target FPSO discussed in this paper.

Extensive research has been conducted on hydrodynamic aspects of turret moored FPSO both experimentally, numerically and theoretically (Luo et al., 2003, Kim et al., 2005, Soares et al., 2005, Oliveira et al., 2014, and Torres et al., 2015), application of disconnectable FPSO in the arctic region (Howell et al., 2001, Duggal et al., 2000), impact of turret location on dynamic behavior of FPSO on waves (Kannah, 2006). Considering multiple coupling effects such as FPSO motion, turret rotation, mooring lines and risers/umbilicals, to accurately predict the FPSO is challenging. In the past studies, the turret cylinder is always assumed to be a rigid body when investigate the FPSO response. Furthermore, incorporating the elasticity of turret cylinder and contact between the elastic lower bearing system and FPSO hull makes the prediction of turret loads a more challenging task. Model testing is one powerful tool in respect of predicting the extreme motion of FPSO. However, compared with numerical simulation, model testing has limitations in the dimensionless parameter discrepancies and difficulty in modeling contact. Fully coupled time domain simulation is adopted in this paper using a vessel/mooring/riser/turret coupling model. The proposed coupled model is capable of modeling the bending and inertia of turret, contact between the lower bearings and FPSO and reaction force at the upper bearings. By virtue of time domain simulation, the snapshot of the most severe loads for turret structure design can be captured.

The objective of this paper is to provide a procedure of obtaining rational and comprehensive turret loads by means of time domain numerical simulation for the subsequent turret structure design and design verification analyses. This paper also presents the methodology to predict turret loads using a coupled turret-FPSO model and demonstrates the criteria of critical loads selection. The work presented in this paper has been adopted to derive various turret loads fed into the structural design of the turret cylinder and gantry of a FPSO. The methodology presented in this paper applies to both disconnetable and permanent internal turret system.

TURRET SYSTEM DESCRIPTION

The turret is a geo-stationary structure located in the bow section of the proposed FPSO, allowing the FPSO to weather vane. Figure 1 demonstrates the proposed FPSO system, which consists of one new built FPSO, 9 evenly spacing mooring legs, and 11 risers/umbilicals. An upper roller bearing and lower self-lubricating bearing arrangement allow the vessel to rotate about the turret. The system is composed of four primary components: the turret gantry, turret swivel, turret structure, and the FPSO turret integration structure.

The turret structure can be broadly divided into the lower, upper and swivel sections. The lower turret includes the following main components:

- Lower bearing pad assemblies;
- Chain table that house the chain stoppers, the I-tubes and flexible risers;
- Mooring/riser installation equipment;

The upper turret consists of the main bearing and the main bearing support structure. The system is designed to transfer loads from the turret through the main bearing to the ship-side main bearing support structure through roller bearing system. The fluid swivel stack assembly is mounted at the top of the upper turret. The turret integration structure includes the lower bearing support structure, the moonpool, the upper bearing support structure and the structural reinforcement necessary to transfer loads into the vessel.

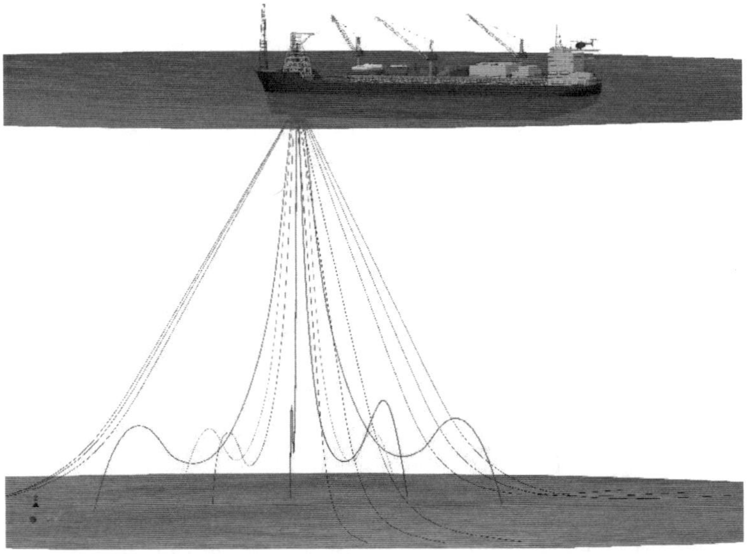

Figure 1 Sketch of FPSO with mooring and riser systems

Figure 2 shows the general arrangement of the turret and its key components, turret cylinder, riser decks, gantry, swivel stack and associated piping system. Figure 3 demonstrates how the turret is integrated into the FPSO hull. The turret is supported by the upper bearing and the lower bearing on the FPSO hull.

Figure 2 3D illustration of turret system integrated into FPSO hull

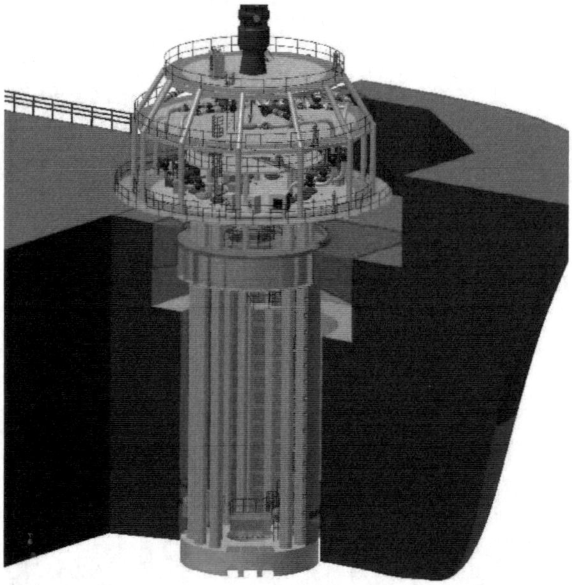

Figure 3 ISO view of turret system with topside and gantry

TURRET LOAD COMPONENTS

Rational identification of critical turret loads is the very first step to further predict the comprehensive turret loads. The major turret loads to be considered in this paper are summarized as follows.

1. Mooring loads

The mooring loads are computed using the time-domain analysis program Orcaflex. In reaction to the mean wind, mean drift wave force and current loads, the mooing system offsets to the mean equilibrium position at the static analysis stage. Then under the wave frequency wave excitation load and the slowly varying loads of wind and wave drift forces, the mooring line tension fluctuates around the mean values. These static and dynamic line tensions are resisted by the turret at the uni-joints installed inside the chain table.

2. Riser loads

Hydrocarbon transfer from the seabed is accomplished with a total of eleven risers and umbilical lines (among them four are designated as future lines). The risers are supported vertically by the internal riser deck and laterally by the bend stiffener that are attached to the bottom flanges of riser I-tubes. In the present analysis, the risers are modeled as flowlines with corresponding mechanical properties and hydrodynamic properties. However, only tensions are simulated since the bend stiffeners are not included. In general, the risers are much more soft, i.e. more compliant to FPSO movement than the mooring lines. The risers are not designed as load bearing components and their horizontal loads are significantly smaller than that of the moorings. Overall, the riser system has limited impact on the turret global load. For the local supporting structural design for risers, the extreme riser loads may be used.

3. Vessel acceleration

The acceleration of the vessel induces inertia forces in the turret. The surge, sway, heave, roll, pitch and yaw accelerations are computed and fed into the overall turret design load predictions. In addition, when the turret system accelerates horizontally and vertically with the FPSO, the fluid surround the turret is excited and this effect is modeled as the added mass in the Finite Element simulation.

4. Gravity loads

The gravity load in the vertical direction is represented by the turret self-weight. However, as the FPSO pitches and rolls, the gravity effect also induces a horizontal component in the ship local axis system. The horizontal component of the gravity load is calculated by weight and tilt angle and included in the turret design load prediction. Both the vertical and horizontal gravity load components are resisted by the interface structures between the FPSO and turret.

5. Wave slam

The turret is assembled inside the FPSO moonpool, so it is protected from the lateral wave slamming load. Wave slam load is not included for the internal turret system.

6. Wind load

Wind loading is applied to components that are elevated and exposed to wind, such as a portion of the upper turret and swivel stacks. It is noted that the direct wind loading is insignificant compared with other load components.

7. Swivel driving post load

The turret interacts with the gantry through the swivel driving posts. It is understood that this load component is small compared with other load components. The driving post loads may be analyzed in the local structural design.

8. Hydrodynamic load

The hydrodynamic load acting on the turret is included as added inertia, i.e. the mass of water that moves with the turret. Between the turret and the FPSO moonpool, there is trapped water, which moves with the FPSO/turret to induce inertia loads in the local structural analysis.

PREDICTION OF TURRET LOADS

1. General approach

The analysis of design loads was conducted using the time domain analysis method. For a defined design seastates, the time domain simulation of the mooring and vessel responses is conducted to derive the mooring line tensions, riser loads and vessel motions including low frequency motion and wave frequency motion.

The key analysis steps are:

- Step 1 - Vessel Hydrodynamics

Hydrodynamic wet surface models of the FPSO at various loading conditions are developed. A three-dimensional diffraction/radiation analysis is performed using the WAMIT to compute the hydrodynamic coefficients, such as added mass, radiation damping, first order wave exciting forces, wave drift force coefficients, and motion RAO's. The wind force coefficient, wave drift force coefficients or QTF's and current profile are input to the Orcaflex for calculation of the steady wind, wave drift and current loads and slow drift motions.

- Step 2 – Define Design Load Cases

The design load cases of environmental conditions and directional combinations of wind, waves and current are defined. The design cases involve the operating condition, the extreme condition and the mooring damaged case.

- Step 3 – Develop a Coupled Vessel/Mooring/Riser System Model

In this paper, the mooring legs and the risers / cables are modeled using Orcaflex, together with the vessel dimensions, first order wave excitation force, wave drift force coefficients and wind/current coefficients. Line composition and actual mechanical properties of the various segments, namely un-stretched lengths, linear weights in air and water, axial stiffness, etc. have been properly defined.

- Step 4 - Simulation of the FPSO Responses and Mooring Line Tensions

These calculations have been carried out using the program Orcaflex which allows to perform time domain simulation of the vessel low frequency horizontal motions (surge, sway and yaw) and six degree of freedom wave frequency motions. Based on the vessel responses, the mooring line tensions were computed using a lumped-mass finite element method which can consider the internal and external pressure of risers.

2. Simplified turret structure

Figure 4 illustrates the simplified turret model. The time domain simulation adopts a coupled model in which the turret is modeled as a beam supported at the lower and upper main bearings. The upper bearings provide horizontal, vertical and bending moment around the horizontal axes. The contact between the turret body and the lower bearing pads can be captured in the proposed turret model. The vertical load is only supported by the upper bearings. The turret is connected with nine anchor legs and risers. The turret follows the motion and acceleration of the FPSO, i.e. no relative movement between the turret and FPSO beside the relative rotation of the FPSO around the turret. In the load model, the lower bearing load is distributed among the contact pads.

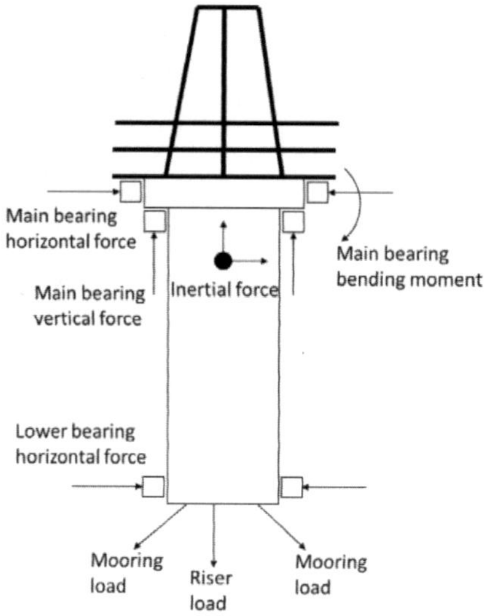

Figure 4 Turret model demonstration

3. Coupled vessel/mooring/riser/turret model

The turret load model illustrated in Figure 5 consists of the vessel, mooring legs, all present risers/cables, the turret body, the chain table, the lower bearing system and properly treated boundary condition at the upper bearing. Figure 6 and Figure 7 present more details of the chain table and lower bearings. Regarding the various coordinate systems involved in the simulation and post-processing of the turret load, seamless communication with the subsequent structural analysis is indispensable for an accurate data interpretation.

Figure 5 Coupled turret load model with all risers and mooring lines

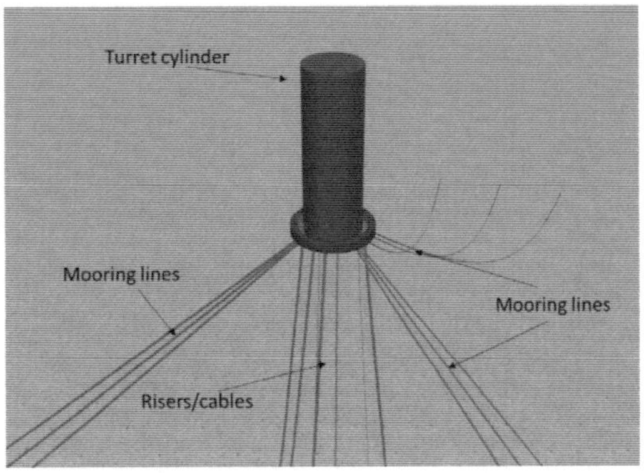

Figure 6 Coupled turret load model with all risers and mooring lines

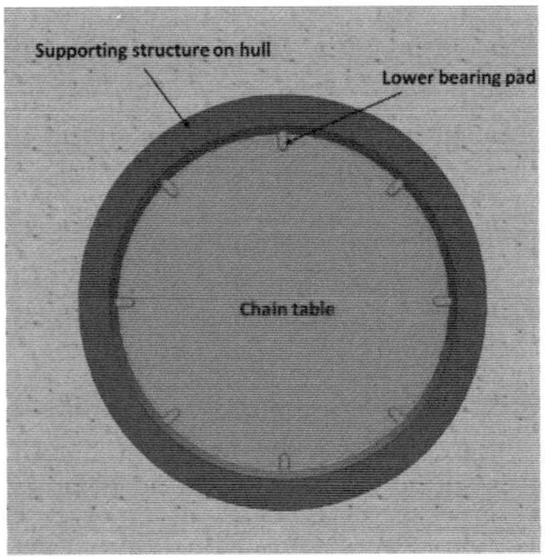

Figure 7 Contact model showing chain table, lower bearing and support structure on vessel hull

4. Environmental load combinations

In case no specific site data on the relative directionalities between wave, wind and current, it assumes wind goes up to 30° to wave and current can travel in the perpendicular direction of wave according to industry experiences. Mooring designs (API RP 2SK, 2005, and API RP 2SM, 2007) for different mooring leg solutions are based on the following direction combination as presented in Table 1. Turret load simulation adopts the same direction combination approach so as to capture the most critical turret load scenarios. When site specified meta-ocean data is available, the dominate wave direction can be determined and used in the time domain simulation.

Table 1 Environmental load combination

Direction Category	Wave Dir (deg)	Wind Dir (deg) wrt wave dir	Current Dir (deg) wrt wave dir
Colinear	any	0	0
Crossed	any	30	45
Transverse	any	30	90

5. Turret design load cases

In order to extract the worst loads for turret structure design, a criteria is needed to setup for purpose of capturing the snapshot of turret loads. Regarding each environmental combination, the turret design load scenarios are illustrated and presented in Table 2. Based on previous project experience, the transverse mooring condition is the most adverse environmental combination for turret structure design.

Table 2 Turret design load

Design Load	Description	Mooring Analysis Case
Max. Main Bearing Horizontal Force	Max Fr=Sqrt(Fx^2+Fy^2)	Transverse condition
Max. Main Bearing Vertical Force	Max Fa	Transverse condition
Max. Main Bearing Bending Moment	Max Mt=Sqrt(Mx^2+My^2)	Transverse condition
Max. Lower Bearing Horizontal Force	Max Fh=Sqrt(Fx^2+Fy^2)	Transverse condition
Max. Turret Horizontal Acceleration	Max Sqrt(Acc.x^2+Acc.y^2)	Transverse condition
Max. Turret Vertical Acceleration	Max Acc.z	Transverse condition
Min. Main Bearing Vertical Force	Min Fa	Transverse condition
Min. Turret Vertical Acceleration	Min Acc.z	Transverse condition

6. Turret load case study

The procedure and coupled model proposed in this paper has been applied to a new built FPSO, planned to be installed in South China Sea in a water depth of 300m. One existing subsea production termination, and two floating production units will be connect with the proposed FPSO. The FPSO is designed specifically to withstand a 500-year typhoon condition including 500-year extreme wave condition with associated wind and current/soliton and 500-year extreme wind condition with associated wave and current/soliton.

The simulation, results and statistical data processing will not be elaborated in detail herein. However, the observations and remarks about coupled time domain simulation using the proposed methodology is more pertinent to the objective of this paper.

The criticalities of each turret control load are summarized as follows.

- Maximum Main Bearing Horizontal Load

Clearly, the maximum main bearing horizontal load occurs in the transverse condition where the FPSO roll motion is close to its maximum. The upper bearing horizontal load also affects the design of main bearing.

- Maximum Main Bearing Vertical Load

The maximum main bearing vertical load is an important design parameter, it governs main bearing system design include the rollers and the frames of the upper main bearing. The maximum load occurs at the instance that the FPSO pitches upward. The mooring vertical load is high at this scenario.

- Maximum Main Bearing Vertical Load

The maximum main bearing bending moment is a critical design parameter for the main bearing system, it affects the structure design of the main bearing for both the frame and the rollers. The ballast condition is more critical since both the maximum horizontal and vertical loads happen in the ballast condition.

- Minimum Main Bearing Vertical Load

The minimum vertical load occurs in the fully loaded condition. The FPSO pitches down, the mooring vertical load is small under this scenario. It can be seen that the minimum vertical load remains positive, i.e. uplift of the turret does not occur. It can also be confirmed by the negative acceleration at the center of gravity of the turret that uplift does not happen.

- Maximum Lower Bearing Horizontal Load

The maximum lower bearing load is driven by the mooring horizontal load. The maximum lower bearing horizontal load occurs in the loaded transverse condition.

- Change of FPSO Relative Heading to Lines

In general, the change of FPSO heading relative to the lines does not affect the turret global loads significantly under the same loading condition.

- Single Mooring Line Damage

It is observed that the single mooring line damage reduces both the horizontal and the vertical load on the main bearing. The reduction on the main bearing bending moment is less than 10%.

CONCLUSIONS

Accurate prediction of turret load determines the confidence of the structure design of turret system. It plays key role in the development of complete turret system including both structure design, mechanical system selection and piping design. A thorough discussion of determination of turret load for internal turret solution is not available in published literature. This work presented in this paper elaborates methodology, procedure, the way of developing a rational coupled model incorporating elasticity of turret cylinder, the way of simulating contact between lower bearing and turret structure and comprehensive criteria to interoperate time series of multiple turret loads. A case study was conducted adopting the proposed approach. The criticality analysis of simulation results reveals the coupling of FPSO motion and the turret control loads, which will also be insightful for similar engineering designs.

ACKNOWLEDGEMENTS

The authors would like to thank CNOOC and COTEC for support of this work and granting permission to publish this paper. The authors are grateful to engineers at COTEC who have contributed to the development of the turret-moored system.

REFERENCES

American Petroleum Institute, "Design and Analysis of Stationkeeping Systems for Floating Structures", API RP 2SK, 2005.

American Petroleum Institute, "Recommended Practice for Design, Manufacture, Installation, and Maintenance of Synthetic Fiber Ropes for Offshore Mooring", API RP 2SM, 2007.

Duggal, A.S., Heyl, C.N., and Poranski, P.F., "Terra Nova FPSO: Integration of Model Tests and Global Analysis", Proceedings of 10th International Society of Offshore and Polar Engineers (ISOPE), Seattle, Washington. 2000.

Luo, Y., and Ye, W., "Mooring and Riser Design for GOM FPSOs in 10000 ft Water Depth", Proceeding of Offshore Technology Conference, Houston, TX. 2005.

Howell, G.B., Duggal, A.S., and Lever, G.V., "The Terra Nova FPSO Turret Mooring System", Houston, TX. 2001.

Soares, C.G., Fonseca, N., Pascoal, R., "Experimental and Numerical Study of the Motions of a Turret Moored FPSO in Waves", Journal of Offshore Mechanics and Arctic Engineering, 2004.

Oliveira, A.C., and Fernandes, A.C., "The Nonlinear Roll Damping of a FPSO Hull", Journal of Offshore Mechanics and Arctic Engineering, 2014.

Luo, Y., and Baudic, S., "Predicting FPSO Responses Using Model Tests and Numerical Analysis", Proceedings of 13th International Society of Offshore and Polar Engineers (ISOPE), Honolulu, Hawaii. 2003.

Kim, M.H., Koo, B.J., Mercier, R.M., and Ward, E.G., "Vessel/mooring/riser Coupled Dynamic Analysis of a Turret-moored FPSO compared with OTRC Experiment", Ocean Engineering 32, pp. 1780-1802, 2005.

COLLISION AVOIDANCE SYSTEM FOR FPSO – SHUTTLE TANKER BERTHING

MARK A. AZADPOUR
AZADPOUR ENTERPRISES

SANAM SAEBI
SBM OFFSHORE

Proceedings of the 21st Offshore Symposium, February 2016, Houston, Texas
Texas Section of the Society of Naval Architects and Marine Engineers

ABSTRACT

In recent years, a number of tankers have run into FPSOs' during offloading cycle due to human error even with utilization of other positioning systems that are currently available. Collision with an FPSO during this cycle is very costly and can put the FPSO out of operation for several months. For instance, a collision in the North Sea on October 8th 2009 caused the FPSO to be out of production at the end of Jan 2010. In order to avoid berthing accidents, we propose a warning system that is separately deployed to provide early collision warning. The system detailed in this paper, assists the berthing process whether human watch person is present or not and is independent of any other system deployed.

Keywords: SNAME, Offshore, Symposium, FPSO, Berthing, Collision Avoidance, Offloading Cycle, Shuttle Tanker

INTRODUCTION

During normal operation of an FPSO, shuttle tankers approach the FPSO during offloading cycle. Collision with an FPSO during this cycle is very costly due to helideck, Hose reel and other expensive equipment located on the aft of the vessel.

In recent years, a number of tankers have run into FPSOs' during offloading cycle caused by human error in conjunction with positioning systems that are available and used today. The berthing process, typically, is monitored by closed circuit cameras and a watch keeper on FPSO, to observe and direct this process manually.

In addition, a number of positioning systems are utilized. DARPS is a system that currently is used to provide distance information from shuttle tanker. These DP systems use various inputs for the determination of position such as ARTIMIS and Fan-beam antennas which are fitted onboard the ship to provide signals to the DP system of the shuttle tanker.

The presence of a tug boat per regulations is required for DP1 systems and is not required with DP2 systems. Thus the lack of human tug watchman and the failure of automation systems have resulted in a number of accidents.

In order to avoid berthing accidents, we propose a warning system that is separately deployed to provide early collision warning. The system detailed in this paper, assists the berthing process whether human watch person is present or not. It is independent of any DP system and does not require input from DP to provide another independent collision avoidance system when a shuttle tanker is approaching an FPSO.

BACKGROUND

The distance maintained between a vessel and the shore, another vessel, or any object is called berthing. Due to the closeness of the vessel during this process, the chances of collision are increased if a safe distance is not maintained. The typical distance between shuttle tanker bow and the FPSO is 250 meters with minimum distance of 150 meters. The minimum distance between tanker bow manifold and the nearest riser line is 40 and the distance to mooring line is 20 meters.

There are two types of collision as shown in Figure 1, the glancing blow or direct hit.

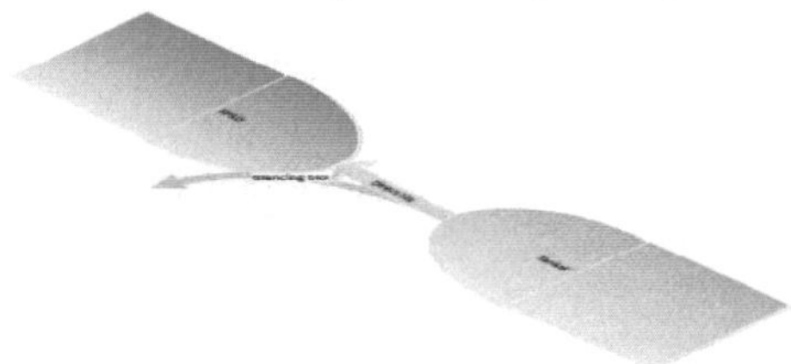

Figure 1 Glancing blow and direct hit are two types of collisions

There are two types of offloading:

- **Indirect:** A small offloading platform is utilized with submerged systems such as OLS (Offshore Loading System), SAL (Single Anchor Loading), STL (Submerged Turret Loading).
- **Direct:** The tanker and FPSO are connected via hawser and hose connections.

There are two types of direct offloading:

- Tandem
- Alongside

In 15 years prior to 2011, there were several accidents in the North sea including DP shuttle tanker collision in tandem offloading occurred on Emerald FSU, Gryphon FPSO, Captain FPSO, Schiehallion FPSO, Norne FPSO, Njord FSU. The highest shuttle speed at the moment of impact reached over 2 knots which translated to an energy transfer in the range of a few Mega Joules (MJ) to 100 MJ. Due to this level of force, major damages have been inflicted on FPSO and in some instance the flare tower at the stern of the FPSO has been damaged.

Characteristic	*Variations (Base case in bold)*
FPSO station keeping capabilities	**Internal turret with 8-12 point mooring system**
FPSO heading keeping capabilities	Without heading control **With heading control**
ST heading and station keeping capabilities	Main propulsion (single or twin screw) No DP system DP1, **DP2** or DP3 systems
Off-loading mode	**DP operated** Taut hawser operated
Interface systems	**With hawser connection** Without hawser connection
Distance FPSO-ST	50 –100 meter **80 meter** 150 meter (not used at present, but considered as a sensitivity)

Figure 2 Variation in FPSO/FSU Field Configurations

In addition to the vessel damages, there is a possibility for hose rupture and oil spillage. For instance, in August 20, 1980, after hawser breakage, the hose was ruptured which caused oil to be pumped to the tanker bow area which subsequently caught fire. In another incident, on December 12th, 2007, a rupture in the hose on Statfjord caused about 4,400 cubic meter of oil to be spilled.

Given the high cost of collisions, there have been many studies conducted on the causes of collisions and the frequency of these close calls. Dr. Haibo Chen, et al. Used the following collision frequency model for practical frequency model between DP shuttle tanker and FPSO:

P(collision) = P(drive_off) x P(failure of recovery | drive_off) (1)

Where:

P (drive_off): Frequency of shuttle tanker drive-off forward which has the potential to cause collision.

P (Failure of recovery | drive_off): Failure probability of recovery initiated from shuttle tanker, given a shuttle tanker drive-off forward.

Using the above model for 1999-2003 case studies, it shows that DP2 shuttle tanker generic drive-off forward frequency is around 0.02 to 0.05 per year (assuming 24 hour operation, and one offloading per week).

Furthermore, in the same study, the main causes of berthing collisions are identified as:

1. Controllable pitch propeller, e.g. failure in pitch control, or feedback
2. Position reference systems, e.g. common mode failure of DARPS
3. Sensors, e.g. failure of one wind sensor, hawser tension sensor, or draught sensor
4. DP software, e.g. software error, hidden bugs
5. Human operators, e.g. human errors by DP operator

To further investigate the causes of collision during off-loading cycle, the main contributors to collision frequency during drive-off were studied. These vulnerable situations during drive-off are:

- surging
- yawing

Tanker is vulnerable to drive-off when relative motions between FPSO and tanker are excessive. Figure 3 depicts each of these situations detailing the position of the shuttle tanker relative to the FPSO.

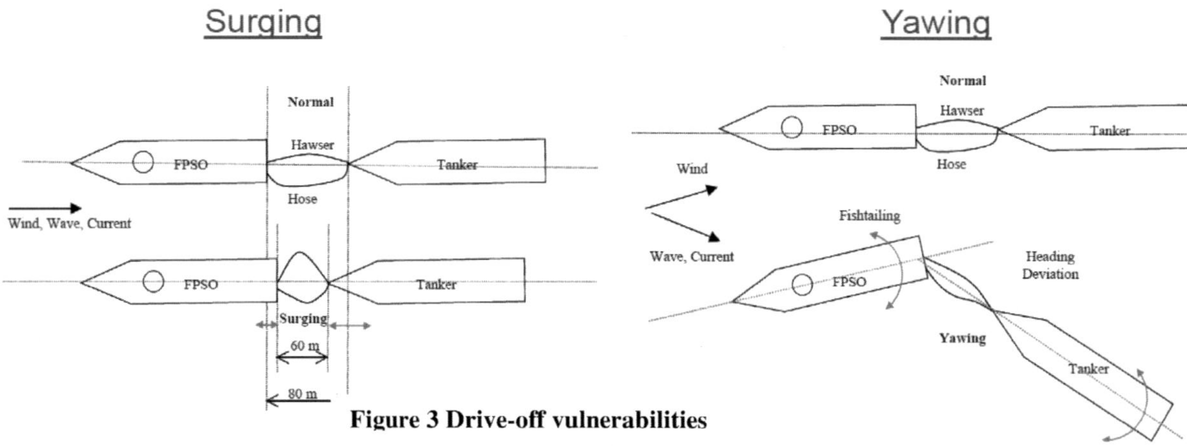

Figure 3 Drive-off vulnerabilities

The significant wave height (Hs) during offloading cycle is determined by the following formula:

Hs = 5.5 m loading / 4.5 m connection (2)

The results of another study of main contributors to collision frequency based on incident experience and expert evaluation is shown in Table 1.

Table 1 Ranking of factors during drive-off

RIF Group	Ranking	Contribution
Technical Dependency alone	4	10%
Human Operation dependency	2	15%
External Condition Alone	7	2%
Technical and Human Combined	1	40%
Technical and External Combined	6	8%
Human/Oper. And External	5	10%
Tech & Human/Oper. & External	3	15%

As Table 1 reveals, human and technical factors account for 40% of collision causes during drive-off and they have the highest contribution percentage. Therefore, addressing them to avoid berthing collisions has the highest Return on Investment (ROI). This is the main focus of this paper and the presented solution as detailed next.

HUMAN INTERVENTION

The natural question to ask during a failure is that why human intervention fails during these situations or it occurs too late to avoid collisions. There have been studies done to determine the root cause of human intervention failure.

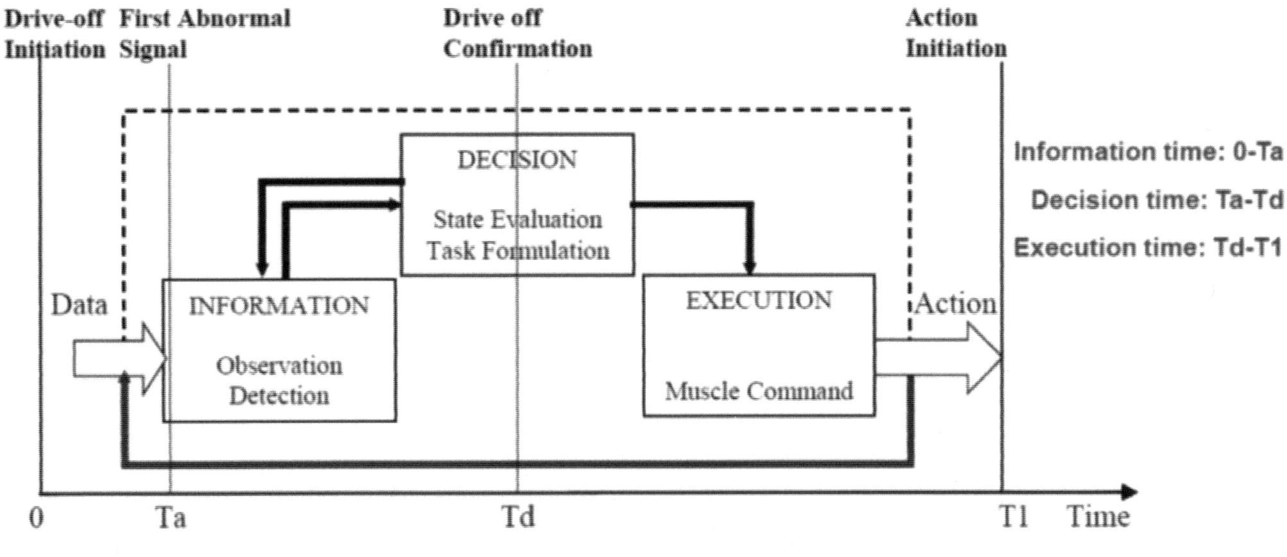

Figure 4 Wickens' model & Step-ladder model for human information processing stages.

Studies have shown that the recovery failure probability is high due to a lack of time for the DP operator to initiate the recovery action in a tanker drive-off scenario.

Figure 4 shows the Wickens' model and step-ladder model for human information processing stages.

Recovery failure reduction can be achieved by:

- Increasing the time window for the operator to react
- Reducing the time it takes for the operator to react.

A substantial increase in the separation distance between FPSO and the tanker, and a possible reduction of the main propeller(s) forward thrust that can be used by the tanker DP system, are two measures to increase the time window for the operator to react.

The effective reduction of operator reaction time can be achieved by a number of measures, i.e., selection, training, procedure, crew resource management, and automation support, which are designed to improve early detection as well as effectively reducing the DP operator's time in diagnosis and situation awareness.

Therefore, this paper focuses on a mechanism by which human error/technical errors are circumvented and the response time is aided by an independent system as described next.

AVOIDANCE SYSTEM

The collision avoidance system consists of three sub- systems as depicted in Figure 5. Step one is the detection of potential collision followed by creating a warning to the operators and concluded by a potential autonomous action to avoid collision by reversing and steering in the opposite direction.

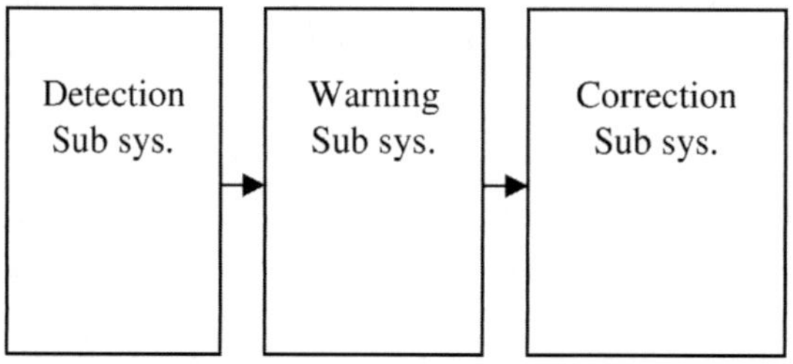

Figure 5 Collision avoidance Sub-systems

Each one of these sub-systems is detailed in the following sections. Optionally, various combinations of these sub-systems could provide users with level of functionality that is meeting the desired functionality/cost target.

DETECTION SUB-SYSTEM

The warning subsystem must detect when the shuttle tanker is getting close to FPSO. The proposed system includes a series of sensors that use already deployed principles including radar, camera, or laser technology to detect close proximity and to alert operator and/or make corrections automatically. Each technology has its own pros and cons. For instance, the radar system is operable in all weather conditions while laser and camera based systems are not operational during poor weather conditions (e.g. in thick foggy conditions). Therefore, the selection of detecting sensors depends on the cost requirement of such system as well as the requirement of deployment environment.

For instance, if radar based detection is chosen, various frequency ranges could be deployed. A radar based detection system could be based on Hughes Electronics radar system running at 77 GHz. The idea is that a series of emitter/detector localized sensor are either mounted to the side of FPSO or are deployed in floating devices with riser as shown in Figure 6. These floating devices are lowered to water level and via suspended mechanism are connected to the side of FPSO in case of alongside offloading and in the aft of the FPSO in the case of tandem offloading.

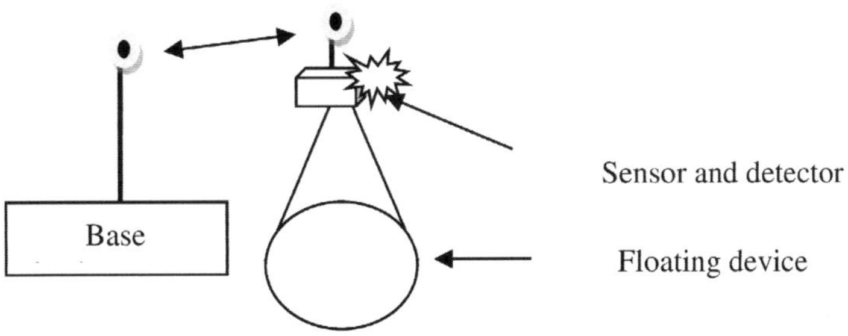

Figure 6 Suspended sensor sub-system

These units would transmit information via radio frequency waves to a base station antenna located on the FPSO. This transmission system could use encryption for security as well as data recovery in case of bad weather condition. These units could be solar powered to be independent and self-sufficient except for maintenance requirement and battery replacement.

WARNING SUB-SYSTEM

The information from various sensor/detector pairs are collected via base station and by use of intelligent software module and in conjunction with GPS system, the exact orientation of tanker and velocity of tanker is calculated and displayed on the operator terminal.

In addition, the warning subsystem includes an operator terminal where persons in charge enter the control limit parameters. These parameters specify the proximity at which an alarm would be entered and operators are notified via wireless devices that the clearances are violated on the tanker remote terminal. For instance, a set of measures at which various alarms are entered is shown in Table 2.

Table 2 Alarm level setting for the warning sub-system

Alarm level	Distance (m)
Red	<=250
Yellow	<=350
Green	> 350

This scenario is referred to as the open loop operation of the system where the manual intervention is required by the operator to avoid collision (Figure 7). In this figure, we designate which portion of the system resides on the FPSO and tanker respectively.

In other words, the detection sub-system detects the distance between FPSO and tanker. This information is fed to the warning sub-system. If the distance violates a control limit entered by the operators, a warning or alarm is generated to alert operators. However, no automatic action is taken and it is up to human intervention to make the manual correction to avoid potential collision.

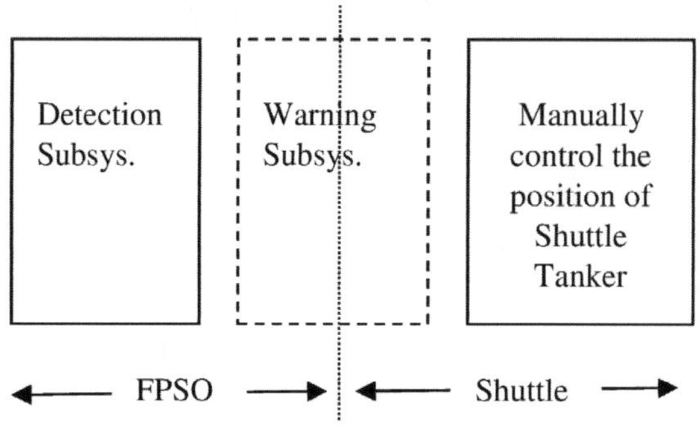

Figure 7 Open loop operation: Dash arrow signifies manual intervention

CORRECTION SUB_SYSTEM

In the case of using correction sub-system, the collision avoidance system is utilized in a closed loop operation where the corrective action is taken automatically by either increasing the distance between the shuttle and FPSO and/or steering the shuttle tanker in the opposite direction. This system is deployed in conjunction with DP2 correction system and interfaces to the same instrumentation to intervene and make corrections via instrumentation as shown in Figure 8.

Information from detection subsystem is directly fed into the correction subsystem.

As shown in Figure 8, in the closed loop configuration, the detection subsystem information is relayed directly to the correction subsystem to allow the software control system make proper adjustments. This software is the DP2 software that is already deployed in the tanker.

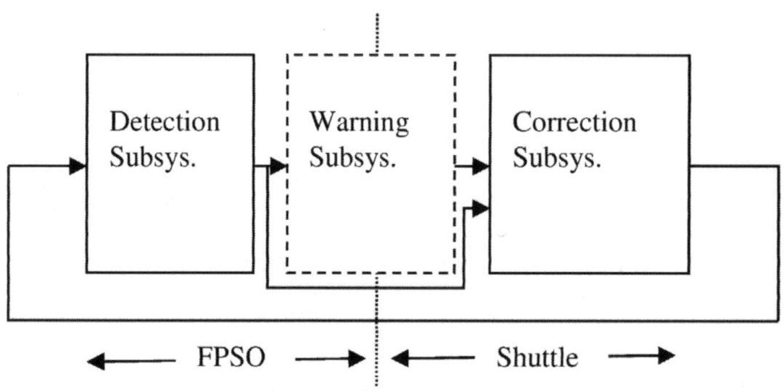

Figure 8 Closed loop operation

The interface requires the use of Application Programming Interface (API) of the DP2 system to send control calls to the DP2 system as shown in Figure 8.1.

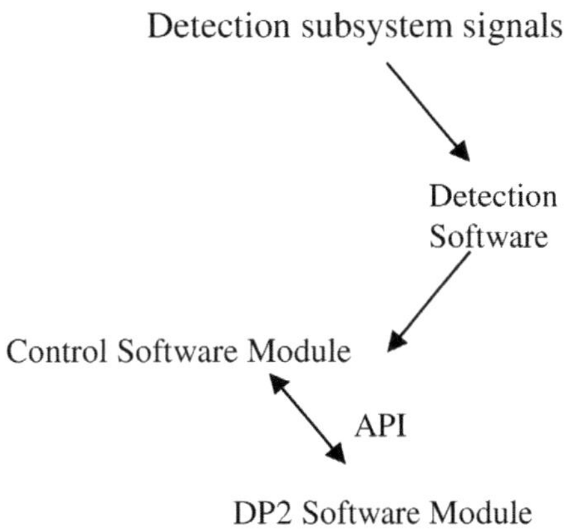

Figure 8.1 API is used to send commands to DP2 software system and to get feedback on actions taken

Furthermore, the adjustment could be refined by the information input into the warning subsystem that is forwarded to the correction subsystem. Upon corrective action taken either manually or automatically, that information is fed to the detection subsystem to verify that indeed the parameters under the consideration are been adjusted properly.

For instance, if the separation distance between the shuttle tanker and the FPSO is of interest, then the detection system should be able to verify the increase or change of direction between the two vessels. This allows for adjustment of alarms, for instance, from high-alert (red) condition to mid-alert (yellow) condition.

In addition to the automatic corrective action, tug boats could be notified via the wireless warning system and to also engage and move the shuttle tanker away from the zone of potential danger as shown in Figure 9 to expedite the corrective action. Since this is a closed loop operation, the movement away from FPSO is detected automatically via the detection sub-system and the proper action is taken in the correction sub-system.

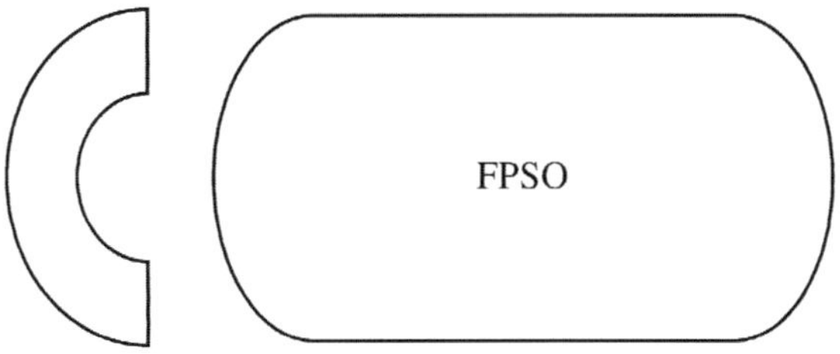

Figure 9 Zone of potential danger

The general closed loop control system is a classic control mechanism which consists of a Proportional, Integral, Derivative (PID) closed loop control system. This loop provides feedback mechanism. The transfer function of a PID controller is given by:

$$\frac{X(s)}{F(s)} = \frac{K_d s^2 + K_p s + K_i}{s^3 + (10 + K_d)s^2 + (20 + K_p)s + K_i} \tag{4}$$

Which contains poles and zeros. In order to have a stable system, poles should be avoided which is part of the software control routine to block such set points.
This loop allows for fine tuning of the amount of correction taken and would adjust the correction amount as the distance increases or to take a more drastic and faster action if vessels move closer to each other. The block diagram of a PID loop is shown in Figure 10.

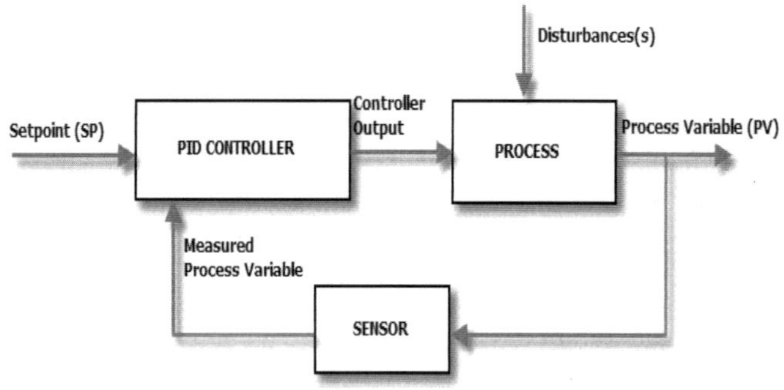

Figure 10- PID Controller with PID incorporated

The PID is tuned per vessel to account for the response time and the dead zone of mechanical system when small corrections are applied to ensure the smooth ramping and avoid over shooting.

These PID set-points are entered via the automatic control software in the case of closed loop configuration. These set-points along with sensor inputs are used to calculate the amount of correction to be applied and this iterative process is continued until desired results are accomplished.

As an example of implementation of such collision avoidance system, consider the OSPREY system by Ocean Tech Systems that has been recently designed to perform the above functionality. The system consists of a base station that is the repository for the information and is permanently installed in an FPSO. The detectors are solar power and installed on the BUOY. There is also a portable unit that is used by the approaching tanker to detect alarms. At the time of this writing there was no information available on increase speed and reduction of probability of collision based on this system.

CONCLUSION

In recent years, a number of tankers have run into FPSOs' during offloading cycle due to human error even with utilization of positioning systems that are currently available. In this paper, the causes and effect of collision between a shuttle tanker and an FPSO during this cycle is detailed. A Collision is very costly and can put the FPSO out of operations for several months. Various types of off-loadings as well as clearances were discussed and human intervention model was portrayed.

In order to avoid berthing accidents, we proposed a warning system that is separately deployed to provide early collision warning. The system detailed in this paper, assists the berthing process whether human watch person is present or not and is independent of any other system deployed. The system consists of three sub-systems to detect potential collisions, to warn and alert responsible personnel, and to take corrective action, if deployed in a closed loop manner. The result is avoidance of such mishaps from occurring and saving substantial costs to the operating companies. In addition, the potential environmental impact and costs due to a collision are avoided.

REFERENCES

SEVAN Marine, "*Significant oil spill risk reduction by introducing a novel scheme for offloading oil from a geostationary FPSO to shuttle tankers* ", 2011.

Journal of Reliability Engineering and System Safety 84, pp 169-186 ,2004.

Chen Haibo, " *Risks in DP Shuttle Tanker Offloading Operations*", http://www.cesos.ntnu.no/attachments/078_S18_Haibo%20Chen.pdf, 2013.

Abbott, J. "1988 Computer-Aided Wheelhouse Systems." *Journal of Maritime Engineering,* 12:3, pp 64-72, 1988.

Tobin, S. and M. Darman. *Elementary Naval Architecture.* New York: McGraw-Hill, 1957.

Fash, Barbara W. and William L. Fash. "Saving the Maya Past for the Future: Copan's New Sculpture Museum." *Peabody Museum of Archaeology,* 1996, www.peabody.harvard.

Jones William," Keeping Cars from Crashing", IEEE Spectrum, http://spectrum.ieee.org/transportation/advanced-cars/keeping-cars-from-crashing

Trav Tiger, "PID en updated feedback - Own work. Licensed under CC BY-SA 3.0 via Commons" - https://commons.wikimedia.org/wiki/File:PID_en_updated_feedback.svg#/media/File:PID_en_updated_feedback.svg

SNAME
THE INTERNATIONAL COMMUNITY
FOR MARITIME AND OCEAN PROFESSIONALS

A PANEL METHOD APPLIED TO THE PREDICTION OF THE PERFORMANCE OF DUCTED PROPELLERS AND THRUSTERS

SPYROS A. KINNAS
PROFESSOR, THE UNIVERSITY OF TEXAS AT AUSTIN

HONGYANG FAN
MASTER'S STUDENT, THE UNIVERSITY OF TEXAS AT AUSTIN

YE TIAN
POST-DOCTORAL FELLOW, THE UNIVERSITY OF TEXAS AT AUSTIN

Proceedings of the 21st Offshore Symposium, February 2016, Houston, Texas
Texas Section of the Society of Naval Architects and Marine Engineers

ABSTRACT

Ducted propellers have been, for a long time, a viable alternative of propulsion; due to their higher efficiency at high thrust coefficients, less sensitivity to the ambient flow, and more robust mechanical layout, than open propellers. Applications of ducted propellers or thrusters can be found in many types of ships, particularly in floating production, storage and offloading (FPSO) and liquid natural gas (LNG) hulls for maintaining a vessel's position.

In the 20th Offshore Symposium of Texas SNAME we presented a hybrid method which coupled a Vortex-Lattice Method (VLM) solver with a Reynolds-Averaged Navier-Stokes (RANS) to simulate the flow around ducted propellers.

In this work, a panel method is applied to the prediction of the performance of ducted propellers. A ducted propeller with a rectangular shape tip and a thruster with a blunt trailing edge duct are considered. The effects of viscosity on the duct are evaluated via coupling the method with an integral boundary layer solver. One significant feature of this method is the application of full wake alignment scheme in which the trailing vortex wake sheets of the blades are aligned with the local flow velocity by also considering the effects of duct and duct wake. Additionally, a procedure of repaneling the duct is simultaneously introduced to improve the accuracy of the method. The results from the improved wake model are compared with those from a simplified wake alignment scheme. At the same time, full-blown RANS simulations of the three dimensional problem are conducted. The propeller thrust and torque, and the blade pressures, predicted by the present panel method, using the improved wake alignment model, show very good agreement with experimental measurements, and with results from a hybrid method developed by the Ocean Engineering Group of UT Austin, as well as with full blown results from RANS solvers.

Keywords: Ducted propeller, Thruster, Panel Method

INTRODUCTION

As a viable alternative of propulsion, ducted propellers can produce higher efficiency than open propellers particularly at high loading conditions or low speeds. The utilization of ducted propellers can be found in various types of marine vehicles as well as offshore structures such as the floating production storage and offloading (FPSO) and liguid natural gas (LNG) hulls. The duct surrounding the blades, in addition to improving efficiency, it can provide extra benefits when compared to open propellers, as the propeller bounded by the duct becomes less sensitive to ambient flow and more resistant to environmental debris.

Satisfactory predictions of the performance of ducted propellers have been a significant concern in the marine industry for a long time. Full blown Reynolds Averaged Navier-Stokes (RANS) simulations, like that of Abdel-Maksoud and Heinke (2003), although can provide reliable results, are considered to be numerically too expensive, especially at the design stage. Additionally, the RANS simulations require significant time in setting the proper grid. Boundary element methods are found to be an ideal alternative since they do not require the discretization of the whole fluid domain and therefore are much more computationally efficient. Hess et al. (1964) published the pioneering work of panel method. Since then, a large number of panel codes have been proposed. Subsequently, Gibson and Lewis (1973), Hess et al. (1985) developed a panel method for the specific application of marine propellers. Morino et al. (1974) introduced a panel method based on Green's formulation in which the primary unknown was the velocity potential. Lee (1987) and Kerwin et al (1987) carried out a comprehensive investigation of panel methods for a more suitable formulation for simulations of open and ducted propellers.

The free vortex sheets shed from the trailing edge of a propeller blade are material surfaces, which have to be aligned with the local flow velocity. The influence of the wake shape on the prediction of propeller performance was initially discussed by Kerwin and Lee (1978). After that Greeley et al. (1982) proposed a fast scheme of wake alignment denoted as PSF-2 wake alignment. In this scheme, the trailing wake shape of a propeller subjected to axisymmetric inflow is investigated through an iterative vortex lattice method. The correct pitch of the wake geometry is firstly calculated at some points in the wake area, after which the wake geometry is updated by adopting the newly computed pitch distribution. The process continues until convergence of the wake geometry is achieved. Kinnas and Pyo (1999) improved this scheme such that it is applicable to the problem of a propeller subjected to inclined inflows. Pyo and Kinnas (1997) developed a high order panel method by adopting biquadratic dipoles on the trailing vortex sheet. This method, though produced good prediction of the roll-up characteristic of the wake and the force performance around design advance ratio, was relatively more time-consuming than the panel method. More recently Lee and Kinnas (2004, 2005) implemented an unsteady wake alignment scheme based on panel method. The method introduced a finite core tip vortex at the tip of the wake sheet and it was applied to the numerical modeling of unsteady wake sheet and developed tip vortex cavitation on hydrofoils and marine propellers. Tian and Kinnas (2012) presented a wake alignment model based on a pseudo-unsteady scheme for a panel method. The scheme is validated with experiments and RANS results in the case of propellers at both design loading condition and high loading conditions.

In the case of ducted propellers there have been hybrid methods, like that by Kerwin et al (1987), where the blade is modeled via a vortex-lattice method (VLM) and the propeller is modeled via a panel method. Panel methods applied on both, propeller and duct, have been developed by Hughes (1997) (who also implemented an orifice equation model to account for the viscous flow through the gap), Lee and Kinnas (2006) (who also extended it in the case of cavitating flows), and Baltazar et al (2013, 2015). Methods which combine a VLM on the propeller with an Euler or RANS solvers on the duct and/or hull have been developed by Gu and Kinnas (2003) (who also implemented a tip gap model), Kinnas et al (2009, 2012a, 2012b) (who also looked at unsteady cavitating flows), Tian et al (2014), and Tian and Kinnas (2015).

In the present work, the PSF-2 type of wake alignment as well as the full wake alignment model of Tian and Kinnas (2012) are extended in the case of ducted propellers, and then applied into a practical case: a ducted propeller with square blade tip. Result from using the PSF-2 type scheme and the full wake alignment scheme will be correlated with experimental measurements and full blown RANS simulations conducted via commercial CFD packages.

METHODOLOGY

1. The Panel Method (Three Dimensional Viscous/Inviscid Interactive Method)

1.1 Inviscid Formulation

A summary of the lower order panel method for three dimensional problems is provided. In order to resolve the effect of viscosity, the panel method is coupled with a two dimensional integral boundary layer solver. Unlike in 2-D problems, in 3-D problems, it would be computationally too intensive to solve for the exact 3-D boundary layer structures through 3-D integral boundary layer equations. For a 3-D geometry like a blade, the present coupling method takes the assumption that the boundary layer growth along the spanwise direction is negligible compared to that in the chordwise direction. To take into account the three dimensional characteristics, the effects from other strips are included.

Given a three-dimensional body such as the duct in Fig.1.1, where the duct surface is divided into N panels in the chordwise direction and the trailing edge wake is divided into N_w panels in the streamwise direction. Both the duct surface and the wake are circumferentially divided into M strips.

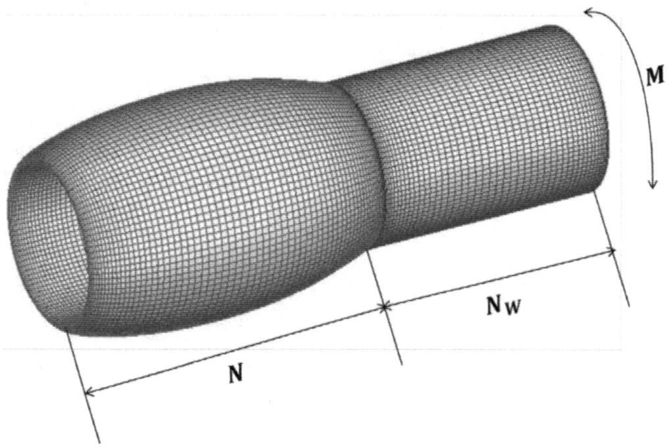

Figure 1.1 Paneling of the duct in the viscous/inviscid interactive method

In 3-D problem, since the interaction between different strips should be taken into consideration, the discretized equation for inviscid formulation could be written in matrix form as:

$$\begin{pmatrix} A^{11} & A^{12} & \cdots & A^{1M} \\ A^{21} & A^{22} & \cdots & A^{2M} \\ \vdots & \vdots & \ddots & \vdots \\ A^{M1} & A^{M2} & \cdots & A^{MM} \end{pmatrix} \begin{pmatrix} \phi^{1,inv} \\ \phi^{2,inv} \\ \vdots \\ \phi^{M,inv} \end{pmatrix} = \begin{pmatrix} S^{11} & S^{12} & \cdots & S^{1M} \\ S^{21} & S^{22} & \cdots & S^{2M} \\ \vdots & \vdots & \ddots & \vdots \\ S^{M1} & S^{M2} & \cdots & S^{MM} \end{pmatrix} \begin{pmatrix} \eta^{1} \\ \eta^{2} \\ \vdots \\ \eta^{M} \end{pmatrix} \tag{1.1}$$

Where A^{PQ} is a matrix of size $N \times N$ which indicates the dipole influence coefficients of panels at strip Q on the control points at strip P.

$$A^{PQ} = \left[A_{ij}^{PQ}\right] = \begin{pmatrix} A_{11}^{PQ} & A_{12}^{PQ} & \cdots & A_{1N}^{PQ} \\ A_{21}^{PQ} & A_{22}^{PQ} & \cdots & A_{2N}^{PQ} \\ \vdots & \vdots & \ddots & \vdots \\ A_{N1}^{PQ} & A_{N2}^{PQ} & \cdots & A_{NN}^{PQ} \end{pmatrix}_{N \times N} \tag{1.2}$$

A_{ij}^{PQ} is the dipole influence coefficient of panel j at strip Q on control point i of strip P.

$\phi^{Q,inv}$ is a vector of size $N \times 1$ indicating the inviscid perturbation potential at strip Q.

$$\boldsymbol{\phi}^{Q,inv} = \left[\phi_j^{Q,inv}\right] = \begin{pmatrix} \phi_1^{Q,inv} \\ \phi_2^{Q,inv} \\ \vdots \\ \phi_N^{Q,inv} \end{pmatrix}_{N\times 1} \quad (1.3)$$

$\phi_j^{Q,inv}$ is the inviscid perturbation potential at panel j of strip Q.

$\boldsymbol{S}^{PQ}$ is a matrix of size $N \times N$ and includes the source influence coefficients of panels at strip Q on the control points at strip P.

$$\boldsymbol{S}^{PQ} = \left[S_{ij}^{PQ}\right] = \begin{pmatrix} S_{11}^{PQ} & S_{12}^{PQ} & \cdots & S_{1N}^{PQ} \\ S_{21}^{PQ} & S_{22}^{PQ} & \cdots & S_{2N}^{PQ} \\ \vdots & \vdots & \ddots & \vdots \\ S_{N1}^{PQ} & S_{N2}^{PQ} & \cdots & S_{NN}^{PQ} \end{pmatrix}_{N\times N} \quad (1.4)$$

S_{ij}^{PQ} is the source influence coefficient of panel j of strip Q on control point i of strip P.

$\boldsymbol{\eta}^{Q}$ is a vector of size $N \times 1$ including the source strength at the strip Q.

$$\boldsymbol{\eta}^{Q} = \left[\eta_j^{Q}\right] = \begin{pmatrix} \eta_1^{Q} \\ \eta_2^{Q} \\ \vdots \\ \eta_N^{Q} \end{pmatrix}_{N\times 1} \quad (1.5)$$

η_j^{Q} is the source strength at panel j of strip Q.

1.2 Viscous Formulation

2-D wall transpiration model is applied in order to include the effects of viscosity. The lower order panel method is coupled with a 2-D integral boundary layer solver through the term of edge velocity.

As shown in Figure 1.2, blowing sources which induce normal velocity on the surfaces are added into the inviscid formulation of the panel method. The surrounding flow in pushed away from the solid surface such that the boundary layer can be simulated.

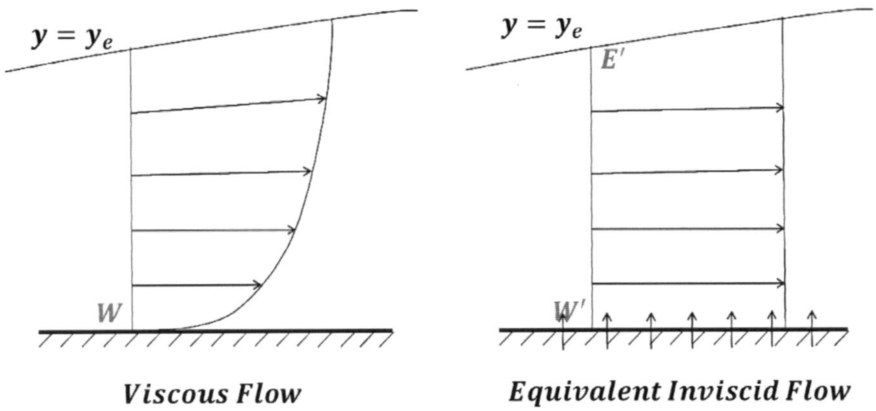

Figure 1.2 Diagram for wall transpiration model, adjusted from Figure 2.1 of Yu (2012)

The blowing sources should be granted with proper strength so that the edge velocity U_e at $y = y_e$ is the same as that in the viscous flow. By applying the steady Bernoulli Equation and the boundary layer theory, it can then be deduced that the pressure at the point W' is the same as that at the point W in the viscous flow. That is, the equivalent inviscid flow shares the same pressure distribution with the real viscous flow. The strength of the boundary layer ("blowing") sources are determined by solving the boundary layer equations via XFOIL, as developed for airfoils by Drela (1989), and later extended to propellers by Hufford et al (1994), Sun and Kinnas (2008), and Kinnas et al (2012).

By taking into account the effects of blowing sources, matrix form of the discretized equation for inviscid formulation, (1.1) is updated into the viscous formulation as follows:

$$\begin{pmatrix} A^{11} & A^{12} & \cdots & A^{1M} \\ A^{21} & A^{22} & \cdots & A^{2M} \\ \vdots & \vdots & \ddots & \vdots \\ A^{M1} & A^{M2} & \cdots & A^{MM} \end{pmatrix} \begin{pmatrix} \phi^{1,vis} \\ \phi^{2,vis} \\ \vdots \\ \phi^{M,vis} \end{pmatrix} = \begin{pmatrix} A^{11} & A^{12} & \cdots & A^{1M} \\ A^{21} & A^{22} & \cdots & A^{2M} \\ \vdots & \vdots & \ddots & \vdots \\ A^{M1} & A^{M2} & \cdots & A^{MM} \end{pmatrix} \begin{pmatrix} \phi^{1,inv} \\ \phi^{2,inv} \\ \vdots \\ \phi^{M,inv} \end{pmatrix} + \begin{pmatrix} B^{11} & B^{12} & \cdots & B^{1M} \\ B^{21} & B^{22} & \cdots & B^{2M} \\ \vdots & \vdots & \ddots & \vdots \\ B^{M1} & B^{M2} & \cdots & B^{MM} \end{pmatrix} \begin{pmatrix} \sigma^{1} \\ \sigma^{2} \\ \vdots \\ \sigma^{M} \end{pmatrix} \tag{1.6}$$

Along a specific strip L, after several steps of reorganization, the above equations can be deduced into total viscous potential which includes inflow velocity potential and viscous perturbation potential with blowing sources terms:

$$\phi_{total}^{L,vis} = \phi_{in}^{L} + \phi^{L,inv} + \textstyle\sum_{m=1,m\neq L}^{M} J^{Lm} \left(\phi^{m,inv} - \phi^{m,vis}\right) + \sum_{m=1}^{M} H^{Lm} \sigma^{m} \tag{1.7}$$

It can be seen that the viscous total potential can be obtained by adding the induced potential by the blowing sources and the interaction term of viscous potential between strips to the inviscid total potential. Differentiating the viscous total potential along the streamwise direction of the body will give the viscous velocity or edge velocity.

Similarly, the viscous potential on the wake can be obtained by summing the induced potential by the source and dipole on the body, the dipole on the wake and the blowing source on the boy as well as wake. The detailed formulation of the viscous/inviscid coupling can be found in Yu (2012), Purohit (2013) and Fan (2015).

1.3 Coupling Procedure

Unlike the coupling procedure in the 2-D problem which is quite straightforward, the coupling algorithm is performed in an iterative manner because of the interactive effects between different strips.

The iterative algorithm is introduced in details as following:

STEP 1: solve for the inviscid velocity as the initial guess of the edge velocity.

STEP 2: calculate the displacement thickness δ^* from the edge velocity distribution on each strip through the two dimensional boundary solver equations.

STEP 3: calculate the blowing source strength σ with the edge velocity and displacement thickness on each strip.

STEP 4: update the edge velocity with the calculated blow source strength.

STEP 5: go to STEP 6 if the converged solution is achieved at each strip. Otherwise go back to STEP 2.

STEP 6: update the viscous potential on the body and wake at all strips.

STEP 7: update the edge velocity at all strips by taking into account the interaction terms between strips and bodies.

STEP 8: continue to STEP 9 if a convergence of edge velocity is obtained. Otherwise return to STEP 1 for new iterations.

STEP 9: save the solutions and exit the iteration process.

2. Wake Alignment Scheme

The free vortex sheets shedding from the trailing edge of a propeller blade are material surfaces, which have to be aligned with the local flow velocity. A lot of research has been carried out on the alignment scheme of the trailing wake sheet since a reasonable representation of the wake is of crucial significance for the calculation of induced velocities on the blade surface and hence the overall loading on the propeller.

2.1 PSF-2 Type Alignment Scheme

The PSF-2 wake model was originally developed by Kerwin (1981) and then extended by Greeley et al. (1982). In this model the trailing vortex wake is composed by the parts of a transition wake and an ultimate wake. The detailed procedure of PSF-2 alignment is presented as follows.

At the first step, the trailing vortex lines at the most inner and outer end of the transition region are aligned with the computed flow. The trailing vortex lines in between are determined through interpolation. The axial and tangential locations of the vortex lines are computed with the induced velocity by the propeller. On the other hand, the radial location of the vortex lines is dependent on the contraction of the transition wake at the tip and the radius of the ultimate wake and the roll-up of the wake is suppressed.

2.2 Full Wake Alignment Scheme

In the model of full wake alignment, the corner points of the wake panels are aligned with the local flow velocity. The details of this scheme may be found in Tian and Kinnas (2012) and Kinnas et al (2015).

In the case of steady, the alignment scheme continues updating the geometry of the wake without solving for the perturbation potential until the criterion of convergence is met or the maximum number of iterations is reached. The dipole influence coefficients by the trailing wake are then reevaluated. If the repaneling of duct, as will be introduced shortly, is involved, the source and dipole influence coefficients by the duct also need to be recomputed. Once the reevaluation of the influence coefficient is finished, the perturbation potentials on the blades, the duct and the wake are solved for. This process will continue until the convergence of wake shape and force performance, normally thrust and torque is achieved.

2.2.2 Repaneling of the Duct

In the improved full wake alignment scheme, the wake geometry is updated within each iteration. The procedure of repaneling the duct is to ensure that the duct panel is aligned with the updated wake after a converged wake shape is achieved at each iteration. As aforementioned, besides the influence coefficients due to the wake, the influence coefficients due to the duct need also to be reevaluated because of the duct repaneling procedure.

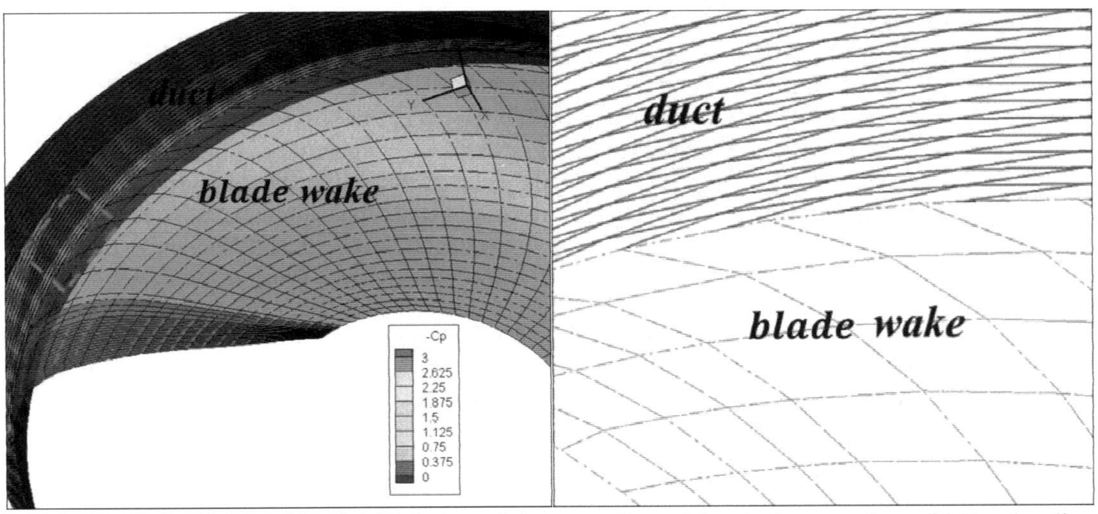

Figure 2.1 Global (left) and local (right) view of the repaneling of duct in the procedure of full wake alignment. The blade wake is at the same pitch with the duct panels, but the panel nodes do not match. From Kinnas et al (2015)

Further improvement related to duct repaneling can be achieved by collapsing the duct panel nodes with the nodes along the blade wake tips, as can be observed in Figure 2.2. It may further help to improve that efficiency of the present panel method, especially in terms of stability of results.

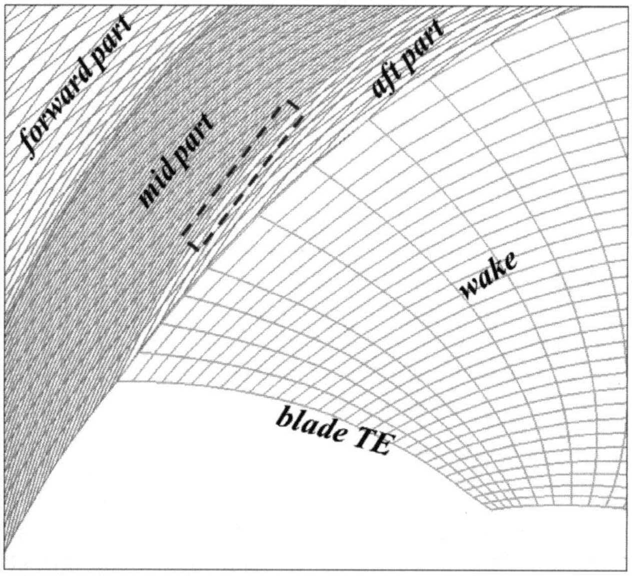

Figure 2.2 Demonstration of collapsing the duct panel nodes with the nodes along the blade wake tips. Note that the panel nodes on the duct and blade wake match. From Kinnas et al (2015)

This procedure, however, requires some more careful treatments. Firstly, the paneling in the mid part of the duct normally needs to follow the blade panel distribution in which cosine spacing is usually applied for better resolution of the blade leading edge and trailing edge. Thus there will be an abrupt transition of panel size from the last panel of the mid part to the first panel to the aft part of the duct. Besides since the panel size in the aft part of the duct is largely decided by the blade wake, there might be panels of large size near the trailing edge of the duct, causing some problem of satisfying the iterative Kutta condition within the present panel method.

2.2.3 Flow Field inside the Duct

Before applying the improved full wake alignment scheme to more complicated problems, it is important to ensure that the evaluation of additional effects of the singularities on the duct and its trailing wake is correctly added into the scheme. The validation is carried out in the problem of a bare duct subjected to uniform inflow as shown in Figure 2.3. The duct shape was adjusted to avoid separation when operating without the propeller. The flow velocity induced by the duct is evaluated at points with fixed axial coordinate but different radial coordinates.

The velocity computed by the present panel method at the space points shown in Figure 2.4 are correlated with that from RANS axisymmetric simulation. It should be noted that the effects of viscosity is involved in the RANS simulation, while it is not in the panel method. Good correlation has been observed at different locations.

On the other hand, it is also important to ensure that the panel method with full wake alignment is grid independent. In Figure 2.5, the calculated velocity at the selected points by the panel method with different panel numbers is shown. The number of chordwise panels is varied while the number of circumferential panels is maintained the same. The results for different panel arrangements agree very well with each other.

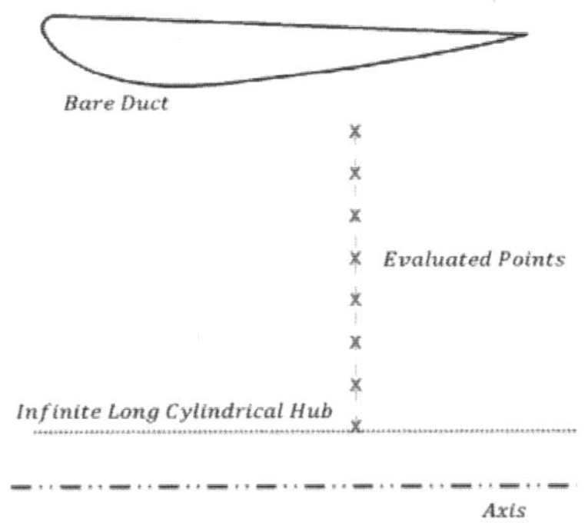

Figure 2.3 Demonstration of the problem of a bare duct subjected to uniform inflow and the points selected for evaluation

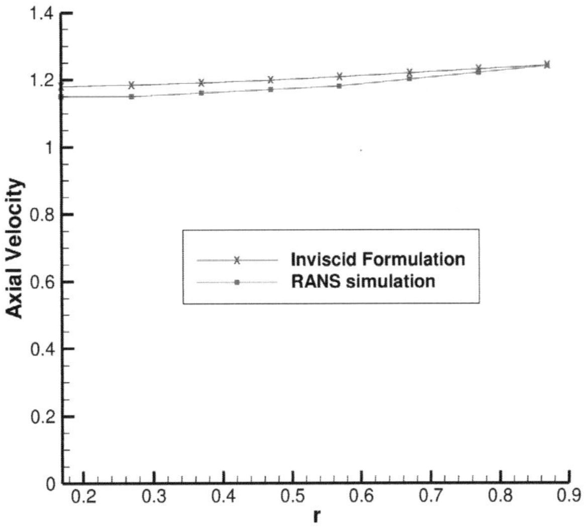

Figure 2.4 The correlation of the axial velocity of the flow at selected points between panel method and RANS (using ANSYS-Fluent) axisymmetric simulation

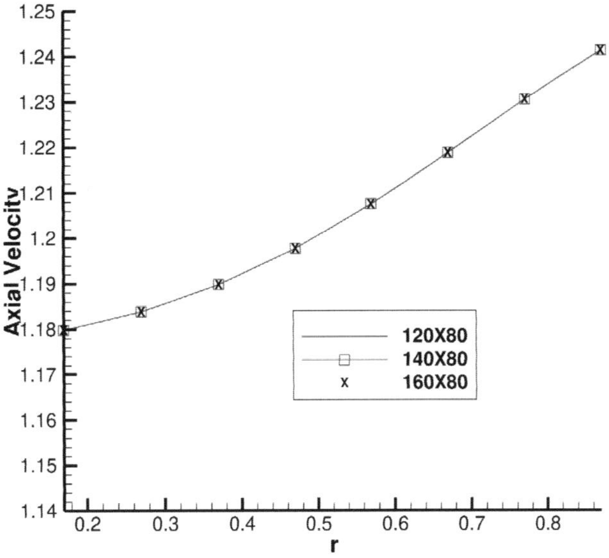

Figure 2.5 Convergence study on the panel method with respect to the evaluated axial velocity

3. Ducted Propeller with Square Blade Tip and Sharp Trailing Edge

The present method is applied to the case of ducted propeller with square blade tip and a sharp trailing edge duct. Two different wake alignment models will be used: PSF-2 type wake model and full wake alignment scheme which has already been introduced in details.

At the same time, full blown RANS simulations have been performed with sufficient spatial resolution in ANSYS Fluent and Star-CCM+, as described in detail in Fan (2015).

The geometry of the ducted propeller and the 2-D section of its duct are indicated in Figure 3.1. It is a 4-blade propeller with square blade tip and bounded within a sharp trailing edge duct 19Am. The duct 19Am is modified by MARIN from duct 19A which has a blunt trailing edge that the current panel method cannot well resolve because of the recirculation region. The design advance ratio J_S of this propeller is around 0.50. Zero gap is used in this case.

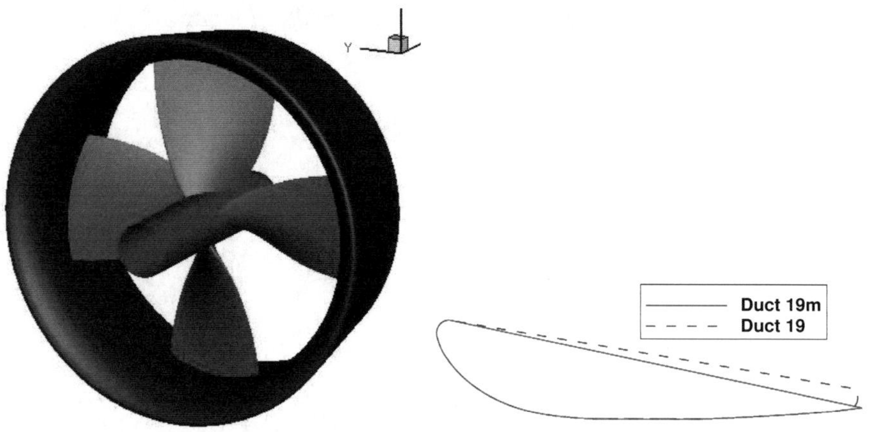

Figure 3.1 Geometry of the ducted propeller with square blade tip, and the section of the sharp trailing edge duct, shown together with the original

3.1 Panel Method

For the panel method, 80× 30 (chordwise×spanwise) and 200×80 (chordwise×circumferential) panels are used to resolve the blade and duct. Two wake models, PSF-2 wake model and full wake alignment (FWA) scheme, are adopted. For all advance ratios, PSF-2 scheme trimmed the blade wakes at 3R downstream with an ultimate wake disk, while the full wake alignment scheme models blade wakes of one revolution. It takes about 30 minutes and 3 minutes for full wake alignment and PSF-2 wake model to complete the run on a single core of Intel Xeon 2.54 GHz CPU. The wake geometry from the two models at J_S =0.4 and 0.6 are shown in Figures 3.2-3.5.

At high loading, the contraction rate of the transition wake is set as zero in PSF-2 wake scheme and the model did not predict the contraction of the transition wake radius. While in full wake alignment model, the rollup of the wake sheets at low advance ratios is well predicted.

As will be shown shortly at all loading conditions, whether contraction of the wake radius has been well predicted (FWA) or not (PSF-2) plays a significant role in if an accurate prediction of force can be obtained.

3.2 Full Blown RANS Simulation

The full blown RANS simulation are performed with ANSYS/Fluent and Star-CCM+ in which periodic interfaces have been applied such that for the present propeller only a quarter of the whole fluid domain is necessary for further simulation. The results from the two commercial codes were very close to each other, and only those from Star-CCM+ are presented in this work. In addition, to better resolve the boundary layer structures on the blades and duct, structured meshing model has been adopted. The same gridding model has been used for the trailing wake of the blades and duct in order to reduce the possible artificial diffusivity.

As also listed in the Table 4.3, *k-ω* SST turbulence model is adopted. QUICK scheme is used for the spatial discretization and SIMPLEC scheme for the pressure correction. Over 6 million polyhedral cells are used to simulate a quarter of the domain with periodic boundary condition. It takes over 32 hours on 32 Intel Xeon 2.54 GHz CPUs for the residuals converged to 1.0E-6. Details of these RANS calculations may be found in Fan (2015).

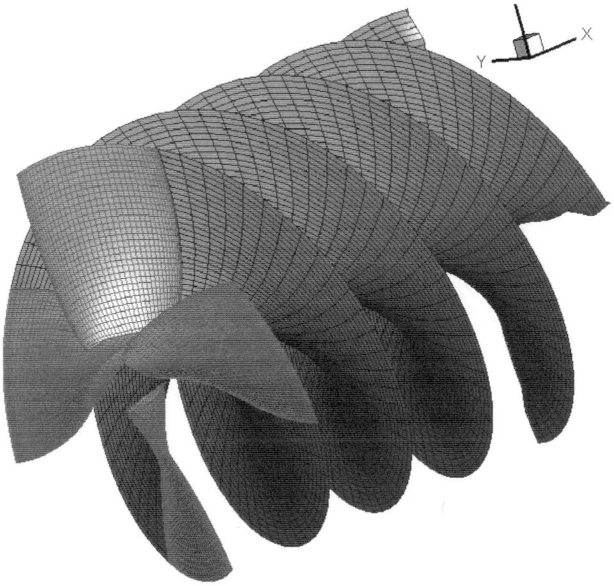

Figure 3.2 Wake geometry generated from PSF-2 wake model at *Js*=0.40

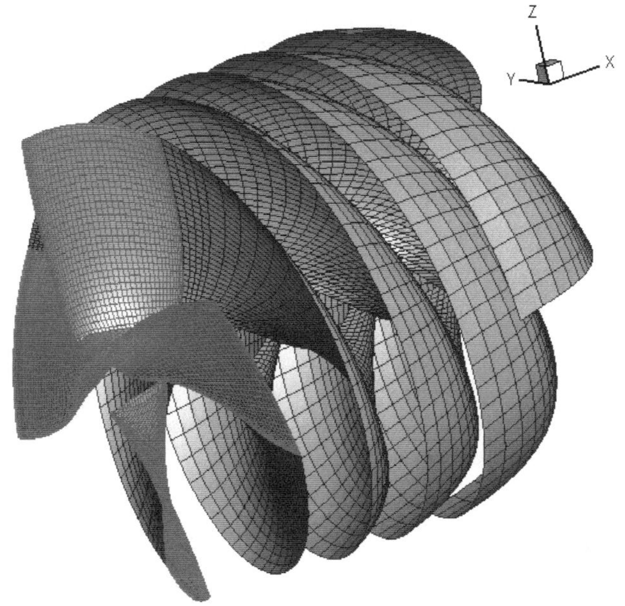

Figure 3.3 Wake geometry generated from full wake alignment scheme at *Js*=0.40

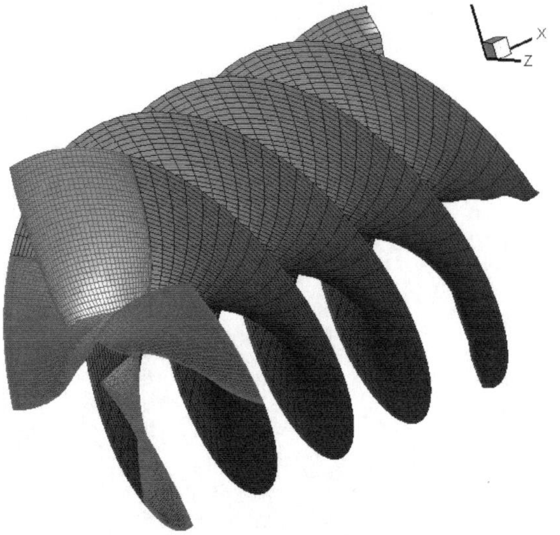

Figure 3.4 Wake geometry generated from PSF-2 wake model at *Js*=0.60

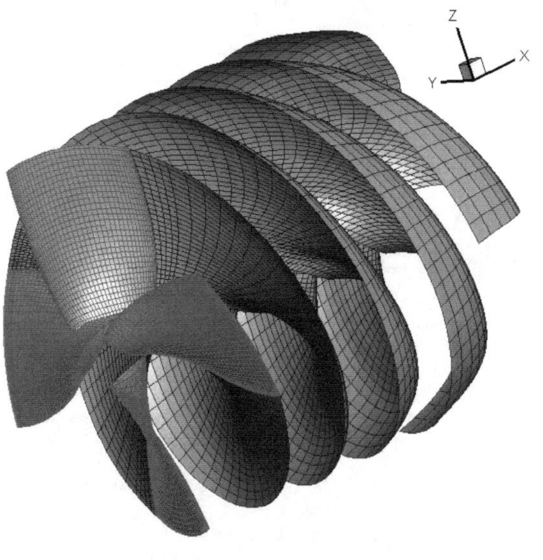

Figure 3.7 Wake geometry generated from full wake alignment scheme at *Js*=0.60

Cell number	Over 6 million
Reynolds number	1.0E+6
Turbulence Model	*k-ω* SST
Pressure Correction Scheme	*SIMPLEC*
Spatial discretization	*QUICK*
CPU Type	Intel Xeon 2.54 GHz CPU
Total time for calculation (32 CPUs)	Over 32 hours

Table 3.1 Parameter setting in the RANS simulation

4 Results

4.1 Propeller Performance

The overall forces, i.e. thrust and torque predicted by the lower order panel method with PSF-2 type scheme and the improved full wake alignment scheme are correlated with that from experimental measurements and the full blown RANS simulation, as shown in Figure 4.1.

It is observed that for this case the panel method with PSF-2 type scheme overestimates significantly the thrust and torque on the propeller at all operating conditions. On the other hand, by applying the improved full wake alignment, the predicted forces agree well with both experimental data and full blown RANS simulations at most advance ratios. Only at very high loading the FWA overestimates the force, even though not considerably.

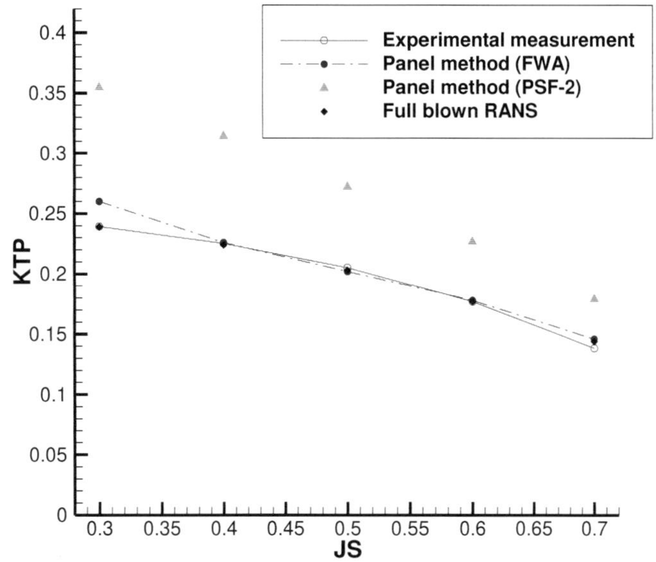

Figure 4.1 Predicted propeller thrust from the present and other methods vs. experimental measurements; from Kinnas et al (2015)

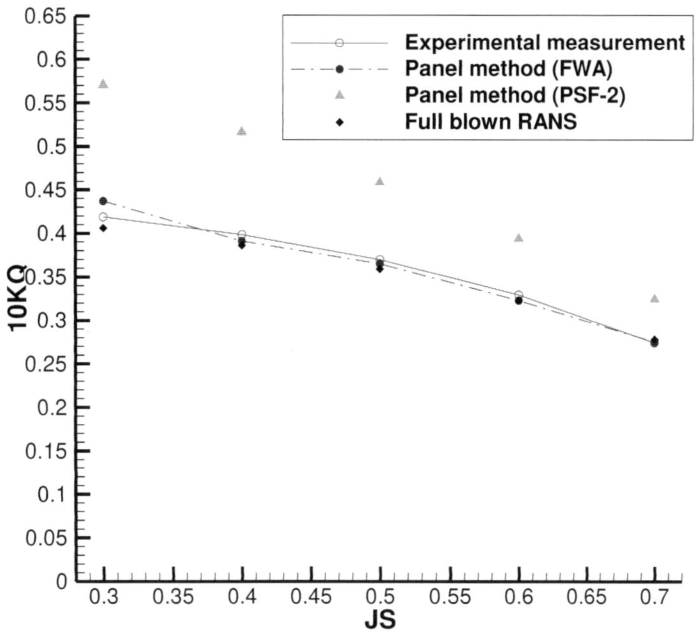

Figure 4.2 Predicted propeller torque from the present and other methods vs. experimental measurements; from Kinnas et al (2015)

4.2 Pressure Distribution on Blades

Figures 4.3 to 4.5 present the detailed correlations of pressure distribution on the propeller blades under three different loading conditions: high loading condition (*Js*=0.40), design loading condition (*Js*=0.50) and low loading condition (*Js*=0.60). For each loading condition, two radial blade sections, 0.65(near the mid of blades), 0.86 (close to the blade tip and duct inner surface) are selected for comparison.

The predictions of pressure distribution by full wake alignment model are in very good agreement with the full blown RANS simulation (using Star-CCM+) at different radial sections and under different loading conditions.

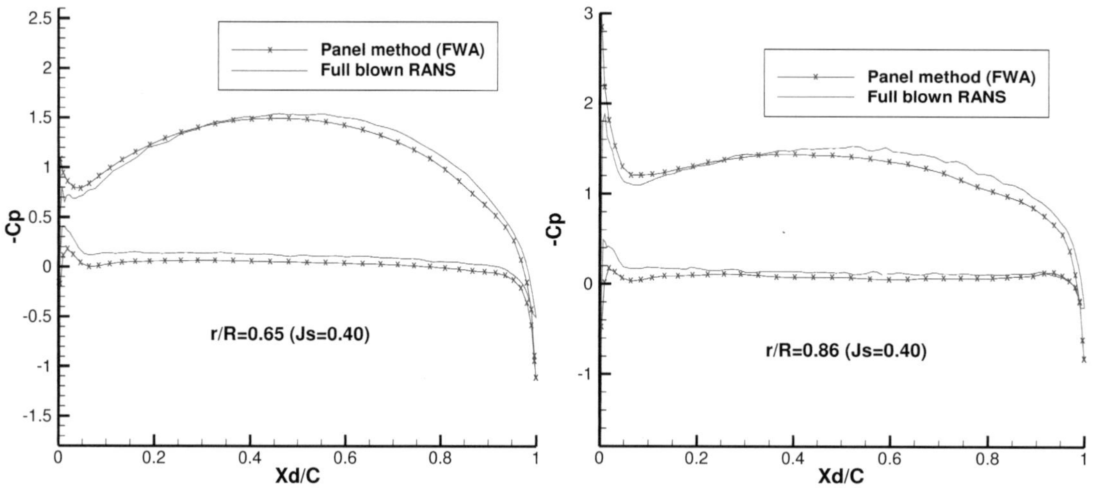

Figure 4.3 Correlation of pressure distribution at blade stations r/R=0.65, 0.86 (Js=0.4)

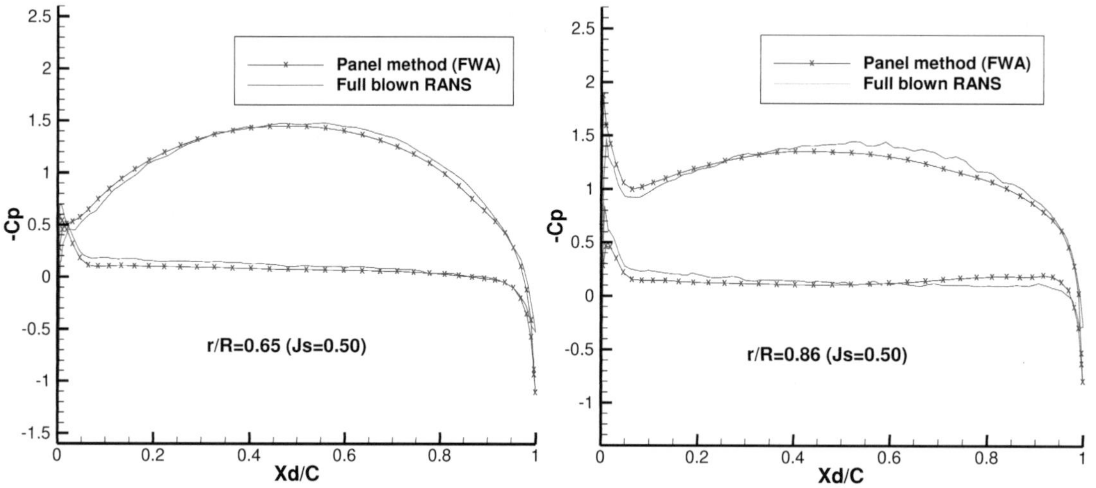

Figure 4.4 Correlation of pressure distribution at blade stations r/R=0.65, 0.86 (Js=0.5)

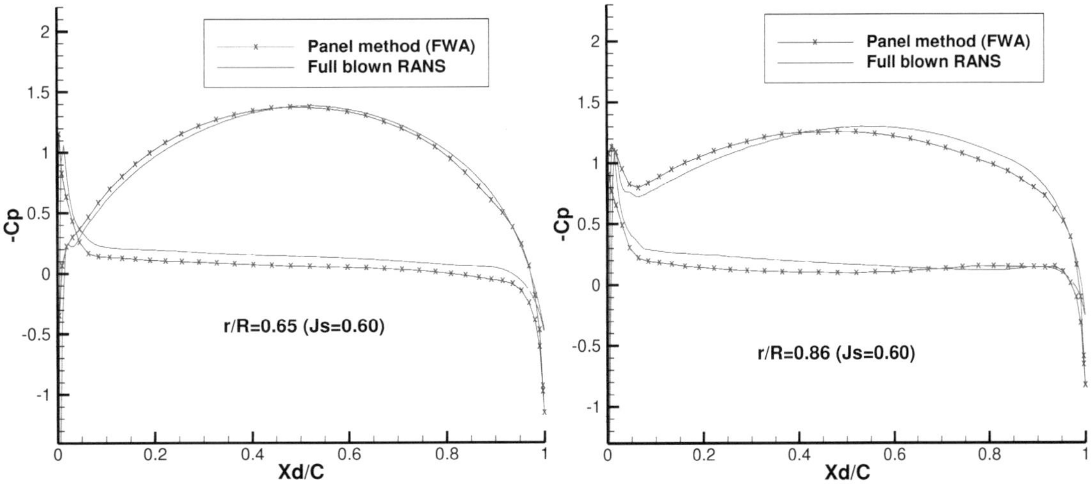

Figure 4.5 Correlation of pressure distribution at blade stations r/R=0.65, 0.86 (Js=0.6)

4.3 Comparison of Computational Efficiency

Since by applying the lower order panel method, it is not required to grid the whole fluid domain, making the panel method much more computational efficient than the full blown RANS simulations. In the current case, the panel method takes only about 1/60 of the CPU time, while using only one processor versus 32 processors used in the case of full blown RANS simulations. In addition, the time devoted in gridding for the full blown RANS can take several days while it is negligible for the panel method.

Method	Full blown RANS	PANEL METHOD
Cell No.	Over 6 million cells	Less than 12K panels (10 iterations for wake)
Total running time	Over 32 hours (32 CPUs)	30 minutes (1 CPU)

Table 4.1 Computational efficiency comparison between the panel method (with 10 iterations to determine the fully aligned wake) and the full blown RANS simulation

5. Application to Thruster with Blunt Trailing Edge Duct

A 4-blade ducted propeller from which the bare duct shape in last section is extracted is investigated in this section with the present extension scheme. An infinite long cylindrical hub is included in the study. The design advance ration of this propeller is around 0.70. The hybrid method that couples a potential Vortex-Lattice Method (VLM) solver with a RANS solver is also applied for detailed correlation. The correlation is carried out at the design advance. The 3-D geometry of the ducted propeller and the section of its duct are shown in Figure 5.1, together with the extention in order to model the separated flow at the duct trailing edge. The method utilized to determine the shape of the extension is described in Fan (2015), Purohit (2013), and Kinnas et al (2013). The coupling with XFOIL and the pressure equality condition in the extension scheme is applied on the circumferentially averaged pressures on the duct.

5.1 RANS-VLM coupling method

The body force within the blade zone of the RANS simulation is shown in Figure 5.2. A reliable body force interpolation scheme has been here applied, as described in Tian et al (2014). Also shown in Figure 5.3 is the axial component of the effective wake calculated at the last iteration. The flow was found to start separating from the pressure side of the duct from 96% duct length from the original trailing edge, which has been used the cut plane location in the calculation of viscous/inviscid interactive method.

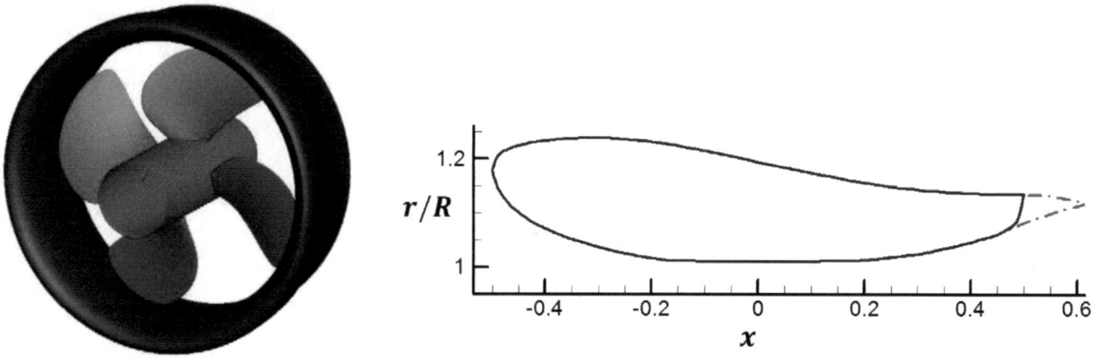

Figure 5.1 3-D geometry of the ducted propeller and the section of its duct with the extension used to model the separated flow

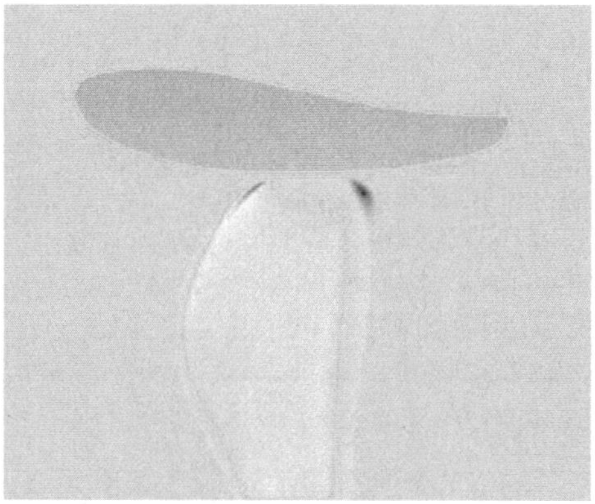

Figure 5.2 Body force within the blade zone of the RANS simulation

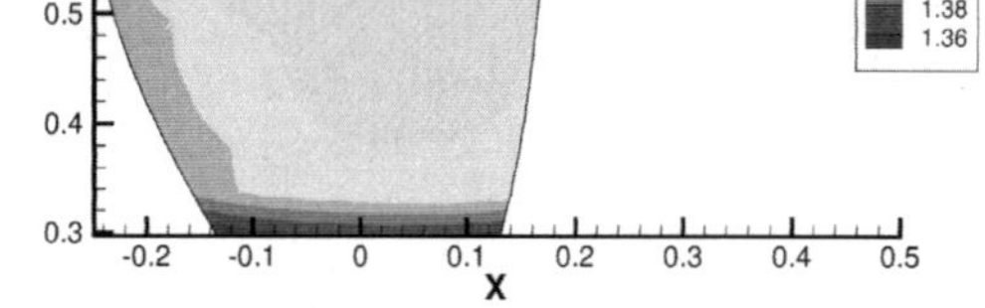

Figure 5.3 Contour plot of the axial velocity component of the effective wake, calculated at the last iteration

5.2 Correlation of results from two methods

The duct shape in the viscous/inviscid interactive formulation is modified through the extension scheme and based on the separation point found in the RANS-VLM run above. The section of the extended duct and the paneling of the ducted propeller are shown in Figure 5.4.

The pressure calculated by the viscous/inviscid interactive method is correlated with that from the RANS/VLM coupling method in Fig. 5.5. The predicted circulation along the blade is shown in Figure 5.6. The two methods agree fairly well except at the blade section near the hub.

Figure 5.4 Extended duct shape and gridding of the ducted propeller in the viscous/inviscid interactive method

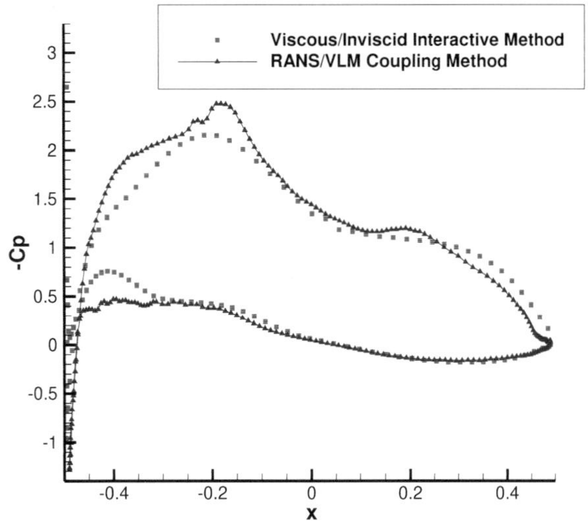

Figure 5.5 Pressure distribution (circumferentially averaged) on the duct, predicted from the two methods

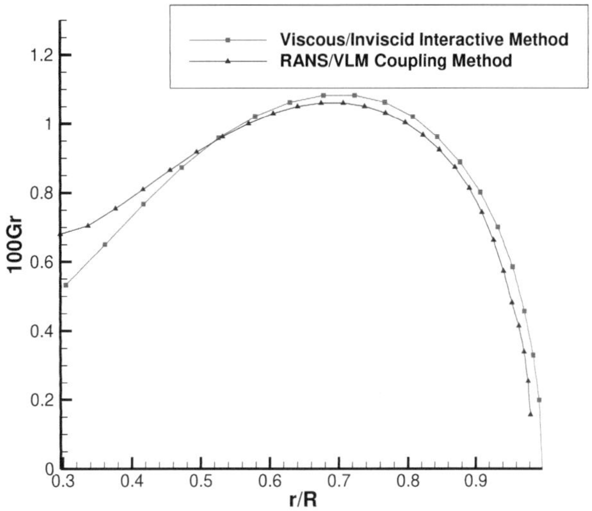

Figure 5.6 Predicted circulation on propeller blade from the two methods

5. Conclusions and Future Work

In this work, an improved full wake alignment scheme has been applied on ducted propellers by aligning the trailing wake with the local velocity. The effects of the singularities on the duct and duct trailing wake, which used to be neglected, are included in the improved scheme. It has been adopted by a lower order panel method for prediction of performance of ducted propellers. It is observed the improved wake model represent well the characteristics of roll-up at the wake tip especially under high loading conditions. In a ducted propeller with square blade tip, it is further confirmed that at various advance ratios, whether contraction of the wake radius has been predicted (full wake alignment) or not (PSF-2 type scheme) will significantly affect the predicted overall performance of the propeller. The predicted forces from the panel method with full wake alignment wake model agree well with experimental data and full blown RANS simulations. However, under the same condition of advance ratios, the force predicted by the PSF-2 type scheme overestimates significantly the thrust and torque on the propeller.

The detailed pressure distribution on the blades has also been investigated. The related correlations have been performed between the present method and full blown RANS simulation. Very good agreement has been achieved by the full wake alignment model at different radial sections and under various loading conditions for a ducted propeller with square blade tip.

The method is also applied in the case of a thruster with duct which has a blunt trailing edge, by handling the separated flow with an extension of the duct shape which is determined by requiring pressure equality at the (actual) trailing edge of the duct. The effects of viscosity have been evaluated via coupling the circumferentially averaged pressure on the duct with an integral boundary layer solver.

In the improved full wake alignment scheme, the process of repaneling the duct has been involved which might cause the following problems. The paneling in the mid part of the duct normally needs to follow the blade panel distribution in which cosine spacing is usually applied for better resolution of the blade leading edge and trailing edge. Thus there will be an abrupt transition of panel size from the last panel of the mid part to the first panel to the aft part of the duct. Also since the panel size in the aft part of the duct is largely decided by the blade wake, there might be panels of large size near the trailing edge of the duct, causing some problem of satisfying the iterative Kutta condition within the present panel method. Efforts must be made to overcome these types of numerical inaccuracies. Additionally, the effects of viscosity on the duct must be examined more extensively, for other geometries and conditions.

6. Acknowledgments

This work was supported by the U.S. Office of Naval Research (Contracts N00014-10-1-0931 & N00014-14-1-0303) and Phases VI and VII of the "Consortium on Cavitation Performance of High Speed Propulsors" with the following current members: American Bureau of Shipping, Kawasaki Heavy Industry Ltd., Rolls-Royce Marine AB, Rolls-Royce Marine AS, SSPA AB, Andritz Hydro GmbH, Wärtsilä Netherlands B.V., Wärtsilä Norway AS, Wärtsilä Lips Defense S.A.S., Wärtsilä CME Zhenjiang Propeller Co. Ltd, and Steerprop Ltd. The authors wish to thank Mr. Seungnam Kim, a graduate student in the Ocean Engineering Group at UT Austin, for his help with writing this paper.

References

Abdel-Maksoud, M. and Heinke, H.-J. (2003). 'Scale effects on ducted propellers'.In Proceedings of the Twenty-Fourth Symposium on Naval Hydrodynamics, Fukuoka, Japan.

Baltazar, J., Rijpkema, D., Falcao de Campos, J. A. C., and Bosschers, J. (2013). 'A comparison of panel method and RANS calculations for a ducted propeller systemin open-water'. In Proceedings of the 3rd International Symposium on Marine Propulsors (smp'13), Launceston, Tasmania,, Australia

Baltazar, J., Falcao de Campos, J. A. C., and Bosschers, J. (2015). "Potential Flow Modelling of Ducted Propellers With a Panel Method," In Proceedings of the 4th International Symposium on Marine Propulsors (smp'15), Austin, Texas, USA

Drela, Mark. "XFOIL: An analysis and design system for low Reynolds number airfoils." In Low Reynolds number aerodynamics, pp. 1-12. Springer Berlin Heidelberg, 1989.

Fan, H. (2015), "An Improved Panel Method for the Prediction of Performance of Ducted Propellers," Masters' thesis, Ocean Engineering Group, CAEE, UT Austin, August, 2015.

Gibson, I., and R. I. Lewis. "Ducted propeller analysis by surface vorticity and actuator disk theory." In Proceedings of the Symposium on Ducted Propeller, RINA, Teddington, England. 1973.

Greeley, David Scott, and Justin Elliot Kerwin. "Numerical methods for propeller design and analysis in steady flow." Transactions-Society of Naval Architects and Marine Engineers 90 (1982): 415-453.

Gu, H. and Kinnas, S.A., "Modeling of Contra-Rotating and Ducted Propellers via Coupling of a Vortex-Lattice with a Finite Volume Method," Propellers/Shafting 2003 Symposium, Society of Naval Architects and Marine Engineers, Virginia Beach, VA, September 17-18, 2003.

Hess, J. L. "Calculation of non-lifting potential flow about arbitrary three dimensional bodies." Journal of Ship Research 8, no. 2 (1964): 22-44.

Hess, John L., and Walte O. Valarezo. "Calculation of steady flow about propellers using a surface panel method." Journal of Propulsion and Power 1, no. 6 (1985): 470-476.

Hufford, Gary S., Mark Drela, and Justin E. Kerwin. "Viscous flow around marine propellers using boundary-layer strip theory." Journal of ship research 38, no. 1 (1994): 52-61.

Hughes, M. J. 'Implementation of a special procedure for modeling the tip clearance flow in a panel method for ducted propellers'. In Proceedings of the Propellers/Shafting '97 Symposium, Virginia Beach, Va, USA, 1997.

Kerwin, J. E. and Lee C.S. Prediction of steady and unsteady marine propeller performance by numerical lifting-surface theory. No. Paper No. 8. 1978.

Kinnas, S., and S. Pyo. "Propeller wake alignment models in uniform and inclined inflow." In Propellers/Shafting'97 Symposium, Virginia Beach, VA. Soc. Naval Arch. & Marine Engnrs. 1997.

Kinnas, S. A., S. H. Chang, Y. H. Yu, and L. He. "A hybrid viscous/potential flow method for the prediction of the performance of podded and ducted propellers." In Proceedings of the Propeller and Shafting Conference, Williamsburg, VA, pp. 1-13. 2009.

Kinnas, S.A., Chang, S.-H., Tian, Y., and Jeon, C.-H. (2012a) "Steady and Unsteady Cavitating Performance of Ducted Propulsors", 22nd International Offshore (Ocean) and Polar Engineering Conference, ISOPE'12, Rhodes, Greece, June, 2012a

Kinnas, S.A., Yu, X., and Tian, Y., "Prediction of propeller performance under high loading conditions with viscous/inviscid interaction and a new wake alignment model," In 29th Symposium on Naval Hydrodynamics, August 2012b

Kinnas, S.A., Jeon, C-H, Purohit, J.B., and Tian, Y. "Prediction of unsteady cavitating performance of ducted propellers." In Third International Symposium on Marine Propulsors (smp2013), 2013.

Kinnas, S.A., Fan, H. and Tian, Y. 2015, "A Panel Method with a Full Wake Alignment Model for the Prediction of the Performance of Ducted Propellers", Journal of Ship Research, Vol. 59, No. 4.

Lee, J.T. "A potential based panel method for the analysis of marine propellers in steady flow." PhD diss., Massachusetts Institute of Technology, 1987.

Lee, H.S. and S. A. Kinnas. "Application of a boundary element method in the prediction of unsteady blade sheet and developed tip vortex cavitation on marine propellers." Journal of ship research 48, no. 1 (2004): 15-30.

Lee, H.S. and Kinnas, S.A., "Fully Unsteady Wake Alignment for Propellers in Non-axisymmetric Flows," Journal of Ship Research, Vol. 49, No. 3, pp. 176-190, September 2005.

Lee, H. S. and Kinnas, S. A. "Prediction of cavitating performance of ducted propeller", in CAV2006: Sixth International Symposium on Cavitation, Wageningen, The Netherlands, September 11-15, 2006

Morino, Luigi, and Ching-Chiang Kuo. "Subsonic potential aerodynamics for complex configurations: a general theory." AIAA journal 12, no. 2 (1974): 191-197.

Purohit, J. (2013), "An Improved Viscous-Inviscid Interactive Method and its Application to Ducted Propellers," Master thesis, CAEE, UT Austin, 2013.

Pyo, S., and S. A. Kinnas. "Propeller wake sheet roll-up modeling in three dimensions." Journal of ship research 41, no. 2 (1997): 81-92.

Sun, H. and Kinnas, S.A., "Performance prediction of cavitating water-jet propulsors using a viscous/inviscid interactive method," *Transactions, SNAME (Society of Naval Architects and Marine Engineers),* Vol. 116, pp. 166-181, 2008

Tian, Y., Jeon, C.-H., and Kinnas, S.A. "Effective Wake Calculation/Application to Ducted Propellers", *Journal of Ship Research*, Vol. 58, No. 2, June 2014, pp. 1–13.

Tian, Y.and Kinnas, S.A. (2012) "A Wake Model for the Prediction of Propeller Performance at Low Advance Ratios", *International Journal of Rotating Machinery,* Volume 2012, Article ID 372364, 11 pages; doi:10.1155/2012/372364

Tian, Y. and Kinnas, S.A., "Thruster and Hull Interaction", Journal of Offshore Mechanics and Arctic Engineering. 137(4), 041801, Aug 01, 2015.

Yu, X. "Three dimensional viscous/inviscid interactive method and its application to propeller blades," Master thesis, Ocean Engineering Group, CAEE, UT Austin, 2012.

THE LOW MOTION FPSO (LM-FPSO); A NOVEL SCR FRIENDLY FLOATER FOR DEEPWATER IN BRAZIL PERSISTENT SWELL ENVIRONMENT

RICARDO ZUCCOLO
INTECSEA, WORLEYPARSONS GROUP

ALAA M. MANSOUR[1]
INTECSEA, WORLEYPARSONS GROUP

CHUNFA WU
INTECSEA, WORLEYPARSONS GROUP

JEFFERSON AZEVEDO
INTECSEA, WORLEYPARSONS GROUP

Proceedings of the 21st Offshore Symposium, February 2016, Houston, Texas
Texas Section of the Society of Naval Architects and Marine Engineers

ABSTRACT

The discoveries made in the pre-salt basin offshore Brazil are among the world's most important in the past decade. The pre-salt province comprises large accumulations of excellent quality, high commercial value light oil, however, the new discoveries are located in ultra-deepwater in areas lacking infra-structure, which makes the Floating Production Storage and Offloading (FPSO) vessel the preferred solution.

Steel Catenary Risers (SCRs) are the preferred solution in wet-tree applications due to their simplicity, robustness and low CAPEX and OPEX compared to other riser options. However, due to its relatively high dynamic motions, FPSO is not a feasible host for SCRs in most environments. Also, for efficient production from such rich reservoirs, facilities supporting high production rate and large diameter risers are necessary. This makes it more challenging to find a robust and commercially attractive riser solution.

In this paper a novel design for an FPSO with the ability to host large diameter SCRs is presented and evaluated for applications in deepwater field developments. The new design, namely, Low Motion FPSO (LM-FPSO), has a hull form with a generally rectangular cross-section. The platform is moored in-place using a conventional mooring system. The LM-FPSO performance is enhanced with the robust low-tech feature, namely, free-hanging Solid Ballast Tank (SBT). The SBT is located at a certain distance below hull keel and connected to the hull through four groups of short tendons. All tendon components are the same as those used in conventional TLPs. Through the mass and added mass of the SBT, the LM-FPSO provides heave, roll and pitch responses better than or comparable to any floating production structure in use today.

The paper presents detailed description of this novel design and its in-service performance. A case study is presented where the LM-FPSO is used in persistent swell in ultra-deepwater offshore Brazil to support large diameter SCRs. The SCR's feasibility is demonstrated and reported. The identified risks and associated mitigations for the new design compared to the conventional FPSO are investigated and reported. The paper concludes with discussions on the project execution plan, cost and schedule benefits when developing fields using this novel design.

Keywords: Low Motion FPSO (LM-FPSO), Solid Ballast Tank (SBT), Steel Catenary Risers (SCR), Wet tree

INTRODUCTION

The development of oil and gas fields located in deep and ultra-deep water in areas with minimal or no infra-structure typically requires the deployment of floating facilities with storage capabilities such as the FPSOs. The FPSO possesses relatively high dynamic motion especially in the heave, roll and pitch modes of motion due to the natural periods located in the wave frequency range in these directions. Typical heave and roll Response Amplitude Operators (RAOs) are presented in Figure 1. Significant platform motion responses are expected in the wave frequency range.

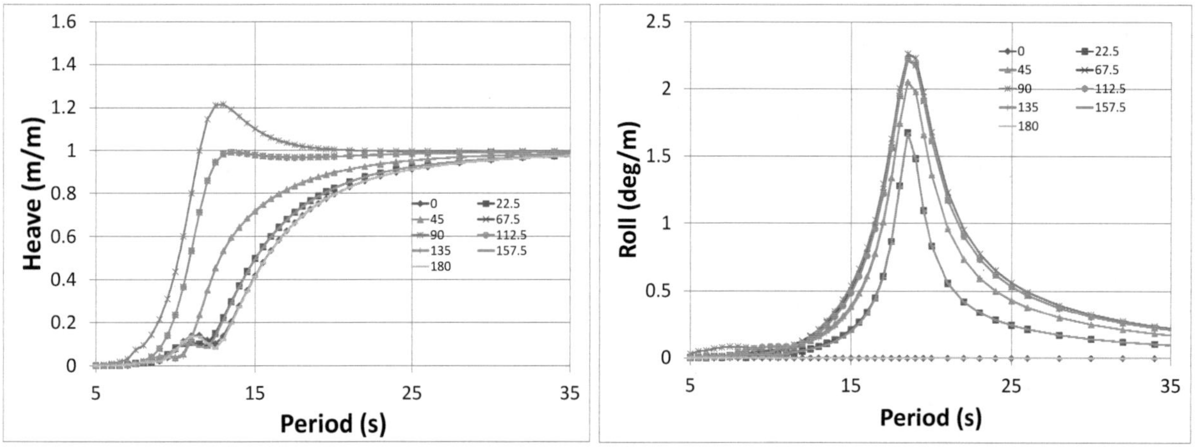

Figure 1 Typical Heave and Roll Motion RAOs for Conventional FPSOs

In wet-tree application, the SCR is considered the most robust riser solution due to its simplicity, long life-cycle and extensive in-service history. It is also the riser solution with the lower combined CAPEX and OPEX cost. However, the high dynamic motion of FPSOs makes them unsuitable to host SCRs. Less robust and more expensive riser system alternative is typically adopted with FPSOs. The problem is even more challenging when large diameter risers are required.

The sail area of a ship-shape FPSO in relatively harsh environment with multi-directional environment mandate the use of turret system to help facilitate the mooring system design, increase the offloading operability and minimize the collision risk between the FPSO and offloading tanker. The turret system comes with significant CAPEX and OPEX costs.

As the development cost increases, it becomes essential to seek novel solution to help reducing the overall cost so that a development business case can be established.

The Low Motion FPSO (LM-FPSO) is a state-of-the-art innovative design developed to maintain the advantages of the conventional FPSOs while offering superior motions and stability performance.

This superior motion response makes the LM-FPSO a suitable host for SCRs and even for Top Tension Risers (TTRs). The shape of the LM-FPSO hull allows the use of a conventional mooring system and eliminates the need for the expensive turret and swivel system.

The LM-FPSO enables not only the use of SCRs on FPSO in deep-water, but the use of large diameter SCRs which meets the field development requirements for high productivity fields such as that in pre-salt basin in ultra-deep water offshore Brazil.

The ability to support TTRs on LM-FPSO presents commercially attractive alternative to the developments with combined FPSOs and wellhead platforms such as those in West Africa since it eliminates the need for the wellhead platform with its associated high CAPEX and OPEX. The advantages and feasibility of using the LM-FPSO to support dry tree applications have been demonstrated by Mansour et. al. (2015).

The LM-FPSO utilizes the Free Hanging Solid Ballast (FHSB) technology to improve the platform response. This technology has been successfully applied to improve the conventional semisubmersible response (see References).

The paper presents an overview of the LM-FPSO design and its superior performance. Case study is also presented where the LM-FPSO is used in persistent swell environment to support SCRs in ultra-deepwater offshore Brazil. The SCR's feasibility is demonstrated. The identified risks and associated mitigations for the new design compared to the conventional FPSO are investigated and reported. The paper concludes with discussions on the project execution plan, cost and schedule benefits when developing fields using this novel design.

THE LM-FPSO DESIGN

The LM-FPSO, (Figure 2), fundamentally consists of a conventional box-shape hull form with generally square or rectangular cross-section. The hull structure is a traditional stiffened plate type structure with easy fabrication. The topside is fabricated in the yard into modules and integrated to the hull at the quayside. The system is moored in-place using conventional spread mooring system.

The LM-FPSO overall performance is enhanced by utilizing a robust and low-tech feature, namely free-hanging Solid Ballast Tank (SBT). The free-hanging SBT is attached to the hull at a certain distance below the keel level away from the wave action by means of a number of short tendons similar to those extensively used in Tension Leg Platforms (TLPs). The hull is fitted with a number of tendon porches at each corner and the SBT is fitted with a corresponding number of bottom tendon receptacles to facilitate the tendon connections at top and bottom, respectively. The tendon porches and the bottom receptacles are identical to those widely used in TLP platforms. The LM-FPSO tendon design follows the API RP 2T requirements, similar to the TLP tendon design. Due to the short tendon length, the expensive coupling equipment is not required. Also, installation by derrick barge offshore is not needed.

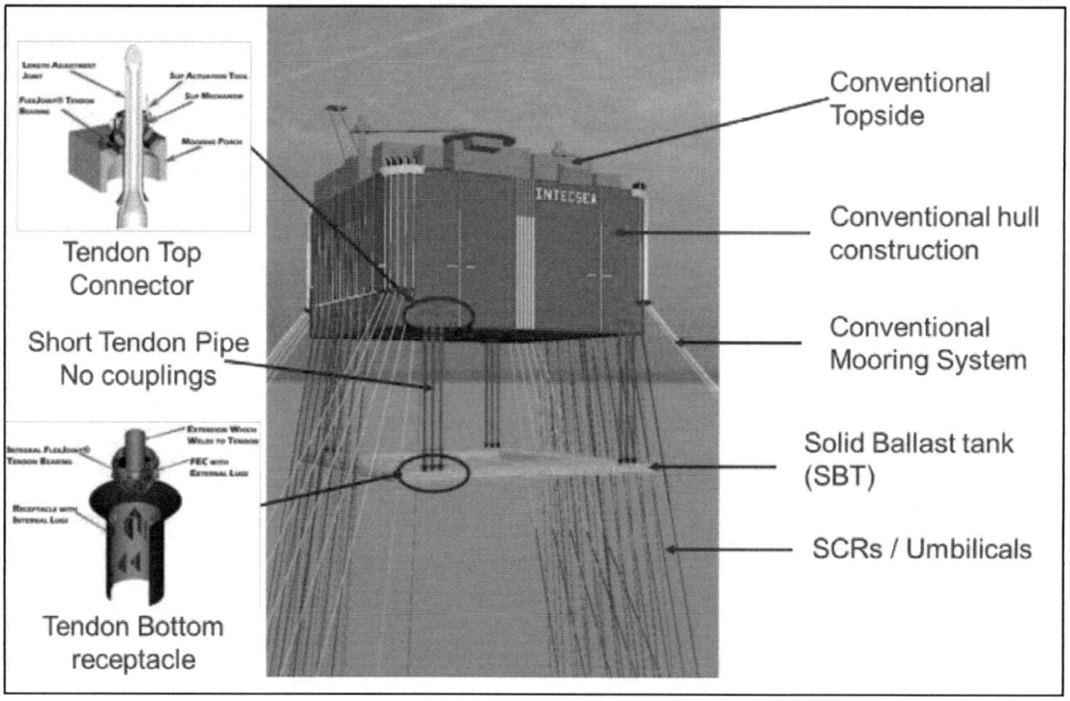

Top connector and bottom receptacle pictures are Courtesy of www.oilstates.com

Figure 2 LM-FPSO Design; a Conventional Hull and More Than 20yrs of Field Proven Components

The SBT is either full or partially filled with solid ballast, depending on the design parameters. The weight of the SBT is adjusted such that the connecting tendons are always kept under tension in extreme and survival hurricane environments and to ensure full coupling especially in the heave, roll and pitch directions between the FPSO hull and the SBT, resulting in the FPSO and SBT to act as a unified system. The introduction of the solid ballast also provides additional system stability by increasing the metacentric height (GM) and system restoring moment.

The addition of the SBT results in a significant increase in the FPSO heave, roll and pitch natural periods, mainly through its own weight and added mass on its large surface area.

During the pre-service conditions, a number of platform mooring chains are used to connect the SBT to the hull and maintain it in a retracted position at close proximity to the hull keel during fabrication and transportation phases of the project. The installed and tensioned chains will ensure no vertical movement will take place during any loading or off-loading operations during the pre-services conditions. Once the platform has arrived at site, the SBT is naturally flooded and the chain jacks are operated to lower the SBT to its desired position.

Once the SBT is positioned at its desired elevation, the short tendons are installed with the help of Anchor Handler type vessel (AHTS) and tendon installation winches on top of the columns. Platform cranes may be used as an alternative to the tendon installation winches.

The chains are then slacked to enable the load transfer from the chains to the tendons and generate the required tendon pretension loads. The platform chains are then disconnected at their lower end (that are attached to the SBT), retracted and reused as part of the platform permanent mooring (station-keeping) system. The pre-tension in the tendons are developed through the in-water weight of the SBT. Figure 2 presents the LM-FPSO in the in-place condition. The solid ballast is added inside the SBT in the dry-dock or at the quayside.

Open tendon porches are adopted to enable the installation procedure to be reversed for the de-commissioning of the platform at the end of its life.

The SBT in the LM-FPSO can be seen as a "suspended" Gravity Base Foundation (GBF). Snorre and Heidrun TLPs are examples of TLPs in-service with GBFs.

THE LM-FPSO CONSTRUCTABILITY AND INSTALLABILITY

The LM-FPSO execution plan strategy is similar to a conventional FPSO, considering the region of installation, project overall execution, and contracting strategy. The presence of the Solid Ballast Tank does not alter the location of fabrication, the hull-topside integration methodology, transportation, installation or hook-up philosophy.

One example for project execution plan (PEP) assumes co-fabrication of the hull, SBT and topside in the same fabrication yard, dry-mating of the hull and topside, and wet-towing of the integrated facility to site. The SBT is fabricated independently in the dry-dock and solid ballast is added. The FPSO hull modules are integrated on top of the SBT in the dry-dock. The topside modules are then mated to the hull. Two mooring platform chains on each corner are used to maintain the SBT in a retracted position during all the pre-service conditions. The dry-dock is flooded and the integrated platform is wet-towed to the installation site (Figure 3).

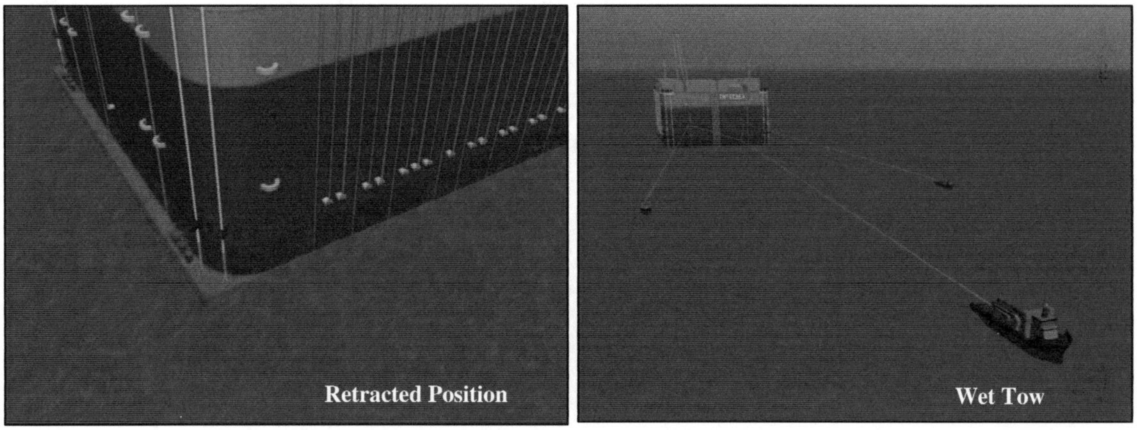

Figure 3 LM-FPSO Fabrication and Wet-Tow

Upon arrival at the site, the towing tugs are used to keep the platform in position, and the chain-jacks are operated to pay-out chains and lower the SBT by about ~5 m below the hull keel. SBT valves are opened to naturally flood the remaining volume of the SBT and develop the pretension on the chains. This operation can be performed via remote controlled valves or through diver operation. The chain-jacks are continually used to lower the SBT to its in-place position bellow keel (Figure 4).

The hull is then ballasted down to the target in-place draft. An AHTS vessel along with platform winches is used to upend and install the short tendons. ROV is used to help guiding the bottom of the short tendons inside the receptacles of the SBT (Figure 5).

The LM-FPSO is then conventionally moored at its desired in-place position with the mooring system.

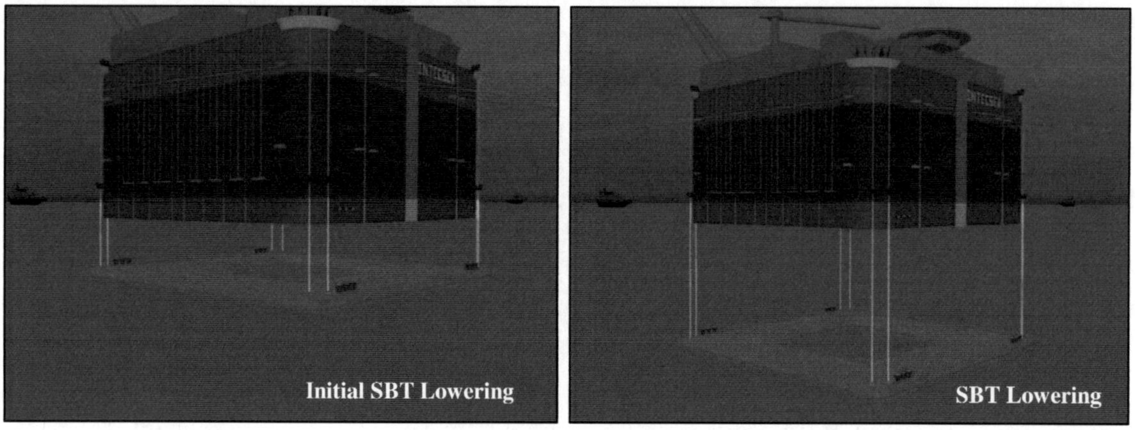

Figure 4 SBT Lowering

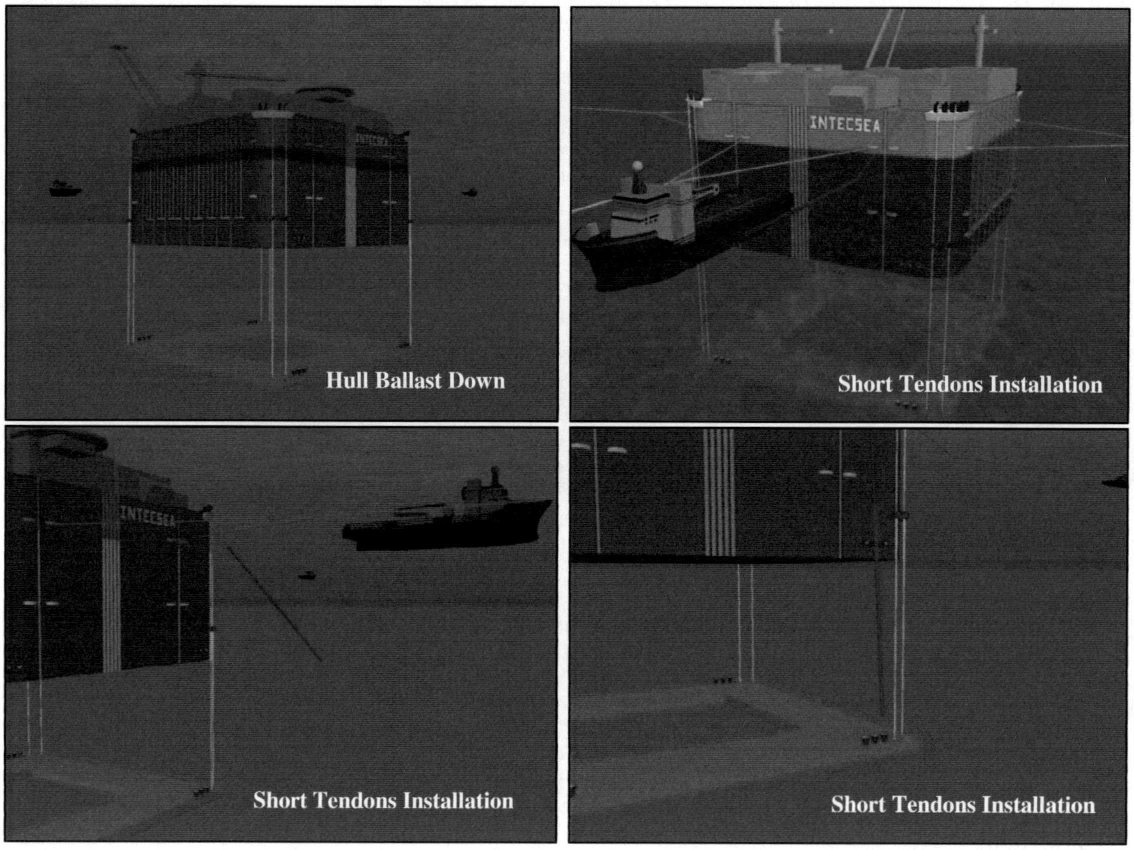

Figure 5 Hull ballasted Down to Target In-place Draft and Short Tendons Upended and Installed

THE LM-FPSO PERFORMANCE

The typical response of the new novel design has been extensively investigated in both frequency domain and time domain using fully coupled simulations.

The LM-FPSO RAOs compared to those for conventional FPSO are presented in Figure 6. It can be seen that the heave and roll responses are significantly reduced compared to the conventional design and the natural periods for the LM-FPSO are now well away from the wave frequency range. Thanks to the mass and added mass on the surface of the SBT.

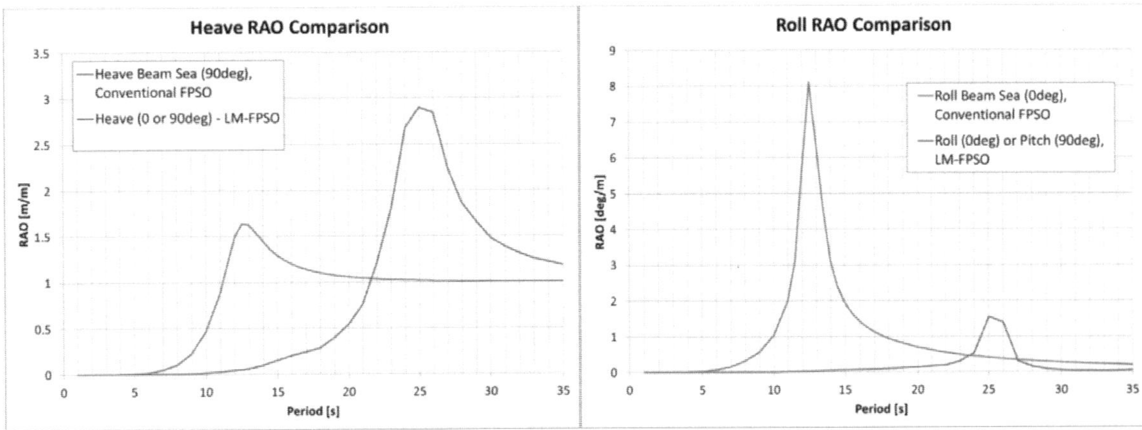

Figure 6 Typical Heave motion RAO (left) and Roll Motion RAOs (right) comparison between LM-FPSO and Conventional FPSO

The typical heave, roll and pitch motion time histories for the LM-FPSO hull and SBT are presented in Figure 7. It is evident that the SBT weight is optimized so that the hull and the SBT are acting as single body in these motion modes.

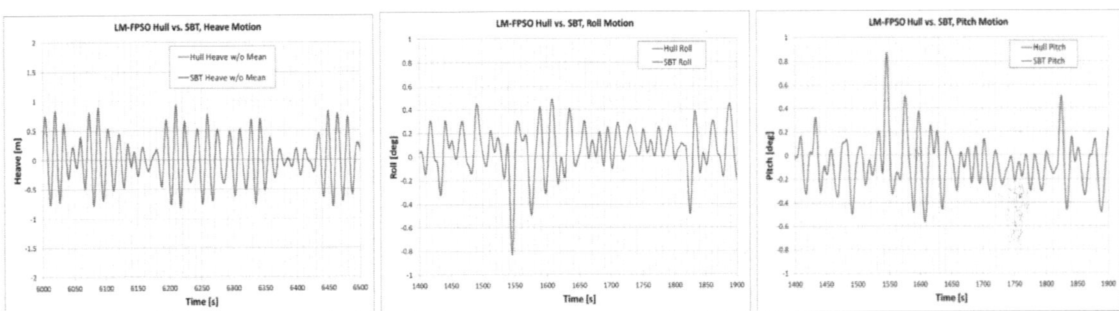

Figure 7 Typical Heave, Roll and Pitch LM-FPSO Hull and SBT Motion Response

Some relative motions between the hull and the SBT are observed in the surge, sway and yaw directions. However, they are limited to first order only and relatively small compared to TLP hull-to-foundation relative motions. Figure 8 presents the hull and SBT motion time histories in surge direction.

The design ensures that positive tendon tension is always maintained even in the survival conditions (Figure 9).

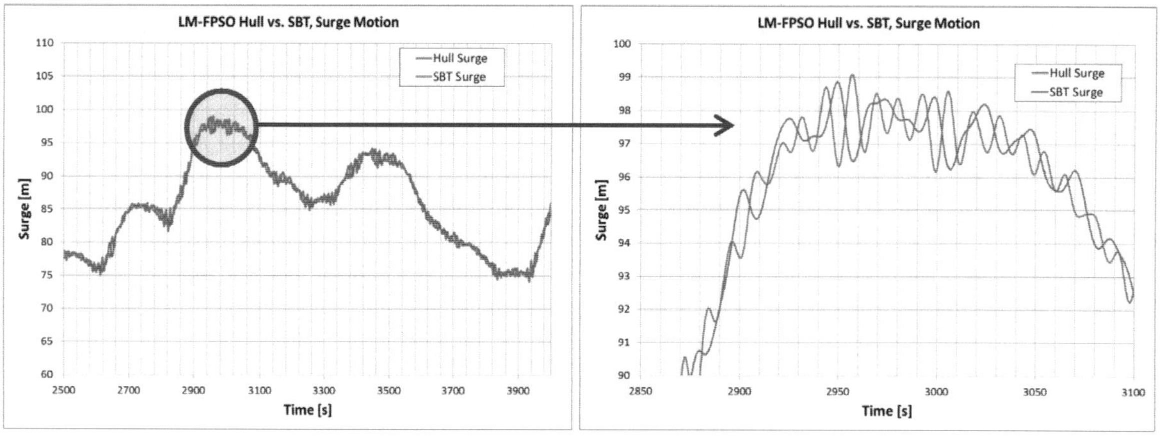

Figure 8 Typical LM-FPSO Surge Motion

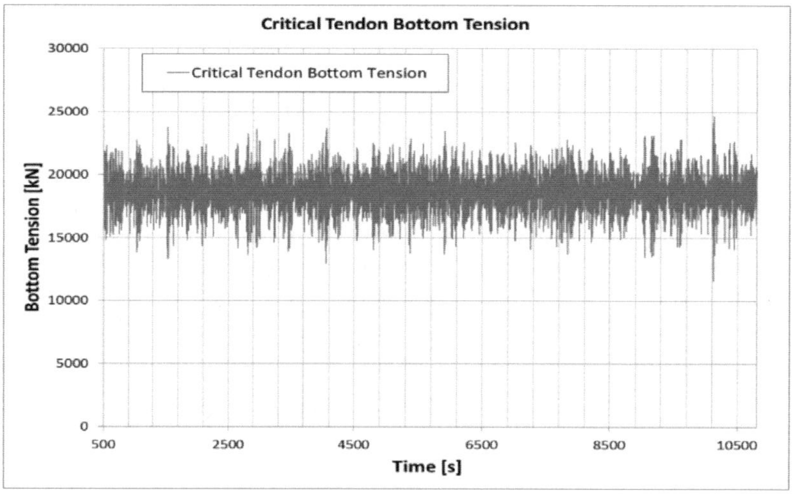

Figure 9 Minimum Tendon Tension at Bottom

CASE STUDY: LM-FPSO AS A WET TREE HOST IN BRAZIL PERSISTENT SWELL ENVIRONMENT

A case study is presented hereinafter where the LM-FPSO is used to support a wet tree application in 3,000 m water depth in Brazil persistent swell environment.

The in-house tools have been used to size the LM-FPSO hull, SBT and mooring system. Stability analyses followed by fully coupled time domain analysis with full Quadratic Transfer Function (QTF) have been performed.

The LM-FPSO is designed to support a topside payload of 40,000 metric tonnes, operating as a wet tree host offshore Brazil supporting 35 SCRs and 18 Umbilicals, with 2MMbbl storage capacity. 14.5-in ID gas export, 6.5-in ID production, 6-in ID gas injection, 4-in ID service, 6.5-in ID water injection are assumed. SCRs layout (Figure 10) have been developed with 3deg separation angle between production risers and 2.5deg between other types of risers.

SCRs departure angles have been assumed: 15deg for gas export, 9deg for production and water injection, 6deg for service and 5deg for umbilical (Figure 11). Recesses at the outer edges of the SBT can be adopted should more clearance between the riser and the SBT is required.

The total lines payload is about 8,233 metric tonnes. Table 1 and Table 2 present payloads, environment and configuration data. Graphical presentation of the configuration is shown in Figure 12.

The mooring is composed by twenty (20) semi-taut mooring legs of chain-polyester-chain, with 120mm R4 platform and anchor chains and 217mm PE (28xMBL EA). The mooring layout has been developed with 4deg separation angle between legs.

The LM-FPSO hull has a square cross section of 101.8x101.8m and integrated facility draft during pre-services of 12.2m (to the bottom of the SBT in the retracted position) which makes the design suitable for construction and integration in a large range of shipyards worldwide, including Brazil. LM-FPSO hull and SBT are connected in-place through four groups of three (3) short tendons of 38-in OD and 1.5-in wall thickness.

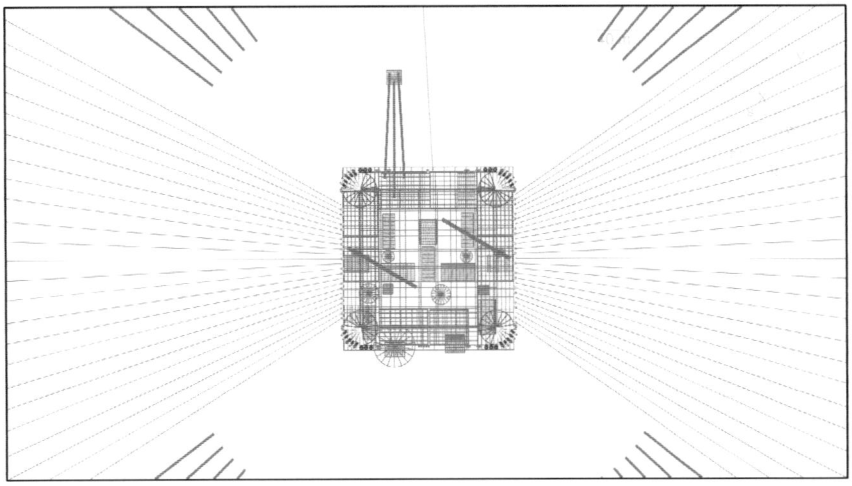

Figure 10 LM-FPSO mooring legs and lines layout

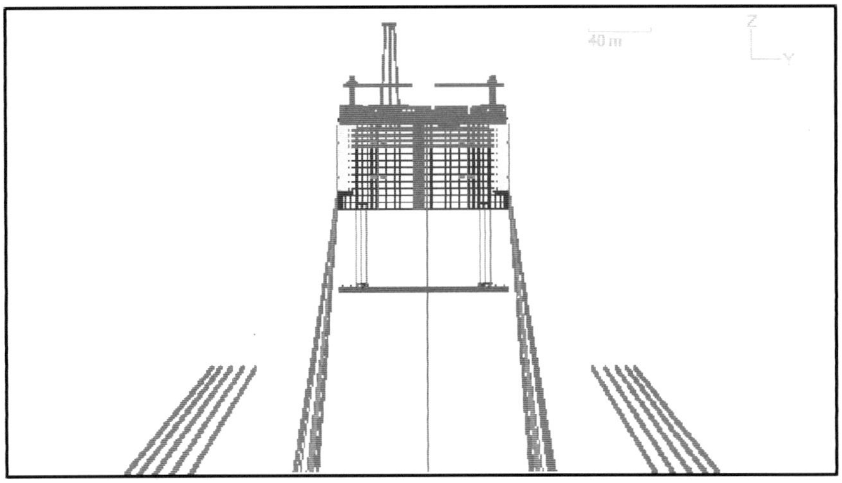

Figure 11 SCRs steep departure angles

Table 1 Payload and environment data for wet tree LM-FPSO

Payload & Env		
Application	(-)	Wet Tree / Ultra Deep Water
Typical Location	(-)	Brazil
		Payloads
Water Depth	(m)	3,000
Topside payload	(Mt)	40,000
Storage	(MMBBL)	2.00
TTRs	(-)	-
SCRs	(-)	35
		Environment
Return period	(yr)	100
Hs	(m)	8.9
Tp	(sec)	12 to 16
Wind speed	(m/s)	25.0
Curr. Speed, surface	(m/s)	1.00

Table 2 Configuration data for wet tree LM-FPSO

Configuration		
Application	(-)	Wet Tree / Ultra Deep Water
Typical Location	(-)	Brazil
		The Hull
L x W x H	(m)	101.8 x 101.8 x 48.6
Corner radius	(m)	10.2
Draft / Freeboard	(m)	39.6 x 9.0
Center Well (L x W)	(m)	None
Hull Lighship Wt	(Mt)	61,613
Displacement	(Mt)	419,338
		The SBT
L x W x H	(m)	107.8 x 101.8 x 3.7
Center Opening L x W	(m)	52.7 x 52.7
SBT steel Weight	(Mt)	3,115
Solid ballast weight	(Mt)	58,961
		The Mooring System
Mooring	(-)	20 Chain-PE-Chain
Platform Chain Dia x L	(mm x m)	120 mm R4 x 150 m
Polyester Dia x L	(mm x m)	217 mm x 4,220m
Anchor Chain Dia x L	(mm x m)	120 mm R4 x 400 m
Tendons	(-)	12 @ 38"OD x 1.5"
Length	(m)	52.0
Pretension	(Mt)	1,895
		System Natural Periods
Heave Np	(sec)	19.00
Roll/Pitch Np	(sec)	25.00
GM	(m)	44.00

As it can be seen in Table 2, the LM-FPSO design has a relatively long heave and roll/pitch natural period of 19 sec and 25 sec, respectively. Unlike the conventional FPSO, while the LM-FPSO offers long roll/pitch natural periods, it maintains very high GM (44 m) and stability. This makes the LM-FPSO less sensitive to the changes in topside vertical center of gravity (VCG). Although it has been reported that very high GM causes uncomfortable motion in the past conventional floaters designs, this is not applicable to LM-FPSO because of the long roll/pitch natural periods.

The overall hull (including the SBT) steel efficiency is found to be 3.5. This means that every 1 metric tonne of hull steel can support 3.5 metric tonnes of payload. That still can be further improved by optimizing the hull steel.

The motion performance of the LM-FPSO has been evaluated in a fully coupled time domain analysis under the effect of wind, wave and current, for 100yr return period. The hull, SBT, connecting tendons, SCRs, and mooring system have been modelled in Orcaflex software and simulated over 3 hours storm conditions for a quartering sea environmental heading (the critical heading). Figure 12 presents the LM-FPSO Orcaflex model.

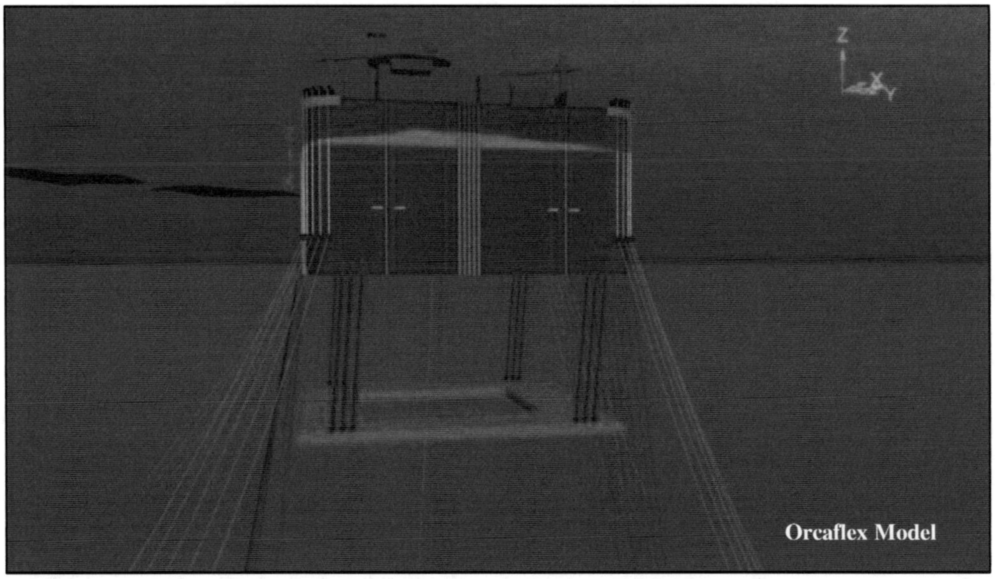

Figure 12 LM-FPSO Hull, SBT, Short tendons and SCRs

The LM-FPSO performance results are presented in Table 3.

Table 3 LM-FPSO Performance Summary

Results		
Application	(-)	Wet Tree / Ultra Deep Water
Typical Location	(-)	Brazil
Return Period	[yr]	100
Max offset	%	4.2%
Max SCR Porch Velocity	(m/s)	0.9
Max heave (SA)	(m)	1.4
Max combined Roll/pitch (SA)	(deg)	1.2
Max Yaw	(deg)	0.3
Max Hull-SBT rel. horizontal motion	(m)	4.9
Minimum tendon tension	(Mt)	1,184
Max Tendon Tension	(Mt)	2,522

The maximum offset for 100yr intact condition is found to be 4.2% of water depth, which indicates that the mooring system can be further optimized. The maximum combined roll and pitch angles are found to be 1.2° for 100yr, which is very low in this extreme environment. Thanks to the high GM and the long roll/pitch natural periods.

The short tendon minimum tensions are found to be positive all the time in 100yr, which satisfy the requirements to avoid snap loads (Figure 13).

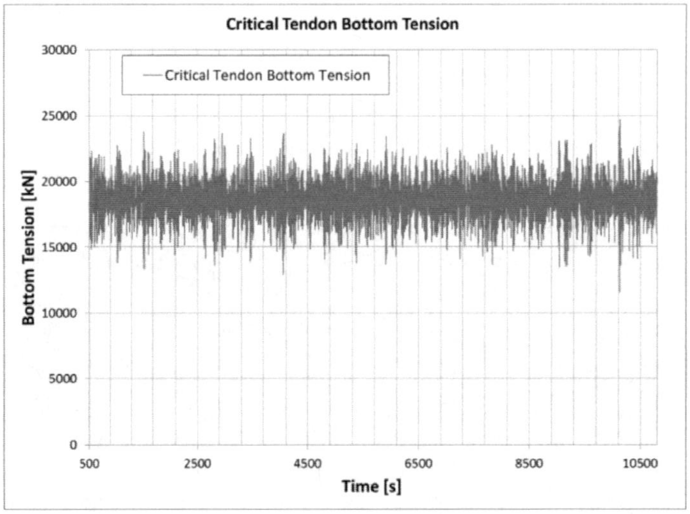

Figure 13 Minimum Bottom Tendon Tension in 100yr Condition

The maximum relative motion between the hull and SBT is found to be 4.9 m for 100yr (Figure 14). The corresponding tendon tilt angle with the vertical is 5.4deg, which is within the tendon flexible joint angle qualification limits of 10 deg.

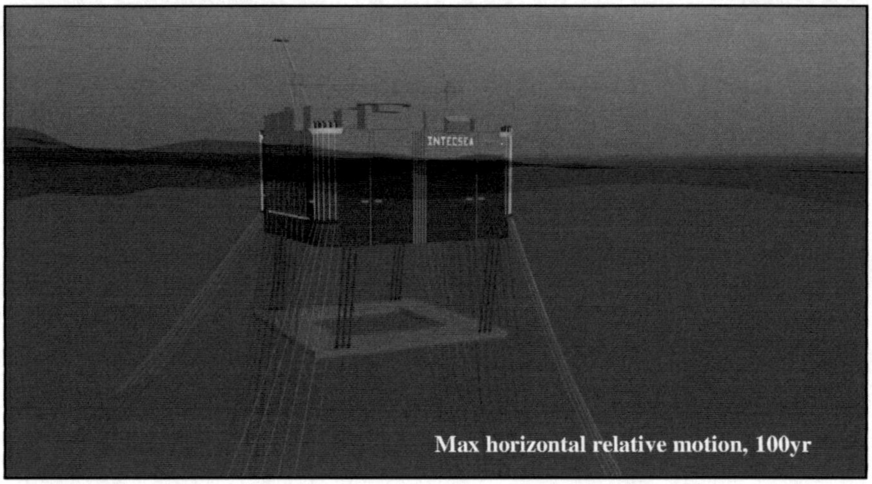

Figure 14 LM-FPSO Hull-SBT Maximum Horizontal Relative Motion in 100yr Condition

Snapshots from the analysis at the instants of maximum heave and maximum combined roll/pitch for 100yr conditions are presented in Figure 15.

Figure 15 LM-FPSO Maximum Heave and Maximum Combined Roll/Pitch Snapshot in 100yr Condition

The SCR porch velocity has been estimated to represent the SCR feasibility limits. Typically SCRs are considered to be feasible when the maximum porch velocity is less than 2 m/sec in 100yr condition. The LM-FPSO maximum riser porch velocity is only 0.9 m/sec in 100yr condition, as presented in Table 3, which indicates a feasible SCR design with sufficient margin.

THE LM-FPSO ECONOMICS

The cost for the LM-FPSO developed above for the Brazilian environment case study has been compared to the cost of a conventional spread moored FPSO. The conventional FPSO has been designed for the same design basis as that used in designing the LM-FPSO.

The LM-FPSO added cost includes the cost of additional hull steel compared to the conventional FPSO, the Total Installed Cost (TIC) for the SBT and the TIC for the short tendons. These items have contributed to a cost increase of about $65 MM.

This cost increase is offset by the cost reduction due to mooring system of $25 MM and the cost reduction of $15 MM per SCR since a conventional FPSO would need to use Steel-Lazy-Wave-Risers (SLWR) instead. The cost delta per each riser will be more should flexible risers are considered. The cost reduction due to these items is estimated to be $550 MM.

Therefore, using the LM-FPSO can result in a cost saving of as high as $485MM compared to the alternative development solution using conventional spread-moored FPSO with SLWR if found to be feasible in this water depth.

ADDED RISKS AND RISK MITIGATIONS

While the LM-FPSO has shown, based on fully coupled numerical analysis, to provide a very benign motion response, three (3) added risks when the LM-FPSO is used in wet tree applications have been identified and mitigated. The first risk is related to snap loads in the short tendons connecting the SBT to the hull, and heave response out of phase. The risk has been mitigated by including sufficient solid ballast in the Solid Ballast Tank to ensure that the tendons remain in tension all the time. Maintaining positive tendon tension in survival conditions is not an API RP 2T requirement but it adds another layer of reliability.

The second risk is dropped objects on the SBT, which has been mitigated by flooding the SBT in the in-service condition.

The third risk is the SBT interference with the SCRs. While this has not been identified as a risk in the case study considered in this paper, should this turn out to be a concern for a particular case study, it can be mitigated by changing the foot-print shape of the SBT to an H-shape and/or using larger SCR departure angles.

CONCLUSIONS

The LM-FPSO is a state-of-the-art innovative design developed to maintain the advantages of the conventional FPSOs while offering superior motions and stability performance. It consists of a conventional box-shape hull that offers easy fabrication and facilitate topside layout.

The LM-FPSO floater has proved to produce significantly improved performance compared to conventional FPSOs through the added feature of free-hanging Solid Ballast Tank (SBT) connected well below the hull keel through four groups of tendons. It is a technology that:

- Enables the use of large diameter SCRs on FPSO in deep-water, instead of expensive and sophisticated riser systems;
- Enables the use of stress joints with SCRs;
- Eliminates the need of a turret and swivel system, using conventional mooring system;
- Has a simple hull form and conventional fabrication;
- Allows quayside integration;
- Has high and flexible hull storage and ballast capacity;
- Provides a large ventilated rectangular deck area;
- Provides improved riser strength and fatigue responses with host platform motion natural periods shifted away from wave frequency energy;
- Is a commercially attractive field development option with substantial savings compared to a conventional FPSO.

The deck is integrated to the hull at the quayside. The integration of the Solid Ballast Tank (SBT) with the FPSO hull is performed also at the quayside. Derrick barges are not required for the short tendons installation. Detailed integration and installation approach has been illustrated and discussed in this paper.

The presence of the solid ballast inside the SBT ensures that the short tendons connecting the SBT to the hull are always in tension. This ensures full coupling between the SBT and the hull in the vertical and rotational directions. The SBT is flooded in the in-place condition to avoid the risk for dropped objects. SBT interference with SCRs has not been identified as a problem based on comprehensive analysis, however, if for a particular case study it turns out to be a concern, it can be mitigated by changing the footprint of the SBT to an H-shape and/or using higher SCR departure angles.

A case study has been presented for the LM-FPSO to host SCRs in 3,000 m water depth in Brazil persistent swell environment. It is found that the new design offers high stability and very benign heave and rotational responses while maintaining high efficiency.

Based on this case study, the estimated cost saving when the LM-FPSO is used compared to a spread-moored FPSO is estimated to be $485MM.

Due to its superior motion, the LM-FPSO can also be used to host Top Tensioned Risers (TTRs) and allow direct vertical access to wells on FPSOs. This would eliminate the need of additional wellhead platform that is typically used in dry-tree applications (Mansour et. al. 2015).

ACKNOWLEDGEMENTS

The authors would like to express their gratitude to INTECSEA Company for sponsoring the preparation of this paper.

REFERENCES

Mansour, A.M., Wu, C., Zuccolo, R., Azevedo, J., "The Low Motion FPSO (LM-FPSO); a Novel SCR and TTR Friendly Floater", Deep Offshore Technology International 2015 (DOT-2015), TX, USA

Mansour, A.M., Wu, C., Kumar, D., Zuccolo, R., "The Tension Leg Semisubmersible (TLS); the Hybrid TLP-Semisubmersible Floater with the Spar Response", Paper 23533, 33rd International Conference on Ocean, Offshore and Arctic Engineering (OMAE2014), CA, USA

Mansour, A.M., Kumar, D., Zuccolo, R., James, J., Heidari, H., "Novel Dry-Tree Semisubmersible Solutions for Deepwater Marginal Field Development", Paper 25351, Offshore Technology Conference Proceedings, Houston, 2014

Mansour, A.M., Kumar, D., Zuccolo, R., James, J., Heidari, H., "Field Development using Semisubmersible Floating Production System with Steel Catenary Risers in Western Australia Harsh Environment", Paper 25265, Offshore Technology Conference Proceedings, Houston, 2014

Mansour, A.M., Dib, M., Zuccolo, R., Bagaria, A., Greiner, B., James, J., Kumar, K., Heidari, H., and Azevedo, J., "A Disconnectable Dry Tree Semisubmersible Design Exposed to Iceberg and Harsh Environment", Paper 24558, Arctic Technology Conference Proceedings, Houston, 2014

Mansour, A. M., Ricardo Zuccolo and Dhiraj Kumar, 2013, "Key Design Considerations for the FHSB Dry Tree Semisubmersible in Harsh Environment", Deep Offshore Technology Conference, Houston, Texas

Mansour, A. M., and Kumar D., 2013, "High Performance Semisubmersible Design for Dry Tree Applications in Harsh Environment," OMAE 2013, Nantes, France.

Yu, J., Mansour, A.M., Yu, A., Zuccolo, R., 2013, "Top Tensioned Riser system design for the FHSB Dry Tree Semisubmersible Platforms," OMAE 2013 Nantes, France.

Mansour, A. M., 2009, "FHS Semi; a Semisubmersible Design Dry Tree Application," OMAE 2009, Honolulu, Hawaii, USA.

Mansour, A. M., and Wu, C., 2008, "New Semisubmersible Design for Ultra-deepwater Dry Tree Application," Deep Offshore Technology Conference, Asia Pacific, Perth, WA.

API RP 2T 3rd Edition, July 2010.

Orcaflex "User Manual Version 9.8d"

SNAME
THE INTERNATIONAL COMMUNITY
FOR MARITIME AND OCEAN PROFESSIONALS

WIND TURBULENCE EFFECTS IN GLOBAL RESPONSES OF A 5 MW WIND TURBINE THREE-COLUMN TLP

IKPOTO E. UDOH
NAVAL ARCHITECT, HOUSTON OFFSHORE ENGINEERING, AN ATKINS COMPANY

JUN ZOU
DIRECTOR OF NAVAL ARCHITECTURE, HOUSTON OFFSHORE ENGINEERING, AN ATKINS COMPANY

Proceedings of the 21st Offshore Symposium, February 2016, Houston, Texas
Texas Section of the Society of Naval Architects and Marine Engineers

ABSTRACT

Wind loads on floating offshore wind turbines (FOWTs) consist of steady and dynamic components and turbulence is the primary characteristic that differentiates their effects. Industry practice has largely focused on steady winds in the design of floating wind turbines, and emphasis on the effects of turbulence has been inadequate. The extent to which turbulence can affect global responses of FOWTs is not sufficiently understood, and is necessary for the design of safe and robust systems. It is known that turbulence can significantly affect the potential for wind turbines to harness required energy from wind fields. For FOWTs, there is valid concern about the extent to which the functionality and safety of the wind turbine system may be hampered, through interactions between the hydrodynamics-induced platform motions, and turbulence-induced motions of the rotor and tower.

Sequel to a prior study addressing the steady-wind aerodynamic effects on the global responses of a 5MW wind turbine three-column TLP, this study investigates the impact of wind turbulence on the global motions, tendon tensions, nacelle accelerations, generated power and tower base responses for the same floating system. The Kaimal and Von Karman wind spectra models (which are some of the most widely used models in offshore wind analysis) are implemented in this study. Results from the steady-wind and turbulent-wind studies are compared and discussed to delineate the effects of wind turbulence. Lessons learned from this study are applicable to any type of FOWT platform and can be effectively incorporated in design with the advantage of achieving safer, robust and optimal FOWT systems.

Keywords: Wind, Turbulence, Turbine, TLP, Responses

INTRODUCTION

Background

Offshore wind energy has acquired global interest owing to its non-polluting, renewable and arguably long term cost-effective attributes. The abundance of sustained winds generally increases with offshore water depths. While relatively higher wind speeds exist offshore, the associated turbulence in the wind fields are also significant and its effects on the response of the wind turbine system must be adequately understood, evaluated and quantified to aid the design of systems for which power generation is consistently assured.

From the perspective of floating platforms, efforts to harness offshore wind energy must overcome two major design concerns: excessive motions which can be induced by environmental loads, and uneconomical cost. These two undesirable factors are related (i.e. floating systems may require expensive designs in order to operate successfully and remain resilient to extreme environments), and their severity typically increases with water depth. With the size and operation of wind turbine systems considered, motions of the supporting platform must be within safe and allowable criteria to ensure optimal operability. Tension leg platforms (TLPs) allow very limited heave, pitch and roll motions (even in extreme seastates) because they are vertically restrained with tendons. Studies such as Tracy (2007) have shown that wind turbines have much less nacelle accelerations with TLPs than they do on floating support structures with slack mooring systems. The potential for TLPs to become significantly utilized in harnessing offshore wind energy is the reason for selecting a TLP concept for this study.

The impact of aerodynamic effects on TLP-type floating offshore wind turbines (FOWTs) has been addressed by several studies, some of which are discussed here. In an experimental study on a single-column TLP concept supporting a 5MW wind turbine, Nihei and Fujioka (2011) report notable effects of wind on platform rotational motions and mooring line dynamics, for cases where wind and waves are considered together. Udoh et al. (2015) studied aerodynamic-induced responses of a three-column FOWT TLP with steady wind fields, showing significant effects of aerodynamics on platform motions, tendon tensions and nacelle accelerations for different seastates. Matha et al. (2009) have showed through load analysis of a wind turbine on TLPs that aerodynamic instability could be severe in platform pitch and roll when the wind turbine is idling in high winds at some yaw misalignments. Although this paper does not address numerous load cases, the impact of high winds when the turbine is in operating condition must be well understood to design FOWT systems with high yield (or consistent power production). Bachynski and Moan (2012) performed parametric studies on single-column TLPs using irregular waves and turbulent winds, and as part of their conclusions, the authors note that motions perpendicular to aligned wind and wave directions were found to be important; in particular, the authors noted that roll motions could be excited by wind turbine and generator torques, and yaw motions could be excited by roll coupling and time-varying aerodynamic loading. It is worth understanding how these effects will manifest in a three-column TLP for operational conditions.

Problem Statement

Although it is known that aerodynamic effects on TLP-type FOWTs can be significant, a clear understanding of the contributions from turbulence on the different parts of the system is needed to adequately design the associated components. In the design of FOWT systems, the response parameters considered typically include platform motions and accelerations, tower-base shear forces and bending moments, tower and blade eigenfrequencies, nacelle accelerations, mooring line (or in the case of TLPs, tendon) tensions and generated power. A fundamental question to address is: how do the magnitudes of these responses correlate with variations in turbulence intensities, for a three-column TLP wind turbine?

RELEVANT THEORY

The mechanics of wind turbines (in operating conditions) primarily includes the effect of rotor thrust, blade drag, lift and deflections, tower deflections and gyroscopic effects. For FOWTs, these effects are coupled to the motions of the support platform. Any significant influence of turbulence on the global responses of the floating system should manifest through some or all of these effects. Each of these mechanisms is briefly described in this section, to set the context for discussions provided in the results section.

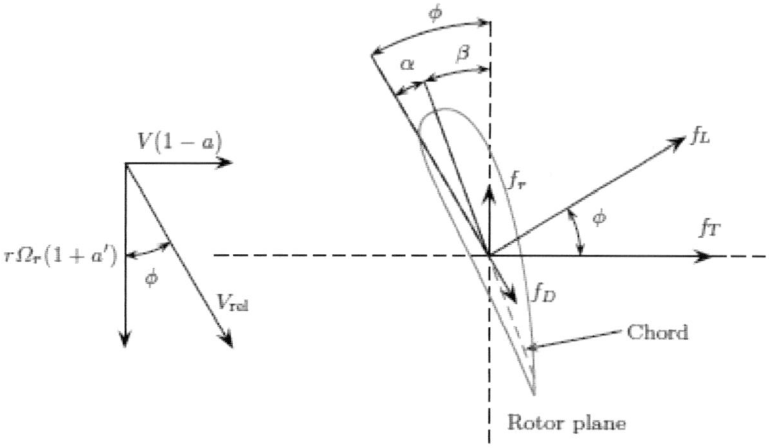

Figure 1: Depiction of Forces on a Blade Element (Bianchi et al., 2007)

Aerodynamic thrust force is the horizontal force imposed by the wind field on the rotor in the downwind direction, when the turbine is operating. To extract kinetic energy from the wind field, the rotor imposes a force in a direction opposite to the thrust force, causing a reduction in the wind velocity (mainly in the horizontal direction). Since the consistency of the wind field is impacted by turbulence, changes in the thrust force, the total useful torque developed by the turbine and the rotor power can be expected due to turbulence. Equations (1) through (3) show the dependence of thrust, torque and power on wind velocity (Bianchi et al., 2007).

$$F_T = \frac{1}{2}\rho\pi R^2 C_T(\lambda,\beta)V^2 \tag{1}$$

$$T_R = \frac{1}{2}\rho\pi R^3 C_Q(\lambda,\beta)V^2 \tag{2}$$

$$P_r = \frac{1}{2}\rho\pi R^2 C_P(\lambda,\beta)V^3 \tag{3}$$

Given the typical dimensions and material compositions of horizontal axis wind turbine blades and towers, their deflections are practically inevitable in operating conditions. The magnitudes of deflections will vary with bending stiffness and length of the members, and may increase with aerodynamic loading. The primary forces acting on blade elements are lift and drag forces (shown in Figure 1), which are a function of the relative velocity between the wind and the blade element. The lift and drag forces are expressed per unit length in Equations (4) and (5), respectively (Bianchi et al., 2007). Under the influence on these loads, flexible blades can experience significant bending stresses and vibrations especially towards the tip. Tower deflections can be induced at the base by motions of the support platform, and by wind forces at the top of the tower. These tower deflections in turn affect blade motions. To avoid resonance in the coupled global responses, the desired scenario is one in which the natural frequencies of the blade and tower deflections are different for each response mode (fore-aft, side-to-side and torsional), and do not intersect with the natural frequencies of the floating support platform either.

$$f_L = \frac{\rho c}{2}V_{rel}^2 C_L(\alpha) \tag{4}$$

$$f_D = \frac{\rho c}{2}V_{rel}^2 C_D(\alpha) \tag{5}$$

A concise explanation of the gyroscopic effect in floating wind turbine platforms, found in Naqvi (2012), an experimental study on a TLP-type and a spar-type wind turbine, is paraphrased here. Essentially, it is possible for revolutions of the rotor (with some angular momentum or torque) about the nominal downwind axis (or X-axis, in the context of this paper) to result in rotations about a second axis (Y-axis in our context, implying pitch motion). Under the occurrence of the rotations in the first two axes, a third component of angular momentum is induced about an axis perpendicular to the first two axes – in our context, an angular momentum in yaw. The support platform therefore experiences coupled pitch and yaw rotations. Naqvi (2012) notes that aerodynamic-induced yaw could have a counteractive effect on the hydrodynamic-induced yaw on the support platform, and that the pitch motion resulting from the gyroscopic effect could either increase or decrease the pitch motions of the platform. This counteractive yaw effect was identified and discussed in the results presented by Udoh et al. (2015).

METHODOLOGY

Coordinate System Definition and System Properties

The coordinate axes (X and Y) used in this study are illustrated in Figure 2 (right); Z = 0 at the waterline. Considering that this paper is a sequel to a related benchmark study (Udoh et al. 2015), the difference between the two coordinate system definitions (i.e. the preceding and current study) is noted here for clarity, and to enhance a comparative interpretation of results between the two studies. Figure 2 (left) depicts the coordinate system used in Udoh et al. (2015) – essentially, the coordinate system used in the current study is obtained by rotating that in Udoh et al (2015) by 90 deg counter-clockwise to align the X-axis with the direction of dominant turbulent wind loads; this was necessary due to limitations in the version of TurbSim (Jonkman and Kincher, 2012) used to develop turbulent wind fields for simulations.

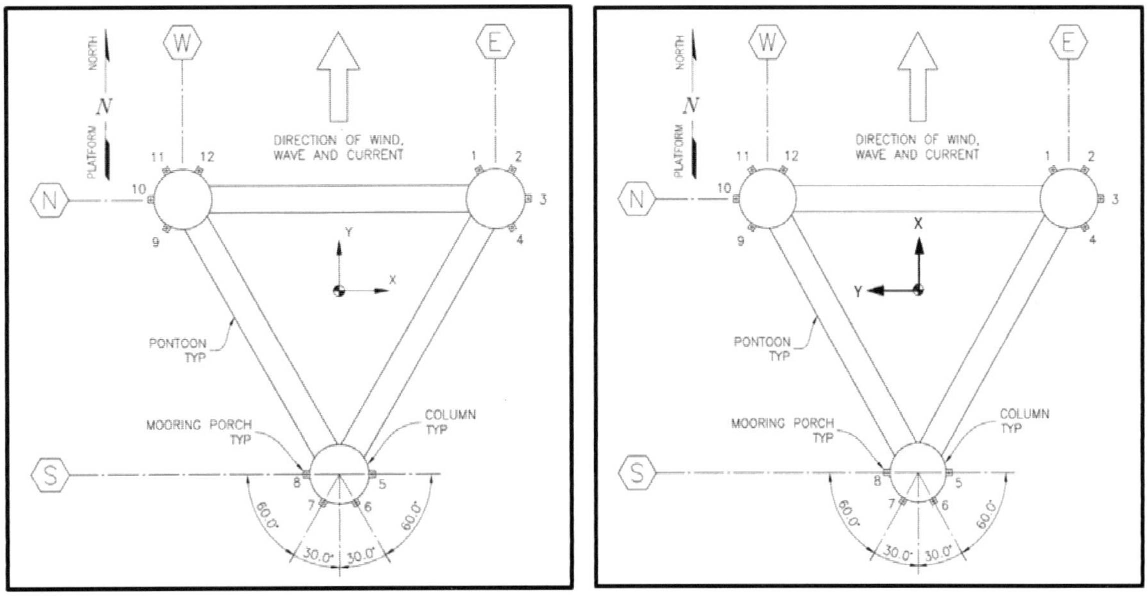

Figure 2: General Arrangement of Three-column TLP (left: Udoh et al. 2015, right: present study)

Nacelle, shaft and tower base coordinates are defined in Figure 3. It is important to note that the nacelle accelerations are reported in the shaft coordinate system (X_s, Y_s, Z_s), while the nacelle heading is defined consistent with the nacelle coordinate system (x_n, y_n, z_n). The nacelle heading is modeled to align with the direction of wind loads, such that the rotor plane is perpendicular to the aerodynamic thrust. The wind turbine used is in this study is the National Renewable Energy Laboratory's (NREL's) 5MW reference wind turbine (Jonkman et al. 2009). Characteristics of the turbine and the three-column TLP are presented in Table 1. Hydrodynamic modeling is performed using the hull mesh shown in Figure 4. A conceptual view of the three-column TLP concepts is shown in Figure 5, and Table 2 shows the natural periods / frequencies of the platform and tower.

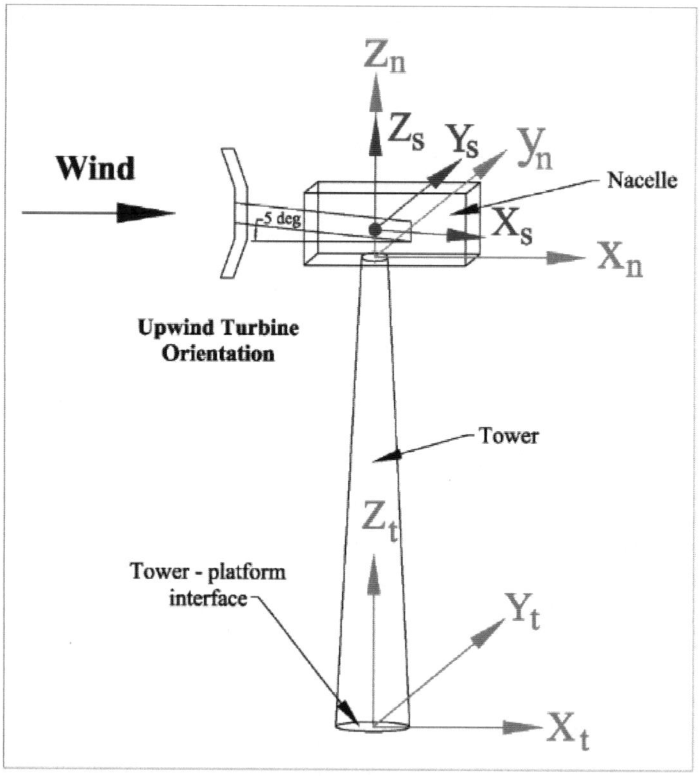

Figure 3: Coordinate System Definitions for Towerbase and Nacelle

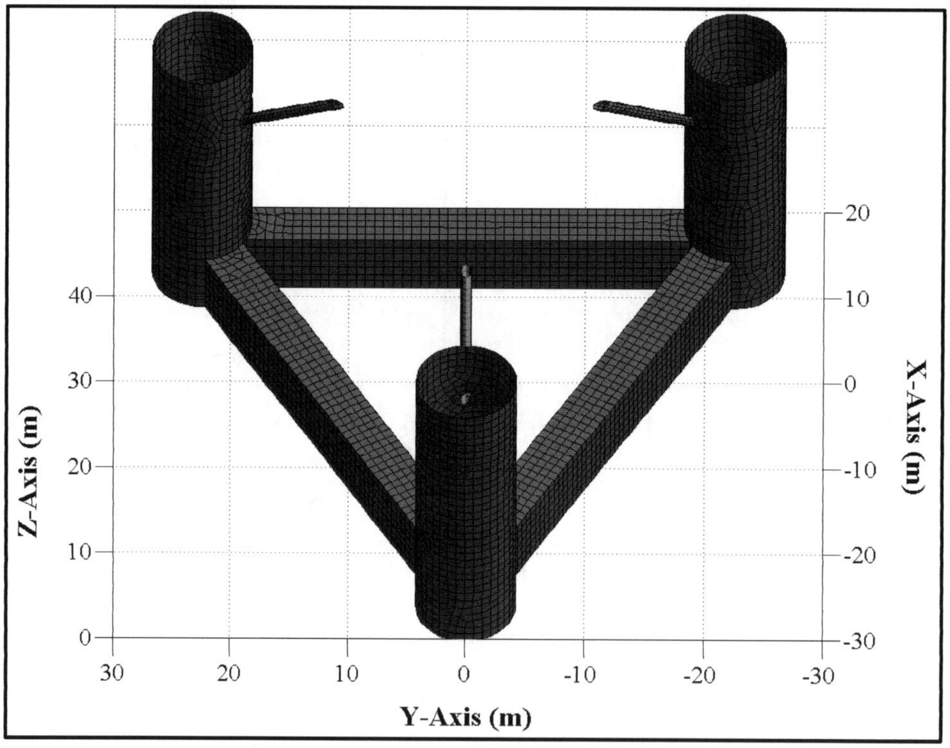

Figure 4: Hull Mesh used in Hydrodynamic Analysis

Table 1: Key Properties of the Three-column TLP wind Turbine

Parameter	Unit	Value
Water depth	[m]	50.0
Hub height from tower base	[m]	90.0
Rotor diameter	[m]	126.0
Lowest rotor to still water level	[m]	27.0
Hub diameter	[m]	3.0
Tower height	[m]	88.2
Tower diameter (top)	[m]	6.0
Tower diameter (bottom)	[m]	10.0
Rotor mass	[mt]	110
Nacelle mass	[mt]	240
Tower mass	[mt]	347
Total wind turbine weight	[mt]	697
Total wind turbine Vcg	[m]	64
Total wind turbine Lcg	[m]	-0.2
Hull draft	[m]	26
Displacement	[mt]	6897.1
Column/column span	[m]	45
Column diamater	[m]	8.5
Pontoon height	[m]	5.5
Pontoon width	[m]	3.8
Column height	[m]	34.0
Column freeboard	[m]	8.0
Hull weight	[mt]	4992.3
Vertical center of gravity	[m]	25.12
Roll radius of gyration	[m]	38.77
Pitch radius of gyration	[m]	38.95
Yaw radius of gyration	[m]	22.37
Tendon mooring pretension	[mt]	1904.8
Number of tendons (chain only)	[m]	12
Total tendon length	[m]	26.0
Chain stopper height	[m]	2.0
Top chain diamter (0.01m corrosion included)	[m]	0.165
Chain grade	[m]	R4S+
Tendon dry weight	[m]	14.20
Tendon wet weight	[m]	12.35
Buoyancy	[%]	13.0

Figure 5: Conceptual view of Three-column TLP Wind Turbine

Table 2: Natural Periods of the Platform and Tower

Mode	Natural Period [sec]	Natural Frequency [Hz]
Surge	25.0	0.04
Sway	25.0	0.04
Heave	0.58	1.72
Roll	1.01	0.99
Pitch	1.01	0.99
Yaw	20.0	0.05
Tower-X (mode 1)	1.30	0.77
Tower-X (mode 2)	0.21	4.87
Tower-Y (mode 1)	1.32	0.76
Tower-Y (mode 2)	0.30	3.32

Computer Programs

Numerical models and other programs applied in analysis have been widely used in the offshore energy industry for successful projects. WAMIT (WAMIT Inc., 2006) is used to calculate hydrodynamic coefficients and response amplitude operators (RAOs) of the floating system. Aerodynamic loads are calculated in FAST (Fatigue, Aerodynamics, Structures and Turbulence – Jonkman and Buhl, 2005); FAST is a publicly available program developed by NREL, with the capability to model the coupled Aerodynamic-Hydrodynamic-Elastic-Control behavior of fixed and floating offshore wind turbines. FAST considers the dynamics of flexible blades, rotor hub, nacelle, and flexible tower of the wind turbine system, as well as the dynamics of the electrical control system, in the calculation of aerodynamic loads. OrcaFlex is a widely used time-domain program, applicable to the analysis of floating structures coupled to slender members (e.g. mooring lines, risers, tendons and cables). In this study, OrcaFlex is used to model the TLP tendons, thus enabling the explicit inclusion of the effects of dynamic tendon tensions in the global responses of the system. Wind and current load coefficients for the TLP hull are calculated using WINDOS (MARIN, 2007), and TurbSim is used to develop turbulent wind fields imposed on the rotor in FAST, and the tower eignemodes are derived using BMODES (Bir, 2007).

A DNV-recommended approach is adopted to estimate the extreme responses of the wind turbine and platform. The expression, which is associated with the probability of 0.9 (or a 90% fractile in the extreme value distribution) is given as (DNV-RP-C103, 2005):

$$\text{Extreme value} = \sigma \times \sqrt{-2In\left(1-0.9^{1/N}\right)} \quad (6)$$

where $N = \dfrac{T_{simulation}}{T_Z}$, σ is the standard deviation, $T_{simulation}$ is the duration simulated and T_Z is the average period between successive zero up-crossings. The T_Z values in Udoh et al. (2015) were inadvertently reported based on the simulation time for one realization, whereas two realizations were used in extreme value analysis. T_Z values reported in this paper are consistent with two realizations for each simulated case. Extreme responses estimated using Equation (6) are used to calculate extreme factors by normalizing using the associated root-mean-square (RMS) values.

Coupled Analysis Methodology

Coupled hydrodynamic-aerodynamic-control (hydro-aero-servo) modeling of the system is achieved by coupling FAST to OrcaFlex such that simulations are stepped in time simultaneously in the two programs, with relevant data exchange. The link between OrcaFlex and FAST is enabled by a dynamic link library named FASTLink.dll (developed through collaborative work by Orcina and NREL).

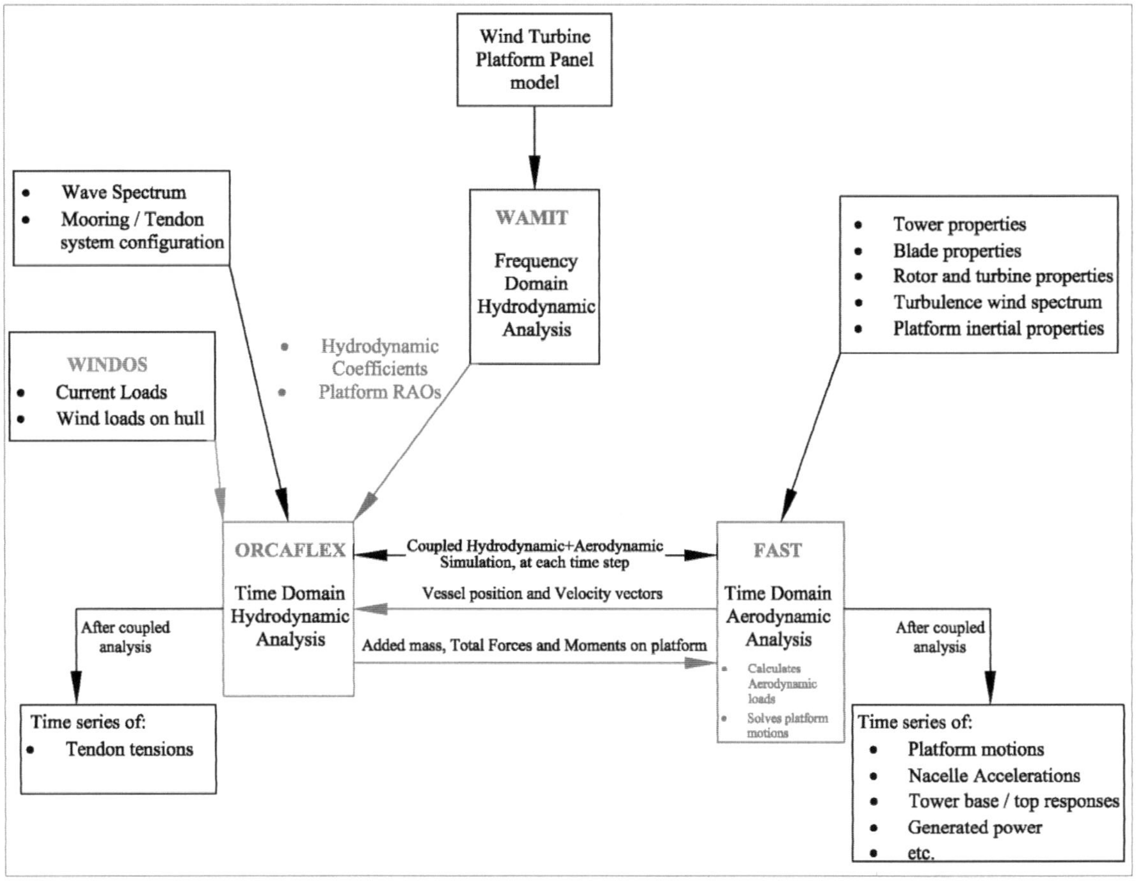

Figure 6: Analysis Workflow

Through FASTlink.dll, FAST passes the floater's instantaneous position and velocity vectors to OrcaFlex which then calculates the total forces and moments on the floater, and in turn passes these loads along with the vessel's added mass back to FAST. The vessel loads are combined with the aerodynamic loads and used in calculating the global motions of the floater (and other results) in FAST; the new vessel positions and velocities are then passed to OrcaFlex for the calculations to be performed in the next time-step, and this process is stepped forward in time through the simulated duration. More details on FASTlink.dll can be found in Masciola et al. (2011). A schematic of the analysis procedure involving the programs discussed in the previous section, and the FAST-OrcaFlex coupling is shown in Figure 6.

Environmental Loads Modeling

Environmental loads modeled in this study are summarized in Table 3. It is important to note that the definition of '50-yrp' and '100-yrp' environments are only representative, and pertain more to the wave characteristics than the wind and current speed magnitudes. The wave spectra for both conditions are shown in Figure 7. A uniform current profile is assumed, and the Norwegian Petroleum Directorate (NPD) spectrum is applied to model dynamic wind loads on the hull and tower. Turbulent winds are applied to the rotor using a full field grid of 2.5 m dimensions in the vertical and horizontal planes. A JONSWAP (Joined North Sea Wave Observation Project) spectrum is used in modeling the wave environments. The wind, wave and current loads are applied collinear along the X-axis (0 deg).

Table 3: Characteristics of the Modeled Environment Loads

Item	Unit	50-yrp	100-yrp
Wave spectrum	[-]	JONSWAP	JONSWAP
Gamma	[-]	2.0	2.0
Significant wave height	[m]	4.11	6.0
Wave peak period	[sec]	12.49	15.09
Wind spectrum (platform)	[-]	NPD	NPD
Mean wind speed (platform)	[m/s]	11.4	11.4
Turbulence wind model (hub/rotor)	[-]	Kaimal and Von Karman	
Turbulence mean wind speed (hub/rotor)	[m/s]	17.0	17.0
Current speed	[m/s]	1.54	1.54
Direction of wind, wave and current	[deg]	0.0	0.0

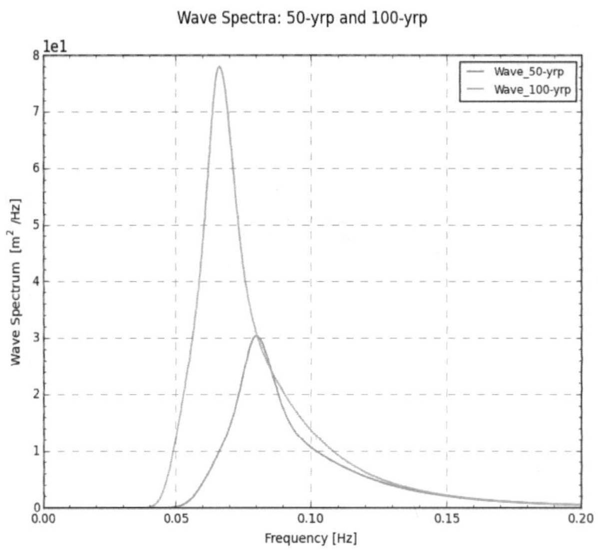

Figure 7: 50-yrp and 100-yrp Wave Spectra

Chaotic fluctuations of wind fields affect the performance of wind turbines through changes in wind velocity. For horizontal axis wind turbines, the horizontal components of wind velocity (especially the one normal to the plane of blade rotation) have significant impacts on the behavior of the system, compared to the vertical component. Spectral models can be used to represent the variation of wind velocity components as stochastic processes. Among the many spectral models in literature, the Kaimal (1972) and Von Karman (1948) spectra are some of the most studied and broadly accepted models in wind energy applications.

The International Electrotechnical Commission (IEC)-defined Kaimal spectrum for the longitudinal (u), transverse (v) and vertical (w) components of wind velocity, as expressed in Jonkman and Kincher (2012), is:

$$S_K(f) = \frac{4\sigma^2_K L_K / \bar{u}_{hub}}{\left(1 + 6fL_K / \bar{u}_{hub}\right)^{5/3}} \tag{7}$$

where K = u, v, w and the associated turbulence spectral parameters applicable to the design of offshore wind turbines (IEC-61400-1, 2005) are summarized in Table 4.

Table 4: Turbulence Spectra Parameters for the Kaimal Model (IEC-61400-1, 2005)

Parameter	Velocity Component Index		
	u	v	w
Standard Deviation, σ_K	σ_u	$0.8\sigma_u$	$0.5\sigma_u$
Integral Scale, L_K	$8.1\Lambda_u$	$2.7\Lambda_u$	$0.66\Lambda_u$

In Table 4, Λ_u is the longitudinal turbulence scale parameter. According to Jonkman and Kincher (2012), the IEC-defined Von Karman spectrum for the longitudinal component of wind velocity is expressed as:

$$S_u(f) = \frac{4\sigma^2_u L / \bar{u}_{hub}}{\left(1 + 71\left(fL / \bar{u}_{hub}\right)^2\right)^{5/6}} \tag{8}$$

and for the transverse and vertical components:

$$S_u(f) = \frac{2\sigma^2_K L / \bar{u}_{hub}}{\left(1 + 71\left(fL / \bar{u}_{hub}\right)^2\right)^{11/6}} \left(1 + 189\left(fL / \bar{u}_{hub}\right)^2\right) \tag{9}$$

where K = v, w. The Kaimal and Von Karman models assume neutral atmospheric stability (i.e. Richardson number = 0), implying that buoyancy is negligible and there is sufficient kinetic energy to homogenize the flow field. As can be deduced from the standard deviations shown in Table 4, the spatial variation of turbulence is anisotropic in the Kaimal model; the Von Karman model on the other hand assumes isotropic turbulence. Figure 8 shows the Kaimal and Von Karman velocity timeseries and turbulence spectra. From visual observation of the timeseries, the X-axis Kaimal wind velocities can be seen to have more low-frequency components than the Von Karman wind velocities in the three categories of turbulence intensity; timeseries of the Y and Z components of turbulent wind velocities are shown in the Appendix. The denotations 'Kaimal-A', 'Kaimal-B', and 'Kaimal-C', represent the data obtained using the Kaimal spectrum with the respective turbulence intensity classes, A, B and C. Similar dentations apply to the Von Karman spectrum as 'Von K.-A', 'Von K.-B' and 'Von K.-C'.

In the modeling of wind loads in FAST for this study, an equilibrium or quasi-steady inflow model is used to simulate the steady wind cases; essentially, the equilibrium model makes the assumption that the aerodynamic wake is always in equilibrium with respect to the inflow and structural response. A dynamic inflow model is used to simulate turbulent wind cases, wherein the model assumes that changes in the inflow or structural response dynamically affect the aerodynamic wake.

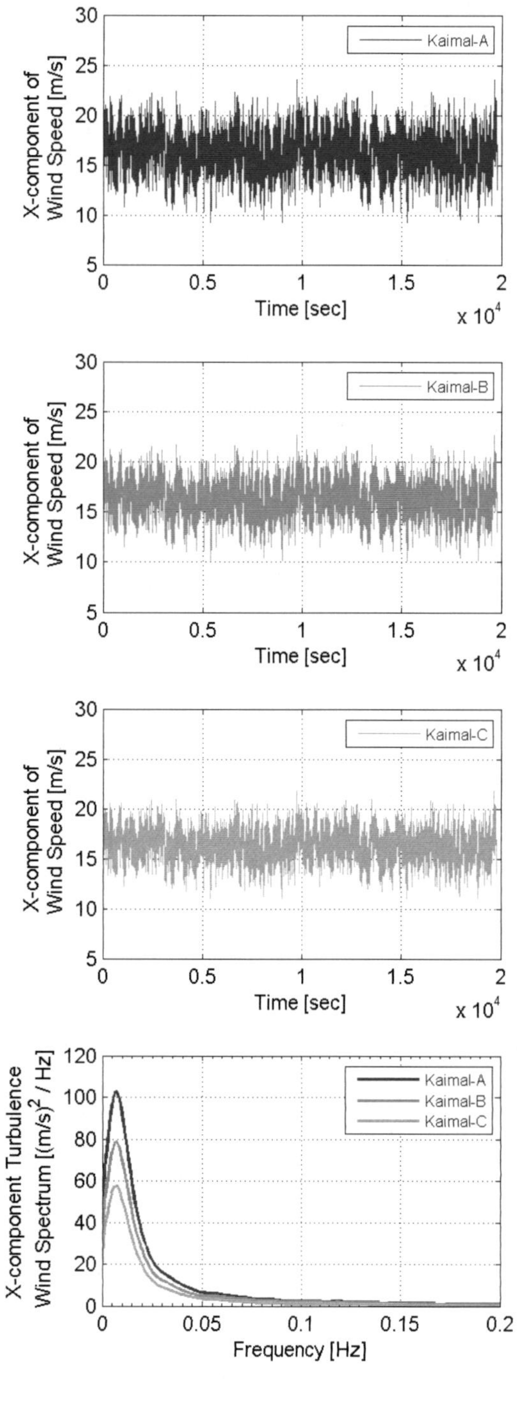

Figure 8: Turbulence Wind Velocity Timeseries and Spectra (left: Kaimal, right: Von Karman)

RESULTS

Majority of the results shown in this paper are spectra of the responses; very limited timeseries data are shown in the main paper for brevity. However, an associated timeseries appendix accompanies this paper as a separate file, but is not published along with the paper. Interested readers may however request this file from the authors, and it will be made available for reference.

Platform Motions in 50-yrp Condition

Motions of the platform in 50-yrp condition indicate significant effects of turbulence, as shown in the response spectra - Figure 9. In all six degrees of freedom, at frequencies where a variation in the spectral energy is seen with respect to turbulence intensity (i.e. A, B and C), the spectra show increase in energy with turbulence intensity. Thus an increase in turbulent wind excitations results in increased response amplitudes. Response spectral peaks in surge occur at the surge natural frequency (0.04 Hz or 25 sec), and the wave peak frequency (0.08 Hz or 12.49 sec). Turbulence-induced responses are clearly evident at the natural frequency, with over 80% increase in spectral energy when comparing the steady wind case to Kaimal-C which has the lowest turbulence intensity considered. No turbulence-induced variation is observed in the response at the wave peak frequency. These observations are consistent with the range of excitations of the Kaimal wind spectra (Figure 8) and the 50-yrp wave spectrum (Figure 7). As expected magnitudes of sway responses are negligible compared to surge considering the direction of environmental loads (i.e. along the X-direction). Heave response spectra correlate very closely to surge responses with peaks also occurring similar frequencies. Some turbulence-induced heave response is seen at the wave peak frequency, with an increase of 20% from the steady wind case to Kaimal-A. This is an interesting observation because there is little or no turbulent wind excitation (and practically no variation of kinetic energy with turbulence intensity) at 0.08 Hz. It implies there is a secondary effect of turbulence that enhances the heave response at the wave peak frequency.

Rotational motions of the platform in 50-yrp show significantly more energy in roll than pitch; this observation is consistent with the gyroscopic effect (explained in the *Relevant Theory* section) on the platform inhibiting the pitch response while enhancing (or at least not inhibiting) the roll response. Although the magnitudes of roll response are low (absolute value less than 0.2 deg), some qualitative findings can still be derived. Roll response due to turbulent wind excitation is seen around 0.01 Hz, where there is zero spectral energy from the steady wind case. A secondary spectral response peak is seen around 0.09 Hz which is similar to the frequency at which a suggested secondary effect occurs in the heave response. The identified similarity indicates some coupling between roll and heave responses. The most significant influence of turbulence in roll response occurs at the yaw natural frequency (0.05 Hz), correlating with the yaw spectral peak response and indicating yaw-induced roll motions of the platform. At that frequency, the steady wind response energies are at least 90% less in roll and yaw.

A comparison of the extreme statistics of motion responses (Table 5 and Figure 10) further shows the influence of turbulence over steady wind effects in each response mode. Noting that the direction of load application is the X-axis, statistics of surge response are indicative of the influence of wind turbulence, whereas sway responses are orders of magnitude smaller and could be considered noise-level responses hence have no significant or practical meaning. Similarly, roll and pitch magnitudes are generally much smaller than yaw amplitudes, hence have less practical significance compared to yaw. Bar plots of the ratios of statistics for 50-yrp are presented in Figure 10. On each bar plot, numeric values of the Kaimal-A ratios are shown. The ratios of statistics are calculated as:

$$Ratio_i = \left(\frac{Statistic_{turbulence}}{Statistic_{steady}} \right)_i \tag{10}$$

For surge, heave and yaw the RMS values in 50-yrp increase by 26%, 31% and 80% due to turbulence, when comparing steady wind and Kaimal-A scenarios; similarly, extreme values of surge, heave and yaw increase by approximately the same percentages due to turbulence between the two scenarios. The extreme factors of the responses between the steady wind and turbulence scenarios are quite similar because the increase in RMS and extreme values for the turbulence scenarios are somewhat proportional.

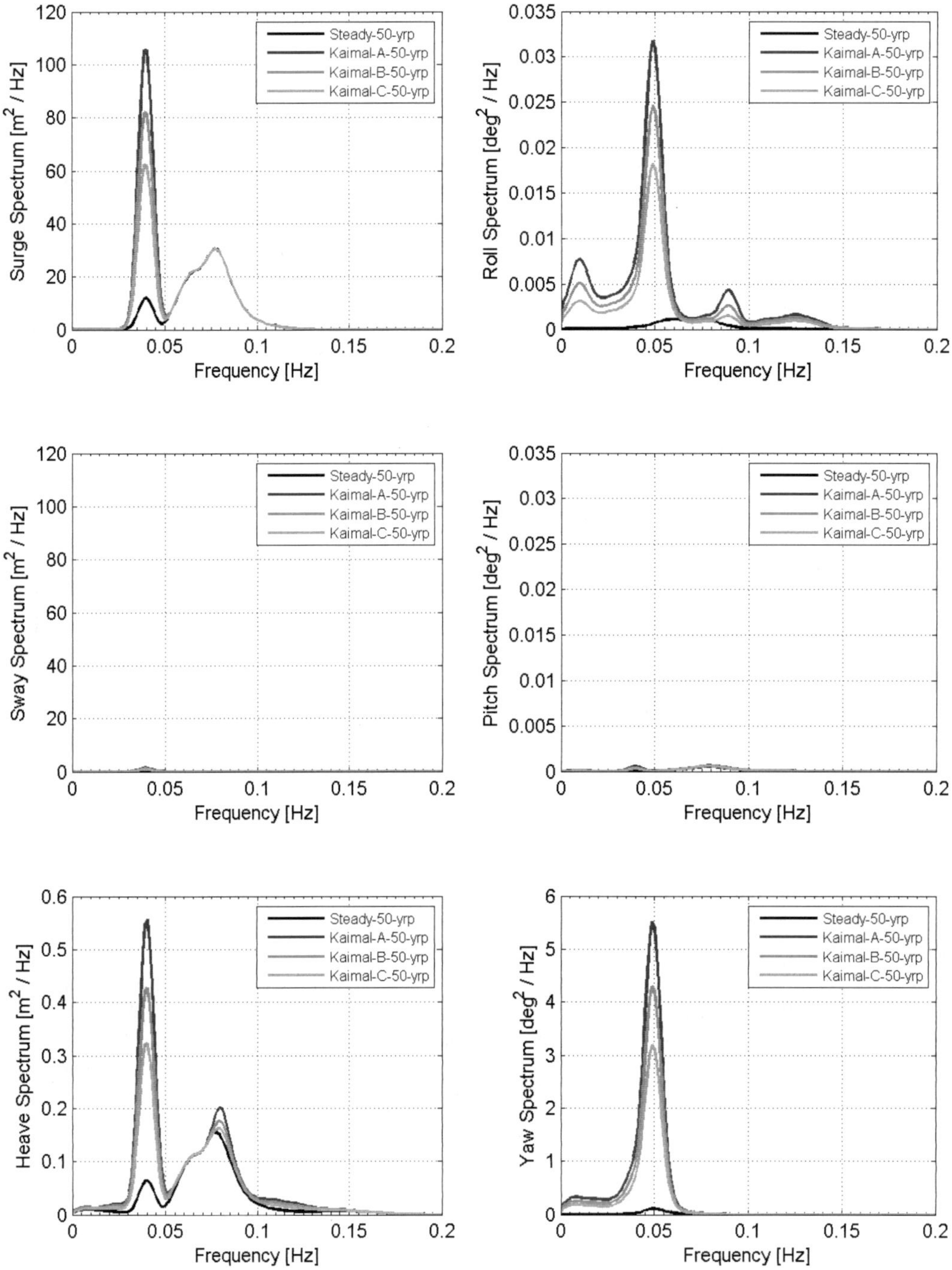

Figure 9: Platform 6DOF Motion Spectra, 50-yrp

Table 5: Platform Motion Statistics, 50-yrp

50-yrp condition							
Steady	Mean	RMS	Tz	90%ile Dynamic Extreme	90%ile Extreme Factor	Maximum	Minimum
Surge [m]	1.844	1.030	13.50	4.497	4.368	6.341	-2.652
Sway [m]	-0.005	0.023	24.90	0.095	4.226	0.090	-0.101
Heave [m]	-0.082	0.078	13.80	0.341	4.363	0.259	-0.423
Roll [deg]	-0.009	0.007	14.70	0.031	4.348	0.022	-0.041
Pitch [deg]	0.012	0.004	11.10	0.018	4.412	0.030	-0.006
Yaw [deg]	-0.129	0.047	19.70	0.200	4.281	0.071	-0.328
Kaimal-A	Mean	RMS	Tz	90%ile Dynamic Extreme	90%ile Extreme Factor	Maximum	Minimum
Surge [m]	1.863	1.403	16.20	6.070	4.326	7.934	-4.207
Sway [m]	-0.018	0.129	25.40	0.545	4.221	0.527	-0.562
Heave [m]	-0.102	0.115	16.80	0.496	4.317	0.393	-0.598
Roll [deg]	-0.003	0.026	13.10	0.115	4.375	0.112	-0.119
Pitch [deg]	0.013	0.006	4.47	0.026	4.614	0.038	-0.013
Yaw [deg]	-0.056	0.283	21.60	1.205	4.259	1.149	-1.261
Kaimal-B	Mean	RMS	Tz	90%ile Dynamic Extreme	90%ile Extreme Factor	Maximum	Minimum
Surge [m]	1.865	1.319	15.80	5.713	4.333	7.578	-3.849
Sway [m]	-0.018	0.113	25.40	0.475	4.221	0.457	-0.493
Heave [m]	-0.097	0.105	16.30	0.455	4.325	0.358	-0.553
Roll [deg]	-0.003	0.023	13.70	0.100	4.365	0.096	-0.103
Pitch [deg]	0.013	0.005	4.71	0.024	4.603	0.037	-0.011
Yaw [deg]	-0.055	0.250	21.60	1.064	4.259	1.009	-1.120
Kaimal-C	Mean	RMS	Tz	90%ile Dynamic Extreme	90%ile Extreme Factor	Maximum	Minimum
Surge [m]	1.865	1.245	15.30	5.401	4.339	7.266	-3.535
Sway [m]	-0.018	0.097	25.40	0.408	4.221	0.390	-0.426
Heave [m]	-0.094	0.098	15.70	0.423	4.334	0.330	-0.517
Roll [deg]	-0.004	0.019	14.50	0.085	4.352	0.081	-0.088
Pitch [deg]	0.013	0.005	5.13	0.022	4.584	0.035	-0.010
Yaw [deg]	-0.055	0.216	21.70	0.920	4.258	0.865	-0.975

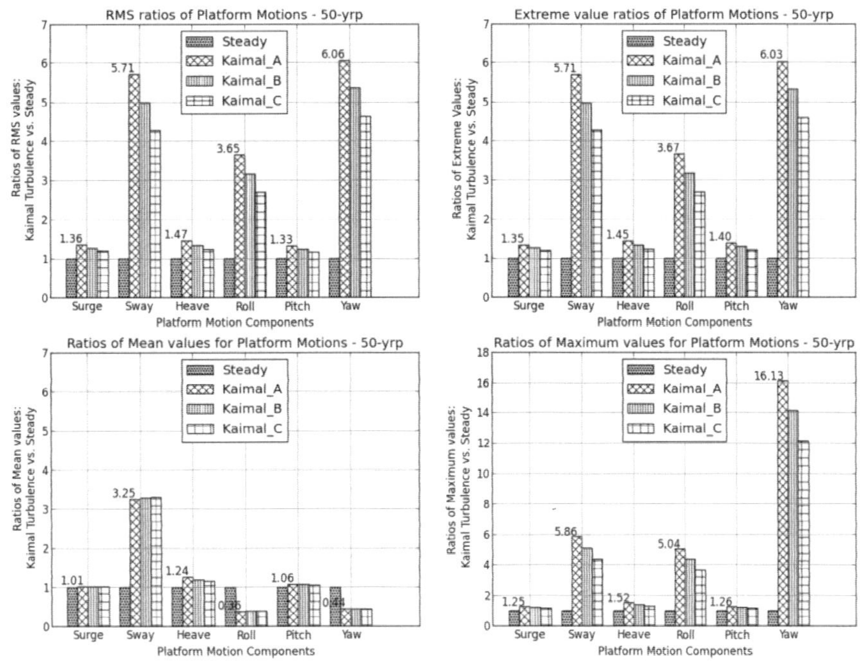

Figure 10: Ratios of Platform Motion Statistics, 50-yrp

Platform Motions in 100-yrp Condition

In the 100-yrp condition, platform motion spectra indicate less variation in energy as a function of turbulence intensity, at the dominant spectral response peaks. In surge response, the influence of the 100-yrp wave environment is quite significant, such that there is barely an 8% increase in the spectral peaks between the steady wind case and Kaimal-A at the surge natural frequency (0.04 Hz); the spectral energies of the steady and turbulent wind cases align over the remainder of the spectral band. A similar observation is made in heave response, with slight increases in spectral peaks due to turbulence at wind spectral peak frequency (0.01 Hz) and the surge natural frequency (0.04 Hz). Similar to the surge response, there is almost no change in heave response peak between the steady wind case and the turbulence cases at the wave peak frequency (0.067 Hz). At 0.09 Hz and 0.11 Hz, interesting responses occur in heave and roll. The platform response amplitude operator (RAO) for heave (Figure 12) shows a peak around 0.13 Hz, and a cancelation at 0.09 Hz. Heave responses at these frequencies show slight increments (7% or less) in energy with turbulence considered, while the increment in roll response energy is over 40% between the steady wind case and Kaimal-C.

Response spectra for roll and yaw for the 100-yrp condition have similar characteristics with those of the 50-yrp condition, except for the peak at 0.11 Hz which is mostly due to coupling between heave and roll – there is some amplitude of heave response (as shown in Figure 12, top left pane) and some wave excitation at this frequency. The distinct peak in roll response occurring at the yaw natural frequency is a clear indication of yaw-induced roll, given the frequency at which the peak occurs and the relative magnitude between the energies in the spectra. Pitch response is again seen to be dampened possibly by gyroscopic effects, causing the dynamic pitch to be much less than the dynamic roll response. This reasoning is supported by the RAOs of pitch and roll (Figure 12), wherein although the magnitudes of pitch response RAOs are at least two orders higher than the roll RAOs over the same frequency band, the platform pitch response is less than the roll response. It is worth mentioning that there is a reasonable amount of energy in the 50-yrp and 100-yrp wave spectra within the frequency ranges where significant peaks are identified in the roll and pitch RAO – this suggests that the influence of the gyroscopic effect on the pitch response must also be significant to dampen the response. For completeness, it is also worth mentioning that much smaller-magnitude RAO peaks exist at the pitch and roll natural frequencies, and are not shown in Figure 12.

The highlighted observations are reflected in the statistics reported in Table 6 and Figure 13. Increments in RMS and extreme values of surge, heave and yaw responses are 2%, 4% and 38% respectively when comparing the steady wind case to Kaimal-A. As mentioned earlier, statistical ratios of sway, roll and pitch have very limited practical significance due to the small magnitudes of the data. It is interesting that the presence of turbulent winds induces a significant reduction in mean platform yaw responses by a factor of 0.44 (in both 50-yrp and 100-yrp) while increasing the dynamic yaw response. However, the maximum platform yaw for the Kaimal-A case (based on extreme value analysis) is about 1.7 degrees which for the particular system studied should not pose any major concern. The coupling between yaw and roll responses is also evident in the reduction of mean roll responses in both 50-yrp and 100-yrp conditions.

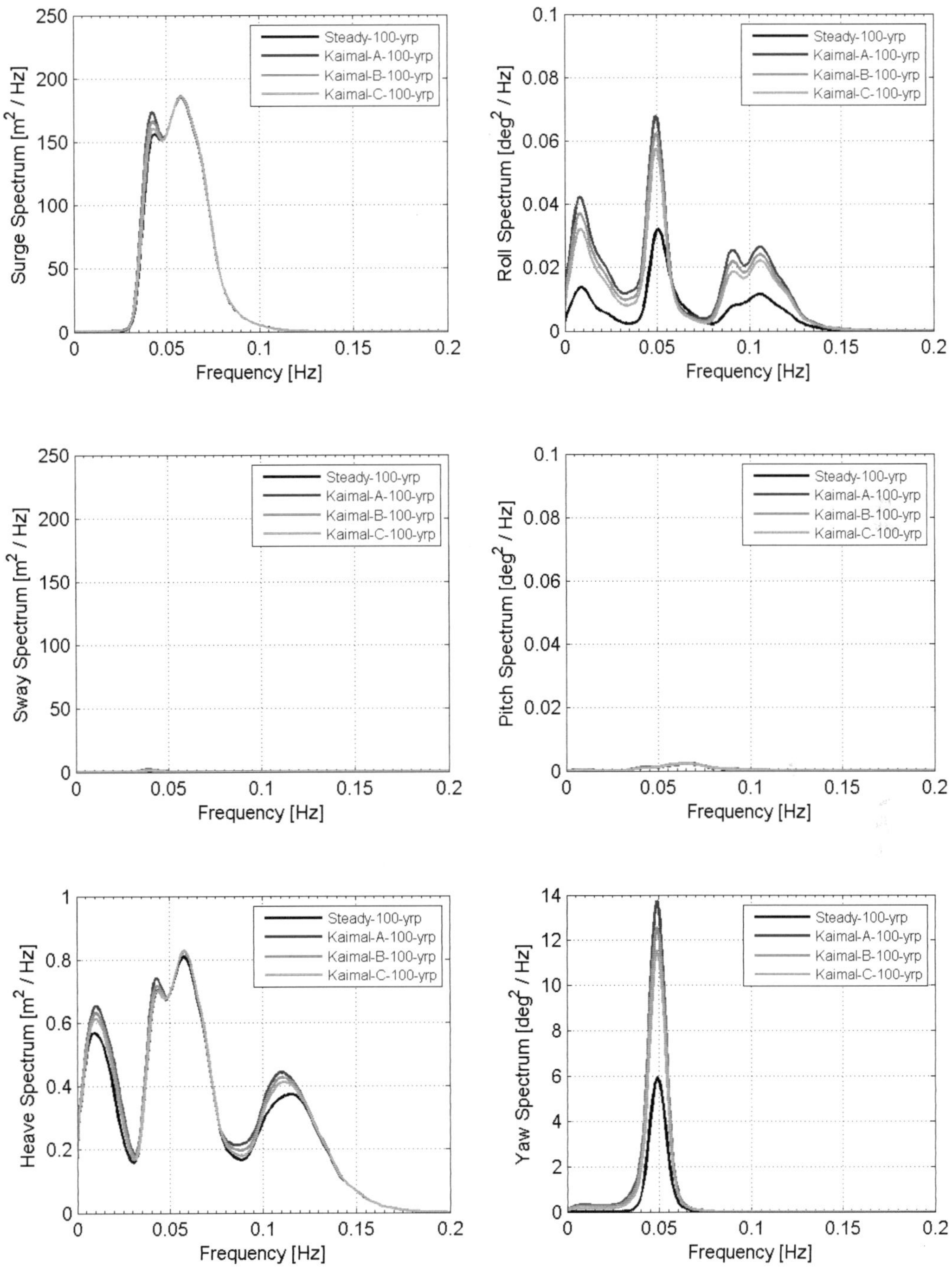

Figure 11: Platform 6DOF Motion Spectra, 100-yrp

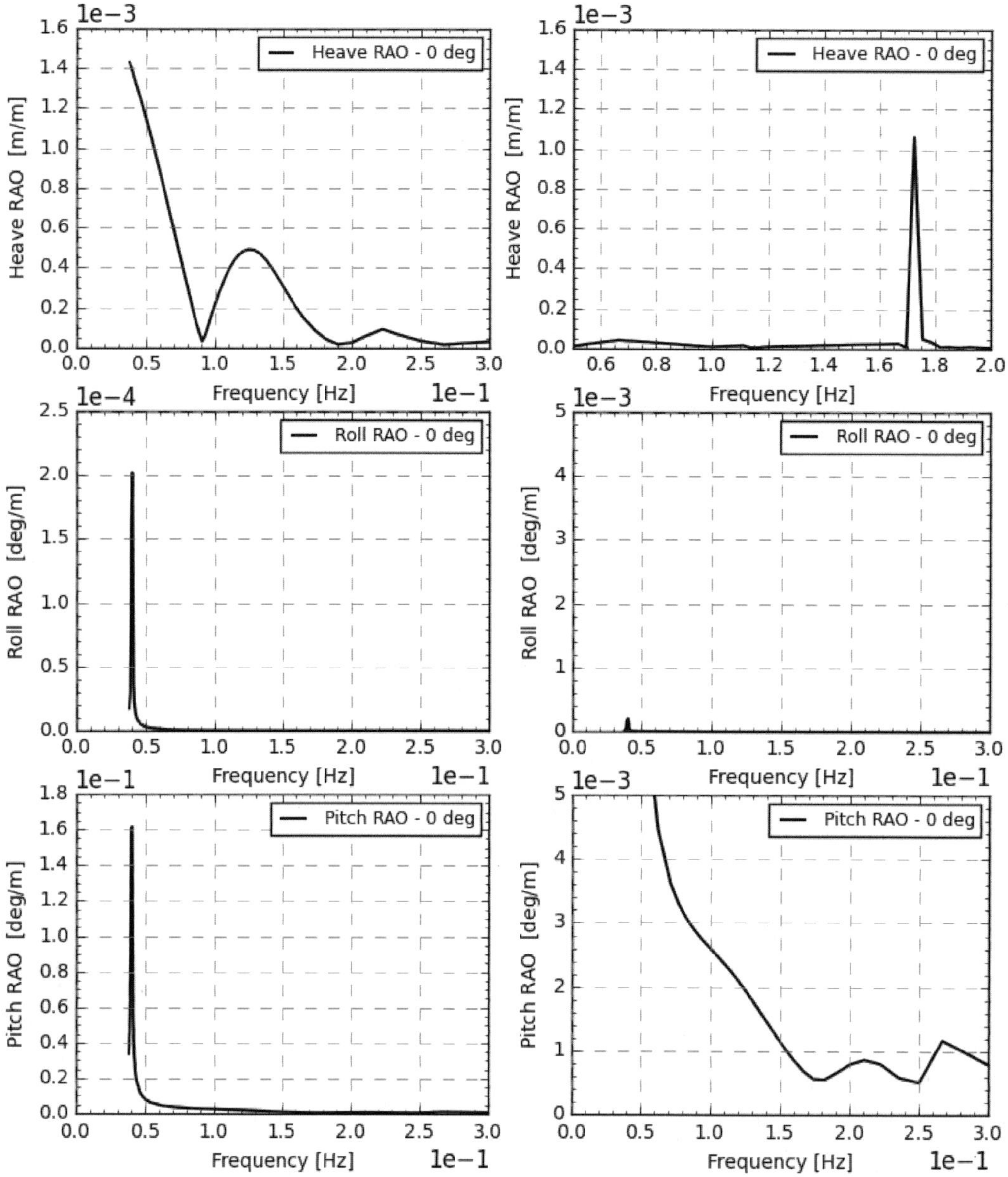

Figure 12: Response Amplitude Operators for Heave, Roll and Pitch

Table 6: Platform Motion Statistics, 100-yrp

100-yrp condition							
Steady	Mean	RMS	Tz	90%ile Dynamic Extreme	90%ile Extreme Factor	Maximum	Minimum
Surge [m]	1.768	2.640	17.10	11.388	4.314	13.157	-9.620
Sway [m]	-0.006	0.071	23.80	0.303	4.236	0.297	-0.309
Heave [m]	-0.194	0.253	15.70	1.098	4.333	0.904	-1.291
Roll [deg]	-0.010	0.035	15.10	0.151	4.342	0.142	-0.161
Pitch [deg]	0.012	0.009	14.50	0.038	4.351	0.051	-0.026
Yaw [deg]	-0.126	0.256	20.20	1.095	4.275	0.969	-1.220
Kaimal-A	Mean	RMS	Tz	90%ile Dynamic Extreme	90%ile Extreme Factor	Maximum	Minimum
Surge [m]	1.793	2.695	17.30	11.619	4.311	13.412	-9.826
Sway [m]	-0.018	0.149	24.80	0.631	4.227	0.612	-0.649
Heave [m]	-0.202	0.263	15.80	1.141	4.332	0.939	-1.343
Roll [deg]	-0.006	0.054	12.70	0.235	4.382	0.228	-0.241
Pitch [deg]	0.013	0.010	6.83	0.044	4.521	0.057	-0.031
Yaw [deg]	-0.056	0.411	21.00	1.755	4.266	1.699	-1.811
Kaimal-B	Mean	RMS	Tz	90%ile Dynamic Extreme	90%ile Extreme Factor	Maximum	Minimum
Surge [m]	1.793	2.676	17.20	11.540	4.312	13.333	-9.747
Sway [m]	-0.019	0.139	24.60	0.586	4.228	0.567	-0.604
Heave [m]	-0.200	0.261	15.80	1.129	4.332	0.929	-1.328
Roll [deg]	-0.006	0.051	13.10	0.221	4.375	0.215	-0.228
Pitch [deg]	0.013	0.010	7.56	0.043	4.499	0.056	-0.030
Yaw [deg]	-0.055	0.390	20.90	1.665	4.267	1.610	-1.721
Kaimal-C	Mean	RMS	Tz	90%ile Dynamic Extreme	90%ile Extreme Factor	Maximum	Minimum
Surge [m]	1.793	2.660	17.20	11.470	4.312	13.263	-9.677
Sway [m]	-0.019	0.129	24.60	0.544	4.228	0.525	-0.563
Heave [m]	-0.198	0.258	15.70	1.116	4.333	0.919	-1.314
Roll [deg]	-0.007	0.048	13.20	0.209	4.374	0.203	-0.216
Pitch [deg]	0.013	0.009	8.26	0.042	4.479	0.055	-0.029
Yaw [deg]	-0.055	0.371	20.80	1.582	4.268	1.527	-1.637

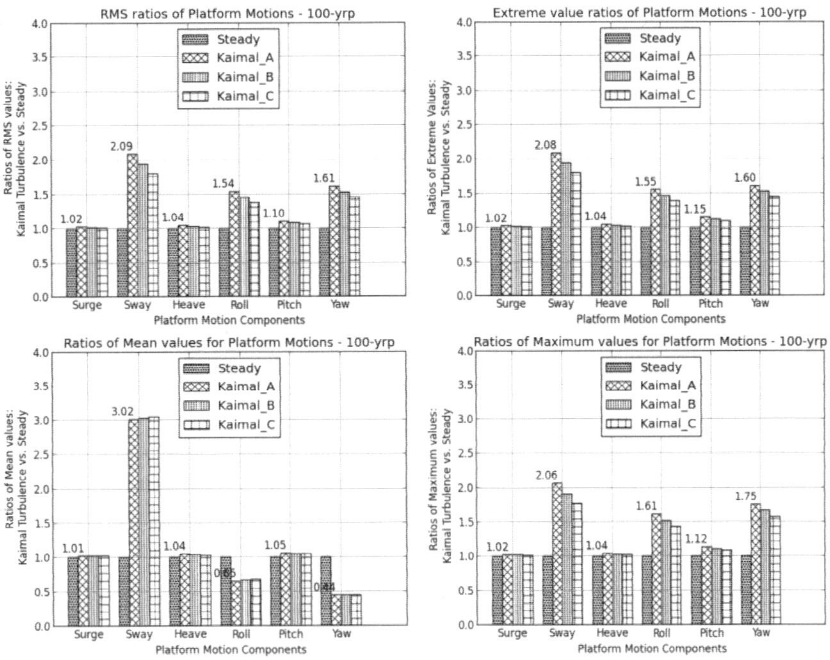

Figure 13: Ratios of Platform Motion Statistics, 100-yrp

Tendon Tensions

We investigated the influence of turbulence on tendon response, namely the top and bottom tensions. Results of bottom tendon tensions are not reported here, but we assessed them to see if any negative tensions (i.e. the chains becoming slack) had occurred. No negative tensions occurred in the simulated cases of 50-yrp and 100-yrp conditions, indicating sufficient pre-tensions in the tendons in regards to the modeled environmental loads. Top tensions are shown and discussed for the 50-yrp and 100-yrp conditions.

Figure 14 shows the spectra of tendon tensions for selected downwind tendons – Tendons 1 and 12, for 50- and 100-yrp. The energy variation at surge natural frequency (0.04 Hz) clearly indicates the influence of turbulence winds in the 50-yrp condition (top panes) for Tendons 1 and 12. Between the steady wind and Kaimal-C cases, the increase in energy at this frequency is over 80% in Tendons 1 and 12. Given the symmetrical locations of Tendons 1 and 12 about the X-axis, it is interesting that there is a difference in the spectral energies of the respective wind cases (steady, Kaimal-A, Kaimal-B and Kaimal-C) between these two tendons. The difference, which seems to grow with turbulence intensity (on close observation), is about 28% for the Kaimal-A case with Tendon 1 having higher energy peaks. This imbalance in energy between symmetrically located tendons is likely due to turbulence-induced yaw motions (especially because roll and pitch magnitudes are much less) of the platform, causing Tendon 1 to experience more dynamic tensions than Tendon 12 around the surge natural frequency. Around 0.6 – 0.7 Hz, higher-frequency responses of the tendons can be seen, although the energy level is much less compared to those at the wave peak frequency and surge natural period. These relatively higher-frequency tendon responses, which are shown in Figure 16 and Figure 17 for downwave and upwave tendons respectively, are likely associated with the blade responses at their flapwise eigenfrequency (0.68 Hz, Ormberg and Bachynsci, 2012). However, the relative broad-banded characteristics of these tensions in the 0.6 – 0.7 Hz range suggest some interference (likely from the tower flexure) on the responses of the blades. This explanation becomes more plausible when the tensions within this frequency band are interpreted in the context of the tower-base reaction loads, which are discussed in the next section.

The conspicuous clarity of turbulence-induced tensions in the 50-yrp conditions is due to the fact that wave-induced responses are not overwhelmingly dominant on the responses. In contrast to the 50-yrp condition, tendon tensions in the 100-yrp condition barely vary due to turbulence across the spectrum as the wave-induced responses are dominant, although some slight variation is seen at the wave peak frequency. Tendon tension spectra and time histories for downwind Tendons 2, 3, 4, 9, 10 and 11 are shown in the time history and spectra appendix file associated with this paper (available on request), and have similar trends to those of Tendons 1 and 12 but with different magnitudes.

Tensions in the upwave tendons (Tendons 5, 6, 7 and 8) are significantly amplified due to turbulence, as can be seen in Figure 15 (at the surge natural frequency, 0.04 Hz) for the 50-yrp condition. By correlating the response peaks of platform motions and tendon tensions in the 50-yrp condition, it is clear that as turbulent winds excite the system at the surge natural frequency, the surge motions are associated with platform setdown or heave motions which in turn induce significant tensions in the tendons at the same frequency. One identified common characteristic of the turbulence-induced responses in the platform motions and tendon tension spectra, is that the energies are relatively narrow-banded compared to wave-induced responses. Since the Kaimal spectrum is fairly narrow-banded, turbulence-induced responses that are narrow-banded imply little or no interference on the excitations that may alter the frequency distribution of the responses. Applying the same interpretation, turbulence-induced responses that are not quite characteristic of the applied Kaimal spectrum (i.e. with fairly broad-banded responses) would suggest the contribution of some interference from other excitations or responses.

Statistics of tendon tensions are presented in Table 7 and Figure 18 for the 50-yrp, and in Table 8 and Figure 19 for the 100-yrp conditions. For the 50-yrp condition, the variation in the RMS and extreme tensions are up to 57% and 60% in the downwave tendons, with Tendon 4 showing the highest increase, when comparing the steady wind case to Kaimal-A (having the highest turbulence intensity). For upwave tendons, the increments in RMS and extreme tensions are similar – 53% and 55% respectively, when comparing the steady wind case to Kaimal-A. The effect of turbulence on the extreme factors, which are indicators of the dynamic responses of the system, show increments of ~4% in the upwave tendons, and up to 5% increase in the downwave tendons when the steady wind case is compared to Kaimal-A. Statistics in the 100-yrp condition show RMS increments of up to 57% in the downwave tendons, and up to 53% in the upwave tendons when comparing the steady wind case to Kaimal-A. For extreme values, turbulence-induced increments are up to 60% in the downwave tendons and 58% in the upwave tendons when the steady wind case is compared to Kaimal-A. Mean values show very small changes (3% or less) due to turbulence across all tendons. Slight increments in extreme factors are also seen in the 100-yrp condition – ~3.8% in the upwave tendons, and up to 5.6% in the downwave tendons when the steady wind case is compared to Kaimal-A. The statistics of tendon tensions indicate that overall (considering all frequencies in each spectrum), the dynamic response of the tendons as indicated by the extreme factors, has approximate increase of 5% due to turbulence. However, the variation of the dynamics as indicated by the RMS statistic increases by over 50% due to turbulence. The implication of these results from a design perspective is that turbulence-induced dynamics are critical for fatigue assessment of tendons (due to high RMS values), and to a much lower degree, increase the loads important for strength analysis of the tendons (based on relatively low percentage increment in extreme factors).

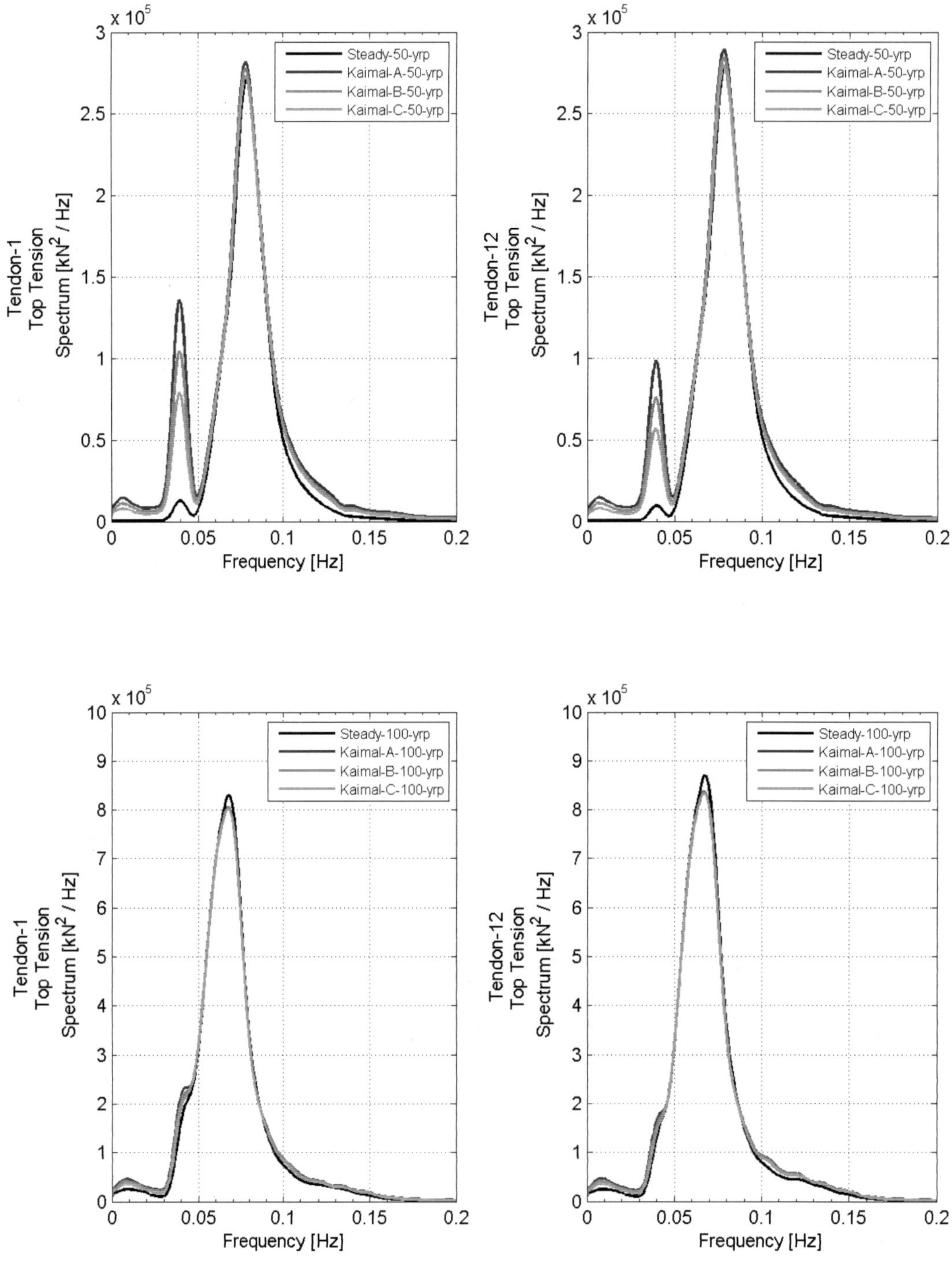

Figure 14: Tendon Tension Spectra for Downwave Tendons (1 and 12), 50-yrp and 100-yrp: 0 – 0.2 Hz

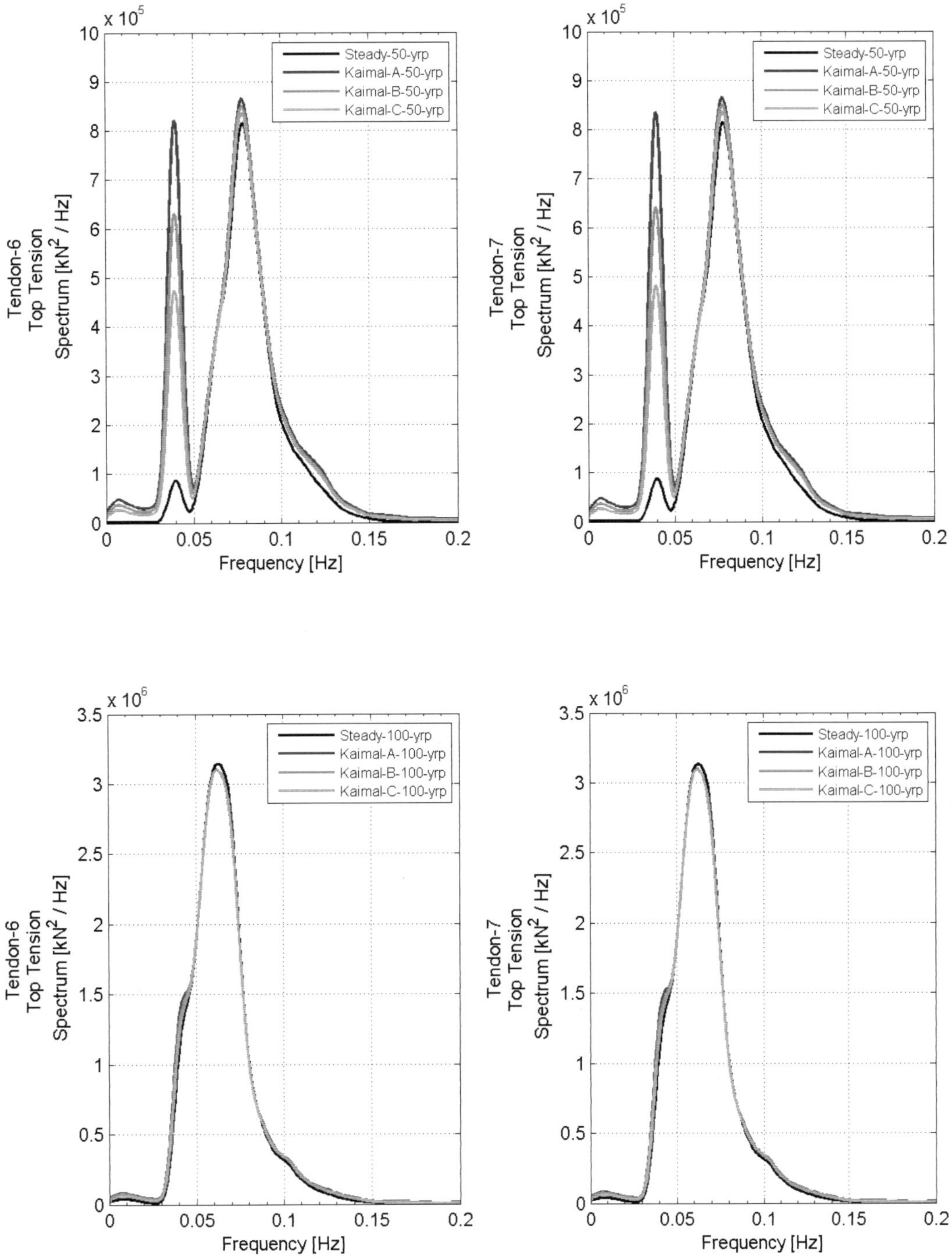

Figure 15: Tendon Tension Spectra for Upwave Tendons (6 and 7), 50-yrp and 100-yrp: 0 – 0.2 Hz

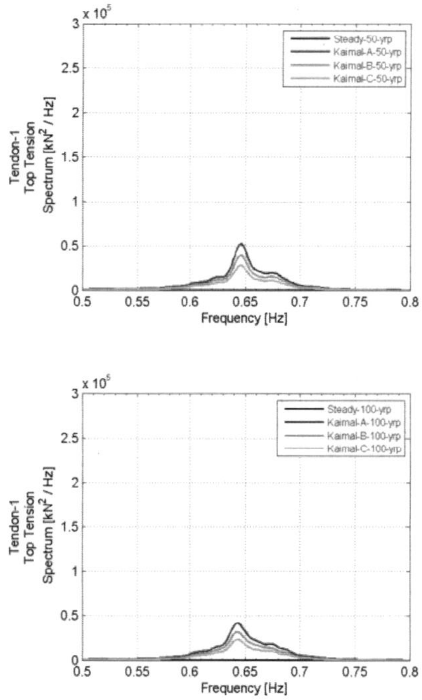

Figure 16: Tendon Tension Spectra for Downwave Tendons (1 and 12), 50-yrp and 100-yrp: 0.5 – 0.8 Hz

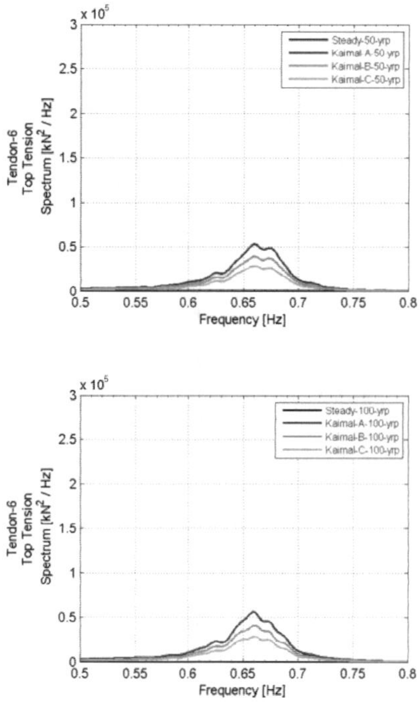

Figure 17: Tendon Tension Spectra for Upwave Tendons (6 and 7), 50-yrp and 100-yrp: 0.5 – 0.8 Hz

Table 7: Tendon Tension Statistics, 50-yrp

50-yrp condition							
Steady	Mean [kN]	RMS [kN]	Tz [sec]	90%ile Dynamic Extreme [kN]	90%ile Extreme Factor	Maximum [kN]	Minimum [kN]
Tendon 1	1281.8	88.4	8.690	394.8	4.468	1676.7	887.0
Tendon 2	1279.7	88.0	8.780	393.1	4.466	1672.8	886.6
Tendon 3	1342.9	70.3	8.110	315.0	4.483	1657.9	1027.9
Tendon 4	1408.8	53.5	7.000	241.4	4.516	1650.2	1167.4
Tendon 5	1981.3	144.2	9.870	640.0	4.439	2621.3	1341.4
Tendon 6	2053.0	166.8	10.000	739.9	4.435	2792.9	1313.1
Tendon 7	2057.0	166.7	10.100	739.3	4.435	2796.3	1317.7
Tendon 8	1989.3	144.0	9.870	639.1	4.439	2628.3	1350.2
Tendon 9	1449.6	55.1	6.910	249.1	4.519	1698.7	1200.6
Tendon 10	1387.6	72.1	7.720	324.1	4.494	1711.7	1063.4
Tendon 11	1320.3	89.8	8.460	401.8	4.474	1722.1	918.5
Tendon 12	1314.5	89.8	8.510	401.8	4.472	1716.3	912.7
Kaimal-A	Mean [kN]	RMS [kN]	Tz [sec]	90%ile Dynamic Extreme [kN]	90%ile Extreme Factor	Maximum [kN]	Minimum [kN]
Tendon 1	1261.8	114.8	3.010	539.7	4.699	1801.5	722.2
Tendon 2	1264.6	117.6	2.830	554.4	4.713	1819.0	710.2
Tendon 3	1329.9	98.9	2.490	468.5	4.739	1798.5	861.4
Tendon 4	1393.3	77.4	2.250	368.6	4.761	1761.8	1024.7
Tendon 5	2001.3	182.1	4.200	842.8	4.628	2844.1	1158.6
Tendon 6	2084.0	210.9	4.190	976.1	4.628	3060.1	1107.9
Tendon 7	2089.6	211.5	4.140	979.3	4.631	3068.8	1110.3
Tendon 8	2012.5	183.3	4.080	849.4	4.634	2861.8	1163.1
Tendon 9	1449.7	74.6	2.380	354.1	4.749	1803.8	1095.6
Tendon 10	1391.5	95.0	2.680	449.0	4.724	1840.5	942.6
Tendon 11	1320.3	114.3	3.080	536.4	4.695	1856.7	783.9
Tendon 12	1306.7	112.2	3.300	525.1	4.680	1831.8	781.6
Kaimal-B	Mean [kN]	RMS [kN]	Tz [sec]	90%ile Dynamic Extreme [kN]	90%ile Extreme Factor	Maximum [kN]	Minimum [kN]
Tendon 1	1262.2	108.9	3.250	509.9	4.683	1772.1	752.3
Tendon 2	1263.3	111.1	3.060	521.5	4.695	1784.8	741.7
Tendon 3	1328.5	92.7	2.670	437.9	4.725	1766.4	890.6
Tendon 4	1393.5	72.3	2.390	343.2	4.748	1736.8	1050.3
Tendon 5	2000.5	173.4	4.550	799.3	4.610	2799.7	1201.2
Tendon 6	2081.2	200.8	4.550	925.6	4.611	3006.7	1155.6
Tendon 7	2086.7	201.2	4.490	928.3	4.613	3015.0	1158.4
Tendon 8	2011.6	174.3	4.430	804.7	4.616	2816.3	1206.9
Tendon 9	1450.1	70.1	2.520	332.2	4.737	1782.3	1117.9
Tendon 10	1390.2	89.8	2.840	422.9	4.711	1813.2	967.3
Tendon 11	1319.1	108.6	3.320	507.9	4.678	1827.0	811.2
Tendon 12	1307.2	107.0	3.530	499.0	4.665	1806.2	808.2
Kaimal-C	Mean [kN]	RMS [kN]	Tz [sec]	90%ile Dynamic Extreme [kN]	90%ile Extreme Factor	Maximum [kN]	Minimum [kN]
Tendon 1	1262.7	103.5	3.530	482.9	4.665	1745.5	779.8
Tendon 2	1262.1	105.1	3.330	491.6	4.678	1753.7	770.5
Tendon 3	1327.3	87.0	2.920	409.5	4.706	1736.8	917.8
Tendon 4	1393.9	67.6	2.590	319.8	4.731	1713.7	1074.0
Tendon 5	1999.7	165.5	4.970	760.0	4.591	2759.7	1239.7
Tendon 6	2078.6	191.7	4.950	880.0	4.592	2958.7	1198.6
Tendon 7	2084.2	192.0	4.900	882.2	4.594	2966.4	1202.0
Tendon 8	2010.8	166.3	4.850	764.4	4.597	2775.3	1246.4
Tendon 9	1450.5	66.1	2.720	311.9	4.720	1762.3	1138.6
Tendon 10	1389.2	84.9	3.080	398.7	4.694	1787.8	990.5
Tendon 11	1318.1	103.4	3.600	481.8	4.661	1799.9	836.2
Tendon 12	1307.7	102.2	3.810	475.1	4.649	1782.8	832.6

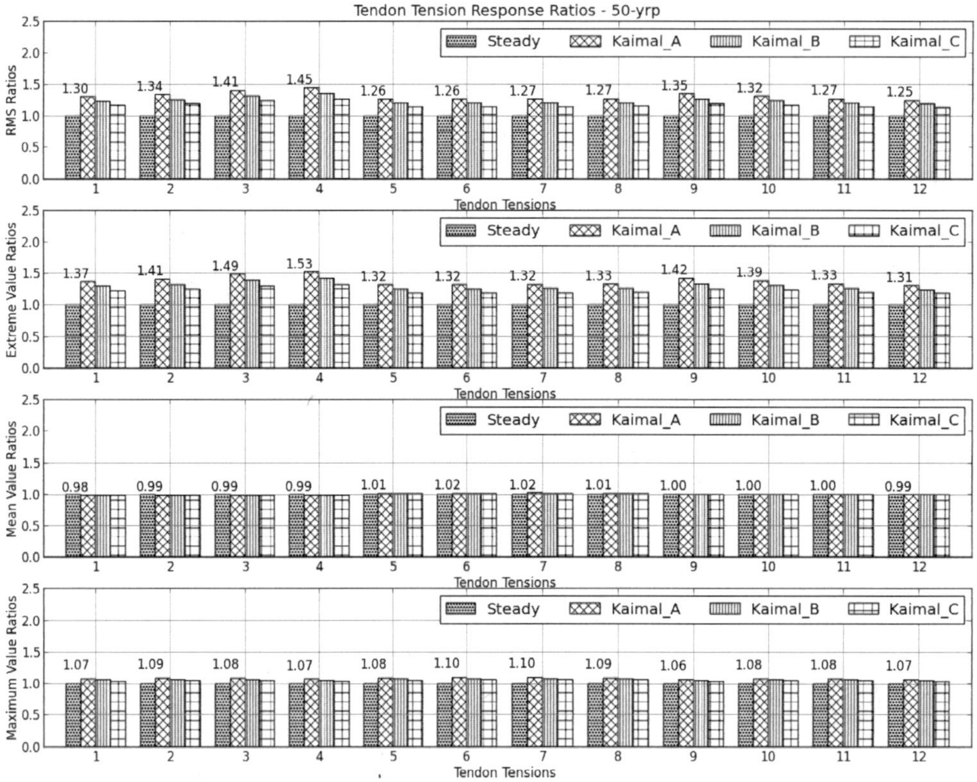

Figure 18: Ratios of Tendon Tension Statistics, 50-yrp

Table 8: Tendon Tension Statistics, 100-yrp

100-yrp condition							
Steady	Mean [kN]	RMS [kN]	Tz [sec]	90%ile Dynamic Extreme [kN]	90%ile Extreme Factor	Maximum [kN]	Minimum [kN]
Tendon 1	1300.7	170.2	13.39	743.8	4.370	2044.5	556.9
Tendon 2	1304.6	170.6	13.30	745.9	4.371	2050.5	558.7
Tendon 3	1365.4	142.1	12.29	623.7	4.390	1989.0	741.7
Tendon 4	1423.7	114.1	11.11	503.6	4.412	1927.4	920.1
Tendon 5	2005.5	296.4	14.24	1291.1	4.356	3296.6	714.4
Tendon 6	2085.6	342.1	14.11	1490.7	4.358	3576.3	594.9
Tendon 7	2089.6	341.8	14.20	1489.1	4.357	3578.7	600.5
Tendon 8	2013.5	296.0	14.30	1288.9	4.355	3302.4	724.6
Tendon 9	1464.6	117.5	10.63	519.5	4.423	1984.1	945.1
Tendon 10	1410.0	145.6	11.72	640.8	4.400	2050.8	769.2
Tendon 11	1345.0	174.3	12.60	764.0	4.384	2109.0	581.1
Tendon 12	1333.3	173.4	12.69	759.8	4.382	2093.1	573.5
Kaimal-A	Mean [kN]	RMS [kN]	Tz [sec]	90%ile Dynamic Extreme [kN]	90%ile Extreme Factor	Maximum [kN]	Minimum [kN]
Tendon 1	1278.1	181.6	4.58	836.8	4.609	2114.9	441.3
Tendon 2	1289.3	184.4	4.31	852.3	4.622	2141.6	437.1
Tendon 3	1352.1	156.3	3.76	726.9	4.651	2079.0	625.2
Tendon 4	1405.2	126.4	3.39	590.6	4.674	1995.8	814.6
Tendon 5	2020.7	309.0	6.77	1397.5	4.523	3418.2	623.2
Tendon 6	2114.2	356.1	6.61	1612.7	4.529	3726.9	501.5
Tendon 7	2119.8	356.3	6.51	1614.8	4.532	3734.6	505.0
Tendon 8	2031.9	309.5	6.54	1402.2	4.531	3434.1	629.7
Tendon 9	1461.6	127.0	3.66	591.4	4.657	2053.0	870.2
Tendon 10	1413.4	156.5	4.13	724.8	4.631	2138.3	688.6
Tendon 11	1344.5	185.2	4.78	851.7	4.599	2196.2	492.8
Tendon 12	1322.5	182.5	5.10	836.9	4.585	2159.4	485.6
Kaimal-B	Mean [kN]	RMS [kN]	Tz [sec]	90%ile Dynamic Extreme [kN]	90%ile Extreme Factor	Maximum [kN]	Minimum [kN]
Tendon 1	1278.6	178.5	5.10	818.7	4.586	2097.4	459.9
Tendon 2	1288.2	180.8	4.77	831.8	4.600	2120.1	456.4
Tendon 3	1351.0	152.8	4.16	707.5	4.630	2058.5	643.5
Tendon 4	1405.7	123.6	3.78	574.7	4.650	1980.4	831.0
Tendon 5	2020.7	305.5	7.44	1375.6	4.502	3396.3	645.1
Tendon 6	2112.4	352.2	7.36	1586.7	4.505	3699.1	525.7
Tendon 7	2118.0	352.4	7.20	1589.2	4.510	3707.2	528.8
Tendon 8	2031.9	306.0	7.29	1379.0	4.507	3410.9	652.9
Tendon 9	1462.2	124.5	4.02	577.5	4.637	2039.8	884.7
Tendon 10	1412.5	153.5	4.54	707.7	4.611	2120.1	704.8
Tendon 11	1343.6	182.0	5.31	833.1	4.577	2176.6	510.5
Tendon 12	1323.2	179.8	5.69	820.3	4.562	2143.4	502.9
Kaimal-C	Mean [kN]	RMS [kN]	Tz [sec]	90%ile Dynamic Extreme [kN]	90%ile Extreme Factor	Maximum [kN]	Minimum [kN]
Tendon 1	1279.2	175.8	5.74	801.8	4.559	2081.0	477.4
Tendon 2	1287.4	177.6	5.40	812.4	4.573	2099.8	474.9
Tendon 3	1350.1	149.7	4.73	689.2	4.602	2039.2	660.9
Tendon 4	1406.2	121.1	4.21	560.3	4.627	1966.6	845.9
Tendon 5	2020.6	302.4	8.21	1355.0	4.481	3375.6	665.6
Tendon 6	2110.7	348.7	8.15	1562.8	4.482	3673.5	547.8
Tendon 7	2116.3	348.8	8.03	1564.6	4.485	3680.9	551.7
Tendon 8	2031.8	302.8	8.06	1357.9	4.485	3389.7	674.0
Tendon 9	1462.8	122.4	4.46	564.6	4.615	2027.4	898.2
Tendon 10	1411.6	150.8	5.09	691.6	4.586	2103.2	720.1
Tendon 11	1342.8	179.2	5.94	815.7	4.552	2158.5	527.1
Tendon 12	1323.9	177.4	6.30	805.2	4.539	2129.1	518.7

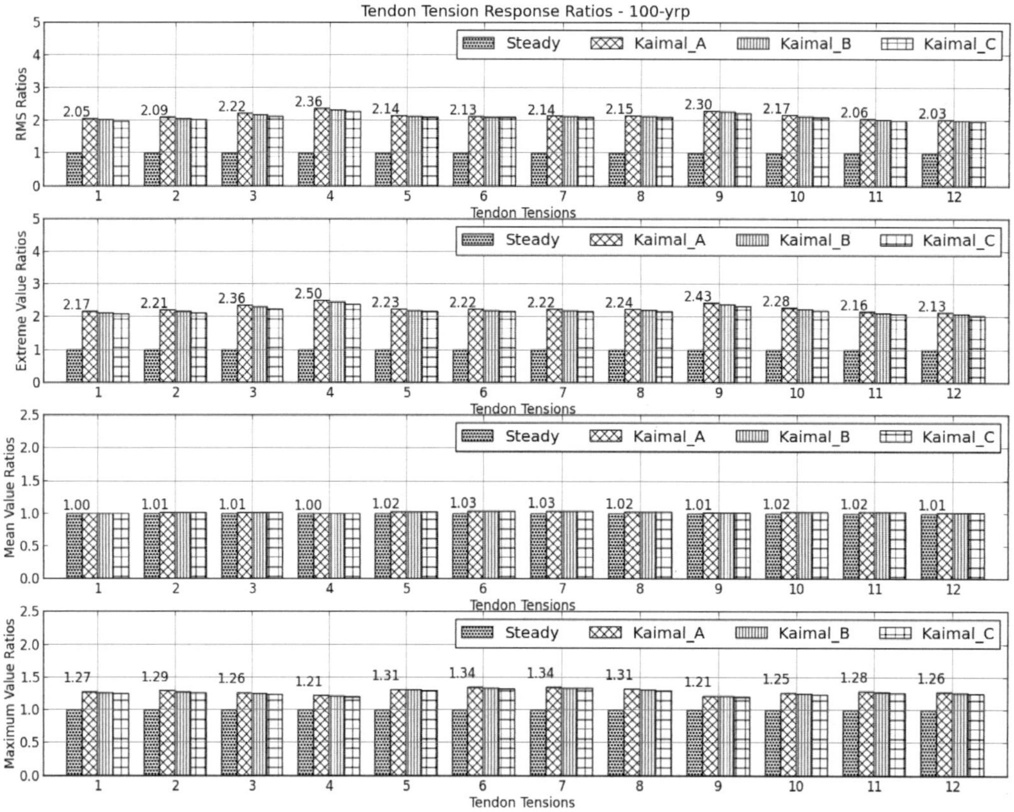

Figure 19: Ratios of Tendon Tension Statistics, 100-yrp

Towerbase Forces and Moments

Reaction loads at the base of the tower are critical to the design of FOWTs, thus understanding the impact of turbulence on their magnitudes and variations is important. We investigated the shear forces in the X_t and Y_t directions, the vertical force in the Z_t direction, and the reaction moments about these three axes at the towerbase.

Figure 20 shows the spectra of towerbase forces and moments for the 50-yrp condition. Towerbase shear forces in the X_t direction are associated with the towerbase moments in the Y_t direction, while the shear forces in the Y_t direction are associated with moments in the X_t direction. Shear forces in the X_t direction, and moments in the Y_t direction have higher magnitudes due to the direction of environmental loading. For easy reference in discussions and bar charts, shear force is denoted S.F. and bending moment is denoted B.M. The S.F. in X_t direction shows the influence of turbulence around the surge natural frequency, and between 0.6 – 0.7 Hz, much similar to the tendon tension spectra. A similar energy distribution is seen in the B.M. spectra of Y_t direction. The energy levels however, are very low indicating almost insignificant variations in dynamics due to turbulence. The S.F. in the Y_t direction, and B.M. in the X_t direction both show some response in the 0.6 – 0.7 Hz range, with energy levels similar to those of the previously mentioned reactions, in the same frequency band. In the towerbase vertical force, turbulence-induced responses can be seen at the peak frequency of the wind spectrum (~0.01 Hz), at the surge natural frequency (0.04 Hz), at the wave peak frequency (0.08 Hz), and the 0.1 – 0.16 Hz range (in which there is some amplification in the heave RAO). Yaw moments at the towerbase clearly show the impact of turbulence; the spectral peaks occur at the peak frequency of the wind spectrum, at which no energy exists for the steady wind case. It is thus very clear that yaw reactions at the towerbase are driven by the low frequency turbulent winds.

Wave-induced responses drive the S.F., B.M and vertical reactions at the towerbase in the 100-yrp condition, as no significant variation is seen in the spectra (Figure 22) due to turbulence. Essentially, the wave loads are much higher than the wind-induced or turbulence-induced loads, that the impacts seem insignificant. In the towerbase yaw reactions however, low-frequency turbulence-induced responses are quite clear (as in the 50-yrp), with peaks occurring at 0.01 Hz.

Statistics of towerbase reactions are reported in Table 9 and Figure 21 for the 50-yrp, and in Table 10 and Figure 23 for the100-yrp. Based on the reaction magnitudes, the forces with practically relevant statistics are the S.F. in the X_t direction and the vertical force, while the moments to consider are B.M. in the Y_t direction (associated with X_t shear force), and the yaw reaction at the towerbase. The ratios of statistics for S.F. in the Y_t direction and its associated B.M. in the X_t direction have little or no practical relevance, as they could be considered noise-level reactions based on their relative magnitudes to the other reactions.

In the 50-yrp, due to turbulence, the RMS of S.F. in the X_t direction increases by 14% and that of the vertical force increases by 27% when the steady wind case is compared to Kaimal-A. The B.M. in the Y_t direction shows an increase in RMS of 19%, while there is a 94% increase in the RMS of yaw moment at the towerbase when the steady wind case is compared to Kaimal-A. Actually, regarding the yaw reaction moment at the towerbase, the practical way to interpret the results is that yaw reactions exist with turbulence, and are almost non-existent in the steady wind case. The extreme values also increase due to turbulence – the S.F. in the X_t direction shows a 19% increase, the vertical force shows an increase of 30%, while the B.M. in the Y_t direction shows an increase of 23% when the steady wind case is compared to Kaimal-A. The extreme value of yaw reaction moment is also overwhelmingly dominant when turbulence is considered. These percentage increments in RMS and extreme values of the towerbase reactions due to wind turbulence are quite significant. The extreme factors of the significant towerbase responses however are in the range of 3.5 – 6.8 %. Statistics of the towerbase reactions in the 100-yrp condition are reflective of the observations made in the spectra – the percentage increments in the reactions due to turbulence is in the range of 4.7 – 10.7 %, except of course for the yaw reaction moment which increases above 80%.

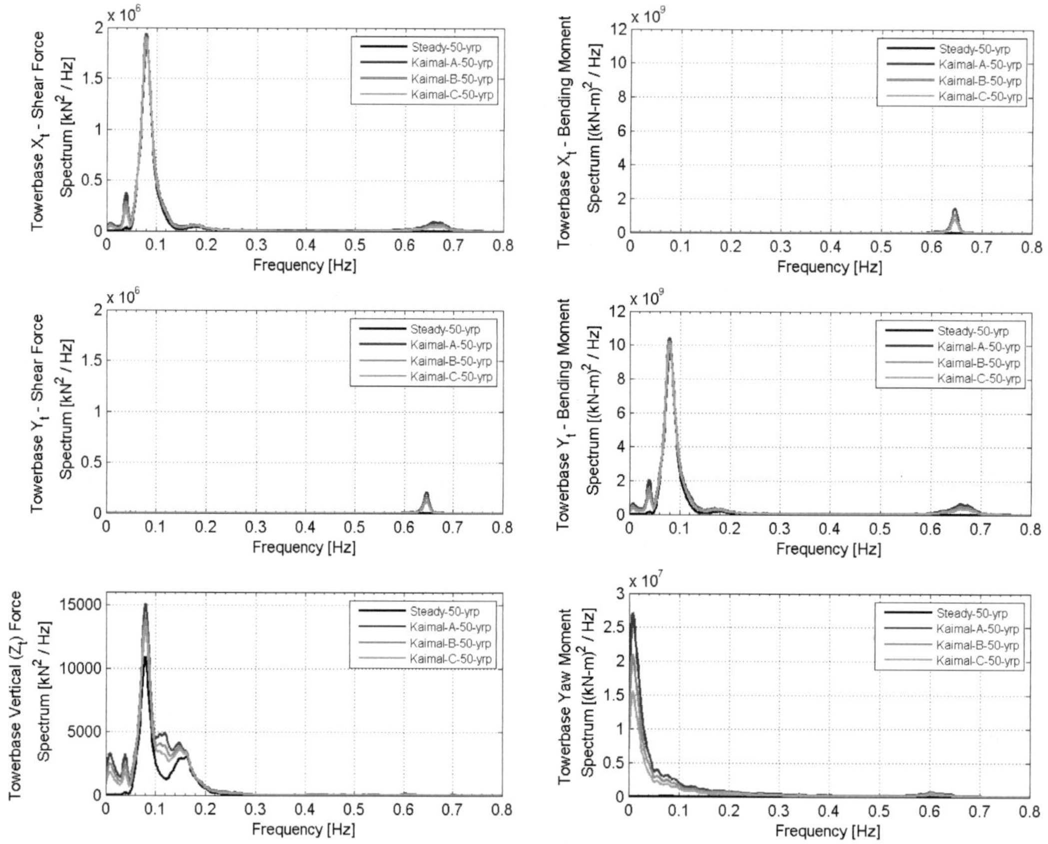

Figure 20: Towerbase Forces and Moments Spectra, 50-yrp

Table 9: Statistics of Towerbase Forces and Moments, 50-yrp

50-yrp condition							
Steady	Mean	RMS	Tz [sec]	90%ile Dynamic Extreme	90%ile Extreme Factor	Maximum	Minimum
Shear Force in $X_t[kN]$	370.9	238.5	10.60	1054.9	4.424	1425.8	-684.1
Shear Force in $Y_t[kN]$	2.7	4.0	3.42	18.6	4.672	21.3	-16.0
Vertical Force, $Z_t[kN]$	-6865.6	22.7	3.70	105.8	4.655	-6759.8	-6971.4
Bending Moment about $X_t[kN-m]$	3960.6	222.7	2.14	1062.6	4.772	5023.1	2898.0
Bending Moment about $Y_t[kN-m]$	32100.7	17141.6	11.10	75655.6	4.414	107756.3	-43554.9
Yaw Moment, about $Z_t[kN-m]$	-763.9	64.2	7.92	288.1	4.488	-475.8	-1052.0
Kaimal-A	Mean	RMS	Tz [sec]	90%ile Dynamic Extreme	90%ile Extreme Factor	Maximum	Minimum
Shear Force in $X_t[kN]$	391.1	278.8	3.61	1299.4	4.661	1690.6	-908.3
Shear Force in $Y_t[kN]$	-9.5	65.3	1.58	315.5	4.834	306.0	-325.0
Vertical Force, $Z_t[kN]$	-6874.5	31.3	1.65	151.0	4.825	-6723.5	-7025.5
Bending Moment about $X_t[kN-m]$	5100.8	5460.6	1.59	26393.0	4.833	31493.8	-21292.2
Bending Moment about $Y_t[kN-m]$	35391.0	21188.7	4.02	98252.9	4.637	133643.9	-62861.9
Yaw Moment, about $Z_t[kN-m]$	-114.7	1162.5	1.72	5599.2	4.817	5484.5	-5714.0
Kaimal-B	Mean	RMS	Tz [sec]	90%ile Dynamic Extreme	90%ile Extreme Factor	Maximum	Minimum
Shear Force in $X_t[kN]$	390.0	269.7	3.86	1252.8	4.646	1642.8	-862.8
Shear Force in $Y_t[kN]$	-9.6	56.9	1.58	275.0	4.834	265.4	-284.6
Vertical Force, $Z_t[kN]$	-6874.5	29.5	1.71	142.0	4.818	-6732.5	-7016.5
Bending Moment about $X_t[kN-m]$	5109.4	4758.1	1.59	22996.6	4.833	28106.1	-17887.2
Bending Moment about $Y_t[kN-m]$	35292.5	20287.3	4.33	93751.1	4.621	129043.6	-58458.6
Yaw Moment, about $Z_t[kN-m]$	-112.0	1020.4	1.66	4922.8	4.824	4810.8	-5034.9
Kaimal-C	Mean	RMS	Tz [sec]	90%ile Dynamic Extreme	90%ile Extreme Factor	Maximum	Minimum
Shear Force in $X_t[kN]$	388.9	261.3	4.07	1211.2	4.634	1600.0	-822.3
Shear Force in $Y_t[kN]$	-9.6	48.4	1.57	233.9	4.835	224.3	-243.5
Vertical Force, $Z_t[kN]$	-6874.4	27.9	1.77	134.3	4.811	-6740.1	-7008.8
Bending Moment about $X_t[kN-m]$	5115.5	4046.1	1.59	19554.7	4.833	24670.1	-14439.2
Bending Moment about $Y_t[kN-m]$	35198.2	19459.2	4.77	89513.5	4.600	124711.7	-54315.3
Yaw Moment, about $Z_t[kN-m]$	-109.3	877.8	1.60	4241.2	4.832	4131.9	-4350.5

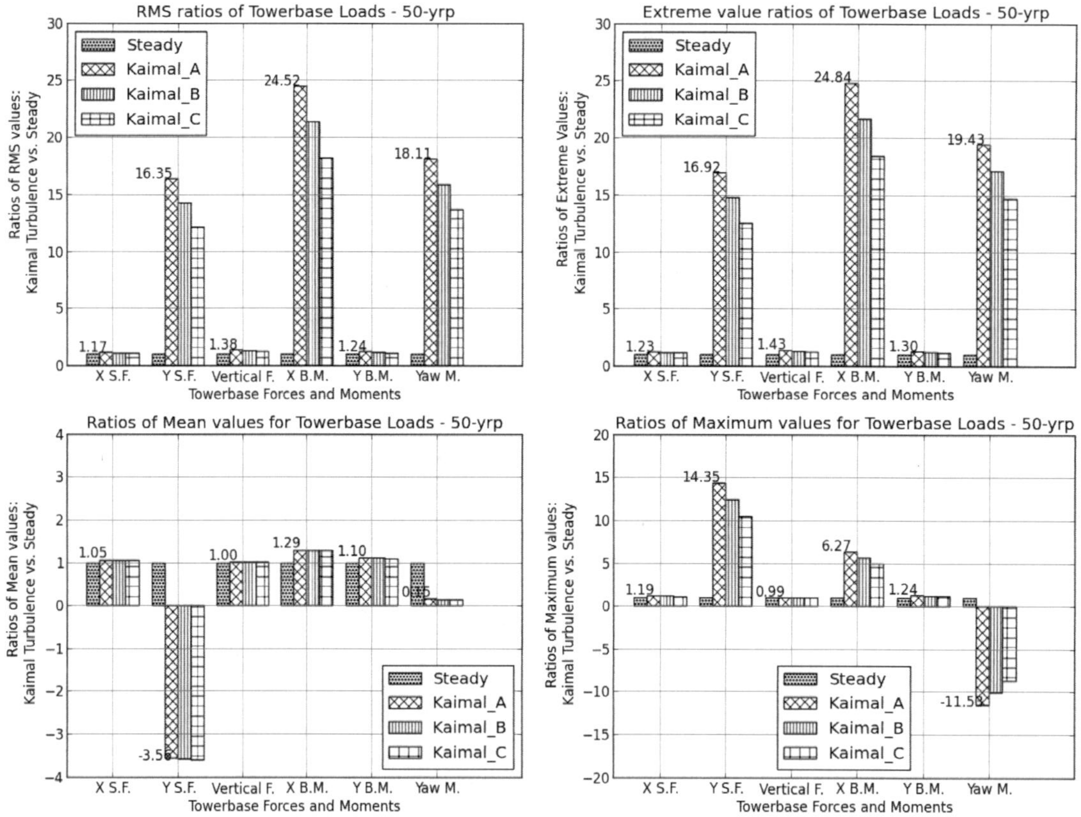

Figure 21: Ratios of Towerbase Forces and Moments Statistics, 50-yrp

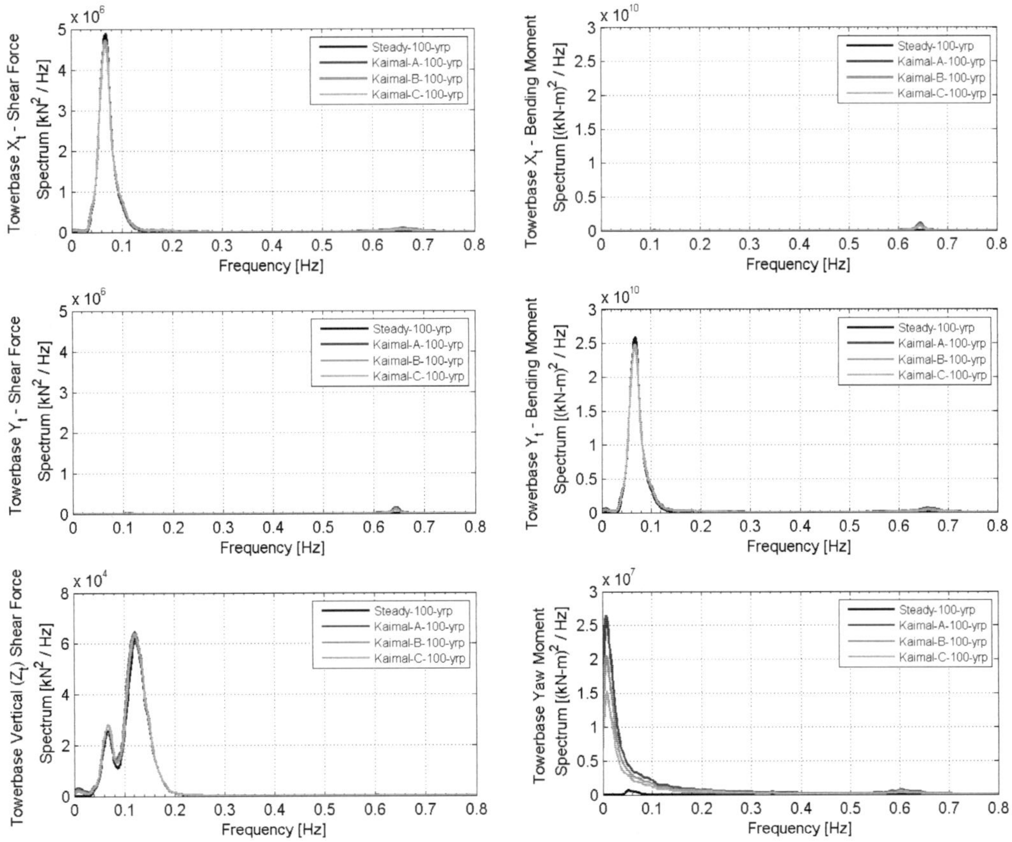

Figure 22: Towerbase Forces and Moments Spectra, 100-yrp

Table 10: Statistics of Towerbase Forces and Moments, 100-yrp

100-yrp condition							
Steady	Mean	RMS	Tz [sec]	90%ile Dynamic Extreme	90%ile Extreme Factor	Maximum	Minimum
Shear Force in $X_t[kN]$	375.4	400.2	12.900	1752.6	4.379	2127.9	-1377.2
Shear Force in $Y_t[kN]$	3.1	16.3	5.280	74.4	4.578	77.5	-71.3
Vertical Force, $Z_t[kN]$	-6866.2	63.8	4.890	293.2	4.595	-6573.0	-7159.4
Bending Moment about $X_t[kN-m]$	3935.1	1115.1	4.440	5147.3	4.616	9082.4	-1212.1
Bending Moment about $Y_t[kN-m]$	32506.7	28892.3	13.400	126244.0	4.369	158750.7	-93737.3
Yaw Moment, about $Z_t[kN-m]$	-740.6	149.0	10.200	660.3	4.432	-80.3	-1400.9
Kaimal-A	Mean	RMS	Tz [sec]	90%ile Dynamic Extreme	90%ile Extreme Factor	Maximum	Minimum
Shear Force in $X_t[kN]$	394.4	418.5	5.140	1918.2	4.584	2312.6	-1523.8
Shear Force in $Y_t[kN]$	-8.7	64.0	1.670	308.4	4.822	299.7	-317.1
Vertical Force, $Z_t[kN]$	-6874.8	67.7	2.460	320.9	4.742	-6553.9	-7195.7
Bending Moment about $X_t[kN-m]$	5028.6	5271.2	1.660	25426.7	4.824	30455.3	-20398.1
Bending Moment about $Y_t[kN-m]$	35667.1	30870.9	5.520	141025.3	4.568	176692.4	-105358.1
Yaw Moment, about $Z_t[kN-m]$	-114.2	1158.3	1.730	5578.3	4.816	5464.1	-5692.5
Kaimal-B	Mean	RMS	Tz [sec]	90%ile Dynamic Extreme	90%ile Extreme Factor	Maximum	Minimum
Shear Force in $X_t[kN]$	393.8	413.7	5.530	1889.6	4.568	2283.4	-1495.8
Shear Force in $Y_t[kN]$	-8.7	56.6	1.690	272.8	4.820	264.1	-281.5
Vertical Force, $Z_t[kN]$	-6874.8	66.9	2.550	316.6	4.734	-6558.3	-7191.4
Bending Moment about $X_t[kN-m]$	5039.3	4648.8	1.680	22413.9	4.821	27453.3	-17374.6
Bending Moment about $Y_t[kN-m]$	35615.1	30367.5	6.090	138075.4	4.547	173690.5	-102460.3
Yaw Moment, about $Z_t[kN-m]$	-112.2	1018.7	1.690	4910.3	4.820	4798.2	-5022.5
Kaimal-C	Mean	RMS	Tz [sec]	90%ile Dynamic Extreme	90%ile Extreme Factor	Maximum	Minimum
Shear Force in $X_t[kN]$	393.1	409.3	6.110	1860.8	4.546	2253.8	-1467.7
Shear Force in $Y_t[kN]$	-8.8	49.3	1.710	237.7	4.817	228.9	-246.5
Vertical Force, $Z_t[kN]$	-6874.8	66.1	2.700	312.3	4.722	-6562.5	-7187.2
Bending Moment about $X_t[kN-m]$	5047.6	4032.4	1.700	19430.9	4.819	24478.5	-14383.3
Bending Moment about $Y_t[kN-m]$	35554.1	29912.0	6.840	135229.8	4.521	170783.9	-99675.7
Yaw Moment, about $Z_t[kN-m]$	-110.1	879.0	1.650	4241.5	4.826	4131.4	-4351.7

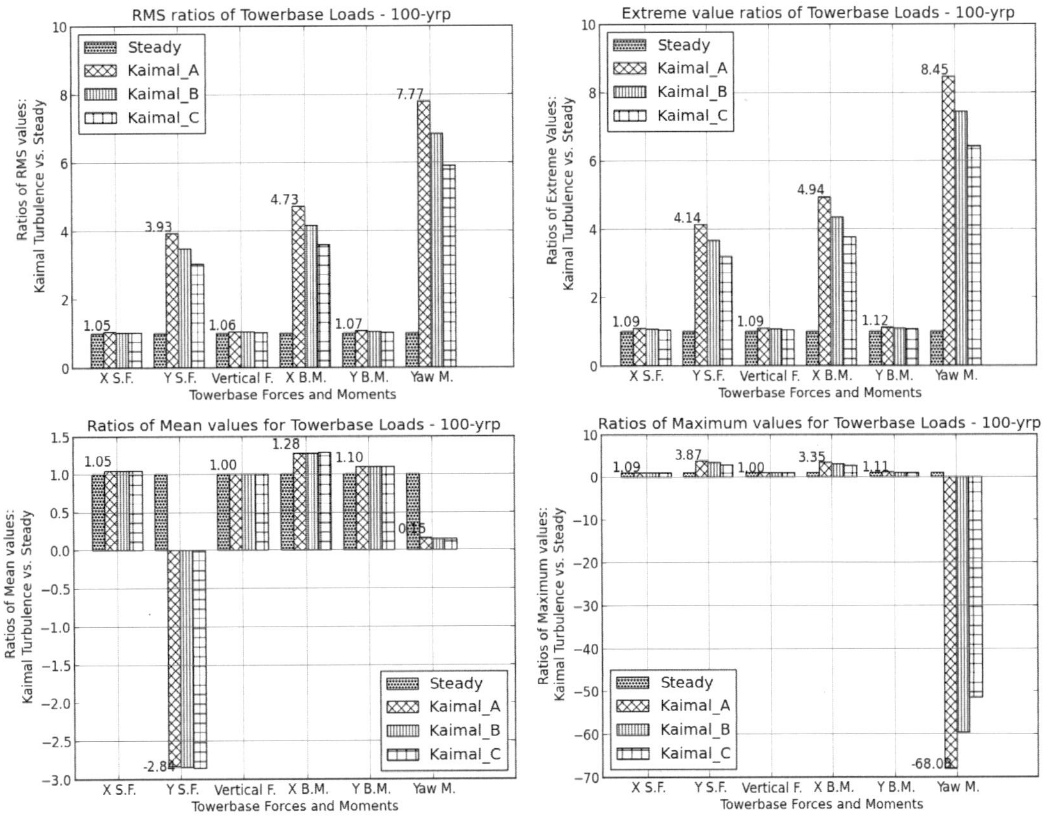

Figure 23: Ratios of Towerbase Forces and Moments Statistics, 100-yrp

Nacelle Accelerations

Spectra of nacelle accelerations are shown in Figure 24. It should be noted that the accelerations reported here do not include the effect of acceleration due to gravity (i.e. g.sinθ); this contribution is however small considering the magnitudes of pitch and roll motions, and would not alter the trends seen in the reported values. As expected, energies in the X_S acceleration spectra are higher than the magnitudes in the Y_S because the X_S axis is in line with the applied environmental loads. In the 50-yrp condition, some influence of turbulence is seen in the X_S acceleration at the surge natural frequency; negligible variations can also be spotted in the Z_S accelerations. There are no changes due to wind turbulence in the nacelle accelerations at the wave peak frequency in the 50-yrp and 100-yrp conditions. This is a direct consequence of the excitations received by the nacelle from the turbulent winds, being much less than wave excitations transferred through the tower to the nacelle. The influence of the turbulence-induced flapwise vibrations of the turbine blades can be seen in the 0.6 – 0.7 Hz band, in both 50-yrp and 100-yrp conditions. At frequencies where the effect of turbulence is seen to increase the energy, the variation of spectral energy with turbulence intensity is very mild.

Statistics of nacelle accelerations are shown in Table 11 and Figure 25 for the 50-yrp condition, and in Table 12 and Figure 26 for the 100-yrp condition. The X_S-axis and Z_S-axis accelerations are the components of importance here; values in the Y_S-axis are very small since this axis is perpendicular to the direction of dominant environmental loads, and the ratios of statistics along this axis are practically insignificant – they are only shown here for completeness.

Statistical ratios in the 50-yrp condition show a 20% and 48% increase in the X_S-axis and Z_S-axis nacelle accelerations respectively, due to turbulence. Extreme values of the accelerations increase by 26% and 49% in the X_S-axis and Z_S-axis, when the steady wind case is compared to Kaimal-A. The RMS increments due to turbulence in the 100-yrp condition are 9% and 8% in the X_S-axis and Z_S-axis respectively, and the corresponding extreme value increments are 15% and 13% - based on comparison between the steady wind case and Kaimal-A. Extreme factors in the 50-yrp condition increase due to wind turbulence by about ~7% and ~3% in the X_S-axis and Z_S-axis respectively, when the Kaimal-A case is compared to the steady wind case; the corresponding increments in the 100-yrp condition are 6.6% and 4.5%.

At least in the 50-yrp condition, the statistics show a significant effect of wind turbulence on the accelerations of the nacelle. The extreme values of nacelle accelerations are significant; for instance in the Kaimal-A case, extreme value of X_S-axis accelerations is approximately 1.54 m/s in the 50-yrp condition, and 2.16 m/s in the 100-yrp condition. These values correspond to 0.16 g and 0.22 g, which are in the region of typically considered nacelle acceleration design criteria limits. Based on these results, FOWT systems with more compliant support platforms could easily pose design challenges in regards to satisfying nacelle acceleration criteria, if wind turbulence excitations are significant.

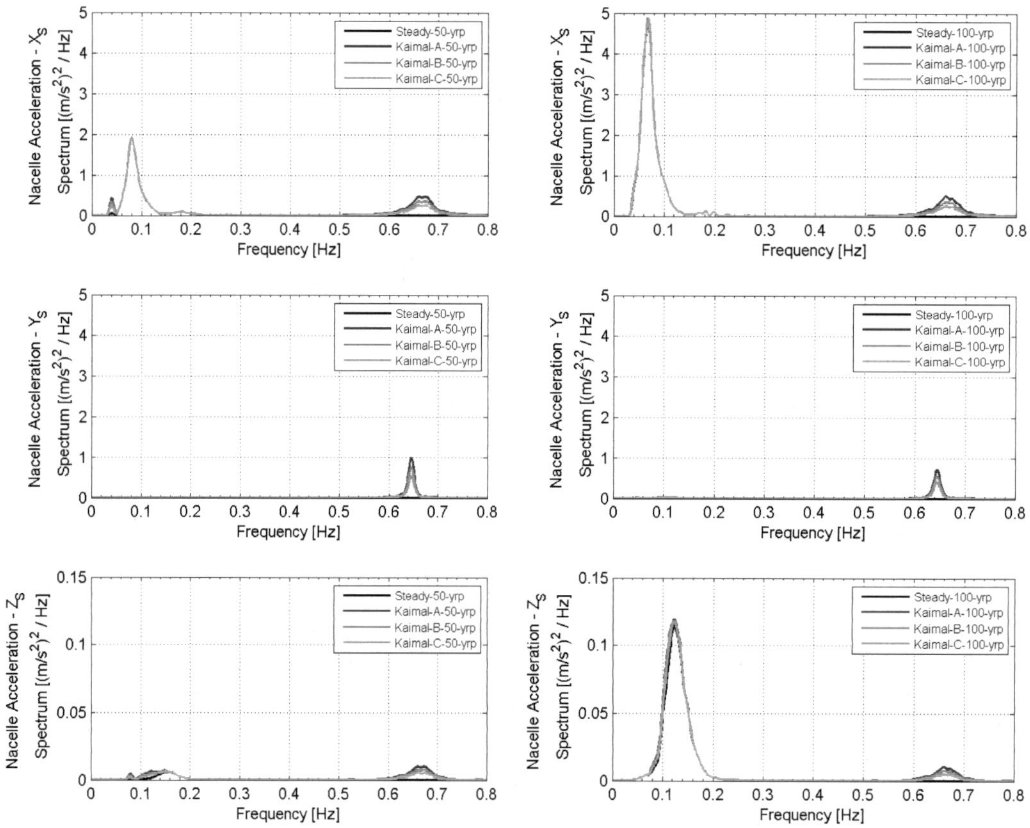

Figure 24: Nacelle Accelerations Spectra, 50-yrp (left panes) and 100-yp (right panes)

Table 11: Statistics of Nacelle Accelerations, 50-yrp

50-yrp condition						
Steady	Mean	RMS	Tz	90%ile Dynamic Extreme	90%ile Extreme Factor	Maximum
Nac. Acc. - $X_S[m/s^2]$	9.863E-05	2.549E-01	9.41	1.134	4.450	1.134E+00
Nac. Acc. - $Y_S[m/s^2]$	-2.824E-04	5.576E-03	1.82	0.027	4.805	2.651E-02
Nac. Acc. - $Z_S[m/s^2]$	-1.076E-04	2.056E-02	0.81	0.102	4.970	1.021E-01
Kaimal-A	Mean	RMS	Tz	90%ile Dynamic Extreme	90%ile Extreme Factor	Maximum
Nac. Acc. - $X_S[m/s^2]$	1.479E-04	3.222E-01	2.10	1.539	4.775	1.539E+00
Nac. Acc. - $Y_S[m/s^2]$	-2.647E-04	1.461E-01	0.49	0.741	5.071	7.407E-01
Nac. Acc. - $Z_S[m/s^2]$	-2.132E-04	3.924E-02	0.41	0.200	5.106	2.002E-01
Kaimal-B	Mean	RMS	Tz	90%ile Dynamic Extreme	90%ile Extreme Factor	Maximum
Nac. Acc. - $X_S[m/s^2]$	1.347E-04	3.074E-01	2.24	1.464	4.761	1.464E+00
Nac. Acc. - $Y_S[m/s^2]$	-2.686E-04	1.280E-01	0.46	0.650	5.082	6.500E-01
Nac. Acc. - $Z_S[m/s^2]$	-1.972E-04	3.554E-02	0.39	0.182	5.113	1.816E-01
Kaimal-C	Mean	RMS	Tz	90%ile Dynamic Extreme	90%ile Extreme Factor	Maximum
Nac. Acc. - $X_S[m/s^2]$	1.242E-04	2.940E-01	2.47	1.394	4.741	1.394E+00
Nac. Acc. - $Y_S[m/s^2]$	-2.717E-04	1.096E-01	0.44	0.558	5.094	5.578E-01
Nac. Acc. - $Z_S[m/s^2]$	-1.832E-04	3.213E-02	0.38	0.165	5.122	1.644E-01

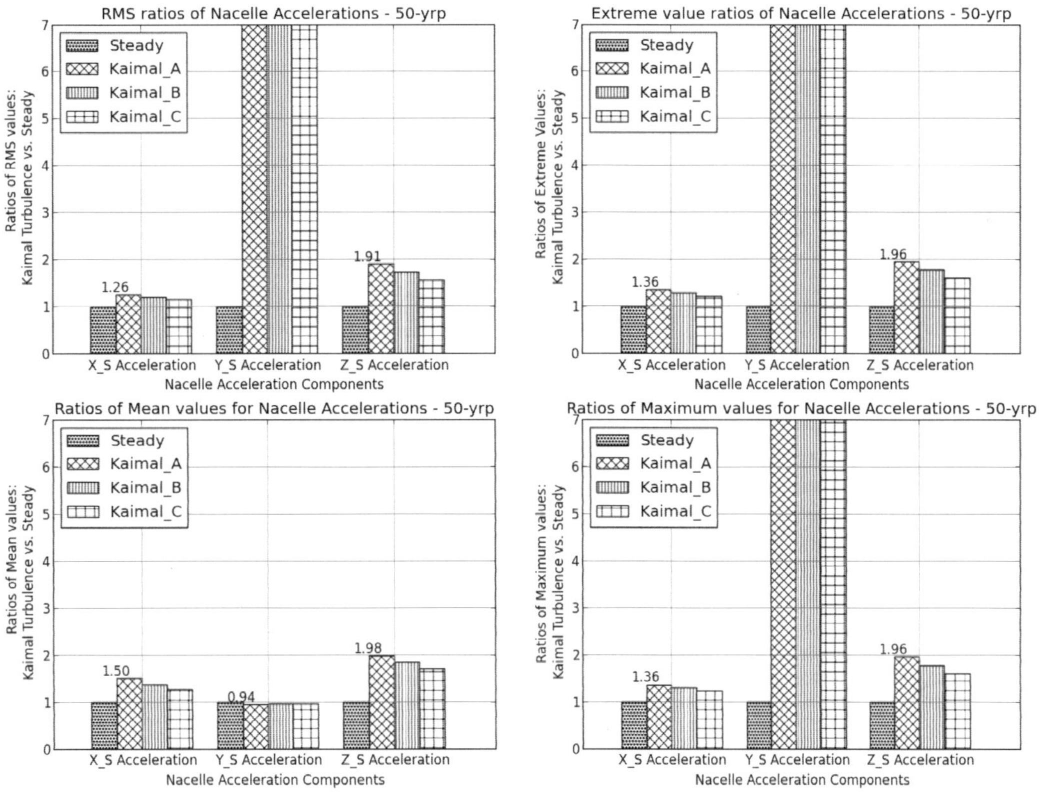

Figure 25: Ratios of Statistics for Nacelle Accelerations, 50-yrp

Table 12: Statistics of Nacelle Accelerations, 100-yrp

100-yrp condition						
Steady	Mean	RMS	Tz	90%ile Dynamic Extreme	90%ile Extreme Factor	Maximum
Nac. Acc. - $X_S[m/s^2]$	1.201E-04	4.173E-01	12.10	1.834	4.394	1.834E+00
Nac. Acc. - $Y_S[m/s^2]$	-4.389E-04	2.289E-02	3.54	0.107	4.664	1.063E-01
Nac. Acc. - $Z_S[m/s^2]$	-3.129E-04	8.005E-02	1.93	0.384	4.793	3.834E-01
Kaimal-A	Mean	RMS	Tz	90%ile Dynamic Extreme	90%ile Extreme Factor	Maximum
Nac. Acc. - $X_S[m/s^2]$	1.668E-04	4.595E-01	2.90	2.163	4.707	2.163E+00
Nac. Acc. - $Y_S[m/s^2]$	-5.437E-04	1.395E-01	0.47	0.708	5.079	7.079E-01
Nac. Acc. - $Z_S[m/s^2]$	-4.142E-04	8.756E-02	0.63	0.440	5.021	4.392E-01
Kaimal-B	Mean	RMS	Tz	90%ile Dynamic Extreme	90%ile Extreme Factor	Maximum
Nac. Acc. - $X_S[m/s^2]$	1.587E-04	4.491E-01	3.16	2.106	4.689	2.106E+00
Nac. Acc. - $Y_S[m/s^2]$	-5.456E-04	1.231E-01	0.45	0.626	5.086	6.256E-01
Nac. Acc. - $Z_S[m/s^2]$	-3.986E-04	8.568E-02	0.63	0.430	5.021	4.298E-01
Kaimal-C	Mean	RMS	Tz	90%ile Dynamic Extreme	90%ile Extreme Factor	Maximum
Nac. Acc. - $X_S[m/s^2]$	1.522E-04	4.405E-01	3.55	2.054	4.664	2.054E+00
Nac. Acc. - $Y_S[m/s^2]$	-5.480E-04	1.068E-01	0.43	0.544	5.095	5.436E-01
Nac. Acc. - $Z_S[m/s^2]$	-3.849E-04	8.399E-02	0.64	0.422	5.019	4.211E-01

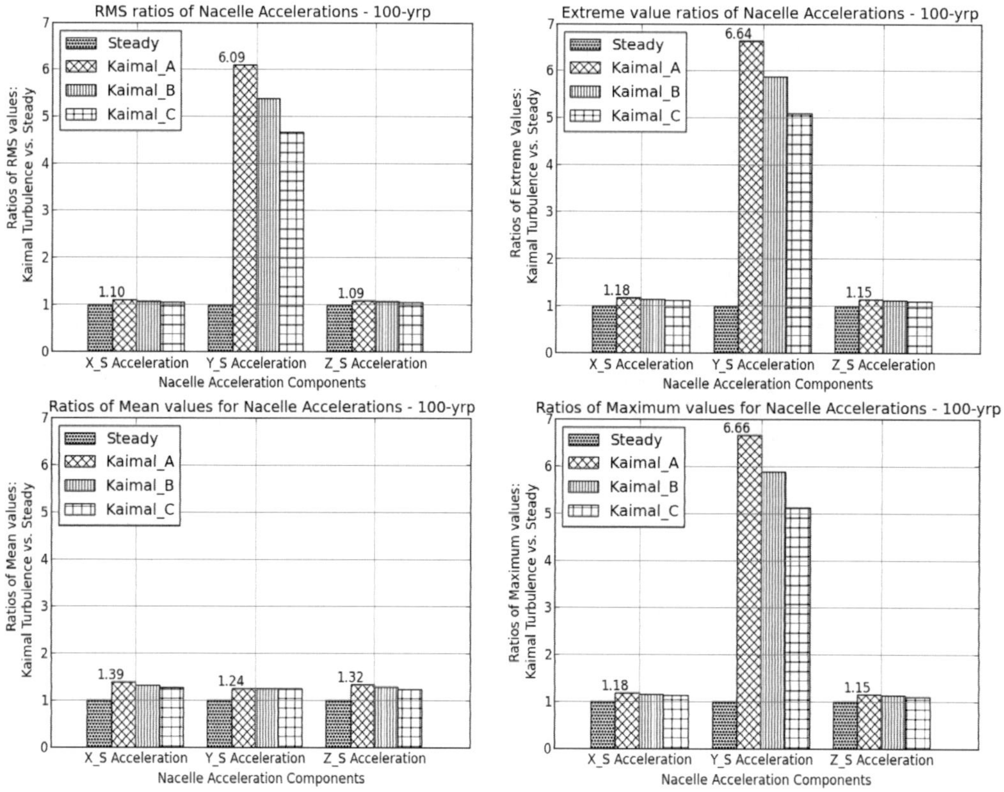

Figure 26: Ratios of Statistics for Nacelle Accelerations, 100-yrp

Towertop Displacements and Rotations

The influences of turbulence on the displacements and rotations of the tower top are consistent with the towerbase reactions and nacelle accelerations. Plots of the towertop displacement and rotational motion spectra are presented in Figure 27 and Figure 28 for the 50-yrp and 100-yrp conditions respectively. Spectral energy distributions show slight variations due to turbulence in the spectral peaks, which occur at similar low and wave frequency regions, as well as in the 0.6 – 0.7 Hz band as observed in the tendon tensions, towerbase reactions and nacelle accelerations. Except for the differences in magnitudes, the characteristics of the spectral responses are quite similar between the 50-yrp and 100-yrp conditions. Energies in the X-axis displacement are higher than those in the Y-axis, as expected. Roll and pitch acceleration responses in the 0.6 – 0.7 Hz band differ in the sense that the pitch acceleration response in this spectral band is more broad-banded than the roll acceleration spectrum which is rather narrow-banded. The total energy levels in pitch and roll accelerations may be quite comparable, but at least qualitatively, the broad-banded pitch acceleration of the towertop in the 0.6 – 0.7 Hz range suggests some interaction between the blade responses and the towertop pitch accelerations. On the other hand, roll accelerations of the towertop within the same frequency band seem to be excited purely by flapwise vibrations of the turbine blades close to their flapwise eigenfrequency (0.68 Hz), as indicated by the very narrow-banded response.

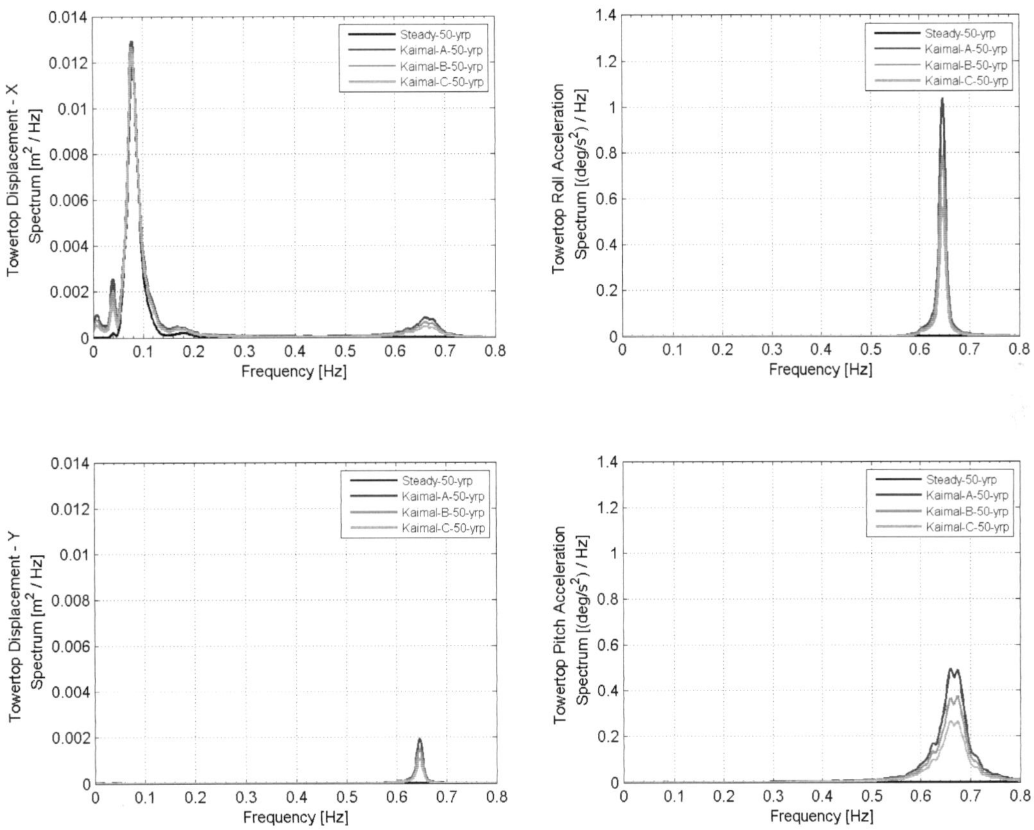

Figure 27: Towertop Response Spectra, 50-yrp

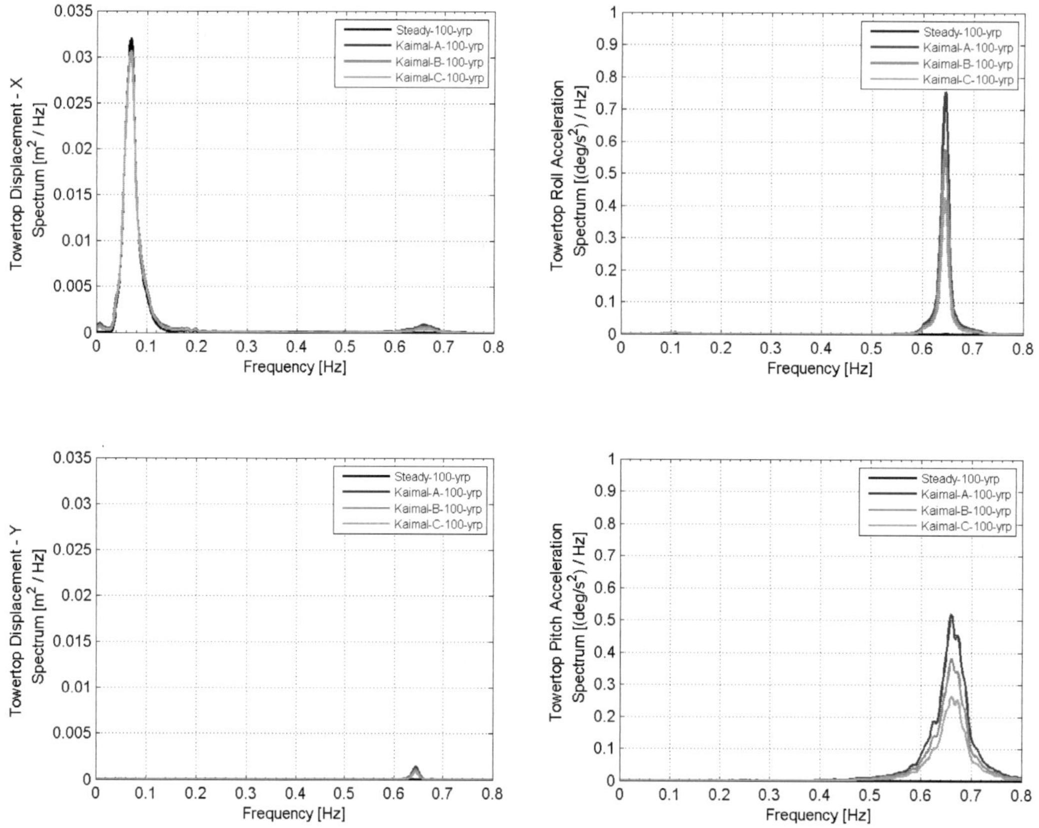

Figure 28: Towertop Response Spectra, 100-yrp

Generated Electrical Power

A comparison of the generated electrical power between the steady wind, Kaimal-A, Kaimal-B and Kaimal-C cases clearly shows significant drops in electrical power as turbulence intensity increases (Figure 29 and Figure 30). Table 13 shows the statistics of generated power; the mean generated power over the simulated 5.5 hours is approximately the same for all cases (~5 MW). The influence of turbulence on the generated power however is seen in the standard deviations, where the standard deviations increase with turbulence intensity. In the 50-yrp condition, there is a 29% increase in standard deviation between the steady wind case and Kaimal-C, and a 36% increase in standard deviation between the steady wind case and Von Karman-C. In the 100-yrp condition, the corresponding percentage increments in standard deviations are 39% and 42% respectively. It is important to keep in perspective the fact that the standard deviations of the steady wind case are already significantly low; as such these increased variances due to turbulence would probably not significantly affect power supply in practice. The low standard deviations (coefficients of variation far less than 1.0) in the generated power time histories, and mean values of approximately 5 MW demonstrate significant dynamic stability of the TLP against the imposed environmental loads. If the platform motions and nacelle responses were excessive, these would reflect in the statistics of generated electrical power.

Besides excessive platform motions, fluctuations in the wind velocity fields due to turbulence could induce reductions in generated electrical power. We investigated the correlation between the X-component of turbulent wind velocity (to the third power), and the generated electrical power for the Kaimal and Von Karman – based simulations. Figure 31 shows the time histories of V_X^3 and generated power for the Kaimal-A case in the 50-yrp condition, and it can be observed as emphasized in the bottom pane that the drops in electrical power occur when there is a preceding reduction in V_X^3. This finding is consistent with Equation (3) which states the dependence of electrical power on V_X^3.

The influence of the severity of sea states on the generated electrical power can be seen in the frequencies and magnitudes of the dips in the generated power time histories between the 50-yrp and 100-yrp sea states. In the 100-yrp, the power fluctuations are due to the combined influence of the higher wave forces on the platform and turbulence effects on the rotor wind fields, hence the higher number of occurrences of dips in the power generated.

Table 13: Statistics of Generated Electrical Power, 50-yrp and 100-yrp

50-yrp condition		
Wind Case	Mean [KW]	Standard Deviation [KW]
Steady	5.00E+03	45.11
Kaimal-A	4.993E+03	106.24
Kaimal-B	4.997E+03	80.53
Kaimal-C	4.999E+03	63.97
Von Karman-A	4.992E+03	108.78
Von Karman-B	4.996E+03	86.50
Von Karman-C	4.998E+03	70.50

100-yrp condition		
Wind Case	Mean [KW]	Standard Deviation [KW]
Steady	4.994E+03	84.68
Kaimal-A	4.971E+03	191.40
Kaimal-B	4.979E+03	162.62
Kaimal-C	4.985E+03	138.46
Von Karman-A	4.968E+03	200.30
Von Karman-B	4.976E+03	172.92
Von Karman-C	4.982E+03	147.18

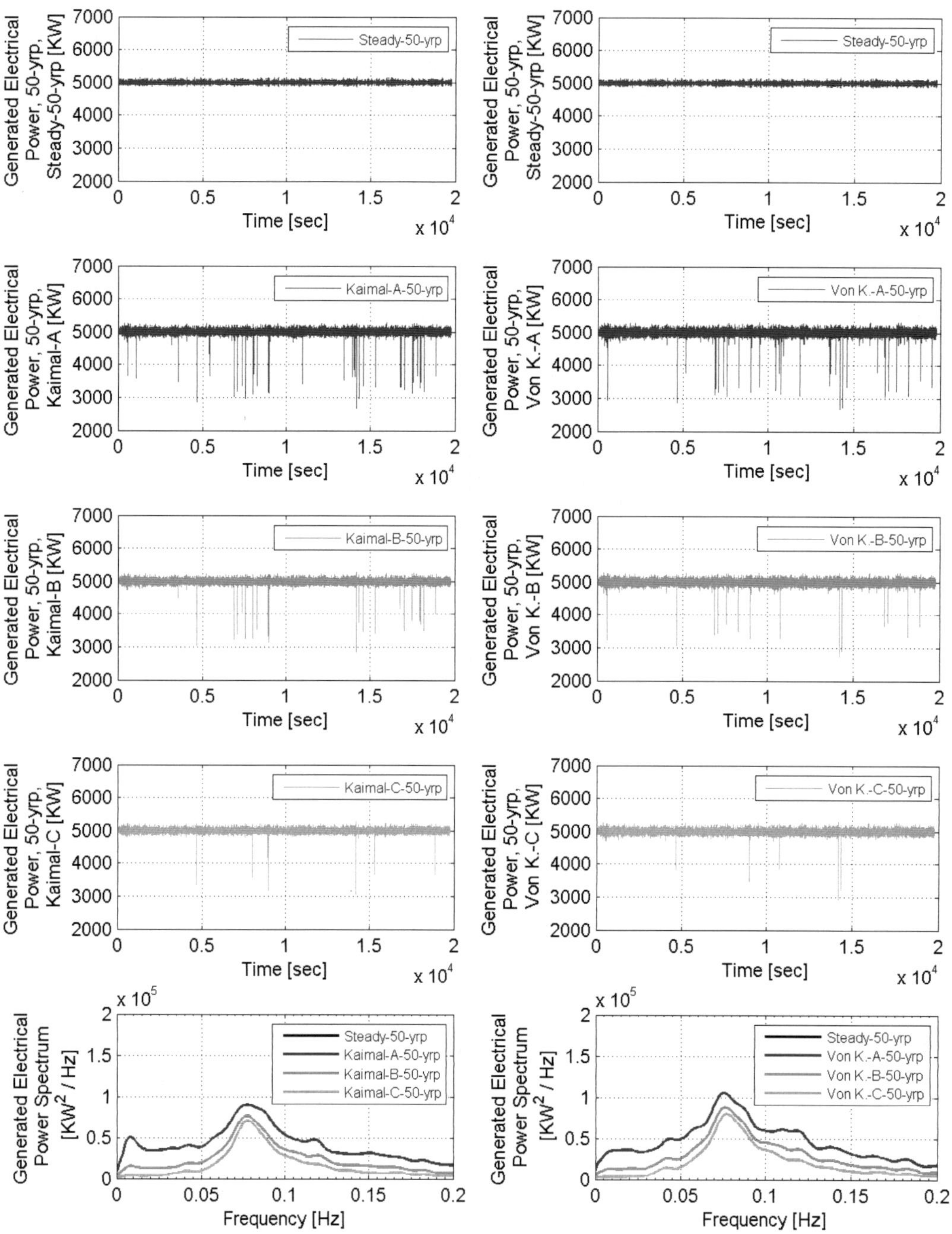

Figure 29: Generated Electric Power Timeseries and Spectra, 50-yrp

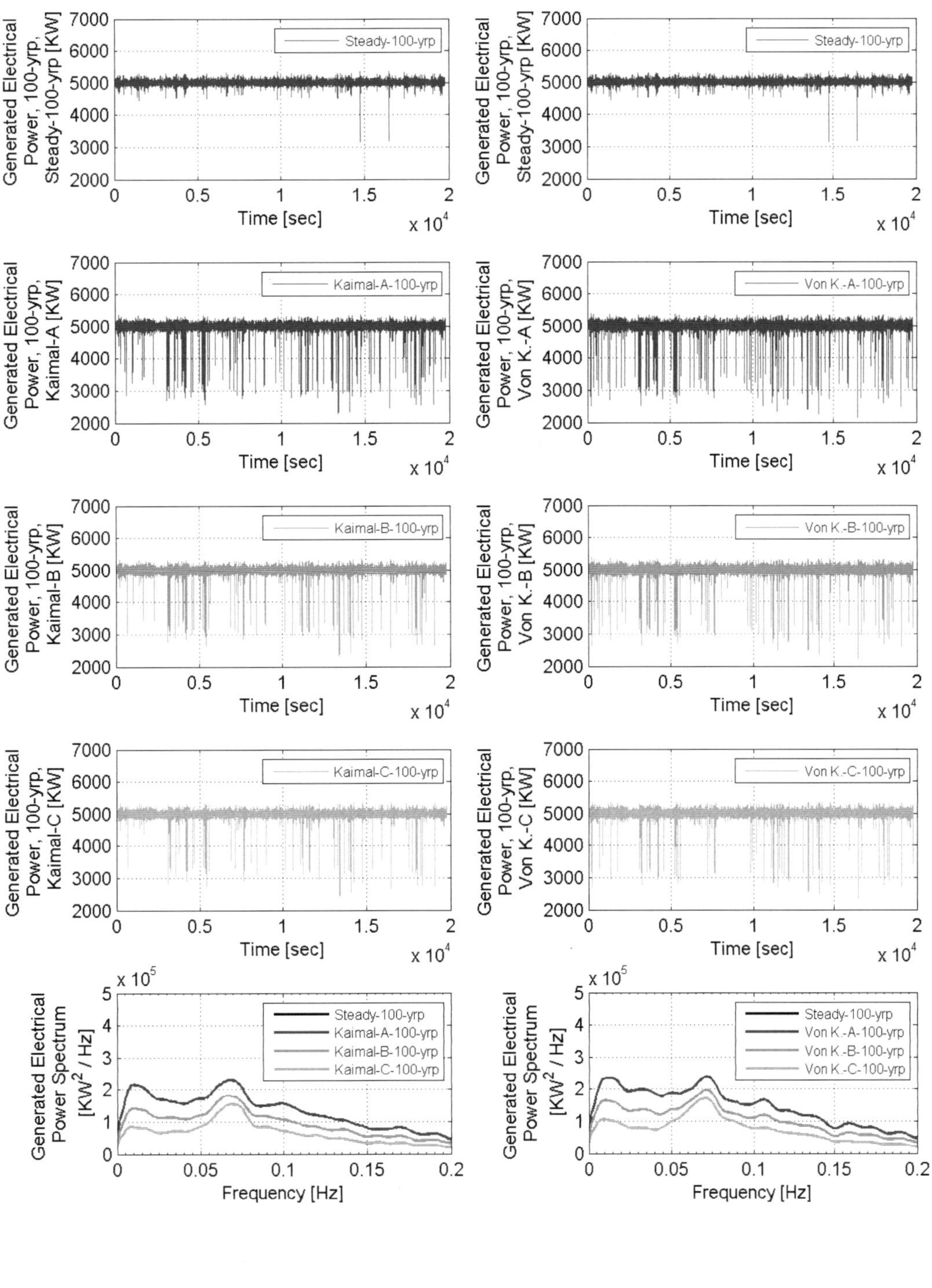

Figure 30: Generated Electric Power Timeseries and Spectra, 100-yrp

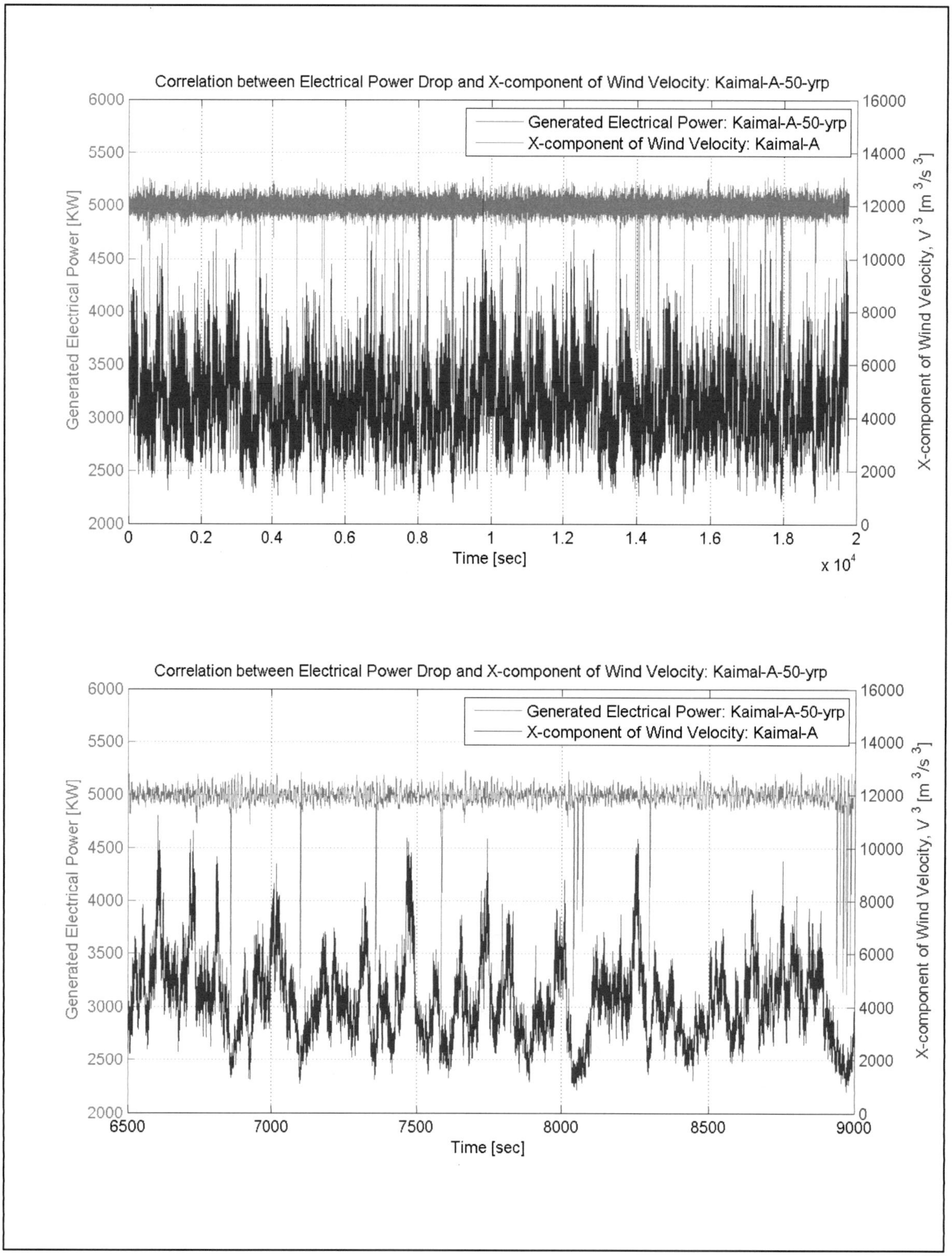

Figure 31: Correlation between Electrical Power Drop and X-component of Wind Velocity

Sensitivity of Platform Motions and Generated Electrical Power to Turbulence Model

Results discussed so far on the influence of turbulent winds have been based on the Kaimal wind spectrum. Acknowledging that there is quite a number of wind turbulence models published in literature, it is important to at least assess more than one model to gain some perspective on how variant the results might be from the previously discussed Kaimal model. Ideally, the best turbulence model to apply to any given project is one which is based on site-specific data and truly models the turbulence loads that the system will experience when installed. However, site-specific models are not always available, and it is common in practice to adopt the most reasonable model for the site, either based on experience or some parameter traits that validate the assumption of a specific model. The Von Karman model is perhaps one of the most commonly studied in the floating wind energy research, and is one of the initial IEC spectral models with characteristics (namely, the component standard deviations and integral turbulence scale) that sufficiently differ from the Kaimal model to provoke responses different from those discussed for the Kaimal spectrum.

We investigated platform motion responses and generated electrical power using the Von Karman model. Only comparisons of spectra are assessed to see the difference in turbulence-induced responses. Figure 32 and Figure 33 show the response spectra for the 50-yrp and 100-yrp platform motions using the Von Karman turbulence wind spectrum. In all degrees of freedom and for both wave conditions, the response spectra track consistently with the Kaimal model in frequency distribution of the energies, but with significantly higher energies for each respective category of turbulence intensity. The Y- and Z-components of turbulence wind velocities (Kaimal and Von Karman) are shown in the Appendix; they show that the amplitudes of the wind speed for the Von Karman model are higher than those of the Kaimal model in the Y and Z components. As shown in their respective wind spectra, the energy content of the Y- and Z-components of the Von Karman and Kaimal winds are still within the same range of frequencies. It is therefore fair to hypothesize, that the higher wind velocities of the Von Karman model in the Y- and / or Z-components impose higher excitations in the wind turbine system to produce higher dynamic responses across all frequency bands where turbulence-induced responses were already seen. It will be interesting to see how the individual components of the system (i.e. blades, nacelle, tower and tendons) respond to this increased turbulence wind excitation (relative to the Kaimal model) from the Y- and / or Z-components – this will be investigated in future work.

The correlation between the dips in the timeseries of generated electrical power and the X-component of turbulence wind velocities (to the third power) also holds true for the Von Karman spectrum – this is shown in Figure 34.

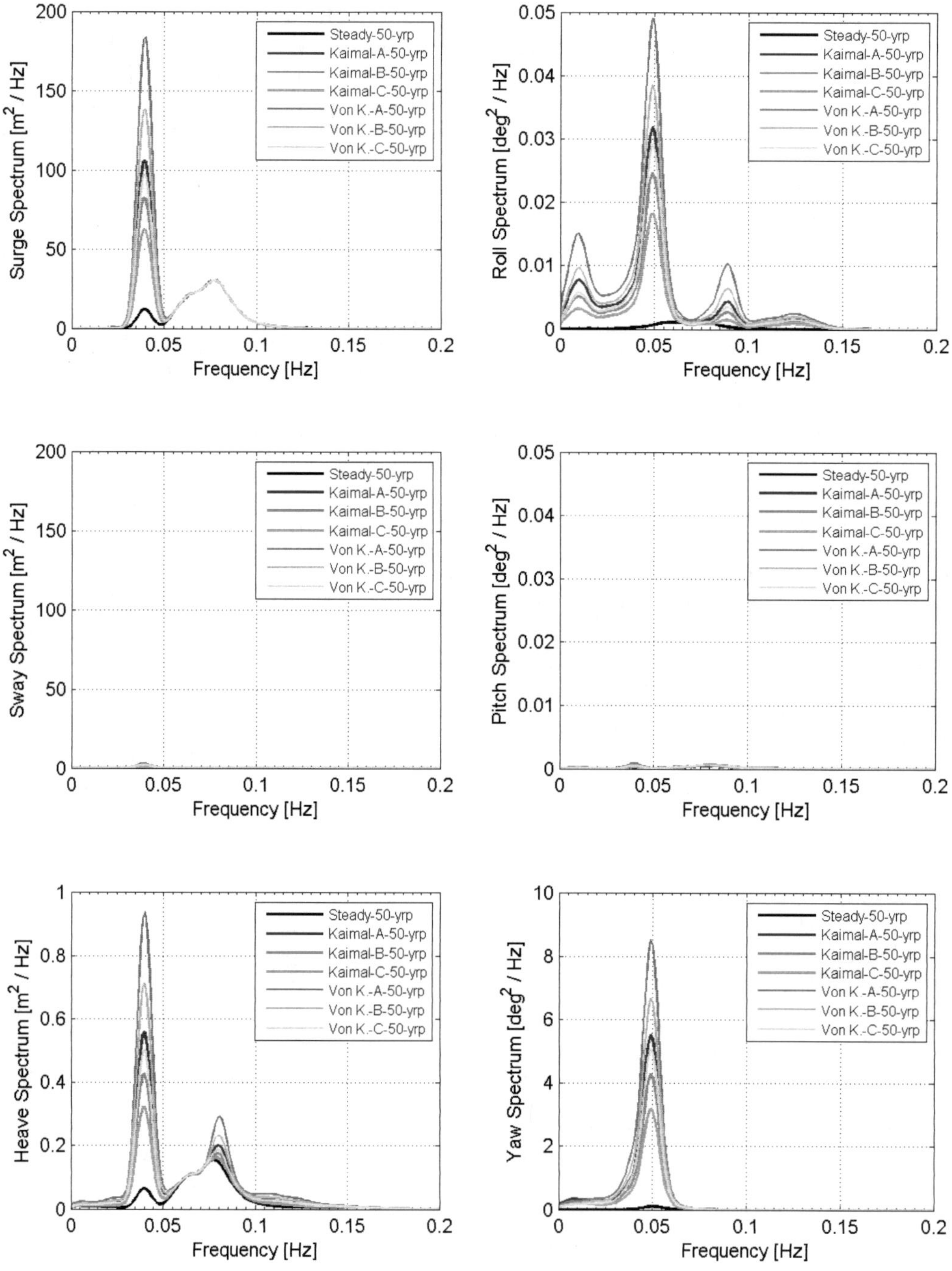

Figure 32: Platform Motions in 50-yrp: Kaimal and Von Karman Turbulence Model

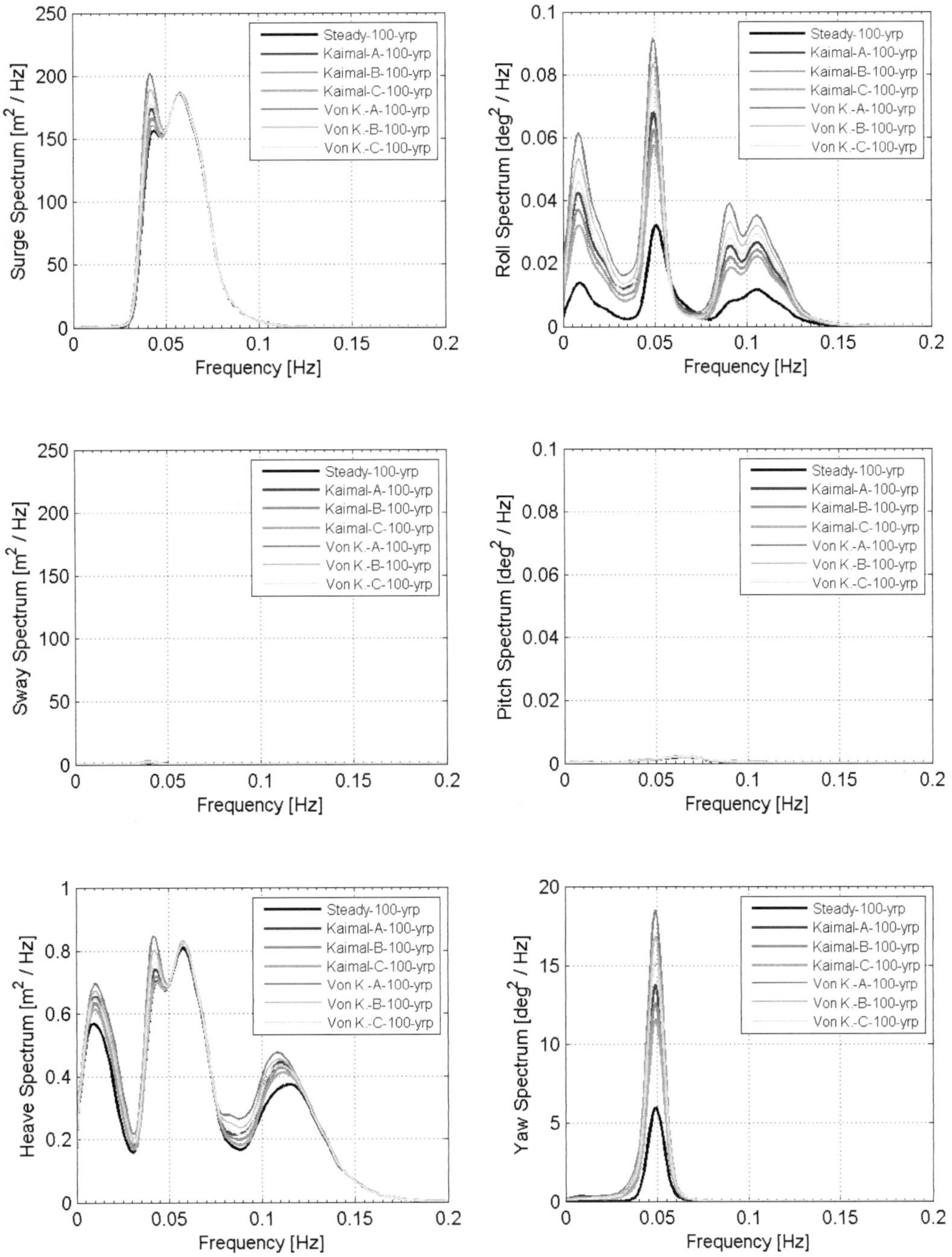

Figure 33: Platform Motions in 100-yrp: Kaimal and Von Karman Turbulence Model

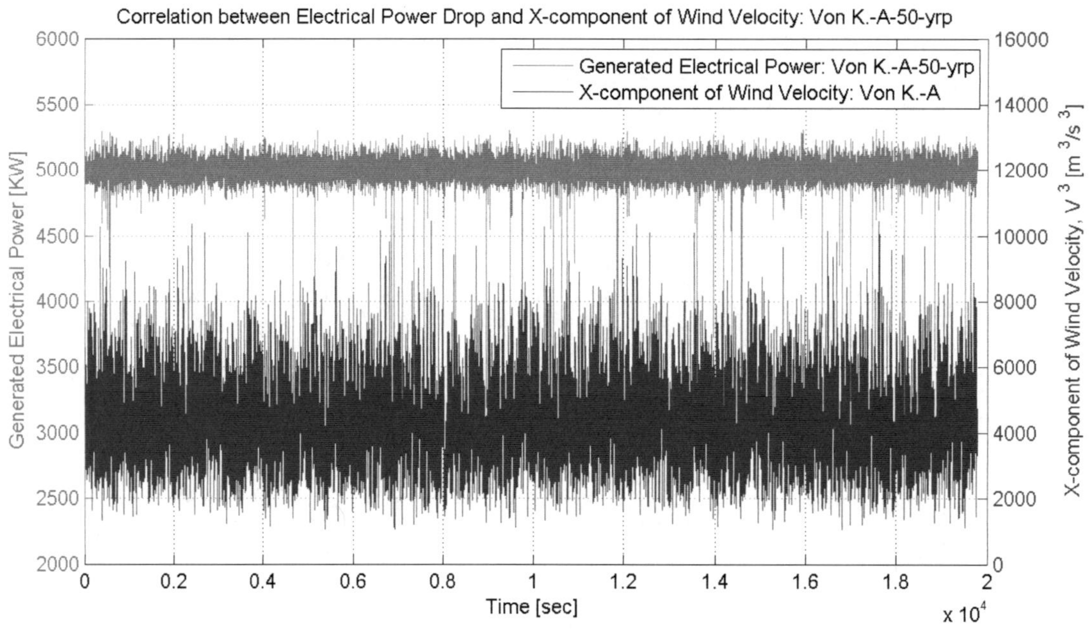

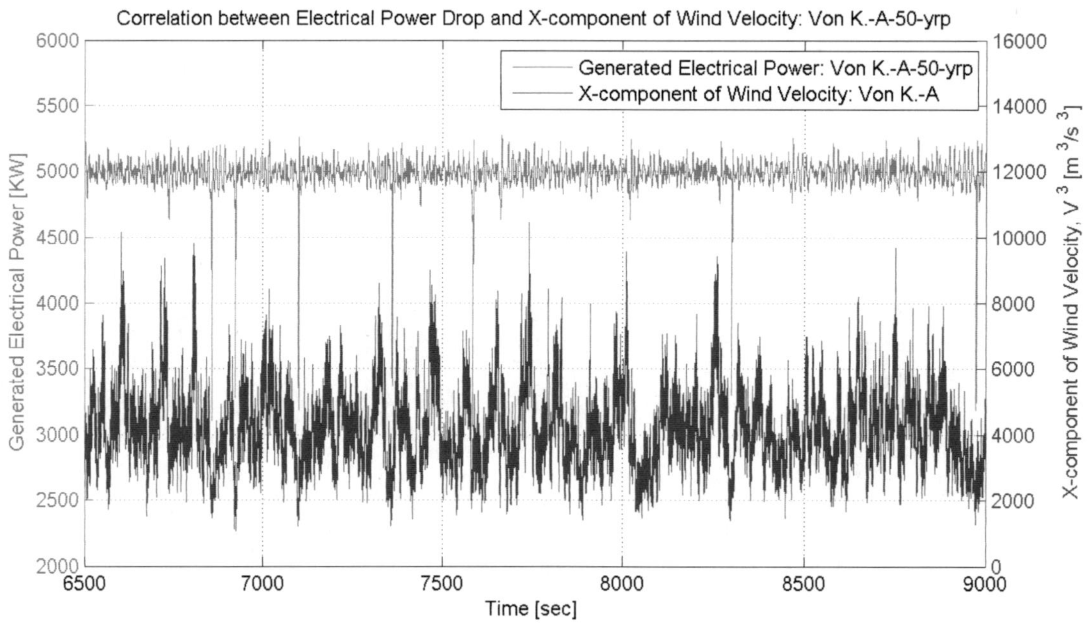

Figure 34: Correlation between Electrical Power Drop and X-component of Wind Velocity: Von Karman Turbulence Model in 50-yrp

CONCLUSIONS

Understanding the influence of wind turbulence on the responses of each primary component in FOWT systems is critical to performing successful designs especially for projects sited in fields with high with turbulence intensities. This study has addressed the characteristics of wind turbulence-induced responses in the platform motions, nacelle accelerations, towertop and towerbase responses, tendon tensions and the electrical power generated by the wind turbine. The three-column TLP designed for this study has overall response attributes of a robust system, in regards to the studied environmental conditions. Some key findings from this study are summarized as follows:

1. Perhaps the most appropriate point to lead the conclusions with is that the extent to which wind turbulence affects or will affect the responses of a given FOWT system depends on the characteristics of the wind field itself. As shown in the last section of the paper, the magnitudes of the responses can differ significantly with wind turbulence spectral models.
2. In the 50-yrp condition the effect of wind turbulence on responses is highest at the surge natural frequency, where there is significant excitation energy from the turbulent winds to drive resonant responses – especially in the platform motions and tendon tension responses. The relevance of this observation needs some emphasis because while sizing floating systems to avoid wave excitations is somewhat common practice, for FOWTs the avoidance of wind turbulence-induced resonant responses is also very important and is more likely to be overlooked in the initial stages of design. For most of the response components, the 100-yrp condition features the dominance of the wave excitations over the turbulent winds, resulting in nearly insignificant variations in the responses due to wind turbulence.
3. RMS and extreme values of tendon tensions increase significantly with wind turbulence, for upwave and downwave tendons. As a result of the RMS and extreme values increasing almost proportionally, the increase in extreme factors are modest, implying modest overall increments in tendon dynamics due to turbulence. However, the large increments in RMS values mean that the wind turbulence can significantly increase the fatigue loads of the tendons.
4. Findings in the towerbase reaction forces and moments correlate consistently with the tendon tensions, and have high increments in RMS values due to wind turbulence. The primary towerbase reactions showing significant effects of wind turbulence are the shear force in the direction of dominant environmental loads, its associated bending moment (in the direction orthogonal to the shear force), the vertical force and the yaw moment. The yaw reaction moment due to wind turbulence completely shows the yaw reaction in the steady wind case to be practically inexistent, and this is seen in the 50-yrp and 100-yrp conditions.
5. Wind turbulence is seen to excite responses at the flapwise eigenfrequency of the blades. This is interpreted to be coupled responses induced by the blades when excited by the turbulent winds, and this effect is seen in the tendon tensions, towerbase reactions, towertop responses and nacelle accelerations.
6. Increments in nacelle accelerations due to wind turbulence are significant, especially in the direction of dominant environmental loads, especially in the 50-yrp condition. The increase in extreme values with wind turbulence suggests that FOWTs with excessive motions will very likely pose difficult design challenges related to satisfying nacelle acceleration criteria, when significant wind turbulence is considered.
7. Fluctuations in generated electrical power increase with wind turbulence. However, the three-column TLP system studied seems to possess favorable response characteristics that minimize impacts on power generation. It has been shown that there is a correlation between the dips in the generated power time series, and reductions in the X-component of wind velocity (to the third power) for both Kaimal and Von Karman-based simulations.

ACKNOWLEDGEMENTS

The authors are thankful to the management of Houston Offshore Engineering, an Atkins company, for providing the approval, support and resources needed to complete this work. We also thank Ms. Ivy Jones of HOE Atkins, for performing an editorial review of the manuscript and providing an objective critique of the text.

REFERENCES

Bachynski, E. E. and Moan, T. (2012), Design Considerations for Tension Leg Platform Wind Turbines, Marine Structures, 29, 89 – 114.

Bianchi, F. D., Battista, H. D. and Mantz, R. J. (2007), Wind Turbine Control Systems, Springer-Verlag London Limited, United Kingdom, ISBN-13: 978-1-84628-492-2.

Bir, G. S., (2007), User's Guide to BMODES, National Renewable Energy Laboratory, Golden, CO. USA.

Det Norske Veritas (2005), Recommended Practice, DNV-RP-C103, Column-Stabilized Units, pp 16.

International Electrotechnical Commission (2005), IEC 61400-1: Wind Turbines, Design Requirements.

International Electrotechnical Commission (2009), IEC 61400-3: Design Requirements for Offshore Wind Turbines.

Jonkman, B. J. and Kilcher, L. (2012), TurbSim User's Guide, National Renewable Energy Laboratory Report, 2008.

Jonkman, J. M. and Buhl, M. L. (2005), FAST User's Guide, NREL/EL-500-38230. National Renewable Energy Laboratory, Golden, CO. USA.

Jonkman, J., Butterfield S., Musial, W. and Scott, G. (2009), Definition of a 5-MW Reference Wind Turbine for Offshore System Development, NREL/TP-500-38060, National Renewable Energy Laboratory, Golden, CO. USA.

Kaimal, J. C., Wyngaard, J. C., Izumi, Y. and Cote, O. R. (1972), Spectral Characteristics of Surface-layer Turbulence, Q.J.R. Meteorological Society, v. 98, pp. 563-598.

Maritime Research Institute Netherlands (MARIN, 2007), Wind Loads on Offshore Structures (WINDOS), User Guide for WINDOS Engineering Tool.

Martha, D., Fischer, T., Kuhn, M. and Jonkman, J. (2010), Model Development and Loads Analysis of a Wind Turbine on a Floating Offshore Tension Leg Platform, NREL/CP-500-46725, National Renewable Energy Laboratory, Golden, CO. USA.

Masciola, M., Robertson A., Jonkman, J. and Driscoll, F. (2011), Investigation of a FAST-OrcaFlex Coupling Module for Integrating Turbine and Mooring Dynamics of Offshore Floating Structures, NREL/CP-5000-52896, National Renewable Energy Laboratory, Golden, CO. USA.

Naqvi, S. K. (2012), Scale Model Experiments on Floating Offshore Wind Turbines, Master of Science Thesis in Mechanical Engineering, Worcester Polytechnic Institute, Massachusetts, U.SA.

Nihei and Fujioka (2011), Aerodynamic Effects on TLP Type Wind Turbines and Predictions of the Electricity They Generate, Journal of Marine Science and Application, 10: 139 – 149.

Orcina Ltd., OrcaFlex User Manual, version 9.7, www.orcina.com.

Ormberg, H. and Bachynski, E. (2012), Global Analysis of Floating Wind Turbines: Code development, Model Sensitivity and Benchmark Study, Proceedings of the Twenty-second International Offshore and Polar Engineering Conference, Rhodes, Greece, June 17-22, 2012.

Robertson A. N. and Jonkman, J. M. (2011), Loads Analysis of Several Offshore Floating Wind Turbine Concepts, International Society of Offshore and Polar Engineers Conference, Maui, Hawaii, June 19-24, 2011.

Tracy, C. (2007), Parametric Design of Floating Offshore Wind Turbines, Master's Thesis, Department of Mechanical Engineering, Massachusetts Institute of Technology, Boston, Massachusetts, USA.

Udoh, I. E. and Zou, J. (2015), Aerodynamic Effects on Global Responses of 5 MW Three Column TLP Wind Turbine, Proceedings of the 20th Offshore Symposium – Texas Section of the Society of Naval Architects and Marine Engineers, Houston, Texas, USA.

Von Karman, T. (1948), Progress in the Statistical Theory of Turbulence, Proceedings of the National Academy of Sciences, v. 34, pp. 530 – 539.

WAMIT Inc. (2006), WAMIT User Manual, version 6.4, Massachusetts Institute of Technology and www.wamit.com.

APPENDIX – TURBLENCE WIND TIMESERIES AND SPECTRA (Y- AND Z-COMPONENTS)

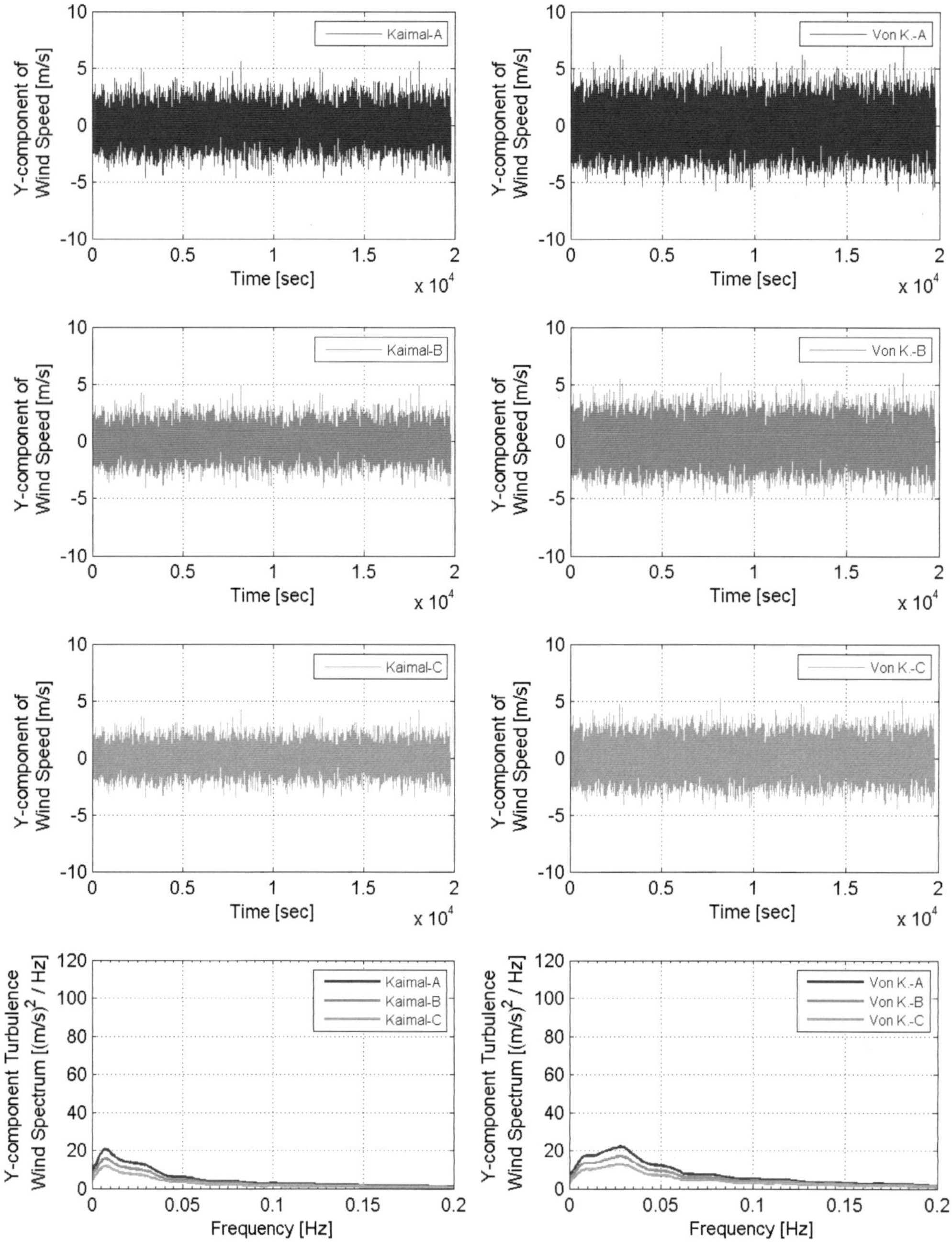

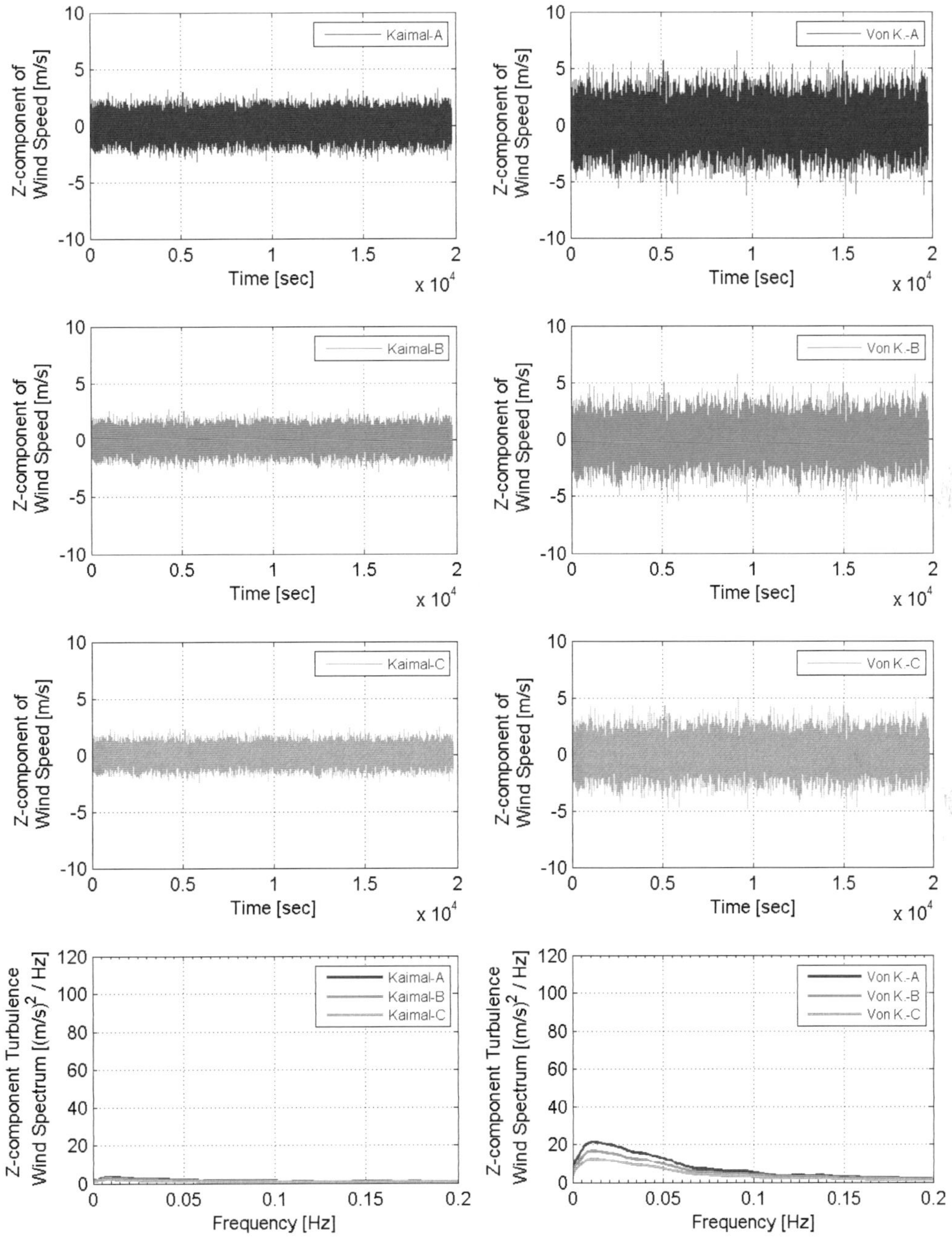
Kaimal-A
Von K.-A
Kaimal-B
Von K.-B
Kaimal-C
Von K.-C
Z-component of Wind Speed [m/s]
Time [sec]
x 10^4
Z-component Turbulence Wind Spectrum [(m/s)^2 / Hz]
Frequency [Hz]

SNAME
THE INTERNATIONAL COMMUNITY
FOR MARITIME AND OCEAN PROFESSIONALS

DESIGN CHALLENGES OF A HYBRID PLATFORM WITH MULTIPLE WIND TURBINES AND WAVE ENERGY CONVERTERS

SUNG YOUN BOO
VL OFFSHORE, HOUSTON, USA
KYONG-HWAN KIM, KANGSU LEE, SEWAN PARK, JONG-SU CHOI, KEYYONG HONG
KOREA RESEARCH INSTITUTE OF SHIP AND OCEAN ENGINEERING, KRISO, KOREA

Proceedings of the 21st Offshore Symposium, February 2016, Houston, Texas
Texas Section of the Society of Naval Architects and Marine Engineers

ABSTRACT

The present paper describes the design challenges of a wind wave hybrid power generation floating platform. The platform is a semi-type which consists of multi columns, pontoons, decks and brace members. The platform with a column span 150m is moored with chain catenary mooring lines at a water depth of 80m to produce power generated from wind and waves. The hybrid system is designed to produce a total of 10MW power from four wind turbines and twenty four wave energy converters (WECs). The platform design is based on industry standards and rules. The wind turbines are installed on four columns located at each corner of the platform while the WECs are placed at the peripheral locations between semi pontoon and deck. The WECs are vertically supported by frames and the vertical linear WEC generators are integrated inside the deck.

Design of the unconventional size of platform faces many design challenges in configuration of the system, structural design, wind turbine wake effects, constructability, loadout, WEC structures, multi-turbine and platform coupled response, mooring system design and power cable and such design challenges are discussed. Brief results of the motion responses, mooring analysis, structural analysis and power cable analysis are also described.

Keywords: Offshore Floating Wind Turbine, Wave Energy Converter, Semi-sub, Floating Structure, Mooring, Hybrid Wind Wave Power Generation, Wind Turbine Wake, Power Cable

INTRODUCTION

There have been considerable attempts to extend the shallow water wind turbine technologies of the bottom mounted structures to the deep water turbines supported by a floating structure. The majority of the floating offshore structure types for the wind application are originated from the oil and gas platforms, for instance, spar, semi-sub and TLP as they are well-proven platforms complying with very strict functioning requirements in harsh environments. Representative prototype floating wind platforms installed are Hywind and Windfloat.

However, the floating wind turbine platform also has numerous deign concerns, for instance, dynamic coupled effects of turbine with platform responses, which is called aero-servo-elasto-hydro couplings. The coupled effects can affect the rotor, structure near the mounting location and mooring system in various aspects.

Most studies on floating offshore turbine have been aimed at wind turbine sitting on a single floater which may enable a designer to apply the existing design tool based on FAST (Jonkman, 2007). More recently, multiple turbine floating systems have drawn interest as they provide a large amount of power in a single platform. Recently, the South Korean government funded a project to develop a wind wave hybrid platform which can facilitate four wind turbines and multiple wave energy converters (WECs) to produce a combined power rate of 10MW (Kim et al., 2015). For this purpose, a semi type floater moored at a water depth of 80m was selected but numerous design challenges were faced upfront along with the conventional design issues. Due to wake effects, consideration of a proper turbine distance is the most critical element in determining the semi dimensions which can raise a series of associated design difficulties. As a result, decided column span of the multi turbine semi is very long compared to the conventional semi, which can then cause construction and loadout difficulties. Additional engineering issues on this structure include design tool to deal with the multi turbine coupling with the semi, the structural design, mooring system, WEC installation and others. In this paper, such design challenges and drivers are discussed.

The present design of the 10MW hybrid platform was evaluated through the analyses in the global strength, motion response, mooring system and power cable, and the results are described briefly.

SITE LOCATION AND METOCEAN DATA

The platform will be installed at the west coast of Jeju island of South Korea depicted in Figure 1. The site water depth is 80m. The location is exposed to persistent winds and waves thoughout the year but strong typhoon passes the site often during the summer and early fall, and typhoon conditions can drive the hybrid system design. As shown in Figure 2, wind directionality is well presented. Metocean conditions for various return periods are summarized in Table 1.

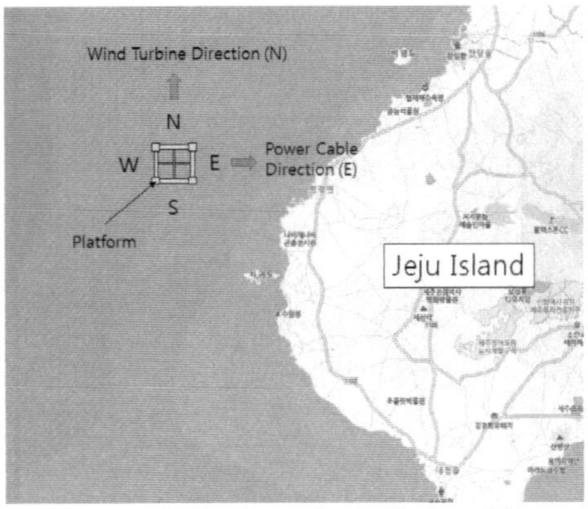

Figure 1 Platform Installation Site

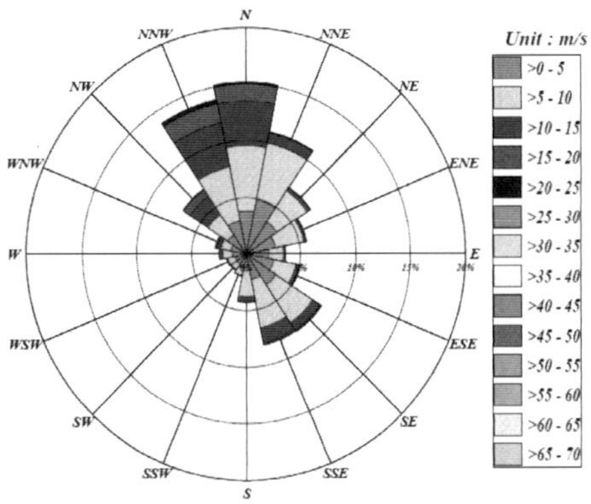

Figure 2 Wind Rose Plot

Table 1 Metocean Data

Items	Unit	1-yr	50-yr	100-yr	500-yr
Hs	m	5.74	9.72	11.32	16.02
Tp	s	10.81	13.98	15.10	17.57
Current	m/s	1.31	1.47	1.57	1.7
Wind	m/s	20.40	43.41	48.41	59.21
Tide	m	4.44	5.59	6.16	N/A

Current: Surface current
Wind:1-hr average at 10m
Tide: Tidal variation

FUNCTIONAL REQUIREMENTS

The wind-wave hybrid power platform produces a combined power rate of 10MW which is 8MW by wind turbines and 2MW by WECs. The WEC systems include a means to park the system during storm events. All the required electrical and marine systems are facilitated in the platform. A single power cable runs toward the coast and is connected to an existing grid. The mooring system is designed to limit the platform offset such that the power cable is in function with no damage. The platform provides facilities to monitor, support and repair the wind turbines and WECs. The wind turbines and WECs are integrated prior to wet-tow to site. Platform will have 25 years of design life.

DESIGN STANRDS AND GUIDELINES

The hybrid power floating platform system design is governed by various design standards, codes and guidelines in hull structure, station keeping, wind turbine, WEC and power cable. The documents for the design are summarized below.

Hull, Mooring and Stability
ABS Rules for Building and Classing Mobile Offshore Drilling Units, 2008
ABS FPI Guide for Building and Classing Floating Production Installations, 2013
API RP 2A Recommended Practice for Planning, Designing and Construction Fixed Offshore Platform – Working Stress Design, 2007
DNV-OS-C201 Structural Design of Offshore Units (WSD method), 2008
DNV-RP-C201 Buckling Strength of Plated Structures, 2010
DNV-RP-C202 Buckling Strength of Shells, 2010
DNV RP C203 Fatigue Design of Offshore Steel Structures, 2010
DNV-RP-F205 Global Performance Analysis of Deepwater Floating Structures, 2010
API RP 2SK Recommended Practice for Design and Analysis of Stationkeeping Systems for Floating Structures, 2nd edition, 2005

Floating Offshore Wind Turbine
ABS Guide for Building and Classing Floating Offshore Wind Turbine Installations, 2013
ABS Global Performance Analysis for Floating Offshore Wind Turbine Installations, 2014
DNV-OS-J101 Design of Offshore Wind Turbine Structures, 2010
IEC 61400-3 Requirements for Offshore Wind Turbines, 2009

Wave Energy Converter
IEC 62600-2 Design Requirement for Marine Energy System (draft), 2014
DNV Guidelines on Design and Operation of Wave Energy Converters, 2005
EMEC Guideline for Design Basis of Marine Energy Conversion System, 2009

Power Cable
API SPEC 17E Specification for Subsea Umbilicals
ISO-13628-5 Petroleum Natural Gas Industries – Design and Operation of Subsea Petroleum Production Systems – Part 5: Subsea Umbilicals
DNV RP F401 Electrical Power Cables in Subsea Applications, February, 2014
DNV RP J301 Subsea Power Cables in Shallow Water Renewable Energy Applications, February, 2014

DESIGN CHALLEGES

There are many challenging design drivers for the hybrid floating platform to support the multiple wind turbines and WECs. Key design drivers are discussed in this Section.

Turbine Distance and Wake Effects

The most notable design challenge in the present platform configuration and sizing is to determine the distance between turbines in order to minimize the wake effects to the downstream turbines. An example photo of the wake clouds behind the turbine in the wind farm is presented in Figure 3. The wake can cause a significant loss of the downstream wind turbine power output and also be a source of fatigue damage to the rotor and structures. The wake effects vary depending on the prevailing wind direction and turbine separation distance. Many empirical and analytical models have been developed and utilized for wind farm layouts although they are not satisfactory in predicting the wake.

Figure 3 Wakes Clouds behind Wind Turbine (Photo courtesy of Vattenfall)

For the present design, the wind turbine distance was decided to be 150m. The distance of 150m will not provide an ideal separation between tandem turbines. However, it may be the maximum distance when incorporating other design drivers discussed in subsequent sections. Wind velocity deficit can be estimated using Jensen model (Jensen, 1983);

$$u_0 - u = u_0\left(1-\sqrt{1-c_t}\right)\left(\frac{D_0}{D_0+2kX_0}\right)^2$$

$$c_t = \frac{2F_t}{\rho\frac{\pi}{4}D_0^2u_0^2}$$

Here u_o and u are speed of free stream wind and wake on the downstream turbine with a sweep diameter of $D_{o.}$ X_o, and C_t and k are separation distance, thrust coefficients and turbulence decay factor.

Figure 4 and Figure 5 present the wind velocities at various downstream locations and power output ratio between downstream (P) and upstream (P_o) turbines aligned in tandem direction, for a 3MW turbine, where a decay factor of 0.04 was assumed. Also CFD results for the case of an upstream speed 10 m/s are shown in Figure 5. It can be observed that power output of the downstream turbine located at 150m behind can be reduced to about 28%. It is seen that the wind velocities from Jensen model are greater than the CFD results but the model reasonably predicts the wake velocities. According to this estimation, the wind turbines may produce approximately a total of 8MW in tandem alignment. However, the wake direction varies with wind direction, which can cause partial or full wake effects on three turbines on downstream, thus total power generated may decrease considerably. This complex phenomenon is under investigation with CFD.

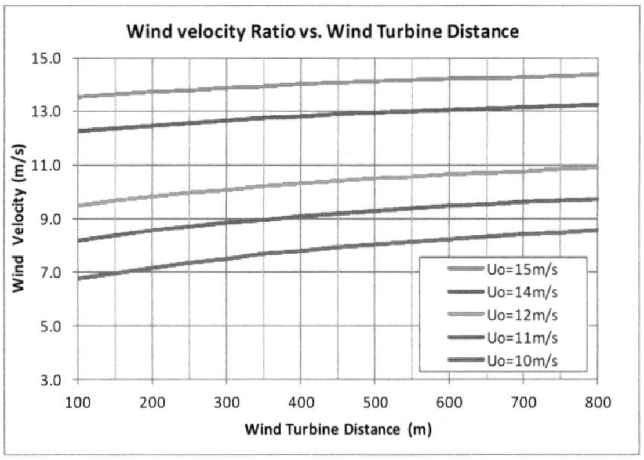

Figure 4 Wind Velocities at Locations behind Turbine by Jensen Model

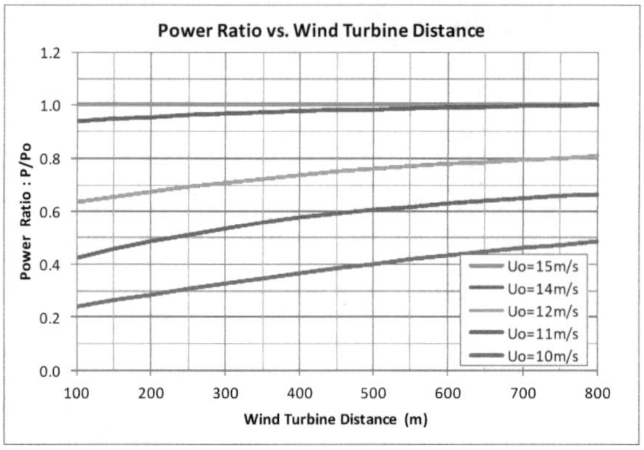

Figure 5 Power Ratio of Upstream to Downstream Turbines

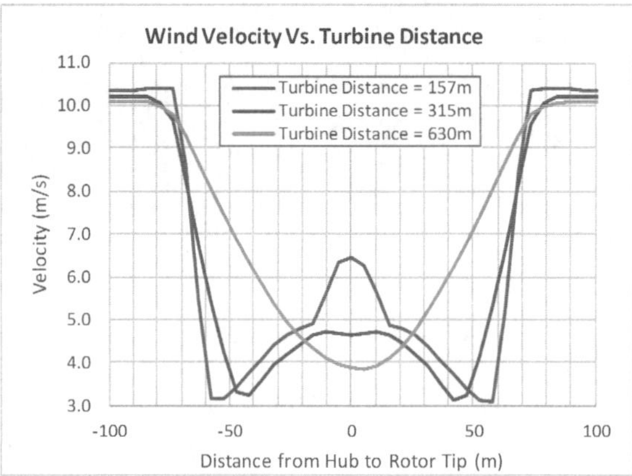

Figure 6 Wind Velocities at Locations behind Turbine from CFD

Floating Platform Type

Among all possible floating types, semi and TLP type of floaters are feasible options for installation in the water depth of 80m. The semi and TLP floaters have pros and cons in several aspects. As presented in Table 1, tidal variation at site is pretty large so that it will affect the floater draft change. The TLP buoyancy and tether tensions are very sensitive to tidal elevation compared to the semi. Thus, the semi type floater was selected for the present application.

The platform consists of multi columns, pontoons, decks and bracing members. The pontoons and decks connect the columns while the bracing members provide the structural support for the columns, pontoons and other structures. Center-to-center column span of the semi is 150m. A wind turbine tower is installed on the top of each column. The WECs are installed between pontoon and deck structure. The deck structure accommodates the WEC generator system.

Constructability and Loadout

Considering the column width, the semi dimensions are now a little greater than 150m in length and breadth which is unconventionally large and also very slender compared to semi rigs used in oil and gas. This size of semi is very problematic in various areas such as yard facility and space availability, construction method, loadout scheme, wind turbine tower and WEC integrations. A dry dock typically provides relatively better construction and easy loadout (or float-out). No dry dock is, however, available to deal with such an unconventional size semi so the following other options can be considered;

- Modular construction on land and welding in water
- Modular construction on land and welding on floating dock or submersible barge
- Entire semi construction on floating dock or submersible barge

- Entire semi construction on land and skid loadout to submersible barge
- Construction of entire lower hull and upper hull (deck) separately, loadout each using heavy lifting crane and integration of both at quayside

The maximum size of floating dock in Korea or commercially available submersible barge has insufficient width to support the whole hybrid semi. A huge overhang of the structure is anticipated during construction or loadout which will cause high bending loads to the semi structure. Thus a new build submersible barge could be optioned for both construction and float-off purposes. Any of the options above affects the structural design, construction, interface, cost and execution.

WEC System Installation

These WECs are installed in the sides of the platform between pontoon and deck. The WEC consists of buoy, shaft, permanent magnet system, coil system and support frames as depicted in Figure 7. Electricity is produced by the vertical oscillation of the magnet system through the linear generator with coil inside the deck. Due to this functional configuration required, penetrations of deck are inevitable. This requires larger size of deck with sufficient strength. Also proper water tight is to be provided to protect the generator system. Many trade-off studies have been carried out to determine the vertical and lateral support frame layout and sizes.

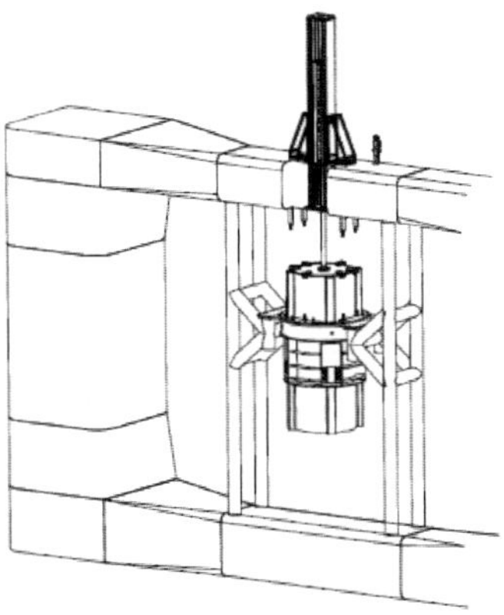

Figure 7 WEC System Installed and Support Frames

Design Tool for Multiple Wind Turbine and Platform Coupled Analysis

Time domain analysis of aero-elasto-hydro coupling with floating structure has been undertaken by many researchers, for instance, Rodier et al. (2000) and Shim et al. (2008). These studies were performed for a single wind turbine mounting on a moored floater, based on FAST (Jonkman, 2007).

Fully coupled analysis of the multiple wind turbines with the floating platform is a very complex and extremely challenging subject. Recently a program to deal with the multiple wind turbine effects was developed (Bae, 2013) and is being extended for the present study.

PLATFORM CONFIGURATION

The wave and wind hybrid power generation platform has four 3MW turbines on the semi columns and twenty four WECs distributed along the four platform sides. The platform is configured by considering several cases of loadout, wet-tow, operating and storm conditions. The total weights of the platform for the in-place operating and storm conditions were estimated considering hull structure, appurtenance, marine growth, ballast, turbines, WECs, marine system, electrical system and reserve, where an appropriate contingency was applied based on the past experiences of the oil and gas rig design. The platform weight differs for each case.

The platform key figures for the storm condition are summarized in Table 2. Also schematics of the platformis illustrated in Figure 8, where the wind turbine rotors are omitted.

Table 2 Key Figures of 10MW Hybrid Semi Platform – Storm Condition

Items	Unit	Value
Displacement	ton	26,848
Draft (design)	m	15.0
Column Span (center to center)	m	150.0
Column Height	m	27.0
Tower + Rotor	ton	1,800
WEC System	ton	2,344
CoG above Keel	m	13.68
Roll / Pitch Radius of Gyration	m	58.56
Yaw Radius of Gyration	m	78.41
Heave Natural Period	s	16.5
Surge / Sway Natural Period	s	103.7
Roll / Pitch Natural Period	s	15.3
Yaw Natural Period	s	115.4

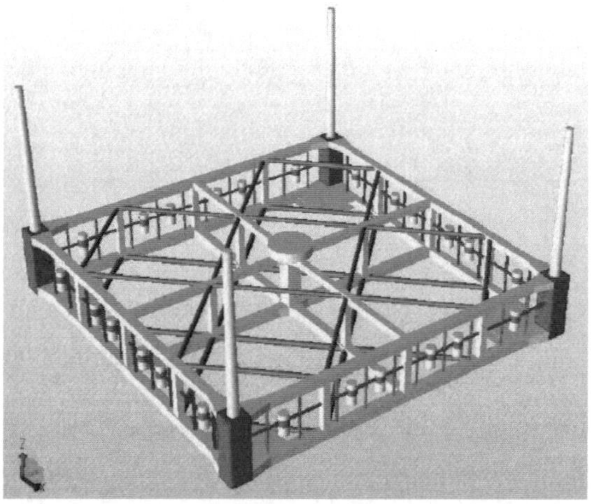

Figure 8 Hybrid Semi Platform Structure Schematics

STRUCTURAL DESIGN

Scantling analysis was conducted on the hull primary structures, deck structures and vertical square members for the initially configured structures presented in Figure 8. It was assumed that the tubular bracings are unstiffened.

The scantling design was verified by a global structure strength analysis using a SACS model. A couple of iterations were carried out by adding bracings and adjusting the structural member sizes to comply with the structural design criteria. One of the modifications is presented in Figure 9 where the added braces to the initial configuration are highlighted in red.

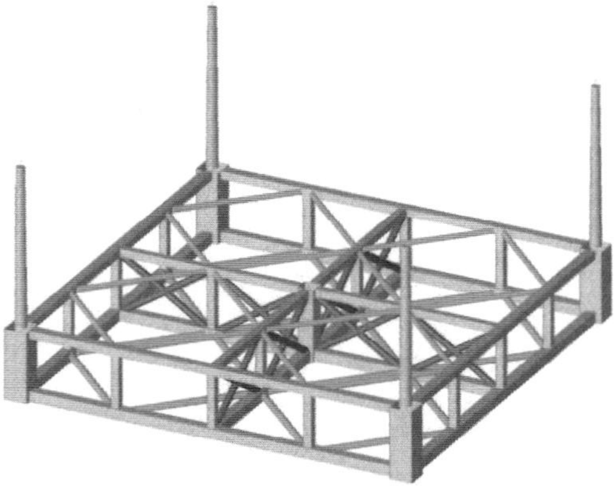

Figure 9 Structure after Modification in SACS Model

MOTION RESPONSE ANALYSIS

One quarter of platform was modeled with panels for analysis. The entire platform panel model is presented in Figure 10. WEC frames were excluded in the present model but their weights and displaced volumes were distributed to adjacent structures modeled.

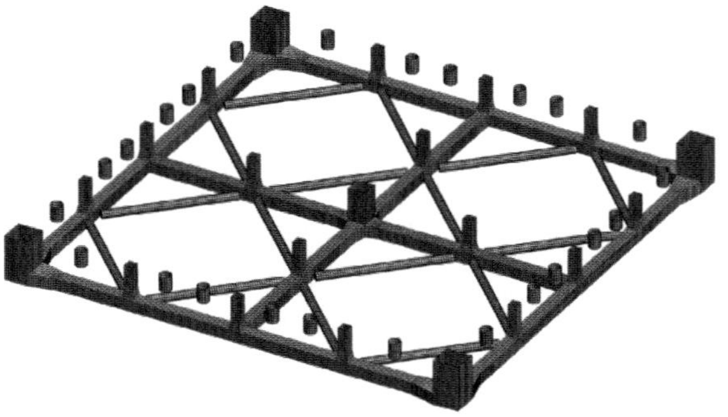

Figure 10 Simplified 3D Panel Model used in Hydrodynamic Analysis

Instead of including hull viscous damping in the WAMIT model, the damping were taken into account by Morison elements with drag coefficient on each element (columns, pontoons, braces, WECs, etc.). The input drag coefficient of the element was assumed initially based on its shapes (DNV-RP-C205, 2007) and then tuned with iterations until the overall current force and moment predicted by OrcaFlex are within acceptable range compared to the values predicted by WINDOS software. The mooring line size and configurations used for the analysis are described in the following section.

Wave tank tests with a 1/50 scale ratio were performed in the KRISO facility. The measured motion RAOs due to white noise and regular wave tank tests are compared with the numerical values by WAMIT (potential damping only) and numerical white noise values (potential plus viscous damping). Also based on the measured decay test results, the damping in the numerical model was correlated and then the simulation was re-run with the adjusted damping. These results are presented as "White Noise: with Correlation" in Figures 11, 12 and 13.

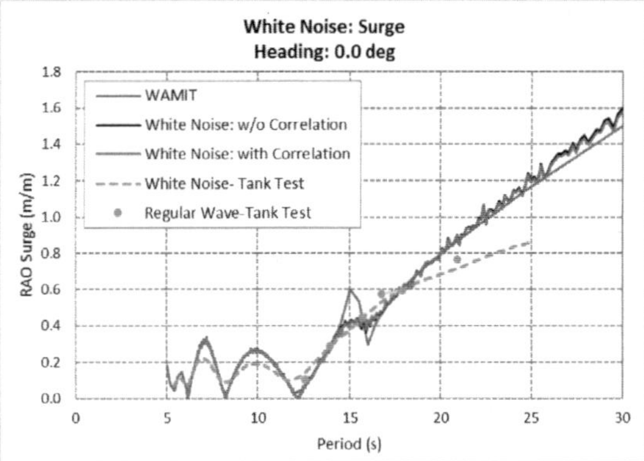

Figure 11 Surge Motion RAO Comparisons

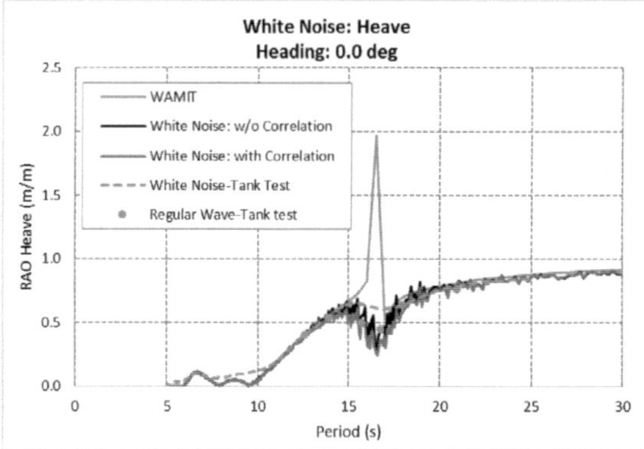

Figure 12 Heave Motion RAO Comparisons

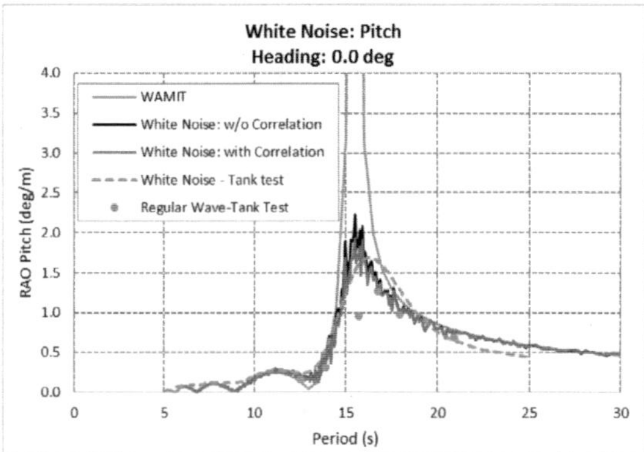

Figure 13 Pitch Motion RAO Comparisons

MOORING ANALYSIS

Several options of mooring layouts along with mooring line length variation have been studied. It is identified that a minimum of 500 ~ 600m long each line with 5 3/8 ~ 6 inch chain is required to limit the offset of the platform for the power cable performance and also to comply with the mooring strength criteria.

Several cases have been studied by changing and combining the number of lines, line length, chain sizes, clump locations and line separation angles. One of these cases was twelve mooring lines with 600m length, where three clumps were utilized. They are in four groups of three and consist of single chain system of R4 studless. Fairleads were assumed located near the keel. Chain diameter for the numerical strength analysis was adjusted accordingly due to the corrosion during the service life of 25 years. The mooring system layout with twelve lines used for the analysis is shown in Figure 14. Analysis was run by OrcaFlex software.

Load cases were selected for DLC 1.6 (operating), DLC 6.1 (50-yr extreme, parked), DLC 9.1 (100-yr survival, parked), based on ABS Guidelines.

Two environmental conditions of COD (codirectional) and MIS (misalignment) were also investigated. Using both Weibull and Rayleigh extreme approaches, the tensions at fairleads and anchors were calculated. It is found that the DLC 6.1 MIS condition for the wave heading 45 degrees governs the mooring strength design. Also the similar trend was observed in platform offset. Additional observation is on the ground chains suspended. The weatherside chains are most likely suspended entirely during the survival event which may require more mooring clumps, longer mooring line length or suction (or pile) anchor rather than a drag anchor. These are now being evaluated to minimize the cost impact.

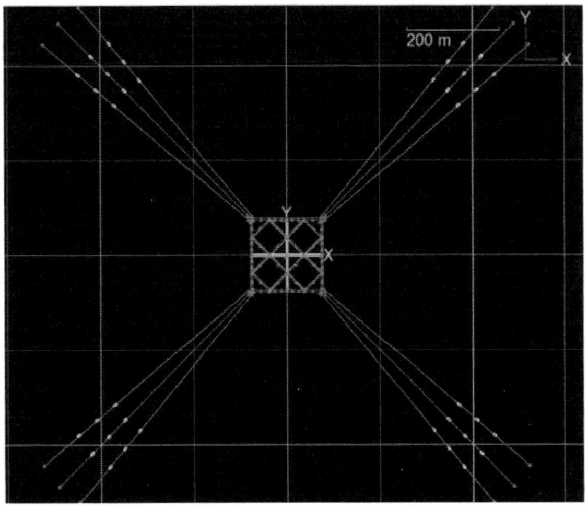

Figure 14 Modeled Twelve Mooring Line Layout

POWER CABLE ANALYSIS

Among numerous options for the power cable installation, a lazy wave configuration was chosen due to water depth at site. A single power cable running toward the platform east was considered. The cable top end was assumed to be connected to the center of the east pontoon keel. Cable diameter used is 105mm.

A screen capture of the numerical model by Orcaflex is shown in Figure 15. By considering the near, far and cross conditions of the platform offset, the lazy wave shape of the cable was determined iteratively by complying with the requirement of tension and minimum bend radius (MBR). The optimized length between the hang-off and first buoy location is identified to be about 62 ~78m with a buoy distribution length of 22m. The buoy OD, length and separation between buoys were 0.4m, 0.4m and 0.5m respectively.

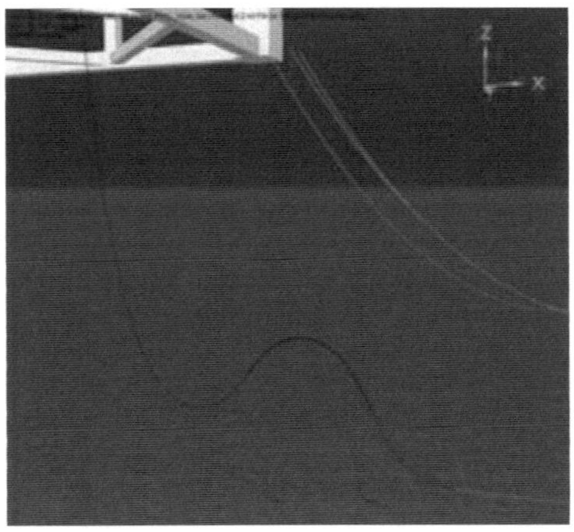

Figure 15 Power Cable Modeling for Lazy Wave Configuration

Figure 16**Error! Reference source not found.** presents the effective tension distribution along the cable arc length for mooring intact, far condition under the extreme environment. The tensions are well below the allowable maximum. MBR for the extreme condition was estimated to be about 4.7m for mooring line intact and 4.2m for one mooring line damage which are greater the allowable MBR. The values were computed with Weibull extreme statistics. Time histories of the bend radius causing the MBR in extreme conditions (mooring intact, one line damage and near case) are plotted in Figure 17.

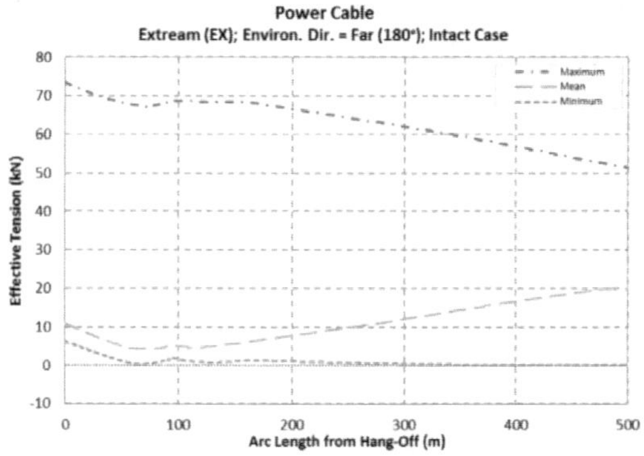

Figure 16 Effective Tensions on Power Cable: Far, Intact, Extreme

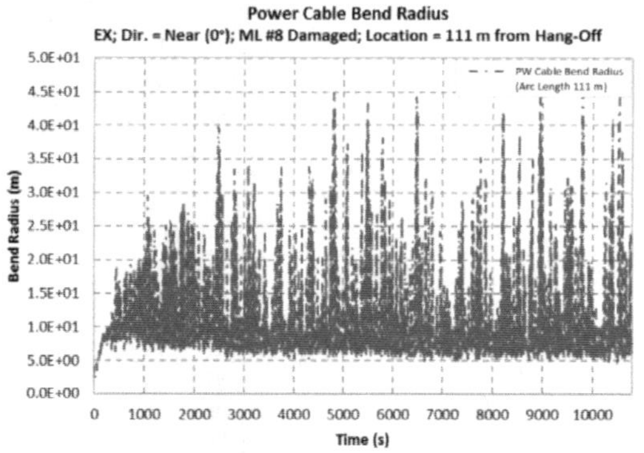

Figure 17 Bend Radius Time Histories: Near, Mooring Line Damage, 111 m from Hang-Off, Extreme

CONCLUSIONS

A wind-wave hybrid power platform has been designed to produce a power rate of 10MW which is a combined power output of 8MW by four wind turbines and 2MW by twenty four WECs. The platform is a four column semi but unconventionally long in length and breadth to provide a minimal turbine separation distance to create less wake effects to the downstream turbines. The wind turbines are mounted on the top of each column while the WECs are installed between pontoon and deck around the perimeter of the platform. WEC power generation systems are integrated inside the deck.

Design challenges of the long slender semi platform were discussed, which encompasses the wake effects, structural design, constructability, loadout, WEC system, and multi-turbine and platform coupled analysis.

The semi structure, mooring and power cable were designed to comply with the industry standards and codes applicable to floating wind turbine along with offshore floating installations. Numerical motion responses were compared with model test data to verify the numerical model. Numerical results on structure, mooring and power cable demonstrate that the current design meets the corresponding design requirements.

ACKNOWLEDGEMENTS

The present work is a result of the project "Development of the design technologies for a 10MW class wave-offshore wind hybrid power generation system" granted by the Ministry of Oceans and Fisheries, Korea. All support is gratefully acknowledged.

REFERENCES

Bae, Y.H., "Coupled Dynamic Analysis of Multiple Unit Floating Offshore Wind Turbine", PhD Thesis, Texas A&M Univ., 2013.

DNV-RP-C205 Environmental Conditions and Environmental Loads, 2007

Jensen, N.O., "A note on wind generator interaction", RISO, 1983

Jonkman, J.M., "Dynamics Modeling and Loads Analysis of an Offshore Floating Wind Turbine", November 2007.

Kim, K.-H., Lee, K., Sohn, J.M., Park, S.-W., Choi, J.-S., Hong, K., "Conceptual Design of 10MW Class

Floating Wave-Offshore Wind Hybrid Power Generation System", Proceedings of the Twenty-fifth (2015) International Ocean and Polar Engineering Conference, 2015

Roddier, D., Cermelli, C., Aubault, A., "WindFloat: A Floating Foundation for Offshore Wind Turbines Part II Hydrodynamics Analysis", June 2009.

Shim, S., Kim, M.H., "Rotor-Floater-Tether Coupled Dynamic Analysis of Offshore Floating Wind Turbines", ISOPE 2008.

IMPROVEMENT OF JACK-UP OPERATING CAPABILITY USING ADD-ON SPUDCANS

JOOST JANSSEN
SENIOR EXPERT ENGINEER, GUSTOMSC

HUGO HOFSTEDE
CONSULTANT, GUSTOMSC

MAAS HOOGEVEEN
ENGINEER, GUSTOMSC

Proceedings of the 21st Offshore Symposium, February 2016, Houston, Texas
Texas Section of the Society of Naval Architects and Marine Engineers

ABSTRACT

Over 200 jack-ups currently operating worldwide are designed for 350 ft water depth and above. In 2016 a large number of 350-ft-plus-class new builds are to be delivered, among which there are 17 CJ46 and 6 CJ50 GustoMSC designs. This paper explores how to extend the operating capabilities of these jack-ups by means of add-on spudcans, e.g. to operate in deeper waters, to withstand higher environmental loads or to carry more variable load.

The effectiveness of add-on spudcans to increase the water depth capability is demonstrated by performing generic Site Specific Assessments (SSA) for the CJ46 and CJ50 jack-ups using generic Gulf of Mexico (GoM) environmental and soil conditions. The presented SSA for these two designs shows opportunities to extend the water depth capability by 50 to 80 ft.

The paper also shows how a CJ50 jack-up can be made suitable to operate in deeper waters as well as to withstand higher environmental loads than originally designed for using typical North Sea harsh-environment and soil conditions. The SSA presented shows that a CJ50 fitted with add-on spudcans can withstand higher environmental loads up to a maximum wave height of 85 ft together with a water depth increase of 110 ft.

Keywords: MODU, Jack-up, add-on spudcans, Site Specific Assessment, SNAME T&R Bulletin 5-5A

INTRODUCTION

Over 200 *350-ft-plus-class* jack-ups are currently operating worldwide. Among a large number of 350 ft plus class new builds to be delivered by the end of 2016, there are 17 CJ46 and 6 CJ50 GustoMSC designs. These units will therefore be ready for employment around the world in the near future. There may be a market for them to operate in larger water depth, more severe environmental conditions or with higher variable loads than currently designed for. This paper shows that add-on spudcans can enhance the operating capability of a jack-up. It aims to quantify the benefits of add-on spudcans by comparing the maximum water depth capability for a specific jack-up with and without add-on spudcans, which is done for two GustoMSC design jack-ups in two regions of operation.

Add-on spudcans are attached underneath the existing spudcan to significantly increase the contact area with the soil. This results in a better support for the jack-up structure and better distribution of loads to the foundation which may contribute to reduce penetrations and increase foundation fixity. Previously, GustoMSC has proven the effectiveness of specifically designed add-on spudcans connected to CJ50 and CJ62 type of legs.

As an alternative to add-on spudcans, leg extensions (new sections on top of legs to extend the length of jack-up legs) could help in cases with limitations due to physical leg length. However, simply adding length may not be a viable solution in case leg strength or foundation bearing capacity are the critical criteria.

The adequacy of a jack-up for a particular site is determined through a site specific assessment (SSA). Its capability to pass such an SSA may be limited by many factors. Some factors are inherent to the platform itself such as for example leg length. Other factors may be related to the environment where it will operate in, such as environmental loads and soil conditions. Cases for GustoMSC CJ46 and CJ50 in the Gulf of Mexico and CJ50 in the North Sea are considered in this paper as presented in Table 1.

Table 1: Cases considered

Jack-up design	**CJ46 and CJ50**	**CJ50**
Area of operation	Gulf of Mexico	North Sea
Water depth	337 – 440 ft	300 – 410 ft
Soil profile	Clay	Sand over clay
Metocean	Tropical Revolving Storm	Harsh environment

GUSTOMSC DRILLING JACK-UPS FOR THIS STUDY

The GustoMSC CJ46 and CJ50 are designed for use in water depths up to respectively 350 ft and 400 ft in moderate environment. These units are equipped with conventional diamond shaped spudcans and provide permanent accommodation for 150 persons. Figure 1 shows a photo of multiple CJ46 jack-ups under construction at the yard of China Merchants Heavy Industry (CMHI).

CJ-series drilling jack-up designs are fitted with three triangular truss legs and fixation systems at each leg chord for optimum load transfer. They are equipped with a retractable X-Y cantilever providing a large drilling envelope. Main particulars and properties of both jack-ups as per design are summarized in Table 2.

Table 2: Jack-up main particulars

Jack-up design	**CJ46**	**CJ50**
Leg type [-]	Open truss; X-braced	Open truss; X-braced
Leg spacing [ft]	151	164
Leg length [ft]	505	541
Hull length [ft]	214	230
Hull breadth [ft]	203	223
Hull depth [ft]	26.2	31.2
Spudcan area [sqft]	1,615	2,150
Total elevated weight* [kips]	24,260	35,065
Variable and combined load* [kips]	6,175	8,800
Preload at seabed [kips]	17,600	25,850

** In survival mode*

Figure 1: CJ46 drilling jack-ups under construction at CMHI yard in China

ADD-ON SPUDCANS

Add-on spudcans can be attached below the existing legs in order to increase operability limits by means of increased foundation fixity and reduced penetrations in the case of softer soils. The add-on is typically connected to the chord ends of an existing leg by means of a dedicated connection system underneath the existing spudcan, without the need for modifications to the existing leg. With the existing spudcans fully retracted, the add-on spudcans thus extend below the base of the hull in transit. These may be designed for a specific purpose or field which is then reflected for example in spudcan area, skirt length and spudcan bottom shape. The leg lower end and spudcan design of each GustoMSC CJ-series are such that it allows mounting of an add-on type of spudcan as well as exchange of add-ons to sister rigs.

The main components of an add-on spudcan are discussed below:

- Spudcan: Steel typically circular shaped with a relatively flat bottom providing foundation support area.
- Skirts: The add-on may be skirted or not, depending on the soil conditions. Generally, skirts may be applied in harder soils such as sands and stiffer clays to provide foundation capacity in excess of preload while add-ons without skirts are typically suited for softer clays. Distinction is made between circumferential skirts and inner skirts. The latter provides the compartmentation which may be needed for structural strength purposes as well for prevention of the soil failure mechanism to go within the volume as enclosed by the skirts.
- Connection system: Specifically designed system that provides load transfer between the legs and add-on spudcan. The connection system comes with the add-on such that these can be used for sister rigs as well.
- Water evacuation pipes: These pipes are open during installation of the jack-up on site to allow for the water to evacuate from the skirt cavity and generally closed during operating conditions of the jack-up.
- Fill pipes: These may be used in case of incomplete penetrations to fill the remaining voids between spudcan bottom and seabed.
- Jetting pipes: These may be used during removal of the jack-up to neutralize the suction pressures underneath the spudcan.

- Spudcan compartments: While the add-on is afloat, these compartments are dry. They are filled to lower the add-ons to the seabed in a shallow and sheltered location and subsequently connected to the jack-up.

Previously add-on spudcans have been successfully used to connect to GustoMSC CJ-series leg designs, which allowed the rigs to operate in larger water depth and harsher environment. Two GustoMSC CJ62 jack-ups have made use of the add-on technique to be able to operate on the fields in the North Sea. A second example is an accommodation unit with CJ50 leg design. They are shown in Figure 2.

Figure 2 Examples of add-on spudcans constructed for use in North Sea

SITE SPECIFIC ASSESSMENT

When a jack-up is intended to operate at a specific site, the specific site conditions may differ from the general design conditions such as loading condition, water depth, airgap, wave height, wind speed, current velocity and soil conditions. A Site Specific Assessment (SSA) addresses and verifies the structural capability of the jack-up taking into account the site specific conditions and loading conditions of the jack-up. Verification is performed on sufficient leg length (reserve), overturning stability, structural strength (leg and holding system) and foundation safety by means of preload check, foundation bearing and sliding capacity. The leg length shall be sufficient for the combination of water depth, airgap and expected maximum leg penetration. The jack-up shall be shown to be capable of resisting the loads during extreme storm conditions as relevant for that particular site reflected by the strength and foundation checks.

Gulf of Mexico

The typical SSAs for the CJ46 and CJ50 in Gulf of Mexico (GoM) conditions are performed in accordance with SNAME-RP T&R Bulletin 5-5A rev 3 (Society of Naval Architects and Marine Engineers, 2008) as well as the GoM Annex (Society of Naval Architects and Marine Engineers, 2007).

Within the GoM it is common practice to evacuate jack-ups prior to arrival of a hurricane so that jack-ups can be considered unmanned during hurricane conditions. Unmanned hurricane conditions for the jack-ups are not elaborated in this paper. The following 3 cases are considered:

1. Assessment Case
   This addresses environmental conditions in accordance with 50 year Tropical Revolving Storm (TRS) independent extremes that will affect the area within 48 hours. Standard load factor (LF=1.15) and resistance factors following SNAME TR5-5A are used.

2. Contingency Case
   This addresses environmental conditions in accordance with 50 year TRS independent extremes that will affect the area within 72 hours. Reduced load factor (LF=1.0) and standard resistance factors following SNAME TR5-5A are used.

3. Winter Storm Case
   This addresses environmental conditions in accordance with winter storm conditions (Dec – May) outside of hurricane season.

The details of metocean conditions are presented in Table 3, based on (Stiff et al., 2006). As the Assessment Case and Contingency Case were found to be governing over Winter Storm Case, the latter is not further addressed here.

North Sea

In the North Sea, as per SNAME TR5-5A (Society of Naval Architects and Marine Engineers, 2008), the SSA is performed for all-year omni-directional metocean data (or seasonal if possible and needed) with a return period of 50-yr as applicable for manned jack-ups which is common practice in the area of the North Sea. The details of typical metocean conditions as used here are presented in Table 4.

SITE CONDITIONS

Metocean conditions

Typical Gulf of Mexico metocean conditions are used for the assessment of the CJ46 and CJ50 as summarized in Table 3 for two relevant water depths (Stiff et al., 2006). For intermediate water depths interpolation is used. The airgap is taken as 62 ft above Lowest Astronomical Tide (LAT) which is applicable for hurricane season and includes a margin for uncertainty in wave crest elevation as well as for settlements in the storm (Society of Naval Architects and Marine Engineers, 2007).

Table 3 Gulf of Mexico metocean conditions

	Gulf of Mexico	
	337ft water depth	440ft water depth
	Winter storm case	
Hmax [ft]	33.0	33.0
Vwind [kn]	56.5	56.5
Vcurrent [kn]	1.12	1.07
	Assessment case	
Hmax [ft]	37.4	37.4
Vwind [kn]	61.0	61.0
Vcurrent [kn]	1.34	1.24
	Contingency case	
Hmax [ft]	41.1	41.1
Vwind [kn]	66.0	66
Vcurrent [kn]	1.43	1.32

Typical North Sea metocean conditions are used for the assessment of the CJ50 as specified in Table 4. The airgap for the lower water depth case is taken as 52.5 ft which is the minimum required and for the larger water depth case 69 ft is used, which allows the 10,000-yr wave crest passes under the hull.

Table 4 North Sea metocean conditions

North Sea		
	Water depth 1 All-year conditions	Water depth 2 All-year conditions
Water depth [ft]	300	410
Hmax [ft]	79	85
Vwind [kn]	82	82
Vcurrent [kn]	2	2

Soil conditions

Typical Gulf of Mexico soil conditions have been selected here. These comprise of a clay top layer with linearly increasing strength from very soft at the seabed surface to stiff at the bottom of this layer. A dense silty sand underlays the clay. The soil profile is presented in Table 5.

Table 5 Gulf of Mexico soil conditions (typical)

Layer no [-]	From [ft]	To [ft]	Cu – top [ksf]	Cu – bottom [ksf]	φ [deg]	Description
1	0	65	0.31	1.04	n/a	Very soft to stiff clay
2	65	160	n/a	n/a	35	Dense silty sand

Typical North Sea soil conditions have been selected here. They consist of a medium dense sand layer at the top which is underlain by a very stiff clay layer of constant strength and of limited thickness. This clay layer is in turn underlain by a competent dense sand layer. The soil profile is shown in Table 6.

Table 6 North Sea soil conditions (typical)

Layer no [-]	From [ft]	To [ft]	Cu – top [ksf]	Cu – bottom [ksf]	φ [deg]	Description
1	0	40	n/a	n/a	35	Medium dense to dense sand
2	40	52	2.40	2.40	n/a	Very stiff clay
3	52	160	n/a	n/a	37	Dense sand

SPUDCAN PENETRATIONS

The preload operation for the CJ46 and CJ50 is executed with full preload using the jacking system. For this operation the hull is jacked-up above the waterline, just clear from the waves. Water ballast is taken in in order to provide a pre-load to the foundation such that the foundation is prepared for a survival storm. After preload the additional ballast water is removed and the hull is jacked up to target airgap. An overview of the preload at the level of the seabed is given in Table 7.

Table 7 Preload per leg at seabed

Preload at seabed [kips]	Conventional spudcan	Conventional spudcan + add-on
CJ46	17,600	18,800
CJ50	25,850	27,500

For the CJ46 and CJ50 in Gulf of Mexico soil conditions with *conventional spudcan* and full preload the penetration may reach about 50 to 55 ft depth. Due to the relatively deep penetration, the soft clay is expected to fill in at least part of the created hole above the spudcan. It is assumed here for the purpose of this paper that this occurs during the preload operation. In case infill occurs after the preload operation the additional load of the clay on top of the spudcan is to be taken into account in the storm conditions as additional vertical load.

By using an *add-on spudcan* (without skirts) with a relatively flat spudcan bottom and increased spudcan area (from 1615 sqft as per conventional spudcan to 3230 sqft for the CJ46 and from 2150 sqft to 4300 sqft for the CJ50), the effective pressure on the foundation is reduced resulting in a penetration after full preload in the range of 20 to 25 ft. As a result, the free leg length under the hull is significantly reduced and the potential of infill of soft clay on top of the spudcan is removed. Figure 4 presents the penetration resistance of both the conventional and add-on spudcan for both jack-ups. It shows that the penetration resistance for the case with add-ons is significantly increased resulting in substantially reduced penetrations. Because of the soil strength increasing with depth no punch through risk is present.

For the CJ50 in typical North Sea soil conditions the penetration with the conventional spudcan results in partial contact of spudcan bottom and seabed. Therefore the maximum preload is applied to the foundation which is done using the fixation system (28,550 kips). A skirted add-on spudcan with increased area from 2150 sqft (as for the conventional spudcan) to 4300 sqft results in full penetration of the skirts and limited contact of the almost flat spudcan bottom with the seabed. However, with filling of the remaining voids between the spudcan and the seabed surface, the calculated foundation capacity can be increased to a value well above the applied preload. Because of the larger area of the skirted add-on which is fully in contact with the soil, the foundation capacity is well above the full preload capacity. Figure 5 shows the penetration resistance of the conventional (relative to spudcan tip) and add-on skirted spudcan (at skirt tip).

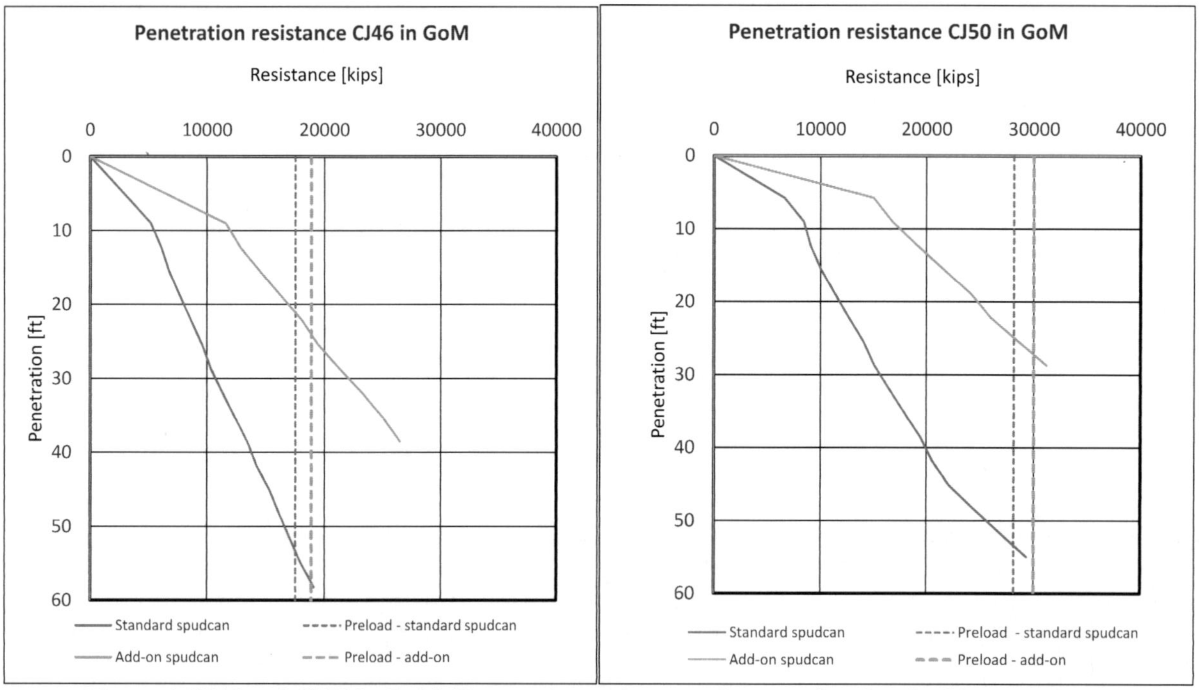

Figure 4 CJ46 and CJ50 in GoM: Penetration resistance of conventional and add-on spudcans

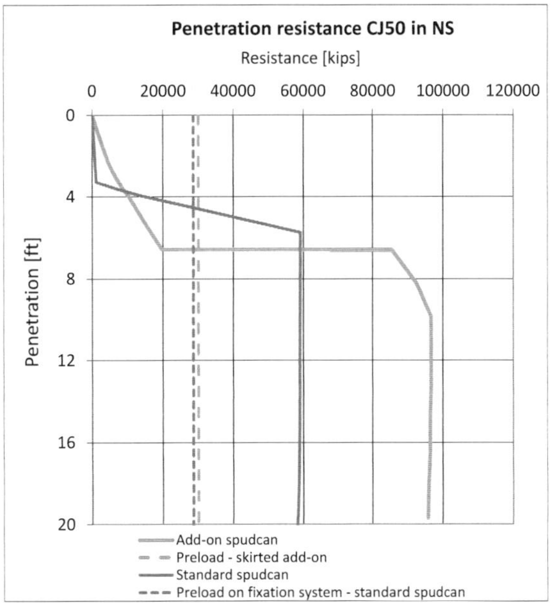

Figure 5 CJ50 in NS: Penetration resistance of conventional and add-on spudcans

RESULTS CJ46 AND CJ50 IN GULF OF MEXICO

Given the typical soil and environmental conditions as stated for the Gulf of Mexico, the CJ46 and CJ50 jack-ups with conventional spudcans may operate in water depths up to respectively 337 ft and 363 ft with the associated deep leg penetrations of 55 to 53 ft respectively. The large penetrations are the reason why these water depths are not the respective rated water depths of the 350-ft-class CJ46 and 400-ft-class CJ50. This is seen from the limiting factor being the available leg length which is fully utilized.

Despite the leg length being fully utilized, the structural and foundation utilizations in extreme storm conditions are found to be acceptable with margin as presented in Table 8. Due to the relatively low utilizations, the variable load was increased compared to the design value.

Table 8 Utilization for CJ46 and CJ50 with conventional spudcan in Gulf of Mexico

	CJ46		CJ50	
	Assessment Case	**Contingency Case**	**Assessment Case**	**Contingency Case**
Water depth [ft]	337		363	
Leg chord [-]	0.62	0.63	0.41	0.41
Fixation system [-]	0.58	0.59	0.42	0.42
Overturning stability [-]	0.29	0.28	0.26	0.27
Preload [-]	0.90	0.88	0.86	0.85
Bearing/sliding [-]	0.67	0.66	0.50	0.50
Fixity percentage	37 %	38 %	47 %	48 %
Leg reserve [ft]	5		5	
Total leg length [ft]	505		541	
Natural period [s]	8.8		8.4	
Leg penetration [ft]	55		53	
Total variable load [kips]	9,500		13,200	

In order to operate in deeper water a leg extension, by adding a leg section on top of the existing leg, may be suggested. An alternative would be to mount an add-on spudcan underneath the existing spudcan which effectively also results in additional leg length, along with other benefits. The assessment and contingency case were re-run for both alternatives and the governing results for these cases are presented in Table 9.

Table 9 Utilization for CJ46 with add-on spudcan or leg extension in Gulf of Mexico

Utilizations CJ46 in GoM - Assessment/Contingency Case		
	With leg extension	**With add-on spudcan**
Water depth [ft]	380	390
Leg chord [-]	0.98	0.63
Fixation system [-]	0.91	0.58
Overturning stability [-]	0.43	0.28
Preload [-]	1.05	0.90
Bearing/sliding [-]	1.00	0.77
Fixity percentage	21%	45%
Leg reserve [ft]	5	5
Total leg length [ft]	548	524
Natural period [s]	9.8	9.1
Leg penetration [ft]	55	21
Total variable load [kips]	9,500	9,500

Table 9 shows that in case of a leg extension alone, the gain in water depth for the CJ46 is limited to 43 ft with the bearing capacity becoming critical. In case add-on spudcans are fitted under the existing spudcans the gain in water depth is approximately 53 ft.

An important benefit of add-on spudcans is reduced penetration depth and therewith also reduced free leg length under the hull. As a result of that, together with the larger add-on diameter, the foundation fixity (indicated as fixity percentage, defined as the ratio of footing moment and total of moments at hull and footing) is maintained while the increase in dynamic loads is limited. This leads to less bending moment in the leg's critical area which enhances the rig's capability to resist environmental loading. A leg extension only results in significantly reduced foundation fixity, substantial increase in dynamic loading and subsequently unacceptable structural and foundation utilizations.

The margin in capability of resisting the storm loading, indicated by strength utilizations, allows for further benefits in the case with add-ons such as a further increase in water depth using leg extensions or a further increase in variable deck load or a combination for the CJ46 in Gulf of Mexico.

Table 10 shows the results for the CJ50 with a leg length increase of 50 ft for the case where this is achieved with a leg extension alone as well as for the case where this is achieved with a combination of add-on spudcans and leg extension. The benefit of the add-on spudcans is that they reduce the penetration such that the maximum water depth can be increased further compared to a leg extension only. The gain in water depth is respectively approximately 47 ft and 77 ft.

Table 10 Utilization for CJ50 with add-on spudcan or leg extension in Gulf of Mexico

Utilizations CJ50 in GoM - Assessment/Contingency Case		
	With leg extension (+ 50 ft)	**With add-on spudcan and leg extension (+ 50 ft)**
Water depth [ft]	410	440
Leg chord [-]	0.47	0.47
Fixation system [-]	0.48	0.47
Overturning stability [-]	0.31	0.31
Preload [-]	0.90	0.90
Bearing/sliding [-]	0.59	0.51
Fixity percentage	42 %	49 %
Leg reserve [ft]	5	5
Total leg length [ft]	590	590
Natural period [s]	9.4	9.4
Leg penetration [ft]	53	24
Total variable load [kips]	13,200	13,200

RESULTS CJ50 IN NORTH SEA

The CJ50 with conventional spudcan was verified for typical soil and all-year metocean conditions for the North Sea. In such harsh environment the as-is jack-up can operate in a water depth up to 300 ft. Structural utilizations are found acceptable with some margin while the foundation bearing capacity is exceeded. However, very limited settlement in the storm results in additional foundation capacity as the spudcan area in contact with the soil increases, therefore this is generally found an acceptable scenario. The results are presented in Table 11.

Table 11 Utilization for CJ50 in North Sea

	Utilizations CJ50 in NS	
	All-year conditions With conventional spudcans	**All-year conditions With add-on spudcans**
Water depth [ft]	300	410
Leg chord [-]	0.93	0.93
Fixation system [-]	0.95	0.92
Overturning stability [-]	0.79	0.76
Preload [-]	0.98	0.95
Bearing/sliding [-]	1.0*	0.46
Fixity percentage	7 %	39 %
Leg reserve [ft]	114	5
Total leg length [ft]	541	541
Natural period [s]	8.2	7.8
Leg penetration [ft]	5	7
Total variable load [kips]	8,800	8,800

* *Bearing check after limited settlement*

In case add-ons are fitted Table 11 shows that an increase of 110 ft of water depth can be achieved. For this case leg length is fully utilized with small margins in structural strength of leg chord and fixation system leaving some room for additional variable load. The foundation is not governing because the add-on spudcans provide foundation capacity well in excess of preload.

SUMMARY OF RESULTS

Both the CJ46 and CJ50 were verified for typical soil and metocean conditions for the Gulf of Mexico area. The following cases were assessed by means of a generic Site Specific Assessment:

- SSA with as-is conventional spudcans
- SSA with leg extension by adding an additional section on top of the leg
- SSA with add-on spudcans connected underneath existing legs

For both the CJ46 and CJ50 the base case with as-is conventional spudcans shows that available leg length is the limiting criterion. Their capability to resist environmental loading in extreme storm is high as the structural strength and foundation capacity are not fully utilized and have margin. The limiting water depth for the CJ46 and CJ50 respectively are 337 ft and 363 ft in combination with the approx. 55 ft leg penetration associated with the typical soil conditions. The large penetrations are the reason why these water depths are not the respective rated water depths of the 350-ft-class CJ46 and 400-ft-class CJ50.

In order to increase the water depth capability additional leg length is needed which can be accommodated for by a leg extension) or by connecting an add-on spudcan underneath the existing conventional spudcan or a combination of both.

For the CJ46 a water depth increase was investigated based on an increase of leg length, either achieved by mounting add-on spudcans underneath the existing spudcan structure or by using a leg extension on top of the existing leg. The latter option resulted in a water depth increase from 337 ft to a maximum of 380 ft where the structural utilizations are the limiting factor. The alternative with add-ons resulted in a maximum water depth of 390 ft where the leg length becomes the limiting criterion again and thus leaving possibilities to further increase the water depth in case the leg is further extended or to allow increase of variable deck load. The add-on spudcan alternative is more effective because it reduces the penetration and maintains the foundation fixity due to the larger diameter.

For the CJ50 a leg length increase of 50 ft was investigated, either achieved by a combination of add-on spudcans and leg extension or by a (longer) leg extension only. The combination with add-on leads to a water depth increase from 363 ft to 440 ft, where the case with a leg extension only results in a water depth increase to 410 ft. The difference is explained by the advantage of add-on spudcan that it results in reduced penetration depth. Both scenarios result in foundation and structural checks with margin while the leg length is governing again. A further extension of the leg may be considered to further increase the water depth capability.

The benefit of add-on spudcans is also shown for harsh-environment areas. For this, the CJ50 was verified for generic North Sea soil and all-year metocean conditions. The as-is jack-up with conventional spudcan was verified to be acceptable for up to 300 ft of water depth. Due to the harsher environment in the North Sea the restriction does not come from leg length but from utilizations in the extreme storm condition. In the case of using add-on skirted spudcans the water depth can be increased to 410 ft. This shows the beneficial effect of the add-on spudcans for the rig being able to withstand more severe environmental loads in deeper water when the structural capacities are close to the limiting utilization for the rig.

CONCLUSIONS

The benefit of add-on spudcans on the operability of jack-ups is demonstrated for the GustoMSC CJ46 and CJ50 designs of which a large number is being constructed in Asia and the majority will become available on the market in 2016. Improved operability, quantified by increased water depth, more severe environmental conditions or increased variable load capability, was explored.

Add-on spudcans are shown to be an effective solution to boost jack-up operating capabilities for both Gulf of Mexico and North Sea. The representative cases used in this paper demonstrate significant water depth capability increase. Besides water depth increase (or a higher airgap), also more severe environmental conditions and more variable load can be achieved depending on site specific conditions.

ACKNOWLEDGEMENTS

We are grateful to David Menzies of MatthewsDaniel Geotechnical Group for the provision of typical soil data for the Gulf of Mexico. We also want to thank Mike White with DNV GL for his help.

REFERENCES

Society for Naval Architects and Marine Engineers, "Recommended practice for site specific assessment of mobile jack-up units", Gulf of Mexico annex, 2007.

Society for Naval Architects and Marine Engineers, "Guidelines for site specific assessments of mobile jack-up units", T&R Bulletin 5-5A, 2008.

Stiff, J.J., Roper, R.R., Cardone, V.J., Cox, A., Lewis, D.R., "Metocean criteria for jack-ups in the Gulf of Mexico", Proceedings of Offshore Technology Conference, Houston, Texas. 2006.

SNAME
THE INTERNATIONAL COMMUNITY
FOR MARITIME AND OCEAN PROFESSIONALS

JACKUPS GOING ON LOCATION – UNDERSTANDING ENERGY PRINCIPLES ON LEG IMPACT LOADS

JOSE H. VAZQUEZ, PH.D.
PRESIDENT – 3DENT TECHNOLOGY

BARTON D. GRASSO, P.E.
DIRECTOR OF ENGINEERING – 3DENT TECHNOLOGY

MARCUS A. GAMINO
ENGINEER I – 3DENT TECHNOLOGY

WEI WANG
ENGINEER I – 3DENT TECHNOLOGY

Proceedings of the 21st Offshore Symposium, February 2016, Houston, Texas
Texas Section of the Society of Naval Architects and Marine Engineers

ABSTRACT

Analyzing self-elevating units in the elevated and afloat modes is fairly well understood, and standards exist for use in the development of limiting environmental criteria, as listed in the Marine Operating Manual (MOM) or as determined by site-specific assessments. However; the transition phase from the afloat mode to the elevated mode, commonly referred to as "going on location" (GoL) is not as well understood. During the GoL phase, the motions of the unit can cause severe loading as the legs impact the seabed when they are lowered. As a safe operating limit, MOMs typically list a single wave height or inclination angle for all conditions, regardless of water depth, soil conditions, wave period and/or direction. While these limits have proven to be safe, they are inherently conservative.

A common approach for establishing safe GoL operating limits is based on the conservation of energy principle. Perhaps due to the expected large pitch/roll response angles near resonance, industry standards usually focus solely on the rotation response at the pitch natural period while ignoring the contribution of heave. It is also common industry practice to equate the pre-impact kinetic energy to the energy absorbed by the leg and jacking system during impact. While these assumptions may generally produce conservative results, the fact that the impact load is directly related to velocity (through kinetic energy) and not motion indicates that the critical period may actually be slightly lower than the rotation natural period. Similarly, it is easy to see that the critical period for impact loads may be one with relatively small rotation and large heave.

From the above, it can be inferred that the parameter of interest for determining GoL leg impact loads is downward spudcan velocity (DSV). It is, therefore, the objective of this paper to illustrate the relationship of DSV and impact load, while answering the question as to whether or not this relationship is independent of oscillation amplitude or wave period. For simplicity, the analyses focus on a single water depth, soil stiffness and direction and are based on a generic jackup suitable for 300 ft water depth.

Keywords: Jackups, Going on Location (GoL), Self-elevating Units, Leg Impact Loads, Spudcan Impact Loads, Downward Spudcan Velocity (DSV)

INTRODUCTION AND BACKGROUND

The analysis of jackups and other self-elevating units in the elevated and afloat modes is fairly well understood, and standards exist for the development of limiting environmental criteria, as listed in the Marine Operating Manual (MOM) or as determined by site-specific assessments. However; the analysis of these units as they transition from the afloat to the elevated mode has been approximated in a number of ways, as the governing factors have not yet been fully isolated. In fact, most jackups are operated under a single wave height limit, which dictates when the legs may be lowered to make contact with the seabed. While the single wave height limit has proven to be a safe operational limit, the fact that it is a single number means that it must be a conservative value since it has to be valid for any wave period, wave direction or soil condition.

Early methodologies for going on location analyses, such as the one described in the DNV guidelines, are based on the principle of energy conservation. For a free-floating unit, subject to harmonic rotational motion, the kinetic energy is given by Equation (1). The potential energy absorbed by a linear spring is given by Equation (2).

$$K.E. = \tfrac{1}{2} I_e \theta'^2 = \tfrac{1}{2} I_e (\omega \theta)^2 \quad (1)$$

$$P.E. = \tfrac{1}{2} k_e s^2 = \tfrac{1}{2} (P_v s) = \tfrac{1}{2} (P_v)(P_v / k_e) = \tfrac{1}{2} (P_v)^2 / k_e \quad (2)$$

where $K.E.$ = Kinetic Energy,

I_e = Effective Rotational Moment of Inertia (actual + hydrodynamic contribution known as added mass),

θ = Rotational Angle, in radians,

$\theta' = d\theta/dt$ = Rotational Velocity, in radians per second,

ω = Radial Oscillation Frequency = $2\pi/T$,

T = Oscillation Period,

$P.E.$ = Potential (spring) Energy,

k_e = Effective Spring Stiffness Constant (combined soil stiffness + leg + jacking system stiffness),

s = Spring Displacement, and

P_v = Vertical component of Impact Load.

Equating kinetic energy to potential energy, and solving for the vertical impact load yields the following equations:

$$\tfrac{1}{2} I_e (\omega \theta)^2 = \tfrac{1}{2} (P_v)^2 / k_e \quad (3)$$

$$P_v = \omega \theta (I_e / k_e)^{1/2} \quad (4)$$

Equation (4) has the same form as the expression in the DNV guidelines for P_v with k_e given as in Equation (5).

$$1/k_e = 1/k_V + 1/k_H (d / h)^2 \quad (5)$$

where k_V = Overall vertical stiffness of the (leg, holding and soil system),

k_H = Overall transverse stiffness of the leg,

d = Horizontal distance from center of rig to the leg (perpendicular to angle of rotation),

h = Vertical distance from seabed to holding system.

Although not explicitly stated, the assumption in the application of the above equation is that the free-floating rotation angle is to be used, and this implies that 100% of the free-floating (pre-impact) kinetic energy due to angular rotation is absorbed by a single leg as potential (spring) energy, and this takes place at the time of first impact. The latter assumption is only valid for initial value problems and is particularly invalid for the case of jacking the legs down, given the out-of-phase relationship between maximum velocity and spudcan elevation. Additionally, the above expression for impact load ignores heave effects altogether.

Furthermore, it is also standard practice to assume that the peak impact load for a given wave height occurs at the natural period, but the peak impact load takes place at the peak rotational velocity, which is usually at a period slightly lower than the natural period. The critical period shifts down further once heave effects are considered. Therefore, while the above approach is assumed to produce conservative results, the fact that it ignores heave response renders that assumption invalid, especially for jackups having pitch/roll natural periods that are noticeably higher than the heave natural period.

Beyond the methodology presented in DNV 31.5, several works have been published in regards to advanced analysis techniques and methodologies for GoL of jackups. Some of the first work published on the subject matter introduced time-domain simulations to account for leg dynamic effects (Lai et al., 1993; Miller et al., 1993; Smith et al., 1994). As is the case in the DNV guidelines, the approach assumes the problem to be directly related to the free-floating rotational response of the jackup, prior to any impact. Further GoL work has been published in recent years. In 2012, Chakrabarti introduced a methodology that uses spudcan velocity with a 3-hour PM spectrum and Rayleigh distribution. In 2014, Kreuzer et al. introduced a methodology that includes wave excitation loads with retardation functions to account for the added mass/damping effects. Also, in 2014, Daun and Olsson introduced a methodology that uses three submodels (hydrodynamic, impact and structural) and an evaluation criterion to establish the operating window for a self-elevating unit. Finally, in 2015, Tan et al. describe a methodology for establishing GoL operating limits for jackup units which solves the non-linear response problem in the time domain using commercial software, accounting for global motions in random seas, leg lowering, transient impact of the spudcan and the seabed, and jackup structural dynamic responses.

It is evident that advancements have been made in the techniques for analyzing jackups in the transient condition of going on location. However; as the methodology becomes more complicated, an understanding of the underlying fundamentals as to the magnitude of the impact load that is calculated is essential, especially if simplified approaches are to be implemented.

The authors contend that a relationship between free-floating response and impact load can be obtained, but instead of simply using rotational velocity for determining leg impact loads, the parameter of interest is downward spudcan velocity (DSV), as it accounts for both rotation and heave effects. Furthermore, the authors assert that the impact load is dependent on the relative contributions of heave and pitch to DSV and as such, the relationship between impact load and DSV varies by period. Therefore, the objective of this paper is to illustrate the relationship of DSV and impact load, while answering the question as to whether or not this relationship is independent or dependent on oscillation period. The analyses focus a generic jackup in regular waves at a single location with 300 ft water depth and very stiff soil conditions. It is noted that while in regular waves the maximum (upward) and minimum (downward) spudcan velocities have the same magnitude, this is not necessarily the case in random waves and it is the downward velocity that induces maximum impact load on the legs - hence the use of DSV.

METHODOLOGY

The authors have independently developed and implemented a methodology for conducting GoL analyses that reflects some of the recent work on the subject. The methodology accounts for soil stiffness, spudcan shape, jacking speed, and compares the calculated loads with the structural capacity of the legs and jacking system to obtain permissible wave height curves. The methodology can be applied to regular waves or random waves in any direction, or combinations of the two. Furthermore, the methodology is based on commercial software and can be implemented in a relatively expedient manner. Wind and current effects are ignored. The GoL analysis methodology is made up of three types of analyses, as outlined below.

- *Frequency-domain diffraction analysis* (using *WAMIT*) to calculate the wave load as well as the added mass/damping coefficients of the jackup at the target depth with its legs lowered near the seabed.
- *Structural analysis* (using *SACS*) to establish the structural capacity envelope of the legs (combinations of vertical force and moment reactions that cause either the pinions, chords or braces to reach their structural capacity).
- *Time-domain response analysis* (using *OrcaFlex*) to calculate the motions and spudcan loads during the jacking operations as the legs make contact with the seabed.

The time-domain response simulation is considered to be divided into four segments, as described in Table 1 and illustrated in Figure 1 (depicting regular waves, but also applicable to random waves).

Table 1 Description of GoL Time-Domain Simulation Segments

Segment	Description
Ramping	Time prior to the actual start of the simulation, taken to be 100 seconds, during which all wave heights are gradually increased from 0 to the target value.
Free-Floating	Time starting at t=0 sec, to the first time there is contact between the seabed and any of the members representing the spudcans.
Transition	Time from the first time there is contact between the seabed and any of the members representing the spudcans, until all legs have made contact and the hull starts to come out of the water.
Elevating	Time after all legs have made and remain in contact with the seabed, resulting in minimal motions except for the elevation of the hull.

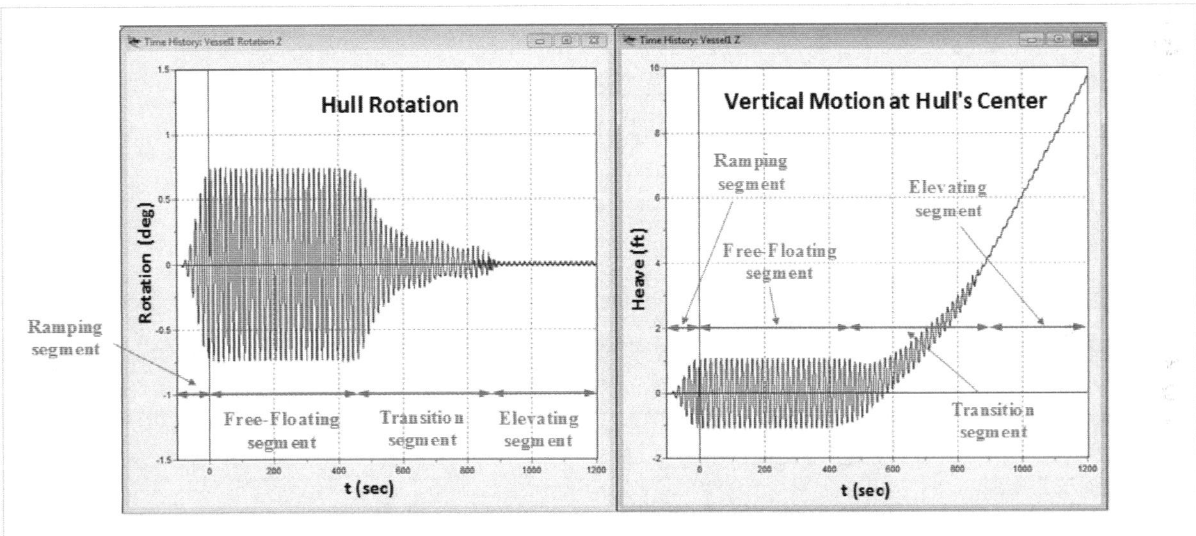

Figure 1 Illustration of GoL Time-Domain Simulation Segments

This GoL methodology is applied herein to test the theory that the parameter of interest for determining leg impact load is DSV, as it accounts for both rotation and heave effects. Since the focus of this paper is on the calculation of the impact load, a structural analyses was not performed to develop the capacity envelope.

For this study, a series of time-domain analyses are conducted to better understand the behavior of the unit as it goes on location. The first part of this study illustrates the balance of kinetic and potential energy and demonstrates how the response (impact load) is noticeably different for heave than it is for pitch. The second part of the study focuses on analyses with a target DSV over a series of periods to illustrate that the peak response is not constant.

JACKUP DESCRIPTION AND FREE-FLOATING RESPONSE CHARACTERISTICS

For the purposes of this study, a generic jackup, suitable for working in moderate environmental conditions without a fixation system is used. The particulars of this jackup are as shown in Table 2, below.

Table 2 Principal Characteristics of Jackup

Hull Length	205 ft
Hull Breadth	204 ft
Hull Depth	25 ft
Loadline Draft	15 ft
Loadline Displacement	28,100 kips
Transverse Leg Spacing	140 ft
Longitudinal Leg Spacing	125 ft
Leg Length	410 ft
Chord Spacing	35 ft
Chord Area	150 in^2
Spudcan Height	18 ft
Spudcan Diameter	45 ft
Total Leg Weight (including Spudcans)	5,805 kips
Total Spudcan Weight	1,500 kips
Jacking Speed	1.5 ft/min
Number of Pinions Per Chord	4
Pinion Stiffness	4,000 kips/in
Mid Elevation of the Pinions Above Main Deck	10 ft

Table 3 shows the mass properties of the unit floating at its loadline draft with the legs near the seabed in 300 ft water depth. The properties are about the origin of the model, located at the water plane, through the center of buoyancy.

Table 3 Inertia and Hydrostatic Properties of Jackup

VCG	-42 ft
Radius of Gyration about Transverse Axis (R_{xx})	147 ft
Radius of Gyration about Longitudinal Axis (R_{yy})	143 ft
Radius of Gyration about Vertical Axis (R_{zz})	73 ft
Heave Hydrostatic Stiffness (k_{33})	1,841 k/ft
Pitch Hydrostatic Stiffness (k_{55})	5.7E6 k-ft/rad

For the time-domain simulations, an equivalent system stiffness value, k_s=37,205 k/ft is used. The value for the equivalent system stiffness is obtained as the combination in series of three stiffness components: a soil stiffness of k_s=500,000 k/ft (i.e., very stiff soil), the leg's axial stiffness k_{leg}=EA/L= 43,200 k/ft and the total pinion stiffness for one leg k_p= 576,000 k/ft. The excitation loads and added mass/damping coefficients, which are input for the time-domain model are obtained from the frequency-domain diffraction analysis. Figure 2 shows the relevant excitation load RAOs. Although not needed for the time-domain simulations, motion response RAOs, including the forward spudcan's downward velocity from the frequency-domain analysis are shown in Figure 3. The motion Response RAOs in Figure 3 correspond to the case with 15% pitch damping.

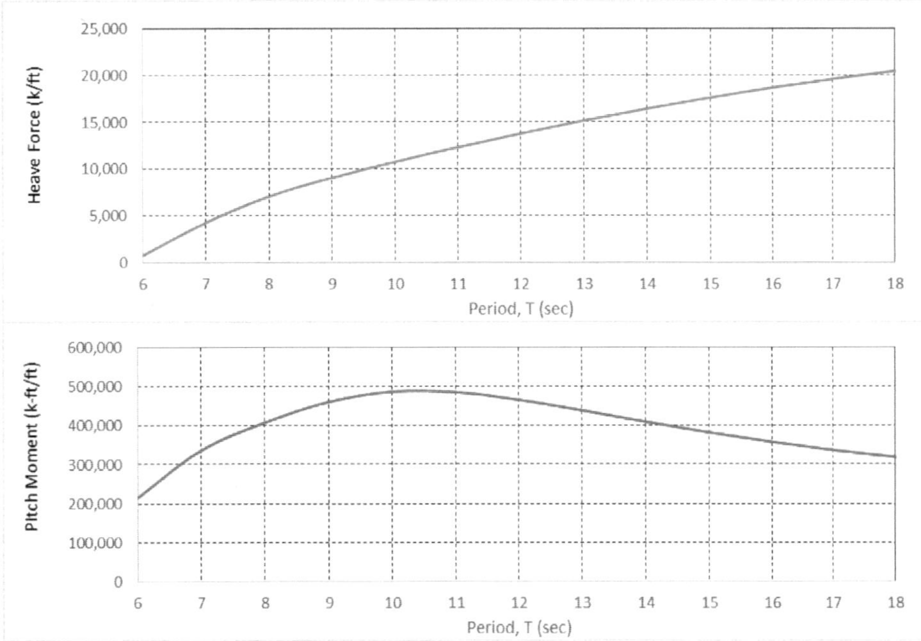

Figure 2 Excitation Load RAOs

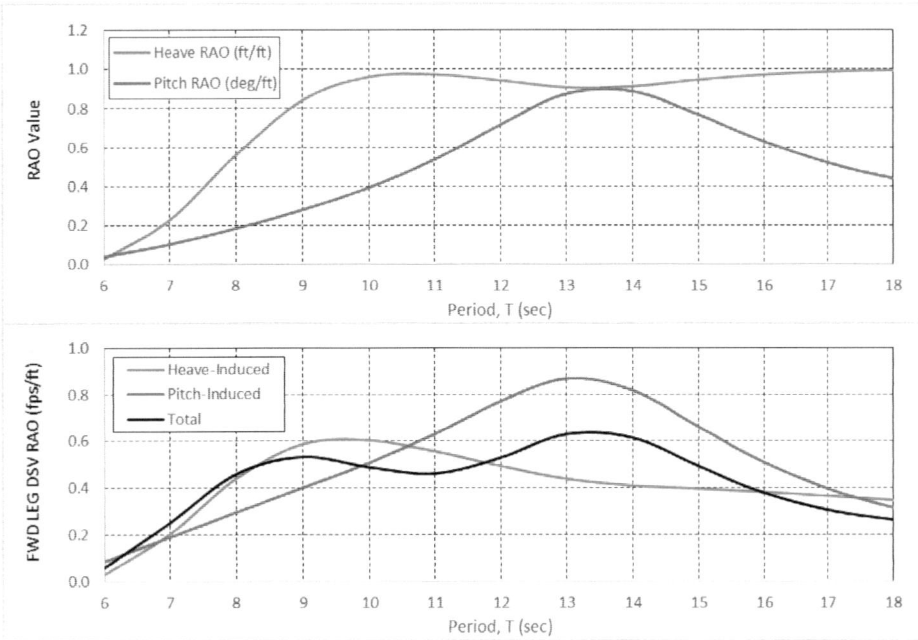

Figure 3 Motion Response RAOs

As can be seen, the natural periods for heave and pitch are 10 and 14 seconds, respectively. Reviewing the forward leg's DSV, it can be seen that for lower periods (6 sec $\leq T \leq$ 8 sec), the total DSV is only slightly larger than the heave-induced DSV. Then, for periods between 8.5 sec and 11.7 sec, as the heave phase angle in the vicinity of the natural period changes, the total DSV is actually lower than the heave-induced DSV. Also noted is that for periods greater than 10 sec, the pitch-induced DSV is greater than the total DSV. The fact that the total DSV is lower than the sum of the pitch-induced and heave-induced DSV is due to the fact that heave and pitch have different phase angles.

PART 1 - ENERGY BALANCE

This part of the study focuses on illustrating the transfer of kinetic and potential energy. For this purpose, four (4) cases are considered, with and without the presence of a seabed. For the cases with a seabed, the seabed is introduced at a time such that the unit has zero pitch so that all spudcans are at the same elevation. Once introduced, the seabed remains stationary. The initial offsets and wave heights are such that the DSV for the forward leg is 2 ft/s at the time the seabed is introduced, for all cases. The four cases considered are as given in Table 4.

Table 4 Cases Considered for Energy Balance Illustration

Load Case	Description
LC1	Heave Decay
LC2	Pitch Decay
LC3	Combined Heave and Pitch Decay
LC4	Response to 12-sec regular waves

All time domain simulations use a time step of 0.01 seconds. The simulations with waves use a ramping segment of 100 seconds. All cases in this part of the study have 15% critical pitch damping. For the calculation of kinetic energy, the added mass values at the natural period are used.

Figure 4 shows the variation of kinetic and potential energy for the cases without a seabed. Figure 5 shows the corresponding energy plots for the cases with a seabed. Figure 6 shows the impact loads for the cases with a seabed. Table 5 shows a summary of the results.

Table 5 Summary of Results from Energy Balance Load Cases Analyzed

Load Case	Time Seabed Introduced (sec)	Maximum Energy (k-ft)	First Impact Load Peak (kips)	Second Impact Load Peak (kips)	Time Interval between First and Second Impact (sec)
LC1	1.92	9,420	7,783 (FWD)	5,863 (FWD)	4.9
LC2	10.68	51,227	9,576 (FWD)	5,668 (AFT)	7.16
LC3	10.69	38,677	9,658 (FWD)	5,468 (AFT)	7.15
LC4	13.99	6,130	9,230 (FWD)	9,569 (FWD)	12.84

It can be observed from the results that the amount of initial potential energy required to achieve the same DSV is quite different for heave and pitch. As expected, the total energy eventually dies-out for load cases 1-3 (due to damping), but that is not the case for load case 4 (due to the continued infusion of energy from the passing wave). For load case 4, the peak energy is seen to eventually become stable a few cycles after introducing the seabed. For this case, the post-seabed energy level is lower than the pre-seabed energy level. It can also be seen that the introduction of the seabed changes the oscillation period of the unit.

From the results of Figure 6, it is evident that the impact loads go from zero to peak in about 0.4 seconds. While the DSV of the forward leg is the same just prior to first impact for all cases, only load cases 2 and 3 have (essentially) the same amount of load on first impact. One more thing that is unique to load case 4 is that the highest load does not take place immediately after the seabed is introduced, but on the second impact. This is due to the significant difference between initial value (decay) problems and forced excitation problems.

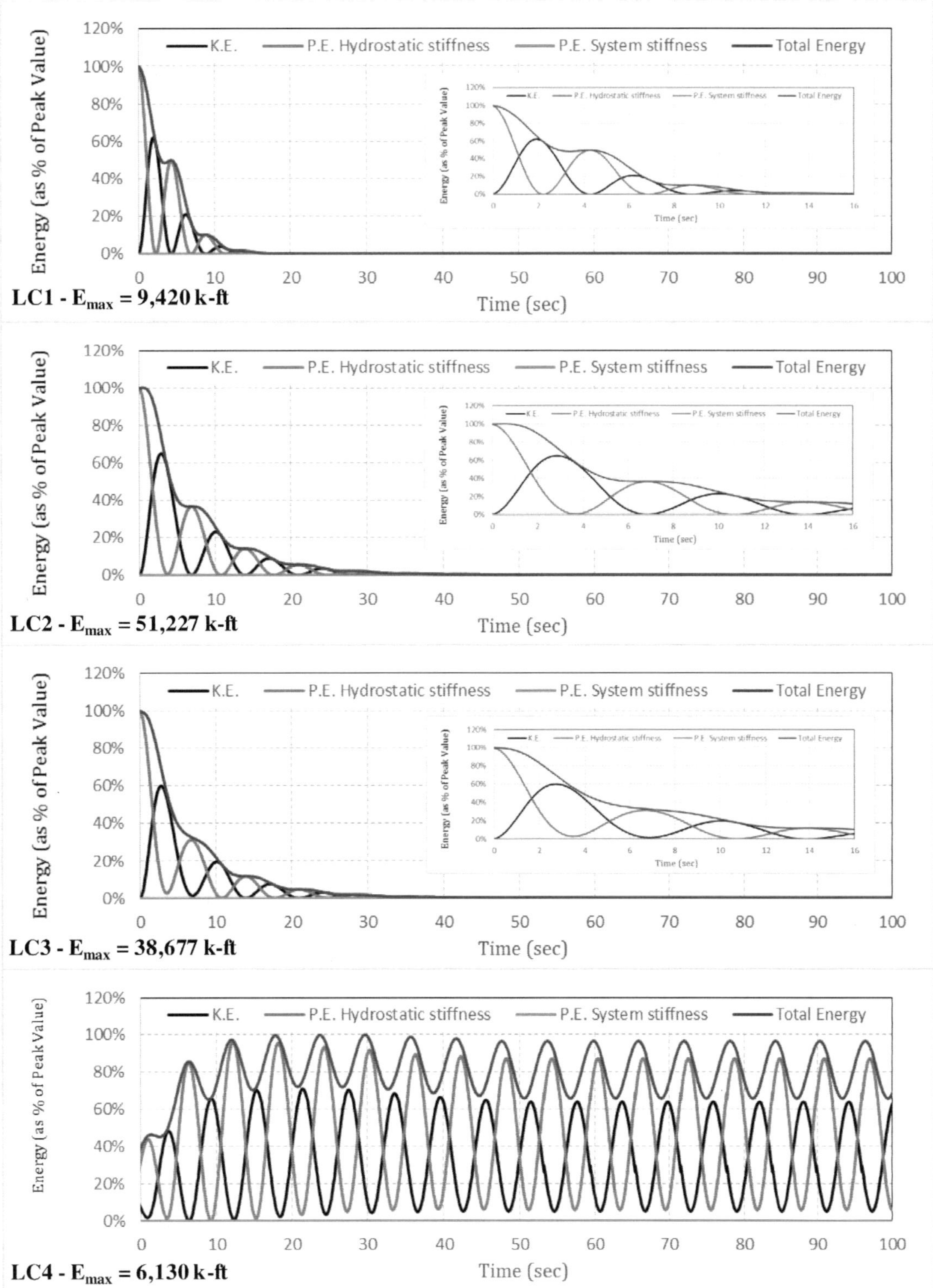

Figure 4 Energy Time-Traces – No Seabed

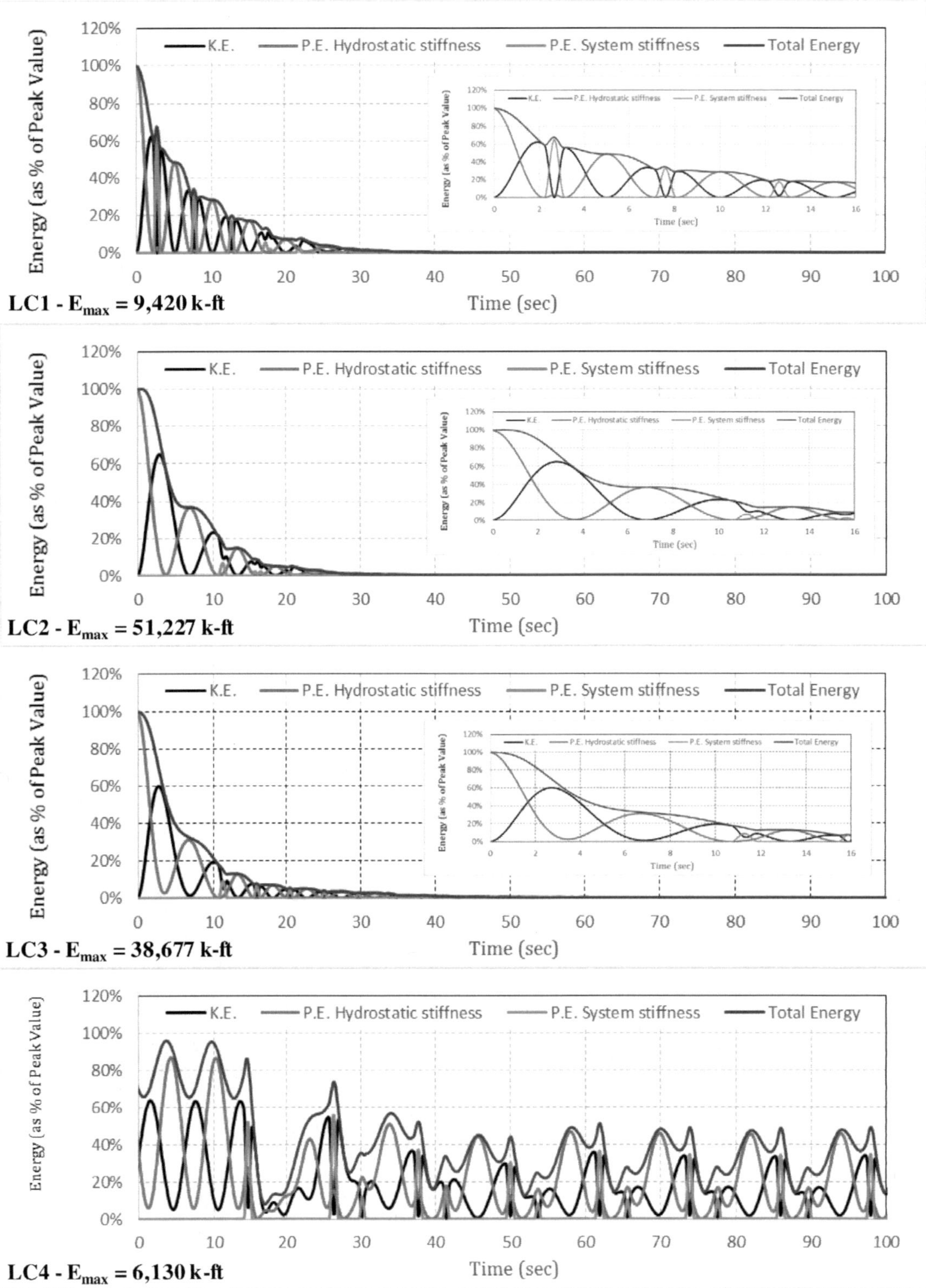

Figure 5 Energy Time-Traces – With Seabed

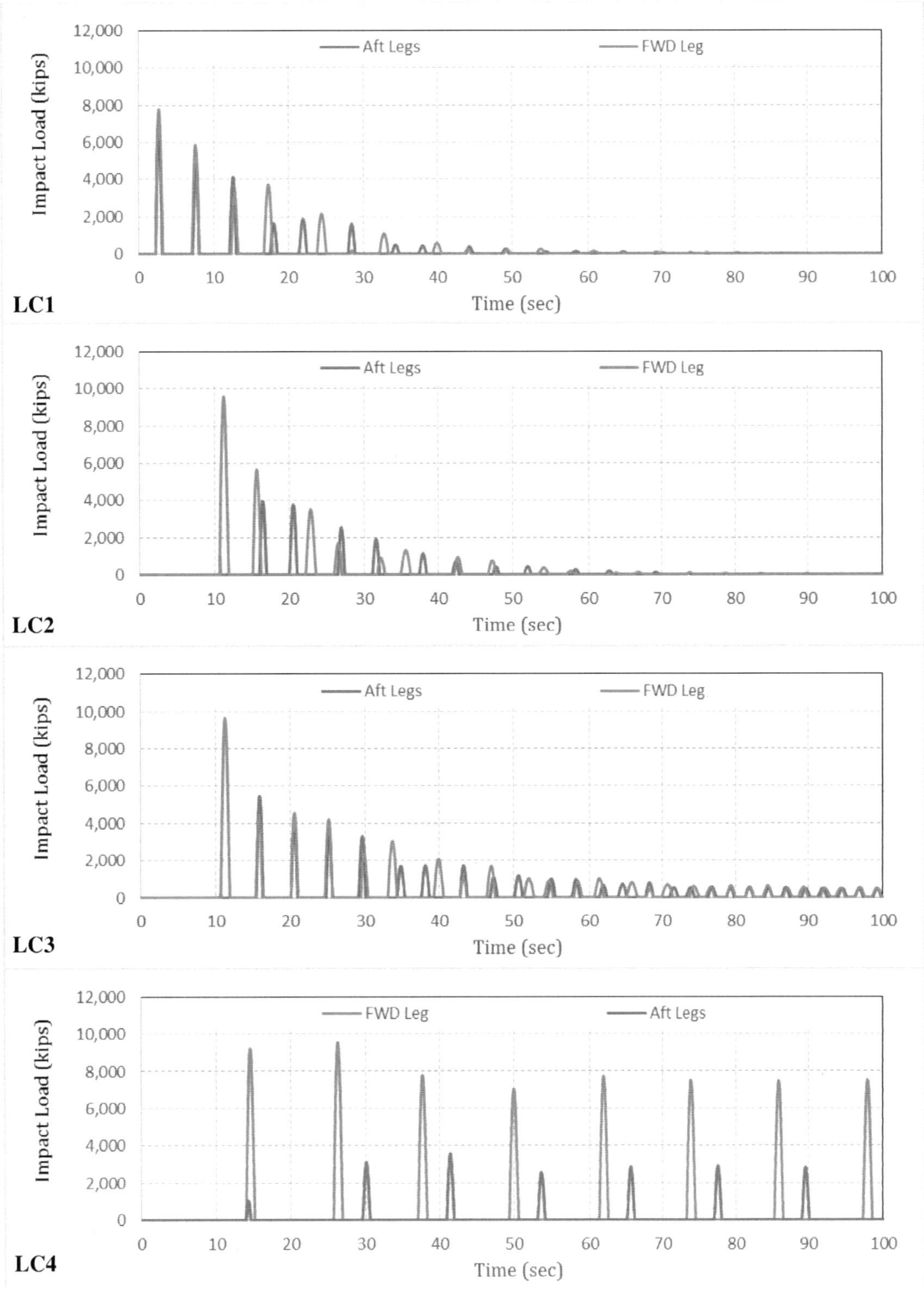

Figure 6 Leg Impact Load Time-Traces

PART 2 –IMPACT LOAD VS VERTICAL SPUDCAN VELOCITY FOR DIFFERENT PERIODS

This part of the study focuses on identifying the peak leg impact load for waves with different periods, but inducing the same free-floating DSV (i.e., peak downward spudcan velocity prior to any contact of the legs with the seabed). Only head seas are considered in the analysis.

Periods from 8 sec to 18 sec are analyzed, in 1-sec intervals. The period range starts before the heave natural period (T_{heave} = 10sec) and ends after the pitch natural period (T_{pitch} = 14sec). While the range of periods could have been expanded, doing so would require the use of disproportionally high wave heights to achieve the target free-floating DSV values. For these simulations, leg jacking is simulated by fictitiously moving the seabed upwards at a speed of 1.5 ft/min.

In order to see if the results are linear with respect to DSV, three speeds are considered (1 ft/s, 2 ft/s and 3 ft/s). For these three cases, 15% external pitch damping value is used. The case with 2 ft/s free-floating DSV is also investigated with 10% and 20% external pitch damping, to assess the effect of the pitch/heave contribution to DSV.

Table 6 provides a summary of the load cases considered, including the wave height values used to achieve the target free-floating DSV values.

Table 6 Cases Considered for Effect of Period on Impact Load Study

	Load Case	LC1	LC2	LC3	LC4	LC5
	Free-Floating DSV (ft/s)	1.0	2.0	3.0	2.0	2.0
	Pitch Damping (%)	15%	15%	15%	10%	20%
Wave Height (ft)	T = 8 sec	3.11	6.22	9.33	6.16	6.30
	T = 9 sec	2.38	4.76	7.14	4.72	4.80
	T = 10 sec	2.19	4.38	6.57	4.36	4.44
	T = 11 sec	2.11	4.22	6.33	4.12	4.34
	T = 12 sec	1.96	3.92	5.88	3.60	4.22
	T = 13 sec	1.85	3.70	5.55	3.02	4.22
	T = 14 sec	1.95	3.90	5.88	3.10	4.52
	T = 15 sec	2.38	4.76	7.11	4.18	5.22
	T = 16 sec	2.97	5.96	8.91	5.80	6.20
	T = 17 sec	3.63	7.26	10.89	7.44	7.26
	T = 18 sec	4.23	8.50	12.75	8.80	8.34

For each period considered, phase angles from 0 to 360 degrees, in 10-degree increments, are investigated to identify the leg impact load. Once the peak impact load and the corresponding time of impact are identified, the peak DSV value is obtained as the maximum DSV immediately prior to the leg making contact with the seabed. Figure 7 shows the peak impact load as a function of both free-floating and peak DSV for the cases with 15% damping (i.e., first three load cases). Figure 8 shows the peak impact load for the cases with 2 ft/s free-floating DSV (i.e., load cases 2, 4 and 5) as a function of peak DSV and as a function of period.

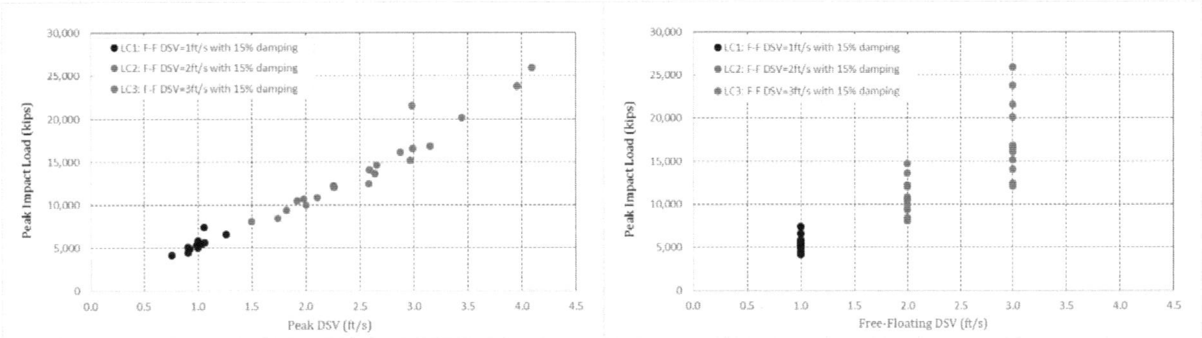

Figure 7 Peak Impact Load vs DSV – 15% Damping

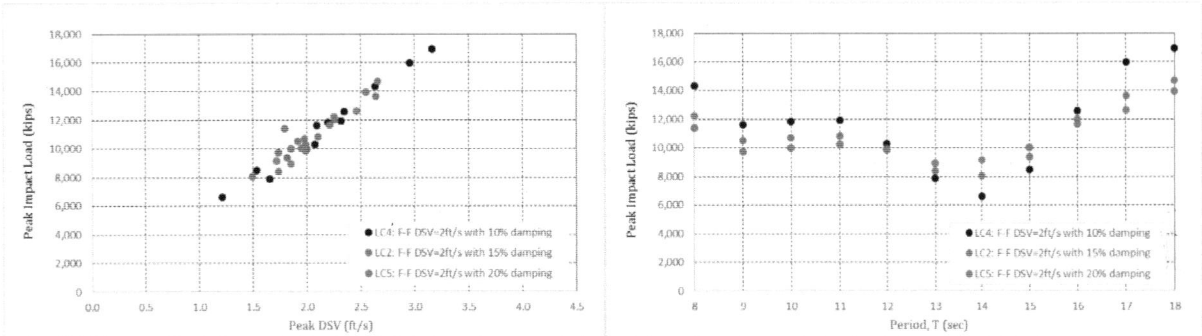

Figure 8 Peak Impact Load vs %Damping – 2 ft/s Free-Floating DSV

Table 6 Impact Load Variation Summary

Load Case	Free-Floating DSV (ft/s)	Pitch Damping (% of Critical)	Peak DSV Range (ft/sec)	Average Peak Impact Load (kips)	Largest Peak Impact Load (kips)	Lowest Peak Impact Load (kips)
LC1	1.0	15%	0.76 - 1.26	5,462	7,410 at T=18 sec	4,140 at T=14 sec
LC2	2.0	15%	1.50 - 2.66	10,945	14,686 at T=18 sec	8,069 at T=14 sec
LC3	3.0	15%	2.26 - 4.09	17,698	25,921 at T=18 sec	12,056 at T=14 sec
LC4	2.0	10%	1.21 - 3.16	11,669	16,924 at T=18 sec	6,611 at T=14 sec
LC5	2.0	20%	1.72 - 2.55	10,682	13,936 at T=18 sec	8,933 at T=13 sec

As is evident from the results presented above, the peak impact load follows a linear trend to the Peak DSV. Also evident is that the peak impact load varies by period, as noted by the separation of the impact load values at different periods for each of the free floating DSV values (1.0 ft/sec, 2.0 ft/sec, and 3.0 ft/sec).

The peak DSV is not equal to the free floating DSV but is instead a range of values centered on the free floating DSV. These values range from 60% up to 158% of the free-floating DSV for the case with the lowest pitch damping. Finally the ratio of the largest peak impact load (for any one phase angle) to the average peak impact load for all phase angles is between 130% and 146%.

CONCLUSIONS

The results of the study show that peak leg impact loads are not only functions of downward spudcan velocity (DSV), but also of the composition of that velocity (i.e., relative contributions from heave and pitch). Furthermore, while the principle of conservation of energy (converting kinetic energy to spring potential energy) is valid in predicting that there is a linear relationship between velocity and impact load, the velocity to be used is not the free-floating DSV, but the peak DSV.

Peak DSV is defined as the maximum downward spudcan velocity immediately before the start of contact between the spudcan and seabed as the impact load builds to the peak value. Figure 9 is included to illustrate the distinction between free-floating DSV and peak DSV.

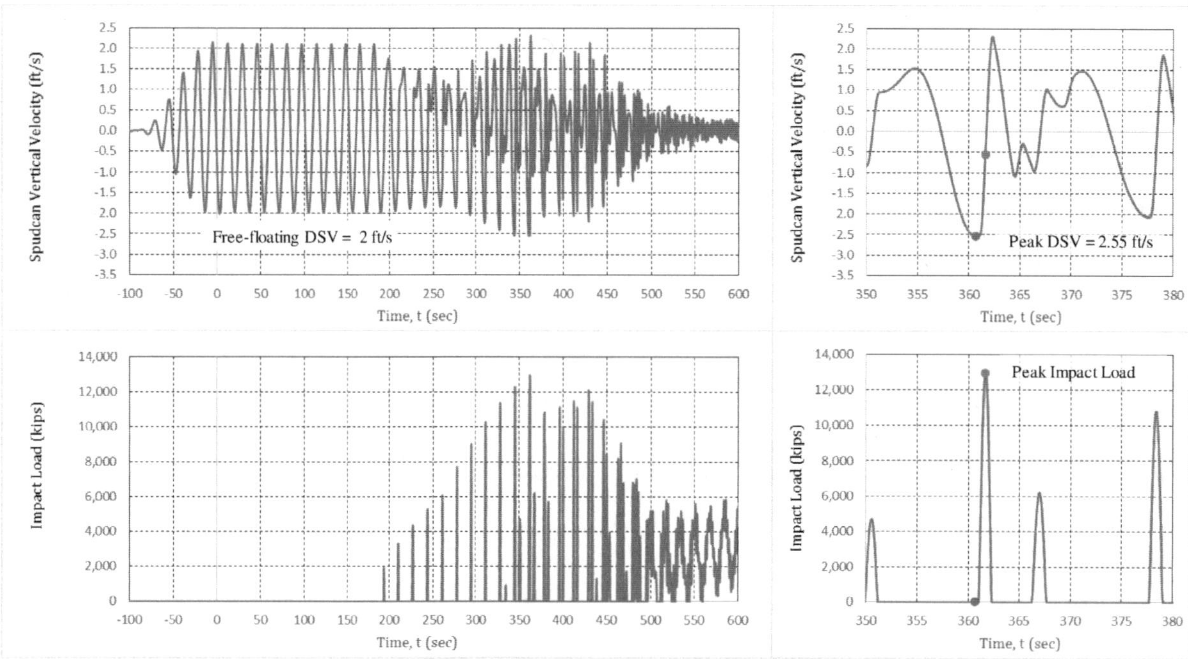

Figure 9 Typical Response Time-Trace for Illustration of Free-Floating vs Peak DSV

The ratio of peak DSV to free-floating DSV was found to change by period, and as such, the peak impact load for a given free-floating DSV varies by period. These results give rise to the concluding statement that when it comes to impact load on jackups going on location, not all velocities are created equal.

ACKNOWLEDGEMENTS

The authors would like to thank the clients and employees of 3Dent Technology who have provided guidance to our understanding of the GoL process. In particular, we thank Richard Roper, Mason Melkowits and Frank Strachan.

REFERENCES

ABS, "Rules for Building and Classing Mobile Offshore Drilling Units," Part 3 Chapter 2 Appendix 2, 2014.

DNV "Classification Notes NO. 31.5 – Strength Analysis of Main Structures of Self-Elevating Units," Section 5.8, 1992.

Chakrabarti P, "Going on Location Study for a Jack-up Rig." In: Proceedings of the 31st International Conference on Ocean, Offshore and Arctic Engineering (OMAE 2012-83034), Rio de Janeiro, Brazil, 2012.

Daun V, Olsson F, "Impact Loads on a Self-elevating Unit during Jacking Operation." Master's Thesis, Department of Shipping and Marine Technology, Chalmers University of Technology, Göteborg, Sweden, 2014.

Kreuzer E, Solowjow E, Qiu G, Hamann T. and Grabe J, "Leg-Seabed Interactions of Jack-up Vessels due to Motions in Irregular Waves," In: Proceedings of the ASME 2014 33rd International Conference on Ocean, Offshore and Arctic Engineering (OMAE2014-24303), San Francisco, CA, USA 2014.

Lai PSK, Lewis, TC, Frieze PA, Miller BL, Smith IAA, "Limiting Motions for Jackup Moving onto Locations." In: Proceedings, The 4th International Conference on the Jack-up Platform Design, Construction & Operation, London, UK, 1993.

Miller BL, Frieze Pa, Lai PSK, Lewis TC, Smith IAA, "Motions and Impact Responses of Jackup Moving onto Location." In: Proceedings of the 25th Offshore Technology Conference (OTC), Houston, Texas, 1993.

ORCAFLEX User Manual, by ORCINA.

SACS User Manual, by Bentley.

Smith IAA, Frieze Pa, Lai PSK, Lewis TC, Miller BL, "Evaluation of Leg Damage Risk for Jackups Going on Location." In: Proceedings of the 26th Offshore Technology Conference (OTC), Houston, Texas, 1994.

Tan P, Chen X, Yu Q, Perry M, Mu H, Chang T, Wong M, Chen, D. "Jackup Going on Location Analysis." In: Proceedings, The 15th International Conference on the Jack-up Platform Design, Construction & Operation, London, UK, 2015.

WAMIT User Manual, by WAMIT INC.

SNAME
THE INTERNATIONAL COMMUNITY
FOR MARITIME AND OCEAN PROFESSIONALS

SUBSEA SYSTEMS RELIABILITY IMPROVEMENTS INSPIRED BY THE AEROSPACE INDUSTRY

OLIVIER BENYESSAAD
OFFSHORE BUSINESS DEVELOPMENT MANAGER, BUREAU VERITAS

PIERRE SECHER
BUSINESS & PROJECT MANAGER ENERGY & INDUSTRY, APSYS

EMMANUEL ARBARETIER
BUSINESS MANAGER AEROSPACE, APSYS

GUSTAVO ROCHA
REGIONAL DIRECTOR OF OPERATION, BUREAU VERITAS

NICOLAS LEGREGEOIS
OFFSHORE GENERAL MANAGER FOR US AND CANADA, BUREAU VERITAS

Proceedings of the 21st Offshore Symposium, February 2016, Houston, Texas
Texas Section of the Society of Naval Architects and Marine Engineers

ABSTRACT

Aerospace industry has been one of the first industries to use and develop reliability techniques before to be followed by numerous industries. The Oil and gas industry has now embraced the idea of reliability analyses for driving design improvements as the potential high costs associated with equipment failures lead to the desire of a highly reliable system and/or a reduction in system uncertainty. Spread of ultra-deep water and developments of recent Subsea processing technologies of this last decade induce the need of state of the art Subsea components as a failure below 1,500m water depth becomes a real challenge, not only in terms of cost but also in terms of environmental safety and reputation for the operating company.

Experience in the highly regulated and disciplined Aerospace business, has shown that the development of system engineering techniques and software tools benefit from finding applications across industries.

The paper will first introduce the main challenges faced by both the Subsea and the Aerospace industries.

The purpose is then to show ways in which safety and reliability techniques from Aerospace already proven across many industries, can add value for the oil and gas organizations.

This paper will also discuss the solutions found for other extreme industries that have relevance to the Subsea application. We will conclude by suggesting ways in which industries can collaborate in order to support and enhance a business model that is not always optimized.

Keywords: SNAME, Reliability, Aerospace, Offshore, Subsea, Symposium

INTRODUCTION

Design for Reliability (DfR) techniques have their origins in the military sector involving the V-1 missile team during World War II (MIL-HDBK-338B, 1998). It was here that it became first understood that an improvement in the reliability of individual components and reconfiguring the system to improve reliability resulted in an increased likelihood of success. Reliability techniques were then quickly adopted in industries where understanding of and improvements in the reliability of complex systems were required, namely by the National Aeronautics and Space Administration (NASA) and the nuclear power industry to prevent such accidents as experienced at a nuclear power plant at Three Mile Island (USA, March 28, 1979) and the explosion of the space shuttle Challenger (USA, January 28, 1986).

Globally, the Oil and Gas Industry has now embraced the idea of reliability approaches such as Reliability, Availability & Maintainability (RAM) and more generally, systems engineering, to drive design improvements. However, it seems that even if Subsea industry is at the edge of reliability-based design, many experts agree on the fact that cross knowledge exercise between other high-tech industries could bring additional knowledge & experience and consolidate current Subsea business model. Over design or unreliable components are not acceptable anymore at less than $30 per barrel. The following will then introduce the main Subsea & Aerospace reliability principles in order to identify the main theoretical and practical gaps. Gaps have been analyzed during dedicated workshops sessions in order to conclude on ways of valuable cooperation between the two industries.

METHODOLOGY

Methodology used to develop the gap analysis between Subsea & aero projects development has been based on the extraction of the most relevant information regarding the best practices as stated in the international standards for instance as well as by including experience of the authors in their respective fields of activity to balance these best practices with their practical applications within the industry. Following review of standards and best practices within Subsea & Aerospace industries, several workshops took place between Bureau Veritas & APSYS companies in order to compare the two approaches and use only the most relevant criteria to discuss the improvement of Subsea development reliability.

1. Definitions and Terminology

Before discussing main reliability & safety issues faced by Subsea & Aerospace industries, it is important to first establish some important definitions and terms.

Subsea Production System (SPS)

A Subsea production system consists of a Subsea completed well, seabed wellhead, Subsea production tree, Subsea tie-in to flowline system, and Subsea equipment including manifold, boosting systems or separation systems as well as control facilities and related power distribution system to operate equipment.

Reliability

The ability of an item to perform a required function, under given conditions of production, environment, and usage, for a required time interval (API RP 17N, March 2009).

Risk

The potential for the realization of unwanted, negative consequences of an event. Risk can be measured by the product of the probability of the event occurrence and the consequence of the event. (API RP 17N, March 2009).

Maintainability

Ability of an item under given conditions of use, to be retained in, or restored to, a state in which it can perform a required function, when maintenance is performed under given conditions and using stated procedures and resources (API RP 17N, March 2009).

Availability

The ability of an item to be in a state to perform a required function under given conditions at a given instant of time, or in average over a given time interval, assuming that the required external resources are provided (API RP 17N, March 2009).

For a repairable system, commonly used to model systems in the Oil and Gas Industry, the inherent availability can be expressed by Eq. 1 (Smith, 1997).

$$A = \frac{\text{MTBF}}{\text{MTBF+MTTR}} \quad (1)$$

Where the MTBF = Mean Time to Failure (MTTF) + MTTR. However, normally the MTTF is much greater than the MTTR and thus MTBF $\approx$ MTTF.

Production Availability

Production Availability is a measure of the actual performance of a production system accounting for production loss due to planned an unplanned outages against the potential production of the system given no outages have occurred within a given system life.

Thus, Production Availability can be expressed as the percentage:

$$\frac{\text{(Actual production allowing for Planned and unplanned downtime)}}{\text{(Potential production with no downtime)}} \times 100 \quad (2)$$

Bath Tub Curve

The failure rate $\lambda(t)$ is a function that describes the number of failures that can be expected to take place over a given period of time. To describe the failure rate, the bath tub curve is used commonly in reliability engineering. The failures exhibited in the first part of the curve, representing a decreasing failure rate, are termed early failures or infant mortality failures. The middle section, representing a constant failure rate, relates to failures that occur randomly during the normal life of the component. The final section, representing an increasing failure rate, is termed wearout failures as increased failures occur towards the end of the components useful life.

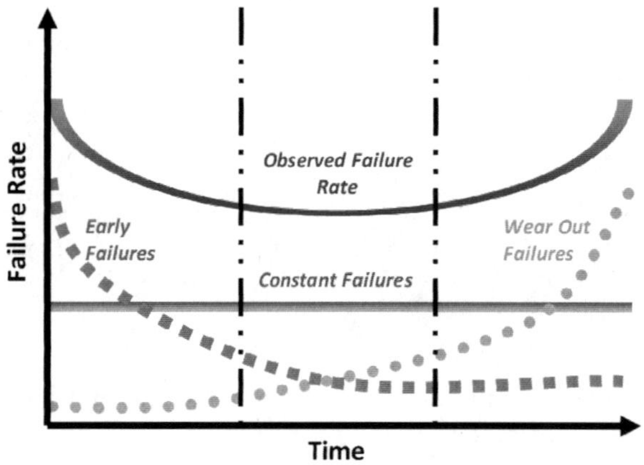

Figure 1 Bathtub Curve (Amstadter, 1977)

Depending on the failure data available, reliability analyses can assume a variety of distributions to describe the system more accurately. Commonly, a constant component failure rate is assumed when maintenance and operations records are not available. Reliability can then be represented by Eq. 3 (Smith, 1997).

$$R = e^{-\lambda t} \quad (3)$$

Where R is the reliability, λ is the failure rate and t represents time.

2. Main challenges faced by Subsea & Aerospace Industries

Even if they are operating in different environments, Subsea and Aerospace systems face some similar challenges.

The first common challenge is related to the increasing complexity of sophisticated systems integration. Automated systems play a key role on operation and safety and software are more and more integrated into these complex systems as aircraft cockpit or Subsea control modules.

A second challenge is the ability to build systems with long life duration. This life duration at some point also needs to be extended for economical reasons. Subsea equipment or aircrafts will remain in activity for decades. An obsolescence management has then to be deployed as an integral part of design, development, production and in-service support in order to minimize the financial and availability impact throughout the product life cycle. Typically, electronic components have been among the most complicated elements to deal with for many industries such as aeronautic but also military industry.

A third common challenge is the environmental conditions in which the Subsea and Aerospace systems are operating. The harsh conditions (mainly temperature & pressure) which are characteristic of deep water or space environments enforce the need of very robust systems.

A forth one would be the condition monitoring of equipment and more precisely Real Time Data Acquisition that is already in place in the aeronautic industry but is still a new field of development within the subsea industry.

Another key challenge linked to the environment is the need to operate in autonomy. Subsea and space systems are operating remotely and reaching high level of reliability is key for their viability. Part of this reliability challenge for the Aerospace industry is the ability for a remote system to reconfigure automatically after a failure. In the aeronautic industry, high reliability targets are provided by the regulation given the safety issues related to people transportation. To be sustainable, there is now no room for over design but only for cost effective & safe design. For both sectors, the need of operating in a sustainable and profitable business model is driven by an optimum design development. The high cost of equipment failure both in terms of intervention and delayed revenue stream from production / operation constraints means that Subsea or space equipment reliability is a paramount aspect. In order to reach high level of reliability in a competitive environment, the aircraft industry has embraced availability issues of the whole supply chain and from design phase to end of life into its business model.

All these common challenges lead then to understand how reliability, availability & safety need to be integrated in Subsea and Aerospace systems engineering to drive design improvements.

Based on these observations and on the recent fall in oil & gas prices, it seems that the offshore industry is ready to move forward with these analogies and use the knowledge developed for aeronautic and some other high tech industries to ensure a more sustainable and profitable business model. As an example, it can be mentioned that the Brazilian operating company Petrobras decided to sign in 2014 a memorandum of understanding (MOU) with the Brazilian aircraft manufacturer Embraer. Responding to the Macondo disaster, Petrobras proposed to adapt the best practices issued from the aeronautic industry to apply it on the design of BOP system.

Another illustration demonstrating the interest of the industry is the "Deep Space, Deep Ocean Forum" held in 2015 in Houston, USA. The first edition of this conference brought more than 500 persons from Aerospace industry such as the NASA as well as Oil & Gas companies like ARAMCO. Conference aimed at demonstrating the different synergies and axes of collaboration to be developed in order to benefit from the cross-experience of both sectors. Slightly before these various initiatives, in 2013, the French testing, inspection and certification company Bureau Veritas signed a MOU with the Airbus group company APSYS in order to investigate the applicability of aeronautic and Aerospace methodology to the offshore industry and more precisely, to the Subsea industry. In June 2015, the MOU has become a cooperation agreement in order to formalize the close relationship and reflection works achieved during about one year and a half.

One of the works performed by the two partners has been linked to the comparison of best practices and standards of both industries and is then partially presented within this paper.

3. Subsea Reliability & Safety Approach

Subsea reliability & safety concerns related to Subsea production systems have been covered by two major sets of international standards: series of API RP 17 and ISO 13 628. These two sets of standards fixed the basis of SPS reliability & safety design both in terms of risk management activities (API RP 17N, March 2009) as well as in terms of guidance for SPS design integrating straight from the beginning all the phases of the project from concept to decommissioning (ISO 13628-1, 2010 & API RP 17A, 2006). It is now well established that reliable design is not only a matter of technical documentations and adequate tests but highly depends on the organizational activities of the project itself. The following sections will summarize the key information presented within the international standards before discussing their practical applicability. It will then be possible to highlight the most relevant reliability concepts to be applied on the whole supply chain and for any step of the life of a Subsea development.

Subsea Development Project Reliability & Technical Risk Management based on API RP 17N

Basically, API RP 17N is based on the same twelve "key" reliability processes as the ones defined by ISO 20815 for production assurance and reliability management (cf Figure 2). However, when ISO 20815 has been developed to cover any assets within the oil & gas industry, API RP 17N has been made specific to cover the SPS Risk & Reliability Management activities. Thus, main trail to ensure that Subsea development projects are performed through a reliable design process is based on the "Define, Plan, Implement & Feedback" loop (DPIF Loop).

First of all, goal, strategy and requirements in terms of risk, RAM and safety are allocated for the whole project. Following this first definition step, the project team is able to develop the appropriate Reliability & Technical Risk Management Plan covering resources, roles & responsibilities, deliverables for each activities and schedule & milestones. Then, during the implementation phase of the defined plan, each project step needs to be checked versus the previously defined performance requirements. The feedback step establishes whether the project stage reliability goals and requirements have or have not been met and includes the success to which the required qualification activities have been implemented. Feedback from operations to projects is also extremely important since lessons learnt from operations regarding reliability & maintainability performance of equipment should always be included as an input to new projects. Figure 2 shows how DPIF loop is applied through all development steps of a project in interaction with the Key Reliability Processes.

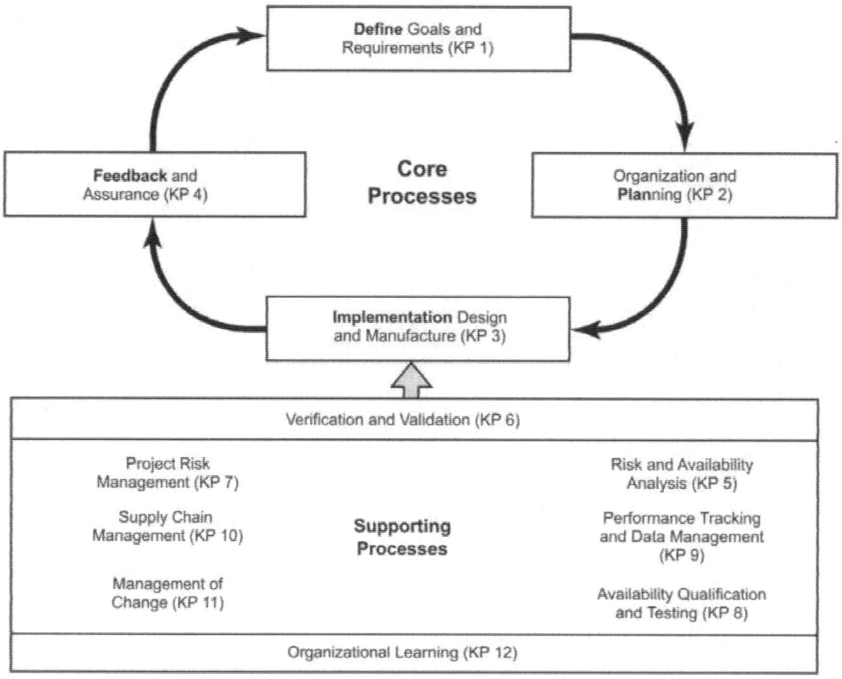

Figure 2 Key Process Interaction with DPIF Loop (API RP 17N, March 2009)

API RP 17N does not give recommendations on the design itself but aims at providing guidance to ensure reliability is considered at each stage of the project. It will help engineering teams to focus on the necessary steps for verifying but also validating the design versus the key indicators defined at the earliest stage of the project. It is worth insisting on the fact that API RP 17N does not cover only technical risks but also any economical & organizational risks a Subsea project can encounter. In this context, Key Reliability Processes are considered as the supporting environment for the reliability activities.

SPS Design & Operation based on ISO 13 628-1

ISO 13 628-1 has been developed to provide general requirements regarding SPS design with regards to all phases of development (design, material selection, testing, manufacturing, operation and decommissioning) and including systems interfaces. ISO 13 628-1 provides a very large, but consistent, definition of SPS as presented in Figure 3.

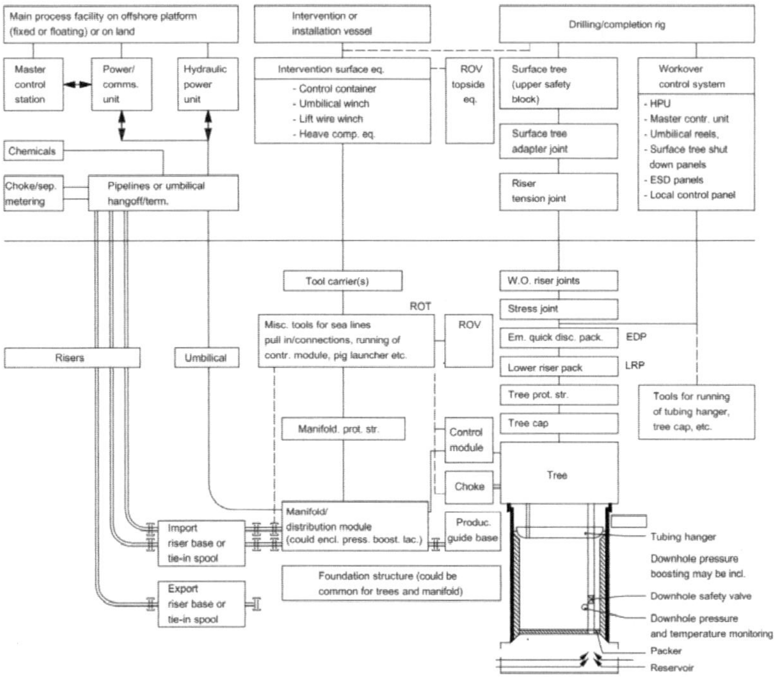

Figure 3 Typical Elements in a SPS (ISO 13 628-1, 2010)

Design recommendations given by ISO 13 628-1 cover design criteria, design loads, material selection, corrosion protection, intervention system design, manufacturing and testing procedures as well as transport, commissioning, operation and decommissioning typical procedures. As stated in most existing standards (ISO, API, NORSOK, etc.), Subsea production systems have to be designed to optimize life-cycle benefit while meeting functional and safety requirements. Thus, system availability needs to be in the central part of the design process straight from beginning. Design strategy should ensure both, reliability of equipment selected (i.e.: inherent availability) by the project team and maximum availability based on the most simple and optimum design and arrangement (i.e.: production availability).

As part of the overall SPS design and in order to ensure safety requirements are met, ISO 13 628-1 recommends to develop a comprehensive barrier philosophy. The barrier philosophy should provide clear and concise guidance on barrier requirements, with the objective of preventing unintentional release of produced/injected fluids which may harm personnel and/or the environment. Given the wide variety of possible field characteristics and equipment configurations, as well as the varying requirements of existing local regulations combined with field operator preferences, standard barrier philosophy is not defined in any standards. However, development of both generic and case-specific barrier philosophies is done through the use of experienced personnel and typically involves the use of risk assessment techniques such as HAZOP, FMEA, QRA, job safety analysis and/or scenario based risk assessment. Where the Safety Case is required, most of these data will be reviewed by the competent authority.

Design Phase

When it comes to a full Subsea field development, design and engineering is generally made system per system (i.e. production / injection wells, separation systems, boosting systems) or even component per component (pumps, manifolds, flowlines). A recent trend within Subsea EPCIs is to join their forces in order to propose pre-FEED, design and engineering of a full Subsea architecture, applying then all the best practices on a full development and not just on a part by part basis.

The practical approach of designing an entire system from reservoir to topside facilities and considering from concept phase, all the requirements and organizational issues of all the phases of the project is called System Engineering (ISO 13 628-1, 2010) and is also often considered as a Life Of Field (LoF) approach.

System engineering discipline activities for Subsea are broadly classified into three primary service areas (Yong Bai & Qiang Bai, 2013):

- Production system design;
- System integration (Layout, Specification, P&IDs, Risk, Reliability & Safety, Economics, etc.);
- Equipment application and development (Power Generation, Boosting, Separation, etc.).

Practically, most of the Subsea designers / EPCs are more and more implementing System Engineering approach to develop safe and reliable Subsea design but also to limit interfaces and installation issues and related additional costs during operation. It is now commonly admitted that gain for industry would be considerable if barriers between the various actors in the Subsea manufacturing / designing industry could fall to let the room to a wider cooperation and coverage of interfaces during design.

A discipline that should benefit even more than the others from System Engineering approach is the flow assurance discipline. Flow assurance issues are defined by ISO 13 628-1 as any "issues that can impact the economic recovery of hydrocarbons in Subsea production systems involving multiphase flow". A reliable and safe Subsea system will then have to be designed with respect to hydrates, wax and asphaltene formations, emulsion or/and sand and water production. Chemical composition of the fluid to be produced is then a real concern as it is one of the driven factors for design. Similarly, as fluid is passing through all process equipment from wells to process topsides / onshore process, a complete understanding of the system is required and it is where System Engineering is intended to bring value.

New Technologies Development

When using technologies already well proven, benefits of a system engineering approach could be limited as return on experience remains the best way to adapt a design even if it can be very costly, especially in case of major accident. When it comes to new technologies or use of an existing system within a new environment, some additional steps are required. Technology Readiness Levels (TRLs) have then been defined within the API RP 17N in order to provide guidance to the step by step approach to be followed with a view to completing the qualification process. Practically, it means that design process and reliability management activities need to be adapted depending on the TRL of the project. The DPIF loop, the KPs definitions as well as the design effort are then defined / detailed differently depending whether the technology is new or not.

In-Service Phase: Data collection, Integrity & Feedback from Operation

It is now among the best Subsea industry practices to ensure equipment reliability data are properly collected during both engineering phases (e.g.: FAT) and operations (e.g.: failure rate, repair time). With respect to this, ISO 14224 has been developed to define the minimum requirements of data to be collected. Practically, only few operators have been able to build their own equipment reliability database and the lack of operational information pushes the whole industry to look at generic reliability database for Oil & Gas equipment such as Offshore REliability DAtabase (OREDA, 2009). Even if well documented, OREDA is limited to anonymous types of equipment, used and maintained differently by various North Sea operating companies and with sometimes a very limited amount of samples or with a large number of failures qualified as "unknown". Most of the Majors have now developed internal specifications regarding capitalization of return on experience but it seems that experience of the project teams or of the certifying authorities remains the best way for a Subsea project to achieve acceptable safety & reliability levels.

Integrity strategy highly relies on the monitoring & inspection systems associated with the Subsea development. Temperature, pressure, flowrate are parameters monitored continuously by sensors. Other parameters such as corrosion and displacement are recorded on a regular basis through ROV, AUV or intelligent pigging system that give key information regarding equipment and system behavior and health. In order to integrate return on experience, a risk based approach is used straight from design phase to (i) improve reliability and maintainability of sensitive items, (ii) ensure that the correct parameters are measured by the sensors and (iii) concentrate inspection efforts on the most relevant equipment items.

4. Aerospace reliability & safety approach

The reliability & safety processes have been defined to support the demonstration of compliance with certification requirements in the context of more complex and integrated system development.

The SAE Aerospace Recommended Practice 4754A provides guidelines for the development of aircraft systems taking into account the overall aircraft operating environment and functions. This document includes validation of requirements and verification of the design implementation for certification and product assurance. It provides practices for showing compliance with the regulations including safety and reliability requirements.

Safety activies are conducted all along the development process, to produce safety requirements from aircraft to item level that will be considered for the development of aircraft level function. The bottom up safety requirement verification process ensures then that system implementation satisfies the validated requirements. The Figure 4 illustrates the different requirement levels.

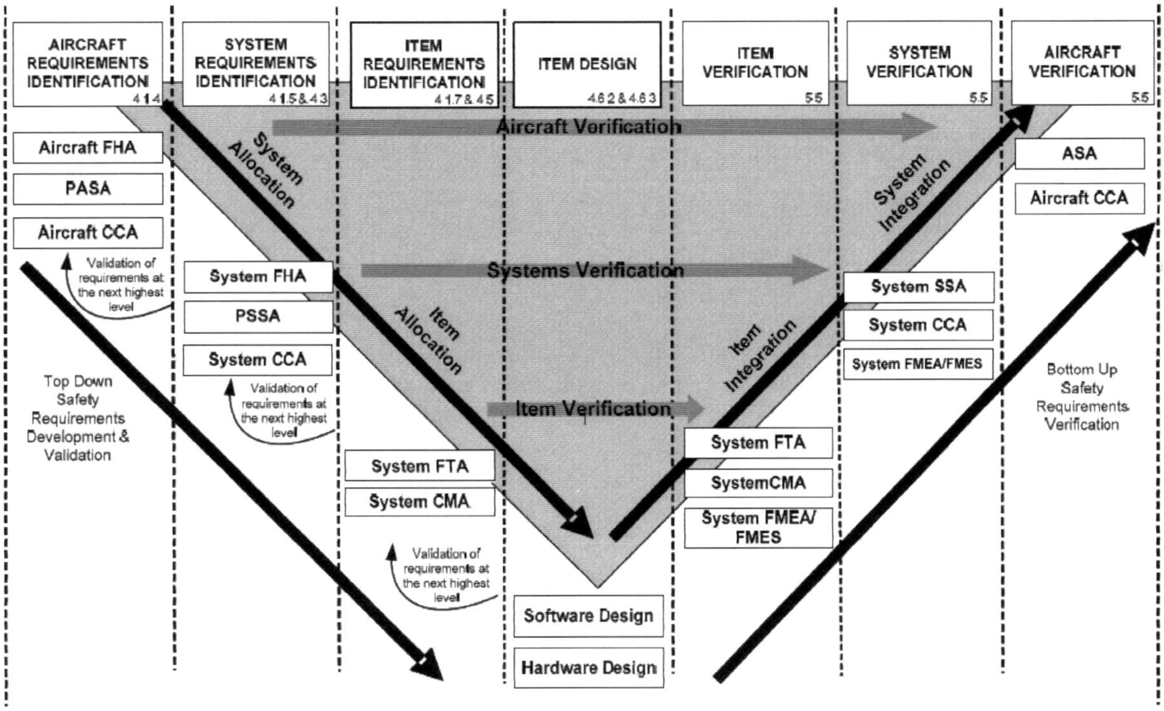

Figure 4 Interaction between Safety & Development Processes

The SAE Aerospace Recommended Practice 4761 provides guidelines for conducting safety assessment process introduced by the ARP4754. Method and results for main ones are described below:

- The Functional Hazard Assessment (aircraft and system FHA) consists of the examination of functions to identify potential functional failures and their effect on the aircraft, the flight crew and the occupants. This process allows the identification of Failure Conditions with their severity level that will be derived into qualitative (independence, Function Development Assurance Level) and quantitative (probability) safety requirements for the design.

- The Preliminary Safety Assessment (PASA/PSSA) is an evaluation of predicted design solution. This process consists of the determination of basic events that can result as single failure or in combination to Failure conditions identified through the FHA process. The evaluation of Failure Condition provides a preliminary indication that the predicted solution can reasonably satisfy the safety requirements allocated to system. This process outputs qualitative and quantitative safety requirements for equipment and interfaces design.

These both activities are updated all along the development process untimely resulting in the Safety Assessment (ASA/SSA). The purpose of this process is to collect, analyze and document verification that aircraft and systems as implemented meet the safety requirements. Part of verification is the consideration of results from (non-comprehensive list):

- Safety activities conducted by system /equipment suppliers (FMEA/FMES) and interfaced systems (SSA)
- Common Cause Analysis
- Human error analysis
- Performances, handling qualities, structure studies
- Maintenance program.

Deviation from validated requirements must be identified and analyzed. When a deviation cannot be accepted due to its effect on aircraft safety, appropriate limitation must be defined to restore an acceptable level of safety.

The Common Cause Analysis (CCA) establishes and verifies that independence requirements identified through FHA and PASA/PSSA processes are satisfied. This includes:

- Particular Risks Analysis, which addresses common cause in relation with events
- Common Mode Analysis, which addresses common cause in relation with the development, manufacturing, installation, maintenance and crew errors
- Zonal Safety Analysis, which addresses common cause in relation with installation

The Development Assurance Level is a measure of rigor applied to the development process to limit, to an acceptable level for the safety, the likelihood of errors occurring during the development process of functions and items. The development Assurance Level is assigned depending on the severity classification of Failure conditions considering the possible independence between development processes that can limit the consequences of development errors. The SAE ARP 4754A provides rules to assign a DAL to functions (FDAL) and items (IDAL), IDAL is then used as an input for the application of applicable standards which are the DO-254 for Electronic Hardware development life-cycle and DO-178 for Software development life-cycle.

Added values of Model Based Safety Analysis on multi systems

The use of Model Based Safety Analysis (MBSA) is particularly beneficial when it comes to the design of complex systems. In Aeronautics, system architectures have reached high complexity level due to large use of shared resources. In "traditional" avionics, one computer (LRU) was dedicated to one function, and the loss of one function involved the loss of items that belong to this function. With recent architectures based on shared resources, loss of one component has consequences on several functions, and indirect effects are sometimes difficult to predict. A cascading and multi effect propagation analysis is needed to project all the consequences at the different levels of the system.

The principle of Model Based Safety Analysis is to extend existing model-based development activities and tools to support safety analysis. It consists of the modeling of the system including mechanical components and software components and it also considers the environment in which the system runs.

The main advantage is that MBSA utilization provides a common reference assuring a more coherent, integrated and cooperative work between safety analysts and designers. This common model ensures that safety analysis outputs are relevant and up-to-date as the system architecture evolves. Additionally, MBSA provides a functional and dysfunctional simulation framework for one system and for multiple interconnected systems and supports exploration of various architectures and design choices by automatically selecting the option satisfying critical safety targets. The use of MBSA computational tools gives the possibility to automate many safety analysis activities, as well as other types of RAMS analysis (e.g.: operational reliability analysis). It enables safety engineer to focus on safety assessments instead of manually handling combinatorial issues. This helps to reduce error risks in the safety assessment regarding missing failure combinations

New Technologies Development

This safety process is generally applied for any new aircraft development or significant change on existing aircraft. The identification of Failure Conditions which result from the assessment of possible failure modes (loss of function, spurious or erratic functioning) of functions and their consequences on the aircraft, is independent from the type of technology used for the implementation of these functions.

However the introduction of new technologies may require additional activities to minimize the risk of failure or error during the development that can affect the aircraft safety. The safety assessment of function or system using new technologies should describe precautions that have been taken to support the demonstration of compliance with safety objectives. Some examples of safety precautions are

- Use of conventional technology as back-up or secondary part for the implementation of critical function
- Dissimilarity to minimize the consequences of development error on the system / aircraft
- Robustness tests to justify reliability figures considered for Failure Condition evaluation
- Specific activities to evaluate potential intrinsic and environmental risk associated with the technology used

In-Service Phase: Data collection, Integrity & Feedback from Operation

Data collected by the aircraft manufacturer is considered at different steps of the safety process. For FHA activity, in service experience is used to confirm the consequences of a failure or a severity level. The consideration of in service experience may also lead to define new hazard not yet identified through the functional failure analysis. For PSSA/SSA, data collected from operation may affect dormancy associated with latent failures such as MTBF or time between two consecutive A/C power up.

In service experience is also considered to improve the development process, the system design & integration (redundancy, monitoring, human machine interface, and installation) and/or technology selection, the procedures…

Data collected by the supplier are mainly used to consolidate reliability figures that will be considered for reliability and safety analysis for new development.

RESULTS

One of the main results of the gap analysis performed is an astonishing number of similarities between standards requirements of both sectors even if level of guidance is often disparate and approaches proposed may be very high level when it comes to Subsea (e.g.: systems engineering approach). The biggest disparities appear with the practical application of the standards and related design approach. This results section will discuss some of them:

- First of all, one of the main differences of language between the Subsea and the aeronautic industries is the notion of "safety". When the aeronautic industry is discussing safety, it has to be understood as "safety & reliability". Indeed, for oil & gas industry, "reliability" in terms of component reliability or process availability and "safety" in terms of human or environmental safety are two different issues that are often treated separately. The failure of an airplane turbine and that of a Subsea compressor can have the same consequence on the performance (i.e. stop of the main function which is "fly" or "produce") but a totally different one on the safety side (i.e. hundreds of life lost for the airplane case vs loss of money for the production installation). Collaboration between aeronautic and Subsea industries would then require the use of the same language to avoid misinterpretation.
- The most significant discrepancy that popped up during the various workshops held between the authors was related to the practical design approach. When Aeronautic design is driven by well-defined standards developed on a global consensus and a model based approach, Subsea approach remains mainly empirical and very conservative. Even if API RP 17N proposes a similar approach to the SAE ARP 4754A, it seems that practical design methods for the Subsea are often much more complicated to put in place and less effective. The main reasons found are the lack of information regarding the component to be selected, the lack of tools to track and assess the objectives allocated and the different use, within the oil & gas industry, of the Systems Engineering method. Development of collaborative tools and methods as well as implementation of a proper organization to support the system engineering activities would improve considerably field development architectures as well as assessment of variation from a base case design.
- As Subsea manufacturers are often different when it comes to equipment such as processing systems, control systems or flowlines, integration of all the components is a real challenge and systems engineering approach is often either limited to an approach at very high level, i.e. covering the full Subsea development but with a little amount of information on the components (e.g.: during concept phase), either on a very limited amount of components i.e. covering a set of equipment proposed by one manufacturer (ex: flowlines, risers and umbilical). Lifecycle cost is then not under control as too many parameters, especially related to the interfaces and installations issues, are unknown. As it is the case within the aeronautic industry, all Subsea package / system suppliers would benefit to be involved through standardized and centralized tools and models to define main input data related to their respective scope of work and provide the project with all the necessary proof to demonstrate their equipment meets the corresponding input value. The operator or an independent third party would then be able to validate all the input data, run independent model calculations and perform a complete verification to ensure there is no lack of consistency or missing information at the interfaces. The models generated would be used at the design phase but also could be updated during operation to apply lesson learnt, to challenge manufacturer input data and then to upgrade future designs.
- Classic engineering approach includes a large number of models: physical analogs, analytic equations, state machines, block diagrams, functional flow diagrams, object-oriented models and computer simulations. Systems Engineering should support the design of a product as well as all the processes to produce it. Then, models should be constructed for both the product and the process. Process models allow, for example, scheduling changes to be studied, dynamic PERT / GANTT charts to be created and sensitivity analyses to show the effects of delaying or accelerating certain subprojects to be performed. Running the process models reveals bottlenecks and fragmented activities, reduces cost and exposes duplication of effort. These models are also used in tradeoff studies and risk management. Systems Engineering Process is not sequential: it is parallel and iterative. However, if Subsea industry is using most of the modeling tools, integration and interaction between all of them is not always properly understood as communication between the various actors is too limited. A more structured approach between the various disciplines involved in a Subsea development based on methods developed for Aeronautic projects would speed up the design process and provide with a better cost

control strategy for the whole lifecycle.

- In Aeronautic, RAM analysis is placed in the center of the decision making strategy. It becomes now more and more common for oil & gas to consider both RAM & CAPEX definition at the same level of importance but it seems that the lack of reliability data and appropriated tools are often a non-negligible limit as confidence in a project depends highly on these inputs. When Aerospace can validate their RAM studies based on a standard and limited set of equipment with a large number of samples, the Subsea industry has to manage with very limited return on experience and a relatively poor database. A strong database for Subsea equipment as well as efficient RAM tools to be used to design systems - and not only to verify them - would bring a strong advantage on a classic design based on team's experience and very conservative specifications. Another interesting aspect is the standardization of Subsea equipment that would reduce considerably CAPEX while simplifying design exercise.

- When it comes to data collection procedure, a well structure approach is defined for both industries (ISO 14224:2006 and SAE ARP 4754). However, practically, it is once again slightly different. For aeronautic, each safety / reliability problem has to be reported and analyzed by the "operating company". Whereas, for instance, a root cause analysis has to be done by Air France and has to be submitted to the manufacturer (ex: Airbus). The process associated is called Experience Design Requirement (EDR). EDR structures the approach of failure analysis to ensure the future design takes into account any necessary modification. In order to ensure a maximum benefit from return on experience, operating companies should provide EPC or manufacturer with all the issues they have met. Assessing properly the incident and associated frequencies as well as communicating efficiently with all the Subsea suppliers would definitely improve equipment reliability while reducing risk. Ultimately, a fair and exploitable return on experience process would allow operators and designers to limit excessive conservatism and reduce in a non-negligible manner the associated costs.

- One reason for Subsea equipment lack of meaningful reliability data is directly linked to the means for data collections. Subsea monitoring and inspections are challenging as it remains complicated in terms of logistics and interpretation. What to look for, how to monitor and how to configure and trend the data are currently some of the most challenging questions as there is a very limited experience with interpretation and trending of such data and very conservative design in Subsea brought down the need of monitoring until now. Maintainability is not established as a priority within the design phase (except when it comes to modularity and basic repair actions) and monitoring has then not been made a priority. Aeronautic has for instance defined the MSG3 approach that aims at adapting the maintenance strategy based on the reliability & safety target. Subsea industry needs then to capitalize on a structured feedback and develop a prognostic approach based on adapted monitoring systems that would be more efficient than a simple diagnostic approach. As an example, a better understanding of the degradation mechanisms and an early discovery of change in behavior of some critical components would allow operator to order Long Lead Items (LLI) and special Offshore Service Vessels way in advance and then limit downtime. It is always worthwhile to remind that every day of production lost could be million dollars.

CONCLUSIONS

Economical viability and safety are the common key drivers of both Subsea and aeronautic developments. Aerospace has been reformed and structured around massive efforts handled by governmental authorities and international cooperation as any critical issues in terms of reliability has an immediate consequence on the lives of any citizen in the world. Railway industry has understood similarly the interest in strengthening its business model based on the best practices issued from space & military industries.

Now, Oil & Gas industry is reaching a critical path where radical changes are required to ensure a maximum availability, reducing development cost (CAPEX), planning in advance OPEX and by optimizing the business model in using return of experience to develop more standardized Subsea projects. Standardization is not just a matter of design itself or manufacturing procedures, it has to be understood as all the organization, competencies, disciplines and any human components that make a project successful. When all the best practices / tools / people that made a project a success are replicable from a project to another, so we can talk about standardization.

Finally, these discussions only set up the first basis of long term cooperation process with industries that went through similar issues decades ago but successfully stepped out of this situation by thinking their industry in a different manner. This paper aims at demonstrating that by simple processes, methodologies and development of some adapted tools, the Subsea industry could benefit from the Aeronautic experience in a straight forward manner.

REFERENCES

MIL-HDBK-338B, *"Reliability Prediction of Electronic Equipment".* Department of Defense, Washington, DC, 1998

American Petroleum Institute "*Recommended practice for Subsea production system reliability and technical risk management*", APIRP 17N, March 2009

Smith, D. "*Reliability, Maintainability and Risk, Butterworth-Heinemann*", Boston, p11, 1997

Amstadter, B. L. *"Reliability Mathematics: Fundementals, Practices, Procedures"*, McGraw –Hill, 1977.

http://www.embraer.com.br/en-us/imprensaeventos/press-releases/noticias/pages/petrobras-e-embraer-assinam-protocolo-para-avaliar-confiabilidade-de-sistemas.aspx, may 2015

Petroleum and natural gas industries "*Design and operation of Subsea production systems – Part 1: General requirements and recommendations*", ISO 13628-1:2010

American Petroleum Institute "*Design and operation of Subsea production Systems – General Requirements and Recommendations*", APIRP 17A, 4th Edition, January 2006, addendum 1, December 2010

NORSOK Z-016 *"Regularity Management & Reliability Technology"*, Rev 1, December 1998, section 6.2 Qualification and application of reliability data, page 22.

Yong Bai & Qiang Bai, "*Subsea Engineering Handbook*", 2010

Petroleum, petrochemical and natural gas industries *"Collection and exchange of reliability and maintenance data for equipment"*, ISO 14224:2006

SINTEF Industrial Management *"Offshore Reliability Data Handbook (OREDA)",* 5th Ed, 2009

SAE ARP 4754-A, *Guidelines for Development of Civil Aircraft and Systems,* 1996

SAE ARP 4761, *Guidelines and methods for conducting the safety assessment process on civil airborne systems and equipment,* 2010

RTCA DO-254/Eurocae ED-80 "*design assurance guidance for airborne electronic hardware*", 2000

RTCA DO-178C "*Software considerations in airborne systems and equipment certification*", 2012

International Council on Systems Engineering, *http://www.incose.org/AboutSE/WhatIsSE*, November 2015

MINI-ROVS FOR UWILD AND TANK SURVEYS OF OFFSHORE INSTALLATIONS

KYLE JACOB SATULA
HULL INSPECTION SUPERVISOR, PROCEANIC

MARK GORDON WALLER
MANAGING DIRECTOR, PROCEANIC

Proceedings of the 21st Offshore Symposium, February 2016, Houston, Texas
Texas Section of the Society of Naval Architects and Marine Engineers

ABSTRACT

For offshore drilling and production facilities, the Under-Water Inspection in Lieu of Dry-dock (UWILD) and other hull inspections can be an expensive, time consuming and disruptive part of vessel maintenance. UWILDs, traditionally performed by divers, raise safety concerns, operational issues, and other restrictions. Some offshore facilities employ a work-class ROV on board which is typically too large and unwieldy to move in close to the hull and perform the detailed inspections required by regulatory and integrity management personnel. Some work-class ROVs cannot operate in the upper water column due to cooling concerns. Divers can work close to the hull, but cannot be in the water with active thrusters, and can only work for a limited duration. Both solutions (divers and a work-class ROV) require a significant support team and a substantial amount of associated equipment. Hull tank inspection presents its own set of hazards, including confined space entry and personnel fatigue.

Many of the constraints are either eliminated or significantly mitigated by use of the mini-ROV. Benefits of mini-ROV inspections stem from the ability to access normally unreachable areas and to inspect tirelessly and indefinitely without risk to human life. Everything necessary for a mini-ROV can be transported in a single helicopter and operated by two ROV pilots. Mini-ROV inspections allow the installation to stay in operation during the inspection.

This article will discuss and compare current options for UWILD and tank surveys of offshore installations and describe in detail the benefits of using mini-ROV technology coupled with experienced structural engineering understanding for the work. This article will also present case studies from Proceanic's extensive track-record, where the use of mini-ROV technology greatly benefited the industry beyond the traditional methods of survey. Challenges and solutions will be discussed, and proven methods will be outlined which draw upon 2+ years of ROV field work.

Keywords: Mini-ROV, ROV, UWILD, Survey, Hull Cleaning, Tank Inspections

INTRODUCTION

As facilities age, particularly as the industry pushes for extending the life of existing assets, integrity management becomes an increasingly important issue. One key area of integrity management is the hull survey, required by class and the flag state. This survey includes bottom survey and internal inspection of all tanks.

Flag and class regulations require that two bottom surveys be carried out in a five year period. In general these surveys are to be carried out in dry dock. However, with the special case of mobile offshore drilling units, which have notably higher day rates than traditional trading vessels, every effort is made to keep these vessels in operation. The Under-Water Inspection in Lieu of Dry-docking (UWILD) Survey was developed to address the desire to operate vessels which either never planned to return to dry dock during their service life (such as an FPSO) or in the case of drilling vessels, would incur significant operational costs to come off station while under contract.

Traditionally, UWILDs have been carried out by divers. Offshore diving has a long history in the oil and gas industry with the earliest reference dating back to the 1940's. Divers are unmatched in dexterity underwater and can more easily manipulate tools when compared to an ROV. However, divers do face a number of operational constraints related to personal safety, physical endurance, and tolerance of environmental conditions.

Hull tank inspections provide their own suite of challenges. Confined space entry is a notoriously dangerous task. According to Nautilus International, more seafarers die in or are injured in confined spaces than in any other onboard work activity.

This paper will present how miniature Remote Operated Vehicles (mini-ROVs) can be used to address the problems for both inspections. Various mini-ROV options will be discussed, and capabilities will be weighed. Finally, two case studies will be presented demonstrating real-world application of the technology and the successes achieved by the equipment.

KEY CHALLENGES DRIVING THE NEED FOR CHANGE

In an industry known for a mentality of, "if it isn't broken, don't fix it," significant drivers are required to encourage the adaptation of a new method for carrying out a task. For the case of UWILD inspections, the following key challenges have been identified, spurring the desire to improve the method:

- Commercial diving is an inherently dangerous task
- Significant amount of equipment involved for a commercial dive team comes with a high logistical and bed-space cost
- Current economic climate is driving a need to reduce costs and reduce downtime
- Environmental constraints such as high currents or significant wave heights when working near-surface leads to considerable diver idle-time when rough weather is present

Larger work-class ROVs address these issues to some degree. A work-class ROV has the benefit of being able to carry out a wide range of tasks from a stable platform, but its large size prevents it from reaching the many tight corners that may be present around critical hull welds, particularly with semisubmersibles. This counteracts the objective of replacing divers in the water, as the goal of using the ROV is to provide an inspection that is of equivalent quality to an inspection carried out by a diver, even if the method of the inspection is different.

Also related to size, work-class ROVs are significantly heavier, posing the risk of vessel damage in the event of a collision. The lack of an aerodynamic design is also an impediment to keeping the ROV on station in an area with significant current and wave forces acting against it. Lastly, many work-class ROVs experience cooling issues when working in shallow waters as they were primarily designed for use in deep water.

To meet these challenges, we must first look in depth at the requirements and intent of a UWILD inspection.

REQUIREMENTS OF THE UWILD INSPECTION

The primary purpose of a UWILD is to establish that the unit's bottom condition is satisfactory and will remain so until the next scheduled inspection. The inspection requirements for a UWILD survey vary by class, vessel type, and age of the vessel but generally consist of the following key parts:

- General visual inspection of all hull structural steel
- Close up inspection of select areas on the bottom and sides of the hull
- Close up inspection of critical hull welds and corners, non-destructive examination may be required
- Inspection of sea chests and discharges
- Close up visual inspection of thrusters and blades, if present

DNV-GL provides a concise handbook on inspection for UWILDs in their offshore technical guidance, OTG-08. The key goal is to sufficiently inspect the structure so that the chances of critical failure before the next scheduled inspection are as low as reasonably possible. Based on these requirements, we investigated what options for mini-ROVs were available that would be capable of carrying out the above requirements while still satisfactorily addressing the challenges that have been identified.

OVERVIEW OF AVAILABLE MINI-ROV SYSTEMS

Based on the key drivers, it was decided to pursue utilization of the mini-ROV technology to accomplish the stated goals. The mini-ROV class contains a wide variety of options, from as small and basic as the build-at-home OpenROV to the larger, full-featured Seabotix vLBV. However, the key attribute of any mini-ROV is that it is able to be launched by hand. This limits the weight of the system to approximately 40 lbs (18 kg).

A variety of mini-ROVs were considered when selecting the optimal system for offshore work. Manufacturers considered include:

- AC-CESS
- Aquabotix
- DeepTrekker
- Outland
- Seabotix
- VideoRay

From these options, the Seabotix vLBV300 and the VideoRay Pro4 were shortlisted for further investigation (see typical arrangements in Figure 1). Key specifications of each ROV are shown in Table 1. We see from the table two key themes:

- The vLBV300 focuses on maneuverability in a standard square-shaped ROV form factor
- The Pro4 is an uncommon form factor which biases towards forward thrust at the expense of some maneuverability

However, one feature elevates the Pro4 ROV: the high power for a small size. This feature was judged very important for the ROV in order to have success in keeping the vessel in operation during the inspection and withstanding the potentially rough environment offshore.

(a) **(b)**

Figure 1 Photos of (a) VideoRay Pro4 and (b) Seabotix vLBV300

Table 1 Specifications of Investigated Mini-ROVs

Item	VideoRay Pro4	SeaBotix vLBV300
Total No. of Carrying Cases	2	5
Total Equipment Weight Including Cases	95 lbs (43 kg)	210 lbs (95 kg)
Power Requirements	110-240 VAC, 800 W	85-265 VAC, 3300 W
Thrusters	2 forward, 1 vertical	4 vectored, 2 vertical
ROV Max Speed	4 kn	3 kn
ROV Size	14.75, 11.4, 8.75 in (37.5, 28.9, 22.3 cm)	24.6, 15.4, 15.4 in (62.5, 39, 39 cm)
ROV Weight	13.5 lbs (6.1 kg)	40 lbs (18 kg)
Max Forward Thrust	22 lbf (10.0 kgf)	49.6 lbf (22.5 kgf)
Thrust to Weight Ratio	1.63	1.24
Net Buoyancy	Adjustable with ballast	Adjustable with ballast
Depth Rating	1000 ft, (305 m)	1000 ft, (305 m)
Automated Navigation Functions	depth, heading	depth, heading, trim (speed)
Tether Options	3 (Performance, Neutral Buoyancy, Negative Buoyancy)	1
Tether Diameter	Negative: 0.30 in (7.62 mm) Performance: 0.38 in (9.6 mm) Neutral: 0.46 in (11.7 mm)	0.35 in (8.9 mm)
Tether Weight	Negative: 0.075 lb/ft (0.112 kg/m) Performance: 0.049 lb/ft (0.073 kg/m) Neutral: 0.073 lb/ft (0.109 kg/m)	0.062 kg/m, 0.042 lb/ft
Tether Net Buoyancy	Negative: FW: -0.0447 lb/ft (-0.0665kg/m). SW: -0.0439 lb/ft (-0.0653 kg/m) Performance FW: 0. SW: +0.0012 lb/ft (+0.0018 kg/m) Neutral: FW: 0. SW: +0.0018 lb/ft (+0.0027 kg/m)	FW: 0. SW: +0.0011 lb/ft (+0.0016 kg/m)
Available Tether Lengths	131 ft (40 m), 251 ft (76.5 m)	820 ft (250 m), 6562 ft (2000 m)
Lighting	3600 lumen LED	2160 lumen LED
Camera Rotation	160°	180°
Camera View Angle	90°	65°
Camera Image Definition	SD 570 line resolution	650 line high resolution color

PERFORMANCE OF THE ROV IN THE OFFSHORE ENVIRONMENT

Most drillships have a draft of not more than 12.0 m and most FPSOs, and most semisubmersibles have an operating draft less than 20.0 m. While working offshore, the wave environment will consist of a combination of short-period wind waves and long-period ocean swell. Due to the long wavelengths of these ocean swell waves, the depth of water affected by the movement of these surface gravity waves is significant and often greater than 20.0 m. As such, both large and small ROVs experience significant movement while near the sea surface. This effect is amplified where enclosed spaces, such as drillship moon pools, create standing waves.

In addition, the mini-ROV's small mass eliminates any risk of damage to the structure being inspected if large wave motions cause significant movement of the ROV. Due to the small inertia of the ROV, it has less risk of damaging itself from impacts against the structure. Since the ROV is launched by hand, often from the structure being inspected, there is no concern over launching the ROV in a heavy wave environment.

These reasons lead us to conclude that wave environment is not a critical factor when determining the limit of the mini-ROV. In fact, it is a requirement that wave environment not be a critical factor. If the mini-ROV were subject to the same operational window as larger ROVs, then the value proposition for them would be much weaker.

However, offshore currents can be a major issue, particularly in the Gulf of Mexico. The Gulf of Mexico Loop Current can produce current as high as 4-5 knots with several eddy currents that spin off the main loop. These eddy currents are temporary but still exhibit surface currents in the range of 2-2.5 knots and can persist for weeks or months. Current in many other areas, particularly near-shore bays and gulfs, tends to be tidal in nature. Thus, even if the magnitude of these currents is strong, they follow a predictable schedule, and inspections can be timed when current is at a minimum during the high and low of each tide cycle.

CASE STUDY I: DRILL SHIP UWILD

The earliest interest for using mini-ROVs for UWILDs came from offshore drilling contractors. Proceanic's first major success in this regard occurred in April 2014. Many drilling contractors had already purchased mini-ROVs for storage on their vessels as a safety measure and to comply with EPA sea chest regulations. In particular, mini-ROVs were identified as resourceful for emergency visual inspection underwater. Such conditions that might warrant a quick look underwater include:

- Anomalous readings from thruster controllers implying something may be fouling the propeller
- Reduced flow rate through a sea chest indicating grating may be obstructed
- Impact or damage while in transit
- Oil or other leakage observed coming from below the vessel

These were intended to only be used occasionally and were to be operated by the rig's officer crew. Because of the infrequency of use and crew turnover, the vessel owner struggled to maintain experienced operators. As a result, usage was limited to short duration inspections away from high risk areas, such as thrusters, where entanglement and damage risk is high.

The owner then approached the Proceanic ROV team about using our mini-ROV crew to operate their equipment to carry out a UWILD inspection for the vessel's 3rd Special Survey. The vessel had recently been re-coated in dry dock; therefore, the hull was expected to be largely free of marine growth. As a result, this vessel was judged a good candidate for UWILD by ROV. As noted above, UWILD by ROV was already an established practice, but with larger inspection and work-class ROVs. The use of mini-ROVs had previously been limited to providing assistance to divers carrying out the primary work.

Particulars of the inspection are as follows:

- Vessel Type: Drillship
- Class: DNV-GL
- Inspection: 3rd Special Survey UWILD
- Vessel Status: Offshore Angola, on-location with dynamic positioning, latched on to well

Various discussions were carried out between the owner and class to formulate the scope and expectations of the survey, and a UWILD plan was produced in accordance with DNV OTG-08.

Upon arrival offshore, the conditions were as expected for offshore West Africa: seas were relatively calm (around 1m wave height) with moderate current (0.5-1.0 knots). Water visibility was considered good, with at least 20 m visibility.

The key success metric of the mini-ROV inspection is that there is no interruption to the on-going operations. This means contending with an active dynamic positioning (DP) system as well as avoiding a myriad of obstructions around the moon pool area; see Figure 2(a).

(a) **(b)**

Figure 2 Photos of (a) typical obstructions coming from drilling moon pool (view from bottom of hull looking at moon pool) during drilling operation and (b) Equipment set up on vessel deck

This is, indeed, where the mini-ROV finds its greatest success. Larger ROVs require launch and recovery systems with a large footprint and high power requirements. Because the setup of these systems takes some time and effort, the ROV is often launched from a single point on the vessel to carry out the inspection. This involves crossing large parts of the vessel with the ROV umbilical, introducing several hazards:

- Passing close by active thrusters
- Snags around drill string and related hoses
- Abrasion from marine growth causing wear or damage to the umbilical

While it is often possible to shut down one thruster of a DP system while maintaining sufficient station-keeping for the drilling operation, it is likely that the umbilical will pass by more than one thruster at some point if trying to inspect the entire underwater portion of the vessel from a single launch position.

However, since the mini-ROV can be launched by hand, the effort required to move the launch station is minimal. The set up for the mini-ROV is shown in Figure 2(b). Thus the ROV can be launched from as close as possible to the item being inspected, minimizing the risk due to excessive umbilical in the water.

Using this method, the inspection went forward smoothly. The mini-ROV demonstrated its ability to maneuver in tight spaces, particularly around the thruster foundation, while resisting the effects of current. Figure 3 shows example snapshots from the ROV camera during inspection.

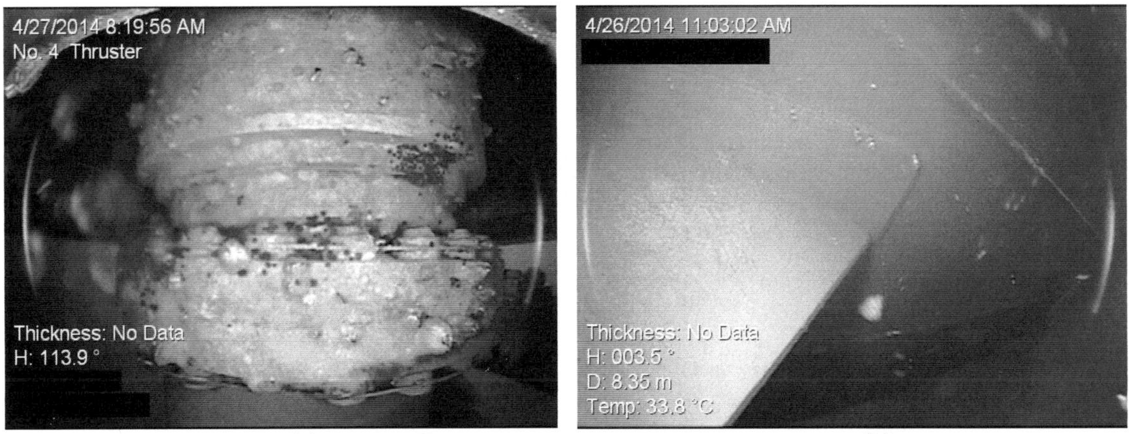

(a) **(b)**

Figure 3 Photos of (a) close-up inspection of thruster foundation in a tight space and (b) close-up inspection of bilge keel termination

The inspection concluded with no anomalies observed. In turn the vessel owner received 3rd Special Survey UWILD credit with no intervention by divers and no need to come off day rate.

CASE STUDY II: TANK INSPECTIONS FOR AN FSO

Using a mini-ROV to carry out a hull tank inspection is a significantly different mode of operation versus use for UWILDs. However, many of the benefits are similar:

- Avoid or eliminate manned entry into confined space
- Increased inspection speed
- No disruption to operations – for example, drillships which are actively drilling are very sensitive to changes in draft, trim, and heel

In many cases, offshore vessels such as drillships and FPSOs are ideal candidates for tank inspection by mini-ROV. In general they have better quality coatings, thus simplifying the inspection process. One requirement put forth by class societies before accepting mini-ROVs for tank inspections in lieu of manned entry is that the tank's coatings must be classified as being in 'good' condition.

This next case study focuses on an FSO situated in the Gulf of Thailand. Key particulars of the inspection are as follows:

- Vessel Type: FSO
- Class: ABS
- Inspection: 3rd Special Continuous Survey
- Vessel Status: Offshore Gulf of Thailand
- Hull Type: Single-skin

The vessel has eight ballast tanks which were due for inspection: six parallel midbody tanks, one fore peak tank, and one aft peak tank. As per the requirements of the 3rd special survey, a combination of general visual inspection

(GVI), close visual inspection (CVI), and ultrasonic thickness (UT) gaugings was required.

Since the ballast tanks were primarily large, open tanks (unlike typical J-shape tanks in a double hull vessel) significant rope access work would have been required. Because of increased field production, the vessel was taking on cargo much faster than originally planned, and offloading cycles were occurring every three days. This rapid loading/offloading meant that keeping an entire ballast tank empty for the duration of the inspection and gauging would be very difficult from an operations perspective.

Using the mini-ROV, ballast tanks did not need to be emptied for inspection and only needed to be briefly topped up while the tank top was inspected. The plan to use the mini-ROV for a tank inspection originally required special approval from ABS. Now, tank inspection by mini-ROV has been adopted as a formal policy.

The primary concern from class was that the mini-ROV would have difficulty navigating within a tank due to limited field of view. As could be seen in Table 1, the mini-ROV exhibits a very wide field of view. This enables better spatial awareness by increasing peripheral vision. Even if the objects at the periphery are highly distorted, they nonetheless increase the pilot's navigational sense. Secondly, pilot selection can have a strong effect on navigating in tanks. Pilots who have a strong background in marine structures or naval architecture are consistently found to be better able to identify their position in the tank and carry out a swifter inspection.

Since the deck of the FSO was reasonably empty and all tank hatches were on the main deck, the team set up a central control station from which to operate the ROV. The ROV was then progressively moved to each tank hatch as the inspection was carried out. Figure 4 shows the control station set up on the main deck.

Figure 4 Photo of equipment set up on deck

First, GVI and CVI were carried out on each tank. The ROV was able to get clear and close-up views of all structure and equipment inside each tank. Since the tank was filled from water offshore, clarity was excellent. Available ROV lighting was more than sufficient to light up the space for visual inspection, see Figure 5.

After visual inspection, gaugings were taken in each tank. The ROV has an externally mounted UT gauge with an articulating arm, thus enabling it to take thickness gaugings at any angle. The probe mounting can be seen in Figure 1(a). The gauging process is a straightforward task. Data is displayed on the video overlay which is then transcribed by the thickness gauging technician. See Figure 6 for screenshots from the ROV while taking thickness gaugings.

The ROV was able to gauge all areas in the ballast tanks required for thickness readings, and the operator received full credit for carrying out the tank inspections. In addition, the survey was carried out in less time than a traditional rope access inspection, resulting in considerable savings for the operator.

(a) (b)

Figure 5 Photos of (a) close-up inspection of bell mouth and (b) close-up inspection deck head

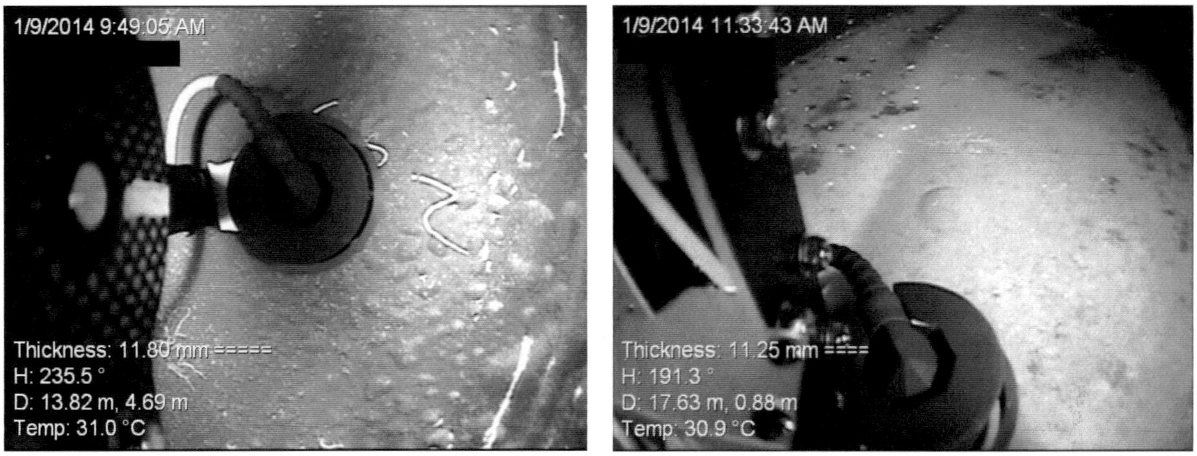

(a) (b)

Figure 6 Photos of (a) taking a gauging on plating and (b) taking a gauging on a side shell longitudinal web

CONCLUSIONS AND PATH FORWARD

Although small in size, the mini-ROV has proven itself capable of addressing some of the problems faced by the offshore industry. The case studies presented are only a small window on the significant track record accrued by mini-ROVs, which have amounted to considerable savings for the operators of the installations.

There are many areas for growth with the mini-ROV. With the exception of ultrasonic thickness measurement, non-destructive examination techniques have yet to be developed for the mini-ROV class. Improvements are also needed in the domain of seachest inspection. Since intake valves are all located within the protected area of the seachest, detailed inspection of the intake valves as per class requirements should be straightforward with a mini-ROV. However, the inherent design of the seachest grating makes opening of the seachest extremely difficult. Better design in the construction stage could enable these gratings to be opened without a diver, thus allowing the mini-ROV to easily inspect inside. Inevitably, the push for increased safety of operation will mean ever more utilization of mini-ROVs for all varieties of inspection.

ACKNOWLEDGEMENTS

The authors would like to thank our early-adopting clients of mini-ROV inspections in the offshore drilling and production industry. The authors would also like to acknowledge Will Boytim and James Fitzgerald for their tireless research and vetting of the mini-ROVs.

REFERENCES

American Bureau of Shipping, "Rules for Building and Classing Floating Production Installations", 2013.

American Bureau of Shipping, "Rules for Building and Classing Mobile Offshore Drilling Units, Part 7 Surveys", 2015.

DNV-GL, "Guideline on UWILD", Offshore Technical Guidance OTG-08, 2014.

DNV-GL, "Guidance on the Use of a Mini-ROV for MOU Ballast Tank Inspections", Technical Update 9-2014, 2014

MMS US Department of the Interior, Minerals Management Service, Gulf of Mexico OCS Region, "History of the Offshore Oil and Gas Industry in Southern Louisiana, Volume I: Papers on the Evolving Offshore Industry", OCS Study, MMS 2008-042, 20008

SNAME
THE INTERNATIONAL COMMUNITY
FOR MARITIME AND OCEAN PROFESSIONALS

CLASS APPROACH FOR LIFE EXTENSION PROCESS OF FLOATING PRODUCTION INSTALLATIONS

SHEWEN LIU, DAVID HUA, CHRISTIANE MACHADO, JER-FANG WU
AMERICAN BUREAU OF SHIPPING

Proceedings of the 21st Offshore Symposium, February 2016, Houston, Texas
Texas Section of the Society of Naval Architects and Marine Engineers

ABSTRACT

In the last five years, during the crest of the offshore industry boom, classification societies have received frequent queries on the feasibility of service life extension for different floating production installations (FPIs). Class society ABS initiated research into its records regarding the status of the oil and gas production fleet, and the sound engineering procedures used to cover demand in specific regions. Results indicate that proper life extension assessments would be required for different FPI types, including column stabilized units (CSUs), spars, tension leg platforms (TLPs), and FPSOs, by region.

In the recent down cycle of the Offshore Industry, life extension of FPIs becomes a more cost-effective solution than decommissioning, for production boosting and startup of new fields. Industry need to extend service life according to up-to-date environmental data and enhanced Classification Rules requires verification that the FPIs to be structurally adequate, and to withstand new design operational loads and additional equipment. ABS has published life extension guidance notes to provide procedures for life extension of aging FPIs. The approach provided in the Guidance Notes is to evaluate the condition of the asset, to verify its adequacy for the extended service based not only on the prescriptive Rule Requirements but also on analysis of unit´s operational behavior at past and present condition, and on comparison with the future predicted behavior for late life. With these verification results, the owner and ABS are able to account for risk and their mitigations and to take appropriate actions to achieve the life extension. The process starts with a baseline survey and a review of the design and in service engineering and survey documentation to establish the current condition, evaluate the asset in "as-is" condition, reassess the strength and remaining fatigue life, implement mitigation and repairs and customize in-service inspection programs or risk-based inspection programs for the extended service. The approach developed was segmented in phases to include decision points to allow owners and operators to interact with Class during the engineering and survey processes on technical issues. These include structural degradation during operation, previous repairs and the most current condition. Particularly, this approach will allow owners/operators to have the chance to make their decisions at certain stages throughout the process and not just at the end of detailed assessments. This paper will summarize the procedure for life extension provided by the guidance notes, and provide an introduction of ABS software HM3D which can be utilized during the life extension process.

Keywords: Life extension, FPI, Strength Assessment, Fatigue Life

INTRODUCTION

The floating production installations (FPI) designed and built to the requirements in ABS FPI rules or guide (ABS, 2015a, and its previous versions) and maintained in accordance with the applicable ABS requirements are typically intended to have a structural design life of 20 years. Many are designed for uninterrupted operation onsite without any dry docking. When a floating installation exceeds its original design life, an evaluation is to be made and appropriate actions are to be taken to extend the life up to the new target operating service life of the installation under the most current data regarding site-specific environmental conditions.

ABS records indicate that about a fifth of the worldwide FPSOs fleet is approaching the end of its designed service life (Machado *et al.*, 2015). While industry have demanded for expanding oil production and new discovery of oil fields, the current down-cycle market condition may prevent spending on expensive new production units. This, aligned with the current low oil prices and down-cycle of the offshore industry regarding investments, makes extending the service life of the existing units more attractive to owners/operators.

Considering the high-demand for safely extending the service life for the existing units, ABS has published Guidance Notes on Life Extension Methodology for Floating Production Installations (hereafter referred as ABS Guidance Notes) (ABS, 2015b) to meet demands from industry and provide comprehensive procedures for industry to follow. This paper gives an introduction of the life extension assessment procedure based on the published guidance notes.

In next section, the life extension process overview flow chart will be presented to show the high-level life extension approach, and more details about each step will be explained in the subsequent sections.

A hull management tool – "ABS Hull Manager 3D (HM3D)" (Kalghatgi, *et al.,* 2009) that converts a 3D CAD model to a database which stores vessel condition information and provides ability to track condition changes due to structural degradation will be introduced in this paper as well. This state-of-the-art system, in conjunction with other ABS software such as EAGLEO FPSO and OSAP, is believed to benefit the life extension process for the considered FPI unit, by shortening the turnaround time for numerical calculations preparation, and allowing the experts to concentrate on investigations of critical areas, scenarios, and risk mitigation actions.

LIFE EXTENSION PROCESS OVERVIEW

The requirements for the life extension of an installation that remains at the same site are given in 1-1-2/5.11.5 of ABS FPI Rules (ABS, 2015a). In general, the classification or continuance of classification of an existing installation for extension of service beyond the design life requires special considerations with respect to documentation reviews, surveys and structural analyses in order to re-verify the adequacy of the installation for its extended services. The following procedure provides more details regarding the steps to be followed in order to assess an installation for a life extension, in particular for those cases where there are changes in the original design parameters and/or arrangements. The life extension process include the following steps listed below and shown in Figure 1.

1. Baseline survey to establish the condition: review original design documentation, plans, modification records and survey reports if any, etc.; survey structure, mooring and machinery to establish condition of the considered unit;
2. Evaluation of the considered unit in "as is" condition: review the results of the structural analysis utilizing results of survey, original plans, MetOcean data and proposed modifications which affect the dead, live, environmental loads on the structures, and original structural analysis;
3. Assessment of the strength and remaining fatigue life trends along service life using the original data and data in the "as-is" condition with proper wastage consideration;
4. Implementation of mitigation and repairs and customize in-service inspection programs or risk-based inspection programs for the extended service.

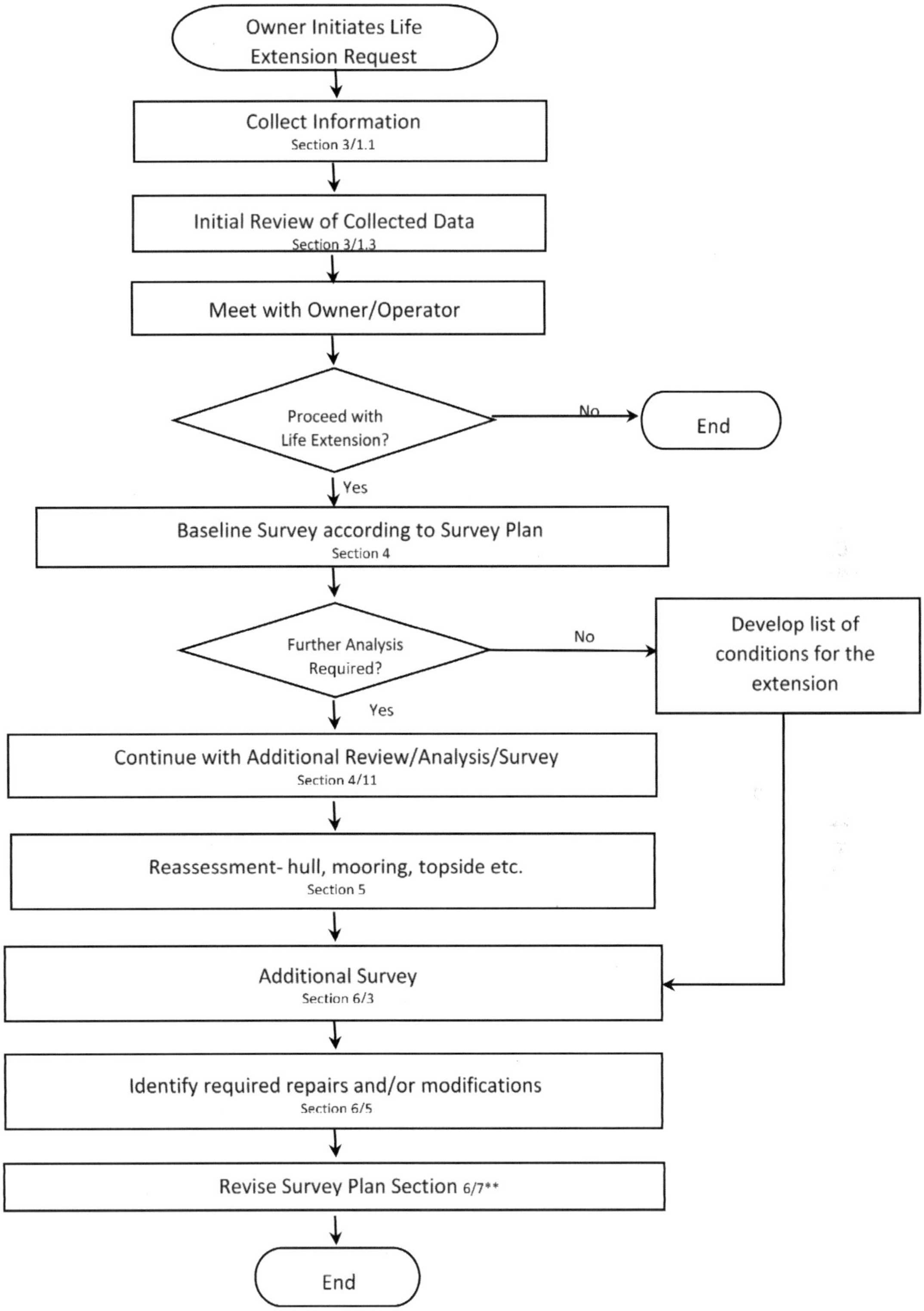

Figure 1. Flow chart for life extension process (*Note, the Sections shown in this chart are referred to ABS Guidance Notes Sections*)

After these processes, ABS records are to be updated to reflect the extended life condition. Class certificate is to be issued for life extension. Where applicable, the existing spectral fatigue analysis (SFA), fatigue life (FL) or remaining fatigue life (RFL) notation with year of maturation is to be implemented or updated accordingly.

It should be highlighted that, even though this process has been established considering a prescriptive engineering approach, it also provides consistent suggestions on engineering analyses and calculation to be prepared as basis and/or support for risk based approaches. On such cases, the impact of modifications of all the design input data mentioned in item 2 and 3 in the behavior of the unit at sea along its service life should be addressed to build a consistent matrix of risk scenarios.

BASELINE INFORMATION AND BASELINE SURVEY

The baseline information such as the design and operational information of the considered FPI, is essential for an engineering assessment of its overall structural integrity. The information includes original design reports, construction/conversion documents, as-is plans, specifications, and survey records during fabrication, conversion, installation and entire service history. The operator needs to ensure that as much information as possible is collected and that it is comprehensive enough to be representative of current conditions at the time of the assessment. If such level of the information cannot be provided, either a prescriptive reassessment of the installation is to be required or extensive investigations are to be conducted in order to estimate consistent behavior of the production unit. Most recent measurements or testing may be utilized for the reassessment.

The documents needed for life extension include: drawings, existing analysis such as hull structure scantling, global and local strength, fatigue and stability analyses; mooring or tendon analysis if applicable, existing survey plan and findings, revision history, survey and gauging reports, repair/replacements/operational history reports, record of modification and updated drawings, updated risk analysis including impact of modification on the risk scenario, MetOcean data and history, measured loads history on mooring, risers, tendons etc.

The following procedure may be applied for Initial assessment of baseline information:

1. Review original design and risk analyses (if applicable) and determine need for further analyses.
2. Assess condition of the FPI from documentation, In-Service Inspection Plan (ISIP) or Survey Plan, survey reports, design basis, existing analyses, etc.
3. Based on review of data, develop additional items for Risk Analysis update, Survey Plan over and above normal Special Survey activity, including the need of weight verification (deadweight survey or equivalent).
4. Agree between Class and Owner/Operator on whether to proceed with the next phase of the life extension. If a decision is made to proceed with the life extension, agree to Survey Plan based on Special Survey requirements and initial assessment of data.

The objective of baseline survey is to inspect the current condition of the considered FPI for life extension. It is the base for reassessment of the integrity of this unit and for determining the baseline condition upon which justification of continued service can be made. The Survey needs to be witnessed and monitored by ABS surveyors. The baseline Survey will include the following activities: Review reports of previous surveys and maintenance; develop an inspection procedure; complete inspection (including underwater) to verify that an accurate assessment of the installation's condition is obtained.

The scope of baseline survey will include corrosion protection system, structures, mooring systems, tendons if applicable, machinery and systems, and other special-considered survey items.

The corrosion protection system is to be re-evaluated to verify that existing anodes are capable of serving the extended design life of the installation.

The structure survey will establish the gauging data which will be the foundation for the next phase of reassessment of integrity of the considered unit.

The machinery and systems survey may contemplate the conditions of design and updates on systems and equipment. It may be used as an important tool for identification of new and updated risk scenario for operations.

REASSESSMENT

If the baseline survey shows the need for further analyses then the reassessment will be required. According to the ABS Guidance Notes (ABS, 2015b), the reassessment of the FPI to undergoing life-extension is to: check if the unit is able to fulfill the requirements for its intended continued service based on the baseline survey information, under considered environment conditions; identify areas to be monitored/repaired and carefully inspected, and possible modifications of structural components to allow continued service.

Each request for Life Extension will be treated on a case-by-case basis and the final extent of the analysis will need to be clarified for each specific project. The decision on scope of further analyses will be based on extent of life extension, extent of reassessment or risk consequence, condition of the vessel/survey history as well as scope of life extension with respect hull, topsides and position mooring systems, etc.

The reassessment process needs to cover hull structure, hull interface, topside structures, mooring system, TLP tendons and tendon connectors, machinery and systems as required by relevant ABS rules. Here we will focus on the hull structure reassessment process. One of the main issue for the reassessment is to better predict the remaining fatigue life for the hull structures, and identify and update critical areas which may need re-inspection, further monitoring and/or repair.

ABS FPI Rules (ABS, 2015a) divide FPIs into two categories: ship-type and non-ship-type installations. The strength assessment for these two categories follows different procedures and requirements in FPI Rules (ABS, 2015a). The strength and fatigue analysis procedures for these two types are summarized below.

Ship-Type Installations

ABS has conducted a comprehensive engineering study on hull structural reassessment in the past 5 years (Machado *et al.*, 2015) for selective representative ship-type installations. The study included the reinvestigation of unit strength, fatigue and stability analyses, including the elaboration of trend lines for strength usage state limits for vessel history, conversion condition and current conditions.

ABS initial scantling evaluation (ISE) and total strength assessment (TSA) (ABS, 2015a) can be used for reassessment of hull structure strength and fatigue analysis. Dynamic loading approach (DLA) (ABS, 2014a) and/or spectral-based fatigue analysis (SFA) (ABS, 2014b) may also be performed in addition or substitution to TSA.

Based on ABS studies, wastages are to be incorporated into the reassessment model. The gross and net scantlings taking consideration of wastages are provided in the Guidance Notes (ABS, 2015b).

A suite of engineering software - ABS Eagle FPSO - can be used for performing ISE and TSA modules for strength evaluation and fatigue analysis. The ISE module represent the current prescriptive Rule requirements investigation and the simplified fatigue analysis for connections of longitudinal stiffeners primary subject to hull girder loads. The TSA module involves the development of the structural analyses. Both modules shall contemplate the necessary updates on environmental conditions, and hydrostatic data, such as GM, minimum and maximum drafts or a scantling reassessment if desired.

On client's decision, a full length ship model, using more complex engineering methodologies such as a dynamic load approach (DLA) and spectral fatigue analyses (SFA), according to 5A-2-1/5.6.1 of the FPI Rules, can be used to reassess the hull structure and fatigue remaining life thoroughly instead of models based on total strength assessment (TSA). Where the full length model is considered, the strength calculation will follow instructions in accordance with ABS Guide for Dynamic Loading Approach for Floating, Production, Storage and Offloading (FPSO) Installations (ABS, 2014a). The fatigue acceptance criteria should follow Section 9 of the ABS Guide for Spectral-Based Fatigue Analysis for Floating, Production, Storage and Offloading (FPSO) Installations (ABS, 2014b). However, correlation of "as-gauged" scantlings instead of gross scantling, and specific corrosion rate still apply for the remaining service life period.

A sample FPSO life extension demonstrate the procedure for reassessment of the structural strength and fatigue analysis. The considered FPSO was built in 1974 as a trading tanker, and was converted into FPSO in 1995 for design life of 20 years. It was considered to have life extension, the number of years additional life needs to be determined by the reassessment results including strength and remaining fatigue life. The reassessment procedure includes the following steps:

1. Determine the Environment Safety Factors (ESF) for strength criteria (β) considering the environmental MetOcean data of conversion period, using ABS SEAS program, which is one of the modules in the ABS EAGLE FPSO suite of program;
2. Determine the ESF for fatigue criteria (α) considering its historic service routes and sites and the

corresponding MetOcean data using ABS SEAS program;

3. Perform structural strength assessment and determine the remaining fatigue life (RFL) using ISE module in ABS EAGLE FPSO program;
4. Perform finite element analysis using the TSA module in ABS EAGLE FPSO program to check critical areas and identify hot spots. Figure 2 shows a web frame shown the typical yielding usage distribution and hot spots in a local FE model as example.

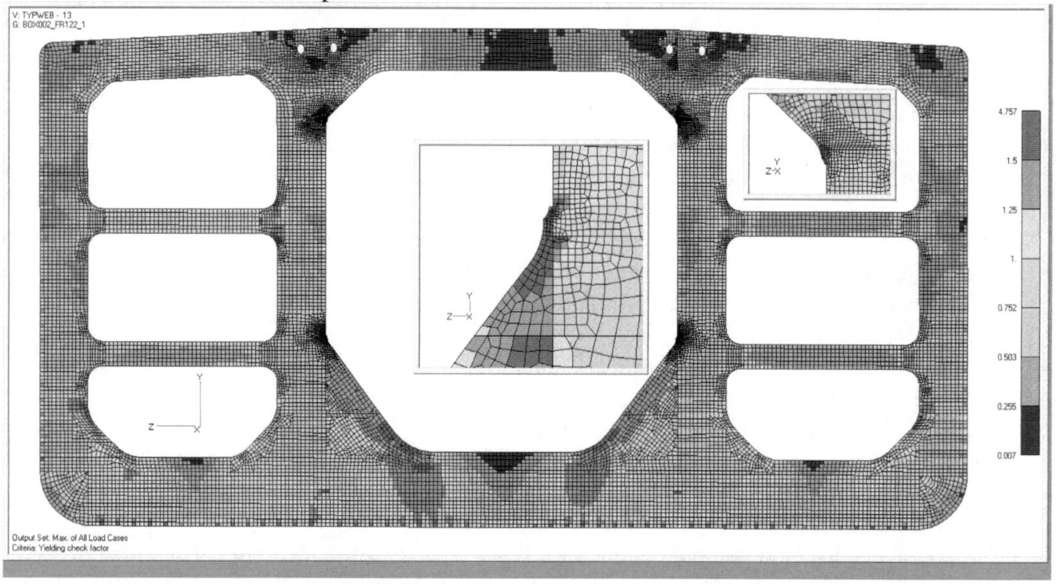

Figure 2. Example of Local Model for Web frame - Usage distribution and hot spots

Non-Ship Type Installations

Reassessment for non-ship type installations may include global strength, local strength and fatigue strength, corrosion protection and stability analysis.

Structural models for assessing the structural strengths are to be updated to incorporate wastages and changes in configuration and weights (as-is conditions). The global strength reassessment of hull structures should be performed for the updated data on environment conditions. Similarly to ship-type installations, decision on the further paths, such as reinforcement, monitoring, or extended in service survey scope shall be discussed and agreed upon based on the results of the structural analyses and their impact on the risk analyses. ABS OSAP program (Offshore Structure Assessment Program) can be used to perform the strength assessment as shown in the FE model in Figure 3.

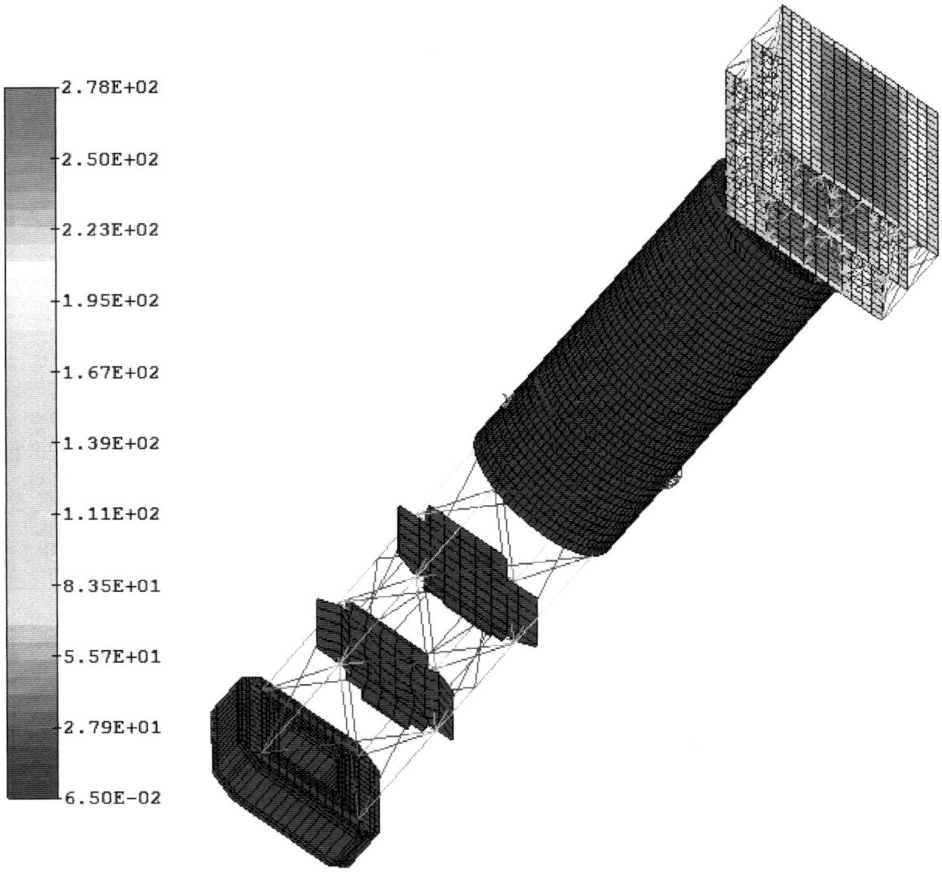

Figure 3. FE Model Example for Strength Analysis in ABS OSAP program (Model is a sample, not actual installation)

A procedure for remaining fatigue strength assessment was developed as described below.

1. Calculate accumulated fatigue damage, that has occurred based on the past service history for operations in transit, past services and site-specific operation at the current environmental criteria, based on a gross (as built/as converted) scantling model;
2. Determine the safety factors for fatigue life of the accumulated damages. The safety factor for fatigue life should be the same as the original design plan, but may be reduced if the current as-is conditions reflect reduction of uncertainty from original design with these conditions: No past findings, Corrosion rate to be within the original design assumption, Reliable load/MetOcean history;
3. Determine the predicted damage for life extension: The predicted annual fatigue damage of the structure in "as-is" condition (as gauged) for the requested extended life is to be calculated at the current site-specific condition using the original wave scatter diagram, or updated scatter diagram if more severe than the original design environmental criteria;
4. Evaluate the remaining fatigue strength. The remaining fatigue strength is evaluated by using one of the following criteria: The total factored damage including the accumulated fatigue damage for the past service and the predicted damage for life extension should be less than 1.0; the requested life extension (L) at the current site is to be less than the remaining fatigue life (RFL).

The fatigue damages can be calculated using the spectral fatigue module offered in the ABS OSAP program.

The same procedure for remaining fatigue strength of hull structures described above can be applied for the remaining fatigue strength assessment of topside structures and other major hull (mainly foundation of heavy equipment, such as mooring, risers, crane, and others) structures.

If the fatigue reassessment is not satisfied for the requested years of life extension, the following mitigation measures may be considered on a case-by-case basis: Fatigue improvements, if feasible, in accordance with ABS requirements or other recognized standards; Increased inspection frequency; Modification of loading conditions, particularly if the original MetOcean data has been updated; Make the originally non-inspectable structure accessible for inspection to reduce the required safety factor.

The possible fatigue-prone areas introduced by modifications are to be identified and assessed.

RE-INSPECTION/REPAIRS

The baseline survey condition and the reassessment can give the basis for determining the extent of repairs/alterations which will be necessary to class the platform for continued operation.

For new observed critical areas, locations with high stress found in the reassessment process, a further survey will be necessary to confirm any alternative solution and/or approach, such as monitoring, or extended in service survey scope. Members and connections found overstressed are to be strengthened. For connections with low fatigue lives found in the reassessment process, their fatigue lives may be improved either by strengthening, or grinding the welds and other approved mitigating methods. If grinding is used, the details of the grinding are to be submitted to ABS for review and approval. Alternative Interval of future periodic surveys may also be determined based on the remaining fatigue lives of these connections.

The new critical areas for inspection are to be identified by comparing the analysis results from the life extension calculation against the critical locations identified in the previous In Service Inspection Plan (ISIP) or Survey Plan.

The additional survey requirements as a result of the reassessment shall be jointly defined by operator and ABS. These will be based on the comparison of the reassessment results and all outstanding items from the baseline survey.

The upgrades, additions and modifications to the marine and industrial machinery and systems will be re-surveyed.

Non-inspectable areas are to be discussed and agreed by ABS and Operator based on the fatigue and strength analysis results.

Where repairs to the floating production installation or its elements identified by life extension reassessment are planned, a complete repair procedure, including the extent of the proposed repair and the need for the surveyor's attendance, is to be discussed among all the participants, such as ABS, operator and executor in order all repairs found necessary by the Surveyor are to be completed to all parties' satisfaction.

RISK ASSESSMENT

Risk assessment may be used to supplement the procedure described in previous sections. The risks addressed are primarily those affecting the safety of an installation, facility or operation, but the methods discussed can also be applied to other categories of risk. Risk assessment may be used to identify the various issues and their risk severity associated with a life extension of the asset. Based on the level of risk assigned, a mitigating approach is to be applied in terms of the level of robustness in the approach. The ABS Guidance Notes on Risk Assessment Applications for the Marine and Offshore Oil and Gas Industries (ABS, 2000) provide the guidelines for defining the concept of risk, describing the methods available to assess risk associated with life extension, and in setting up and conducting successful risk studies.

However, one important consideration to be included in the risk assessment is the comparison between the risk study at conversion scenario and the proposed scenario for life extension. It should be demonstrated that risks and degradation of the barriers are not escalating with extension of service life.

Risk assessment techniques may be used to demonstrate that alternatives and novel features provide acceptable levels of safety in line with current offshore and marine industry practice. The ABS Guide for Risk Evaluations for the Classification of Marine-Related Facilities (ABS, 2003a) provides guidance to ABS clients on how to prepare a risk evaluation to demonstrate equivalency or acceptability for a proposed life extension of floating production installation.

Risk evaluations for the justification of alternative arrangements or novel features may be applicable either to the installation as a whole, or to individual systems, subsystems or components. ABS will consider the application of risk evaluations for alternative arrangements and novel features in the life extension of the floating production installations, verification surveys during construction, and surveys for maintenance of class.

Risk-based inspection (RBI) is a risk assessment and management process that is focused on risks associated with failure modes initiated by material deterioration, and controlled primarily through equipment and structural inspection, as described in ABS Guide for Surveys Using Risk-Based Inspection for the Offshore Industry (ABS, 2003b). Requirement for risk-based inspection plan development and execution are provided in the ABS Guidance Notes (ABS, 2015b).

ABS STRUCTURAL INTEGRITY MANAGEMENT TOOLS (HULL MANAGER 3D - HM3D)

ABS has a tiered set of Structural Integrity Management software tools that allow the operator to manage the condition data of their asset. The toolset includes Hull Manager (HM), Hull Manager 3D (HM3D) and HM3D-FE Toolkit. Hull Manager 3D (HM3D) is a state-of-the-art tool that enable ABS to offer lifecycle condition management services for ship and offshore installations using 3D virtual structural model to manage, analyze and trend future condition. While the purpose of HM3D is to provide lifecycle management, this tool will help and benefit the life extension process for the unit under consideration, since it contains asset specific information required for baseline information and survey, reassessment and re-inspection for hull structures.

HM3D utilizes a 3D CAD model that is converted into a relational database that stores vessel/FPI's condition information and provides the ability to track condition changes due to structure degradation. The baseline information about hull structure required in the life extension procedure can be easily extracted from the database managed by HM3D. HM and HM3D manage inspections, track condition, plan repairs, and interface with selective structural analysis tools for anomaly treatment and life extension planning, with ability to track and trend condition of the asset for life/extended life. HM3D can export the structural model along with topographical data like material grade and ultrasonic thickness measurements to the HM3D-FE Toolkit where a finite element (FE) model can be created which helps to determine the remaining life, to provide the as-is condition of the hull required by the baseline survey. The generated finite element models with the as-gauged condition provides the reassessment of the hull structure to determine the hull structure strength and fatigue assessment. The HM3D tool has the ability to create gauge plans based on Class Rule requirements for Special Surveys. It has the ability to assign more gauge points to substantially corroded areas, critical areas and any structure nominated by user as a suspect area. HM3D is capable of applying various thickness methodologies to ungauged structures and even to extrapolate the thickness of structures in the future based on measurements done in the past. This enables the structural engineers to age forward the structures based on latest available gauging utilizing corrosion rate (manual or historical) to get a desired FE model for analysis.

FE model created by HM3D-FE Toolkit may be used by ABS engineering software such as EAGLE FPSO (for ship-shaped units) and OSAP (for non-ship shaped units) along with loading scenarios and boundary conditions. The ABS SIM software provide Class and its clients an good tool for life extension of any FPI under consideration to create models as shown in Figure 3 with gauged data.

SUMMARY

To meet the industry demand for life extension of existing floating production units, ABS has performed comprehensive engineering study on the reassessment of aging FPIs for a safe extended service life and developed the state-of-the-art methodology for evaluation of life extension of the aging units. These effort led to the publication of the guidance notes for life extension methodology for floating production installations, which give the industry a clear and easy-to-follow procedure for life extension.

It is worth noting that with the integration of the well-developed ABS structural integrity management tools (HM, HM3D, HM3D-FE Toolkit) and the ABS Life Extension procedure, the necessary information for a typical life extension process; such as the design, inspections, condition tracking, repair plans, gauging information, and interfaces with other structural strength assessment tools to track and trend the condition, will be readily available for the engineering review, as such, it will benefit the life extension process when necessary.

REFERENCES

ABS, 2000, Guidance Notes on Risk Assessment Application for the Marine and Offshore Oil and Gas Industries,

ABS, 2003a, Guide for Risk Evaluations for the Classification of Marine-Related Facilities,

ABS, 2003b, Guide for Surveys Using Risk-Based Inspection for the Offshore Industry,

ABS, 2014a, Guide for Dynamic Loading Approach for Floating, Production, Storage and Offloading (FPSO) Installations,

ABS, 2014b, Guide for Spectral-Based Fatigue Analysis for Floating, Production, Storage and Offloading (FPSO) Installations,

ABS, 2015a, Rules for Building and Classing Floating Production Installations,

ABS, 2015b, Guidance Notes on Life Extension Methodology for Floating Production Installations,

Kalghatgi, S. G., Serratella, C., Hagan J. B., 2009, Hull Inspection and Maintenance Systems, WMTC 2009

Machado, C., Wu, J & Huang W., 2015, Classification Society Studies on FPSO Life Extension Requirements, OTC-26202-MS, Brazil Offshore Technology Conference, 2015

SNAME
THE INTERNATIONAL COMMUNITY
FOR MARITIME AND OCEAN PROFESSIONALS

INTEGRITY MANAGEMENT SERVICES FOR FLOATING UNITS FROM DESIGN TO DECOMMISSIONING

JONATHAN BOUTROT
OFFSHORE BUSINESS DEVELOPMENT MANAGER, BUREAU VERITAS

NICOLAS LEGREGEOIS
OFFSHORE GENERAL MANAGER FOR US AND CANADA AT BUREAU VERITAS

Proceedings of the 21st Offshore Symposium, February 2016, Houston, Texas
Texas Section of the Society of Naval Architects and Marine Engineers

ABSTRACT

By reviewing the process of Asset Integrity Management applied from the design phase, continued through the in-service life and obsolescence considerations for decommissioning, this paper presents a view of managing the safety and operational efficiency of floating units.

The paper will go through the presentation of a didactic process ensuring the Asset Integrity of the structure and equipment in answering a few questions:
How to ensure a correct knowledge of the offshore facilities' condition?
What are the main degradation processes threatening the facilities?
What will be the response of a facility to the main threats over the years?
How to implement a cost-effective Inspection Strategy?
Answers to those questions will involve various services embedded into a global Integrity Management process such as risk analysis, CFD and hydrodynamics, structural FEM calculations, condition monitoring, Risk-Based Inspection strategy and 3D geometrical models for visualization and reporting.

Established at design phase and updated during the in-service life of the Unit, this global Asset Integrity process for the Unit will support and optimize the inspection programme based on risk considerations by focusing on the most critical elements to the safety.

Besides an OPEX cost reduction by optimizing the inspection efforts, the engineering services that are parts of the Asset Integrity Management are fully embedded into that global process. This brings another cut in costs by saving time and thus cost when performing engineering analysis.

Keywords: Structural Integrity Management, Risk-Based Inspection, Hazards, Risk Assessment

INTRODUCTION

The growing demand for production efficiency and sustainability are part of the main requirements in the Oil & Gas Industry. Offshore installations, pipelines and platforms need to be built to work in a hostile and highly regulated environment, and they need to keep on working safely and efficiently throughout their design life and often beyond into life extension.

At the same time it is necessary not to compromise the operational safety and quality. Based on that, the Integrity Management, which is considered as an integral part of engineering design, is developed in order to ensure a reasonable balance among facility's efficiency, sustainability, safety, quality and costs.

The disciplines and services of Asset Integrity Management deliver cost savings and improved uptime over the entire life cycle of a facility. Asset life cycle costs optimization finds the right balance between CAPEX and OPEX.

AIM services identify and prioritize degradation mechanisms and hazards and set in place appropriate monitoring, inspection and mitigation actions, from design review to obsolescence planning and life extension for all offshore and onshore assets. They integrate design review, risk assessment, naval architecture, structural engineering, process engineering, maintenance and inspections.

On top of that, an overall management web-based system is established and maintained, creating synergies between the different AIM services and serving as an 'integrity dashboard' allowing the Operators to monitor the safety of their assets, remotely and on a daily basis.

Such required system is a modern Asset Integrity Maintenance System (AIMS) which means a planned and systematic way of building the data required to understand the condition of the Asset and having in place an inspection and modelling system to keep that up to date.

It sounds simple, but in practice, because there are many contractual and other parties involved, a full AIMS built up from the as-built condition is rare.

For floating Units being built today, Operators can benefit from experience and begin to integrate a comprehensive AIMS process from the design phase, involving all parties. Then when they face in-service issues, they will have the information they need to make sound decisions.

As Units age and new technologies and regulations come into play, AIM services provide the knowledge, data and software solution to upgrade or maintain the asset to meet the new environment and service needs.

OFFSHORE FLOATING UNITS' MAIN THREATS AND DEGRADATION PROCESSES

Several levels of hazards are entailed to the activities performed in Offshore Oil & Gas platforms and they may result in severe consequences to the life and health of workers, pollution of the environment, economic and production losses. Physical effects which can occur in case of a major accident event in Offshore Oil & Gas units include:

- Spillages: large amounts of hydrocarbons and pollutants discharged uncontrolled in the installation or in the sea in a short period of time
- Pool Spread and Evaporation: The occurrence of a spillage followed by the generation of a pool spread and evaporation can be the source for subsequent fire and explosion.
- Dispersion: The dispersion process depends on the humidity and heat transfer from the surface. Several safety researches as well as computer modeling have been developed to better understand release events followed by dispersion, and eventual subsequent fire and explosion scenarios.
- Occurrence of a fire after ignition of released hydrocarbons;
- Occurrence of explosions after gas release, formation and ignition of an explosive cloud.
- Rapid Phase Transition events (RPTs): A RPT can eject large amount of liquid into the air, which will evaporate during the fall down.
- Vapor Cloud Explosion: After long time of researches, experts concluded that it is necessary the vapor cloud is generated in a congested or confined region of the installation to establish the mechanism of the flame acceleration so a blast may occur.
- Boiling Liquid Expanding Vapor Explosions (BLEVES): This physical effect can present catastrophic results, although it can be avoided when pressurized storage and transport are not used.
- Escalation: Nowadays this is one of the main concerns regarding the offshore gas projects such as Floating Storage Regasification Units (FSRUs) and Floating Liquefied Natural Gas (FLNGs) units. Although proven technologies are being deployed in these new projects, there are still several technical challenges to be overcome. For instance, the space available between the process units in a gas floating unit is quite limited and this could increase the severity of an explosion, for instance. So the minimum safety distances between the process units have to be carefully established.

Regarding the hull structure, there are two main degradation mechanisms that could be threatening the asset integrity: cracks – due to fatigue or strength issues – and corrosion. They are linked because corroded structure is weakened which increases the load on fatigue-prone structural details which then leads to fatigue cracking.

We know a lot about fatigue today, but we know much more about corrosion because it has been studied and worked on for much longer. Fatigue issues need deep structural analysis to forecast and correct, corrosion is visible and simple to combat.

The risk of structural failures and the resulting potential consequences to the asset integrity, life and health of the workers, pollution of the environment and direct economic damage can be assessed qualitatively and/or quantitatively by an Asset Integrity Management System as presented next.

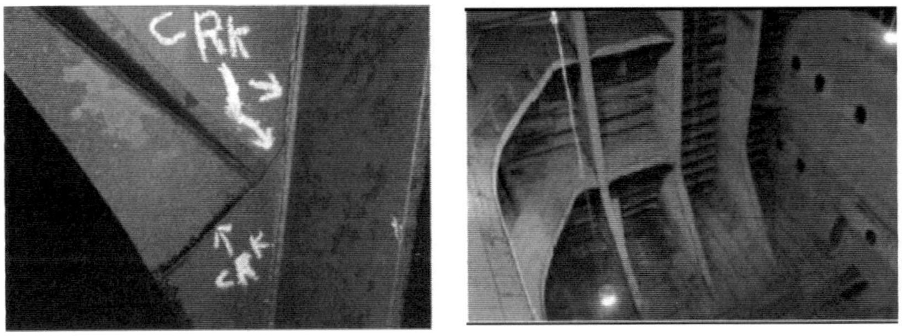

Figure 1 – Illustration of degradation mechanisms

ASSET INTEGRITY MANAGEMENT – GENERAL OVERVIEW

The difficulty for operators, especially of older units, is that indicators of degradation mechanisms do not present themselves in a uniform or coherent manner, and the consequences of these indicators may fall into the different areas of responsibility of different areas of the company or project. The data about the indicators may be available, but it may not always be available to the people who can make sense of it or who need it to plan how to avoid an incident or downtime.

There are tools and systems available to enable operators to know the condition of their asset and to monitor and predict how that condition is changing over time. They are grouped under the title of AIM Systems. The most important single thing about AIMS is that they are not simply engineering tools for engineers. They are a cluster of software tools, inspection techniques, data management and data sharing platforms and good organizational practices. Each part alone provides only part of the solution, and to be effective all parts have to be embraced by the management of all the parties involved.

Thus, at the heart of Asset Integrity Management is a continuously updated assessment loop that can be divided into 4 main steps: Data Management, Evaluation, Strategy, and Program.

- Data Management: all available documents describing the asset are reviewed and relevant information is used as input for the Evaluation step
- Evaluation: using as input the information selected from the Data review, a risk evaluation is carried out
- Strategy: the risk assessment results are reviewed and organized in order to provide a realistic, practical inspection strategy giving methods and intervals
- (Inspection) Program: application of the recommended strategy. Inspection results (condition data) will be the feedback for a new evaluation

And as described in the next chapters, a complete portfolio of tools and services is integrated in this loop, in order to support offshore operators facing the Integrity Management challenges of their floating units.

For instance, structural integrity for hull, topsides and mooring, process integrity, electrical and instrumentation devices, HSE and data management are all parts of the integrated AIM solution, set in place to establish and maintain a Risk-Based Inspection plan through the Asset life.

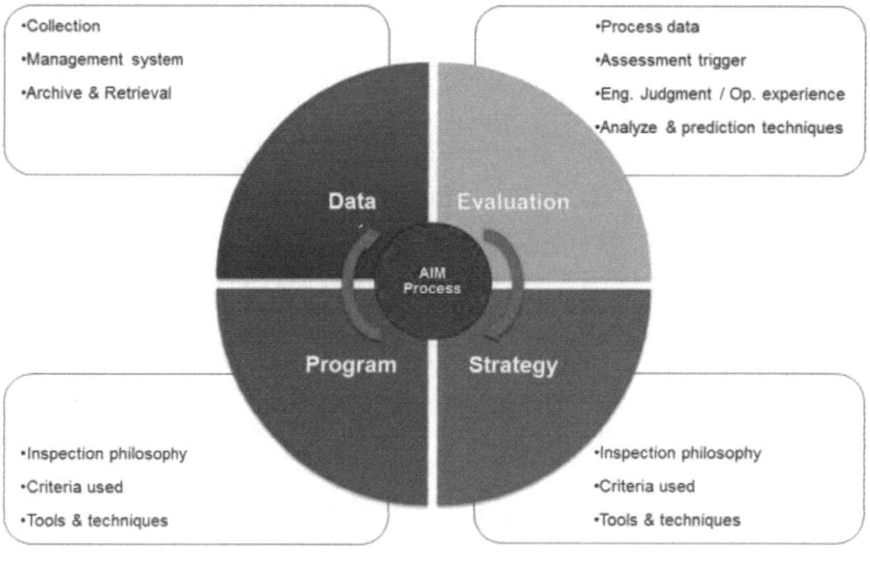

Figure 2 – AIM loop

Another crucial aspect is the right timing to establish the AIM methodology. The ideal AIM starts at the design phase and continues through the in-service phase of the asset. Currently many FPSOs are built by one project team then passed to an operational team with no link over the AIMS. When a project team passes a new build over to the operations department a lot of useful information for integrity management is not properly handed over. One such item is an adequate inspection plan. That will only be generated if the actual inspection and maintenance teams are involved in the loop during the creation of the hull and topsides inspection program. Their input is necessary to make sure the final deliverable will be used.

This may involve several different departments in the operating company alone, such as the project team, the safety department, marine operations, maintenance, and all these need to liaise with several external parties and regulators. That means close attention to data management, and data input and accessibility.

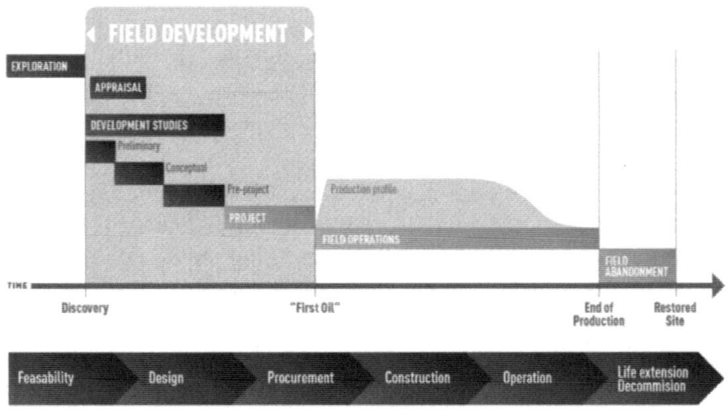

Figure 3 – Asset Life cycle

One example of solution for the Asset Integrity overall management is to use an open web-based system where controlled access allows AIM participation worldwide. The system is designed to improve inspection management and information control for different types of facilities. Information is stored and secured in a unique database and accessed in real time through the internet or intranet. Users' profile can be defined according to the roles of the different people participating in the AIM activities, such as Asset Managers, Surveyors and Administrative staff.

Such system shall bring together data collection and management, risk-based inspection and analysis and state of the art maintenance planning and monitoring to deliver a system which avoids unexpected or unplanned stoppages and outages and protects the asset value, combined with easy reporting and global web-based management.

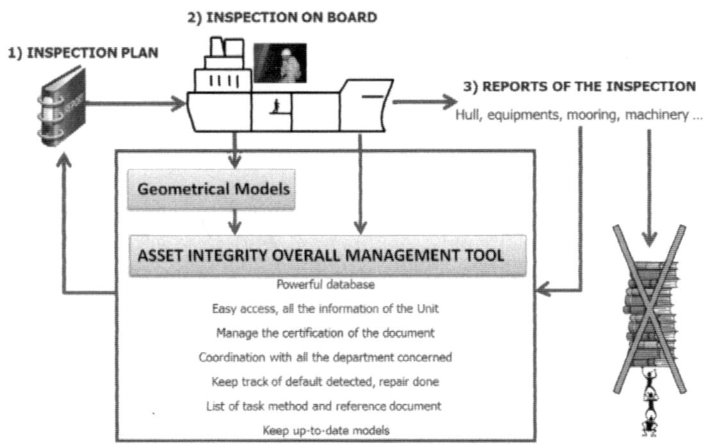

Figure 4 – Illustration of Asset Integrity overall management tool

AIM – DATA MANAGEMENT: ENSURING THOROUGH KNOWLEDGE OF THE ASSET

Data management is a key element of any AIM system and it involves not only design and construction data but the natural feedback from the life cycle. How will you build in a learning process so that continuous improvement is part of the system? It is at the assessment and evaluation stage that a decision needs to be made. Data management consists in a managed system for archive and retrieval of asset data and other pertinent records.

It has to be kept in mind that the AIM approach is built on a Risk-Based Inspection (RBI) process which efficiency and accuracy is highly dependent on the data used in input. Hence, it is necessary to implement a robust system that can store and process a huge quantity of information.

As an example, the following gives a non-exhaustive list of data used in the context of AIM and RBI:

- Structural drawings of the installation: up-to-date drawings reflecting the 'as-is' configuration of the asset has to be available and shared between all AIM actors for multiple purposes: prepare and report inspections, input for modeling (FE models, risk analysis, etc), analyze the topside layout configuration

- Analysis reports covering a wide range of domains: naval architecture such as fatigue and strength analysis, risk analysis such as HAZID, HAZOP, QRA, Fire & Explosion Risk Analysis (FERA) or Quantitative Risk Analysis (QRA). Those analysis are usually first performed during the design phase of the project by the EPC, the Shipyard and/or a third party and they may be updated during the in-service phase after a design modification (e.g. new topside module)

- Survey reports: at delivery of the asset, a baseline is laid-down. Extensive inspections and measurements campaigns are launched for the hull structure internal spaces, outer shell, underwater areas, ballast and cargo tanks and all the associated systems. During the in-service phase of the asset life, inspections continue allowing for a continuous comparison between the as-is state and the as-built state. The findings from inspection activities are used to facilitate an understanding of the causes of potential failure and to priorities management actions.

By storing all this data in a unique system also used for the overall management, the ways in which facilities may fail to deliver their intended performance can be identified and appropriate, effective and efficient plans to prevent those failures can be implemented. Analyses revealing the weakness of the asset are further described in the next chapter.

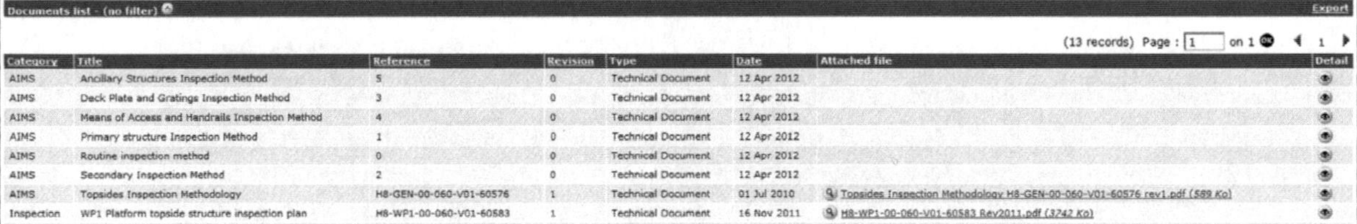
Project Follow Up | Tree | Documents | Models | Planned Inspections | Non Planned Surveys | Comments | Corrective Actions

Documents list - (no filter) — Export

(13 records) Page : 1 on 1 — 1

Category	Title	Reference	Revision	Type	Date	Attached file	Detail
AIMS	Ancillary Structures Inspection Method	5	0	Technical Document	12 Apr 2012		
AIMS	Deck Plate and Gratings Inspection Method	3	0	Technical Document	12 Apr 2012		
AIMS	Means of Access and Handrails Inspection Method	4	0	Technical Document	12 Apr 2012		
AIMS	Primary structure Inspection Method	1	0	Technical Document	12 Apr 2012		
AIMS	Routine inspection method	0	0	Technical Document	12 Apr 2012		
AIMS	Secondary Inspection Method	2	0	Technical Document	12 Apr 2012		
AIMS	Topsides Inspection Methodology	H8-GEN-00-060-V01-60576	1	Technical Document	01 Jul 2010	Topsides Inspection Methodology H8-GEN-00-060-V01-60576 rev1.pdf (589 Ko)	
Inspection	WP1 Platform topside structure inspection plan	H8-WP1-00-060-V01-60583	1	Technical Document	16 Nov 2011	H8-WP1-00-060-V01-60583 Rev2011.pdf (3742 Ko)	
AIMS	WP1 Structure Integrity Management plan	H8-WP1-0-V01-60586	0'	Technical Document	10 Dec 2010	H8-WP1-0-V01-60586 WP1 Structure Integrity Management plan.pdf (2003 Ko)	
AIMS		Survey report sheet	0	Technical Document	09 Feb 2012	WP1 Survey Report Sheet.doc (86 Ko)	
AIMS		CVI Inspection method	0	Technical Document		CVI Inspection Method.doc (28 Ko)	
AIMS		GVI Inspection method	0	Technical Document		GVI Inspection Method.doc (33 Ko)	
AIMS		NDT/NDE Inspection method	0	Technical Document		NDT-NDE Inspection Method.doc (29 Ko)	

Documents attached to the database: reports, Inspection sheets, procedures

Figure 5 – Illustration of Data Management tool

AIM – EVALUATION: ASSESSMENT OF INTEGRITY FAILURES

Two main categories of integrity failures have to be evaluated separately due to their nature: equipment/process failures resulting from operations and structural hull failures, mainly induced by fatigue cracks or corrosion.
The first part of this chapter treats the assessment of integrity failure with regards to equipment & process.

Part of an AIM state of mind is the importance of Risk Assessment studies, starting by risk identification. But there is a danger at this stage that the model and the process obscure the vital reality. This process is based on the data that goes in, on the equipment, its use, its history and its previous failures. If that data is flawed, then the whole process going forward is weakened, because the wrong decisions on inspection programmes and Condition Based Maintenance will be made.

Also, risk analysis will establish the consequences and seriousness of any structure failure, which can be quantified with the probability and frequency of occurrence to determine the risk level. All the data analysis and modelling leads to a very detailed reliability centred maintenance plan for the asset and equipment and a risk-based inspection plan for the unit structure intended to detect early signs before a small crack becomes a major problem.

Risk Assessment aims at identifying, understanding and evaluating the threats which could compromise the safety and health of all the parts involved in an Offshore Oil & Gas project. The Risk Assessment can help the companies to identify which risks need particular attention and what control measures could be put in place.

The following will present a typical process of Risk Assessment methodology that can be performed to ensure that all the risks linked to an offshore installation dealing with oil & gas are understood and under control. Additionally, historical data of accidents have demonstrated that a significant part of the accidents are caused by human factors like procedures not followed, poor interfaces, etc. So it is important not to focus solely on technical factors that could lead to an accident during the Risk Assessment.

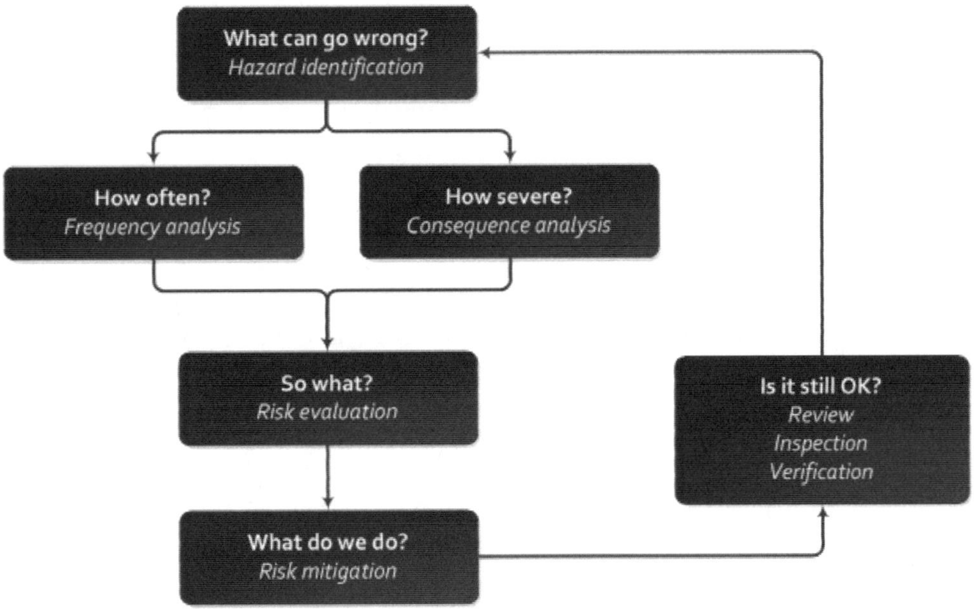

Figure 6 – Risk Assessment Methodology

Hazard identification is the first step of the risk assessment and it corresponds to a qualitative review of likely accident scenarios and it includes a range of techniques that study the magnitude of these accident events on an installation and suggest appropriate prevention, mitigation and risk control measures.

Through the Hazard Assessment results, it is possible to select a list of scenarios that should be modelled quantitatively during the consequence analysis step. HAZID and HAZOP are examples of techniques commonly used to identify which scenarios in an offshore unit could expose workers to injury and assets to damage.

Hazard Identification (HAZID) Study

The Hazard Identification (HAZID) is one of the most important and commonly techniques used to identify which scenarios in an offshore unit could expose workers to injury and assets to damage.

The HAZID method is a combination of identification, analysis and brainstorming based on the hazards identified in a prepared checklist, which is then divided into main sections containing hazard categories. During a workshop session, a team of expert analyses whether the existing controls are appropriate to eliminate the hazards or mitigate the consequences. Also, the risk of the identified scenarios is qualitatively estimated and classified according to a risk matrix which lists different levels of expected likelihood and impact of occurrence.

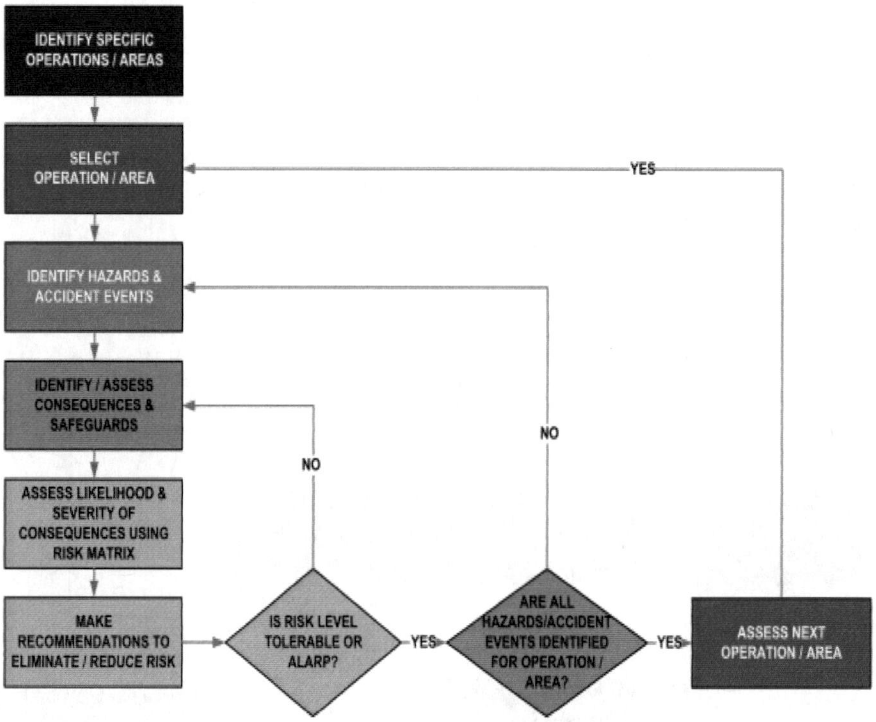

Figure 7 – HAZID Methodology

The HAZID takes the opportunity to ensure hazards such as overpressure in cargo tanks, cryogenic spill as well as loss of production caused by piping damage have been considered and confirm safety systems will be implemented. HAZID does not aim to provide technical solution but it will prioritize the key issues to be investigated quantitatively in the Quantitative Risk Assessment (QRA).

Hazard and Operability (HAZOP) Study

Hazard and Operability Study (HAZOP) aims to find out how deviations in process parameters (flow, pressure, level, temperature and other) could compromise safety, environment, operational continuity, final product characteristics or the assets integrity.

A HAZOP is performed by, firstly, dividing systematically the installation in systems, sub-systems and finally in "nodes" (smallest part of system or sub-system composed by process equipment and streams). Then, technical workshops aims at searching, for each process parameter, all possible causes and consequences for deviations from normal operating conditions, using "guide-words" like "no flow", "lower level", "higher pressure", etc. Following the discussions, the team identifies, for each deviation, the associated existing safeguard and detection possibilities.

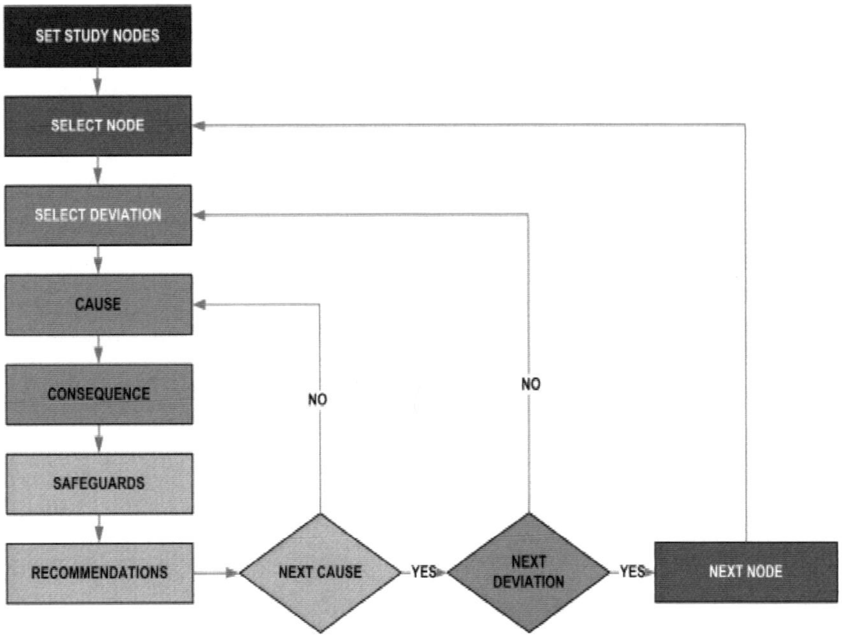

Figure 8 – HAZOP Methodology

The next steps of the risk assessment, after the hazard identification, will be the calculation of risk levels based on probability of occurrence and consequences of each major accidental event. Function of the calculated risk level, a risk-based inspection programme will be established.

The frequency analysis aims at estimating the occurrences of each major accident event identified during the Hazard Assessment stage in a given time period. In other words, the hazard event frequency is calculated by determining the quantity of certain type of an accident event per unit time.

The consequence analyses evaluate the resulting effects and impacts on the personnel in case an accident occurs. In a quantitative basis, the consequence is defined as per the following equation:

$$\text{Consequence} = \text{Occupancy} \times \text{Vulnerability}$$

- Occupancy: proportion of time (over a year) exposed to work hazards; and
- Vulnerability: probability that exposure to the hazard will result in fatality.

The Risk Quantification combines the frequencies and consequences estimated for each event identified during the qualitative step to determine the risk. It is possible to calculate different types of risk including loss of life, impairment of safety functions, property damage, business interruption, environmental pollution.

Regarding the assessment of hull structural integrity failures, fatigue and corrosion were identified as the main source of degradation.

The objective of the assessment of structural failure is to provide an as-is condition assessment of the Asset's hull which will identify weakness points in the structure, estimate the remaining fatigue life of the hull, estimate steel renewal and repair needs and prepare the forward inspection and maintenance plans.

Finite Elements Analysis of the as-is structure will look in detail at yielding, buckling and spectral or deterministic fatigue criteria using partial or full-length 3D models. A simpler 2D analysis can be used for hull girder analysis. More detailed FEA can zoom in on crack propagation, impacts and past damages.

None of this analysis will be useful if the inputs are not clear, correct and complete. They have to include:

- Drawings, hydrodynamics & heading analysis
- Mooring analysis
- Capacity plan
- Trim & stability booklet
- Weight distribution
- Mooring loads
- Tanker trading routes history for conversions
- Towing/transit phases
- Site metocean conditions
- Loading conditions / Load cases
- Loading/Unloading sequence
- Repair history
- Corrosion condition
- Detected failures, damages, particular findings: cracks, dents, pitting, collision…

Fatigue analysis works best with a preliminary fatigue screening analysis to scan fatigue weaknesses, and assessment of typical longitudinal connections. This is then supplemented by detailed assessment of typical locations determined by experience and the output of the fatigue scan. A full analysis of previously detected cracks on board has to be made and special attention given to areas such as the turret interface and topside mountings in case of turret-moored FPSO.

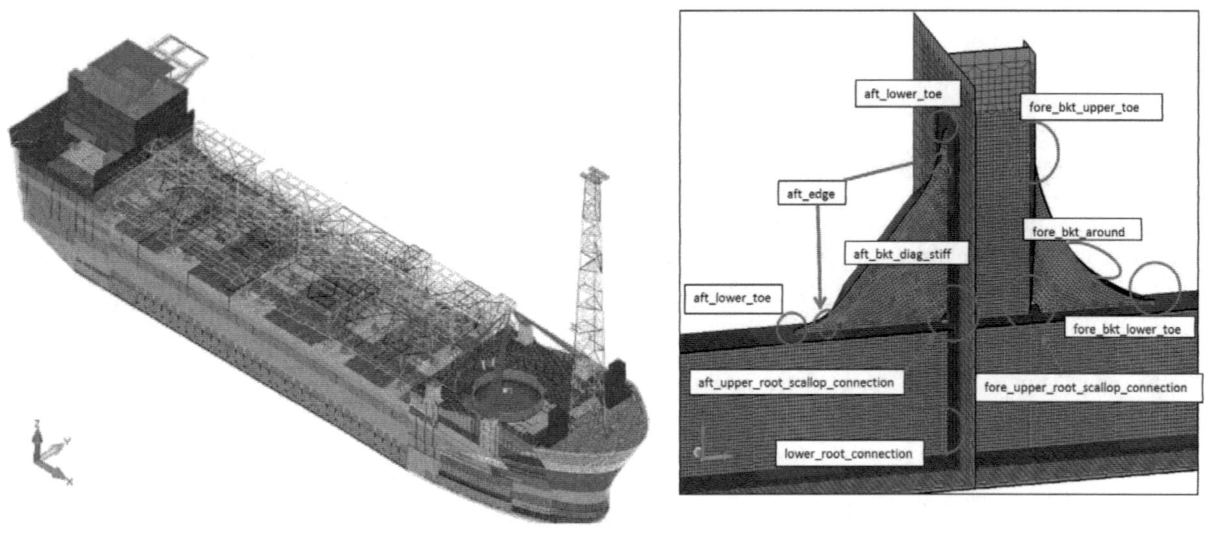

Figure 9: FPSO – FE complete ship model and refined details for fatigue analysis

Analysis of the input data, particularly the comparison between the as-is state and the as-built state, if available, will produce a corrosion status report. It should show a measured thickness diminution comparison with maximum allowance criteria based on class or any other standard.

3D modeling tool delivers corrosion assessment and steel renewal plans using a geometrical model on which we can populate the as-gauged plate thicknesses. Corrosion rates from class Rules or user defined are applied to deliver corrosion prediction and steel renewal prediction for the required period ahead.

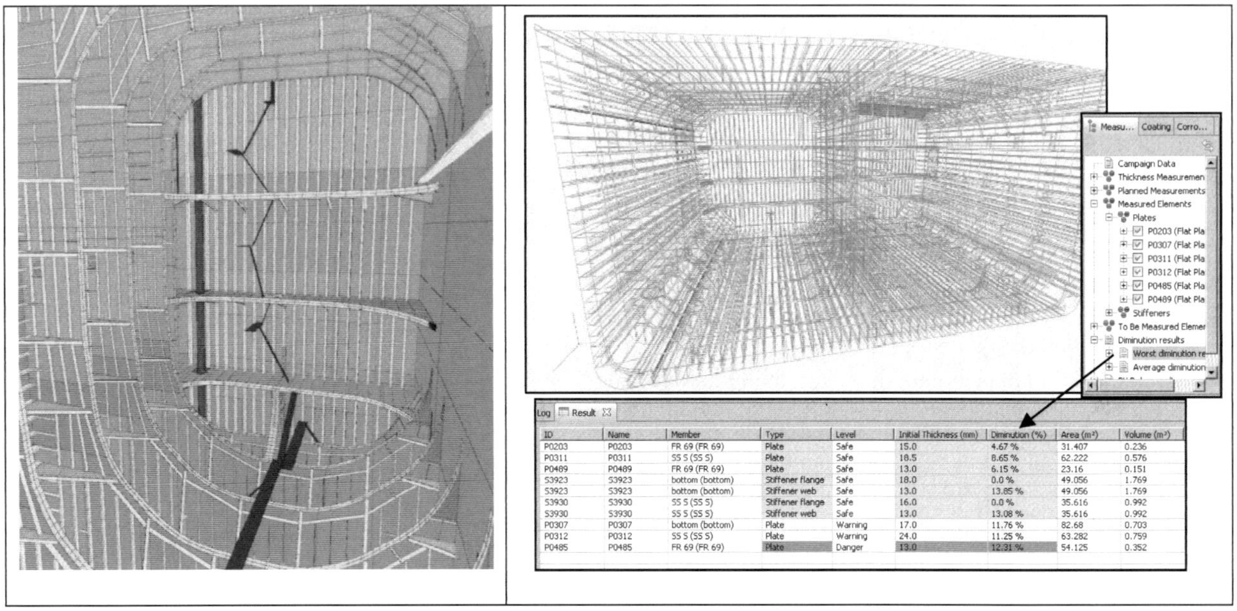

Figure 10: 3D geometrical models for corrosion analysis

Such software is useful for processing and for visualization, but it depends on a good inspection regime which collects relevant data with no ambiguity. The inspection plan is vital, it has to use Risk-Based techniques to define what to inspect, how to inspect and when to inspect.

But that in turn is only useful if there is in place a common and simple system for recording and transmitting and storing the inspection data, in other words it needs an overall management tool as described before.

AIM – STRATEGY: RISK-BASED INSPECTION PLAN

Summarizing the different steps of the AIM methodology, data should be collected, analyzed and processed; then specific analysis – HAZID, HAZOP, fatigue analysis, corrosion assessment – should be performed to identify all hazards and threats to the integrity, if not already covered by the data managed. Based on all this information, the risk assessment of the identified accidental scenarios can be performed and an inspection programme can be established based on risk considerations.

Risk-Based Inspection (RBI) applicable parameters must include coating degradation, corrosion, fabrication quality control, fatigue and crack propagation, and well-thought through risk characterization, which measures each risk in terms of safety, the environment and the business impact.

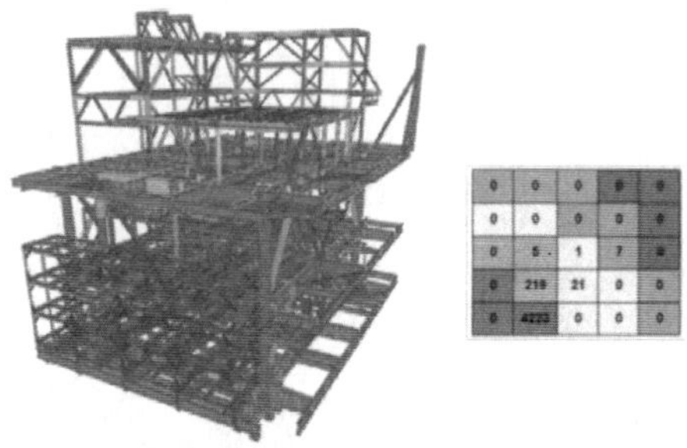

Risk Level	Inspection Interval
V	Mitigation Required
IV	1 years and NDT
III	2 years and CVI
II	3 years and CVI
I	5 years and CVI

Figure 11: RBI programme visualized on geometrical model

Information management is critical to properly order the inspections of elements and to properly report the inspection that has been performed. It is then highly recommended to use geometrical models – as previously described – to report the RBI program detailing location of inspection, inspection conditions/method and findings (if any) of previous inspection campaigns.

Geometrical models including the RBI programme and inspections feedback shall then be uploaded into the overall management system to keep the 'AIM loop' updated and closed.

Benefits include minimized operation downtime, inspections focused on critical areas and components, not only prescriptive, implementation of preventive programs and planned maintenance activities, inspection program accounting for operating conditions and accessibility to structural areas. It is a demonstration of compliance for regulators and internal/external stakeholders that the integrity management program meets expectations. It is in line with similar frameworks proposed by recognized industry recommended practices and standards, such as the proposed API RP 2SIM.

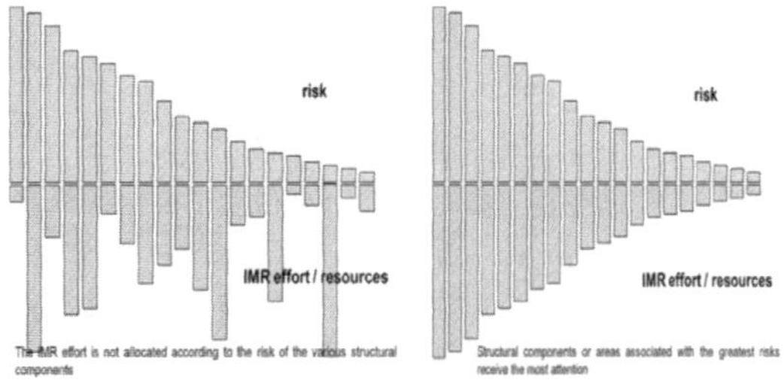

Figure 12: RBI objective: inspection effort optimization

AIM – PROGRAM: INSPECTION ACTIVITIES & OFFSHORE EXECUTION

Based on the selected path the actual strategy is drafted followed by the actual operational implementation on the unit. The integrity manager has to set the availability and safety criteria for the asset, and risk ranking and prioritizing flow from that. The class, state, verification and company requirements have to be married up with an inspection scheme and a system which is accessible to all the partners and departments involved.

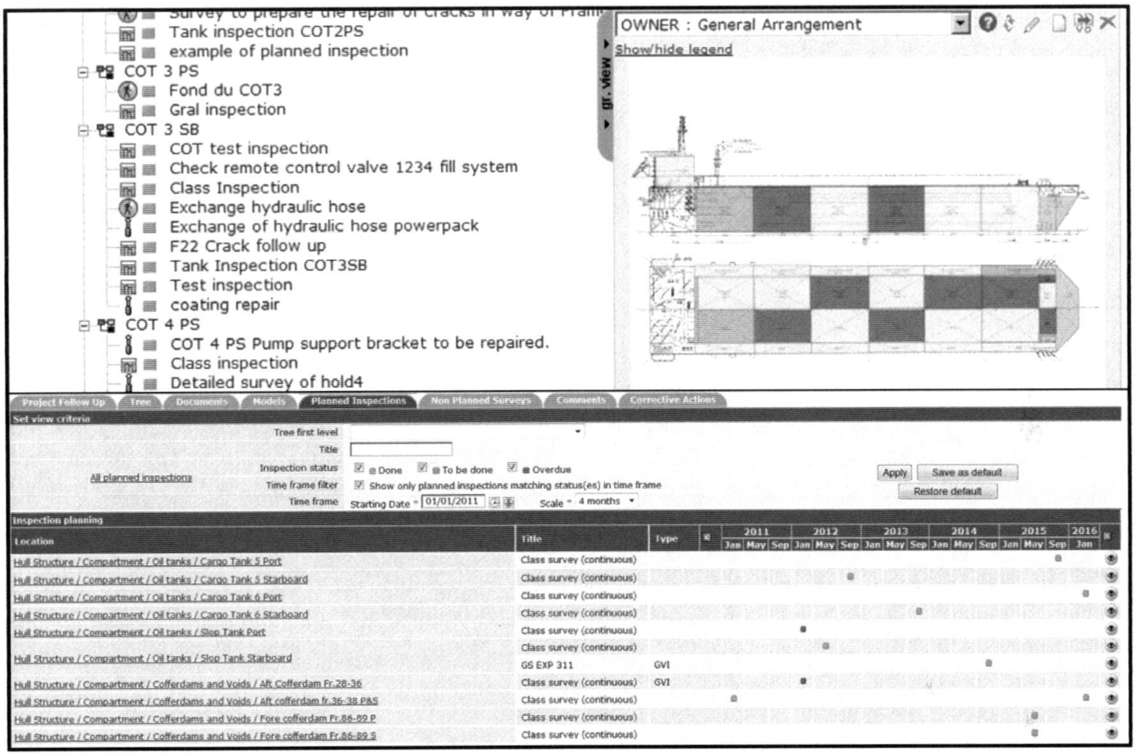

Figure 13: Inspection schedule

The Asset Integrity overall management system will allow the surveyor to retrieve the inspection scope of work with detailed information thanks to the embedded data center. After its survey, reports will be uploaded into the system, processed and stored. Corrective actions for instance will be automatically tracked.

Finally, it has to be reminded that the Asset Integrity Management has to be continuously updated, and inspection results will be the feedback for a new evaluation so as to close the loop!

CONCLUSIONS

This paper presents the main steps to develop an effective Asset Integrity Management methodology ensuring safety and reliability. It shall integrate all Asset Integrity specific services previously described, including corrosion assessment, RBI, inspection schedule and fatigue hot spot identification. At its heart is a continuously updated assessment loop: data collection, RBI analysis, inspection schedule, survey reports, data update…

Benefits include minimized operation downtime, inspections focused on critical areas and components, implementation of preventive programs and planned maintenance activities, inspection program accounting for operating conditions and accessibility to structural areas. It is a demonstration of compliance for regulators and internal/external stakeholders that the AIM program meets expectations. It is in line with similar frameworks proposed by recognized industry recommended practices and international standards.

All those different analyses methodologies were presented to demonstrate how important it is to select the most suitable methodology according to the main goals of the project, considering as well the project phase and quantity of information available at the time of the analysis performance. In order to develop a successful Asset Integrity Management plan, key members of the workforce, including senior management and other key decision makers, must be involved. The outputs of the Asset Integrity Management process will provide the key workforce members with the information needed to make inspection prioritization and optimization.

On this basis, it can be concluded that Integrity Management enables the oil & gas companies to justify the investment to improve the safety of the installation and to create a greater confidence in performing process and transfer operations in Offshore Oil & Gas units.

REFERENCES

American Petroleum Institute, "Recommended Practice for Structural Integrity Management of Fixed Offshore Structures", API RP 2SIM

UK HSE Books 2006 – Plant Ageing, prepared by TWI Ltd, ABB Engineering Services, SCS Ltd and Allianz Cornhill Engineering

Bureau Veritas. Safety Studies Workshops Methodology Statement. Bureau Veritas Internal Guide-line. 2010

Health and Safety Executive (HSE). Reducing Risks, Protecting People. HSEs Decision-making Process. HMSO, London. 2001b

International Maritime Organization (IMO) risk matrix – IMO recommendations. In Guidelines for Formal Safety Assessment (FSA) (MEPC/Circ.392 MSC/Circ.1023). International Maritime Organization (IMO). 2002

TNO. Guideline for Quantitative Risk Assessment (The Purple Book). Netherlands: Directorate General of Labour.

ULTRA-DEEPWATER GAS FIELD DEVELOPMENT OPTIONS: A STUDY ON FLNG AND 'SUBSEA-TO-BEACH' OPTIMAL APPLICABILITY

LUIS ALEJANDRO BARRERO
SR. TECHNICAL ADVISOR, GRANHERNE, A KBR COMPANY

FRANCESCO BELTRAMI
SR. TECHNICAL ADVISOR, GRANHERNE, A KBR COMPANY

Proceedings of the 21st Offshore Symposium, February 2016, Houston, Texas
Texas Section of the Society of Naval Architects and Marine Engineers

ABSTRACT

Multiple development options are currently available for the production of LNG with the gas extracted from ultra-deepwater gas reservoirs. The main are (a) Subsea-to-Beach coupled with onshore LNG - with or without late-field-life gas compression; (b) Dry-tree/ Wet-tree Host coupled with onshore LNG; and (c) Floating LNG.

Subsea-to-Beach and Dry-tree/ Wet-tree Hosts are field-proven options, within their own respective technological limits and financial applicability boundaries, and are being used for major capital projects. FLNG is now in the commercial phase with several worldwide projects under execution, and is becoming a more and more popular alternative development 'building block' to be adopted in some specific cases.

Several factors affect the decision on the 'best' development option for ultra-deepwater gas reservoirs. The readiness status of subsea technologies is one of these factors due to the fact that a subsea production system is required in all the development options discussed here, as well as location and features of the reservoir(s) to be produced, the recoverable reserves, and the size of the LNG train(s) and of the entire offshore and onshore facilities.

In the light of the increasingly wider interest of the industry in FLNG developments for offshore gas fields, it is worth creating simplified, yet reliable and robust, work methods to confidently discuss the Subsea-to-Beach and FLNG alternatives for a given project.

The present paper is a contribution to this discussion on Subsea-to-Beach versus FLNG, especially to the identification of the applicability limits for these development options.

Relevant Subsea-to-Beach and FLNG projects are selected and reviewed; the technological limits of the enabling subsea technologies are investigated; the key technical and decision factors in the Concept Selection are assessed and then discussed with respect to non-technical factors such as HSE, risks, economic viability and technology readiness status.

Keywords: Deepwater Gas Field, Offshore Field Development, Subsea-to-Beach, FLNG, Subsea Production System

INTRODUCTION

This paper was conceived to continue a multi-disciplinary investigation on the selection criteria to be adopted while developing deepwater and ultra-deepwater gas fields, especially with respect to the choice between Subsea-to-Beach and FLNG. Two studies were conducted in late 2014 and early 2015. Their results were presented at international conferences but, due to the relevance of the subjects discussed, the work continued during the second half of 2015. This paper was written to document the progress made by the industry over the past months and to add to the discussion the floating Host concept (with either dry trees or wet trees). For these reasons, the main objectives of this paper are:

- Projects Review: an updated review of public domain information on selected projects is presented, showing typical examples for the different options
- Technologies Investigation: the main technologies required by the different options are revisited and presented; technologies needed by future projects are discussed
- Technical Factors and Decision Factors: the paper presents the investigation of the most critical factors affecting the preliminary design of production systems for deepwater gas field developments, and
- Non-Technical Decision Factors: a brief qualitative review of the non-technical factors affecting concept selection and preliminary design.

METHODOLOGY

Projects Review

Sixty-nine (69) projects have been reviewed during this study. They are listed in Table 1 to Table 7 and refer to six (6) long subsea tie-backs to a pre-existing host; seventeen (17) Subsea-to-Beach projects (past and present); seven (7) Subsea-to-Beach projects (future); fourteen (14) present and future FLNG projects; eight (8) fixed and floating Hosts for deepwater gas development, not necessarily coupled to LNG production; six (6) shallow water hosts connected to large onshore LNG plants in different countries; and eleven (11) deepwater GoM floaters designed to handle large quantities of gas in addition to oil processing. This last group is considered as a sort of term of reference to show where the industry capabilities stand at this time. Similarly, the inclusion of shallow water gas projects, and of other deepwater projects not associated with LNG production, provides a benchmark for the deepwater and ultra-deepwater gas field developments that are the main focus of this paper.

Over the past two decades the industry has looked with increasing interest at Subsea-to-Beach projects as an effective way of developing deepwater gas fields. Several operators have adopted this approach in different geographic areas, in mature fields and new provinces, to produce large individual gas reservoirs or group of smaller reservoirs. Large Subsea-to-Beach projects designed to feed onshore LNG plants are currently under execution: in the near future, the world LNG market is likely to change as a result of the new multi-MTPA (million ton per annum) LNG production capacity associated to Subsea-to-Beach, especially due to large projects such as Gorgon in Australia (15.6 MTPA) and to other projects in Mozambique (50 MTPA). The current global LNG production is approximately 240 MTPA (stable over the past four years), while the overall liquefaction plants capacity is around 300 MTPA across the world. At this time of writing is not clear how the LNG market will develop in the near future and if the expected growth (up to possibly 500 MTPA) will happen for real.

As the Subsea-to-Beach, the floating Host is another field-proven option within its own respective technological limits and financial applicability boundaries, and is being used for major capital projects. During the present investigation, examples of deepwater Dry-tree Hosts and Wet-tree Hosts were searched in the public domain information to determine if these concepts have been used in conjunction with onshore LNG. The partially unexpected results of this specific investigation are discussed in the next section of the paper.

The industry has also seen the potential of FLNG developments both in shallow water and deepwater. FLNG is becoming more and more popular as an alternative development 'building block' to be considered in some specific cases. FLNG is now in the commercial phase with several international projects either in execution or in early study stage. The success of these projects will open a new market enabling the commercial exploitation of abundant stranded gas reserves. FLNG may be also used to produce landlocked gas reserves (for example shale gas): this application has not been investigated in this paper as the focus is here on deepwater gas field developments.

These are the main development options that are currently available for the production of LNG with the gas extracted from ultra-deepwater gas reservoirs. Some projects from these different developments options for ultra-deepwater gas fields were selected from public domain information and were reviewed analyzing their main parameters.

Technologies Investigation

To better discuss advantages and disadvantages of the different developments options in a given project, it is necessary to preliminarily investigate the readiness level, qualification status and market availability of the main components of the required subsea production system (SPS). This was carried out in a previous paper (Beltrami, 2015): the results of that study have been revised and updated here, while the main findings are confirmed.

The main technology gaps and most critical challenges are discussed with respect to field development and as a part of a wider discussion on how to exploit deepwater gas reservoirs. Technology readiness needs to be considered when sanctioning an offshore project, and it plays a critical role while assessing the feasibility of a given project within a pre-ordered timeframe. Being aware of the technologies currently available on the market will help operators to confidently narrow down the number of initial concepts created during concept selection. Looking at the future market trends will enable the subsea manufacturers to effectively develop the technologies required by the upcoming projects.

Key Technical Factors and Decision Factors

A large number of different technical factors and decision factors have to be taken into account. Since it may be difficult to deal with the abundant data provided by different disciplines that is subject to modifications during the initial stages of a project, it is worth attempting to determine which factors have the most relevant impact on the development decision. Focusing on a limited number of factors will help Concept Definition and Selection, and will likely increase the confidence of the decision makers while selecting the developments options to be progressed to Pre-FEED. Key technical and decision factors for deepwater gas field development have been extensively studied in the past (see Beltrami and al., 2015). That paper is the starting point of the present discussion on the key factors that affect the practical applicability of the different concepts under examination (namely Subsea-to-Beach, FLNG, and floating Host): after reviewing public domain information on selected projects, a case study was proposed to determine how technical feasibility and SPS capital expenditure (CAPEX) were affected by the main different factors previously short-listed. The case study was conceived adopting a multi-discipline approach based on offshore field development; flow assurance; subsea engineering; pipeline engineering; and cost engineering. The cases with high subsea CAPEX were then investigated as possible candidates for FLNG application.

Non-Technical Factors

Examples of non-technical decision factors to be considered are: geopolitics considerations; offshore lease terms and conditions; local regulations; contractual strategies and market conditions; decommissioning requirements; risk mitigation and management for technical and non-technical risks. Non-technical factors are very important but are also difficult to quantify. For this reason, it is advisable to identify the sources of uncertainty at the beginning of a project. A brief overview of these non-technical decision factors is presented in this paper.

RESULTS

This section presents and discusses the results obtained through the Projects Review and the Technologies Investigation. The key technical and decision Factors and the non-technical factors are also discussed.

Projects Review

The main results are summarized in Figure 3. This chart shows each reviewed project with two key factors that affect the preliminary design of the field development, i.e. the step-out distance from shore (km) and the prevailing water depth (m) of the subsea wells. Four main Sectors have been identified in the chart. They will be used to simplify the discussion in the sections that follow. The red horizontal line (corresponding to 300m WD) represents the boundary of shallow water as defined in this paper.

Sector I mainly comprises present Subsea-to-Beach project, but other concepts are present too. Sector II contains future Subsea-to-Beach and long subsea tie-backs, plus different types of floaters. Sector III shows examples of fixed and floating hosts and some FLNG applications. Only one present project populates Sector IV: Independence Hub appears to be the unique example of ultra deepwater gas field with a distance from shore exceeding 200km. The other possible project in Sector IV is the FLNG vessel considered for Santos Basin in Brazil: this project is understood to be postponed at this time. To better explore Sector IV, selected GoM floaters (Wet-tree Hosts and Dry-tree Hosts) were studied as examples of deepwater and ultra-deepwater development concepts. The results of this additional study are presented in Figure 4.

The long subsea tie-backs were carried out to explore the limits of the subsea technologies and to produce gas that otherwise would not be exploitable, see Table 1. In this table, the step-out distance is the distance between the wells and the pre-existing host. Some of these fields are very far from shore and reporting the distance between the wells and the mainland might be misleading. Long subsea tie-backs are mentioned here because they remain a technical benchmark for the following Subsea-to-Beach projects. Examples of Subsea-to-Beach, FLNG, and floating Host applications are discussed in the notes that follow.

Subsea-to-Beach

The typical Subsea-to-Beach project is shown in Figure 1 - subsea wells in the bottom left corner - together with the subsea tie-backs to a shallow water host (fixed platform) and to a deepwater floating Host. The main difference between these development concepts is that Subsea-to-Beach requires multiphase export to the onshore processing plant while in the subsea tie-backs the gas is processed offshore. Treated, single-phase gas is then exported to the onshore plant. Subsea-to-Beach is also represented in Figure 2.

A selected list of present and future Subsea-to-Beach projects is presented in Table 2 and in Table 3, including the cases where the produced gas is fed to an onshore LNG plant. Notable existing Subsea-to-Beach projects coupled to onshore LNG production are Snohvit in the Barents Sea and the West Delta Deep Marine (WDDM) concession in Egypt. Notable future projects are for example Gorgon and Jansz-lo (under construction in Western Australia). Similarly, Subsea-to-Beach and onshore LNG are going to be adopted in East Africa, a new, promising gas province.

The known applicability limits of Subsea-to-Beach are shown in Figure 3 (water depth to step-out distance). The SPS technical feasibility extends up to approx. a 200km step-out distance from shore and up to 1,500m water depth (WD). It is understood that the step-out distance in future projects may not exceed these current applicability limits, while the water depth is going to exceed 2,000m. Subsea-to-Beach projects are unique because of their:

- Lack of offshore host (fixed platform in shallow water, floater in deeper waters) and of production risers
- Multiphase flow, wet gas with liquids (not processed offshore)
- Natural flow, no offshore or subsea compression (though may require compression at a later stage of field life)
- No pipeline insulation; need to constantly inject mono-ethylene-glycol (MEG) to prevent hydrate formation
- Long step-out distances from subsea wells to onshore terminal
- Large volumes of gas to be handled: great number of subsea wells in some cases and very large productivity of individual subsea wells in other cases
- Need for engineered shore approach.

Ultra-deepwater Subsea-to-Beach projects (1,500m WD and more) also require:

- Transition from deep offshore to continental shelf affecting pipelines design
- Power and communication distribution over long distances.

Subsea-to-Beach projects share with long gas subsea tie-backs the step-out distance range, within 200km, see Figure 3, but at present their water depth is limited below 1,500m. Looking at Figure 3, it can be implied that future Subsea-to-Beach projects will go to deeper waters. Examples of these projects are understood to be Block 2 from Statoil in Tanzania (2,500m), and some other fields from BG also in Tanzania (2,000m).

Dry-tree/ Wet-tree Host

Figure 1 shows a typical shallow water production host (fixed platform) and a typical deepwater production host (floater). In this case the floater is connected to subsea wells (wet trees). In principle, a Dry-tree Host may be possible too. This concept is not shown in Figure 1. A floating host might be also required for the late-field-life compression of the produced gas, especially if the subsea wells feed an onshore LNG plant. This concept is presented in Figure 2. It should be noted that also in this case, the floater is a Wet-tree Host. This is due to the fact that this host is an addition to a project that initially starts as a Subsea-to-Beach development.

Relevant deepwater gas fields that have been developed, or are being developed, by means of shallow water fixed hosts or deepwater floaters are presented in Table 5. Not all these projects are finalized to LNG production; nevertheless the proposed examples properly reflect the current status of the offshore industry.

The floating Host option is being used for major capital projects, for example in Inpex Ichthys. This was the largest discovery of hydrocarbon liquids in Australia in more than 40 years. The Ichthys LNG project ranks among the largest and most significant energy projects in the world. The project includes some of the world's biggest and most advanced offshore facilities, massive onshore processing facilities near Darwin in the northern Territory of Western Australia and an 889km-pipeline to unite them. The unusual length of this pipeline is due to environmental constrains that dictate a relatively distant LNG plant location. The Ichthys Field covers an area of around 800 km^2 in water averaging depths of around 250 meters. The estimating production starting year is 2017, see Table 5.

The Shtokman gas and condensate field is located in the central part of the Russian sector of the Barents Sea shelf, about 600km northeast of Murmansk, where sea depth varies between 320m and 340m.
Field operations will be performed using deepwater production systems and special floating production units (FPU) equipped with quick-disconnect system and able to swiftly move off icebergs. The produced gas will be conveyed via the offshore trunklines to the coast near the Teriberka settlement where the LNG plant will be located. Gas will be produced by twin four-slot templates and then it will be conveyed by flexible production risers from a template to the FPU. Gas and condensate separation, and gas processing operations will be performed aboard the FPU. After that, processed gas will be exported from the FPU by flexible risers. A special device will connect risers to two trunklines that will carry gas to onshore facilities.

The Malampaya field is located 80km off the coast of Palawan Island, in the Republic of Philippines. The platform consists of a deck, supported by a concrete gravity sub-structure (CGS). The platform is located in 43m WD while the deepwater subsea wells are at depth of 850m. The processed gas is compressed and exported through a 504km pipeline to the onshore facility. The field has been developed using a manifold cluster of subsea wells tied back to the platform.

The Pluto facility operates as "not normally manned" and is controlled from the onshore Pluto Gas Plant near Karratha in Western Australia. Operations include production of hydrocarbons from the reservoirs, receiving hydrocarbons at the Pluto platform, located in 85m WD, and transport via the trunkline to the onshore facility for processing. Gas is piped to shore along a 180km, 36-in pipeline to the new Burrup LNG Park, located between the North West Shelf Venture and Dampier Port. Also in this case, the pipeline routing is understood to reflect environmental constrains. The Burrup LNG park has been designed to aggregate gas from the Carnarvon Basin to support future LNG growth and create another domestic gas supply hub for Western Australia.

Located approximately 240km southeast of Venice, Louisiana in more than 2,400m WD, the Independence Hub is the deepest offshore gas platform ever installed in the world (the current world record for an oil-producing floater is Cascade & Chinook FPSO.) The platform is a deep-draft, semi-submersible platform that is connected initially to eight (8) natural gas fields. Production from the hub is transported to shore via the 216km Independence Trail Natural Gas Pipeline. This project is mentioned here because it is an important example of deepwater gas field developments, even if not directly related to onshore LNG production.

Notable examples of fixed hosts used to develop shallow water gas fields to produce LNG in nearby onshore locations are listed in Table 6. Sakhalin is an oil and gas development, but is mentioned here because it represents a valid way of recover and monetize hydrocarbons that otherwise would be wasted.

It can be noted that fixed and floating Host projects are mainly located on Sectors I, II and III of Figure 3, with the unique exception of the above mentioned Independence Hub floater. Hosts projects cover a great range from step-out distances, going from 35km to 889km and also cover a great WD range, from 75m to 2,440 m. However, Host projects are not found for great step-out distances combined with high WD.

It has also to be noted that only Wet-tree Hosts have been encountered in the present investigation, although it is understood that various Dry-tree/ Wet-tree Hosts were initially considered for Ormen Lange (see Gustavsson, 2003). The possible lack of examples of Dry-tree Hosts selected for deepwater gas field development may be due to the fact that only a limited number of large deepwater and ultra-deepwater gas fields have been developed in practice so far.

FLNG

A typical illustration for ultra-deepwater FLNG is shown in Figure 2. Floating LNG has a relatively short history: some projects have been sanctioned and are currently in the execution phase, others have been temporarily postponed due to the present market conditions, see Table 4. The success of these projects will open a new market enabling the commercial exploitation of abundant stranded gas reserves.

The Prelude FLNG development in Western Australia is an example. It is located in 250m WD and 200km from shore. FLNG is also currently being developed for deepwater gas fields. It is worth noting that FLNG was initially considered for an ultra-deepwater development in the Brazilian offshore (Santos Basin). Even if that project appears to be on hold at this time, it indirectly confirms the potential of FLNG in ultra-deepwater.

Subsea Technologies Investigation

A Subsea-to-Beach project may require an extensive qualification program for technologies that are not yet field proven. Technology readiness issues are less frequently encountered for the Subsea Production Systems associated to FLNG. It can also be expected that flow assurance challenges will be easier to manage with the smaller size SPS required by FLNG than by the large Subsea-to-Beach SPS.

Regarding Subsea-to-Beach and subsea tie-backs, Figure 3 shows an interesting fact: some of the subsea technologies required for future ultra-deepwater Subsea-to-Beach projects have been already employed by past and existing long gas subsea tie-backs. Despite the challenges to produce gas reservoirs in 1,500m to 2,000m WD at a 200km distance from the coast, technologies and operational lessons learned are available to the industry as a valid starting point.

Nowadays, subsea hardware and controls are designed for 3,000m WD even when used in shallower waters. Technology qualification programs, according to API or DNV, are carried out as applicable to ensure that the subsea technologies that are not field-proven for a given project undergo a complete test program to confirm their fitness for purpose.

Flowlines, pipelines and trunklines are designed according to API or DNV codes and standards to meet the project requirements. Terminations are provided with hubs to connect jumpers, and valves to be used to isolate a line if required during the field life. The design of long pipelines and trunklines can be challenging especially if the wells are far from the coast and there is a transition from the deep offshore seabed to the continental shelf. Another challenge in pipeline design is the shore approach and crossing, as this requires specialized technologies to minimize the impact on the environment. Pipelines need to be piggable with smart pigs. Even if the design and the installation of these ultra-deepwater pipelines may be challenging, the current technologies permit deeper waters and longer step-out distances. A special pipeline is the MEG pipeline used to transfer MEG from the onshore terminal to the offshore location(s). This smaller size pipeline is usually piggy-backed to a larger production pipeline. The MEG leaving its pipeline enters an infield control umbilical: for this reason, MEG cleanliness may be an issue.

Control Umbilicals contain multiple hydraulic and chemical tubes to supply control fluid and chemicals respectively; power and communication (on copper) quads and fiber optics (for large data volume communications). Considering the long step-out distances of these projects, umbilical costs are relevant. Electro-hydraulic multiplexed control systems are mature technologies even if the exchange of data and power distribution over long distances is still challenging. Subsea transformers will be needed to ensure the required electric power is effectively available on the seabed. Large umbilical cross sections will be required for high power distribution over long distances. Another challenge for control umbilicals is the maximum length that can be produced without splicing (in some cases it is not possible to manufacture the required length in one run and splicing cannot be avoided). The size, weight and complexity of some deepwater umbilical terminations have increased over time: it might be necessary to re-think these terminations, and also how subsea distribution units are designed. The main technical challenges for ultra-deepwater Subsea-to-Beach gas developments appear to be:

- Possible need for offshore (host) or subsea gas compression when the subsea wells start depleting. Offshore compression is technically feasible while, at this time, subsea gas compression is not yet a completely mature technology. Ormen Lange subsea compression project has been temporarily suspended, see (Rijnbeek, 2014), while Asgard and Gullfaks subsea compression projects started their operations in summer 2015. The operational results of these two new projects will help understanding if in the future subsea gas compression becomes viable for ultra-deepwater Subsea-to-Beach projects

- Subsea power distribution equipment and power cables (large cross sections), especially in case of subsea gas compression where tens of MWs are required at the seabed
- Reliable high-speed subsea communications over long distances capable of handling increasingly larger data volumes due to intelligent well completions and subsea sensors (such as WGFM, sand detector, etc.)
- MEG cleanliness control after regeneration (including possible subsea filtration systems); MEG dosing to ensure proper hydrate formation prevention; cocktailing of MEG and other chemicals, if possible, to limit the number of chemical tubes in the control umbilicals
- Obsolescence of electronic components and backward compatibility of new controls equipment added to the subsea production system: this is an issue for all SPS's but is here more relevant because the design life of Subsea-to-Beach projects is shifting from the usual 20-25 years to 30 years and above: for example, it is understood that the design life for Gorgon pipelines and subsea controls is 50 years.

Marine Technologies Investigation

The undeniably limited presence of ultra-deepwater gas fields developed with a Dry-tree/ Wet-tree Host appears to be motivated mainly by historical and commercial reasons, and not by technical reasons. Designing ultra-deepwater floaters is well within the current industry capabilities. To better understand this fact, a list of GoM floaters, designed to produce oil and relevant amounts of gas, has been selected, see Table 7. The water depth of each Host has been searched in the public domain, while the minimum distance from shore has been retrieved from information made available by BOEM. In the end, the applicability of these Hosts has been compared to the applicability range of the other field development options. Most of the fields listed in Table 7 have been developed with Wet-tree Hosts. The only examples of Dry-tree Hosts are TLP's: Magnolia, Marlin and Ursa.

The results shown in Figure 4 seem to confirm that, broadly speaking, Subsea-to-Beach and Hosts share the same applicability range, although Hosts may be used for step-out distances exceeding the 200km technology limit existing for Subsea-to-Beach.

Key Technical Factors and Decision Factors

As discussed in (Beltrami et al., 2015) there are several technical and decision factors at play in the Concept Selection for a deepwater gas field, but only a limited number of real 'key' factors need to be taken into account. It is worth remembering that technical factors are those that cannot be modified, for example the metocean conditions of a given offshore location. The main technical factors that were shortlisted in that study are:

- Recoverable reserves
- Step-out distance from shore
- Water depth at wells location.

On the contrary, the decision factors are the results of an intentional design choice, for example between specifying single or dual flowline(s) or defining the number and distance of different drill centers. In the above mentioned paper, these main decision factors for deepwater gas field development were studied:

- Number of wells per drill center.
- Well flowrate
- Overall production flowrate
- Drill center count
- Drill center distance.

The most critical factors identified on the basis of the SPS CAPEX, still applicable to the Subsea-to-Beach applications discussed here, were the step-out distance from shore (technical factor, T); the overall production flowrate (decision, D) and the drill center distance (D). Less critical factors were the recoverable reserves (T), the water depth at wells location (T) and the drill center count (D). Somehow intermediate factors appeared to be the well count (D) and the well flowrate (D).

Repeating the same exercise taking into account the cost of drilling and completion for the required subsea wells, the most critical factors become the overall production flowrate (D), the well count (D), the step-out distance from shore (T) and the water depth at wells location (T). In this paper, the focus was on the step-out distance from shore and on the water depth at the wells location. Being technical factors, these parameters cannot be changed and appear to be critical in orienting the decision on how to develop a given offshore gas field. The overall production flowrate and the well count are decision factors. They may be modified and optimized to improve the project economics. They are similarly very important, but in the end may not be as critical as the step-out distance from shore and the wells water depth.

Non-Technical Factors

Non-technical risks will act as a go/ no-go regulator in making the final choice between possible development options. Examples of project uncertainties are:

- Local HSE requirements, including decommissioning
- Oil and gas market conditions
- Global geo-political situation
- Relationship with governments, authorities, trade unions, other parties including local content requirements
- Relationship with partners and other stake-holders
- Offshore lease terms and conditions
- Contractual strategy.

Examples of possible financial risks are:

- Project delays
- Cost escalation during project execution
- Plant(s) non performing as expected
- Need for de-bottlenecking and brownfield work
- Late life compression requirements for Subsea-to-Beach.

Taking into account these non-technical factors in a worst-case scenario, it can be concluded that Subsea-to-Beach coupled to onshore LNG offers these flexibility measures:

- SPS reconfiguration and expansion
- LNG train additions on the mainland
- Increased safety: no people working offshore.

At the same time, in the same worst-case scenario, FLNG and floating Hosts offer these advantages:

- No need for late life compression
- Smaller environmental footprint
- Reduced decommissioning costs.

Floating Hosts have a longer history than FLNG, but this is likely to change when the planned shallow water and deepwater FLNG vessels start operations in the near future.

As a concluding remark of this discussion, the importance of the 'right' sizing of production facilities should not be underestimated. Offshore projects are sanctioned only in presence of robust economics supporting the field development. Identifying the right size of the project may be challenging and risky: larger facilities require great investments to be repaid, but at the same time offer the largest rewards. Smaller facilities are less capital intensive, but the profitability of the development may be limited by this size.

Concepts Applicability Ranges

Table 8 presents the main results of the present investigation with respect to the development concepts that have been already used or are being adopted to develop deepwater gas fields producing LNG. The concepts proven at this time of writing are Subsea-to-Beach for deepwater and ultra-deepwater application and fixed Host for shallow water and for deepwater application: in this last case the wells are in deeper waters than the Host.

Several development concepts are not field proven yet, but projects are under execution. The experience with the first generation of FLNG vessels will be particularly interesting to the industry to determine if this development concept can be used in different geographic areas and in various environmental conditions. Wet-tree Hosts may be a valid alternative to FLNG and to Subsea-to-Beach but this has to be necessarily discussed on a case-by-case basis.

Considering both Table 8 and Figure 3 it appears that FLNG and Hosts have - in principle - the widest applicability ranges and that Hosts may be valid candidates for Sector IV in the same Figure 3. This appears to be somehow confirmed by Figure 4: Lucius spar is a recent example of ultra-deepwater host located very far from the shore.

In the light of the same results, Subsea-to-Beach may be easier to justify if the offshore field(s) location is close to the shore, and if a smaller number of drill centers is required (resulting in smaller production flowrates.) FLNG might work better in presence of a large reservoir or a cluster of reservoirs located far from the shore and requiring distant drill centers, preferably in shallower waters. A Dry-tree/ Wet-tree Host may work better in deeper waters and in cases where environmental conditions or geopolitics suggests not going with FLNG.

Possible Evolution Lines

Observing Figure 5 it is possible to identify two (2) trends, one more noticeable than the other, and a possible trend too undefined to be considered as a real trend at this time. Moving from Sector I to Sector II evidently appears to be a natural evolution of the offshore projects from shallower water to increasingly greater water depths (trend 1). It can be expected that the current 1,500m WD limit will be exceeded in the near future more and more frequently. At the same time, the 200km step-out distance remains, at this time, difficult to cross. No future Subsea-to-Beach projects appear to be studied or carried out for such step-out distances.

The current limitations to the subsea technologies may explain why beyond 200km step-out distance only FLNG and Hosts have been/ are considered. The number of deepwater gas developments in Sector III is still limited but it can be expected that the evolution line from Sector I to Sector III, still not too evident, will become more important in the future (trend 2.)

In the light of the information gathered for this paper, it is not possible to conclude if and when the third trend will develop and when Sector IV will be populated with ultra-deepwater gas fields developed with FLNG or floating Hosts: at this time only limited examples are known and only Independence Hub is operational. Chances are, however, that the floater required will be a Wet-tree Host. A possible development using subsea technologies in Sector IV will require real technological breakthrough in many disciplines including subsea power distribution and control, subsea communications, electrical actuations, HIPPS and late-life subsea wet gas compression.

CONCLUSIONS

Subsea-to-Beach is a development concept that has been successfully used in shallow water and deepwater. It is now moving to deeper waters, especially in new provinces lacking production infrastructure. The current 200km step-out distance limit might be crossed in the future, but this requires intensive technology development work to ensure that the enabling technologies, including subsea wet gas compression, are qualified and ready for an ultra-deepwater environment.

FLNG is a relatively new, but versatile, development concept. It is expected to work in deepwater and ultra-deepwater and also for step-out distances exceeding the subsea technology limit (200km). Future FLNG projects already appear in Sector II and Sector III in Figure 3: it is expected that in the future more FLNG projects will populate these two Sectors. FLNG has also been considered for Sector IV: it has to be seen if an FLNG vessel will be installed in ultra-deepwater at great distance from the shore.

Several Wet-tree Hosts have been encountered during this investigation. A floater connected to subsea wells will process and compress the produced gas in order to ensure its export over long, or even remarkably long, distances. Some of the projects mentioned in Figure 3 are currently under execution. Their successful start-up and operation will hopefully confirm the validity of this development concept, whose applicability is limited to very large reservoirs due to the cost of the facilities (subsea, floater and host.)

No examples of Dry-tree Hosts have been identified in conjunction with the development of deepwater and ultra-deepwater gas fields. If confirmed, this is an interesting result. Possible reasons for not selecting this type of host may be the large size of the deepwater gas reservoirs, requiring multiple and distant subsea drill centers, and the number and size of required wells: several conventional wells or a more limited number of large-bore wells. Also the selected GoM examples show that in deepwater the preference appears to go to Wet-tree Hosts: only a limited number of Dry-tree Hosts have been encountered in the investigation, see Figure 4.

The same figure also confirmed that the technologies required to design, install and operate floating hosts in ultra-deepwater and very far from the shore are currently available; see the Lucius spar that recently started its operations. Floaters appear to be the best candidates to populate Sector IV in Figure 3. At present, this Sector only contains two projects (Independence Hub and Lucius) that are not finalized to LNG production but this may change in the future. There may be technical and non-technical reasons to explain why Sector IV is partially populated, in particular the fact that there might be a limited number of ultra-deepwater gas reservoirs discovered beyond the 200km distance from shore and in waters deeper that 1,500m. Another reason might be the fact that such reservoirs, if any, may be in international waters: this would add some bureaucratic complexity to their development.

ACKNOWLEDGEMENTS

Thanks to Javier Gonzalez for the illustrations in the Appendix and to the Granherne colleagues who have reviewed this paper providing their highly appreciated suggestions and comments. The feedback provided by Dr. S. Basu during the final review of the manuscript was also greatly appreciated.

REFERENCES AND BIBLIOGRAPHY

Beltrami, F., "Subsea Technologies for Ultra-Deepwater 'Subsea-To-Beach' Gas Field Development: A Critical Review of Current Achievements and Future Challenges", 26th World Gas Conference, Paris, France, 2015.

Beltrami, F., Olugbode, F., Barrero, L., and Subramanian, K., "FLNG vs 'Subsea-To-Beach': A Study on the Critical Factors governing Concept Selection in Deepwater Gas Field Developments", SPE Annual Technical Conference and Exhibition (ATCE), Houston, Texas, 2015. Paper SPE-174927-MS.

Boekhorst, A., Steenson, B., and van der Velde, H., "FLNG: Applying Advanced Technology to Bring More Natural Gas to Market", Offshore Technology Conference. Houston, Texas, 2015. Paper OTC-25907-MS.

Dekker, M., and Reid, D. "Deepwater Development Strategy", Offshore Technology Conference, Houston, Texas, 2014. Paper OTC-25135-MS.

Eisbrenner, K., Lee, J., and Choi, D., "Stranded Gas Field Development with Cluster LNG Technology", Offshore Technology Conference, Houston, Texas, 2013. Paper OTC-24205-MS.

Gharaibah, E., and Parsons, R. "Flow Assurance Solutions and Control Systems for Long Step Out Subsea Production Systems", Offshore Technology Conference, Houston, Texas, 2014. Paper OTC-25458-MS.

Gustavsson, F., Eriksen, R., Brekke, G., and Kraggerud, A. G., "Selecting the Field Development Concept for Ormen Lange", Offshore Technology Conference, Houston, Texas, 2003. OTC Paper-15307-MS.

Gyllenhammar, E., Boireau, C., Riviere, L., and Ulvestad, A., "Deep Offshore Gas Fields: A New Challenge for the Industry", Offshore Technology Conference, Houston, Texas, 2015. Paper OTC-25802-MS.

Hartell, W. D., and Greenwald, J. R., "Deepwater Development Production and Export Options", Asia Pacific Oil and Gas Conference and Exhibition, Jakarta, Indonesia, 2009. Paper SPE-123415-MS.

Huaiyin, L., Xuebo, D., and Kai, Z., "Review and Outlook on Arctic Offshore Facilities & Technologies, Offshore Technology Conference, Houston, Texas, 2015. Paper OTC-25541-MS.

International Gas Union, "World LNG Report - World Gas Conference Edition", 2015.

Irmann-Jacobsen, T. B., Hop, B., Awolola, T. K., Ferrario, R., and Vanzan, D., "Subsea Gas Compression for a Greenfield Long Distance Tie-Back to an Onshore LNG Facility", Offshore Mediterranean Conference and Exhibition, Ravenna, Italy, 2015. Paper OMC-2015-435.

Maddahi, M., and Mortazavi, S. J., "A Review on Offshore Concepts and Feasibility Study Considerations", SPE Asia Pacific Oil and Gas Conference and Exhibition, Jakarta, Indonesia, 2015. Paper SPE-147875-MS.

"Platforms-Nearest-Distance-To-Shore.xlsx", file retrieved on 11-Dec-2015 from the website: www.boem.gov/Platforms-Nearest-Distance-To-Shore/

Reid, D., Nunez, D., and Dekker, M., "Deepwater Development: Wet or Dry Tree?", Offshore Technology Conference Brazil, Rio de Janeiro, Brazil, 2013. Paper OTC-24517-MS.

Rijnbeek, R., "Sub Sea Compression at Ormen Lange", International Petroleum Technology Conference, Doha, Qatar, 2014. Paper IPTC-17649-MS.

Riviere, L., Decrin, M. K., Sibaud, E., and Anres, S., "Long Tie-Backs Development in Deep Waters", Offshore Mediterranean Conference and Exhibition, Ravenna, Italy, 2011. Paper OMC-2011-145.

Xia, J., and D'Souza, R. B., "Floating Production Platform Selection for Developing Deepwater Gas Fields off North West Australia", SPE Asia Pacific Oil and Gas Conference and Exhibition, Perth, Australia, 2012. Paper SPE-158717-MS.

ABBREVIATIONS

API	American Petroleum Institute
BOEM	Bureau of Ocean Energy Management
CAPEX	Capital Expenditure
CGS	Concrete Gravity Sub-Structure
DNV	Det Norske Veritas
FEED	Front End Engineering Design
FLNG	Floating LNG
FPSO	Floating Production Storage and Offloading
FPU	Floating Production Unit
GoM	Gulf of Mexico
HIPPS	High Integrity Pressure Protection System
HSE	Health, Safety and the Environment
JDA	Joint Development Area
LNG	Liquefied Natural Gas
MEG	Mono-Ethylene-Glycol
mmscfd	Million Standard Cubic Feet per Day
MTPA	Million Ton Per Annum
MW	Mega Watts
SPS	Subsea Production System
TBD	To Be Determined
TLP	Tension Leg Platform
WD	Water Depth
WDDM	West Delta Deep Marine
WGFM	Wet Gas Flow Meter

APPENDIX: TABLES AND ILLUSTRATIONS

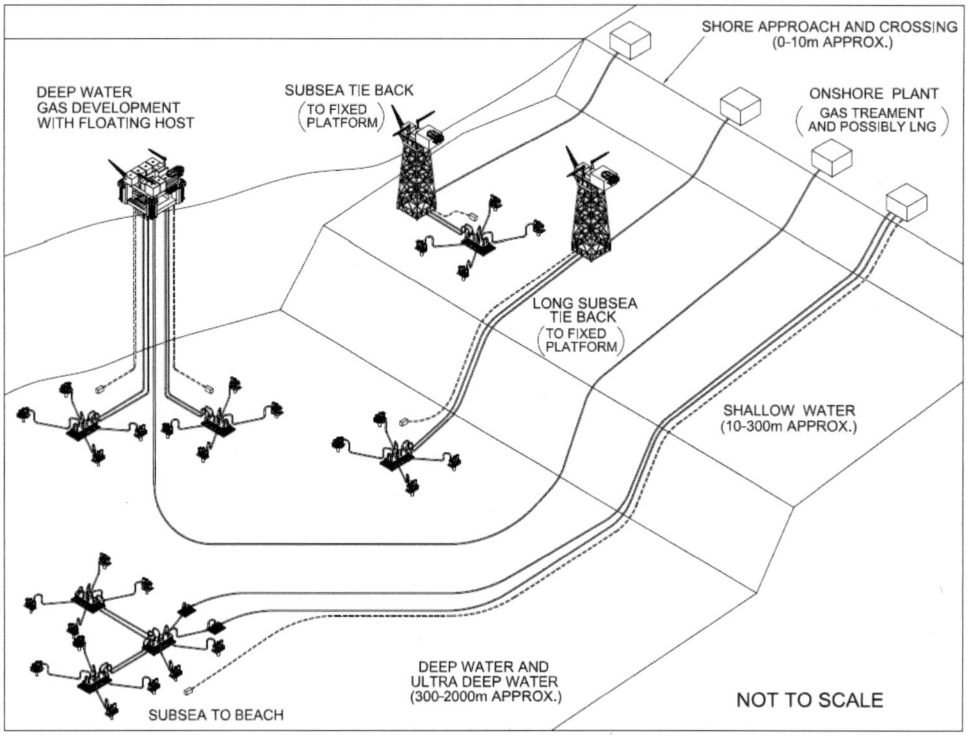

Figure 1 Offshore Gas Field Development Concepts

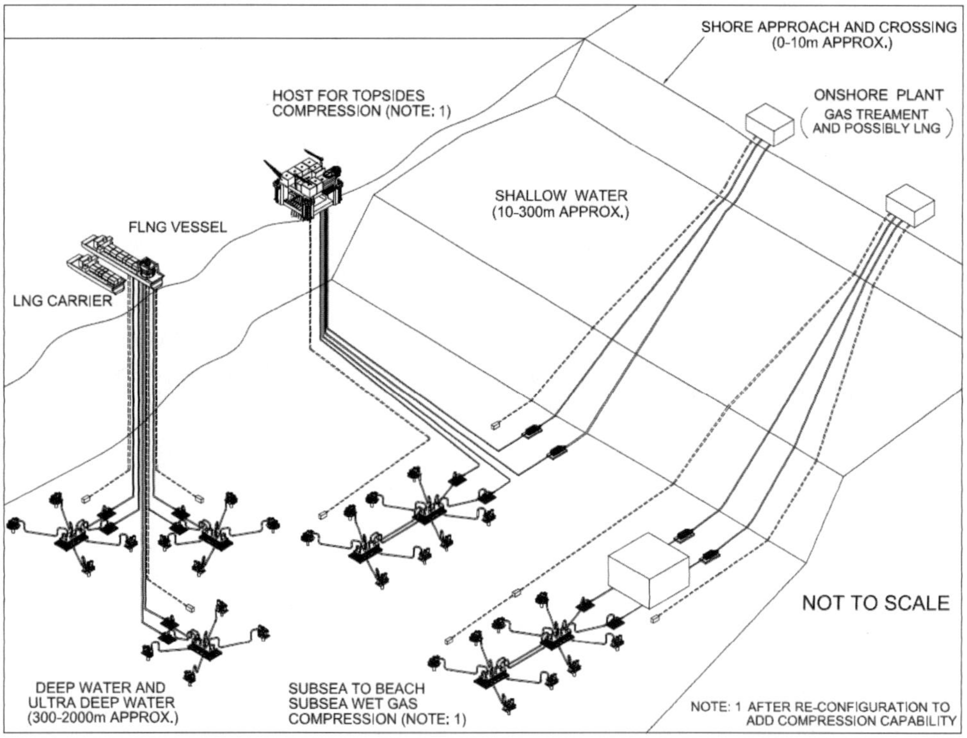

Figure 2 Offshore Gas Field Development Concepts (continued)

Table 1 Selected Long Gas Subsea Tieback Projects

Year	Operator	Field	Location	Water Depth (m)	Step-out (km)	Notes
2019E	Noble Energy	Leviathan	Mediterranean Sea	1,500	130	
2015E	Noble Energy	Tamar South West	Mediterranean Sea	1,700	148.8	
2014	Husky Oil	Liwan	South China Sea	1,500	75	
2013	Noble Energy	Tamar	Mediterranean Sea	1,700	148.8	
2002	Total	Canyon Express	Gulf of Mexico	2,195	100	
1997	Shell	Mensa	Gulf of Mexico	1,615	110	

Note: 'Step-out' distance here refers to the distance between the subsea wells and the pre-existing host.

Table 2 Selected Current and Ongoing Subsea-to-Beach Projects

Year	Operator	Field	Location	Water Depth (m)	Step-out (km)	Notes
2016E	Chevron	Gorgon	Western Australia	200	65	LNG
2016E	Chevron	Jansz-lo	Western Australia	1,350	180	LNG
2016E	Total	Laggan-Tormore	Atlantic Ocean	600	143	Phase I
2015	Shell	Corrib	Atlantic Ocean	355	83	
2013	Gazprom	Kirinskoye	Sea of Okhotsk	90	28	
2013	BHP	Macedon	Western Australia	180	75	
2013	Tullow Oil	Kudu	Atlantic Ocean	170	170	
2010	Santos	Vic/ P44	Western Australia	70	55	
2009	BG Group	Sequoia	Mediterranean Sea	570	120	LNG, WDDM VI
2008	BP	Taurt	Mediterranean Sea	108	72	
2007	Statoil	Snohvit, Albatross	Barents Sea	345	178	LNG
2007	Shell	Ormen Lange	North Sea	1,100	120	
2005	BHP	Minerva	Victoria (Australia)	60	11	
2005	BG Group	Sapphire	Mediterranean Sea	1,100	120	LNG, WDDM III
2005	BG Group	Simian, Sienna	Mediterranean Sea	1,100	120	LNG, WDDM II
2003	BG Group	Scarab, Saffron	Mediterranean Sea	850	90	WDDM I
1981	Eni	Emilio	Adriatic Sea	84	32	

Table 3 Selected Future Subsea-to-Beach Projects

Year	Operator	Field	Location	Water Depth (m)	Step-out (km)	Notes
TBD	Statoil	Block 2	Tanzania	2,400	80	LNG
TBD	BG Group	Various	Tanzania	2,000	100	LNG
TBD	Total	Laggan-Tormore	Atlantic Ocean	600	143	Phase II
TBD	Chevron	Greater Gorgon	Western Australia	300	188	LNG
2022E	Eni	Mamba	Mozambique	1,580	40*	LNG
2019E	Anadarko	Prosperidade	Mozambique	1,500	50	LNG
2019E	Anadarko	Golfinho	Mozambique	1,500	100	LNG

E = Expected
* = Estimated

Table 4 Selected FLNG Projects

Year	Operator	Field	Location	Water Depth (m)	Step-out (km)	Notes
2024E	Inpex	Abadi	East Indonesia Offshore	800	150	See Note 1
2021E	ExxonMobil/ BHP	Scarborough	Western Australia	1,000	200	
2021E	PTTEP	Cash Maple	Western Australia	130	680	Under Evaluation
2019E	Ophir	Fortuna	Equatorial Guinea, Block R	1,900	140	
2019E	Woodside	Browse	Western Australia	600	425	
2019E	Eni	Coral South	Mozambique	2,260	60*	
-	Engie (GDF)	Petrel, Tern, Frigate	W. Australia - Bonaparte	100*	250	Suspended
2018E	Noble Energy	Leviathan	Mediterranean Sea	1,600	130	Suspended
2018E	Noble Energy	Tamar, Dalit	Mediterranean Sea	1,700	148.8	Suspended
2018E	Woodside	Greater Sunrise	JDA Papua/ Australia	600	450	See Note 2
-	Petrobras	Santos Basin	Brazilian Offshore	2,200	360	Suspended
2018E	Petronas	Rotan	Malaysia Offshore	1,150	130	PFLNG2
2017E	Shell	Prelude	Western Australia	250	200	
2016E	Petronas	Kanowit	Malaysia Offshore	200	180	PFLNG1

E = Expected
* = Estimated

Note 1: FLNG capacity upgraded from 2.5 to 7.5 MTPA. New start-up date announced.

Note 2: FLNG option to be abandoned. Tie-back to Timor-Leste (150km) currently considered.

Table 5 Fixed and Floating Hosts

Year	Operator	Field	Location	Water Depth (m)	Step-out (km)	Notes
2020E	Chevron	Gendalo Gehem	Indonesia	1,829	80*	Postponed
2017E	BP	West Delta Nile	Egypt	800	75*	
2017E	Inpex	Ichthys	Western Australia	170	889	LNG
2016E	Gazprom	Shtokman	Barents Sea	340	600	LNG, Postponed
2012	Woodside	Pluto	Western Australia	850	190	LNG
2009	Reliance Ind.	KG-D6	East Coast of India	900*	35*	
2007	Anadarko	Independence Hub	Gulf of Mexico	2,440	216	
2001	Shell	Malampaya	Philippines	850	500	

E = Expected
* = Estimated

Table 6 Shallow Water Host plus Onshore LNG

Year	Operator	Field	Location	Water Depth (m)	Step-out (km)	Notes
2016E	Chevron	Wheatstone	Western Australia	260	225	LNG
2009	BP	Tangguh	Indonesia	60	130	LNG
2006	ConocoPhillips	Bayu-Undan	W. Australia - Darwin	80	502	LNG
1999	Gazprom	Sakhalin	Russia	50	15	LNG
1999	RasGas	North Field	Qatar	75	80	LNG
1989	Woodside	Northwest Shelf	Western Australia	125	135	LNG

E = Expected

Table 7 Selected GoM Deepwater Floaters with Relevant Gas Processing Capacity

Year	Operator	Field	Location	Tree	Water Depth (m)	Step-out (km)	Gas (mmscfd)
2015	Anadarko	Lucius	GoM	Wet	2,164	350*	450
2014	BP	Nakika	GoM	Wet	1,933	96	550
2004	Anadarko	Marco Polo	GoM	Wet	1,310	187	300
2004	ConocoPhillips	Magnolia	GoM	**Dry**	1,432	237	150
2003	Anadarko	Gunnison	GoM	Wet	945	248	200
2002	Anadarko	Nansen	GoM	Wet	1,120	187	200
2002	Anadarko	Boomvang	GoM	Wet	1,053	183	200
2000	ExxonMobil	Diana	GoM	Wet	1,463	220	325
1999	Eni Petroleum	Allegheny	GoM	Wet	1,005	151	195*
1999	McMoRan	Marlin	GoM	**Dry**	972	103	250
1999	Shell	Ursa	GoM	**Dry**	1,306	90	400

* = Estimated

Table 8 Applicability of Development Concepts: Present Status and Future Trends

No	WD Type	Well Completion	Host Type	LNG Plant	Status	Examples
01	Shallow water	Dry Tree	Fixed Host (Steel jacket or other type of platform)	Onshore	Proven	Qatar North Field Tangguh Sakhalin Bayu-Undan North West Shelf
02	Shallow water	Wet Tree	Fixed Host (Steel gravity base platform)	Onshore	Developing	Wheatstone (future)
03	Shallow water	Wet Tree	Floating Host (Semi-submersible)	Onshore	Developing	Ichthys (future)
04	Shallow water	Wet Tree	FLNG vessel	Offshore	Developing	Prelude (future) Kanowit (future)
05	Shallow water	Wet Tree	No Host	Onshore	Developing	Gorgon (future)
06	Deepwater	Dry Tree	Floating Host	Onshore	-	None
07	Deepwater	Wet Tree	Fixed Host (Steel jacket platform)	Onshore	Proven	Pluto
08	Deepwater	Wet Tree	Floating Host (FPU)	Onshore	Postponed	Shtockman
09	Deepwater	Wet Tree	FLNG vessel	Offshore	Developing	Abadi (future) Browse (future) Scarborough (future) Rotan (future)
10	Deepwater	Wet Tree	No Host	Onshore	Proven	Snohvit Simian, Sienna, Sapphire, Sequoia (WDDM) Jansz-lo (future)
11	Ultra-deepwater	Dry Tree	Floating Host	Onshore	-	None
12	Ultra-deepwater	Wet Tree	Floating Host (Semi-submersible)	None	Proven	Independence Hub
13	Ultra-deepwater	Wet Tree	FLNG vessel	Offshore	Developing	Coral South and North (future) Fortuna (future) Santos Basis (future, postponed)
14	Ultra-deepwater	Wet Tree	No Host	Onshore	Developing	Prosperidade, Golfinho (future) Mamba (future)

Note: 'Shallow water' here refers to 0-300m WD; 'Deepwater' to 300-1,500m WD; and 'Ultra-deepwater' to 1,500m+ WD.

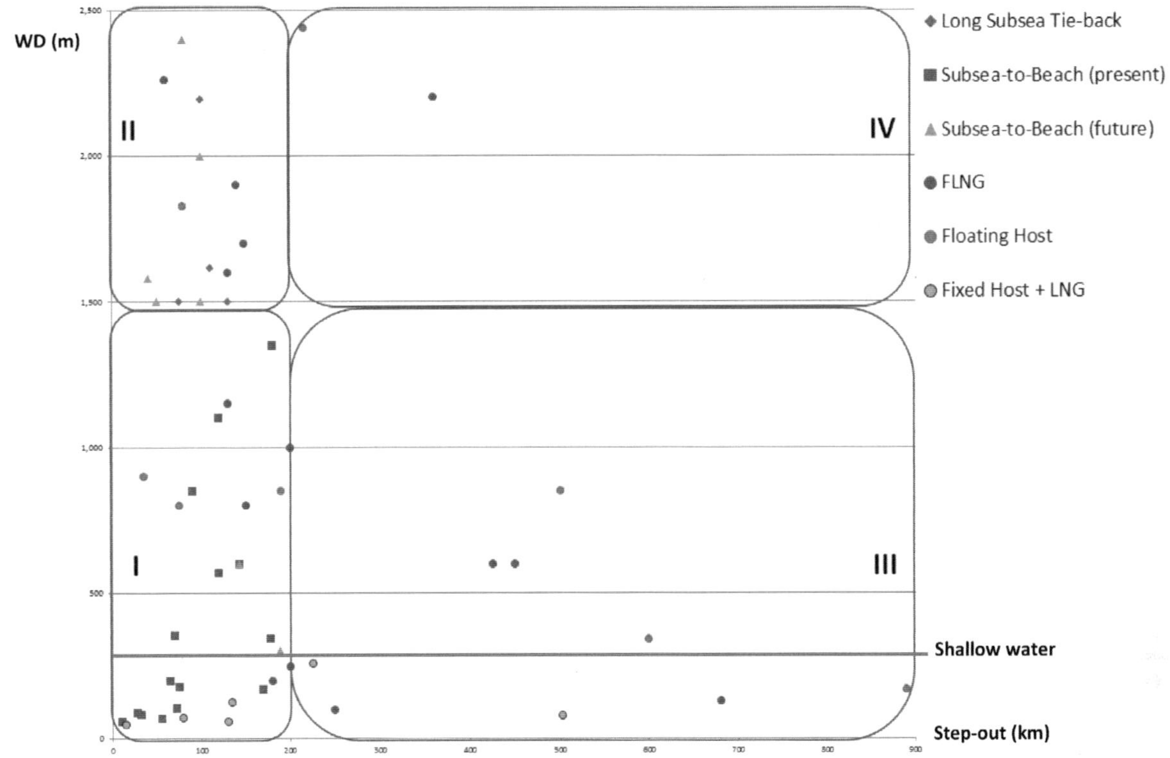

Figure 3 Applicability Ranges for Subsea-to-Beach, FLNG and Hosts

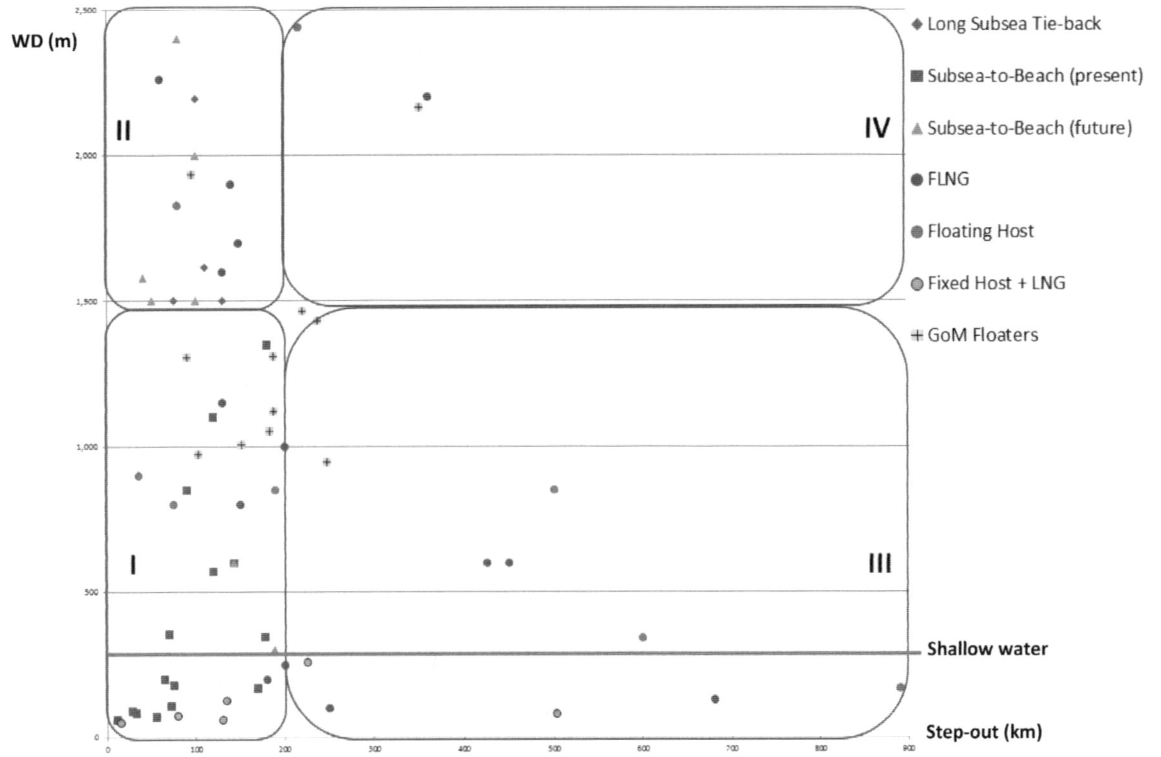

Figure 4 Applicability Ranges for Subsea-to-Beach, FLNG and GoM Hosts

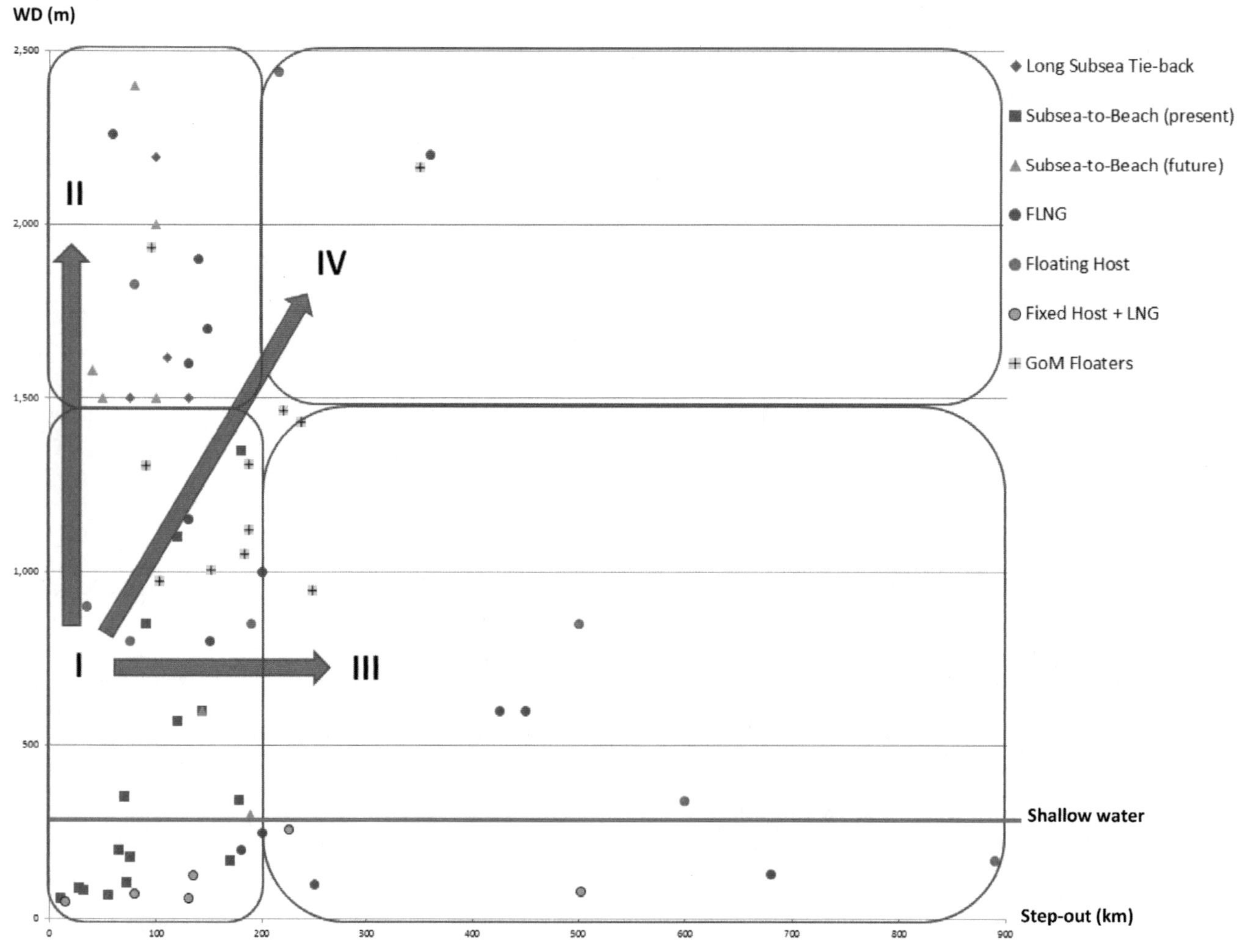

Figure 5 Evolution Lines for Concept Applicability

SNAME
THE INTERNATIONAL COMMUNITY
FOR MARITIME AND OCEAN PROFESSIONALS

OPTICAL FIBER SENSORS FOR SUBSEA AND TOPSIDE ASSET INTEGRITY MONITORING APPLICATIONS

VICTOR SERVETTE
OPTICAL FIBER ENGINEER, CEMENTYS, HOUSTON, TEXAS

VINCENT LAMOUR
CHIEF TECHNOLOGY OFFICER, CEMENTYS, HOUSTON, TEXAS

Proceedings of the 21st Offshore Symposium, February 2016, Houston, Texas
Texas Section of the Society of Naval Architects and Marine Engineers

ABSTRACT

Asset integrity monitoring is becoming a critical activity as Oil&Gas operators needs to secure their existing production assets and extend their service life. The last decade happened to be a turning point for sensing technologies as deeper subsea oil fields and intelligent oil wells required the extensive use of optical fiber sensors. As a result, optical fibers sensor technologies are now matured and off-the-shelf solutions for offshore and subsea asset integrity monitoring.

Fiber optics differs from other monitoring technologies by being totally passive: this means there is no need for subsea electronics or power batteries, making it an ideal choice for long term offshore and subsea asset integrity monitoring. Added to easy real-time communication, high multiplexing capabilities and distributed sensing, optical fiber brings new possibilities to existing traditional sensing systems.

Optical fibers are also extremely robust and, when deployed correctly, has a lifetime of more than 25 years, such as in the telecommunication industry.

Different optical fiber sensor technologies have been developed for Oil& Gas applications, creating a wide diversity of sensing capabilities, for both punctual (Fiber Bragg Gratings) and distributed sensing (Distributed Temperature Sensing, Distributed Strain Sensing, Distributed Acoustic Sensing, Distributed Pressure Sensing).

In this paper we present the latest optical fiber sensor monitoring solutions and applications for offshore and subsea asset integrity monitoring. The assets studied include Hulls and Topside Structures (Structural Health Monitoring), Risers (fatigue monitoring on Touch Down Zone and Hang-Off point), Mooring Chains and Turrets (load monitoring by load cells or load shackles), and subsea tiebacks and umbilicals (distributed monitoring of both temperature and strain for subsea leak detection and life of field monitoring).

Keywords: SNAME, Offshore, Symposium, Optical Fiber, Fiber Bragg Grating, DTS, DAS, DSS, Raman, Brillouin, Rayleigh, Life of Field, Asset Integrity Monitoring, Structural Health Monitoring

INTRODUCTION

Optical fiber sensor is bringing a disruptive revolution to the sensor market, by allowing new measurement capabilities. Its intrinsic properties make it very well adapted to the offshore and subsea environments:

- Passive sensors: the sensors do not need any electronics or batteries to send out measurements
- No inline power : well adapted for topside Class 1 Div 1 environments
- Good for static and dynamic measurements : 25 years without drift and up to 25 kHz for FBG sensors
- Not affected by ElectroMagnetic (EM) field: sensors can be implemented near high voltage or strong electromagnets
- Very high multiplexing possibilities: typically more than 50 000 sensors on a single fiber for distributed sensing
- Non-intrusive: a fiber is 125µm thick, only one fiber is needed
- Frequency measurements: high signal to noise ratio, insensitive to inline losses or de/reconnections

For comparison, a classic resistive strain gage is very sensible to moisture ingress and EM fields, needs to be recalibrated (offset drift) and needs typically 4 wires per gauge.

In this paper, we first present the two major technologies used for optical fiber sensor monitoring : Fiber Bragg Gratings "FBG" for discrete measurements and Optical Time Domain Reflectometry for distributed measurements. Then we present the different applications of optical fiber sensors in the Oil&Gas offshore and subsea industries.

TECHNOLOGIES

A wide range of optical fiber sensors were developed for different industries. In this paper, we focus on the two major technologies that has been used for the last decade in the Oil&Gas industry: Fiber Bragg Grating based sensors and Optical Time Domain Reflectometry based sensors.

1. What is an optical fiber?

In order to understand optical fiber sensors, one must first look at an optical fiber. An optical fiber is a flexible string made of extruded glass capable of transmitting light from one end to the other. It is extremely small (125µm, thinner than an hair) and totally passive: it does not transport energy other than light and does not need any power supply. The core fiber can be protected within a cable, making an optical fiber cable just as strong as a conventional electric cable.

In order to confine the light, an optical fiber is made out of different layers of silica, with different optical indices (see Figure 1). The difference of indices between the core and the cladding ensures that the light will stay within the core of the fiber. The size of the core can change, making the fiber either single-mode (core of typically 9µm) or multi-mode (typically 50 or 62.5µm). The fiber is then protected by a buffer and inserted inside a cable, within different thickness and protection depending on the application.

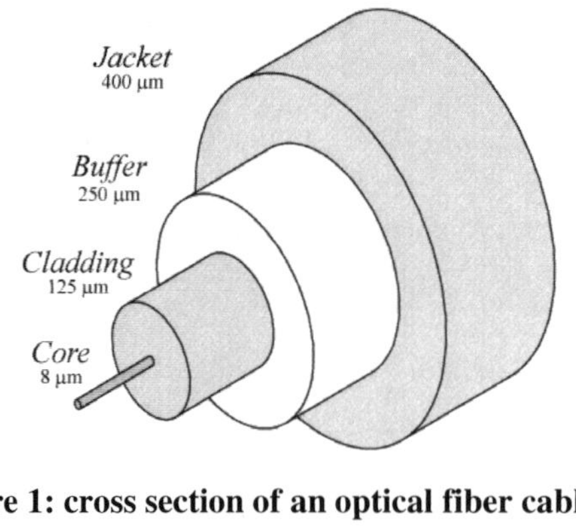

Figure 1: cross section of an optical fiber cable

2. Fiber Bragg Grating Sensors

Fiber Bragg Grating (FBG) Sensor is the most common type of optical sensors, used for a wide variety of applications. A FBG is an optical element that can be implemented locally inside a fiber. It has the property of reflecting only one wavelength while transmitting all the others. The reflected wavelength depends on the properties of the fiber locally (strain and temperature).

It is composed of a small periodic variation of the optical index inside the core of the fiber. The frequency of the reflected wavelength is directly linked to the period of the grating ($\lambda B = 2n_e\Lambda_B$ (*1*), Figure 2), so one can easily see that a change in the period of the grating (due to either mechanical or thermal effects) will change the frequency of the reflected wavelength. They are similar to electrical strain gauges in their response, this is why they are often referred to as optical strain gauges.

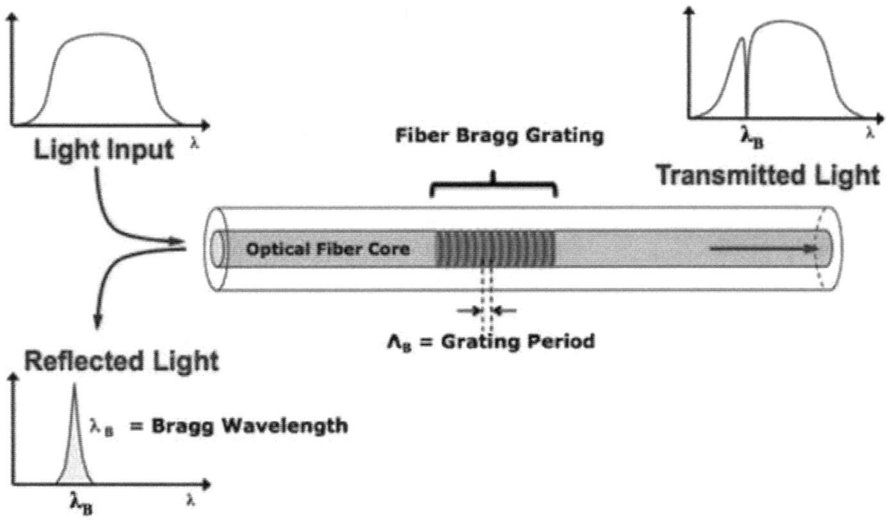

Figure 2: Fiber Bragg Gratings

$$\lambda_B = 2n_e\Lambda_B \tag{1}$$

Where λ_B is the wavelength of the reflected light, n_e the effective index of the fiber and Λ_B is the grating period.

3. Distributed Fiber Optic Sensors by Optical Time Domain Reflectometry measurements

Distributed Sensing led a major breakthrough in the optical fiber sensing technologies: a single telecommunication fiber is able to provide measurements typically every foot for several miles ! No punctual sensors are necessary, the fiber itself reacts to its environment and sends the information back to both ends, it acts both as sensor and as a communication device.

This technology is based on Optical Time Domain Reflectometry, meaning that it analyses the light backscattered from a fiber (Figure *3*). A laser pushes a short single frequency light pulse through the fiber, that pulse is backscattered in the fiber due to different physical phenomena (Rayleigh, Raman, Brillouin backscatterings), this backscattered light is then analyzed. The location of the information is known thanks to the time of flight of the backscattered light (in the same way a radar computes distance).

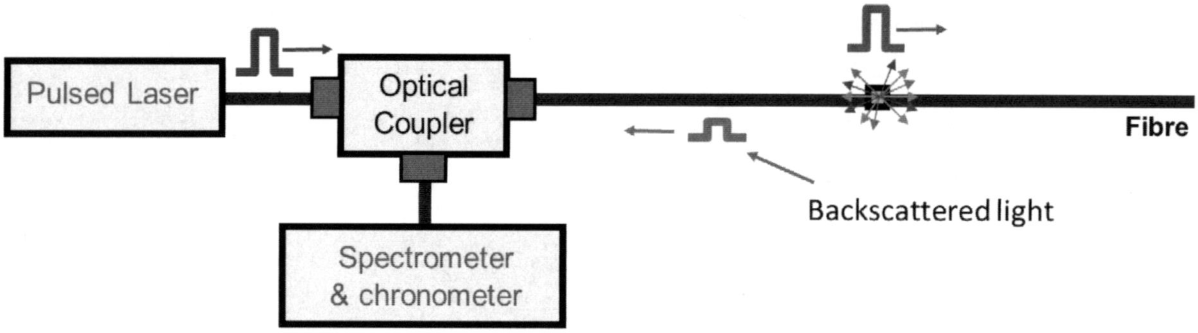

Figure 3: basic layout of a reflectometry measurement

The backscattered light is analyzed in frequency thanks to an optical spectrometer. The received spectrum contains interaction peaks due to different physical phenomena (see Figure 4):

- The Rayleigh peak has the same frequency as the laser pulse. Its intensity will vary with local vibrations and temperature
- The Brillouin peaks will see their frequency shifted due to both local temperature and strain changes
- The Raman peaks intensity will vary with temperature

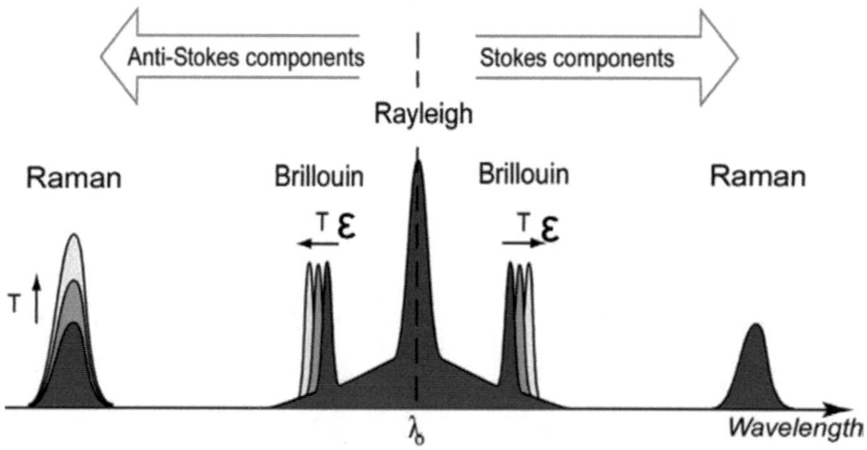

Figure 4: Typical spectrum of backscattered light

By analysing the different peaks, a single fiber can serve as a distributed sensor for temperature (DTS), strain (DSS), vibration (DVS) and acoustics (DAS).

4. Case study : comparaison between optical and electrical strain gauges for pipeline fatigue analysis

In collaboration with Stress Engineering Services, several tests were performed to compare optical strain gages to classic electrical strain gauges. These tests were conducted at Stress Engineering Facility in Houston. A fatigue test sample of pipe was instrumented by Stress with several electrical strain gauges using their regular procedures. We then installed 2 FBG StrainLux gages: one for strain and one for temperature measurements.

vic

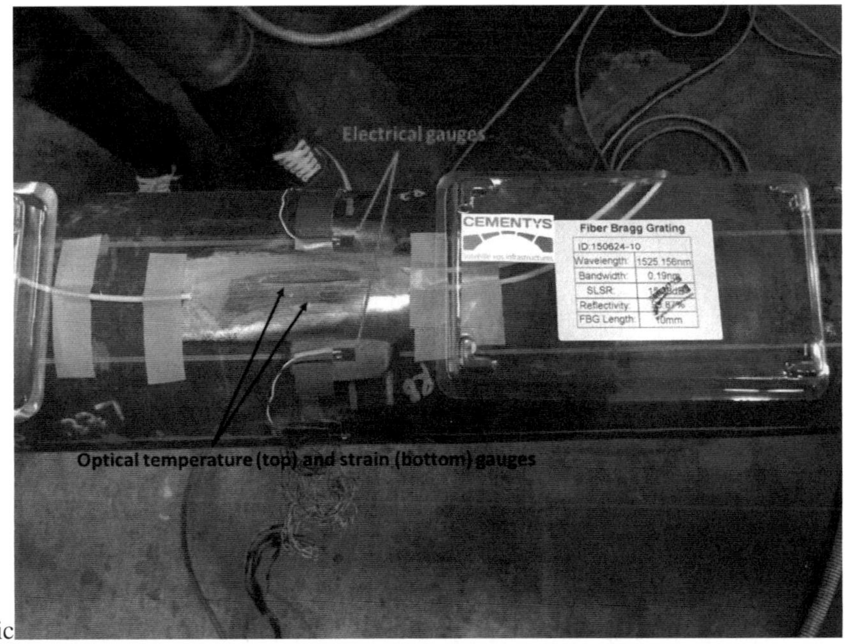

Figure 5: Picture of the gauges installed on the fatigue test sample before being protected

This test setup allowed us to make the sample vibrate at the desired value of cyclic strain.

Multiple tests were done on the sample as to isolate different measurement artefacts on the gages. The first test was a simple baseline measurement, to measure the noise level of each gage. During this base line measurement, the sample is at rest. The results are given on the Figure 6

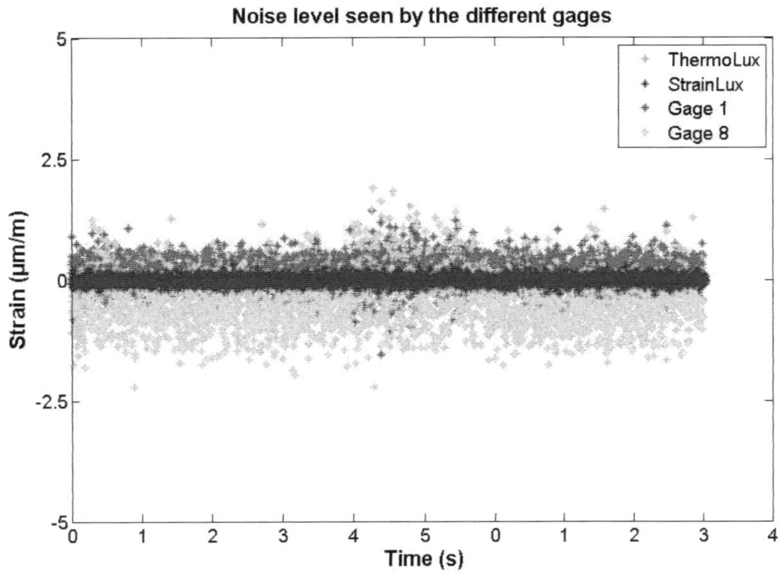

Figure 6: Noise level of optical and strain gages with time

One can clearly see that on this setup, optical gages have a lower noise level than the electrical gages (around 0.3µstrain versus 1.5 µstrains).

Several strain tests were done, for different values of enforced oscillating strain, by vibrating the sample close to its natural frequency. Figure 7 gives an example of results obtained:

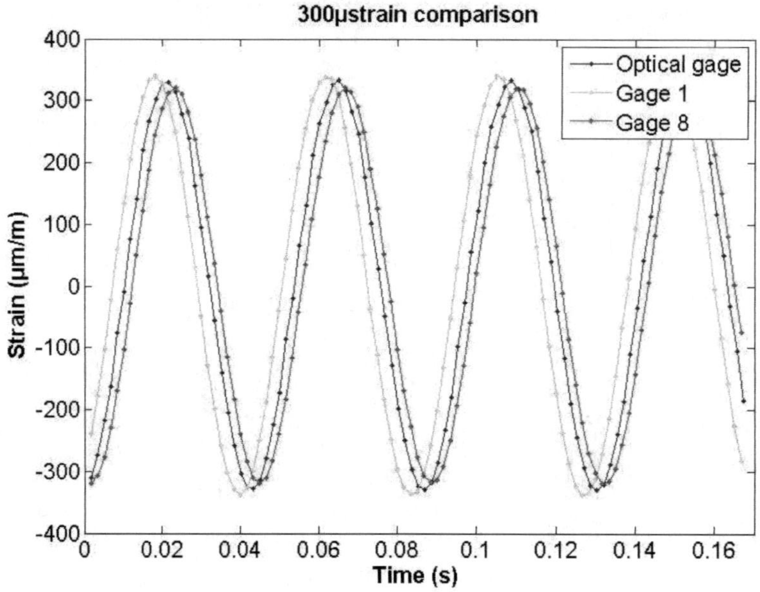

Figure 7: Strain measurement comparison

The measurments obtained with the optical gage is in accordance with the strain values measured by the electrical gages.

The final test presented in this paper is the influence of electromagnetism (EM) noise on the measurements. For this, a powerful magnet was moved close to the gages while the test sample is a rest. Figure 8 shows the influence of EM fields for both technologies.

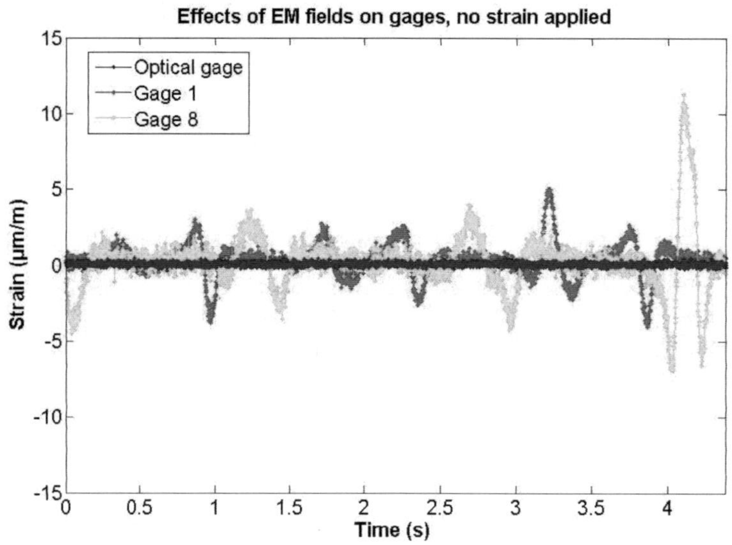

Figure 8: Strain measured by gages due to EM fields

Figure 8 clearly shows the effects of EM fields near electrical gages: the noise level drastically increases and creates eratic measurements. On the other hand, we can see that the optical gage isn't inflenced. This makes optical fiber sensor, the technology of choice when measurements are need in strong EM fields, such as close to electric pumps, valve or generators.

APPLICATIONS

In this section, we will detail how optical fiber sensors can be used for Asset Integrity Monitoring in the Offshore industry. For the sake of clarity, we present the field applications from the subsea reservoir to the offshore platform (see.Figure 9)

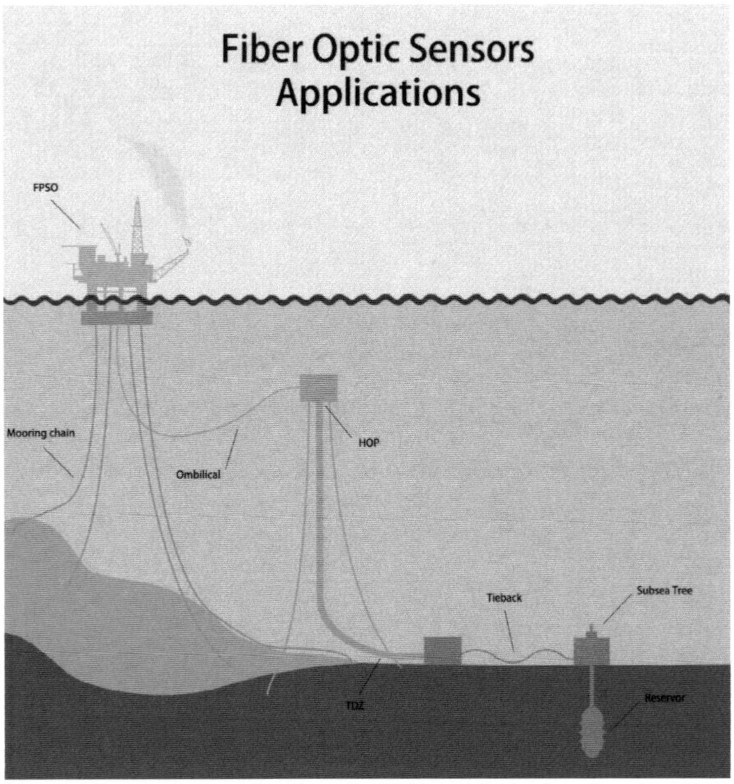

Figure 9: Fiber optic sensors applications in Offshore Industry (source: Cementys.com)

1. Downhole Applications

In the reservoir, several parameters are important to monitor in order to optimize the product flow and to enhance oil recovery. Temperature and pressure are the most common measured parameters. Today, the majority of pressure and temperature cells use embedded electronics and quartz technologies. They give good results but are their use may be limited by several factors: they cannot work permanently in high temperatures (>200°C or 390°F) and are difficult to multiplex on a single line.

For this reason, several companies developed pressure and temperature cells working on FBGs. These cells allow measurements in high temperatures with direct communication. The FBG is used as an optical strain gauge, attached to a sensitive element subject to pressure.

Distributed Fiber Optic Sensor measurements are also commonly performed in the downhole fields. Both temperature and acoustics can be monitored using an optical fiber cable deployed in the well by using DTS and DAS interrogators. The results were found to be extremely helpful during well completion and operation phases, this is why more than 10 000 wells (mostly onshore) have been equipped with distributed Fiber Optic Sensing measurements systems (Jacobs 2015).

The major advantage for using optical fiber sensors in subsea downhole applications relies on measuring remotely and permanently passive downhole sensors with very long existing umbilical lines (over 10 miles). The relatively high cost of optical wet mate connectors is still a limiting factor for Optical Fiber Sensor deployment on Subsea Wells.

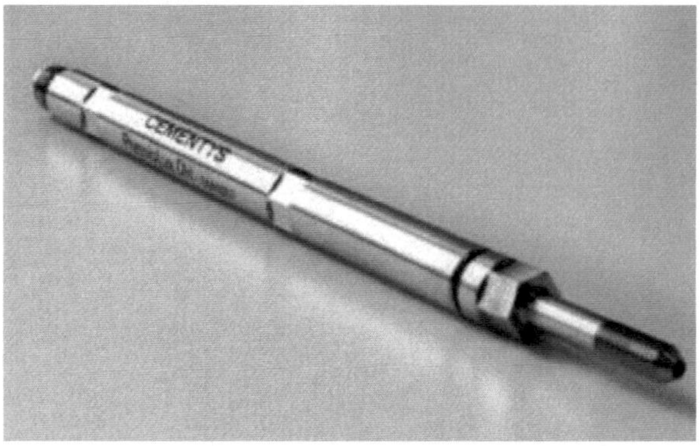

Figure 10: Example of a ½'' Optical Fiber Downhole Pressure/Temperature gauge for High Temperature / High Pressure Oil Fields.

2. Subsea tree applications

Subsea trees, BOP stack and modules are large structures, subjected to important loads and vibrations from the wells and tiebacks. A failure of those critical structures can cause great safety and environmental issues. Structural Health Monitoring can be typically performed through 3 longitudinal strain sensors allowing vibration and stress analysis. If needed, a "wet mate" connector, allowing subsea connection and disconnection of the optical line, can be installed so the sensors can be interrogated by a ROV, for periodic survey readings.

Other Process sensors such as Pressure sensors, Flowmeters, thermometer or accelerometers can use the FBG technology to be integrated to existing optical telecommunication subsea networks without any power.

3. Subsea tiebacks and flowlines.

Subsea tiebacks and flowlines are critical assets that are difficult to inspect. They need to be engineered as to make sure the pipe will not be obstructed by the transported product. The classic solution today is to double the line, to be able to send pigs for cleaning and be sure that one will be always available if the other gets clogged due, for example, to hydrate formation.

To remedy this issue, new pipes have been developed to increase the product temperature, either passively (insulated layers on the pipe) or actively (electrically trace-heated pipes). In both cases, it is important to measure the distribution of the temperature along the pipe: it will help production by reducing the risk of clogging and, in the case of heated lines, it will make sure the system is working properly.

Distributed Temperature Sensing is very well adapted to this application: the fiber cable can easily be implemented on the pipe before its installation, to be connected to existing communication lines once laid. An interrogation box topside will then give out temperature data all along the line by connecting it to the fiber (DNV 2009).

Such a system can also be used to see and prevent the apparition of lateral buckling: by interrogating the fiber with a Brillouin based interrogator for Distributed Stress Sensing, one can see the progressive apparition of stresses inside the line. Those stresses being linked to the lateral buckling of the tieback, proactive measures can be put in place to make sure the design stress are never exceeded.

Figure 11: Example of a fiber optic cable inside a concrete weight coating on a pipeline, before offshore towing and S-Lay in shallow waters (Etang de Berre 2015 project)

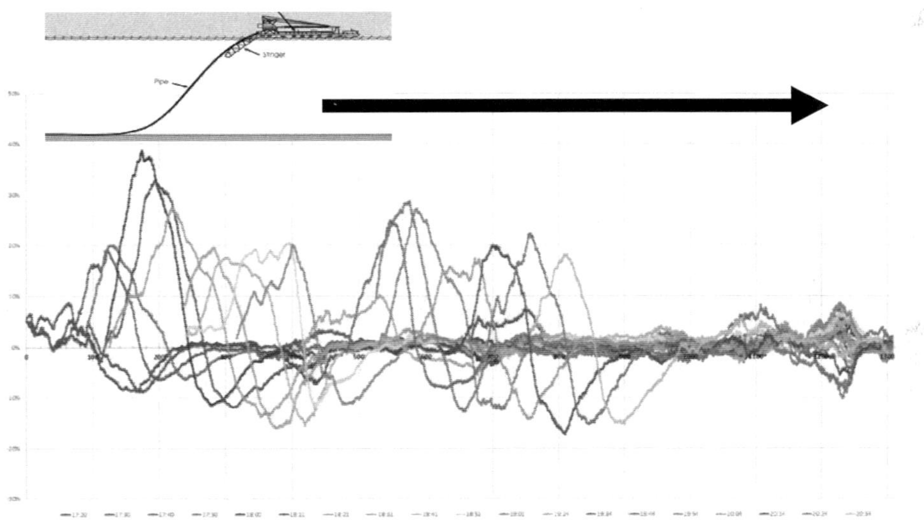

Figure 12: Strain distribution over the 1,3km pipeline during the S-Lay process (each line corresponds to an instantaneous distributed strain measurements, during the step-by-step S-Lay process)

4. Corrosion

Corrosion is omnipresent in the subsea and offshore environments. It is important to control it to ensure the durability and reliability of all the equipment installed. Today, several solutions exist to monitor the corrosion progress, for example cathodic protection monitoring or the use of coupons. Problem being that those element need an inspection to know if they need to be replaced. This can be costly, especially if the coupon is hard to reach (for example a coupon installed on a production tubing, where the whole line needs to be brought up to inspect it).

FBG sensors can be used for corrosion monitoring thanks to their very low drift in time (no effect of moisture ingress or electromagnetic noise). If engineered and installed correctly, the instrumented coupon will relieve stress as its section is getting thinner due to corrosion, this stress relief can be monitored thanks to an optical strain gauge. This type of measurements allows conditioned based maintenance approach, instead of a preventive maintenance strategy, and thus a lower maintenance cost as well as increasing the safety of the system.

5. Risers

Rises are equivalent to a field's highway: they link all the small oil routes at the seabed to structure topside. All the oil produced from a field will go through this complex structure (rigid or flexible risers), its health is thus critical. Today, most of the riser service lives are computed through cumulative fatigue models. Additionally, there are also sensors on the market to monitor them (mainly the hang of point), using either vibrating wire technology, strain gages or LVDT extensometers. They have been widely installed and give good results but have limitations: being electronic based sensors they need subsea power, meaning the batteries need to be changed by an ROV periodically for long term permanent operation. The technologies can also be limited in dynamic to a few Hertz, making the measurement of Vortex Induced Vibration difficult (DNV 2008).

Different companies are working on optical sensors for riser monitoring to avoid those limitation. An FBG based sensor installed at the hang-off point would not need any subsea power and electronics, and could give real time measurements up to 25kHz. As existing sensors, they could be clamped to an installed riser or fitted in the engineering design.

Even though they have a lot of advantages, solutions still need to be developed before an installation can take place: in order to function, the sensors need a continuous optical line to the platform. This can be done either by installing a dedicated umbilical, or by connecting the line to an existing optical line used for telecommunication. Both solutions have their pros and cons and need to be developed and properly installed.

Another critical part of the riser is the Touch Down Zone (TDZ): with tides, waves and other environmental factors, the location of the touch down point may vary up to several tens of meters. This can mean strong localized stresses and possible fatigue points. An FBG solution could be well adapted, thanks to its high multiplexing capabilities: an array of optical gauges can be deployed along the TDZ, allowing a complete stress analysis of the zone with low intrusion to this critical part, all the sensors can be multiplexed on a single fiber. The array can be connected to the Subsea Umbilical Termination Assembly (SUTA) via a patch-cord and wet mate connectors once the riser is installed. There, an existing communication optical fiber umbilical line is used to link the sensors to the topside.

At last, we can monitor the shape of the riser with time. Currents and temperature gradients can greatly affect the riser and create localized weak points. Distributed Strain Sensing (DSS) can be used to know the strains the all along the riser. These distributed strain values could be computed to displacement by integration and thus giving the wanted shape information. Here-again solutions need to be qualified before an installation can be done. The main issue would be the deployment of the optical fiber on an existing riser. Two solutions are thought-of, either by clamping a cable strait down the structure (piggy-back configuration), or twisting the cable regularly down the riser (helix configuration). The length of the riser could also be an issue: displacement being computed by integration of strain, the calculation error will increase rapidly without reference points.

6. Mooring chains and turrets

There are few sensors available on the market for mooring chain load or tilt monitoring applications. They can mainly be divided into three main categories: smart shackles, load cells or tilt sensors.

Smart shackles work by monitoring the stress inside a bended pin of the shackle to come back to load. They have to be installed on the last chain link and are responsible for the full load of the chain. Fiber Optical Shackles will give an absolute value of the load on the mooring chain without any moisture effect or electronic drift. This type of smart shackle has to be planned in advance before the chain installation.

The load cells differ from shackles by being installed on the chain stopper structure. A typical optical Load Cell will perform permanently in offshore environments without recalibration. Another possibility is to install optical strain gages on several links of the chain. Clamped to the link, they measure the variations in load with time.

Both smart shackles and load cells already exist on the market, but working with classic electrical gages. This means an important drift with time and a high dependency on environment (humidity, temperature…). These issues make the items unreliable after a few months of measurements. Transforming those items by including optical fiber gages would help this reliability issue. This is why several companies are working on the development and installation of such sensors.

A local measurement of tilt also gives valuable information on the load of the chain: most of the mooring systems use a catenary disposition, where the tilt of a known point is directly linked to the load of the line. This system can be either in 1 or 2 directions, to see the influences of currents. Those measurements can give an absolute value of the load, as well as information in load change with time. Once again, fiber optical technology has the advantage of not having any battery or electronics subsea: the sensor is totally passive and the measurements are taken real time thanks to an telecom optical cable connected to the platform.

In all applications, most sensors used in Offshore applications use classic electronic gages. This means that the measurements will deteriorate rapidly due to drift of the sensors and to their dependence on their environment (humidity, temperature, loses on the line) after only few months of operation.

This issue can be avoided by switching the electronic technology to optical FBG technology. The sensible elements need only very little changes to be adapted to fiber optics.

7. Structural Health Monitoring

For safety reason, it is important to monitor the long term asset integrity of the topside structures. To do so, Sensors can be implemented on specific regions of the structure to get additional data and make sure the mechanical behavior is as it was predicted by the initial model.

In the case where service life of a platform needs to be extended, how can you be sure it is in good health? The initial model cannot be used anymore, and the lack of data concerning the structure makes it very difficult to ensure its integrity from simulation only.

Optical sensors can be helpful in such a case : by installing several sensors at key positions, one can compute the frequency of resonance of the different modes of vibration, thanks to a vibration analysis. These sensors can be implemented at the end of the initial life expectancy of the structure, and make sure the frequencies do not vary with time: a change in frequency means a behavioral change of the structure, thus aging. This method doesn't give absolute integrity on the structure, but it is monitoring the changes over time (slow aging process).

Sensors using different technologies can be used for this application, but FBG based sensors are well adapted: being totally passive, they can be implemented on the structure without major changes. Other technologies might have trouble adapting to class 1 div 1 environments (highly explosive environments). All the sensors can be multiplexed on the same line, making the cable line non-intrusive. Lastly, they are long term sensors capable of measurements for several years without drift, and with high frequency measurements adapted to vibration analysis.

CONCLUSIONS

A multitude of industrial sectors will benefit from the new optical fiber sensor technologies. From the oil well to the topside structure, optical fiber sensor can bring complementary and previously impossible data to improve safety and production of Oil&Gas assets. Such long term monitoring solutions can provide near-term and long-term cost-saving benefits such as:

- Supervised lifetime extension for existing assets, which are candidates for replacement
- Asset Integrity Monitoring during installation, commissioning and early life, reducing Vessel Time when inspection / expertise is needed.
- Risk Based Management and Condition Based Maintenance during service life

Although the optical fiber technology is fully developed and already in place in other industries (civil engineering, nuclear, health sciences…), installation challenges slowed the implementations of optical fiber sensors in the Offshore industry. Most of those challenges are getting resolved (wetmate connectors, optical pass-through assemblies, optical umbilicals and jumpers…), meaning that the offshore industry will soon benefit from this Asset Integrity Monitoring enabling technologies.

REFERENCES

Trent Jacobs, "Downhole Fiber-Optic Monitoring: An Evolving Technology", Journal of Petroleum Technology 66 (08), pp. 44-53, 2015;

DNV, "Recommended Practice DNV-RP-F206 - Riser Integrity Management", April 2008.

DNV, "Recommended Practice DNV-RP-F116 - Integrity Management of Submarine Pipeline Systems", October 2009.

API, "Recommended Practice 2SIM: Structural Integrity Management of Fixed Offshore Structures", API G2SIM01, November 2014.

NEW GUIDELINES FOR CERTIFICATION OF OFFSHORE ACCESS SYSTEMS USING MOTION COMPENSATED GANGWAYS

BENJAMIN EUSTACHE
OFFSHORE RULES MANAGER, BUREAU VERITAS

ROLAND WERKHOVEN
CONVERSION MANAGER, BUREAU VERITAS

LAURA-MAE MACADRE
MARINE RENEWABLE ENERGY SPECIALIST, BUREAU VERITAS

GIJSBERT DE JONG
OSV & TUG MARKET SEGMENT MANAGER, BUREAU VERITAS

NICOLAS LEGREGEOIS
OFFSHORE GENERAL MANAGER USA & CANADA, BUREAU VERITAS

Proceedings of the 21st Offshore Symposium, February 2016, Houston, Texas
Texas Section of the Society of Naval Architects and Marine Engineers

ABSTRACT

Motion compensated Offshore Access Systems (OAS) are emerging technologies offering substantial advantages to offshore support vessels. Improvements in terms of personal safety during transfer from a ship to an offshore asset and extension of operating weather windows compared to conventional methods are examples of these benefits.
Indeed, the conventional personal transfer systems (lifted basket, bump-and-jump from a crew boat or helicopter transfer) present some risk regarding safety, limited operating time frame or high cost that the new OAS significantly help to improve. Even though the first models of compensated gangways were installed 10 years ago, neither a design standard nor a certification procedure for these systems has been published at the time this paper was written.
In recent years, many different technologies have emerged in the industry, which may nevertheless be divided into two categories: the passive transfer gangways, which are securely connected to the offshore asset and then put in free-flow for personal transfer, and the active transfer gangways, which remain motion compensated during transfer of personal. Undoubtedly the safety issues would substantially differ from one type of system to the other, as well as the critical components.
Based on experience accumulated over the past years with the Dutch developers and manufacturers of OAS, major challenges and issues are highlighted in this paper. As several different designs exist, the safety principles and certification framework proposed in this paper are generic and not design-specific. Due to the inherent novelty of motion compensated gangway systems, particular attention is paid to the assessment of new features. Consequently,

a risk-based Guidance is provided to describe a method for the qualification of new technology to be used for the most innovative parts of the OAS.
This paper presents the main OAS technologies on the market, the related safety issues and the developments made by Bureau Veritas in relation with the publication of the new Guidance Note NI629 Certification of Offshore Access Systems. It develops safety principles for both active and passive OAS, design loading conditions, requirements for structural assessment, machinery and control system design as well as testing procedures.
Keywords: offshore access system, motion compensated gangway, personal transfer, Walk-to-Work, W2W.

INTRODUCTION

An Offshore Access System (OAS) is a piece of equipment intended for the safe transfer of personnel between a ship and an offshore facility. It usually includes a pedestal and a gangway which can be motion compensated, and a connection system at the gangway tip, eventually secured on the offshore facility.

This piece of offshore equipment is rather a new concept and at the time this paper was written, no dedicated standard on design or certification of motion compensated gangways were published. Nowadays, the reference standards used as a basis for certification of offshore access gangways are technical standards applicable for lifting appliance certification, e.g. EN 13852, API SPEC 2C or other rules from Classification Societies. One of the issues when using these standards is that, for instance, they explicitly require the motion compensation system to be switched off during man ridding operations. This is of course unacceptable for systems which precisely rely on such features to provide a safe means of transfer.

Having a reliable technical standard is essential for the manufacturers, operators and national authorities but also banks and insurance companies. It creates a common core of knowledge and an acceptable level of safety shared between all parties. By building on its established leadership in inspection and certification services as well as its recognized expertise in marine equipment and systems, Bureau Veritas is indisputably well placed to develop such document. These are the reasons which have preceded the writing of the new Guidance Note NI629 "Guidelines for Certification of Offshore Access Systems", to be published in 2016. It should also be noted that the development of the Guidance Note was conducted in parallel to several certification projects, including the Z Bridge OAS developed by the manufacturer Z Technologies, in the Netherlands (see Figure 1).

When developing new standards the aim of certification bodies is to produce rules and tools that can be applicable to a wide range of designs. Consequently, trying not to be design-specific, a proposal of a generic certification process is made in this paper. Existing certification principles from more mature sectors such as offshore lifting appliances, drilling equipment or ever fairground machinery, have been adapted to the specificities of OAS. Their respective application and interactions are developed. In addition, particular attention is paid to the novelty induced by offshore access systems. Consequently, a risk-based approach based on qualification of new technology is developed to address the most innovative parts of motion compensated transfer systems.

Figure 1 - Z Bridge – ***Courtesy of Z Technologies***

WHAT IS AN OFFSHORE ACCESS SYSTEM?

Early developments

Offshore Access Systems based on motion compensated gangways are relatively new in the offshore market. The first system was developed in 2007 by the company Offshore Solution. It started as an answer to the demand from the Oil & Gas (O&G) industry to efficiently maintain assets.

Improving safety became more and more important as several accidents had happened during personnel transfer operations. Indeed, transferring personnel is known as one of the more risky marine operations, together with offshore lifting and diving.

Industrial personnel, till that time, used to be transferred to platforms by crew boat or by helicopter. But not all platforms were equipped with helideck platforms, rendering them unreachable for helicopters. Alternatively, crew boats were and still are deployed. In that case, transfer is performed by jumping over (bump-and-jump), or by using a sling rope or a transfer basket. To date those techniques are considered to be not ideal with respect to safety.

The limiting environmental conditions with the above technics are mainly the significant wave height, swell, wind speed and fog. Depending on the location, the above could lead to an overall operability as low as about 70 percent.

On the other hand, Offshore Access Systems deployed from Dynamically Positioned (DP) vessels would increase considerably the operability (see Figure 2). At the same time, a DP vessel with an OAS could stay for longer periods at sea with the capability to accommodate personnel at considerable distance from shore. Compared to transfer using crew boats or helicopters, another advantage of using a larger vessel with an OAS is the possibility to carry and transfer large spare parts.

It appeared soon that the operability of a DP vessel using an OAS might be more than 80 percent and as high as 95%. Although in the beginning the O&G industry was cautious about this technology, with the simultaneous development of the Marine Renewable Energy (MRE) sector which at the time was less mature and more open to new concepts, the success of OAS was rapidly recognized. As safety is of paramount importance in the offshore O&G and MRE industries, the need for independent assessment and certification of these new transfer systems is shared by all parties involved.

Figure 2 – DP2 W2W Multi-Purpose Support Vessel Kroonborg - ***Courtesy of Wagenborg***

It is also no surprise to see new manufacturers proposing different technologies every year in this fast growing market.

Technical features

An OAS is a system intended for the safe transfer of personnel - and potentially also equipment - between a ship and an offshore facility. It usually includes a pedestal and a gangway - which can be motion compensated - , and a connection system at the gangway tip to be secured on the offshore facility. Motion compensated systems used on OAS are intended to compensate the ship motions while the OAS is in connection phase or in personnel transfer phase.

Offshore Access Systems fall within the following two categories:

- Active OAS: which is a fully motion compensated system which remain activated both in connection phase and in personnel transfer phase
- Passive OAS: which is in "free-flow", i.e. not driven, once it is connected to the offshore facility but may be motion compensated during the connection phase.

The active OAS needs power and control all the time, from connecting to an asset, transferring personnel to disconnecting. Redundancy of the systems during all phases is critical and subject to careful review.

The passive OAS does not need power and control during transfer of personnel. Once connected to the platform, the power motion control system is switched off and the gangway let in free-flow mode, whereby the gangway is allowed to follow the relative motions between the vessel and the offshore facility (within design limits).

In case of power failure, whether of electrical or hydraulic origin, the connection of the OAS is ensured and will not jeopardize the personnel safety.

Most OAS are equipped with a ship motion compensating system. These systems can compensate from a single motion (1 degree of freedom), i.e. heave compensation, up to full motion compensation (6 degrees of freedom: heave, roll, pitch, surge, sway and yaw).

An important aspect is also the combination of the ship's station keeping ability with the OAS capability. The footprint of the DP vessel is an essential feature. If the vessel's footprint exceeds the geometrical limits of the OAS, the system as a whole will not perform and thus will not deliver the expected operability.

The height relatively to the waterline, from which an OAS is deployed, is another important issue. For O&G platforms this height is generally between 8 and 16 meters, which is directly associated with the luffing angle of the gangway. Some operators for example take an angle of +/- 10 degrees as a maximum for safe walking.

Now with the upcoming offshore wind industry, this is becoming an issue. Indeed, the height of a landing platform on a windmill is at least 20 meters, not taking into consideration tidal differences.

An OAS could of course be positioned at a higher level by installing a higher pedestal or positioning the equipment on a higher superstructure. But the higher the position, the higher the excursions that may be expected. To cope with this, designers have recently developed new features, for instance lighter construction, counter balance, fast responding electrical devices instead of hydraulics and, most importantly, variable height deployment. This latest feature is being trialed at sea at the time of writing this paper.

Most OAS used in the industry are designed with gangway lengths typically between 20 and 40 meters, with up to 60 metres for the largest ones. These systems are normally used in a harsh environment.

Smaller OAS designed for accessing facilities on the continental shelve also exist. These systems are normally used on wind farm support vessels and crew boats operating in mild weather conditions with a maximum significant wave height of about 1.50 meters and wind speed not exceeding 20 knots, corresponding to approximately 5 Bf.

With the success of the OAS for transfer of personnel, the industry is now requiring transfer of equipment and small cargo packages, typically between 500 and 1500 kg, which poses additional challenges. To date the typical load on gangways is 300 kg, corresponding to two men carrying a third on a stretcher. Despite the fact that an OAS is not a crane, vertical and horizontal transfer of loads will have to be considered in the total concept by the designer and the certification body.

Safety principles

Mechanical and control systems should remain operational in the occurrence of a failure of a single component or a failure of the power supply. For control systems, the use of an Uninterrupted Power Supply (UPS) is already common practise (back-up by batteries). For mechanical systems, technical requirements would depend on the technology used and whether the OAS is driven by hydraulics, electrical actuators or a combination of both. In case of hydraulics, back-up is provided by hydraulic accumulators. For electrically driven OAS an emergency power source is provided.

Another important component in motion controlled OAS is the Motion Response Unit (MRU). Even though this component is not an essential feature of passive OAS, it is of the utmost importance in active OAS. Due to its safety criticality the MRU is to be certified by a certification body.

The maximum motion capacity of an OAS, i.e. its operating envelope, shall be correlated with the ship motions foreseen in the intended most severe sea states and also with the station keeping capability of the DP system. An alarm system with traffic lights shall be fitted for the personal using the gangway, whereby a green, amber and red light code indicate respectively whether the transfer of personnel is safe, the personnel must evacuate the gangway or the OAS must be disconnected (see Figure 3).

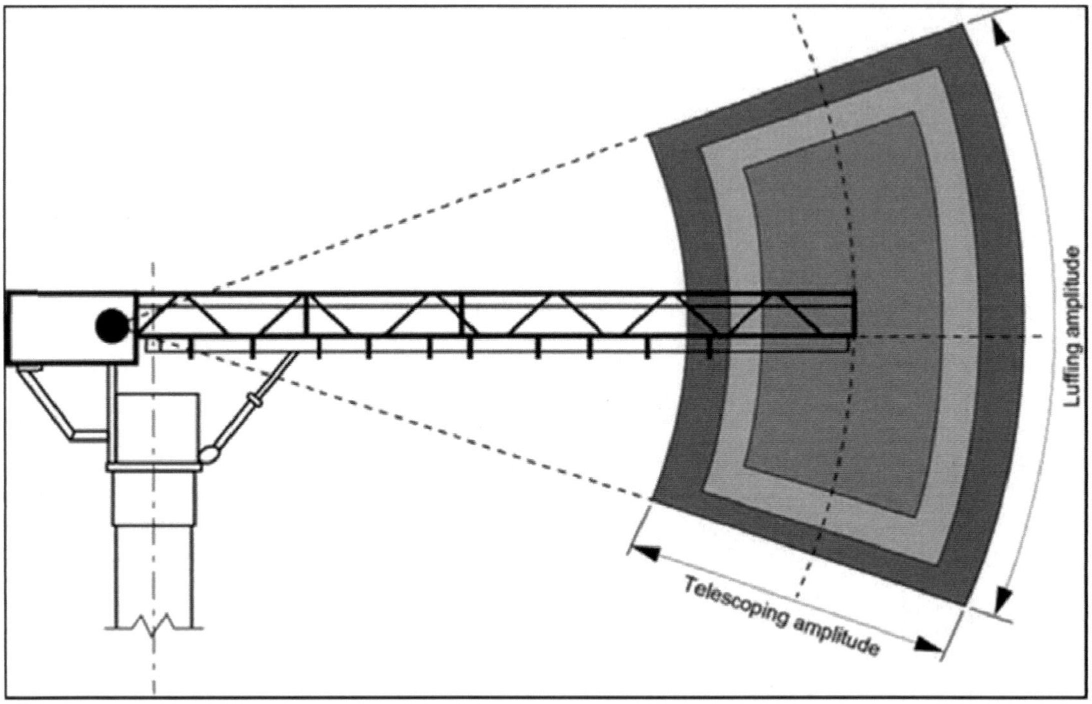

Figure 3 – Example of an operating envelope and alarm system

DEVELOPMENT OF NEW GUIDELINES FOR CERTIFICATION

The certification considerations presented here will be detailed in Bureau Veritas new Guidance Note NI629 "Guidelines for Certification of Offshore Access Systems", to be published in 2016. In this document, Bureau Veritas provides a generic (not design-specific) certification methodology dedicated to offshore access systems. As motion compensated gangways are highly innovative technologies, the development of certification procedures should be considered as an iterative process carried out jointly between certification bodies, technology developers, project operators and professional organisations.

To support the development of OAS, codes and standards are necessary in order to ensure the safety, quality and performance of these innovative technologies. These documents aim at gathering best practices as well as design and operational requirements, forming a reference basis for certification. A robust certification system is of great help for building up confidence between the various OAS stakeholders, including insurers, investors, designers, project developers and final users.

The development of a reference standard is necessary for the certification process in order to check compliance of the OAS against a set of requirements. However, as mentioned in the introduction, the motion compensated gangway is still a new technology and the existing record of in-service information and analysis of past experience is insufficient to provide a comprehensive set of dedicated rules. As a result, almost no guidelines neither standard dedicated to OAS have been published so far. Bureau Veritas NI629 would surely be one of the first dedicated documents.

Consequently, the first recommendations are based on existing rules and standards from related sectors such as offshore handling systems and lifting appliances but also from fairground and amusement park machinery and structures (EN 13 814) for instance. Indeed, some analogies can be made in terms of safety principles between a fairground ride simulator and a motion compensated personnel transfer system. Of course the standards from related sectors are to be adapted to the actual context, bearing in mind cost-benefit considerations. In addition, an alternative approach based on qualification of new technologies is required to address the innovative features of the OAS that are not covered by any standard, as detailed below.

Qualification of new technologies

When neither code nor standard are available for the certification of a specific component or system, the qualification method described in Bureau Veritas Guidance Note NI 525 "Risk Based Qualification of New Technology" is implemented to ensure safety of the considered component.

First, a Technology Assessment step will be carried out to identify for each component the maturity and potential similarities of the application conditions (see Table 1). For each component that is not "proven" in "similar conditions of application", the potential hazards induced by the component failure will be identified and assessed in terms of frequency and severity of the consequences. A FMECA can be used to conduct this study. Finally, associated mitigation measures and qualification activities (calculations or physical tests) will be recommended to ensure the safety of the component or system.

Part of this analysis will be performed during a dedicated risk analysis workshop involving a group of experts having complementary expertise in the relevant fields for the addressed component (mechanics, electronics, hydraulics, etc.).

The component is considered "safe" once all the individual qualification activities have been performed, reviewed and validated by a certification body.

Only few components are likely to be concerned by this total lack of standards. Very innovative technologies might be concerned, but in most other cases, existing standards from related sectors should be available as a basis.

Table 1 - Technology assessment scale

	Application conditions	
Technology maturity	Similar	Different
Proven	0	1
Limited field history	1	2
New	2	3

CERTIFICATION PROCEDURE

Certification or conformity assessment can be defined as a procedure by which a third party gives written assurance that a product, process or service conforms to specified requirements. Certification of equipment generally involves several stages as detailed here after (see Figure 4).

Risk assessment

The risk assessment stage is essential to systematically identify the hazards associated with accidental situations, failure of systems, human error etc. A HAZard IDdentification (HAZID) or equivalent risk assessment method can support the implementation of this study.

Furethermore, the failure of components within the system is to be studied, i.e. failures modes, their possible causes and associated consequences for personal safety, the environment or the integrity of asset studied. Failure Modes and Effects Analysis (FMEA) or equivalent risk assessment method can be used to conduct this study. These risk assessment studies are to be conducted with regard to the safety principles such as the single failure principle.

FMEA and HAZID would then be used as a means to describe the systems and its operating issues and should be updated through the development stages of the system.

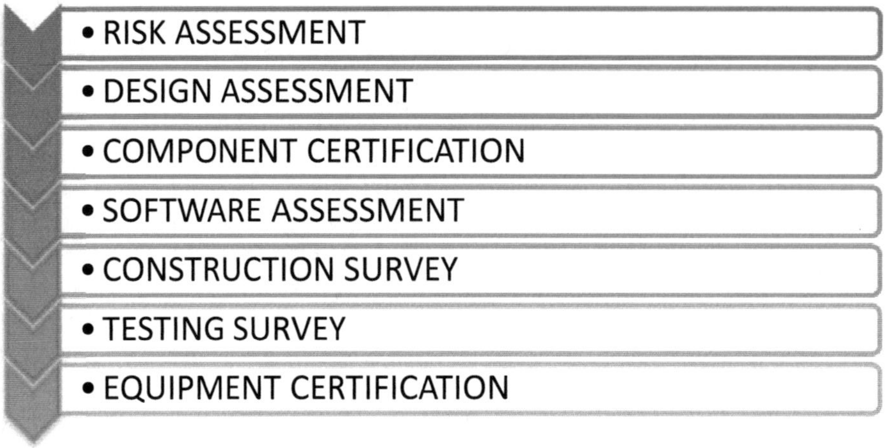

Figure 4 – Certification scheme

Design assessment

The purpose of the design assessment is to verify the conformity of the safety arrangement, the structural scantlings, the electrical installations and the control systems, as described in the equipment documentation, with the rules requirements.

The design documentation should be submitted to the certification body, including:

- the general description of the system,
- the reference standards,
- additional requirements or deviations from these codes and standards when required, and additional requirements which are not covered by existing codes and standards,
- the design parameters, assumptions, methodologies and principles,
- the drawings and reports regarding:
  - the structural arrangement and scantlings,
  - the mechanical systems,
  - the electrical installations,
  - the control systems,
  - the safety features.

The purpose of structural assessment is to verify the structural integrity of the OAS before it is manufactured and tested. Independent analytical and/or numerical calculations may be performed by the certification body, typically for major load carrying structural components such as the gangway and the associated supporting structure.

Component certification

The purpose of component certification is to confirm that major components - such as a hydraulic power units or cylinders - are designed, documented and manufactured in conformity with design assumptions, specific standards and other technical requirements.

Software assessment

As the safety of active OAS relies on the inherent reliability and dependability of its control system, the assessment of the logic controller is to be performed against Bureau Veritas Guidance Note NI425 "Software Assessment for Shipboard Computer Based System".

The increasing complexity, sophistication and integration of shipboard systems such as OAS induce risks. As it is no longer possible to ensure operational safety without looking into the software, testing and verification of control systems and systems integration are needed. Systems hazard analysis techniques can be applied to define risk mitigation measures, which enable to resolve issues in the early project stage with a view to prevent defects, avoid time loss and control cost. In fact, a structured, systems engineering approach throughout the life-cycle (FEED/design, construction, operation) is needed to ensure continuous safe and efficient operation.

To that end Bureau Veritas and Athens Group (Houston, TX) have developed a joint service offering for the integrated systems lifecycle verification and validation of the control software. The service combines early lifecycle software quality assurance (the "white box" approach) with later lifecycle verification and validation of functionality (the "black box" approach) through comprehensive acceptance testing including hardware-in-the loop (HIL) testing.

The white box approach is particularly important in the earlier lifecycle of a project. The methodology establishes traceability of requirements and change management, tests execution paths inside the code and focuses on detecting of software coding defects.

The black box approach is important during the later stage of the project lifecycle. The methodology consists of the development of functional test plans traceable to user requirements and testing of functionality at the interfaces (especially the hardware control interface). The focus is on validating the functionality of the software and ensuring compliance with user and regulatory requirements.

Construction survey

The construction survey aims at ensuring that the manufacturer has built the OAS in conformity with the documentation verified during the design assessment. This evaluation includes the following elements:

- review of the manufacturing plan and procedures;
- quality system evaluation: conformity with ISO 9001 or ad-hoc evaluation of the quality system in place;
- survey during construction including verification that requirements regarding critical components and manufacturing processes are correctly implemented in production and assembly.

Testing survey

Testing attendance by the certification body is essential in order to verify the compliance of the system within the required performance. Before attending testing the test protocol is to be reviewed by the certification body.

Testing should include:

- functional testing in order to verify that the OAS displays the behaviour predicted in the design
- load measurements in order to validate design calculations and to determine the magnitude of loads under specific conditions
- overload testing i.e. proof load of the load bearing parts of the system

CONCLUSION

As collection and analysis of past experience is required over a long period of time in order to build a consensus-based certification scheme there are currently neither international standard nor certification procedure dedicated to Offshore Access Systems. Taking the benefits of the experience gained by performing design review and certification of OAS from several manufacturers, Bureau Veritas proposes a generic certification methodology which is further detailed in the new Guidance Note NI629 "Guidelines for Certification of Offshore Access Systems", to be published in 2016. To take into account for the novelties induced by OAS, qualification of new technology is recommended for the most innovative parts of the device, which are not fully covered by existing codes and standards.

These new guidelines are intended to support the growth of OAS fitted on offshore support vessels - in particular for the dynamic Marine Renewable Energy market - by offering a reference basis for certification with an acceptable level of safety shared between the various OAS stakeholders, including manufacturers, insurers, investors, project developers and end users.

ACKNOWLEDGEMENTS

The authors would like to thank ZTechnologies for granting the right to use the picture of the Zbridge prototype, and more generally for entrusting Bureau Veritas to providing assistance for their new developments.

The authors would like to thank Athens Group for reviewing the paper and providing valuable input.

REFERENCES

Bureau Veritas NI525, Qualification of new technologies, Methodological Guidelines, December 2010

Bureau Veritas NI425, Software Assessment for Shipboard Computer Based System, July 2008

EN 13 814: 2004 - Fairground and amusement park machinery and structures

Bard, J. and Thalemann, F., "Offshore Infrastructure: Ports and Vessels", a report of the Offshore Renewable Energy Conversion platforms – Coordination Action (ORECCA), Fraunhofer IWES, not dated.

Robert P., "Series production and design optimisation", proceedings from WAVEC Seminar, November 2013

A DYNAMIC SUBSTRUCTURING APPROACH TO IMPROVED GLOBAL STRUCTURAL DYNAMIC AND STRESS ANALYSIS OF TOPSIDE/HULL SYSTEMS

DR. ARYA MAJED
TECHNICAL ADVISOR, ADVANCED ANALYSIS, INTECSEA INC., HOUSTON, TX (USA)

DR. ALAA MANSOUR
MARINE ENGINEERING MANAGER, INTECSEA INC., HOUSTON, TX (USA)

MR. LUCA CHINELLO
SENIOR ENGINEERING SPECIALIST, INTECSEA INC., HOUSTON, TX (USA)

MR. YAMING WAN
EXECTIVE STRUCTURAL ENGINEER, INTECSEA INC., HOUSTON, TX (USA)

Proceedings of the 21st Offshore Symposium, February 2016, Houston, Texas
Texas Section of the Society of Naval Architects and Marine Engineers

ABSTRACT

Current practice in global structural dynamic analysis of topside/hull systems involves executing a global finite element model for wave pressure inputs at specific periods. The main purpose of this global model is to generate loads for detailed local models to compute stresses; the global model itself lacks the fidelity and detailed modeling necessary for direct stress computation. The main driver behind executing separate global and local analyses is a computational constraint - each local model can be several million degrees of freedom making an upfront direct integration of a more accurate combined global model for dynamic analyses infeasible. The known issues with this traditional process include: 1) global model lacking the fidelity required for accurate computation of system modal properties (especially for shorter wave period regime), impacting the loads and stresses, 2) local analysis boundary conditions and loads mapping effecting the stress accuracy, and 3) long turnaround schedules preventing sensitivity and trade-off studies for risk/cost reduction and design optimization.

This paper introduces a dynamically substructured global structural dynamic solution which is a significant improvement over the traditional solution in terms of accuracy, computational speeds, and schedules. The improved solution enables a significantly higher fidelity global model, computes its structural dynamic response for the entire wave regime, and directly computes stresses as a function of wave period - all in a few minutes of computation. The improved process eliminates the lower fidelity global loads model utilized in the traditional process. Therefore, the inaccuracies in the lower fidelity structural dynamic loads analysis are removed. Furthermore, the improved process eliminates the local analyses and associated schedules/costs. This eliminates inaccuracies due to local model boundary conditions and loads mapping. In addition, the entire wave spectrum is analyzed in a single analysis, generating complete high resolution dynamic stress spectra as a function of wave period. This ensures that all wave periods important to the stress response are automatically accounted.

Finally, the very fast computation allow for trade-off and sensitivity studies to optimize structure and reduce overall risks, schedules and costs. Therefore, with dynamic substructuring solution, all of the deficiencies in the traditional global analysis are removed. Given the accuracy and extreme computation speeds afforded by the dynamically substructured solution, the next step will be the utilizing the irregular wave pressures as forcing functions. This more realistic representation of the forcing functions will result in the most realistic stresses possible. A proprietary nonlinear dynamic substructuring (NDS) solver, FLEXAS, is utilized to demonstrate this capability on a topside/hull FEM.

Keywords: Component-Mode Synthesis (CMS), Finite Element Model (FEM), Frequency-Domain, Dynamic Substructuring, Regular Wave, Irregular Wave, Topside/Hull

INTRODUCTION

The integrated topside/hull global structural dynamic analysis is a critical task in the design of offshore platforms. The structural designs and finite element models (FEMs) evolve in an iterative manner driven by the global loads analyses. The input forcing functions include wave pressures, assumed as regular waves. Typically, the global FEM does not include the structural details necessary to directly extract accurate stresses; their main use is to generate loads for the detailed local models. The major driver behind this separate global and local FEMs process is limitations in computational power. The local models can be several million degrees of freedom (DoFs) each. Directly integrating them into the global model to improve accuracy is not computationally feasible.

There are multiple known issues with the above global/local process that impact the accuracy of stresses and the overall design robustness. First, since this is a structural dynamics analysis, the accurate capture of the system natural frequencies and mode shapes in the range of regular wave period inputs is of critical importance to accurate loads. The lower fidelity global FEM is not ideal for accurate frequency/mode-shape calculations especially for the shorter wave period regimes. This will have an impact on the loads accuracy. Second, the local FEMs boundary conditions will play an important role in the accuracy of the stresses. It is often a difficult matter to accurately replicate the deformed shapes of the local model as it would have operated in the global system especially for over-constrained boundaries. Third, the global analysis is done at certain discrete wave periods deemed important; it is not executed as an incremental period sweep through the entire wave period range so that all potential resonances important to stress generation may be exercised. The reason behind this practice includes analysis schedule considerations driven by computation times. The consequence from this user wave period selection can include missing certain wave period inputs that can result in high stresses at critical regions. Finally, the above-mentioned process requires long schedules and is therefore incongruent with executing trade-off and sensitivity analyses to optimize design and reduce risks, costs, and schedules.

This paper introduces dynamic substructuring and component-mode synthesis (CMS) to significantly increase accuracy and computational speeds. The CMS process involves transforming detailed local FEMs into dynamic math models (DMMs). A frequency domain CMS procedure follows where the impedances at the connection degrees of freedom (DoFs) are utilized to couple the DMMs and form the integrated global system model. The CMS process couples the substructure modes to form system modes guaranteeing accurate system modal properties for the entire range of wave periods. Therefore, the dynamically substructured global system model is of equal fidelity to direct integration of the detailed local FEMs, within the user selected input wave period range.

The dynamic substructuring process overcomes all the stated deficiencies in the traditional global/local process. The dynamically substructured global model is not only of significantly higher fidelity but also executes at speeds that are a small fraction of the traditional global model. The benefits of the improved process include:

1. A higher fidelity global model resulting in a more accurate structural dynamic response. This removes the traditional process inaccuracies due to a lower fidelity global model.

2. A direct computation of stresses from the global dynamic response eliminating the need for separate local analyses. This removes the traditional process inaccuracies due to local model boundary conditions and loads mapping along with the long schedules.

3. A complete sweep of all wave periods in a single fast analysis resulting in high resolution stress amplitudes as a function of wave period (stress spectra). This removes the traditional process risk of not executing wave periods that may be critical to stress response.

4. Enabling a complete sweep through the wave spectra resulting in high resolution stress spectra (stress amplitude as a function of wave period). This removes the traditional process risk of not exercising wave periods that may be critical to the stress response.

5. Enabling trade and sensitivity studies to better assess the design, reduce risks, and optimize the coupled structure. Traditional process long analysis schedules are not congruent with efficient design process.

It is difficult to provide an exact apples to apples comparison of computation times between the traditional and dynamically substructured global solutions since 1) the traditional solution does not execute a sweep through the entire wave period regime and 2) the traditional solution requires additional local analyses to compute stresses. However, it is worth noting that just the execution of the traditional global FEM for frequencies and mode-shapes in the range of wave periods of interest took several hours in a commercial FEA package utilizing a fast eigensolver on a multi-core computer. From that, one can extrapolate that the dynamic response solution would have taken significantly longer computation times to which the additional schedules for the local analyses computations would still have to be added. In drastic contrast, the dynamically substructured global model, which is of significantly greater fidelity than the traditional global model, took just <u>2 minutes</u> to execute the dynamic response analysis for the entire wave regime and compute stresses!

As an added note, the solver utilized for these analyses is FLEXAS, a legacy nonlinear dynamic substructured (NDS) solver utilized in many complex applications for major operators including nonlinear dynamic simulations of flexible risers, dropped objects, and mechanical connectors. The solver has been validated/benchmarked by major operators and DeepStar against both full scale and bench test experiments.

METHODOLOGY

With dynamic substructuring, a substructure's DMM is derived directly from the high fidelity FEM via a coordinate transformation. This coordinate transformation is the means to significant computational efficiencies via the reduction of the FEM DoFs to a selected number of modal and physical DoFs. External force application DoFs need not be retained since the coordinate transformation may also serve to generalize the forces to the DMM DoFs. Furthermore, the coordinate transformation provides an accurate and direct path for the calculation of internal loads and stresses via load and stress transformation matrices (LTMs and STMs). Once the DMMs are created, the dynamic analysis process is composed of the following steps: (1) couple the DMMs to form the integrated global assembly, (2) execute dynamic analysis cases and recover generalized response, and (3) for each case, recover the desired local stresses.

Three classes of coordinate transformations exist: fixed-boundary, free-boundary, and mixed-boundary. The nomenclature comes from the method's requirement for normal mode boundary conditions. Each class of transformation has its own unique properties, applicability, and efficiencies. The coordinate transformation utilized for these analyses is the so called Residual Flexibility Mixed Boundary (RFMB) also referred to as the Majed-Henkel transformation. The coordinate transformation is given as follows:

$$\begin{Bmatrix} x_b \\ x_c \\ x_o \end{Bmatrix} = \begin{bmatrix} I_{bb} & 0 & 0 \\ 0 & I_{cc} & 0 \\ \psi_{ob}^{c} - g_{oc}^{R} g_{cc}^{R\,-1} \psi_{cb}^{c} & g_{oc}^{R} g_{cc}^{R\,-1} & \phi_{ok}^{N} - g_{oc}^{R} g_{cc}^{R\,-1} \phi_{ck}^{N} \end{bmatrix} \begin{Bmatrix} x_b \\ x_c \\ q_k \end{Bmatrix} \tag{1}$$

The derivation of the above coordinate transformation is beyond the scope of this paper and is given in Majed et al (2005). It is worth pointing out that the transformation is a mix of the following submatrices: 1) constraint modes, ψ, due to unit displacements on the b-set boundary DoFs that remain fixed during the eigenvalue problem, 2) residual flexibility, g, due to unit forces at the c-set boundary DoFs that remain free during the eigenvalue problem, and 3) a truncated set of normal modes, ϕ, computed with the b-set DoFs constrained. It can be shown that the transformation retains full flexibility at the DMM physical (connection) DoFs and retains the full dynamics of the FEM up to the user selected truncation frequency for the normal modes. The reduction of DoFs and hence the computational efficiency arises from the number of kept modes, k, being significantly less than the number of FEM DoFs. The RFMB transformation is then utilized to form the DMM's stiffness, mass, and damping matrices in the set of reduced-order coordinates (b+c+k) ready for substructure coupling.

The frequency domain CMS procedure utilized for these analyses is an impedance coupling technique where the impedances of DMMs 1 and 2 are coupled via the following equation:

$$\begin{Bmatrix} F_k^{(1)} \\ F_t \\ F_k^{(2)} \end{Bmatrix} = \begin{bmatrix} Z_{kk}^{(1)} & Z_{kt}^{(1)} & 0 \\ Z_{tk}^{(1)} & Z_{tt}^{(1)} + Z_{tt}^{(2)} & Z_{tk}^{(2)} \\ 0 & Z_{kt}^{(2)} & Z_{kk}^{(2)} \end{bmatrix} \begin{Bmatrix} \dot{q}_k^{(1)} \\ \dot{x}_t \\ \dot{q}_k^{(2)} \end{Bmatrix} \tag{2}$$

In the above equation, F is the vector of external forces and Z denotes mechanical impedance, a complex-valued matrix which is a function of the DMM's stiffness, mass, and damping matrices. The index t denotes the physical partitions, which for this case, is the connection DoFs (b+c). Equation (2) represents the integrated system of DMMs, accurate for dynamic loads up to the user-defined truncation frequency. For this case, 1.5 times the highest wave frequency of interest was retained to make certain all system modes in the range of wave frequencies are accurate. Once the frequency sweep of equation (2) is completed, the stresses are directly computed from the eq. 2 response vector via the stress transformation matrix.

EXAMPLE OF APPLICATION

Figures 1 and 2 show the traditional global topside/hull FEM and a typical local detailed column FEMs (Column A), respectively. The traditional global FEM is ~1M DoFs and the column A detailed FEM is ~2M DoFs.

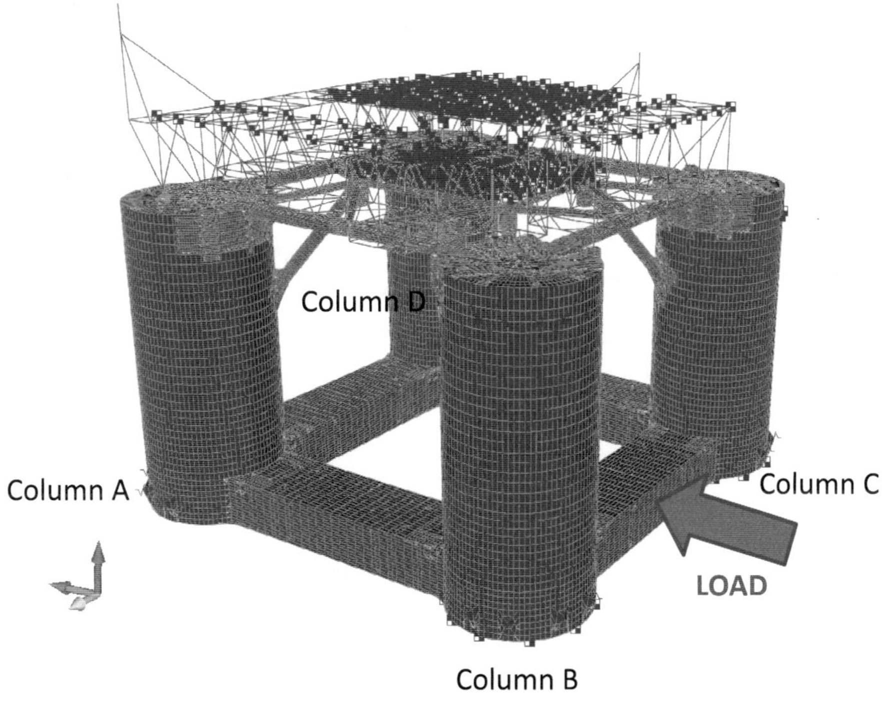

Figure 1 Traditional global FEM

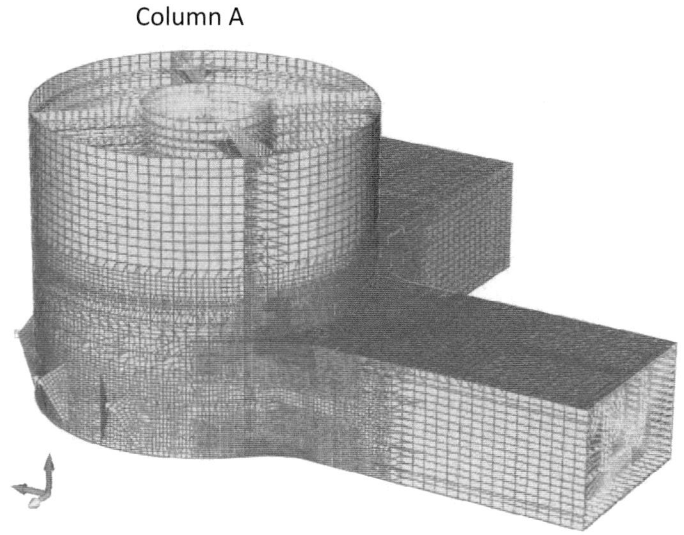

Figure 2 Detailed local FEM of a column structure

Figure 3 depicts the dynamic substructuring process utilized in this paper. The FEMs in the upper part of Figure 3 are transformed to DMMs via eq. (1). A frequency domain CMS is then executed, per eq. (2), to couple the DMMs. FLEXAS solver is utilized to execute the wave inputs and solve both the traditional global model (transformed to a single DMM) as well as the dynamically substructured global model depicted at the bottom of Figure 3. The direct integration of the FEMs in Figure 3 would have resulted in a 3M DoF FEM. The dynamically substructured system, per eq. (2) is just 2,500 DoFs and possess the exact same accuracy as the direct integration of Column A detailed FEM and the global FEM (minus the Column A FEM) as depicted in Figure 3.

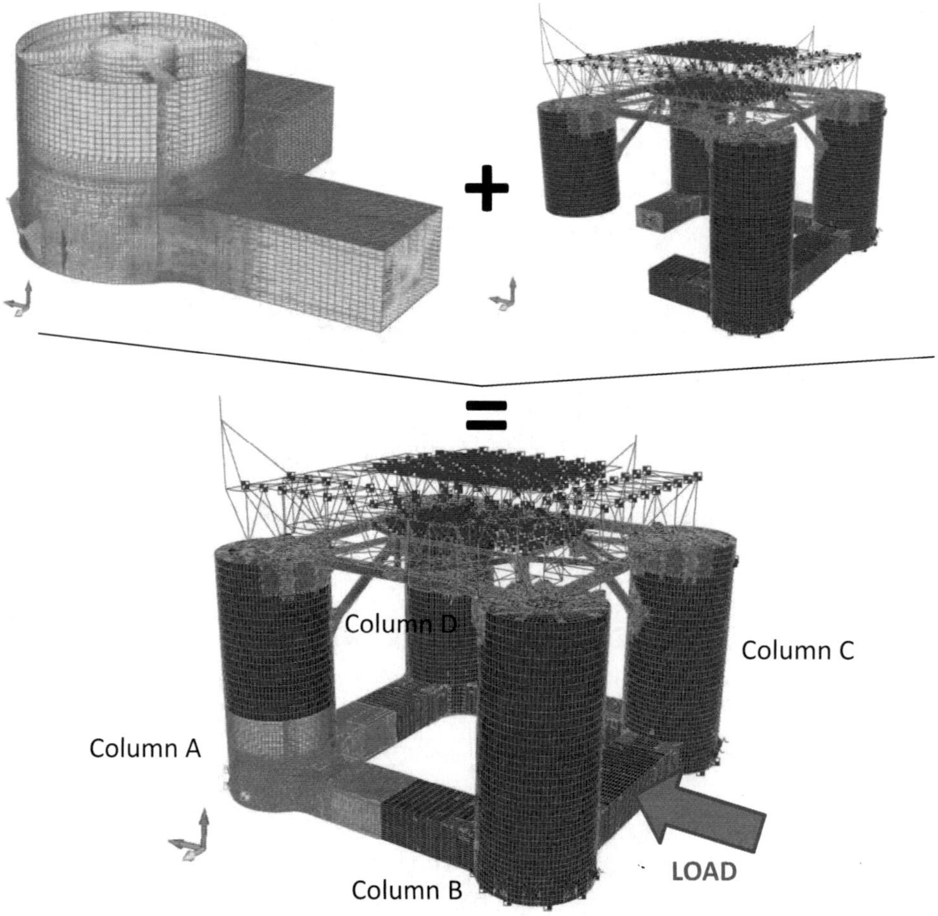

Figure 3 - Dynamically substructured global model

Stresses are computed in Column A and C elements for both the traditional global and dynamically substructured models. The stress recovery locations are depicted in Figures 4 and 5. It is anticipated that the stresses in Column C for both analyses will be similar given the distance of the Column C from the dynamically substructured detailed Column A. It is also anticipated that the stresses at Column A between the two analyses will be different, especially for the shorter wave periods where the deficiencies of the traditional global FEM will become more apparent.

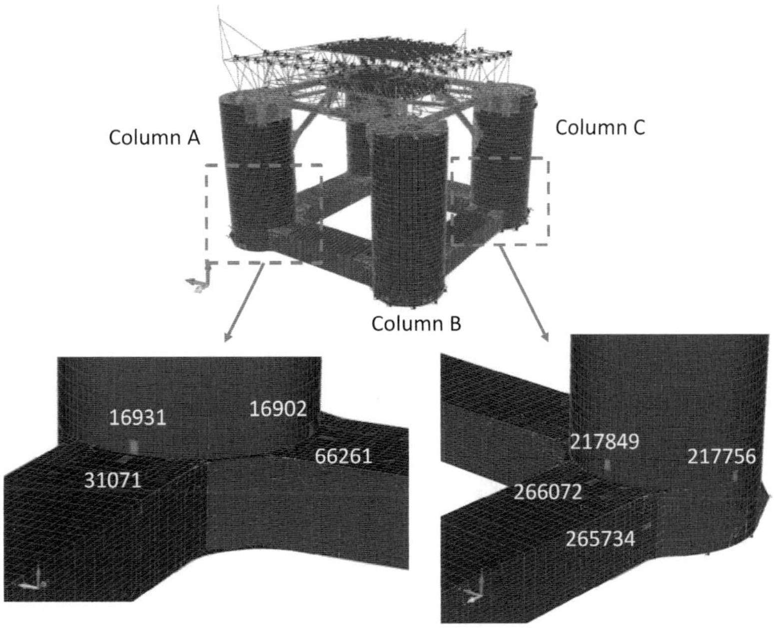

Figure 4 Elements for stress output – Traditional global model

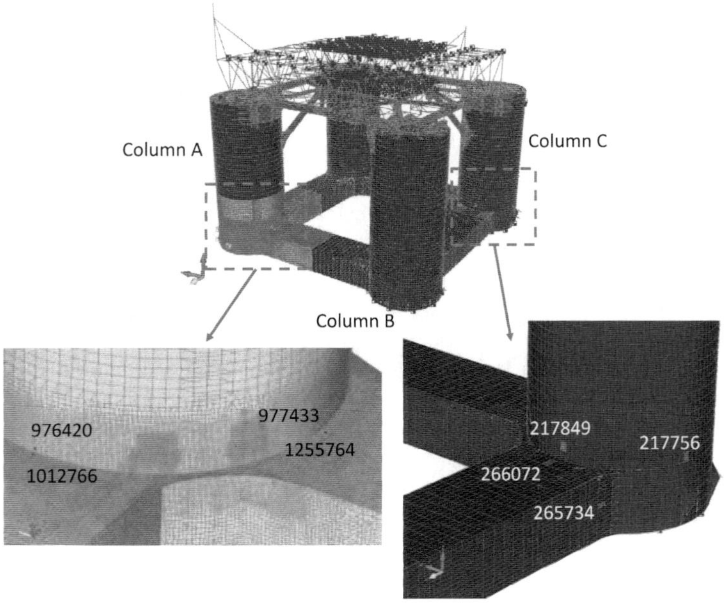

Figure 5 Elements for stress output – Dynamically substructured global model

RESULTS

Figures 6 and 7 present the von Mises stresses, at the element locations defined in Figures 4 and 5, from the FLEXAS traditional global and dynamically substructured global analyses, respectively. Please note that the stresses are computed against the entire wave period of regime of interest, from 3 to 15 seconds, for both FLEXAS solutions.

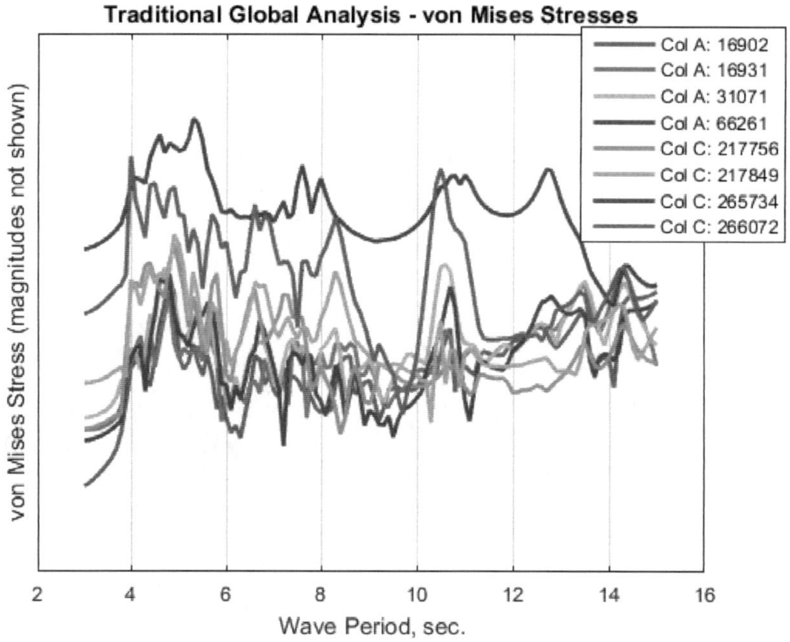

Figure 6 - Traditional global analysis – von Mises Stresses

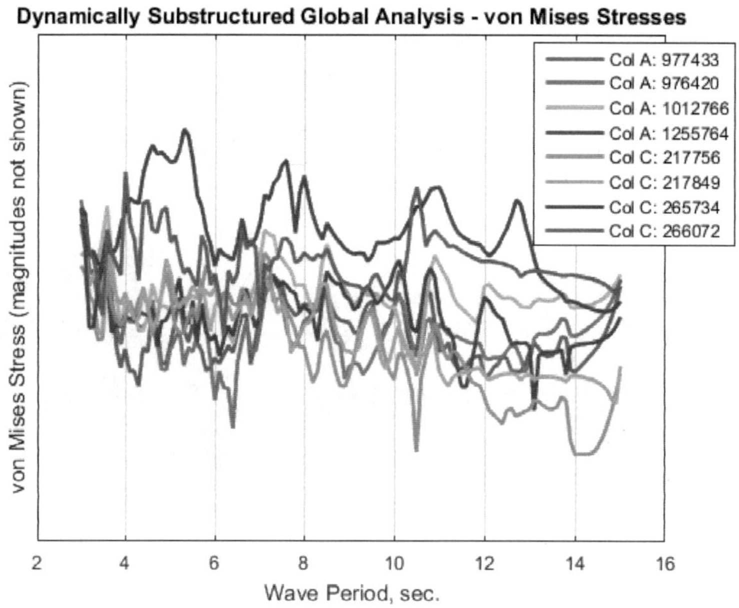

Figure 7 – Dynamically substructured global analysis – von Mises stress

Figure 8 shows the comparison of Column C stresses between the traditional and dynamically substructured global analyses. As expected, the stresses compare well between the two analyses for this location since it is far removed from the location of the detailed Column A. However, the added resolution of the dynamically substructured global system is easily evident even in this result. In addition, this result serves to benchmark the dynamically substructured global solution.

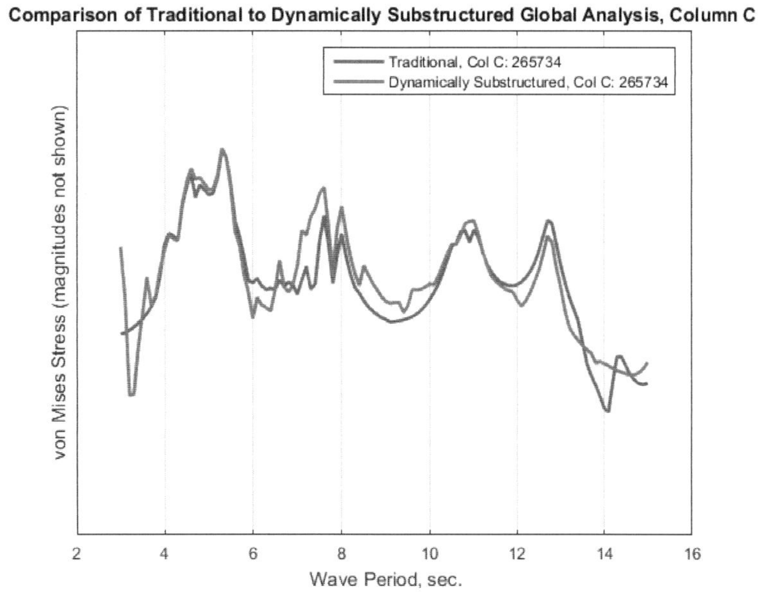

Figure 8 – Comparison of Traditional to Dynamically Substructured Global Analysis, Column C

Figure 9 shows the comparison of Column A stresses between the traditional and dynamically substructured global analyses. Here, we expect major differences between the global and dynamically substructured global solutions particularly in the shorter wave period regime where the fidelity of the traditional global model increasingly tails off.

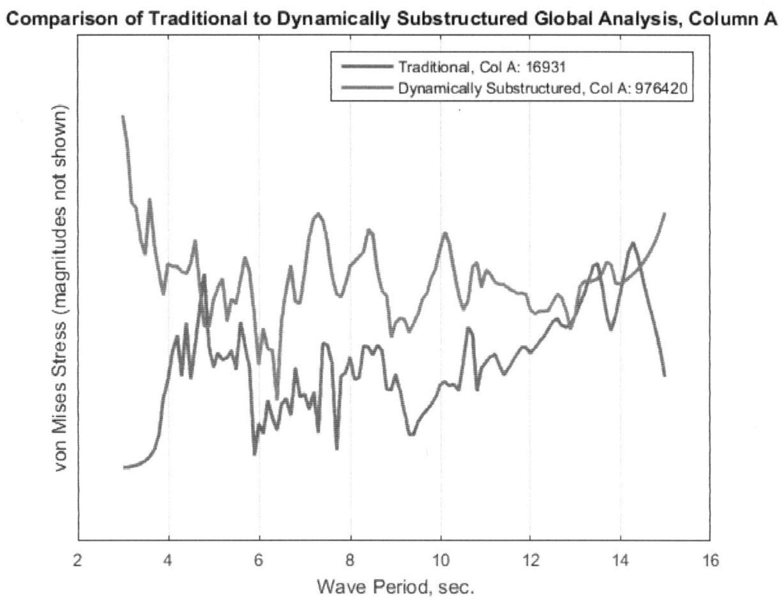

Figure 9 – Comparison of Traditional to Dynamically Substructured Global Analysis, Column A

CONCLUSIONS

It was demonstrated that the dynamically substructured global structural dynamic solution is a significant improvement over the traditional solution in terms of accuracy, computational speeds, and schedules. The improved solution enables a significantly higher fidelity global model, computes its structural dynamic response for the entire wave regime, and directly computes stresses - all in a few minutes of computation.

The improved process eliminates the lower fidelity global loads model utilized in the traditional process to satisfy computational constraints. Therefore, the inaccuracies in the lower fidelity structural dynamic loads analysis are removed. Furthermore, the improved process eliminates the local analyses and associated schedules/costs. This eliminates inaccuracies due to local model boundary conditions and loads mapping. In addition, the entire wave spectrum is analyzed in a single analysis, generating complete high resolution dynamic stress spectra as a function of wave period. This ensures that all wave periods important to the stress response are automatically accounted.

Finally, the very fast computations allow for trade-off and sensitivity studies to optimize structure and reduce overall risks, schedules and costs. Given the accuracy and extreme computation speeds afforded by the dynamically substructured solution, the next step will be the utilizing the irregular wave pressures as forcing functions. This more realistic representation of the pressure field will result in the most realistic stresses possible.

ACKNOWLEDGEMENTS

A word of thanks to the INTECSEA's structure group for providing the FEMs to carry out this work. An additional word of thanks to the INTECSEA's global technology for supporting this task.

REFERENCES

Majed, A., and Bhat, S., "Dynamics of a Vertically Dropped Flexible Jumper in Deep Waters", Proceedings of the 20th Offshore Symposium of the Society of Naval Architects and Marine Engineers, Houston, Texas, February 2015.

Majed, A., Chinello, L., and Cooper, P., "Computationally Efficient Flexible Pipe Bench Test Simulations Using Nonlinear Dynamic Substructures", Proceedings of International Society of Offshore and Polar Engineers (ISOPE), Kona, Hawaii. 2015.

Majed, A., Henkel, E., and Wilson, C., "Improved Method of Mixed-Boundary Component Representation in Structural Dynamic Analysis", AIAA Journal of Spacecraft and Rockets, Vol. 42, No. 5, 2005.

SNAME
THE INTERNATIONAL COMMUNITY
FOR MARITIME AND OCEAN PROFESSIONALS

DRIFT-OFF LIMITS FOR THE DRILLING RISER OF DRILLSHIP IN ENVIRONMENTS FROM NORMAL OPERATION TO STORMS AND LOOP CURRENT

SHIYU CHEN
SR. ENGINEER, P.E., ENSCO PLC, HOUSTON, TX
HONG YI
PROFESSOR, SHANGHAI JIAOTONG UNIVERSITY, SHANGHAI, CHINA

Proceedings of the 21st Offshore Symposium, February 2016, Houston, Texas
Texas Section of the Society of Naval Architects and Marine Engineers

ABSTRACT

For offshore drillship, a riser and drift off analysis is usually performed for each well to find out the operability, tension settings, operating limits (percentage of offsets) of the riser for certain target mud weights and environment (such as 1 year winter storm and 10 year winter storm including wave, current, and wind). Based on the results of riser and drift off analysis, drilling operation can be done for the target mud weight in environments up to the operable environment which an operating offset limit is obtained and shown from the riser analysis (for example, 1 year winter storm wave and current). For the drift off analysis, we usually do an analysis for a typical environment (such as 1 year winter storm) to get watch circles for that environment.

But in reality, the sea state is changing every day and is not the same as the typical environment used for the riser and drift off analysis. Most of the time (normal operation), the environment is calm sea conditions (with small waves, low wind speed, and low or high currents). Because watch circles change with environment, the watch circles for normal operation are much different from that of the typical environment. Drillers may need to know drift off limits in other environments (such as normal operation environment) to get a real picture for drift off and watch circle. Also, drillers want to know the drift off limits in high loop current environment to make operational decisions.
Keywords: Drillship, Drift-Off Limits, Watch Circles, Normal Operation Environment, Winter Storms, Loop Current

INTRODUCTION

For the drilling riser drift off analysis, we usually do an analysis for a typical environment (such as 1 year winter storm) to get watch circles for that environment. In reality, the sea state is changing every day and is not the same as the typical environment used for the riser and drift off analysis. Most of the time (normal operation), the environment is calm sea conditions (with small waves, low wind speed, and low or high currents). Normal operation environment (wave height and period, wind speed, current profile in certain time range such as a week.) is measured by drilling operator in the field. Because watch circles change with environment, the watch circles for normal operation are much different from that of the typical environment. Drillers may need to know drift off limits in other environments (such as normal operation environment and in high loop current environment) to get a real picture for drift off and watch circle, and make operational decisions.

METHODOLOGY

This paper presents some studies and drift off analysis to find drift off limits for more environments (normal operation, winter storm environment, and loop current environment) for the riser system of drillship. MCSKenny DeepRiser software version 3.2 was used to do the drift off analysis. Through drift off analysis, drift off limits for drillship in normal operation conditions were obtained. Also, drift off limits for drillship in high loop current environment were obtained. Those drift off limits will help driller to know watch circles in cases/environments which may happen most of the time.

DRIFT OFF ANALYSIS OF A DRILLSHIP FOR GC825-2 WELL (DEEP WATER) AND FOR MC842-1 WELL (SHALLOW WATER) IN DIFFERENT ENVIRONMENTS

The first drift off analysis was done for an Ensco drillship at 4950ft water depth for GC825-2 well.

Information for the well, metocean data, and riser spaceout are the following:

Vessel: An Ensco drillship (as shown in Figure 1), Well name: Green Canyon 825-2 in GOM, Water Depth: 4950ft, Targeted mud density: 12.7ppg, Wellhead Stick Up Above Mudline: 10ft.

Figure 1 Ensco Drillship

The riser stack-up (spaceout) is shown as in Table 1 (expected TJ stroke is 25.9, targeted mud density: 12.7ppg)

Riser Stack-Up & Particulars								
Joint Name	No.	Main Tube Inner Diameter (in)	Joint Length (ft)	Elevation to Bottom of Joint (ft)	Dry Weight of Section (kips)	Wet Weight of Section (kips)	Section Length (ft)	Depth to Bottom of Section (ft)
BOP	1	18	27.61	10	565.08	491.3	27.61	4940.00
LMRP	1	18	11.3	37.61	349.84	304.16	11.3	4912.39
LFJ Oilstates Subsea 6.0 KSI (94	1	0	8.86	48.91	30	29.9	8.86	4901.09
Slick 0.875in WT	12	19.5	75	57.77	507.86	344.29	900	4892.23
10kft buoy 55inX0.875in	6	19.5	75	957.77	366.97	12.19	450	3992.23
8Kft buoy 55inX0.875in	17	19.5	75	1407.77	1005.41	3.69	1275	3542.23
5Kft buoy 55inX0.875in	14	19.5	75	2682.77	806.15	-12.52	1050	2267.23
3Kft buoy 55inX0.875in	4	19.5	75	3732.77	217.75	-18.14	300	1217.23
2Kft buoy 55inX0.875in	10	19.5	75	4032.77	525.97	-57.45	750	917.23
Slick 0.875in WT	1	19.5	75	4782.77	42.32	28.69	75	167.23
18.75ft Pup x 0.875in WT	2	19.5	18.75	4857.77	26.45	20	37.5	92.23
Telescopic_OB	1	22	78.2	4895.27	70.78	61.54	78.2	54.73
Telescopic_IB	1	19	71.2	4973.47	20.2	17.56	71.2	
UFJ Oilstates Diverter II-3 (13 ki	1	0	9.04	5010.71	9	8.9	9.04	
Total Weight (LFJ-TJ)					3569.66	382.29		
Total Weight (Ring to LMRP)						749		
Weight (Ring to BOP)						1228.05		

Buoyancy Ratio	89%
Expected TJ Stroke (ft)	25.9
Water Depth (ft)	4950

Table 1 Riser Stack-Up

Metocean Conditions are provided in Table 2

Return Period	1 Winter	10 Winter	Normal Operation	
Hs (ft)	16.1	21.0	6.0	
Tp (s)	10.7	12.0	4.0	
Shape	Jonswap		Jonswap	
Gamma	1	2	1	
WS (knots)1hr, 10m	33.1	36.9	19.4	
Current				
Depth (ft)	Speed (knots)		Depth (ft)	Speed (knots)
0	0.82	0.91	0.00	0.50
98.4	0.62	0.70	98.4	0.28
196.9	0.29	0.33	196.9	0.10
295.3	0.00	0.00	295.3	0.00
4950	0.00	0.00	4950	0.00

Loop Current and Associated Wind and Wave			
Hs (ft)	6.56		
Tp (s)	6.3		
Shape	Jonswap		
Gamma	1		
WS (ft/s) - 1hr @ 10m	32.8		
Current (ft/s)			
Depth (ft)	High Shear	Mod Shear	Low Shear
0	2.6	4.1	3.2
164	2,6	4.1	3.2
197	2.4	3.9	3.2
230	2.2	3.7	3.2
262	2.1	3.5	3.2
295	1.9	3.2	3.1
328	1.7	3.0	2.9
492	0.9	2.2	2.7
656	0.5	1.7	2.4
820	0.3	1.4	2.2
984	0.3	1.2	1.9
1312	0.2	1.0	1.5
1640	0.2	0.9	1.3
1968	0.2	0.7	1.1
2296	0.2	0.7	1.0
2624	0.2	0.6	0.8
2952	0.2	0.5	0.7
3280	0.2	0.5	0.5
4950	0.5	0.5	0.5

Table 2 Metocean Conditions (1yr and 10yr winter storms, Normal Operation, 1yr loop current)

Drift-Off Results for Normal Operation, 1yr Winter Storm and 1yr Loop

The following criteria were used as limits for the riser system during the Drift-Off analysis.

Operational Data

Water Depth (ft)	Draft (ft)	Mud Weight, (ppg)	Recommended Top Tension (kips)
4,950	39.3	8.56	1007
		12.7	1420

Environmental Data

Return Period	Wave Height Hs (ft)	Wave Period Tp-c (sec)	Wind Speed (knots)	Surface Current (knots)	Soil curve
~~95% Non-Exc~~	~~8.0~~	~~7.6~~	~~38.1~~	~~3.1~~	~~N/A~~
Normal Operation	6	4	19.4	0.50	P-Y
1 Year WS with 90% wind*	16.1	10.7~~8.3~~	29.8	0.82~~1~~	P-Y~~N/A~~
1 Year Loop	6.56	6.3~~8.3~~	19.4	2.43~~1~~	P-Y~~N/A~~
~~10yr Loop~~	~~4.9~~	~~6~~	~~20.4~~	~~5.13~~	~~N/A~~

* Wind speed reduced to 90% to get a min watch circle of 20ft.

Drift Off Criteria

Item	Available	Limit	Reference
TJ Stroke (ft)	65~~.0~~	65	Vendor spec
Tensioner wire travel (ft)	50.0	50	-
Upper Flex (degree)	15.0	15.0	-
Lower Flex (degree)	10.0	10	-
Casing Bending Stress	56 ksi Yield	80%*(Yield) = 44.8 ksi	API RP 17G
Other Component Stress	80 ksi Yield	2/3*(Yield) = 53.3 ksi	API Spec 16A, 16F
Wellhead Bending Moment	4000 kips-ft	4000 kips-ft	Vendor spec
Time from red alert to EDS (sec)	N/A	63	-
Time from yellow to red alert (sec)	N/A	60	-
Heave (ft)	N/A	Per RAO and Hs	Vessel RAO
Tide (ft)	N/A	2	Standard GOM
Safety Margin (ft)	N/A	2	-

Two sets of results at 5 degree and 45 degree headings (relative to the collinear environment direction), in terms of watch circles and alert time, are recommended for setting the WSOC (well specific operating criteria). They are listed below.

Drift Off Limit Violations (Normal Operation)

Normal Operation First

Mud (ppg)	Heading (deg)	P.O.D. Time (sec)	P.O.D. Dist. (ft)	Red Time (s)	Red Dist. (ft)	Yellow Time (s)	Yellow Dist. (ft)	Limiting Criteria
8.56	5	446.5	362.3	383.5	267.6	323.5	185.4	Tensioner Stroke
	45	336.0	368.5	273.0	271.4	213.0	174.8	Tensioner Stroke
12.7	5	420.5	300.4	357.5	214.2	297.5	145.5	Max. Casing Bending Stress
	45	301.0	305.4	238.0	209.8	178.0	122.1	Max. Casing Bending Stress

Drift Off Limit Violations (1yr Storm with 90% wind)

1yr W Storm First Limits:

Mud (ppg)	Heading (deg)	P.O.D. Time (sec)	P.O.D. Dist. (ft)	Red Time (s)	Red Dist. (ft)	Yellow Time (s)	Yellow Dist. (ft)	Limiting Criteria
8.56	5	253.5	303.6	190.5	158.9	130.5	72.4	Tensioner Stroke
	45	184.0	302.3	121.0	134.7	61.0	32.0	Tensioner Stroke
12.7	5	269.0	324.6	206.0	175.7	146.0	84.1	Tensioner Stroke
	45	196.0	330.4	133.0	161.5	73.0	46.8	Tensioner Stroke

Drift Off Limit Violations (1yr Loop)

1yr Loop First Limits:

Mud (ppg)	Heading (deg)	P.O.D. Time (sec)	P.O.D. Dist. (ft)	Red Time (s)	Red Dist. (ft)	Yellow Time (s)	Yellow Dist. (ft)	Limiting Criteria
8.56	5	298.5	364.0	235.5	245.5	175.5	149.1	Max. Casing Bending Stress
	45	210.0	377.5	147.0	206.0	87.0	79.5	Tensioner Stroke
12.7	5	275.5	303.2	212.5	197.4	152.5	110.7	Max. Casing Bending Stress
	45	190.0	316.5	127.0	157.9	67.0	48.2	Max. Casing Bending Stress

Conclusion: The yellow watch circles in normal operation environment are much larger than those of 1yr Winter Storm and 1yr loop environment.

The second drift off analysis was done for the Ensco drillship at 2135ft water depth for MC842-1 well (shallow water). Targeted mud density: 16ppg, Wellhead Stick Up Above Mudline: 12ft.

The riser stack-up (spaceout) is shown as in Table 3 (expected TJ stroke is 27.7ft, targeted mud density 16ppg)

Riser Stack-Up & Particulars								
Joint Name	No.	Main Tube Inner Diameter (in)	Joint Length (ft)	Elevation to Bottom of Joint (ft)	Dry Weight of Section (kips)	Wet Weight of Section (kips)	Section Length (ft)	Depth to Bottom of Section (ft)
BOP	1	18	27.61	12	565.08	491.3	27.61	2123.00
LMRP	1	18	11.3	39.61	349.84	304.16	11.3	2095.39
LFJ Oilstates Subsea 6.0 KSI (94	1	0	8.86	50.91	30	29.9	8.86	2084.09
Slick 0.875in WT	12	19.5	75	59.77	507.86	344.29	900	2075.23
3Kft buoy 55inX0.875in	3	19.5	75	959.77	163.31	-13.61	225	1175.23
2Kft buoy 55inX0.875in	10	19.5	75	1184.77	525.97	-57.45	750	950.23
Slick 0.875in WT	1	19.5	75	1934.77	42.32	28.69	75	200.23
32.5ft Pup x 0.875in WT	1	19.5	32.5	2009.77	20.5	14.43	32.5	125.23
18.75ft Pup x 0.875in WT	2	19.5	18.75	2042.27	26.45	20	37.5	92.73
Telescopic_OB	1	22	78.2	2079.77	70.78	61.54	78.2	55.23
Telescopic_IB	1	19	71.2	2157.97	20.2	17.56	71.2	
UFJ Oilstates Diverter II-3 (13 ki	1	0	9.04	2195.71	9	8.9	9.04	
Total Weight (LFJ-TJ)					1357.19	397.89		
Total Weight (Ring to LMRP)						764.6		
Weight (Ring to BOP)						1243.65		

Buoyancy Ratio		71%
Expected TJ Stroke (ft)		27.7
Water Depth (ft)		2135

Table 3 Riser Stack-Up

Metocean Conditions are provided in Table 4

Return Period	1 Winter	10 Winter	Normal Operation		90% Non-Exceedance		75% Non-Exceedance	
Hs (ft)	10.8	19.4	6.0		7.1		4.4	
Tp (s)	8.0	10.3	4.0		6.7		5.4	
Shape	Jonswap		Jonswap		Jonswap		Jonswap	
Gamma	1.6	2	1		1		1	
WS (ft/s)1hr, 10m	48.4	62.1	37.5		33.2		26.3	
Current								
Depth (ft)	Speed (ft/s)		Depth (ft)	Speed (knots)	Depth (ft)	Speed (knots)	Depth (ft)	Speed (knots)
0	0.50	0.60	0.00	0.50	0.00	0.50	0.00	0.50
200	0.50	0.60	98.4	0.28	98.4	0.28	98.4	0.28
300	0.30	0.30	196.9	0.10	196.9	0.10	196.9	0.10
400	0.30	0.30	295	0.00	295	0.00	295	0.00
2135	0.30	0.30	2135	0.00	2135	0.00	2135	0.00

Loop Current and Associated Wind and Wave		
Hs (ft)	7.78	
Tp (s)	6.93	
Shape	Jonswap	
WS (ft/s) - 1hr @ 10m	37.5	
Current (ft/s)		
Depth (ft)	1yr	10yr
0	3.72	5.39
82	3.72	5.38
164	3.69	5.34
246	3.29	4.77
328	2.76	4.00
410	2.40	3.47
492	2.14	3.10
574	1.91	2.77
656	1.71	2.48
838	1.54	2.22
820	1.42	2.06
902	1.31	1.89
984	1.19	1.73
1066	1.11	1.61`
1148	1.03	1.50
1230	0.99	1.43
1312	0.92	1.34
1394	0.86	1.25
1476	0.82	1.18
1558	0.79	1.14
1640	0.74	1.07
1722	0.70	1.01
1804	0.67	0.96
1996	0.66	0.92
1968	0.66	0.88
2051	0.66	0.86
2133	0.66	0.81

Table 4 Metocean Conditions (1yr and 10yr winter storm, normal operation, 1yr loop current)

Drift-Off Results for Normal Operation, 1yr Winter Storm and 1yr Loop, 90% Non-Exc, 75% Non-Exc

The following criteria were used as limits for the riser system during the Drift-Off analysis.

Operational Data

Water Depth (ft)	Draft (ft)	Mud Weight, (ppg)	Recommended Top Tension (kips)
2,135	39.3	8.56	1027
		16	1346

Environmental Data

Return Period	Wave Height Hs (ft)	Wave Period Tp-c (sec)	Wind Speed (knots)	Surface Current (knots)	Soil curve
~~95% Non-Exc~~	~~8.0~~	~~7.6~~	~~38.1~~	~~3.1~~	~~N/A~~
Normal Operation	6	4	22.2 (17.8* for 30deg heading)	0.50	P-Y
1 Year WS	10.8	8.0~~8.3~~	28.7	0.30	P-Y~~N/A~~
1 Year Loop	7.78	6.93~~8.3~~	22.2	2.2~~1~~	P-Y~~N/A~~
90% Non-Exc	7.1	6.7	19.7	0.50	P-Y
75% Non-Exc	4.4	5.4	15.6	0.50	P-Y

* Wind speed reduced to 80% to get a min watch circle of 20ft for 30deg heading.

Drift Off Criteria

Item	Available	Limit	Reference
TJ Stroke (ft)	65.~~0~~	65	Vendor spec
Tensioner wire travel (ft)	50.0	50	-
Upper Flex (degree)	15.0	15.0	-
Lower Flex (degree)	10.0	10	-
Casing Bending Stress	56 ksi Yield	80%*(Yield) = 44.8 ksi	API RP 17G
Other Component Stress	80 ksi Yield	2/3*(Yield) = 53.3 ksi	API Spec 16A, 16F
Wellhead Bending Moment	4000 kips-ft	4000 kips-ft	Vendor spec
Time from red alert to EDS (sec)	N/A	63	-
Time from yellow to red alert (sec)	N/A	60	-
Heave (ft)	N/A	Per RAO and Hs	Vessel RAO
Tide (ft)	N/A	2	Standard GOM
Safety Margin (ft)	N/A	2	-

Two sets of results at 5 degree and 45 degree headings, in terms of watch circles and alert time, are recommended for setting the WSOC. They are listed below.

Drift Off Limit Violations (Normal Operation)

Normal Operation First Limits:

Mud (ppg)	Heading (deg)	P.O.D.		Red		Yellow		Limiting Criteria
		Time (sec)	Dist. (ft)	Time (s)	Dist. (ft)	Time (s)	Dist. (ft)	
8.56	5	274.5	161.2	211.5	97.3	151.5	51.8	Max. Casing Bending Stress
	30*	246.0	152.9	183.0	86.4	123.0	38.6	Max. Casing Bending Stress
16	5	264.5	138.9	201.5	82.8	141.5	41.9	Max. Casing Bending Stress
	30*	229.5	131.1	166.5	70.5	106.5	28.6	Max. Casing Bending Stress

* Wind speed reduced to 30ft/s (80%) to get a min watch circle of 20ft

Drift Off Limit Violations (1yr Storm)

1yr W Storm First Limits:

Mud (ppg)	Heading (deg)	P.O.D.		Red		Yellow		Limiting Criteria
		Time (sec)	Dist. (ft)	Time (s)	Dist. (ft)	Time (s)	Dist. (ft)	
8.56	5	206.5	163.9	143.5	77.5	83.5	25.7	Max. Casing Bending Stress
	30	151.5	159.4	88.5	49.8	28.5	3.9	Max. Casing Bending Stress
16	5	199.5	141.4	136.5	64.5	76.5	18.6	Max. Casing Bending Stress
	30	141.0	135.6	78.0	37.7	18.0	1.3	Max. Casing Bending Stress

Drift Off Limit Violations (1yr Loop)

1yr Loop First Limits:

Mud (ppg)	Heading (deg)	P.O.D.		Red		Yellow		Limiting Criteria
		Time (sec)	Dist. (ft)	Time (s)	Dist. (ft)	Time (s)	Dist. (ft)	
8.56	5	194.5	162.7	131.5	82.4	71.5	25.9	Max. Casing Bending Stress
	30	145.0	158.7	82.0	53.2	22.0	3.2	Max. Casing Bending Stress
16	5	184.5	141.1	121.5	66.9	61.5	17.0	Max. Casing Bending Stress
	30	134.5	136.8	71.5	40.5	11.5	0.6	Max. Casing Bending Stress

Drift Off Limit Violations (90% Non-Exc)

90% Non-Exe and Normal Current First Limits:

Mud (ppg)	Heading (deg)	P.O.D.		Red		Yellow		Limiting Criteria
		Time (sec)	Dist. (ft)	Time (s)	Dist. (ft)	Time (s)	Dist. (ft)	
8.56	5	316.5	158.0	253.5	104.6	193.5	64.3	Max. Casing Bending Stress
	30	221.5	154.4	158.5	79.1	98.5	29.6	Max. Casing Bending Stress
16	5	306.0	136.7	243.0	90.3	183.0	54.2	Max. Casing Bending Stress
	30	207.0	132.7	144.0	64.3	84.0	21.1	Max. Casing Bending Stress

Drift Off Limit Violations (75% Non-Exc)

75% Non-Exe and Normal Current First Limits:

Mud (ppg)	Heading (deg)	P.O.D.		Red		Yellow		Limiting Criteria
		Time (sec)	Dist. (ft)	Time (s)	Dist. (ft)	Time (s)	Dist. (ft)	
8.56	5	426.5	151.7	363.5	116.3	303.5	87.7	Max. Casing Bending Stress
	30	283.5	150.9	220.5	95.6	160.5	51.5	Max. Casing Bending Stress
16	5	415.0	130.5	352.0	101.0	292.0	76.4	Max. Casing Bending Stress
	30	264.5	129.5	201.5	78.9	141.5	39.7	Max. Casing Bending Stress

CONCLUSIONS

The result shows drift off limits during normal operation and high loop current environment are much different from that of the typical environment.

1. During normal operation, the watch circles are much larger than that of 1yr winter storm (This is because the wave height, wind speed of normal operation environment are much less than that of 1yr winter storm).
2. In shallow water (2135ft) case, even the yellow watch circle is less than 20ft for 30 deg heading in 1yr storm environment, the yellow watch circles are larger than 20ft in Normal Operation environment.
3. Watch circles increase as the wave height, wind speed and current speed decrease. Depending on the targeted rig utilization rate, 90% Non-Exc or 75% Non-Exc probably is a better environment to be used to perform the drift-off analysis to find out if the (yellow) watch circles meet the requirement.
4. The drift off limit in Normal Operation environment will help driller to know watch circles in cases which may happen most of the time, and help them to make operational decisions.
5. Currently, API and the industry have not specified how to use the drift-off analysis (what kind of

environment to use) to make sure the watch circles for offshore drilling are good enough or not. It is part of the risk control and management (Lost power or DP system failure happened only occasionally). This paper showed a way to do the drift-off analysis and get watch circles more practically (such as using 90% Non-Exc or 75% Non-Exc environment).

ACKNOWLEDGEMENTS

Thanks for Ensco plc for supporting this research and study.

REFERENCES

American Petroleum Institute, "Recommended Practice for Design, Selection, Operation and Maintenance of Marine Drilling Riser Systems", API RP 16Q, First Edition, Nov. 1, 1993.

American Petroleum Institute, "Specification for Marine Drilling Riser Equipment", API Specification 16F, First Edition, Aug. 2004.

American Petroleum Institute, "Recommended Practice for Completion/Workover Risers", API RP 17G, Second Edition, Jul. 2006.

Kenneth Bhalla and Yusong Cao, "Watch Circle Assessment of Drilling Risers During a Drift-Off and Drive-Off Event of a Dynamically Positioned Vesse'", Proceedings of the Dynamic Positioning Conference 2005, Houston, USA, Nov. 15-16, 2005

Lew Weingarth, "Refining the DP Watch Circle", Proceedings of The Dynamic Positioning Conference 2006, Houston, USA, Nov. 17-18, 2006.

SNAME
THE INTERNATIONAL COMMUNITY
FOR MARITIME AND OCEAN PROFESSIONALS

A SOLUTION TO ACCURATE OFFSHORE DIMENSIONAL CONTROL

TIMOTHY DAVID GREESON
ENGINEERING MANAGER, PROCEANIC

MARK GORDON WALLER
MANAGING DIRECTOR, PROCEANIC

Proceedings of the 21st Offshore Symposium, February 2016, Houston, Texas
Texas Section of the Society of Naval Architects and Marine Engineers

ABSTRACT

Accurate offshore dimensional control is essential for a variety of offshore projects including setting of equipment, damage assessments, modification, refurbishment, and integration of structure, piping, and other components.

Traditional offshore measurement methods include use of physical tape measures and drawings with field verifications or as-built drawings. 3D point cloud scanning with lasers is now commonly used throughout the industry but does not provide real-time feedback, requires powerful computers for post-processing, and is only accurate to +/- 2mm (varies depending on the machines used).

A more precise and accurate method of measurement is use of a total station device with auto-leveling features turned off. This method is accurate, portable, and versatile, and it is also more cost effective than traditional 3D point cloud scanning. Placing control point targets around the objects of interest allows an engineer with a standard computer to quickly take measurements from multiple locations and then reference them into a single global coordinate system. Advantages of this method include real time feedback of measurements, means to measure dangerous areas from a safe position, increased accuracy, and increased precision. Total stations are a common sight in ship yards, but they are usually operated by a field technician (not a surveyor and not an engineer). Total stations are not commonly used offshore since they require special consideration for use. They also require a small amount of post-processing in a spreadsheet to be relatable to global coordinates.

This article will discuss and compare current options for offshore dimensional control and describe the use of the total station method for offshore work. The article will also present several case studies, drawing from offshore dimensional control projects, where use of a total station benefited the project beyond the traditional methods of offshore dimensional control.

Keywords: Dimensional Control, Total Station, Theodolite, 3D Cloud, Damage Assessment, Offshore Equipment Installation, Surveying

INTRODUCTION

Accurate offshore dimensional control is essential for a variety of offshore projects including setting of equipment, damage assessments, modification, and refurbishment and integration of structure, piping, and other components. Although quay-side modifications are almost always preferred, some situations call for construction / modification work to be performed offshore. During any fabrication work, dimensional control is critical to the success of the project. Dimensional control can be performed in a number of ways including use of traditional measuring tapes, use of laser measuring tapes, and use of levels / plumb bobs, etc.

3D cloud scanning involves use of laser instruments to take thousands of measurements and generate a 3D point cloud for use in CAD modeling. The results of individual scans can be compiled into one large overall model for use in project planning and detail design stages of the project. Although a very powerful tool, the resolution is limited to +/- 2mm depending on the equipment being used. Furthermore, results are not instantly usable and require post processing (usually shore based).

THE SOLUTION

A more precise and accurate method of measurements is use of a total station device. This method is accurate, portable, versatile, and more cost effective than methods of near-comparable accuracy, such as 3D point cloud scanning.

Total stations are made by several different manufacturers and are widely used in traditional onshore surveying. Total stations are also a very common tool used in shipyards to perform dimensional control, again using traditional surveying methods. Most of these instruments have sophisticated sensors and internal processors to allow them to be used by non-technical personnel. They have an auto-level feature that enables the processor to establish a new station within a known coordinate systems using only two points. However, on floating facilities, the ship or facility is constantly in motion, and the vertical axis of the total station is constantly changing compared to the coordinate system of the facility. Even small motions from a fixed-jacket platform or jack-up drilling rig will cause the coordinate system to rotate enough to cause error.

The solution is to use a total station with the auto-level feature turned off. Turning off this feature will allow the instrument to be used in any level of vessel motions, but it will also disable several key features, including the ability to use two points to re-establish a new station within an existing coordinate system. This function is what allows measurements to be taken of anything and from anywhere on the vessel into a single coordinate system, regardless of line of sight. This critical ability can be re-attained by measuring a series of unchanging control point targets around the objects of interest or around the entire vessel. An engineer with a standard computer can quickly take measurements from multiple locations and then reference them into a single global coordinate system using a simple coordinate transformation.

ADVANTAGES

Using a total station for dimensional control provides instant feedback and yields very accurate results. The total station is far less involved and tedious than other methods for measuring and data recording. Manual measuring methods such as tape measures usually involve a coordinated effort between at least two people for each measurement. This introduces multiple sources of human error for each measurement. Combined with the fact that the error built-in to the equipment is much greater than a total station, accuracy of measurements obtained with a total station is orders of magnitude higher. Furthermore, measurements are recorded electronically automatically, saving time and limiting unnecessary personnel, and shots can be taken across the rig from non-dangerous and non-intrusive locations, which saves permitting time and reduces risk.

Depending on the application of the equipment, 3D scanning can be an effective tool for use in dimensional control. Scanners can quickly amass a large amount of data over an entire area, which can then be used to generate 3D models. This is an extremely useful technology for creating full models of facilities and for planning construction and modification projects. Also, by eliminating human error, the accuracy of measurements is improved over manual methods. However, in many project applications, 3D models are unnecessary and an inefficient way to perform the required dimensional control. When only specific dimensions are desired, the total station is a much faster and more accurate method. With 3D scanning, many scans typically have to be compiled and merged via control points. This post-processing can lead to days or weeks of lag before any useful dimensions can be obtained. When the total station is used, post-processing can be performed immediately on site.

Most crucially, use of a total station will ultimately yield cost savings (without compromising accuracy or precision) in most dimensional control applications. The time saved over manual measuring methods can significantly reduce personnel and time on board. And the equipment overhead cost of a total station is typically much less than 3D scanners, and the post-processing costs are negligible in comparison.

Although the published tolerance for the most machines is +/- 0.001' (0.3 mm), this value is given for the distance from the machine to the point in question. While the distance from the machine is important in traditional surveying, this type of use of the machine is only interested in the relative distances from points located relatively close together. The resulting accuracy of the relative dimensions between two adjacent points is better than the published tolerances.

INSTRUMENT SELECTION

Instrument selection is very important. This paper is written based on experience with a Nikon 3.M instrument, but other resolutions and other brands are available. The smaller the angular measurement, the more accurate the machine will be. Any selected instrument must have the following key features: ability to turn off the auto-level feature, resolution to at least 3 seconds, ability to acquire reflector-less targets, and they must be light and portable for use offshore.

Using a traditional survey tripod, the bottom is placed onto a surveyor's triangle (commonly used for surveying on concrete). Magnets can be added to the base for increased stability during rougher seas.

USES AND LIMITATIONS

This method of dimensional control can be used to perform on site damage assessments, planning for refurbishments, project pre-planning, lift pre-planning, and many other offshore fabrication projects.

During damage assessments, personnel can take readings from non-hazardous or unaffected areas. The operator can also attain real-time, pre-processed results and adjust his assessment zone as needed.

As with all methods of dimensional control, this method does have limitations. This method is meant for surveying a relatively small area or small number of targets.

Like all laser based measuring devices, the instrument is not able to attain a return signal at high angle of incidences nor from dark colors. A combination of these two limitations (for example, during a post-fire inspection), is particularly difficult.

The instrument is water resistant but is not meant to be used in the rain. During field tests, wet steel has been difficult to survey at angles of incidence beyond 45 degrees. Towel or flame drying the steel allows the instrument to attain a return signal.

CASE STUDY I: OFFSHORE SETTING OF ANCHOR WINCHES

A client desired to add a second anchor winch on each column of a semisubmersible. The new set of anchor winches was to be powered using a common drive shaft between the existing winches and the new ones.

To minimize shipyard downtime, the foundations were scheduled to be installed offshore. To further complicate the problem, the new anchor winch foundations had to be aligned to within +/- 1mm of the existing anchor winch foundations to allow the drive shaft to align properly. The use of shims was discouraged and very limited.

To accomplish this tight fabrication tolerance, the foundations were fabricated onshore with 2 inches excess material that would need to be trimmed offshore. An engineer with a total station was sent with an offshore fabrication team. Upon arrival to the rig, the engineer then went to the first corner to have its foundation installed and began taking measurements of the existing foundation and the location where the new foundation would be installed. Several control points were established in the immediate surrounding area to use periodically for verification that the station had not moved within a set of measurements.

Once the exact desired height of the new foundations was determined, the fabrication team began trimming the foundations. Once the foundations were trimmed, they were placed onto the deck, and the total station was again used to verify the height and location of the foundations. After making any necessary adjustments, the foundations were welded into position using welding methods intended to not distort the structure. Several times during the welding process, the total station was used to verify that the foundations remained within tolerance. All remaining winch foundations on the other columns were installed using the same method. During this project, the engineer was able to support up to two fabrication teams working simultaneously at two different locations on the rig.

The final result was that the foundations were placed within the final tolerances. The tolerance of the foundation was controlled by the fabrication methods used. A small set of shims pre-made for final adjustments based on the as-built dimensional control work.

CASE STUDY II: RIG DAMAGE ASSESSMENT

A jack-up drilling rig caught fire while drilling offshore. The drilling package was reported to be heavily damaged, and the derrick was claimed to be warped from the heat of the fire. An engineer was requested to perform a visual and dimensional control survey of the damage. The remainder of the discussion of this case study focuses solely on the dimensional control work.

The rig was brought into a shipyard for assessments and repairs. Upon arrival at the rig, the engineer performed a visual survey to determine the scope of the dimensional control work. During this initial survey, several bulkheads were found to be warped, and several beams were observed to be sagging. The derrick did not appear to have any deformations.

To assess the bulkheads, a station was established immediately in front of the bulkheads, with the coordinate system as orthogonal as possible. Using this technique allows the engineer to interpret the results, although post-processing is required for the most accurate results. Shots were taken on all four corners of the bulkhead and then on a grid pattern within the bulkhead. The grid density was increased in way of the affected / deformed areas. A plane was established using a best-fit plane through points considered to be outside of the affected areas. The results indicated significant deformities in some of the bulkheads. Normal fabrication tolerances and deflections due to static loads were taken into account when interpreting the final results.

To quantify the deflections in the sagging beams, a station was established underneath the beams. A 90-deg eye piece attachment as used to allow the instrument to take measurements from overhead. Once post-processed, the results indicated that the beams were sagging well beyond the original fabrication tolerances.

Finally, the derrick was investigated using the total station. The area immediately surrounding the derrick was too close to be able to get a measurement up the entire derrick. To be able to attain all of the necessary measurements, the station was established on the helideck, across the rig from the derrick but well within the range of the total station. Ten measurements were taken along the height of each derrick leg: five on the column flange to measure deformations in one member plane and five on the column web to measure deformations in the other member plane. The results indicated that the derrick was not deformed beyond the original fabrication tolerances.

CONCLUSIONS

The use of a total station in onshore fabrication is a common practice. However, use of the total station offshore has been limited due to instrument and personnel limitations. Allowing an engineer to operate the total station with consideration given to the instrument's limitations and the ability to post-process data in the field allows greatly expanded applications of this method of dimensional control for offshore use.

Other uses include but are not limited to the following: Jack-up leg damage assessments, general project pre-planning, B.O.P. dimensional control, Distance measurements for multiple platform complexes, etc.

Proceanic is a leader in innovative offshore engineering and inspections, using new and existing technology in a variety of applications to provide a safer, more cost effective, and more accurate inspections.

REFERENCES

[1] Total Station Nivo Series Instruction Manual, Version A 1.0.0, July 2009.

[2] Faro Laser Scanner Focus Manual, October 2011.

SNAME
THE INTERNATIONAL COMMUNITY
FOR MARITIME AND OCEAN PROFESSIONALS

BEST PRACTICES FOR CFD ANALYSIS OF VIM AND VIV FOR OFFSHORE STRUCTURES

MUSTAFA C. KARA
ENGINEER, DNV GL

JAN KAUFMANN
SR. PROJECT ENGINEER, DNV GL

VOLKER BERTRAM
SR. PROJECT MANAGER, DNV GL

ROBERT B. GORDON
SR. PRINCIPAL ENGINEER, DNV GL

PARTHA SHARMA
HEAD OF SECTION: RISERS & FLOWLINES, DNV GL

Proceedings of the 21st Offshore Symposium, February 2016, Houston, Texas
Texas Section of the Society of Naval Architects and Marine Engineers

ABSTRACT

CFD is increasingly used as an alternative to model tests in the assessment of fluid dynamics aspects of offshore structures. Applications are found both in design and operation. This paper presents guidelines for applying CFD to the investigation of vortex-induced motions of moored floating structures and to vortex-induced vibration of subsea structures such as rigid jumpers.

A summary of main techniques is provided, along with short-comings and capabilities. Recommended procedures for quality assurance are also presented. To illustrate the best practices, the paper includes a case study of VIM of a deep draft semi-submersible.

Keywords: CFD, VIM, VIV, Offshore Structures

INTRODUCTION

Growing availability of high-performance computers and robust computational fluid dynamics (CFD) software have enabled the use of CFD for the solution of many problems in Offshore Engineering that were previously addressed only experimentally. One such application is vortex-induced vibrations (VIV) of offshore structures such as moored platforms, marine risers and subsea pipelines containing free-span sections.

The cyclic motion of a moored platform induced by vortex shedding is referred to as vortex-induced motion (VIM). VIM may cause significant fatigue damage to moorings and risers. In fact, the economic viability of deepwater projects may depend on the accuracy of VIM predictions during the design phase.

Full-scale measurements of vortex-induced motions have consistently indicated that model tests over-predict the full-scale response (e.g., Ma, et al., 2013). Recently, CFD has been used as a design tool to provide VIM data at full-scale conditions. So far, the strategy has been to validate CFD solutions against model tests and then used the validated CFD model to obtain solutions at full scale, e.g. Kim, et al. (2015); Vinayan, et al. (2015); Antony, et al. (2015a,b); Kara, et al. (2016).

CFD simulation of VIM is challenging due to factors such as bluff-body flow separation, fluid-structure interaction, mooring characteristics (including viscous damping effects) and Reynolds Number dependence. VIM is a transient phenomenon and must be modeled as such in CFD simulations. Good resolution of the boundary layers and wake region is important as is turbulence modeling. Typically, a detached-eddy type turbulence model is required to adequately represent wake turbulence.

These requirements result in very large simulations with grids on the order of 10 million cells. Periods of motion are around 100 s and time steps are typically less than 0.1 s. For stable motion statistics, simulations must run 20-30 motion cycles. Each simulation can take days to execute on a modern cluster computer using 1000 cores.

Validation against model tests is relatively straightforward. However, published full-scale data are few and uncertain, as ambient conditions (current, wind, waves), as-built geometry with assorted appendages and marine fouling introduce complications. Questions remain concerning spatial and temporal resolutions in full-scale CFD simulations.

This paper provides guidelines for CFD simulations of VIM and VIV. DNV GL has as a long-goal to evolve these guidelines into a Recommended Practice.

ELEMENTS OF COMPUTATIONAL FLUID DYNAMICS

Basic Elements

Basic elements dealing with the governing flow equations, numerical methods for solving these equations and the associated errors and uncertainties are described here. The important topics of verification and validation are also discussed.

Flow Classification: Flows can be classified in many ways. The flow model selected should be as simple as possible, but not any simpler. If there is no experience for a given flow, initial studies with more elaborate flow models may elucidate whether simplifications are justifiable or not. VIV and VIM problems for offshore structures are usually modeled as unsteady, viscous, turbulent, single-phase, incompressible flows. Practical engineering problems usually require a 3d solution.

Flow Equations: Numerical flow simulations are based on field equations and boundary conditions. The Navier-Stokes (N-S) equations together with the continuity equation suffice to describe the physics for most VIM/VIV problems. In practical solutions for VIM/VIV, turbulence must be represented using a semi-empirical model such as detached eddy simulation (DES). DES combines Reynolds-averaged Navier Stokes equations (RANSE) in the near-wall region with Large Eddy Simulation (LES) in the wake region (Spalart, 2009). The RANSE part typically is modeled using the relatively simple one-equation Spalart-Allmaras model (Spalart, 2009) or Menter's shear stress transport k-ω model (Mentor, 2012). This approach often yields accuracy levels comparable to pure LES with remarkably reduced computational costs.

Numerical Techniques: CFD comprises methods that solve the basic field equations subject to boundary conditions by approaches involving a large number of finite elements. These approaches lead automatically to a large number of unknowns. Methods used include Finite Element, Finite Difference and Finite Volume. While the majority of modern, commercial CFD codes are based on Finite Volume Method, in principle, any of these approaches is acceptable to solve viscous flow problems.

Errors & Uncertainties: All models are an approximation of reality and thus contain errors (difference between reality and computation). Errors come from various sources and can be classified in hierarchal order as follows:

Round-off error: Limited machine accuracy in representing real-valued numbers in computers (e.g. 64-bit representation) and error propagation in arithmetic operation introduce errors. As long as error propagation does not play a significant role, round-off errors are considered to be negligible compared to the other errors.

Iteration errors: CFD iteratively solves the governing nonlinear differential equations. Iteration errors are then defined as the difference between the converged solution and the solution obtained after a limited number of iterations. Iteration errors can be controlled by monitoring the residuals.

Discretization model errors: CFD uses assorted schemes to approximate differential and integral expressions. Theoretically, errors due to discretization will vanish as the grid size goes to zero. However, this is not a practical option, as required computational resources would go to infinity at the same time. Discretization errors can be reduced effectively and efficiently by using local grid refinement in regions with high discretization errors. Higher-order schemes may also reduce discretization errors. Generally, the grid should be finer in regions with high flow gradients.

In order to estimate the magnitude of discretization error, two methods are possible: using different discretization schemes or using different meshes. The second option may employ e.g. Richardson extrapolation (see Wikipedia). This is recommended to estimate grid-independent solutions with the advantage that it can be used for field variables and integral values alike. For complex industrial problems, limited storage space and computational time can pose practical barriers for this technique. Then comparing results from different discretization schemes is often an attractive alternative. Time steps should be chosen sufficiently fine; e.g. the Courant-Friedrichs-Lewy (CFL) number (see Wikipedia) in the flow regions of interest should be less than 1. Typically, variations of time steps are investigated separately from variations of spatial resolution. For flow oscillations, at least 200 to 400 steps per period are recommended.

Model errors: The real world is modelled first by a physical model. Simplifications in the assumptions introduce high-level errors. For example, assuming incompressible flow and thus constant density is a simplification. Another example is uniform inflow at an inlet as boundary condition. Truncating the model at unphysical domain boundaries (respectively imposing wrong or simplified boundary conditions at domain boundaries) is a frequent source of error. Validation is used to ensure that model errors for a given application remain acceptably small.

Model errors, e.g. using too simplified basic equations or boundary conditions, are the most critical source of errors in practice. Modeling strategies should thus be based on (validated) experience with similar flows. Variations of model parameters may help to establish confidence in the model, but does not constitute validation.

Verification and Validation: Verification and validation are procedures used for quality assurance in CFD (e.g. Oberkampf and Trucano, 2002; ASME, 2009). The PMBOK (2013) Guide provides the following definitions:

Verification: "The evaluation of whether or not a product, service, or system complies with a regulation, requirement, specification, or imposed condition. It is often an internal process."

Validation: "The assurance that a product, service, or system meets the needs of the customer and other identified stakeholders. It often involves acceptance and suitability with external customers."

CFD verification focuses on proper implementation of equations and numerical techniques. It strives to identify and eliminate programming errors. Often analytical test cases are used to test software in part or in total. CFD users should obtain and review verification documentation from CFD code suppliers. Commercial CFD codes with large user base may be regarded as implicitly verified for practical purposes. This still leaves model parameters to be verified as suitable for the application. For academic problems, convergence orders can be determined in rigorous mathematical methods. For complex geometries and complex flows, verification of schemes is less straight-forward. For example grid refinement may not extend all the way to the boundary layer as wall functions require minimum distances from walls. Users should then adhere at least to the following measures for quality assurance:

Examine iterative convergence with documented thresholds for residuals: "A first indication of the convergence to the solution is the reduction of the residuals [...]. However, different types of flow require different levels of residual reduction. [...] As a result, it is also necessary to monitor the solution during convergence and to plot pre-defined target quantities of the simulation as a function of the residuals. A visual observation of the solution fields at different levels of convergence is recommended." (Zigh and Solis, 2013).

Examine basic consistency (conservation of mass, momentum): It is recommended to monitor global conservation of mass and momentum (and, if applicable, energy).

Examine flow plausibility, especially at domain boundaries

In case of doubt, it is recommended to perform variations of spatial and temporal resolution, as well as variations of key control parameters of the model, and examine resulting changes in results.

CFD validation focuses on the physical and numerical model. It strives to identify applicability of a chosen model and associated errors of model parameters (grid parameters, turbulence models, etc.). Validation employs experimental data. It is important to quantify uncertainty in experimental data as well.

Validation should look at integral numbers (such as drag force), but also at flow details. For single scalar quantities, agreement between computation and target value may be due to fortunate error cancellation or parameter manipulation to obtain a given target. Blind tests and international validation test cases are generally the best option to validate a code.

"Validation cannot be understood as a binary ('yes' or 'no') problem. From an engineering viewpoint, validation is an estimation problem: What is the measure of agreement between the computational result and the experimental result, and how much is the measure affected by numerical error in the computational solution and by the experimental uncertainty? The answers depend clearly on application and user. Acceptance criteria are in most cases determined very vaguely […]," (Zigh and Solis, 2013).

If CFD codes have been validated for a given application, proven model parameters (grid type and fineness, control parameters, turbulence model, etc.) can be employed and standard quality assurance measures suffice. In many cases, CFD codes have been validated for a broad spectrum of flow problems, but not the given application at hand. In this case, more attention has to be paid to quality assurance including sensitivity studies for key parameters.

Key Techniques

Key techniques include turbulence modeling, free-surface modeling, rigid body motion, and fluid-structure interaction. For VIM/VIV problems, free-surface modeling is not typically considered.

Turbulence Models: Turbulence is in general not fully understood. All turbulence models are semi-empirical, combining general shape functions from physical insight with empirically determined constants. No turbulence model used for offshore or maritime flows has been investigated for its suitability at the free surface.

Turbulence models generally use a reference length scale and reference velocity scale. Alternatively, the velocity scale can be expressed as the fraction of length scale and a time scale. To obtain the proper dimension, the eddy viscosity is expressed proportional to the product of length scale and velocity scale. The length scale is characteristic for the larger turbulence structures which are mainly contributing to the momentum transfer in the fluid. The velocity scale is characteristic for the intensity of the turbulent fluctuations.

Turbulence models relevant to VIM /VIV problems include:

k-ε model: The most popular turbulence model for ship flow applications in practice remains the standard k-ε model. This model solves two additional transport equations, one for the turbulent kinetic energy k and one for its dissipation rate, ε. These equations contain four empirical constants which were determined (in a best fit approach) for very simple flows in physical and numerical experiments. The applicability to other turbulent flows (e.g. around maritime geometries) was never explicitly validated.

The k-ε model cannot be applied directly at a wall (ship hull) as it assumes inherently high (local) Reynolds numbers. If a no-slip condition (zero relative speed at the hull) is to be enforced directly at the wall, the ε differential equation must be substituted by an algebraic equation near the wall. This is the so-called low-Re k-ε model. However, more popular is the introduction of a wall function coupled to the standard k-ε model. The wall function is empirically determined for two-dimensional flows.

The k-ε model appears suitable for flows with a predominant boundary-layer character. The wall function makes the approach numerically efficient, but the model is in principle not capable of predicting flow separation for curved surfaces.

k-ω model: k-ω models can be seen as a further evolution of k-ε models. ω can be interpreted as a 'turbulence frequency'. k-ω models yield better agreement with experiments in many cases; however, they react more sensitively to grid quality.

LES/DES models: Large-eddy simulations (LES) may eventually solve the current debate over turbulence modeling for engineering applications, but for high-Reynolds number flows computational requirements exceed largely available computing resources. Many LES solutions in practice have insufficient grid resolution. Then numerical diffusion on the grid acts as a hidden turbulence model and results show disturbingly high grid dependency. DES (detached eddy simulations), as described in the sub-chapter, "Flow equations" offer a more promising compromise between required computational resources and achievable accuracy.

All commonly used turbulence models are plagued by considerable uncertainties. The applicability of a turbulence model can only be evaluated in benchmark tests for similar flows.

Rigid-Body Motion: In many maritime and offshore applications the motion of the investigated body must be taken into account in order to give an accurate representation of the physical problem. Structural deformation of the body can be neglected when the elastic deformations are small compared to the body motion. They then can be assumed to have no influence on the flow. This is called rigid-body motion.

In general, rigid-body motions can be categorized as follows:

1. Cases where the motion is known and the body motion is the key driver of the flow, e.g. propeller motion at given rpm
2. Cases where the motion is driven by the flow, e.g. ship or offshore platform motion in waves or vortex-induced motions of a floating offshore unit
3. A combination of the above two cases, e.g. a self-propelled ship in waves

Fluid forces acting on the body surface, body position and motion affect the flow. This leads to a coupling of fluid flow and rigid-body motion:

Loose or explicit coupling: In each time step, fluid and body motion equations are solved using the values of the previous time step for the boundary conditions. This method is frequently unstable or requires very small time steps for converging solutions. Therefore it is rarely used in practical applications.

Strong or implicit coupling: In each time step, fluid and body motion equations are solved iteratively. For numerical stabilization, often under-relaxation techniques are applied.

The mesh must reflect the body motions. Different mesh motion strategies can be applied:

Entrainment: The whole grid follows the body motion. The cells do not change their shape and therefore the cell quality remains. This comes with almost no additional computational cost, but can lead to the requirement of large refinement areas in free-surface flows.

Sliding mesh: This is a special form of the whole grid motion for rotatory motion. The body-fixed part of the mesh moves with the body while the rest of the mesh is unchanged. The flow values at the interface of both meshes must be interpolated.

Mesh morphing: The mesh vertices are morphed, which allows arbitrary mesh motions. The mesh topology is kept during the simulation (i.e. neighbor cells do not change) but cell shapes change during the simulation; this can affect cell quality. This is more general than entrainment, but involves higher computational cost.

Special forms of morphing: E.g. spherical linear interpolation (SLERP). If applicable, this approach combines advantages of entrainment (in terms of computational costs) and morphing (in terms of flexibility).

Overset grid: Here a background mesh remains fixed while a body-fixed mesh follows the motion. Flow variables are interpolated between both meshes. This allows a very general setup and therefore a wide range of applications but can be computationally expensive due to the interpolation, especially in parallel computing.

The selection of mesh motion strategy depends both on the flow problem and the flow solver.

In addition to an appropriate meshing strategy, simulation parameters have to be chosen carefully to ensure reliable results. In particular, it must be ensured that:

- Convergence is reached during each time step, i.e. there are a sufficient number of iterations per time step, where the mesh is moved.
- The time step is small enough to accurately resolve the motion (e.g. number of time steps per wave period, per propeller revolution or per period of motion of interest).

Fluid-Structure Interaction: For rigid-body motions, the elastic deformations of the investigated structure are not of interest or can be assumed to have no influence of the flow. However, in many engineering scenarios large-amplitude vibrations raise the need for fluid-structure interaction (FSI) simulations. Such vortex-induced vibration or motion problems may appear for bridges, chimney stacks, marine cables, and offshore riser pipes subjected to flows of air or water.

For the solution of FSI problems, one can combine the fluid and the structure representation to a single problem (closed or monolithic formulation) and solve both parts simultaneously. In a separate (partitioned) formulation for the fluid and structural part, a sequential solution procedure is needed with an underlying coupling strategy. Hou et al. (2012) survey various approaches with comments on accuracy and efficiency, emphasizing the difference between conformal and non-conformal meshing (i.e. immersed boundary) methods. Non-conformal mesh methods have almost no relevance for industrial applications and commercial software. They are therefore not addressed in the guideline.

The advantage of the partitioned formulation over the monolithic approach is that tailor-made solvers can be used for the fluid (CFD) and the structure (CSD) parts (e.g. a finite-volume method for CFD and a finite-element method for CSD). This gives more flexibility in practical applications. Both formulations have in common that in most cases non-conformal boundaries require special care for the information transfer.

Fluid-structure coupling (i.e. coupling between fluid forces and structural deformation) may be "weak" or "strong". Weak coupling is not recommended for most applications due to stability problems. The user must ensure during the simulation that enough iterations per time step are conducted to ensure the desired level of convergence. For stability reason it is often helpful to employ under-relaxation. One-way coupling (i.e. pure mapping of CFD forces to the CSD model) can only be used for problems where the structural deformation has no influence on the flow.

For conformal meshes, the CFD mesh has to be updated to follow the structural deformation (e.g. mesh morphing). In most cases the CSD representation does not match the CFD mesh (typically the CSD mesh is coarser than the CFD mesh). Then special interpolation schemes must be used to ensure sufficient CFD mesh quality at the FSI interface.

SOLUTION PROCESS & QUALITY MEASURES

This section discusses geometry modeling, grid generation, initial and boundary conditions, computation, and post processing.

Geometry Modeling: CFD analyses start generally with a geometric CAD (Computer-Aided Design) representation of the investigated bodies. The CAD model must closely represent the actual geometry. Any gaps between surface patches in the CAD surface must be significantly smaller than the local mesh size to ensure proper meshing.

For meshing, compromises between level of geometric details and mesh quality (with reasonable cell count) are often necessary. Geometric detail can be important for vortex generation. Mesh quality is important for numerical stability of the simulation. Often it is appropriate to omit small geometrical details for computational efficiency. Such decisions should be guided by physical insight on whether these details may be significant for the investigation at hand.

The CAD model must have the scale and origin as intended for the analysis. Especially for validation studies, orientation and origin must be consistent if e.g. forces and moments are compared with model tests.

Grid Generation: Starting from a CAD representation, the flow domain has to be discretized into a numerical grid. Meshing strategy and type depend on the flow problem, the used meshing software (i.e. the underlying algorithm), and the flow solver. Therefore only general advice can be given.

The following grid types may be employed:

- Structured meshes (single-block or multi-block)
- Unstructured meshes with tetrahedral or polyhedral cell shapes, often with high-quality prism layer cells near the hull surface
- Hybrid meshes with a combination of the above

During mesh generation, the user should check grid quality parameters, such as:

- Orthogonality
- Cell skew
- Cell aspect ratio
- Cell volume jumps

In areas with high flow gradients, grid resolution must be fine enough to capture these gradients. To avoid unnecessarily high cell counts, grids are refined mostly using local refinement (isotropic or anisotropic).

Checking grid quality before the simulation is important, as grid quality is decisive for quality of results. Time is better spent on generating a good mesh than on adopting solver settings to get converging solutions on a bad mesh.

During simulation, it should be checked whether desired y+ values are obtained, i.e. whether the height of the first prism layer cell was chosen appropriately. If possible, the use of wall functions should be avoided and the user must ensure that y+ values near the wall boundary are close to 1. If this is impractical and logarithmic wall functions are used y+ values should be in the range of 30 to 100.

During simulation it is also important to check if the initial grid refinement is in accordance with the calculated flow field, e.g. areas with high gradients are resolved properly. Mesh refinement zones should be adopted accordingly if necessary.

Initial and Boundary Conditions: Due to the limited (finite) size of the simulation domain, appropriate boundary conditions have to be set at the domain boundaries. Two types of boundary conditions and combinations of them are most commonly encountered:

1. Dirichlet conditions specify a flow variable (pressure, velocity component, etc.).
2. Neumann conditions specify a first derivative of a flow variable.

In general, boundary conditions should be imposed at flow regions where the flow is well defined, see e.g. Ch. 6.10 in MARNET (2002). Most important physical boundary conditions are:

Inflow (inlet) condition: For pressure and velocity, Dirichlet conditions specify magnitudes depending on flow direction. Velocities are often prescribed as uniform or following a profile (based on a theoretical function or interpolated from discrete data points). E.g. a wind profile may follow from empirical approximations or measurements from wind tunnel experiments.

Inflow conditions should be applied far enough upstream to ensure that the flow is not dominated by the inlet. If there is no experience for similar problems, systematical sensitivity studies should be performed to ensure that the inlet is "far enough" upstream, i.e. further shift of the inlet position has no significant effect on the flow at the region of interest.

Turbulence properties require special care. Most codes allow specifying:

- turbulent kinetic energy k and turbulent dissipation rate (ε or ω), or
- turbulent intensity and length scale
- eddy viscosity ratio

If available, these values should be set according to experimental test data. Otherwise, they should be set based on verifications and (own or published) experience with similar problems. When neither model test data nor comparable simulation data are available comprehensive sensitivity studies should investigate the influence of the turbulence quantities specified at the inlet. As a rule of thumb, for external flows (around ships, submarines, or offshore platforms), we recommend turbulence intensities of 0.5 – 1% and eddy viscosity ratios μ_t/μ of 5 – 10. Specification of the length scale affects the decay of the turbulence level. High values lead to slower decay of the turbulence and a high eddy viscosity. High eddy viscosity values at the inlet can in turn dominate the flow solution and lead to over-predicted viscosity at the body.

Outlet condition: The outlet must be far enough downstream to have only weak influence on the upstream flow.

Typically, Neumann conditions for velocity and turbulence properties are imposed, and a Dirichlet condition for static pressure. Extreme care is needed when specifying flow velocities and directions on the outlet plane.

Wall boundary condition at the body: For moving bodies, velocities at the wall boundary must account for the body motion. This is handled in most codes automatically.

Wall boundary conditions may be imposed with or without wall functions, depending on the near-wall grid resolution and the chosen turbulence model. For most applications, wall roughness can be neglected and the wall can be assumed to be hydraulically smooth; if wall roughness must be taken into account, detailed measurement data should be used to specify more advanced modelling near the wall.

Symmetry: Symmetry is often used in order to reduce the domain size. However, even for symmetric bodies and symmetric (e.g. uniform) inflow, the flow itself can be asymmetric, as vortex formation (e.g. von Karman vortex street) may introduce asymmetries. If in doubt, at least one flow simulation without symmetry conditions may elucidate whether the symmetry conditions may be imposed for numerical efficiency.

Periodic (spatial cyclic) conditions: As the flow field within the solution domain is not known, an approximate initial distribution must be specified which is in most cases just a uniform flow field based on the values from the boundary conditions. In free-surface seakeeping simulations, the velocity and phase of harmonically oscillating flow can be initialized based on analytical values. For transient simulations, one may speed up convergence by conducting a preceding steady-state calculation.

Computation: Most flow problems in industrial applications need mesh sizes which require distributed processing in order to get reasonable response times. For the parallelization, a good compromise has to be found between number of processors and efficiency of each processor. As a rule of thumb, for most CFD codes and hardware architectures at least 50,000–100,000 cells per processor are needed to avoid excessive message passing between processors.

During the simulation, the iterative convergence should be monitored. Two types of convergence checks should be made:

Residual convergence check: Dependent on the used code, normalized residuals of various solution equations are given. Usually these residuals should reduce by several orders of magnitude.

Flow variables check: For steady state simulations, monitored values of interest (e.g. drag or lift) must converge to constant values. For periodic problems (e.g. motions in harmonic waves), periodicity vs. time can be checked. Plausibility checks (e.g. for natural period) are recommended. For transient problems (e.g. unsteady flow separations), we recommend to check how mean values are affected by different sampling windows.

Post-Processing: Post-processing here refers to text and graphical information. It involves extracting desired flow properties from computed flow fields. Flow properties may concern scalar values (thrust, lift, drag, etc...) or distributed variables (pressures, velocities, etc.).

The documentation should give the employed software including version number. For commonly used commercial software, a simple reference, possibly with link to a website suffices. For other codes, references should be given for theory and validation. Preference should be given to references that are widely available.

All important model features should be described, including:

- geometry and coordinate systems
- domain extent
- initial and boundary conditions
- grid type, cell count, cell expansion rates, etc. Suitable figures of grids with global and detailed views should be included as appropriate.
- turbulence model and other associated models as appropriate (e.g. cavitation models, combustion models, etc.)
- control parameters such as time step size, time functions for blending factors, numerical schemes, etc.

Quality assurance measures shall be documented, including convergence of time histories of residuals of the solved equations (e.g. mass and momentum, turbulence). For flow simulations, where the CFD user has insufficient experience, spatial and temporal sensitivity studies shall be performed and the essential results of these studies shall be documented (grid independence, sensitivity of control parameters, turbulence model, etc.).

If quality assurance measures are omitted, justification for this decision must be given.

Wherever feasible and appropriate, computed flow properties should be compared to results from analytic, experimental or other numerical approaches to validate & verify the simulations. Distributions of key flow variables shall be visualized for plausibility checks and for communicating findings and recommendations. Key flow variables my include velocity and pressure distributions in regions of interest, wall shear stress distributions on wall boundaries at specific time steps, free-surface elevation, etc. In particular, regions with large gradients and/or regions with low cell quality should be inspected for quality assurance.

VIM & VIV PROBLEM DESCRIPTION

The topic of fluid–structure interaction (FSI) and specifically VIM or VIV is of critical importance in a number of engineering scenarios. Moored platforms and offshore risers represent a selection of engineering structures that are known to undergo vibrations of significant amplitude when subjected to incident flow, often resulting in fatigue damage with an associated risk of catastrophic failure (e.g., Kara et al., 2015).

VIM in offshore applications is found for moored platforms excited to large planar motion (surge, sway and yaw) due to flow separation. The flow around such platforms in currents (e.g. loop currents in the Gulf of Mexico) can become highly unstable due to unsteady vortex generation. The vortices lead to fluctuating exciting forces which can result in large motion amplitudes due to inherently small damping for planar motions. Accurate prediction of such VIM amplitudes is important especially for design aspects such as riser and mooring fatigue.

VIV in offshore applications can be found e.g. for subsea jumper systems under the influence of sea bottom currents. Most widely used subsea jumper systems are planar, two-dimensional configurations such as M-shaped jumpers. Though this shape brings increased flexibility over horizontal free spanning systems, it leads to unique challenges due to VIV for both in-line and cross-flow conditions.

Vortex-Induced Motions

Flows which lead to VIM are highly unsteady and feature strong coupling of flow and body motion. For such flows, the use of classical RANSE methods is not suitable. To predict the pronounced flow separations leading to VIM, at least the larger turbulent scales must be resolved. Therefore, for VIM analyses, the use of Scale-Resolving Simulations (SRS) (following the definition in Menter, 2012) is highly recommended. For most industrial applications, LES methods are still impractical due to the high computational effort especially due to the need of very fine meshes for a sufficient near-wall resolution. Here hybrid approaches such as DES (Menter, 2012), have become popular alternatives. However, these hybrid models require special attention to various details (Menter, 2012) such as: turbulence model selection, grid generation, and control parameter settings. Consequently, care is also required for post-processing, solution interpretation and quality assurance. The following gives some guidelines:

Turbulence model selection: RANSE models are too coarse. Turbulence models with finer scale resolution must be used. If LES methods are not affordable for the application at hand, hybrid models (such as DES or SAS) should be chosen. Ideally, both DES and SAS models should be compared, especially if no experience for similar problems is available.

Grid generation: In regions, where the RANSE model is applied, a suitable RANSE mesh should be modelled, e.g. appropriate near-wall resolution with $y+<1$, enough cell layers to resolve the boundary layer flow with adequate stretching (typically 10 – 15). In the SRS region, at least 20 cells per characteristic diameter are necessary (Spalart, 2001; Menter, 2012). The characteristic diameter of an offshore platform in this context is the column diameter. In some cases, a much finer spatial resolution is needed. Therefore mesh sensitivity studies are indicated.

Steps to determine appropriate mesh size are (Menter, 2012; Antony, 2015a,b):

- Define the wake zone where the vortex shedding will have most influence.
- Initially refine the wake with node size of D/15, where D is the diameter of the cylinder.
- Determine the turbulent length-scale based on the SST k-ω model from a steady state calculation.
- Calculate the characteristic length of the grid in the wake zone.
- If the ratio of turbulence length-scale to characteristic length-scale is larger than 1, the mesh refinement in the wake is sufficient to use DES or LES models in the region of interest. Otherwise the mesh needs further refinement.
- Once a mesh size has been determined, an appropriate time-step can be selected. The time step should be chosen such that Courant numbers<0.1 are obtained. This value is recommended based on systematic sensitivity studies.

Numerical Setting: For the convective terms, second-order schemes are recommended. For industrial applications with complex geometries, it is often impossible to use pure central differencing schemes. Then blended schemes are used to overcome instabilities and unphysical oscillations. In some cases with lower mesh quality (e.g. hybrid hexahedral and polyhedral meshes), filtered schemes can help to reduce grid induced distortions. But in general, this should be avoided. It is preferable to spend more time on trying to generate a high-quality mesh. For the turbulence equations, often first-order upwind scheme are sufficient. For hexahedral meshes with lower quality (e.g. high skewness) and for polyhedral meshes, a least-square method (LSM) for gradient calculation is recommended. SAS should always be used with LSM (Menter, 2012). For time integration, a second-order scheme is recommended, e.g. backward Euler or Crank-Nicolson, sometimes blending with a slight amount of first-order explicit Euler scheme. Crank-Nicolson is less robust (Ferziger and Peric, 2002). Time steps size should be chosen such that CFL numbers < 0.1 are obtained in the LES region.

Post Processing: Visual inspection of results contributes to quality assurance. Iso-surfaces of the Q-criterion (Hunt et al., 1988) should be visualized at various time steps showing the evolution of turbulence structures. The Q-criterion is a more suitable measure than the vorticity, as it avoids "steady shear layers". Q-values should not be normalized, as then plots "can elevate very weak turbulence structures to the same level as the strong ones and thereby produce an incorrect picture" (Menter, 2012). The iso-surface should have a positive value, as "a coherent vortex is defined as a region where Q > 0, Green et al. (2007). It is often helpful to color the iso-structure by some flow variable distribution, such as velocity magnitude, Courant number, eddy viscosity ratio, etc.

The following checklist may assist in judging the result:

- Check if unsteady turbulence develops at the expected locations.
- Check large-scale symmetries/asymmetries of the flow.
- Check the solution for numerical wiggles (odd-even decoupling)
- Check the size of the resolved eddies and see if they are as one would expect from the grid resolution.
- Check that the Courant number on these eddies is < 1. Check the eddy-viscosity ratio. It should be much smaller than in the RANSE region.
- For global SRS turbulence models, check if the turbulence structures develop early in the separating shear layer or if a noticeable delay is observed.
- For unsteady inlet conditions, check if synthetic turbulence is reasonable and does not decay.
- Check the progress of the simulation towards a "statistically converged" solution: The resolved turbulence requires some time until it has developed and has been transported through the domain. Time-averaging has to wait until that stage has been achieved.

Further general guidelines for post-processing in VIM analyses include:

- Include pictures of turbulence structures in reports on VIM applications.
- Animation help understanding the flow physics in VIM applications.
- Add monitoring points at interesting locations in the flow and plot time histories of flow quantities to demonstrate statistical convergence.

Averaging is an important step in analyzing and post-processing of scale resolving simulations. In most cases, average integral values (e.g. drag or lift values) or average motion amplitudes are of interest. For averaging, the start-up phase (until turbulent structures are fully developed) should be excluded. This start-up phase will depend on the flow problem and the quality of the initialization (e.g. restart from results of a RANSE simulation). The sensitivity of average values to different sampling windows should be checked. Also it should be checked if the average value is fairly constant; drifting values indicate poor convergence.

Vortex-Induced Vibration

This paper mainly focuses on VIM. For the sake of completeness, some general comments are given here for VIV. There is not much experience with VIV simulations in industry projects, as the field is relatively new. In general, the same recommendations as for VIM apply also for VIV regarding turbulence modelling, grid generation, numerical settings and post-processing. Due to the highly unsteady nature, most VIV simulations will require strong fluid-structure interaction coupling. Both CFD and structural parts require proper modelling, i.e. high-quality grids and correct mass and stiffness parameters.

CASE STUDY

The case study considers a dry-tree, paired column (PC) semi-submersible platform. Antony, et al. (2015a,b) applied three different commercial CFD solvers for VIM analyses for this case (FLUENT™, ACUSOLVE™, and STAR CCM+™). All simulations were for a scaled down model used in the model tow tests. Horizontal stiffness matched that provided in the model tests by four lateral springs. Model details are provided in Antony, et al. (2015a).

We compare here CFD and model-test results. CFD results were obtained using OpenFOAM™, Version-2.4.0. The flow was modeled as single-phase flow. The free surface was neglected, which is justified by the very low Froude number. The simulation model used a second-order implicit scheme (backward differencing) for time derivatives. For spatial derivatives, a second-order upwind scheme was used, stabilized for convective terms. For turbulence quantities, a second-order TVD limited linear scheme was used. Pressure-Velocity coupling followed the PISO and a merged PISO-SIMPLE (PIMPLE) algorithm (Weller, 2005). The outer (PIMPLE) iterations per time step were fixed to 10. For the motions, we used the DNV GL in-house 6DoF solver with non-linear coupling of accelerations and velocities (El Moctar et al., 2010). Degrees of freedom were suppressed consistently with the model test. The 6DoF solver uses an iterative formulation for implicit two-way coupling, a Newmark time integration scheme, and has an interface for generalized external forces (such as the spring forces for horizontal stiffness). Time steps were chosen such that CFL numbers <0.1 were ensured in the wake region.

Three turbulence models were compared:

- URANS simulation using the k-ω SST model
- DES using the Spalart-Allmaras improved delayed detached eddy (SA-IDDES) model
- Scale-Adaptive Simulation using the k-ω SAS model

A low initial turbulence level was assumed, with turbulent intensity set to 1% and eddy-viscosity ratio set to 10.

The CAD model of the submerged portion of the platform is shown in Fig. 1. The CFD domain is shown in Fig. 2. A trimmed hex mesh was used with prism layers and isotropic refinement near the hull and in the wake region. Ten prism cell layers were used at the hull, with the first cell height fixed to ensure y+ < 1. Three meshes were generated, refining systematically the scaling base with factor √2. This resulted in cell counts of 2 million (m1), 4.5 million (m2) and 10 million (m3). For the medium grid (m2) (Fig. 3), this gave a cell size in the wake region of 6% of D. Mesh sensitivity studies indicate sufficient grid resolution for the medium mesh.

Figure 1: CAD surface for CFD mesh generation

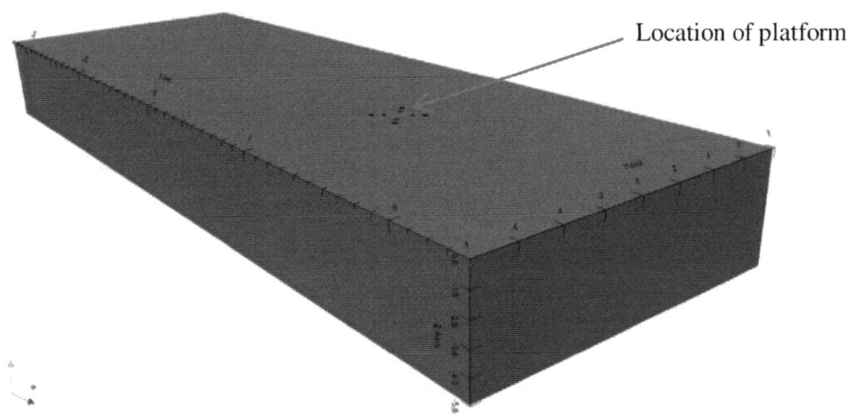

Figure 2: CFD domain extent

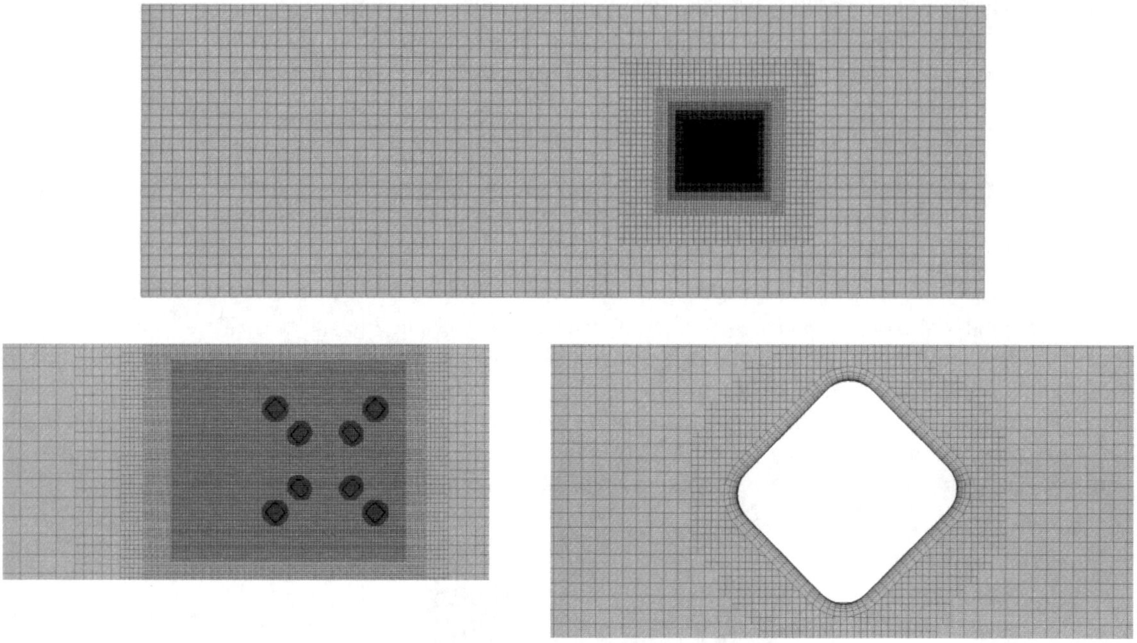

Figure 3: Details of mesh

For vortex-induced motions, the platform was investigated in heading 0° and for reduced velocities U_r = 4, 6, 8, and 10. For VIM the reduced velocity U_r is a commonly used non-dimensional parameter based on the current velocity (U), natural period (T_n) and platform column diameter (D); $U_r = (U\ T_n)/D$. For this case, only calculations with DES were conducted, as URANS failed already at the simpler drag problem and the SAS model showed no improvements.

Figure 4 compares results of the VIM simulations with experimental results. In general the results agree very well for the nominal average normalized sway amplitudes and the average zero crossing sway periods. For all VIM simulations, the first 100 s (corresponding to 6-8 VIM cycles) were excluded from the evaluation to account for the start-up phase and to let the turbulence structures develop. For the statistical evaluation, at least 15 cycles were taken.

The biggest discrepancy between experimental and CFD results occurs at reduced velocity U_r = 6. For U_r = 5 there is a large spreading of the experimental results of the three different runs which indicates the beginning of lock-in, Antony (2015a,b), known to be very unstable, especially for VIM tests at lower mass ratios. In these tests the mass ratio was less than one. The lock-in phase is not fixed to one single reduced velocity but rather a continuous process; a larger variance in the experimental results can also be seen for U_r = 4 and slightly for U_r = 6. At higher reduced velocities (i.e. system is completely locked-in) both the scatter in the experimental results is reduced and model tests and CFD simulations agree very well.

SUMMARY

CFD is increasingly being used as an alternative to model tests in the assessment of fluid dynamics aspects of offshore structures. This paper attempts to provide a best practice guideline for applying CFD to the investigation of VIM and VIV problems of offshore structures. DNV GL has as a long-goal to evolve these guidelines into a Recommended Practice.

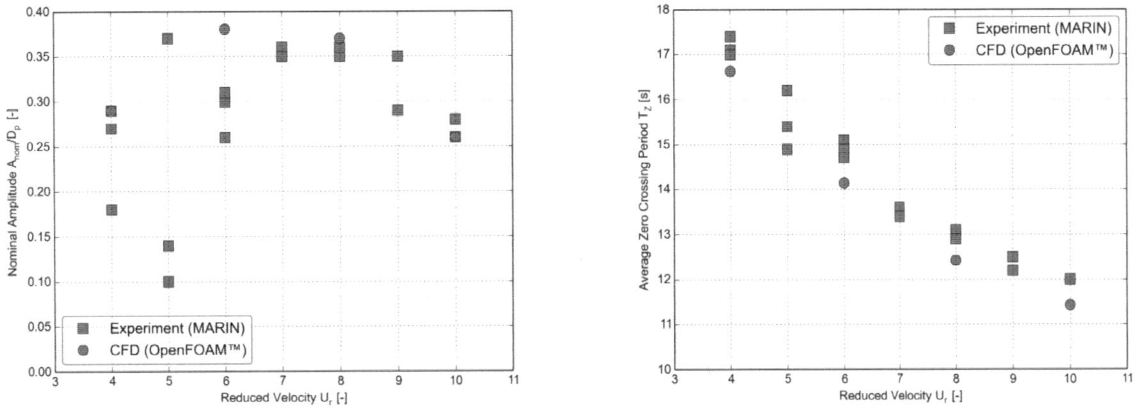

Figure 4: Experimental and CFD results for nominal amplitude (left) and average zero-crossing period (right)

ACKNOWLEDGEMENTS

We thank our DNV GL colleagues Jan Oberhagemann and Tobias Zorn for their expert contribution to the guidelines summarized in this paper. The CFD example used here is part of DNV GL's in-kind contribution to the "Vortex Induced Vibration Study for Deep Draft Column Stabilized Floaters", a study funded by RPSEA, Project Manager–William J. Head and executed by Houston Offshore Engineering (HOE), Principle Investigator–Arun Antony. DNV GL would like to acknowledge the input (CAD geometry and CFD analysis report) and generous guidance to execute this work from Arun Antony and Vimal Vinayan of HOE.

REFERENCES

ANTONY, A. (2015a), Vortex Induced Motion Study for Deep Draft Column Stabilized Floaters, RPSEA CFD Analysis Report, DNV GL, Houston

ANTONY, A. (2015b), Vortex Induced Motion Study for Deep Draft Column Stabilized Floaters, RPSEA Existing VIM Data and Design Practices Summary Report, DNV GL, Houston

ANTONY, A.; VINAYAN, V.; KIM, S.J.; HOLMES, S.; SPERNJAK, D. (2015a), A CFD based analysis of the vortex induced motion of deep-draft semisubmersibles, 25th Int. Ocean and Polar Eng. Conf. (ISOPE), Kona

ANTONY, A.; VINAYAN, V.; HOLMES, S.; SPERNJAK, D.; KIM, S.J.; HALKYARD, J. (2015b), VIM study for deep draft column stabilized floaters, Offshore Technology Conf. (OTC), Houston

ASME (2009), Guide for Verification and Validation in Computational Fluid Dynamics and Heat Transfer, American Society of Mechanical Engineers

EL MOCTAR, B.; KAUFMANN, J.; LEY, J.; OBERHAGEMANN, J.; SHIGUNOV, V.; ZORN, T. (2010), Prediction of ship resistance and ship motions using RANSE, Gothenburg 2010—A Workshop on Numerical Ship Hydrodynamics, Gothenburg

FERZIGER, J.; PERIC, M. (2002), Computational Mechanics for Fluid Mechanics, Springer

GREEN, M.A.; ROWLEY, C.W.; HALLER, G. (2007), Detection of Lagrangian coherent structures in three-dimensional turbulence, J. Fluid Mech. 572, pp.111-120

HOU, G.; WANG, J.; LAYTON, A. (2012), Numerical Methods for Fluid-Structure Interaction — A Review, Commun. Comput. Phys. 12, pp. 337-377.

KARA, M.C.; STOESSER, T.; McSHERRY, R. (2015), Calculation of fluid-structure interaction: methods, refinements, applications, Eng. and Comp. Mech., Inst. Civil Eng.

KARA, M.C.; KAUFMANN, J.; GORDON, R.; SHARMA, P.P.; LU, J.Y.; (2016), Addressing the pressing needs of offshore ultra-deepwater floating facilities and risers: Application of CFD for computing VIM induced fatigue damage of SCRs and moorings, Offshore Technology Conf., Houston (in press)

KIM, S.J.; SPERNJAK, D.; HOLMES, S.; VINAYAN, V.; ANTONY, A. (2015), Vortex-induced motion of floating structures: CFD sensitivity considerations of turbulence model and mesh refinement, 34th Int. Conf. Ocean, Offshore and Arctic Engineering (OMAE), St. John's

MA, W.; WU, G.; THOMPSON, H.M.; PRISLIN, I.; MAROJU, S. (2013), Vortex Induced Motions of a Column Stabilized Floater, Proceedings of the D.O.T. International Conference.

MARNET (2002), Best Practice Guidelines for Marine Applications of Computational Fluid Dynamics, MARNET CFD, Final report

MENTER, F.R. (2012), Best Practice: Scale-resolving simulations in ANSYS CFD, ANSYS Germany

OBERKAMPF, W.L.; TRUCANO, T.G. (2002), Verification and Validation in Computational Fluid Dynamics, Report SAND2002 – 0529, Sandia Nat. Lab., Albuquerque

PMBOK (2013), A Guide to the Project Management Body of Knowledge (PMBOK Guide), Project Management Institute, Newtown Square

SPALART, P.R. (2009), Detached-Eddy Simulation, Ann. Rev. Fluid Mech. 41, pp.181–202

VINAYAN, V.; ANTONY, A.; HALKYARD, J.; KIM, S.J.; HOLMES, S.; SPERNJAK, D. (2015), Vortex-induced motion of deep-draft semisubmersibles: A CFD based parametric study, 34th Int. Conf. Ocean, Offshore and Arctic Engineering (OMAE), St. John's

WELLER, H.G. (2005), Pressure-Velocity Solution Algorithm for Transient Flows, Technical Report, OpenCFD Ltd.

ZIGH, G.; SOLIS, J. (2013), Computational Fluid Dynamics Best Practice Guidelines for Dry Cask Applications, Report, US Nuclear Regulatory Commission (US NRC), Washington, DC

NOMENCLATURE

6DoF	Six **D**egrees-**o**f-**F**reedom
CAD	**C**omputer-**A**ided **D**esign
CFD	**C**omputational **F**luid **D**ynamics
CFL	Courant-Friedrichs-Lewy condition
CSD	**C**omputational **S**tructural **D**ynamics
DES	**D**etached **E**ddy **S**imulation
FSI	**F**luid-**S**tructure **I**nteraction
LES	**L**arge **E**ddy **S**imulation
N-S	**N**avier-**S**tokes Equations
PIMPLE	Merged **PISO-SIMPLE** algorithm
PISO	**P**ressure **I**mplicit with **S**plit **O**perator algorithm
PMBOK	**P**roject **M**anagement **B**ook **o**f **K**nowledge
RANSE	**R**eynolds **A**veraged **N**avier-**S**tokes **E**qs.
SA-IDDES	**S**palart-**A**llmaras
SAS	**S**cale **A**daptive **S**imulation
SLERP	**S**pherical **L**in**E**ar Inte**RP**olation
SIMPLE	**S**emi-**I**mplicit **M**ethod for **P**ressure **L**inked **E**quations
SRS	**S**cale-**R**esolving **S**imulations
VIM	**V**ortex **I**nduced **M**otion
TVD	**T**otal **V**ariation **D**iminishing

D	Characteristic length
T_n	Natural Period
U	Free-stream velocity
U_r	Reduced velocity
k	Eddy length scale
ε	Dissipation scale
μ	Viscosity
μ_t	Turbulent viscosity
ω	Vorticity scale
y+	Dimensionless wall distance

SNAME
THE INTERNATIONAL COMMUNITY
FOR MARITIME AND OCEAN PROFESSIONALS

DEVELOPMENT OF SECONDARY COLUMN ENHANCED TLP CONCEPT FOR CENTRAL GULF OF MEXICO

JUN ZOU
DIRECTOR OF NAVAL ARCHITECTURE
HOUSTON OFFSHORE ENGINEERING, LLC
AN ATKINS COMPANY

Proceedings of the 21st Offshore Symposium, February 2016, Houston, Texas
Texas Section of the Society of Naval Architects and Marine Engineers

ABSTRACT

Recent crude oil prices were dropped more than 50% and new norm price may be fluctuated around $40 to $80 per barrel in many years. In addition, metocean criteria have been increased significantly after Hurricane Katrina and Rita which have imposed considerable impacts on field development in GoM, prior to steep declination of crude oil prices. Under multifactorial pressures on economics of field development in deep water of central GoM, it imposes extraordinary difficulty in sanction of deep water projects and calls for innovative solutions to significantly reduce cost and meet challenges. This paper is intended to present one development scenario of TLP concept in response to the market calls.

One study scenario has been carried out based on same design basis and topsides payloads for conventional TLP (CTLP) and secondary column enhanced TLP (SCE-TLP) for central GoM in 4,000 ft water depth. Preliminary global performance analysis has been undertaken to ensure satisfaction of design criteria, such as extreme tendon tensions and minimum airgap, etc. In addition, a high level hull structural scantling of SCE-TLP was performed in order to yield reliable hull steel weight estimate. Preliminary comparison of deck/hull structure weights and tendon/foundation weights of CTLP and SCE-TLP is highlighted and major advantages of SCE-TLP are identified and discussed.

Keywords: Conventional TLP (CTLP), Secondary Column Enhanced TLP (SCE-TLP), Shear Forces, Pry/Squeeze Loads, Central GoM, Field Economics

INTRODUCTION

Recently, oil and gas prices have nasty downward move and will stay in new norm $40 ~ $80 per barrel for many years which brings more pain for oil and gas producers. It has been reported that worldwide projects worth around $1.5 trillion remain uneconomic at current oil prices. There are extremely strong pressures on offshore industry to improve field economics and keep deep water floating platform project afloat. In the Gulf of Mexico, there is another key driver, post Katrina metocean criteria, to push the overall project cost increasing significantly. Under multifactorial pressures on economics of field development in deep water of central GoM, it imposes extraordinary difficulty in sanction of deep water projects and calls for innovative solutions to significantly reduce cost and meet challenges.

To achieve major capital cost reductions, it calls for entire industry collaborations to accumulate savings on drilling, topsides, hull, moorings/tendons, risers, foundations and construction, transportation and installation. This paper presents one example of TLP design in central GoM through innovation hull arrangement to achieve savings on deck, hull, tendon and foundation.

Secondary column enhanced TLP (SCE TLP), as illustrated in Figure 1, has 8 columns, 4 inner columns and 4 outer columns (or called secondary columns); 8 rectangle pontoons and conventional wellbay. Inner column spacing is decoupled of stability requirements from pre-service conditions, such as load out, quayside integration, float on/off, wet tow and hull/tendon ratcheting. Inner column span will sorely depend on best supporting location for optimizing deck steel and adequate space of well bay arrangement for TTRs risers to ensure not having TTR riser interference with pontoon or column.

Figure 1 Secondary Column Enhanced TLP

SCE TLP development philosophies are summarized as a) Maintaining good features of CTLP, e.g. quayside integration, moderate draft around 90 ft ~ 120 ft; b) Conventional deck/hull structures; c) Conventional well bay and TTR arrangement; d) More efficient deck/hull design; e) More efficient tendon/foundation design; f) SCRs friendly; g) Maintaining constructability, transportability & installability and h) Maintaining high HSE (Heath, Safety and Environment) standard.

SCE TLP concept features, as depicted in Figure 2, are highlighted as 1) Maintain quayside integration & de-couple dependencies between a) column spacing requirement for pre-service stability and best span for deck supporting; b) Column spacing and best locations for tethers; c) topsides & hull design cycles since topsides can proceed independently layout and design after inner column span and size is settled down as long as topsides payloads and CGs within topsides not to exceed weight and CG envelop; d) improve deck and hull structure efficiency; e) Reduce tendon dynamic responses to down size tendon and foundation; f) More transparent to waves in well bay; g) Damage Tolerance; h) High flexibility, scalability and expandability.

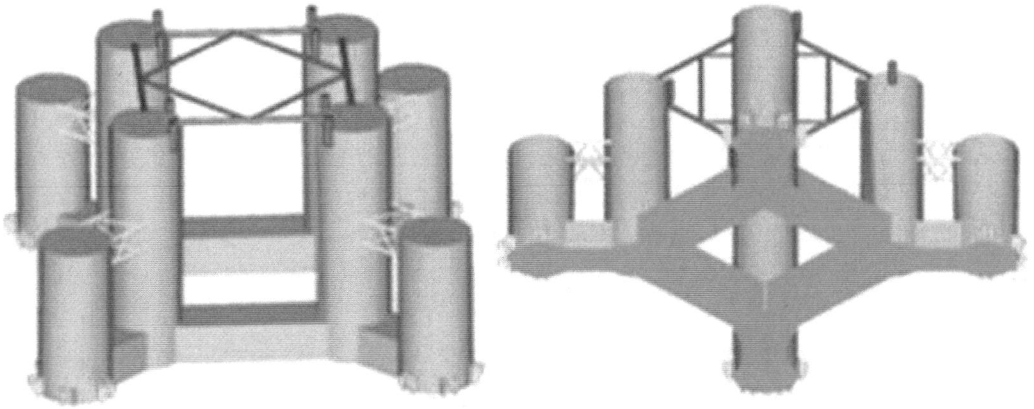

Figure 2 SCE TLP Hull Configuration

In order to have a better understanding of the hull form features, one study scenario has been carried out based on same design basis and topsides payloads for conventional TLP (CTLP) and secondary column enhanced TLP (SCE-TLP) for central GoM in 4,000 ft water depth.

Preliminary global performance analysis has been undertaken to ensure satisfaction of design criteria. Characteristic responses to identify the hull form features are 1) wave upwelling RAOs; 2) tendon tension RAOs; and 3) shear forces and pry/squeeze RAOs.

The outline of the paper is as follows. First, descriptions of the study case including topsides payload, TTRs, SCRs and metocean criteria are summarized. Second, CTLP and SCE TLP configurations are sized and highlighted. Third, the numerical results of three characteristic responses are presented and compared. Fourth, major savings on deck, hull, tendon and foundation are given. Finally, main conclusions are drawn.

DESCRIPTIONS OF THE STUDY CASE

Hull Sizing Basis

Hull sizing basis has been summarized in Table 1. In this table, topsides dry weight includes drilling rig, facilities and utilizes and living quarter weights but excludes deck structural weight; topsides operating weight includes contents and drilling fluids besides topsides dry weight. TTR risers tensions and SCRs and Umbilicals vertical loads were accounted separately. Ship channel width and water depth constrains were also specified by assuming quayside integration site at Kiewit.

Table 2 Hull Sizing Basis

Topsides dry weight	st	15000
Topsides operating weight	st	20000
Deck dimension		
Width (E-W)	ft	260
Length (N-S)	ft	275
Height (BOS to TOS)	ft	45
Number TTR slots (4x6)		24
Number TTRs		20
Number of SCRs		14
Number of Umbillicas		8
Total TTRs tension	st	9000
Total SCRs + Umbilicals loads	st	1000
Max TLP width (edge to edge)	ft	345
Ship channel tow draft	ft	42

Metoceam Criteria

Based on API-RP-2MT [ref. 1], 100-yr and 1,000-yr wave and wind dominant hurricanes have been summarized in Table 2.

Table 3 Central GoM 100-yr and 1,000-yr Hurricanes

Central GoM	100-yr Hurricanes		1,000-yr Hurricanes	
	Wave Dominant	Wind Dominant	Wave Dominant	Wind Dominant
Waves				
Significant wave height (ft)	49.2	51.8	61.7	65.0
Peak spectral period (s)	15.0	15.4	16.8	17.2
Max wave crest height (ft)	58.1	61.0	71.9	75.5
Max wave heigt (ft)	86.9	91.5	108.9	114.5
Period of max waves (s)	13.5	13.9	15.1	15.5
Winds (at 10m elevation)				
1-hr mean wind speed (ft/s)	157.5	149.6	196.9	187.0
Currents				
Speed (ft/s) | depth (ft)	5.91 | 0	5.91 | 0	8.20 | 0	8.20 | 0
	4.43 | 125	4.43 | 125	5.55 | 154	5.55 | 154
	0.33 | 249	0.33 | 249	0.33 | 312	0.33 | 312

TLP CONFIGURATIONS

CTLP Hull Configuration

Based on hull sizing basis presented in Table 1, metocean criteria highlighted in Table 2 and common practice per API-RP-2T [ref. 2], a Conventional TLP (CTLP) hull with upper column frame has been sized. The principal hull dimensions have been summarized in Table 3 and hull configuration is illustrated in Figure 3. From A-A view in Figure 3, this CTLP has 4 round columns, 4 rectangle pontoons and 16 tendons. Upper column frame arrangement is shown in B-B view and elevation view in Figure 3. Displacement 95,560 st as given in Table 3 is similar to the Shell URSA TLP.

Table 1 CTLP Hull Configuration

Draft	(ft)	106.0
Displacement	(st)	95,560
Column span (Center to Center)	(ft)	235
Column diameter	(ft)	80
Pontoon dimension "width x height"	(ft)	42 x 31.5
Column freeboard	(ft)	95
Column height	(ft)	201
Deck post height	(ft)	10
Still water airgap	(ft)	105
Deck dimension (width x length x height)	(ft)	260 x 275 x 45
Total Weight	(st)	60,700

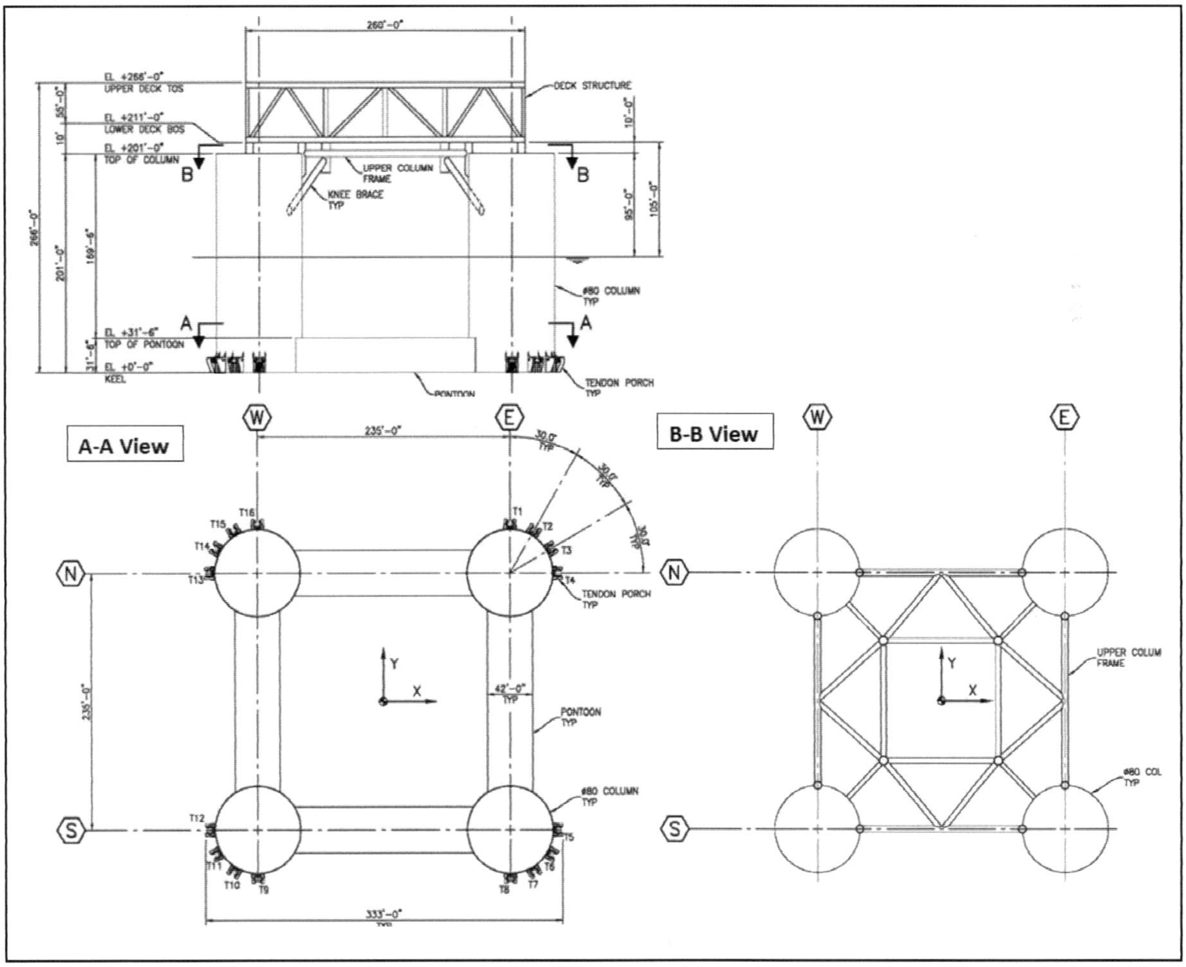

Figure 3 CTLP Principal Dimensions

SCE TLP Hull Configuration

Based on same hull sizing basis in Table 1 and same metocean criteria in Table 2, a Secondary Column Enhanced TLP (SCE TLP) hull without upper column frame has been sized. The principal hull dimensions have been summarized in Table 4 and hull configuration is illustrated in Figure 4. For this sizing exercise, diameters of inner column and outer column were assumed to be same for simplification. Diameter of inner column and outer column can be different and altered for different purposes if needed.

Table 2 SCE TLP Hull Configuration

Draft	(ft)	106
Displacement	(st)	87,500
Inner column spacing	(ft)	155
Inner column diameter	(ft)	56
Inner column height	(ft)	192
Pontoon dimension "width x height"	(ft)	38 x 31.5
Outer column diameter	(ft)	56
Outer column height	(ft)	198
Deck post height	(ft)	10
Still water airgap	(ft)	102
Total weight	(st)	54,200

From A-A view in Figure 4, this SCE TLP has 8 round columns, 8 rectangle pontoons and 16 tendons. From Tables 3 and 4, the SCE TLP inner column span is 155 ft while CTLP column span is 235 ft; SCE TLP column diameter is 56 ft while CTLP column diameter is 80 ft. SCE TLP has 8 shorter and smaller pontoons while CTLP has 4 larger and longer pontoon. Structural efficiency of SCE TLP hull configuration benefits through this innovative arrangement. Top of column layouts is presented in B-B view of Figure 4. Outer of column and inner column connection is illustrated in C-C view of Figure 4. Deck structure was connected to inner columns and all tendons are connected on outer columns. Loads/weights passing have been separated and re-directed with respect to CTLP arrangements which will yield more efficient structural designs.

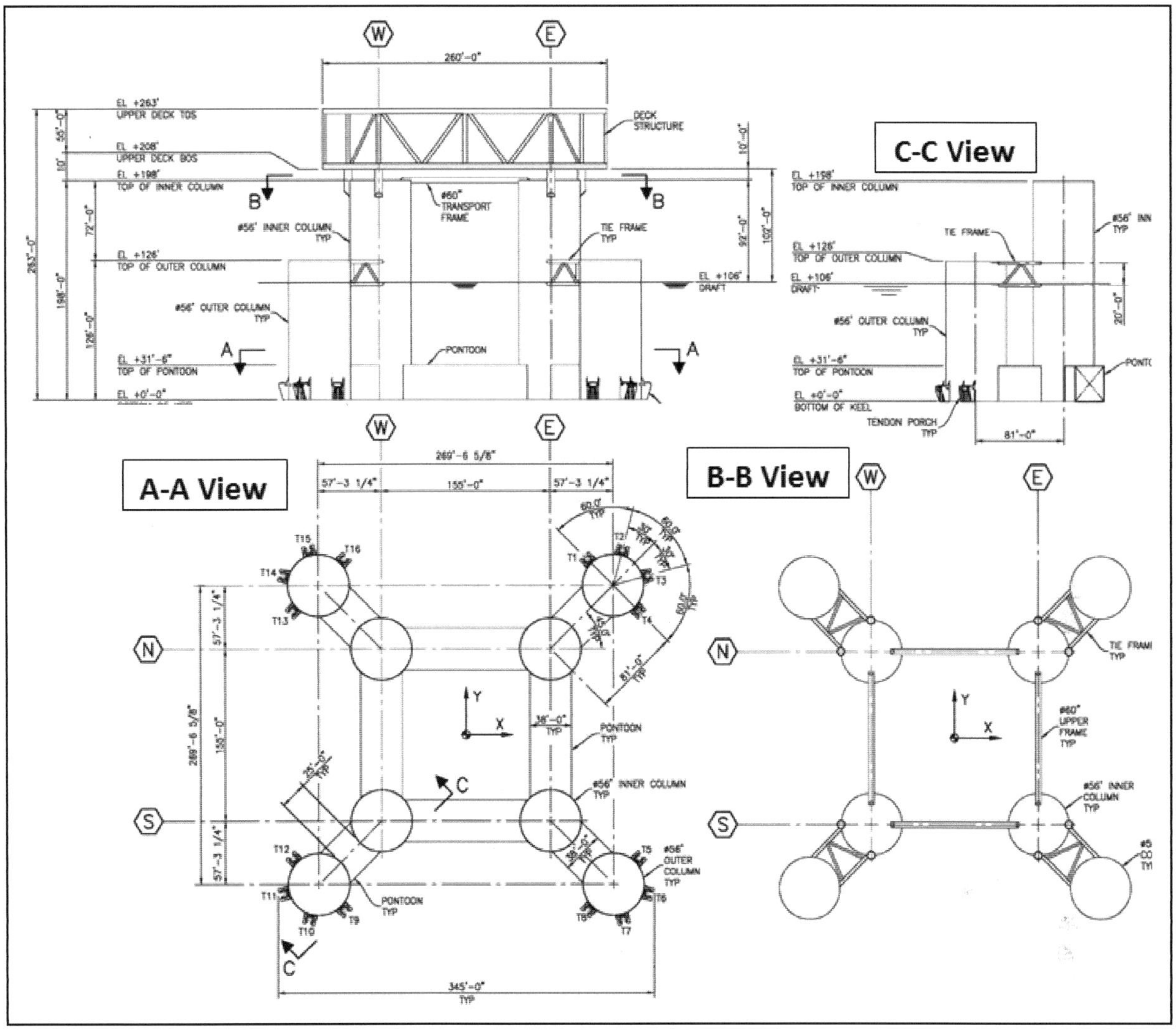

Figure 4 SCE TLP Principal Dimensions

NUMERCIAL RESULTS

In order to demonstrate the unique features of SCE TLP, some selected response characteristics, such as wave upwelling, tendon tension, hull splitting force, shear force and pry/squeeze loads, are presented, compared and discussed in this section.

Wave Upwelling RAOs

The most critical locations for 0 deg and 45 deg of both CTLP and SCE TLP have been selected, wave upwelling RAOs were computed and depicted in Figure 5.

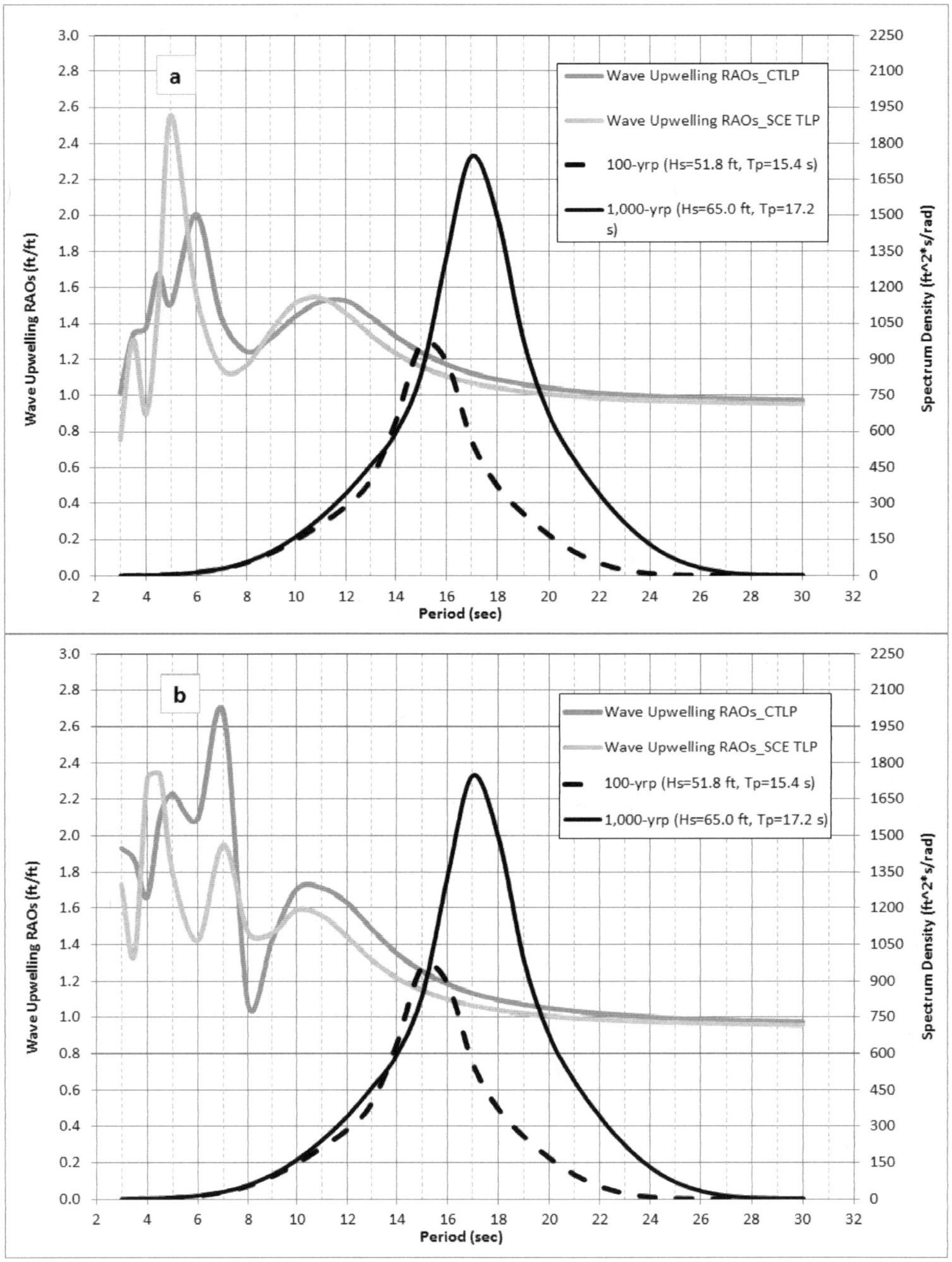

Figure 5 Comparison of wave upwelling RAOs, a) 0 deg heading; b) 45 deg heading

100-yr hurricane and 1,000-yr hurricane wave dominant conditions were considered and standard deviations of wave upwelling of CTLP and SCE TLP in 0 deg and 45 deg headings are given in Table 5.

Table 5 Comparisons of wave upwelling of CTLP and SCE TLP in 100-yr and 1,000-yr Hurricanes

Wave Upwelling		100 yr Hurricane			1,000 yr Hurricane		
		CTLP A	SCE TLP B	Percentage (A-B)/B*100%	CTLP C	SCE TLP D	Percentage (C-D)/D*100%
Stdev (0 deg)	(ft)	17.1	16.3	5%	20.3	19.4	5%
Stdev (45 deg)	(ft)	18.5	16.8	10%	21.6	19.8	9%

As a high level comparison, the extreme value equation used for this exercise is of the form recommended by DNV (2005), see [ref. 3] as following,

$$\text{Extreme value} = \sigma \times \sqrt{-2In\left(1-0.9^{1/N}\right)} \qquad (1)$$

where $N = \frac{T_{simulation}}{T_Z}$, σ is the standard deviation, $T_{simulation}$ is the 3-hr duration (10,800 seconds) and T_Z is the average period between successive zero upcrossings. The dynamic extreme factors were determined as the ratio of the extreme value to the corresponding standard deviation.

From Table 5, it has been found that wave upwelling of 45 deg heading is more critical for both hull forms. By considering 45 deg max wave upwelling, extreme factor is about 4.3 based on Equation 1 and max wave upwelling of CTLP and SCE TLP are 92.9 ft and 85.1 ft respectively in 1,000-yr hurricanes. The max wave upwelling of CTLP is about 7.8 ft higher than that of SCE TLP. In other word, SCE TLP column freeboard can be 7.8 ft less to have equivalent min airgap. Form Tables 3 and 4, it can be seen only 3 ft column freeboard was considered for SCE TLP sizing.

Tendon Tensions RAOs

45 deg up-wave and down tendon tension RAOs are illustrated in Figure 6a and 6b respectively. Standard deviations of up-wave and down-wave tendon tensions of CTLP and SCE TLP in 45 deg headings are given in Table 6.

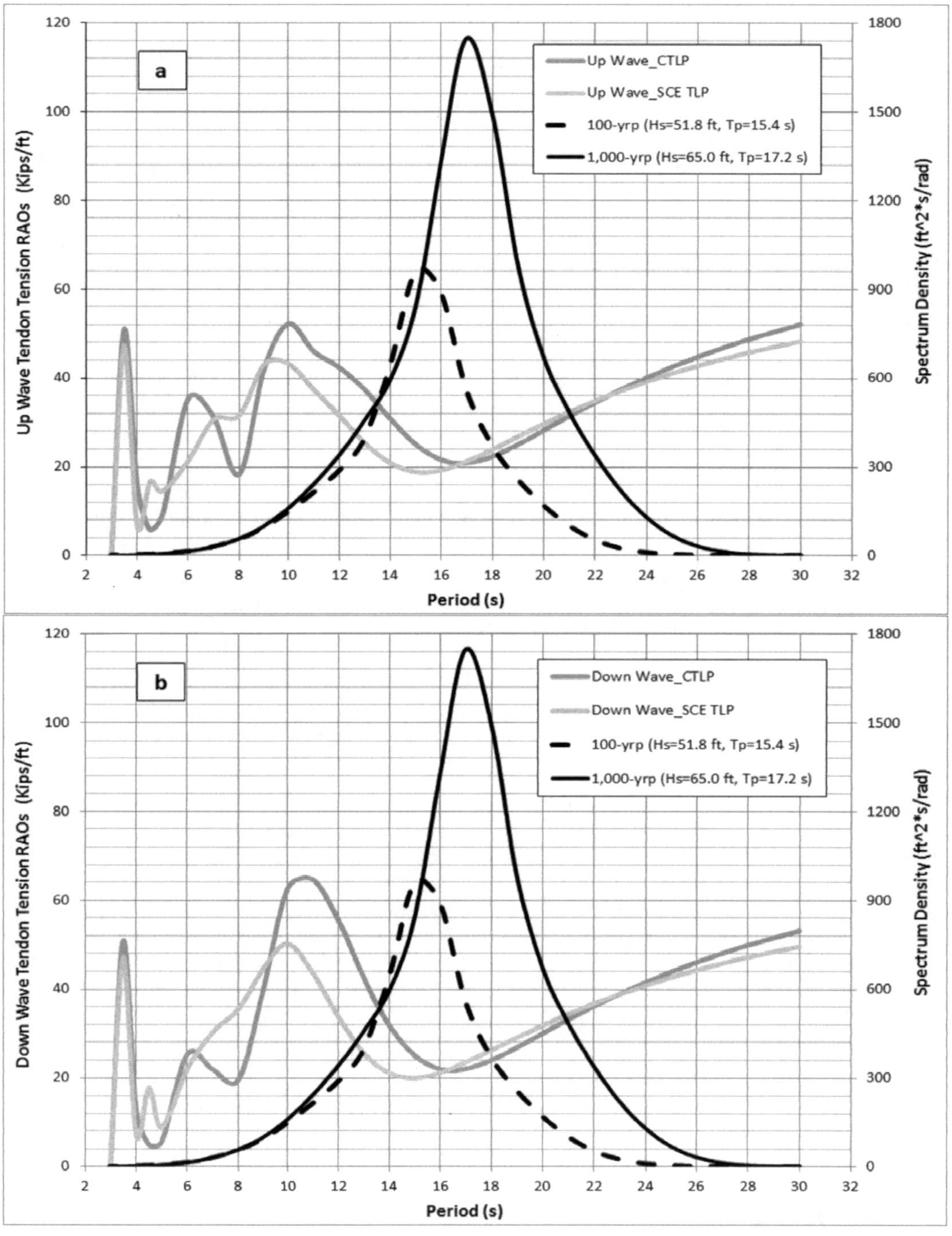

Figure 6 Comparison of tendon tension RAOs: a) up-wave tendon; b) down-wave tendon

Table 6 Standard deviation of wave frequency up-wave and down-wave tendon tensions

Wave-frequency tendon tensions	100 yr Hurricane			1,000 yr Hurricane		
	CTLP A	SCE TLP B	Percentage (A-B)/B*100%	CTLP C	SCE TLP D	Percentage (C-D)/D*100%
Up-wave stdev (45 deg) (kips)	423	355	19%	498	441	13%
Dn-wave stdev (45 deg) (kips)	484	385	26%	564	478	18%

From Table 6, it has been observed that down-wave wave-frequency tendon tension of 45 deg heading is more critical for both hull forms. By considering 45 deg max down-wave wave-frequency tendon tension extreme factor is about 4.3 per Equation 1 and max down-wave wave-frequency tendon tensions of CTLP and SCE TLP are 2,425 kips and 2055 kips respectively in 1,000-yr hurricanes. The max down-wave wave-frequency tendon tension of CTLP is about 370 kips per tendon higher than that of SCE TLP. In other word, SCE TLP tendon pre-tension can be 370 kips less to have equivalent min tendon tension in 1,000-yr hurricane. Other design parameters, such as max offsets and min airgap, shall be considered and final tendon pretension may not be reduced by 370 kips per tendon. However, this did indicate tendon for SCE TLP can be down sized considerably to reduce tendon and foundation cost.

Hull Split Forces

The split force is defined as the difference in net force along X-axis direction, between the two halves of the hull as illustrated in Figure . Figure presents the hull split force RAO for the two hull forms for a wave heading of 45 deg. The extreme hull split forces in 100-yr and 1,000-yr hurricane events is compared in Table .

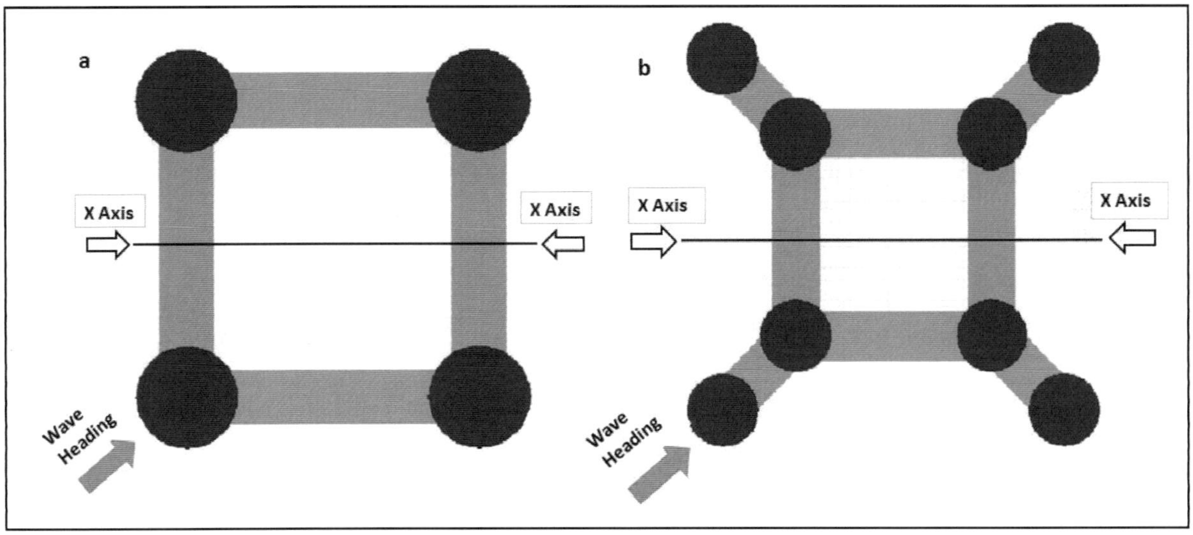

Figure 7 Hull split force definitions for 45 degree wave heading: a) CTLP; b) SCE-TLP

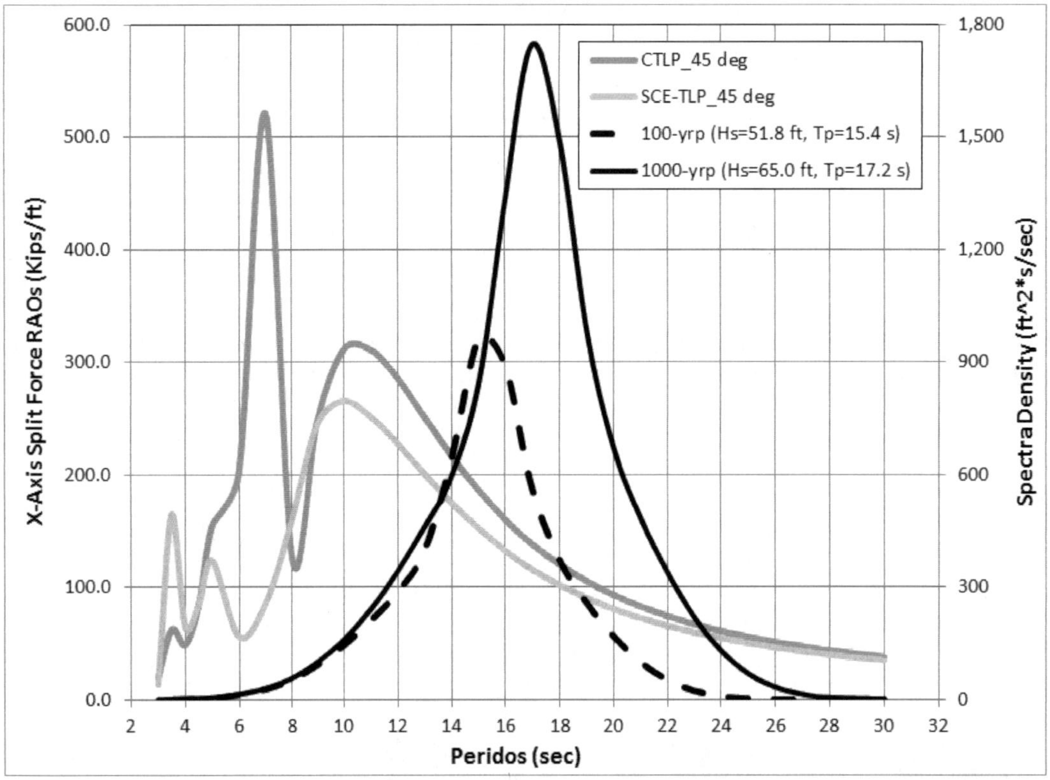

Figure 8 Hull X-axis split force RAOs at 45 deg wave heading for CTLP and SCE-TLP

Table 7 Standard deviation of hull split forces of CTLP and SCE TLP

Wave-frequency hull split force	**100 yr Hurricane**			**1,000 yr Hurricane**		
	CTLP A	**SCE TLP B**	**Percentage (A-B)/B*100%**	**CTLP C**	**SCE TLP D**	**Percentage (C-D)/D*100%**
Along x-axis stdev (kips)	3249	2434	33%	3567	2706	32%

Based on Equation 1, max extreme factor is about 4.3. From Table 7, max wave-frequency x-axis hull split forces of CTLP and SCE TLP are 13,971 kips and 11,636 kips respectively in 1,000-yr hurricanes. The max wave-frequency hull x-axis split forces of CTLP is about 2,335 kips higher than that of SCE TLP.

Shear Forces between Columns and Pontoons

An important consideration for offshore platform structural design is the shear forces at connections between hull members. Figure defines two types of shear forces for the CTLP and SCE TLP hulls, (a) between the column and pontoon, and (b) between inner column and main pontoon.

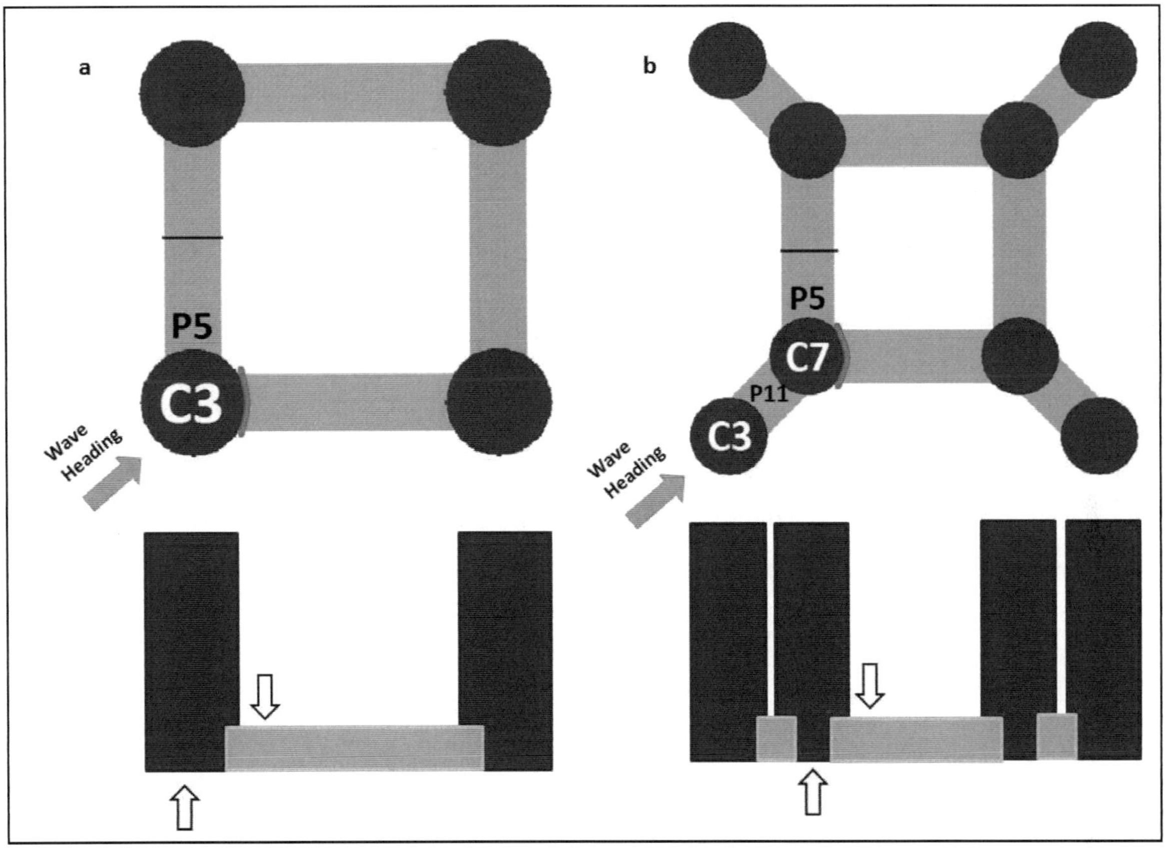

Figure 9 Vertical shear force definitions: (a) At CTLP column and pontoon; (b) At SCE-TLP inner column and main pontoon

Figure plots the shear force RAOs at the defined locations of the CTLP and SCE TLP as shown in Figure 9 in 45 deg heading. The standard deviation of shear forces in 100 yr and 1,000 yr hurricane waves are presented in Table .

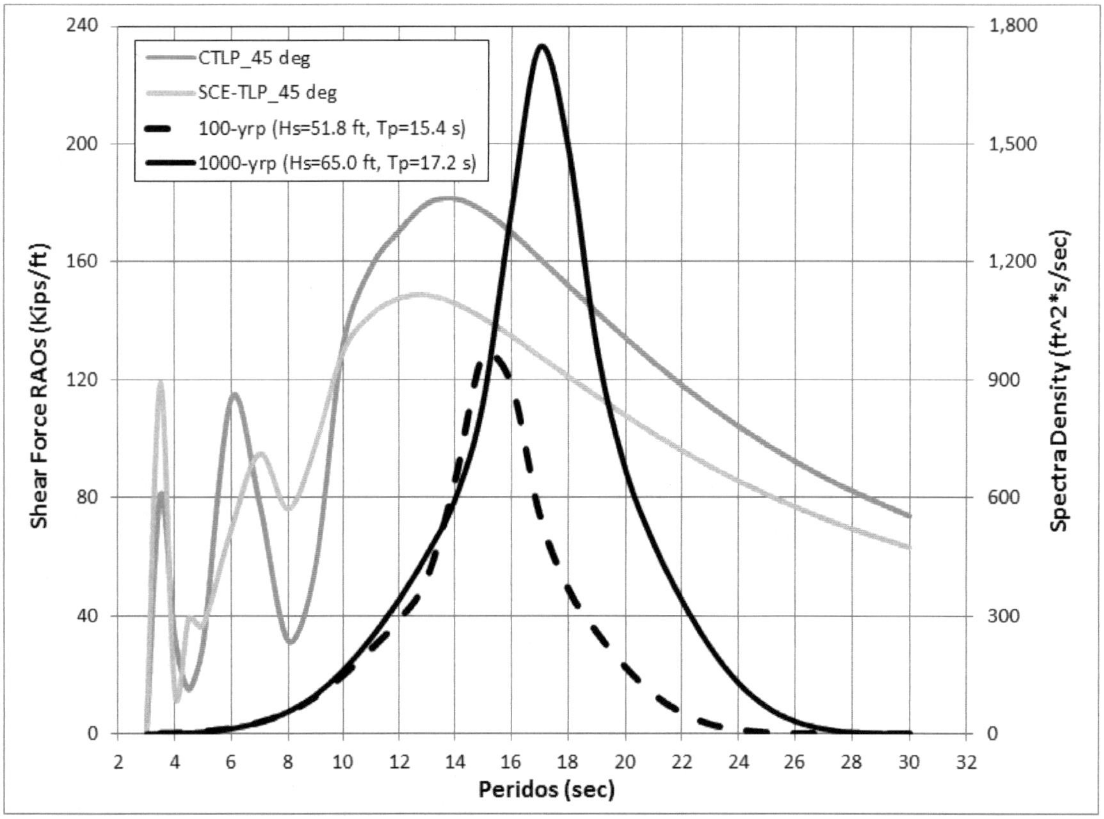

Figure 10 Vertical shear force RAOs of CTLP and SCE-TLP

Table 8 Standard deviations of shear forces of CTLP and SCE TLP

Wave-frequency shear forces	100 yr Hurricane			1,000 yr Hurricane		
	CTLP A	SCE TLP B	Percentage (A-B)/B*100%	CTLP C	SCE TLP D	Percentage (C-D)/D*100%
stdev (kips)	2167	1817	19%	2649	2194	21%

Based on Equation 1, max extreme factor is about 4.3. From Table 8, max wave-frequency shear forces of CTLP and SCE TLP are 11,391 kips and 9,434 kips respectively in 1,000-yr hurricanes. The max wave-frequency shear forces of CTLP is about 1,957 kips higher than that of SCE TLP.

Pry/Squeeze Forces between Columns

The characteristic pry/squeeze forces between opposite columns are another key factor for the hull and deck structural design. For SCE TLP, both the opposite outer columns and opposite inner columns are considered as illustrated in Figure 11. The definitions of pry/squeeze loads for CTLP and SCE TLP are presented in

Figure . **Figure** 12 plots the pry/squeeze force RAOs as defined in Figure 11 at a 45 degree wave heading. **Table** summarizes the standard deviations of pry/squeeze forces of CTLP and SCE TLP outer to outer columns in 100-yr and 1,000-yr hurricanes. **Table** presents the standard deviations of pry/squeeze forces of SCE TLP outer to outer columns and inner to inner columns in 100-yr and 1,000-yr hurricanes.

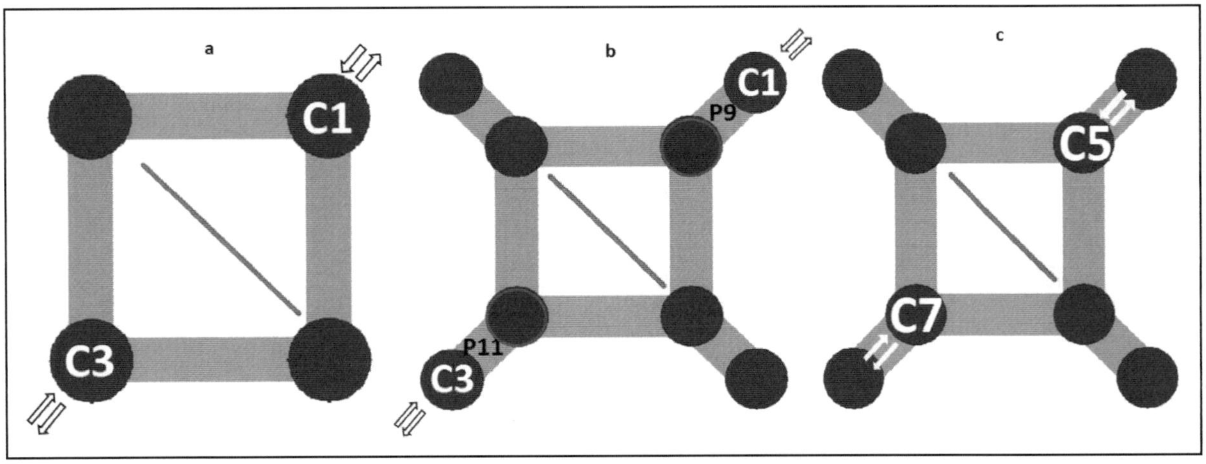

Figure 11 Definitions of pry/squeeze loads in 45 deg heading: a) CTLP, column to column; b) SCE-TLP, outer column to outer column; c) SCE-TLP, inner column to inner column

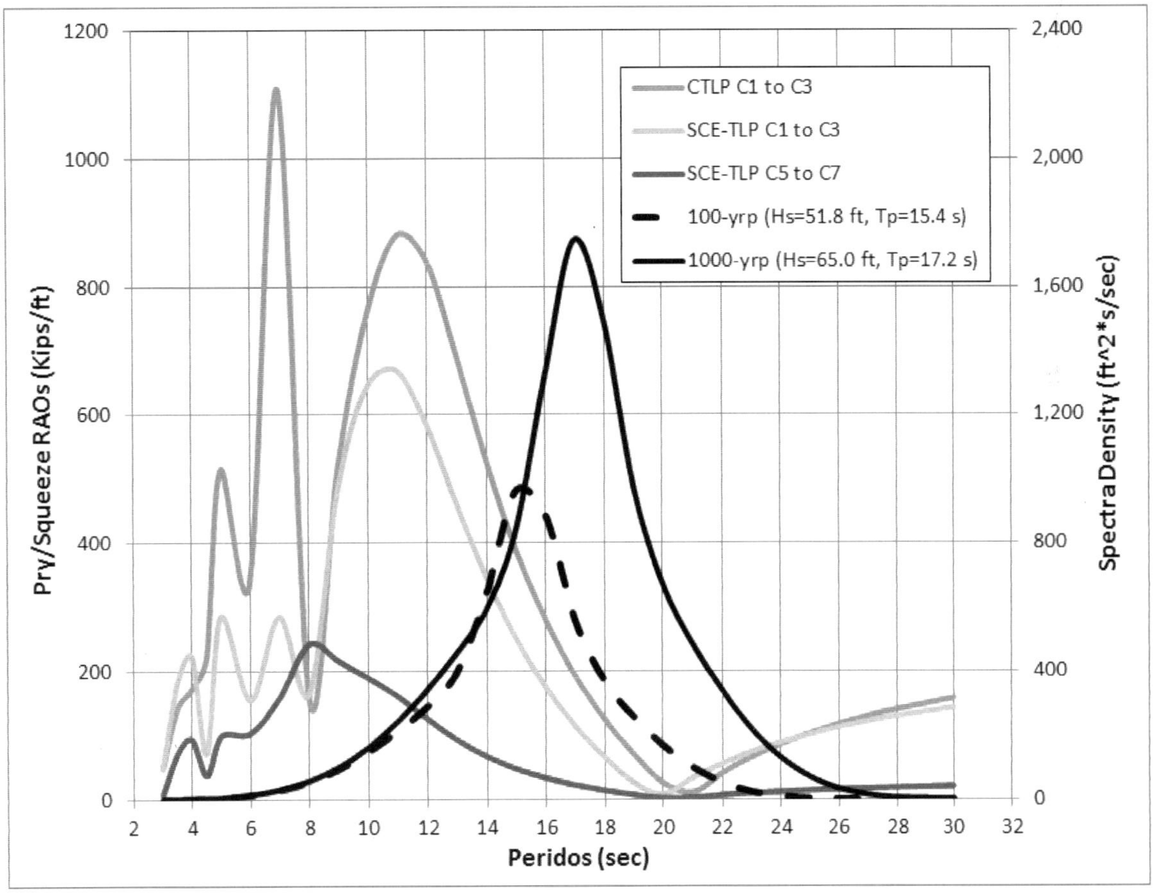

Figure 12 Pry/Squeeze RAOs of CTLP and SCE-TLP as defined in Figure 11

Table 9 Standard deviations of pry/squeeze loads of CTLP and SCE TLP outer to outer columns

CTLP and SCE TLP outer/outer	100 yr Hurricane			1,000 yr Hurricane		
	CTLP A	SCE TLP B	Percentage (A-B)/B*100%	CTLP C	SCE TLP D	Percentage (C-D)/D*100%
stdev (kips)	7542	5200	45%	7979	5504	45%

Table 10 Standard deviations of pry/squeeze loads of SCE TLP outer to outer and inner to inner columns

SCE TLP outer/outer and inner to inner	100 yr Hurricane			1,000 yr Hurricane		
	Inner A	Outer B	Percentage A/B*100%	Inner C	Outer D	Percentage C/D*100%
stdev (kips)	2992	5200	58%	3128	5504	57%

The pry/squeeze forces at SCE TLP outer columns are 45% of those at the CTLP columns. In addition, the pry/squeeze forces at the SCE TLP inner columns are 58% and 57% of those at the outer columns in 100-yr and 1,000-yr hurricanes respectively.

MAJOR SAVINGS

After preliminary global performance analysis, hull and tendon configurations have been settled down. Preliminary deck and hull structural analyses have been carried out. Deck primary steel weights of both hulls per different column spacing have been analyzed. Similarly hull structural scantling was undertaken and hull structural weights of both hulls have been derived.

Major savings are summarized in Table xx. There are a few key notes highlighted as follows:

- Upper column frame weight was included in deck primary steel which can be grouped into hull steel weight for CTLP and no UCF was considered for SCE-TLP due to much smaller column spacing
- Tendon weight saving on Item 4 in Table xx does not include foundation weight saving. Foundation saving shall be anticipated at similar percentage of tendon weight which will be obtained in future study.

Table 11 Summary of major savings

Items	TLP type	Unit	CTLP A	SCE-TLP B	Delta 1 A-B
1	**Main deck steel**	st	**8100**	**6300**	**1800**
2	**Secondary deck steel**	st	**2960**	**2960**	**0**
3	**Hull steel**	st	**20850**	**17300**	**3550**
4	**Tendon**	st	**19600**	**15800**	**3800**
5	**Total (Items 1+2+3+4)**	st	**51510**	**42360**	**9150**

By a rough calculation/estimation, the savings in Table 11 are worth about 100MM USD.

CONCLUSIONS

Two TLP hulls and tendons were sized to carry same topside payload (excluding deck structure), and TTR and SCRs tensions. Preliminary global performance analysis as well as high level deck and hull structural work have been undertaken in this exercise. In this paper, the selected characteristic responses were presented to display SCE TLP hull unique features. The main conclusions are summarized and highlighted as follows:

- **Wave upwelling**. As illustrated in Figure 5, wave upwelling RAOs of the SCE TLP at the most critical location are consistently lower than that of corresponding value of CTLP due to smaller column more transparent to incident wave trains. Column freeboard of the SCE TLP hull can be cut down future by approximately 5 ft which will benefit pre-service stability, quayside integration, and result in reduced wind loads and hull steel. Overall, it will result in a more efficient hull and tendon/foundation design, and ease of transportation and installation.
- **Tendon tension RAOs.** As depicted in Figure 6, up-wave and down-wave tendon tension RAOs of the SCE TLP for 45 deg heading are consistently lower than that of corresponding value of CTLP. Wave-frequency tendon dynamic tensions in 100-yr and 1,000-yr hurricanes are significantly lower which indicate tendon size of the SCE TLP can be smaller and tendon pretension can be reduced. Thus hull displacement can be less which will result in less hull steel. It can be reasonably anticipated that similar reduction on foundation piles which haven't reported in this paper. Overall, it will result in a more efficient hull and tendon/foundation design.
- **Hull split forces.** From
- Table and Figure 8, hull split forces of the SCE TLP are considerably lower than those of the CTLP in 100 yr and 1,000 yr hurricanes. This implies the innovative hull column/pontoon arrangement is a considerable advantage for reduction split forces. The hull structure will result in a more efficient design.
- **Shear forces.** From Table 8 and Figure 10, shear forces of SCE TLP are dramatically lower than those of CTLP in 100-yr and 1,000-yr hurricanes. Thus, the SCE TLP will have a more efficient column/pontoon structural design.
- **Pry/Squeeze loads.** It is well known that pry/squeeze loads affect hull and deck structure design. From Tables 9 and 10 and Figure 12, pry/squeeze loads of SCE TLP are dramatically lower than those of CTLP in 100-yr and 1,000-yr hurricanes. Thus, a more efficient hull and deck structural design can be anticipated.

Since SCE TLP development is still at early stage and validation of wave basin model tests on extreme responses is still pending and shall be carried out in near future to mature the concept and improve the concept readiness.

ACKNOWLEDGEMENTS

The author would like to acknowledge the management of Houston Offshore Engineering, LLC, an Atkins Company for their support of the research on development of SCE-TLP and for allowing the publication of the study results. The author is also grateful for kind assistances of HT Chen and Dongxing Jia on numerical computations.

REFERENCES

[1] API RP 2MET, Derivation of Metocean Design and Operating Conditions, First Edition, American Petroleum Institute, November, 2014.
[2] API RP 2T, Recommended Practice for Planning, Designing and Constructing Tension Leg Platforms, Third Edition, American Petroleum Institute, July, 2010.
[3] Det Norske Veritas (2005), Recommended Practice, DNV-RP-C103, Column-Stabilised Units, pp 16.

SIMULATING TURBULENCE FOR OCEAN CURRENT TURBINE

PARAKRAM PYAKUREL
FLORIDA ATLANTIC UNIVERSITY

JAMES H. VANZWIETEN
FLORIDA ATLANTIC UNIVERSITY

PALANISWAMY ANANTHAKRISHNAN
FLORIDA ATLANTIC UNIVERSITY

WENLONG TIAN
NORTHWESTERN POLYTECHNICAL UNIVERSITY

Proceedings of the 21st Offshore Symposium, February 2016, Houston, Texas
Texas Section of the Society of Naval Architects and Marine Engineers

ABSTRACT

This paper describes an approach for incorporating oceanic turbulence effects into numerical simulations of Ocean Current Turbines (OCT)s. This approach uses a combination of analytic expressions and numerical methods that can be called by variable time step integration routines without requiring the generation of a-priori ambient flow time histories. Ambient turbulence intensity and mean flow speed are user input parameters and Kolmogorov's five-thirds law is utilized to obtain the frequency spectra upon which the algorithm is based.
Keywords: Turbulence, Simulation, Inertial subrange, Marine renewable energy

NOMENCLATURE

C = Coherence decay constant
Coh = Coherence function
ϵ = Turbulent kinetic energy dissipation rate
f = Frequency
$\bar{u}$ = Mean velocity component in flow direction
u_f = Fluctuating velocity component of turbulent flow
u_k^* = Resultant fluctuating component of flow speed associated with mid-frequency component k
H= Velocity weighting factor
θ= phase angle
u = water speed at any instant
S = Power Spectral Density
w = Wave number
σ = Standard deviation

TI = Turbulence intensity

INTRODUCTION

Electricity can be produced from river, tidal, and ocean currents by converting the kinetic energy in these flows into electrical energy using in-stream hydrokinetic turbines. Ocean energy has a great potential to reduce carbon emissions in the long term because ocean energy technologies do not generate Green House Gases (GHG) in operation and have low lifecycle GHG emissions, providing the potential to significantly contribute to emissions reductions [1]. The ocean energy resource off the east coast of the United States has an estimated 19 GW of technically extractable energy [2], which is equivalent to 4.07% of electricity consumed by the U.S. in 2014 [3]. Average kinetic energy densities off Florida above 3.0 kW/m^2 and off North Carolina above 2.0 kW/m^2 have been calculated from ocean current measurements [4]. Similarly, a total power of around 50.78 GW has been estimated as being technically recoverable from tidal streams [5], and the total recoverable riverine hydrokinetic energy resource estimated for the continental United States is 13.7 GW [6]. River and tidal turbines have been installed in multiple locations globally, device development is currently underway for ocean current systems, and research is being carried out to transition single device installations into turbine arrays [7].

Ambient turbulence affects both wind and in-stream hydrokinetic turbines. Turbulence affects wind turbine structural response, power production, and wake evolution [8]. Interaction among turbines in marine turbine array systems are also found to be significantly affected by ambient turbulence intensity [9]. The turbulent wakes generated from the upstream turbines have significant impact on fatigue loads of downstream turbine [10]. In addition, the far wake decay behind turbine is also strongly dependent on ambient turbulence intensity [11].

The inertial subrange is the part of the wave-number spectrum where both the supply and the dissipation of turbulent energy are small compared with the rate of energy transfer through the spectrum [12]. In the inertial subrange, energy is transferred from larger scale eddies to smaller scale eddies with negligible viscous dissipation [13]. Turbulence energy spectrum in the inertial subrange can be generalized to be inversely proportional to the five third power of wave number [14] and analysis of turbulence at two tidal energy sites have been carried out by considering inertial subrange [15].

The US National Renewable Energy Laboratory (NREL) has developed a numerical simulation tool known as TurbSim to simulate turbulent flow. TurbSim has used a statistical model to numerically simulate time series of three-component wind-speed vectors at points in a two-dimensional vertical rectangular grid that is fixed in space. In this method, spectra of velocity components and spatial coherence were defined in the frequency domain, and an inverse Fourier transform was used to generate time series. Stationary process was assumed by TurbSim to generate time series that can be integrated into numerical simulation packages [16].

This paper develops a method to simulate turbulence that can be integrated into turbine performance simulation. Turbulent effects on ocean current turbines will be characterized by applying this approach to a turbine simulation developed by Southeast National Marine Renewable Energy Center (SNMREC) [17]. An approach similar to TurbSim is adopted in that statistical model is used and stationary process is considered. The simulation built in this paper is for inertial subrange. Sets of analytical expressions are derived (section 2) to calculate instantaneous velocity so that entire time series do not have to be generated and variable time steps can be used. While only the velocity component in the direction of flow is simulated in this paper, this method can be extended to simulate all 3 components of velocity.

METHODOLOGY

The section presents an approach for generating a set of analytic expressions that can be used to define flow speed at a set of fixed grid-points as a function of time. This allows for a numerical simulation to request this water velocity component at the pre-defined locations for any time, eliminating the need for generating time histories of water velocity prior to running numerical simulations. Analytic expressions written for flow velocity in the mean flow direction at each spatial point are obtained that are a function of time. Turbulence velocity is simulated using the following user input parameters: mean velocity component in flow direction $\overline{u}$, Turbulence Intensity TI, as well as the minimum and maximum frequencies f_{min} and f_{max} considered. Any type of spatial grid (circular or rectangular) that contains location points over which turbulent velocity shall be generated can be considered.

- **Turbulence and Turbulence Spectra for the Inertial Sub-range**

In this section a statistical approach for characterizing turbulence is discussed and a turbulence spectrum is defined for the inertial sub-range as a function of the limits of the frequency range being considered, and the turbulence intensity over this frequency range. This formulation is valid for frequency ranges that can be approximated as existing within the inertial sub-range, with similar approaches required if the frequency spectra deviates significantly from this assumption.

In general, a turbulent flow can reach a statistically stationary state if the data time history is collected for long period of time i.e. the statistics can be considered independent of time for large sampling period [18]. A velocity component u at any time t can be expressed as [19]:

$$u(t) = \bar{u} + u_f(t), \tag{1}$$

where $\bar{u}$ is the mean velocity component and $u_f(t)$ is fluctuating component of velocity associated with this direction with zero mean. The root mean square value u^* of the fluctuating components $u_f(t)$ for a total of N_p measurements is then [20]:

$$u^* = \sqrt{\frac{1}{N_p}\sum_{l=1}^{l=N_p}\left(u_{f,l}\right)^2}, \tag{2}$$

where $u_{f,l}$ represents l^{th} measurement of the fluctuating component $u_f(t)$. This equation is valid for discrete data sets where measurements are recorded at a fixed time step.

The power spectral density (PSD) calculated from water velocity measurements made at a point in a turbulent flow gives an indication of the turbulence energy content at different frequencies [13]. The PSD for wave number w has following proportionality in inertial sub-range [21]:

$$S(w) \propto \epsilon^{2/3} w^{-5/3}, \tag{3}$$

where $S(w)$ is PSD expressed as function of wave number and ϵ is turbulence kinetic energy dissipation rate. For a given frequency, f, the wave number, w, is directly proportional to frequency [13], which implies the power spectral density $S(f)$ as a function of the f can be calculated as:

$$S(f) \propto \epsilon^{\frac{2}{3}} f^{-\frac{5}{3}} \left(\frac{\bar{u}}{2\pi}\right)^{2/3}. \tag{4}$$

For a constant TI, which is an input parameter in this algorithm, the dissipation rate ϵ is also a constant. Therefore, since mean water velocity component $\bar{u}$ is also constant, the above equation can be re-written as:

$$S(f) = Af^{-\frac{5}{3}}, \tag{5}$$

where A is constant of proportionality that is equivalent to $\epsilon^{\frac{2}{3}}\left(\frac{\bar{u}}{2\pi}\right)^{2/3}$.

The PSD defines how the power of signal (u^{*2}) is distributed in frequency space [22]. Hence, the standard deviation σ and the integral of one sided PSD $S(f)$ in frequency domain can be related by [16]:

$$u^{*2} = \sigma^2 = 2 \times \int_0^\infty S(f)df. \tag{6}$$

The turbulence intensity, TI, is defined by [23]:

$$TI = \frac{u^*}{\bar{u}}. \tag{7}$$

From Equation 1 it can be seen that $u_f(t) = u(t) - \bar{u}$. Therefore, u^* is equal to the standard deviation σ of

$u(t)$. Consequently, TI can be expressed as $TI = \sigma/\bar{u}$.

Plugging $u^* = TI \times \bar{u}$ and $S(f) = Af^{-\frac{5}{3}}$ into Equation 6, the constant A can be obtained by integrating the right hand side of Equation 6 from f_{min} to f_{max}:

$$A = \frac{\bar{u}^2 \times TI^2}{3\left[\frac{1}{f_{min}^{\frac{2}{3}}} - \frac{1}{f_{max}^{\frac{2}{3}}}\right]}. \tag{8}$$

Therefore, the frequency spectra in the inertial sub-range can be defined as follows

$$S = \frac{\bar{u}^2 \times TI^2}{3\left[\frac{1}{f_{min}^{\frac{2}{3}}} - \frac{1}{f_{max}^{\frac{2}{3}}}\right]} f^{-5/3}, \tag{9}$$

assuming the TI is derived from this same frequency range. Five minutes window has been used to measure TI experimentally for tidal sites [15].

- **Single Location Turbulence Expression**

This section develops an analytic expression for the turbulent velocity at a single fixed location. This expression is defined as a function of the turbulent spectra listed in the previous section, the frequency range being considered under the assumption that this is within the limiting frequencies of inertial sub-range, and time.

The frequency range defined from f_{min} to f_{max} is discretized into N equal parts such that there are f_1 , f_2 ....f_N, f_{N+1} frequency points defining the edges of individual frequency elements within this range. The difference between any two consecutive frequencies, df, is therefore $df = (f_{max} - f_{min})/N$. A mid-frequency point f_k^* between any two consecutive frequencies f_k and f_{k+1} is found as $f_k^* = (f_k + f_{k+1})/2$. Thus, there are an N number of mid-frequencies.

Using the value of A calculated from Equation 8, the amplitude of the resultant fluctuating velocity u_k^* for mid-frequency f_k^* is determined as:

$$u_k^{*2} = \int_{f_k}^{f_{k+1}} S(f)df,$$

$$u_k^{*2} = \int_{f_k}^{f_{k+1}} Af^{-\frac{5}{3}}df,$$

$$u_k^* = \sqrt{\frac{3A}{2}\left(\frac{1}{f_k^{\frac{2}{3}}} - \frac{1}{f_{k+1}^{\frac{2}{3}}}\right)}. \tag{10}$$

Using this approach the amplitude fluctuating water velocity component associated with every mid-frequency can be calculated. To calculate the total fluctuating water velocity, $u_f(t)$, as a function of time, a combination of fluctuating velocities at every frequency is considered:

$$u_f(t) = \sum_{k=1}^{k=N} u_k^* \sin(2\pi f_k^* t + \theta_k), \tag{11}$$

where θ_k is the random phase angle that has equal probability of occurrence between 0 and 2π and is associated with mid-frequency component f_k^* .

Equation 11 is then plugged in Equation 1 to define the turbulent water velocity at any time t. The final expression can therefore be written as:

$$u(t) = \bar{u} + \sum_{k=1}^{k=N} u_k^* \sin(2\pi f_k^* t + \theta_k). \tag{12}$$

- **Spatially Correlated Turbulent Flow Field for a Fixed Grid**

This section describes approach to develop partially correlated flow field for multiple locations on a fixed grid. Correlation between two signals at nodes i and j as a function of frequency can be calculated using the cross spectral density $S_{ij}(f)$ [24]. The cross spectral density between nodes i and j is defined as suggested by [25] according to:

$$S_{ij}(f) = Coh_{ij}(f)\sqrt{S_{ii}(f)S_{jj}(f)}, \tag{13}$$

where $S_{ii}(f)$ is PSD at node i, $S_{jj}(f)$ is PSD at node j, and $Coh_{ij}(f)$ is a coherence function between nodes i and j. This coherence function can be estimated as a function of frequency, distance between the two points and the mean flow velocity. The expression below has been suggested for Coh_{ij} [25]:

$$Coh_{ij} = \exp(-\frac{C\Delta r_{ij} f}{\bar{u}}), \tag{14}$$

where Δr_{ij} is distance between nodes i and j, and C is a coherence decay constant.

The frequency f in Equation 14 can be replaced by mid-frequency f_k^* to make Coh_{ij} a constant between nodes i and j for frequency points f_k and f_{k+1} by defining Coh_{ij}^k as:

$$Coh_{ij}^k = \exp(-\frac{C\Delta r_{ij} f_k^*}{\bar{u}}). \tag{15}$$

The cross spectral density $S_{ij}(f)$ between any two nodes i and j is found by assuming auto-spectrum to be same for every location, i.e. $S_{ii}(f) = S_{jj}(f)$. Therefore, Equation 13 can be written as:

$$S_{ij}(f) = Coh_{ij}^k A f^{-\frac{5}{3}}. \tag{16}$$

If velocities at M number of locations on a fixed grid are to be simulated, an M x M lower triangular matrix of velocity weighing factor H can be determined for any frequency f using Choleksy's decomposition as suggested in [25]:

$$\begin{aligned}
&H_{11} = S_{11}^{1/2},\\
&H_{21} = S_{21}/H_{11},\\
&H_{22} = (S_{22} - H_{21}^2)^{1/2},\\
&H_{31} = S_{31}/H_{11},\\
&\cdot\\
&\cdot\\
&\cdot\\
&H_{ij} = (S_{ij} - \textstyle\sum_{l=1}^{l=j-1} H_{il}\, H_{jl})/H_{jj},\\
&H_{jj} = \left(S_{jj} - \textstyle\sum_{l=1}^{l=j-1} H_{jl}^2\right)^{1/2}.
\end{aligned} \tag{17}$$

In this algorithm it is assumed that any node j is affected by only other preceding nodes (nodes $<j$). Therefore, output velocities at any node j is not dependent on velocities inputs of all nodes, but only on velocities inputs from nodes preceding j, as the correlation with node numbers > j will be guaranteed at the time of their calculation. This artifact is also used in TurbSim [16].

Random phases θ_{kl} for mid-frequency component f_k^* are generated for all locations l. These random phases have equal probability of occurrence between 0 and 2π. Velocity u_{kj}^* at node j as a function of mid-frequency component f_k^* is then obtained as:

$$u_{kj}^* = \sum_{l=1}^{j} H_{jl}(f_k^*) e^{i\theta_{kl}}. \tag{18}$$

It is seen that Equation 18 gives a complex value for u_{kj}^* whose resultant phase angle θ_{kj}^R can be determined as:

$$\theta_{kj}^R = \tan^{-1}\left(\frac{imag(u_{kj}^*)}{real(u_{kj}^*)}\right), \tag{19}$$

where $imag$ indicates that the imaginary part of the signal is selected and $real$ indicates that the real part of the signal is selected.

Velocity $u_j(t)$ in time domain at node j is now obtained by slight modification of Equation 12 as:

$$u_j(t) = \bar{u} + \sum_{k=1}^{k=N} u_{kj}^* \sin(2\pi f_k^* t + \theta_{kj}^R). \tag{20}$$

Time history can also be generated if required by plugging different values of t in Equation 20. If a period of simulation run time T to generate time history is chosen instead of an instantaneous time t, T should be greater or equal to inverse of f_{min} in order to resolve the lowest frequency component i.e. $T \geq 1/f_{min}$. The minimum sampling frequency f_s can be found from Nyquist criteria as $f_s = 2 f_{max}$. If N_p is the number of data points/measurements such that the total run time T is discretized to N_p number of points, then $f_s = N_p/T$, and the minimum value of N_p required is $N_p = 2 f_{max} T$.

RESULTS

The simulation approach described in this paper is coded in MATLAB to simulate turbulent flow. A circular grid with 40 points is utilized when creating these results, where azimuthal angle is discretized to 8 equal parts and radius is discretized to 5 equal parts. The radius of the grid is 10 m, which is a likely rotor radius for an ocean current turbine. For mean flow velocity $\bar{u} = 2\,m/s$, ambient turbulence intensity of 0.15 ($TI = 15\%$), $f_{min} = 0.01\,Hz$, $f_{max} = 5\,Hz$ and instantaneous time $t = 5\,s$, simulation results for various coherence decay constants C of 0.01, 5, and 10 are shown in Figures 1-3 respectively. For wind turbine, values of C ranging from 2 to 27 has been suggested [25]. The frequency range considered includes the frequencies in inertial subrange as measured in [15]. The plots on the left side of these figures (labeled as "(a)") show that velocity calculated for each of the 40 nodes at $t = 5\,s$. These plots show that the for a smaller decay constant, C, and therefore larger correlation, the flow field varies less between grid points with variations in these figures of about 0.05, 1.1, and 1.3 m/s respectively. This axial flow velocity variation can be seen in the color plots on the right where color change represents velocity change over the circular grid. Since $C = 0.01$ represents highly coherent state with Coh_{ij} defined by Equation 15 approximating to value 1, the velocities at all the location points are nearly same as seen in Figure 1. However, as C increases, Coh_{ij} decreases and velocities are less coherent and more random as seen in Figures 2 and 3. The horizontal dashed lines in Figures 1-3 represent mean flow velocity.

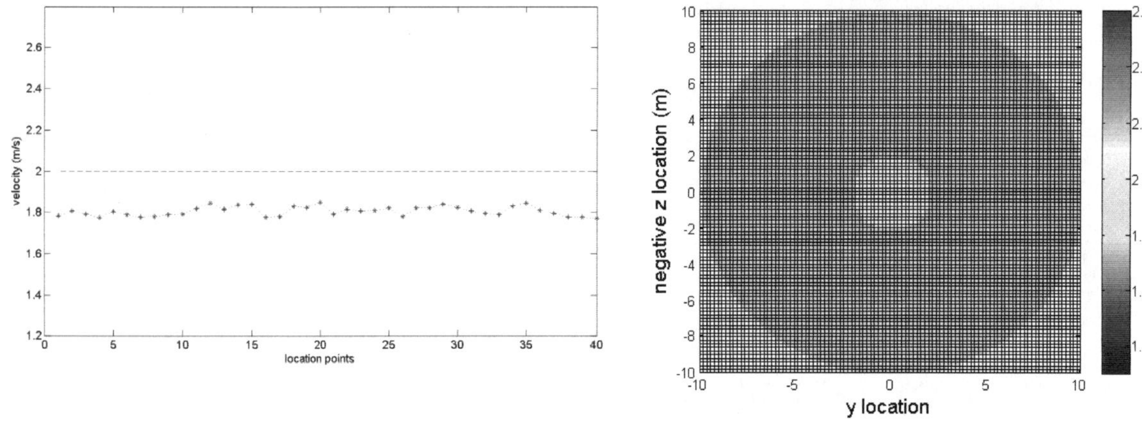

(a) (b)

Figure 1: Simulated turbulent velocity for C=0.01

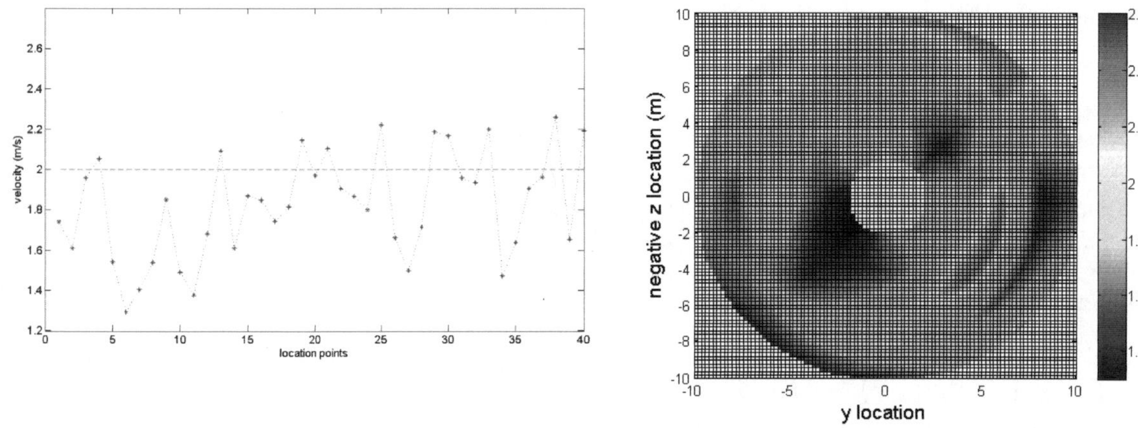

(a) (b)

Figure 2: Simulated turbulent velocity for C=5

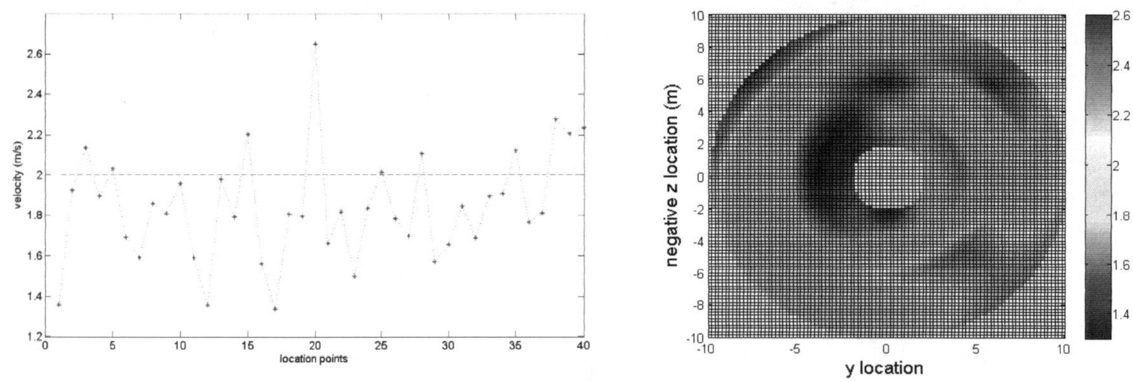

(a) (b)

Figure 3: Simulated turbulent velocity for C=10

Simulation results for velocity time histories are presented in Figures 4 and 5 for $\bar{u} = 2\ m/s$, $C = 5$ and TI of 5% and 15%. Site measurements in tidal channels in USA and Chile have indicated that TI varies between 5% and 20% [26]. It is seen that as TI increases the velocity fluctuation from mean value also increases. These results highlight how much higher peak velocities can be obtained for more turbulent locations.

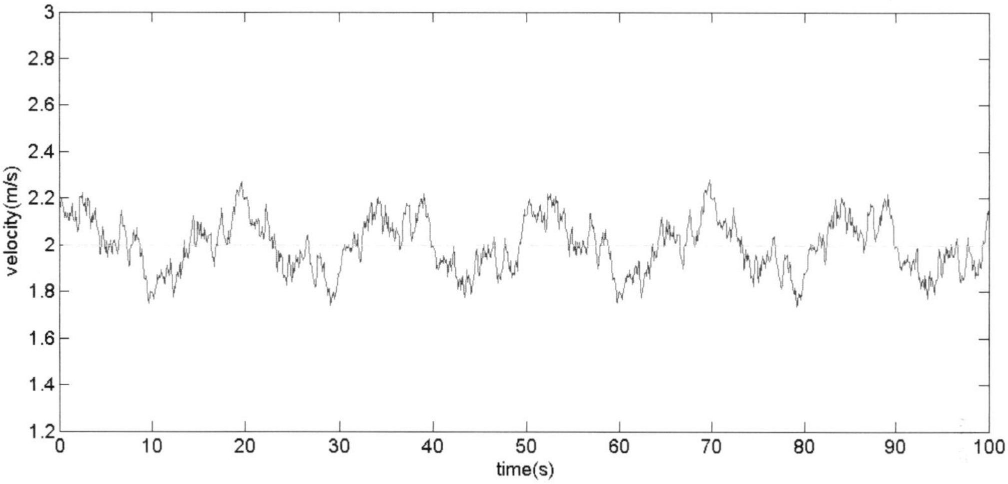

Figure 4: Velocity time history at location 1 for TI=5%

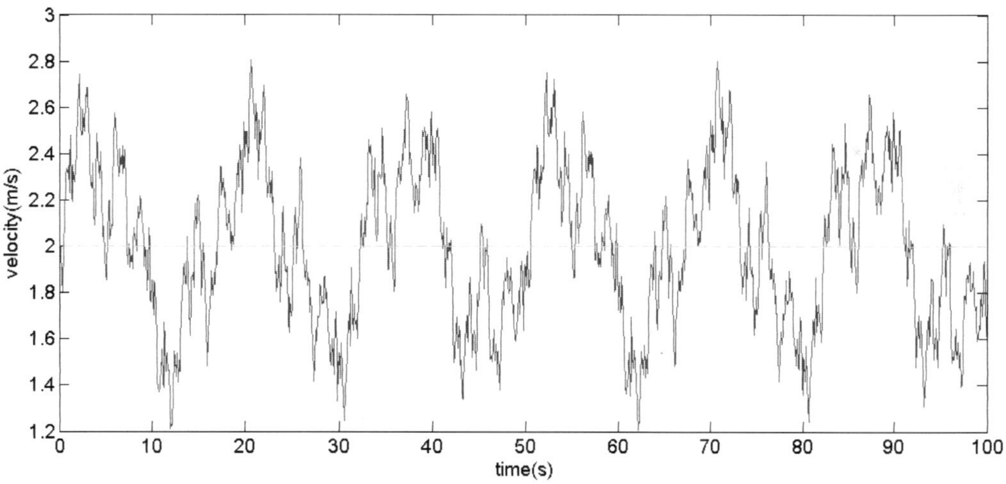

Figure 5: Velocity time history at location 1 for TI=15%

CONCLUSIONS

An approach to simulate turbulent flow speed is presented for frequencies in inertial sub-range by assuming Kolmogorov's five-thirds law to hold true. Turbulent flow can be simulated over a user defined fixed grid by selecting appropriate correlation among locations using this method. Analytical expressions are used along with numerical methods to allow simulation of velocity at any instant of time. In addition, velocity time history can also be generated if required. The approach discussed in this paper can be integrated into numerical simulations of ocean current turbine to evaluate effects of turbulence on turbine performance.

ACKNOWLEDGEMENTS

We would like to thank the National Science Foundation (NSF) and specifically the Energy, Power, Control and Networks (EPCN) program for their valuable ongoing support in this research within the framework of grant ECCS-1308168 'Collaborative Research: Optimized Harvesting of Hydrokinetic Power by Ocean Current Turbine Farms Using Integrated Control'. The SNMREC is supported by the State of Florida and the U.S. Department of Energy. Wenlong Tian would also like to thank the Doctorate Foundation of Northwestern Polytechnical University.

REFERENCES

[1] Lewis, A., Estefen, S., Huckerby, J., W. Musial, Pontes, T., and Torres-Martinez, J., "Ocean Energy-Special Report on Renewable Energy Sources and Climate Change Mitigation," Intergovernmental Panel on Climate Change (IPCC), 2011.

[2] Georgia Tech Research Corp., "Assessment of energy production potential from ocean currents along the United States coastline," Georgia Tech Research Corp., Atlanta, GA, 2013.

[3] U.S. Energy Information Administration, "Annual Energy Review," 2015. < http://www.eia.gov/>

[4] Machado, M.C.P.M, VanZwieten, J.H., Pinos, I., "A Measurement Based Analyses of the Hydrokinetic Energy in the Gulf Stream," Accepted to the Journal of Ocean and Wind Energy, no. JOWE AS-R-06, In press.

[5] Georgia Tech Research Corp., "Assessment of Energy Production Potential from Tidal Streams in the United States," Georgia Tech Research Corporation, 2011.

[6] EPRI, "Assessment and Mapping of the Riverine Hydrokinetic Resource in the Continental United States," Technical Report, Electric Power Research Institute, 2012.

[7] Myers, L.E, Bahaj, A.S., "An experimental investigation simulating flow effects in first generation marine current energy converter arrays," Renewable Energy, 37 (2012) 28-36, 2011.

[8] Churchfield, M.J., Lee, S., Michalakes, J., & Moriarty, P.J, "A numerical study of the effects of atmospheric and wake turbulence on wind turbine dynamics, Journal of Turbulence," 13, N14, 2012.

[9] Mycek, P., Gaurier, B., Germain, G., Pinon, G., & Rivoalen, E., "Experimental study of the turbulence intensity effects on marine current turbines behaviour. Part II: Two interacting turbines," Renewable Energy 68 (2014) 876-892, 2014.

[10] Lee, S., Churchfield, M., Moriarty, P., Jonkman, J., and Michalakes, J., "Atmospheric and Wake Turbulence Impacts on Wind Turbine Fatigue Loading," National Renewable Energy Laboratory, 2011.

[11] Maganga, F., Germain, G., King, J., Pinon, G., and Rivoalen, E., "Experimental characterisation of flow effects on marine current turbine behaviour and on its wake properties," IET Renewable Power Generation, Volume 4, Issue 6, pp. 498–509, 2010.

[12] Bradshaw, P., "Conditions for the Existence of an Inertial Subrange in Turbulent Flow", Reports and Memoranda No. 3603, AERONAUTICAL RESEARCH COUNCIL, London, 1969.

[13] Johnson, R.W., The handbook of fluid dynamics, CRC Press and Springer, ISBN 3-540-64612-4, 1998.

[14] Chasnov, J.R., "Simulation of the Kolmogorov inertial subrange using an improved subgrid model," Phys. Fluid A, Vol. 3, No. 1, 1991, pp. 188-200, 1991.

[15] Thomson, J., Polagye, B., Durgesh, V., and Richmond, M. C., "Measurements of Turbulence at Two Tidal Energy Sites in Puget Sound, WA", IEEE Journal of Oceanic Engineering, Vol. 37, No. 3, July 2012.

[16] Jonkman, B., and Kilcher, L., "Turbsim User's Guide: Version 1.06.00," NREL, 2012.

[17] VanZwieten Jr, J. H., Vanrietvelde, N., and Hacker, B.L., "Numerical Simulation of an Experimental Ocean Current Turbine," IEEE Journal of Oceanic Engineering, Vol. 38, No. 1, pp. 131-143, 2013.

[18] Pope, S. B., Turbulent Flows, Cambridge University Press, ISBN 0 521 59886 9, 2000.

[19] Chen, C.J., Jaw, S.Y., Fundamentals of Turbulence Modeling, Taylor & Francis, ISBN 1-56032-405-8, 1997.

[20] Massachusetts Institute of Technology, 7. Basics of Turbulent Flow, http://www.mit.edu/course/1/1.061/www/dream/SEVEN/SEVENTHEORY.PDF <Accessed January 25, 2016>

[21] Hunt, J.C.R., and Vassilicos, J.C., "Kolmogorov's Contributions to the Physical and Geometrical Understanding of Small-Scale Turbulence and Recent Developments", Royal Society Publishing, Proceedings: Mathematical and Physical Sciences, Vol. 434, No. 1890, Turbulence and Stochastic Process: Kolmogorov's Ideas 50 Years On, Jul. 8, 1991, pp. 183-210, 1991.

[22] Ferro, M., "Experimental study on turbulent pipe flow", Technical Reports from Royal Institute of Technology, KTH, Sweden, 2012.

[23] Camp, T.R., Shin, H.W., "Turbulence Intensity and Length Scale Measurements in Multistage Compressors", ASME, International Gas Turbine and Aeroengine Congress and Exposition, The Hague, Netherlands, June 13-16, 1994.

[24] Crocker, M.J., Handbook of Acoustics, John Wiley & Sons, Inc., 1998.

[25] Veers, P.S., "Three-Dimensional Wind Simulation," Sandia National Laboratories, 1988.

[26] Thomson, J., Kilcher, L., Richmond, M., Talbert, J., deKlerk, A., Polagye, B., Guerra, M., Cienfuegos, R., "Tidal turbulence spectra from a compliance mooring", Proceedings of the 1st Marine Energy Technology Symposium, METS2013, April 10-11, 2013, Washington, D.C.

REMOTE OPERATED UNDERWATER WELDING VEHICLE

KARTHIK S
CENTRE FOR MARITIME RESEARCH, AMET UNIVERSITY, CHENNAI, TAMILNADU, INDIA

Proceedings of the 21st Offshore Symposium, February 2016, Houston, Texas
Texas Section of the Society of Naval Architects and Marine Engineers

ABSTRACT

The underwater welding is the emerging technology of the present and the future. Many offshore platforms and ships need the underwater welding process during emergency. There exist many research institutes globally in developing techniques for underwater welding. A dire need for developing the underwater vehicle for welding processes in marine applications. The recent developments in the underwater vehicles have led to a revolutionary change in the design and system development for coastal surveillance, sea bed inspection and ocean bed exploration. This paper presents a review in implementation of a robot arm for an autonomous underwater welding vehicle. It deals with review of various underwater welding techniques for marine applications and structured such a manner all aspects of underwater welding process with robot manipulator as per the classification requirements. A brief review of its characteristics, applications and the risk mitigation factors of underwater welding techniques for marine applications. The end result is a model that provides accurate predictions for the joint torques of a two-link robot arm in a simple form implemented with algorithms for precision planning and control during underwater welding application.
Keywords: Underwater vehicle, welding, application, robot arm, laser weld, development.

INTRODUCTION

The first ever underwater welding was carried out by British Admiralty Dockyard for sealing leaking ship rivets below the water line. Underwater welding is an important tool for underwater fabrication works. In 1946, special waterproof electrodes were developed in Holland. In recent years the number of offshore structures including oil drilling rigs, pipelines, and platforms are being installed significantly. Some of these structures will experience failures of its elements during normal usage and during unpredicted occurrences like storms, collisions. Any repair method will require the use of underwater welding vehicle. The recent developments in the marine world have led to a revolutionary change in the design and development of various systems. Developments in underwater technology are one of such changes. Welding processes have been used extensively as a joining technique, used in design and fabrication of various structures like naval ships, airplanes, automobiles, bridges, pressure vessels, etc. Underwater welding vehicle has emerged as a better option in contrast to other the underwater vehicle in underwater environment.

Another emerging area in marine application welding systems is the underwater welding technique. Underwater Welding has been significantly used. However, its use has not reached to a significant level in a welding environment due to number of factors. The underwater welding process came into existence with the development of water-proof electrodes in 1940's. It is the process of welding at elevated pressures, normally underwater. It may be carried out in the water itself or dry inside a specially constructed positive pressure envelope, thereby providing a special environment. Under water is also known as "hyperbaric welding" when used in a dry environment, and "underwater welding" when in a wet environment. The application areas of this underwater welding technique include a wide variety of structures, such as repair ships, offshore oil platforms, and pipelines. Steel is the most common material welded. Figure 1 show the area to be weld under water.

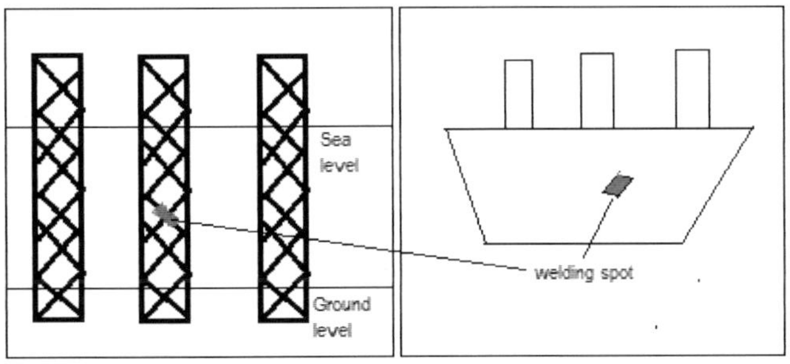

Figure 1. Pointing the welding spot

Various researchers have defined the concept of underwater welding in different ways challenge is to create an environment for welding. These are discussed in the below in this article. One of the main advantages of this underwater welding vehicle is the savings in terms of time and human. For repair and maintenance jobs in marine applications, the equipment vessel need not be taken out to carry out the welding operation. The main difficulties in underwater welding are the presence of a higher pressure due to the water head under which welding takes place, chilling action of the water on the weld metal (which might change the metallurgical structures and properties), the possibility of producing the arc mixtures of hydrogen and oxygen in pockets, which might set up an explosion, and the common danger sustained by divers, of having nitrogen diffused in the blood in dangerous proportions. Hydrogen and oxygen are dissociated from the water and will travel separately as bubbles Oxygen cutting is about 60 percent efficient a above river beds, especially in mud, because trapped methane gas in the proper concentrations can explode. There is a risk to the welder or worker of electric shock. There is a risk that defects may remain undetected. The other main area of risk is to the life or health of the welder or diver from nitrogen introduced into the blood stream during exposure at increased pressure. Furthermore, complete insulation of the welding circuit is an essential requirement of underwater welding vehicle.

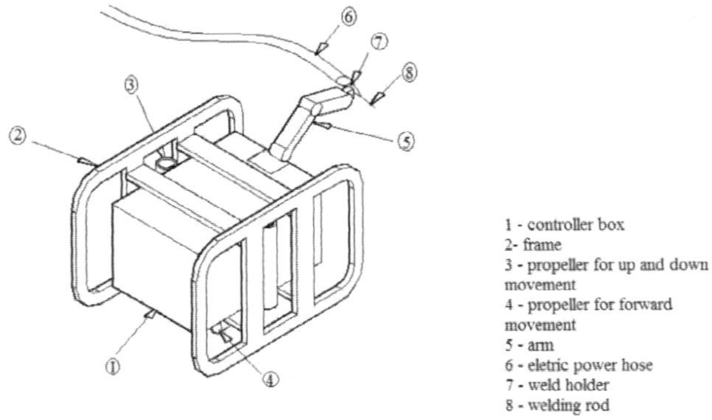

Figure 2. 3D structure for underwater welding vehicle

A remote-controlled repair welding vehicle which uses laser welding technology in underwater environment was developed. This is an underwater vehicle technology combined with a laser welding technology. This report will describe the structure and performance of this vehicle, and the welding test results. The repair welding vehicle consists of two parts. The one is driving equipment, and the other is welding unit. It can glide in water, move around the defected area, and stay on the welding area. After that it starts YAG laser repair welding. The target of this technology is outer surface repair of some pipeline, ship made some steel. Degradation by General Corrosion and so on might be occurred at outer surface in underwater. If the damaged area is wide, repair welding works are done.

Underwater Welding in Marine Applications

There are various ways in which underwater welding vehicle can be performed. In underwater welding vehicle, the main challenge is to create an environment for welding. One of the main advantages of this welding technique is the savings in terms of time. For repair and maintenance jobs in marine applications, the equipment/vessel need not be taken out to carry out the welding operation. In practice, the use of underwater wet welding for offshore repairs has been limited mainly because of porosity and low toughness in the resulting welds. With appropriate consumable design, however, it is possible to reduce porosity and to enhance weld material toughness through micro structural refinement. Hence, welding in offshore and marine application is an important area of research and needs considerable attention and understanding where, many problems are still unsolved. In the present review, a brief understanding of the problems in underwater welding vehicle will be discussed in context to the existing welding techniques. Detailed description of a few advanced welding techniques has also been made.

Laser Welding

Laser as a source of coherent and monochromatic radiation, has a wide scope of application in materials processing. Laser assisted welding, because of the sheer volume proportion of work and advancement over the years, constitutes the most important operations among the laser joining processes the front view of the schematic set up for laser underwater welding with a filler rod. The focused electron beam is made to irradiate the work piece or joint at the given level and speed. A shroud gas protects the weld pool from undue oxidation and provides with the required oxygen flow. Laser heating fuses the work piece or plate edges and joins once the beam is withdrawn. In case of welding with filler, melting is primarily confined to the feeding wire tip while a part of the substrate being irradiated melts to insure a smooth joint. In either case, the work piece rather than the beam travels at a rate conducive for welding and maintaining a minimum Heat Affected Zone (HAZ). There are two fundamental modes of laser welding depending on the beam power/configuration and its focus with respect to the work piece:

- Conduction welding
- keyhole or penetration welding

Conduction limited welding occurs when the beam is out of focus and power density is low insufficient to cause boiling at the given welding speed. In deep penetration or keyhole welding, there is sufficient energy/unit length to cause evaporation and hence, a hole forms in the melt pool. The 'keyhole' behaves like an optical black body in that the radiation enters the hole and is Underwater laser assisted welding compared to the other underwater welding methods has the advantages of low heat input, easy to transfer energy and control adaptability. The low heat input is of significance for reducing of the sensitivity of stainless steels to Stress Corrosion Cracking. Underwater LBW has not been used in application, however, because a series of problems have not been solved yet. The plasma induced by the interaction of the laser beam and the metal vapor or the shielding gas in CO_2 laser assisted welding has shielding effect on the laser beam, but the plume induced in YAG laser assisted welding has not such shielding effect on laser energy transferring. No matter CO2 or YAG laser assisted welding, the optical emissions induced in the welding process indicate the basic characteristics of the keyhole and the variation of welding parameters.

ULBW Compared to GTAW

The filler metal used for Underwater Laser Beam Welding (ULBW) is the same as that for Gas Tungsten Arc Welding (GTAW), and is selected for suitability for the application and base material being joined. It is introduced into the pool created by the laser beam in a manner very similar to that seen in during welding with a gas metal arc welding machine. However, ULBW is a completely automatic welding process. In this respect, ULBW differs from GTAW performed with a machine, where the underwater vehicle makes adjustments during welding.

Methods

When laser radiation interacts with the work piece, secondary radiation is generated. This radiation contains information about the process stability and can thus be used to detect process errors. The secondary radiation is recorded with one or several high speed CMOS cameras (with up to 40000 fps) a shown in figure 2. Thus a high temporal correlation between the recorded frames is achieved. In state of the art laser welding process control systems the temporal dependencies are not fully exploited. Therefore we propose the use of Hidden Markov Models (HMM) to be able to model the temporal dependencies and increase the classification performance.

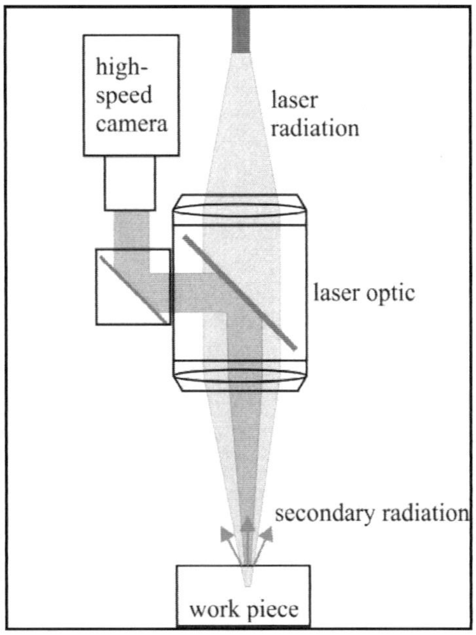

Figure 3. Method of laser beam welding

Laser Sensing

The use of robotics in arc welding applications serves the purposes of decreasing cycle time and improving the quality and repeatability of welds. This is only possible when the weld joints themselves are consistent and repeatable. Unfortunately, due to inadequate tooling and part preparation, weld joints often deviate greatly from the programmed weld path. The use of advanced tools such as TAST, Touch Sensing, 2D or 3D Vision Systems, and Laser Vision Tracking helps to minimize problems caused by part and tooling imperfections and allows for an expansion of welding opportunities that were previously impossible.

Touch Sensing

A low cost, software based, joint location sensing system is Touch Sensing. Touch Sensing works by using the welding electrode, nozzle, or other sensing pointer to make electrical contact with the part. The robot stores the position data and then makes adjustments automatically to the entire weld path before the weld start. Touch Sensing allows for adjustments in one, two, or three dimensions. One disadvantage to using Touch Sensing is the increase in cycle time due to performing the search routines. Added time typically ranges from 3–5 seconds per joint. In addition, for Touch Sensing to be most effective, the part must have defined edges for the sensor to find.

Another relatively low cost software based method used to track weld joints is Through Arc Seam Tracking (TAST). TAST uses welding current feedback to determine the vertical position of the torch. As contact tip to work distance increases, current gets smaller, and as the contact tip to work distance shortens, the amount of current increases. TAST corrects variations in the torch's vertical position to maintain a constant stick-out. Similarly, TAST uses welding current feedback and the weave function of the robot to determine the lateral position of the torch in the weld joint. At the center of the joint, arc current is at a minimum. As the torch reaches the edge of its weave cycle, arc current peaks. If the value of the peak current at the edge of the weave cycle increases, the torch is moving away from the weld joint, and TAST makes the necessary corrections.

While TAST offers a low cost solution to correcting for imperfect weld joints, the set-up of the system variables is challenging and requires a thorough understanding of the weld process. In addition, TAST cannot be used to weld aluminum, because the change in current due to change in stick out is not significant enough to track. Once the set-up is complete, the weld process must remain stable to prevent erroneous current feedback data from going back to the robot. TAST is best used in applications with material thickness greater than 2 mm and travel speeds between 35–50 inches (inches per minute).

2D and 3D Vision System

2D Vision Systems can be used in the same capacity as Touch Sensing for improving cycle time and quality. Vision is used for joint location tracking, robot guidance, and error proofing prior to establishing an arc. In robot guidance, for each part or weld seam, the operator calibrates the camera and teaches the weld path on an ideal part. This reference image is stored in the robot's memory. On each part thereafter, the camera takes a picture before an arc is established and the robot performs a pattern match between the reference image and the new image. The robot then calculates any offsets and adjusts the entire weld path accordingly.

Robo Pipe Torch While Touch Sensing can take 3-5 seconds to perform the same calculations for each joint, using a 2D camera reduces this time to tenths of a second. In addition, the use of a 2D camera makes it possible to track seams on material thinner than 1 mm since the robot does not have to actually touch the part to find the seam. To minimize the potential for error, the camera takes a picture to verify correct part orientation and presence of safety features. This error detection function either prevents an error in process or detects it before performing next operations. This eliminates the need for costly proximity sensors and reduces the risk of damaging tooling or the robot.

Lighting is critical in 2D Vision Systems. Any ambient light including sunlight, overhead lights, or other welding arcs within close proximity can distort the image and decrease the accuracy of the pattern match. This can make 2D Vision Systems cumbersome to use on shiny material such as aluminum. Also, the harsh environment created by a welding can hinder the camera's accuracy. If the camera is torch mounted, the lens must be protected from heat and spatter from the weld. 3D Vision Systems employ the same technology as 2D Vision Systems with the addition of a laser. While 2D Vision Systems detect part movement in the X and Y directions and part rotation, 3D Vision Systems add detection for changes in height, pitch, and yaw.

Laser Vision Tracking

Laser Vision Tracking provides a three-dimensional real time method for tracking weld seams on material thickness less than 1 mm or joints with less than 1 mm gap. The laser sensor is attached to the robot arm approximately one inch ahead of the welding torch. A laser diode projects a line onto the part, and the camera detects the joint seam based on the distortion in the laser line. The camera's software then provides feedback to the robot regarding gap and fit up variations. This data allows the robot to make necessary adjustments to torch position, travel speed, weave amplitude, and other weld parameters such as voltage and current. Laser Vision Tracking does not hamper overall travel speed, as weld speeds of 100ipm are still attainable. Since the laser operates on a single wavelength, it is not susceptible to interference from ambient light. Software algorithms that reduce noise and unwanted feedback also make it possible to use Laser Vision Tracking on shiny material such as aluminum and stainless steel. In addition, the camera is protected from weld spatter and fumes and intense heat by being mounted at the one inch look-ahead distance from the torch. The move from analog to digital technology greatly improved the capability of Laser Vision Tracking. Digital technology allows for the sensor to have a wider range and clearer signal. Compared to their analog counterparts, digital sensors are roughly four times more precise and have faster scanning rates. One of the main disadvantages still associated with Laser Vision Tracking. In addition, since the sensor is mounted to the torch, it can limit torch access in some applications.

For any robotic application, there are several available options to help improve weld quality and repeatability, including TAST, Touch Sensing, 2D Vision, and Laser Vision Tracking. These options have their own benefits and limitations, and only with a thorough evaluation of each application can a user determine which system can provide the most return on investment in terms of increased productivity and improved quality.

Robot Arm

The robot arms are normally stationary. In most cases, they are bolted at the shoulder to a specific position and the robot arm can move with great speed and accuracy to perform repetitive tasks such as spot welding and painting. Manipulator robots are very much unlike the autonomous mobile robots whereby the intelligence provided to them does not make them adapt to the environment in which they are. In most cases, most manipulator robots are capable of handling many end-effectors in order to increase the versatility of their use. These various end-effectors can be used for several purposes such as welding, painting, screwing and assembling. Although manipulator robots can be very versatile, they suffer from a fundamental disadvantage, which is lack of mobility. A fixed manipulator robot has a limited range of motion that depends on where it is bolted down, in contrast to a mobile robot that is capable of moving about.

Implementation of robot arm

In the design of systems, there are generally two methods of approach namely

- Top-down method
- Bottom-Up method

The top-down method is usually applied in designing a system from the scratch while the down-top method is used for reverse designing of an already existing system or functional design as in software engineering. We employed the top-down approach in the design of the robot arm project and how the various modules were integrated to arrive at the entire system. A block diagram model of the robot arm control is shown in figure 3.

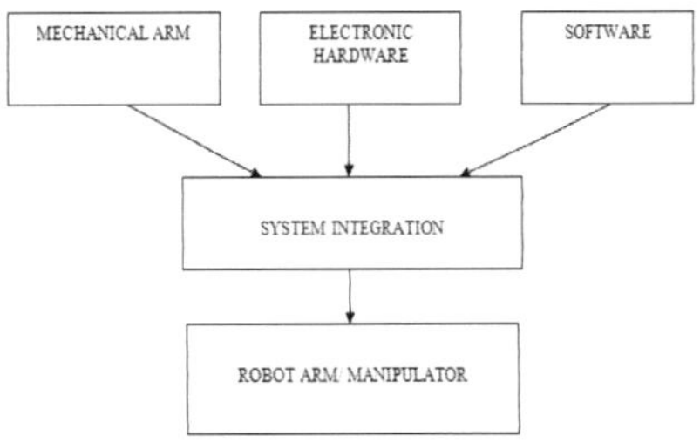

Figure 4 Robot arm control chart

The feedback sensors are optic-couplers located on the robot joints which send feedback information to the microcontroller on the robot's position. The actuators are the stepper and dc motors at each joint.

Hydrodynamic Modeling

There are many possible types of flow characteristics for two-link arm motion that can occur, depending on what flow regime the manipulator is operating in. This research has focused on the types of motion used when the manipulator is welding typically small acceleration slews over angular distances that are large enough that at least part of the manipulator has reached steady state. For these cases, the changes in added mass parameters are negligible and drag is the dominant hydrodynamic effect to be modeled. The average Reynolds number for each link was assumed to be within the range 100 < Re < 105. In this research, it was beneficial to use flow visualization to determine the important regions of the fluid flow to include in the model. A technique that has been used to analyze human swimming motion was modified for this purpose using small sections of heat shrink linked together in a chain and attached along each link. This provides a nearly neutrally buoyant stream behind the trailing edge which indicates what the dominant flow characteristics are. A video camera mounted above the manipulator captures the dynamic images for post analysis. See Figure 4, for a description of the notation used for referencing the 2 DOF manipulator and its motion.

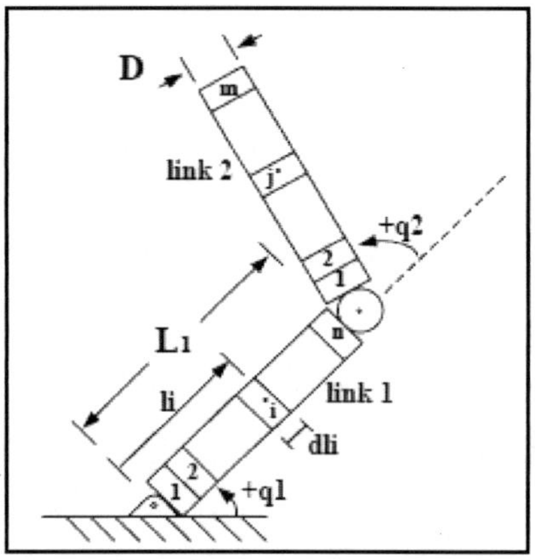

Figure 5. Notation for two link arm

For positive shoulder rotation, the flow visualization for various joint trajectories will be described for cases where the elbow joint is held at a fixed angle. When q2 is positive but still at small angles, the flow around the tip of the second link is increased, and there is still pressure gradient induced flow from the first link onto the second one toward the tip of link 2. As q2 increases, however, the fluid motion over the second link is in the direction of its free stream angle of attack, flowing from the tip of link 2 toward the elbow joint. This type of fluid motion is defined as geometric flow. Consequently, the flow up the first link and down the second link meets slightly below the elbow at the top of link 1. For negative values of q2, the geometric flow is in the same direction as the pressure gradient induced flow towards the tip of the second link. There is no flow around the tip instead, the flow bluffs of the end. When q2 is close to -90 degrees, the flow up from the tip of the first link appears to bounce of the elbow and back down onto itself instead of continuing up the second link.

CONCLUSION

The Welding process is implemented in underwater vehicle for marine application. It's controlled by the remotely operated and welded by using the optic camera path tracking. The improved methods presented with the goal of improving the model for underwater welding process with a two-link robotic arm for underwater vehicle. It is formulated for a two-link arm with a fixed elbow angle, includes separate drag coefficients on both links. Depending on the joint angles and trajectories, the flow can be dominated by various 3-D effects. The representation can be used both in the modeling of an arm vehicle system for preliminary analysis of control system designs, as well as control and planning algorithms for the arms themselves in order to achieve adequate endpoint placement.

ACKNOWLEDGEMENT

I am grateful to Dr.K.N.G.Reddy Director- centre for maritime research and AMET University for all the encouragement and excellent support received while working on this paper, and I thank to my family for motivation and support towards the research with me all the time.

REFERENCES

Karthik S, "Underwater vehicle for surveillance with navigation and swarm network communication", Indian journal of science and technology" Vol 7(S6), 22–31, October 2014.

Jyotsna Dutta Majumdar, "Underwater welding present status and future scope", Journal of Naval Architecture and Marine Engineering, June 2006.

R.T. Brown and K. Masubuchi, " Fundamental Research on Underwater Welding", Book from Effect of water environment on metallurgical structures of welds .

A. Anand and A. Khajuria, "Welding process in Marine application: A review", International journal of mechanical engineering and robotics research", January 2013.

Masataka et al, "Development of Underwater Laser Cladding and Underwater Laser seal Welding Techniques for Reactor Components", International Conference on Nuclear Engineering, July 2006.

David Hoffman et al, "Welding" book published on 2010,

David J.Hall, "Robotic Sensing Device", Department of electrical engineering, Carnegie-Mellon University.

Wei Xu et al, "Research on Laser Ranging System Based on Time Delay Estimation", Journal on Photonics and Optoelectronics" 2011.

X.Q.Chen et al, "Mobile Robots – Past Present and Future", Department of mechanical engineering, University of Canterbury, 2008.

SNAME
THE INTERNATIONAL COMMUNITY
FOR MARITIME AND OCEAN PROFESSIONALS

TIME DOMAIN SIMULATION OF LARGE AMPLITUDE MOTIONS IN SHALLOW WATER

Amitava Guha, Abhilash Somayajula and Jeffrey Falzarano

Marine Dynamics Laboratory, Department of Ocean Engineering, Texas A&M University, Texas, USA,77843-3136

Proceedings of the 21st Offshore Symposium, February 2016, Houston, Texas
Texas Section of the Society of Naval Architects and Marine Engineers

ABSTRACT

A finite depth 3D Green function has been developed to estimate wave loads on the floating body in frequency domain which is then coupled with a time domain motion simulation tool. Nonlinear hydrostatic and Froude-Krylov forces have been incorporated considering the instantaneous wetted surface and the nonlinear equation of motion is then solved considering the large-amplitude rotations of the body. However, the option of using a consistent second order forcing model is also retained in the code where the linear diffraction forces and moments are augmented with the drift forces calculated from Newman approximation of difference frequency quadratic transfer functions (QTF).

The main advantage of the method lies in its capability to simulate large-amplitude motions of ship shaped and non-ship shaped vessels with or without mooring and with or without forward speed in deep or shallow water depths. This allows investigation of various nonlinear effects and ensures safe operation. It is particularly useful in modelling the motions of a ship in a port channel or towing of offshore platforms from fabrication yard to deep water installation site.

In this article, a methodical approach in validating the numerical results against published and other established numerical programs is presented for the developed finite depth Green function and hydrodynamic coefficient predictions. Also the drift forces and moments obtained by different forms of the Newman approximation are compared and validated against commercially available software. The ability of the time domain simulation program to capture nonlinearities is also shown by simulating the Mathieu type instability in heave and pitch modes of a Single Column Floater.

Keywords: Time domain, Shallow Water, Forward Speed, Large Amplitude, Hydrodynamics, Motion Simulation, Parametric Excitation, Single Column Floater

1 INTRODUCTION

Hydrodynamic wave excitation force and motion prediction using potential theory is a common practice in the marine and offshore industry. With the growing interest in renewable resources, the offshore wind and hydro kinetic energy extraction methods are being engineered applying similar techniques. Consideration of the finite depth effect and nonlinear forces are of critical importance in designing mooring systems and determining safe operational conditions which directly impacts the downtime.

A 3D Green function based potential theory method has been developed to obtain hydrodynamic coefficients and motion in frequency domain (Guha, 2012; Guha and Falzarano, 2015a). This method facilitates a quick analysis using discretized wetted surface of a vessel without modeling the water surface or other boundaries. A new finite depth Green function is developed and incorporated in this potential code to assess the effect of water depth on hydrodynamic coefficients. As the computational time for a numerical finite depth Green function has significant effect on the frequency domain analysis time, efficient algorithms bases on Newman (1990), Linton (1999), Pidcock (1985), Li (2001) and Liu et al. (2015) are implemented. In addition, the forward speed approximations using encounter frequency is also incorporated to simulate vessels traveling with nonzero forward speed in finite water depth.

Based on the frequency domain tool a time domain program has been developed to simulate the nonlinear equations of motion of a generalized floater. The developed tool includes the calculation of the nonlinear force vector including the nonlinear Froude-Krylov and hydrostatic forces and also is capable of handling large amplitudes of motion using an Euler angle consideration. In this paper two approaches of calculating the nonlinear force vector are described and their relative advantages and limitations are highlighted. The paper also demonstrates the application of both methods to two examples to demonstrate the importance of both methods. The first method is the calculation of nonlinear Froude-Krylov and hydrostatic force vector by integrating the pressure over the instantaneous underwater area under the incident wave. The second method is the more common approach of consideration of 2nd order forces using a quadratic transfer function (QTF) which is consistent upto second order. The second method is applied to the KCS container ship and its results are validated against the results from a commercially available software. Later the significant advantage of the first method is demonstrated by simulating the Mathieu type instability for the single column floater (SCF) which otherwise cannot be captured using a standard second order analysis. The SCF was originally suggested as a better alternative to classic spar (Huang et al., 2000) due to the advantages of wet towing to site and not requiring an upending for installation. However, it was later found that the Mathieu type instability was a major concern and the project was abandoned (Haslum, 2004). To avoid such scenarios, automated optimization techniques using evolutionary algorithms (Guha and Falzarano, 2015c) may be used to mitigate instability in the early design stage.

The time domain simulations also allow for a further statistical analysis which can help predict the extreme responses (Guha et al., 2013) and characterize its non-Gaussian probabilistic nature (Somayajula and Falzarano, 2014b).

2 FORMULATION

2.1 Estimation of wave excitation forces in finite water depth with or without forward speed

The fluid structure interaction forces on a floating body in a random sea is predicted by means of the principle of superposition. The random or irregular waves can be represented as a sum of constitutive regular waves and the linear forces on the floating body due to each regular wave can be summed together to obtain the total hydrodynamic force on the body. This leads to the frequency domain potential flow analysis where the wave induced forces are obtained for particular wave frequencies in an ideal fluid (incompressible, inviscid and irrotational).

To formulate the boundary value problem, a body (floating or fully submerged) is considered to be traveling with steady forward speed U or equivalently fixed in a current of equal magnitude and opposite direction. Regular wave of amplitude A, incidence frequency ω_I with heading angle β traveling in uniform water depth h is considered. The effect of finite water depth on wave frequency is expressed in terms of the dispersion relation $\omega^2 = kg\tanh(kh)$, where k is the wave number. An efficient code to solve the dispersion relation applying higher order iterative technique following Newman (1990) is developed.

The forward speed changes the frequency at which the body encounter the waves (Doppler effect), which is expressed as the encounter frequency $\omega_e = \omega_I - \frac{\omega_I^2}{g}U\cos\beta$. The unsteady fluid potential around the body is given by:

$$\Phi(\mathbf{x},t) = [-Ux+\phi_S(\mathbf{x})] + \left[\phi_I(\mathbf{x},\beta,\omega_I)+\phi_D(\mathbf{x},\beta,\omega_I)+\sum_{j=1}^{6}\eta_j\phi_j(\mathbf{x},U,\omega_e)\right]e^{i\omega_e t} \tag{2.1}$$

where ϕ_S is the steady disturbance potential due to forward motion of the body in calm water. ϕ_I is the potential due to incident wave, ϕ_D is the diffraction potential and ϕ_j is the radiation potential. η_j is the motion amplitude of the body in 6 degrees of freedom. Continuity condition in the fluid domain gives the Laplace equation:

$$\nabla^2\Phi = 0 \tag{2.2}$$

where the potential can be solved by applying following boundary conditions:

1. Combined dynamic and kinematic free surface boundary condition: $\left[\left(i\omega_e - U\frac{\partial}{\partial x}\right)^2 + g\frac{\partial}{\partial z}\right](\phi_I,\phi_D,\phi_j) = 0$ on $z=0$
2. The bottom boundary condition: $\frac{\partial}{\partial z}(\phi_I,\phi_D,\phi_j) = 0$ on $z=-h$
3. The Sommerfeld radiation condition: $\lim_{kr\to\infty}\sqrt{kr}\left(\frac{\partial}{\partial r} - ik\right)(\phi-\phi_I) = 0$
4. The body surface no-penetration boundary condition:
   (a) Radiation : $\frac{\partial\phi_j}{\partial n} = i\omega_e n_j + Um_j$ on S
   (b) Diffraction: $\frac{\partial\phi_I}{\partial n} + \frac{\partial\phi_D}{\partial n} = 0$ on S

where simplified $m_j \approx (0,0,0,0,n_3,-n_2)$ terms with body normal $\mathbf{n} = (n_1,n_2,n_3)$ are applied. The linear incident wave potential satisfying above boundary conditions is given by:

$$\phi_I = \frac{igA}{\omega_I}\frac{\cosh(k_I(z+h))}{\cosh(k_I h)}e^{-ik_I(x\cos\beta + y\sin\beta)} \tag{2.3}$$

The radiation potential in forward speed can be obtained using the zero speed potential as:

$$\phi_j = \phi_j^0 \qquad \text{for } j = 1,2,3,4 \qquad\qquad \phi_5 = \phi_5^0 + \frac{U}{i\omega_e}\phi_3^0 \qquad\qquad \phi_6 = \phi_6^0 - \frac{U}{i\omega_e}\phi_2^0 \tag{2.4}$$

Analytical solution for the radiation potential ϕ_j and the diffraction potential ϕ_D are generally not available for arbitrary shaped bodies, which necessitates the use of numerical techniques such as the source distribution method.

2.1.1 The source distribution method

To obtain the diffraction and radiation velocity potentials, the body surface is discretized into N quadrilateral panels of area ΔS_j and an uniform source of unknown strength is distributed uniformly over each panel. The zero speed velocity potential can be written in terms of the unknown source strength σ as:

$$\phi^0 = \frac{1}{4\pi}\int_S \sigma(\mathbf{x}_s)G(\mathbf{x},\mathbf{x}_s)ds \tag{2.5}$$

where $\mathbf{x}$ is the field point where the potential is being evaluated due to a source at $\mathbf{x_s}$. Here, $G(\mathbf{x},\mathbf{x}_s)$ is the finite depth Green function which satisfies the free surface, bottom and radiation boundary conditions. Applying the body boundary condition gives:

$$-\frac{1}{2}\sigma_i + \frac{1}{4\pi}\sum_{j=1}^{N}\sigma(\mathbf{x}_s)\frac{\partial G}{\partial n}(\mathbf{x},\mathbf{x}_s)\Delta S_j = \begin{cases} i\omega_e n_j & : \text{Radiation} \\ -\dfrac{\partial \phi_I}{\partial n} & : \text{Diffraction} \end{cases} \tag{2.6}$$

This system of linear equations are solved using efficient matrix inversion algorithms to obtain the source strength σ_i on each panel.

2.1.2 The finite depth Green function

In obtaining the source strength on each panel of the body, the most computational burden is because of the Green function evaluation. For a body with N panels, the Green function must be evaluated for at least $(N-1)\times(N-1)$ times and 4 times this more in the case of a higher order integration method such as Gauss quadrature, which consumes a large portion of the CPU time. To overcome this, an efficient finite depth Green function has been developed. The analytical form of the finite depth Green function is given in John (1949, 1950) which is also summarized in Wehausen and Laitone (1960). In terms of numerical stability and the computational efficiency the finite depth Green function is divided into two separate functions which are used in the domain $R/h < 0.5$ and $R/h > 0.5$, where $R = \left[(x-\xi)^2 + (y-\eta)^2\right]^{1/2}$ is the horizontal distance from the source $q(\xi,\eta,\zeta)$ to the field point $p(x,y,z)$.

Integral form of the Green function

The integral form of the Green function is suitable for $R/h < 0.5$ and found to be unstable for $R/h > 7$. The analytical form of this function is given by Wehausen and Laitone (1960) as:

$$\begin{aligned} G(p;q) =& \frac{1}{r} + \frac{1}{r^*} \\ &+ 2PV\int_0^\infty \frac{(\mu+K)e^{-\mu h}\cosh(\mu(\zeta+h))cosh(\mu(z+h))}{\mu\sinh(\mu h) - K\cosh(\mu h)}J_0(\mu R)d\mu \\ &+ i\frac{2\pi(k+K)e^{-kh}\sinh(kh)\cosh(k(\zeta+h))\cosh(k(z+h))}{Kh + \sinh^2(kh)}J_0(kR) \end{aligned} \tag{2.7}$$

where

$$\begin{aligned} K =& \frac{\omega^2}{g} = k\tanh(kh) & r =& [(x-\xi)^2 + (y-\eta)^2 + (z-\zeta)^2]^{\frac{1}{2}} \\ r^* =& [(x-\xi)^2 + (y-\eta)^2 + (z+2h+\zeta)^2]^{\frac{1}{2}} & R =& [(x-\xi)^2 + (y-\eta)^2]^{\frac{1}{2}} \end{aligned} \tag{2.8}$$

The solution of the Rankine part for both source and the image source is obtained using the method described by Hess and Smith (1964). The full numerical implementation details for this is given in Guha (2012). The imaginary part of the Green function can be directly calculated. The numerical difficulty is presented mostly by the principal value integral part of the Green function. Using method of substitution the singularity of the function is removed and a form of that can be numerically integrated is obtained. The Gauss-Laguerre quadrature is implemented to obtain the integral value efficiently.

Series form of the Green function

The series form of the Green function is applicable for the rest of the domain $(R/h > 0.5)$. John (1949, 1950) derived the infinite series expansion for the finite depth Green function as:

$$\begin{aligned} G(p;q) =& 2\pi \frac{K^2-k^2}{k^2h-K^2h+K}\cosh(k(z+h))\cosh(k(\zeta+h))\,[Y_0(kR)+iJ_0(kR)]] \\ &+4\sum_{n=1}^{\infty}\frac{k_n^2+K^2}{k_n^2h+K^2h-K}\cos(k_n(z+h))\cos(k_n(\zeta+h))K(k_nR) \end{aligned} \tag{2.9}$$

where k_n denotes the set of corresponding positive real roots of equation: $k_n\tanh(k_nh) = -K$. Again, the function can be broken into real and imaginary parts. The imaginary part can be rewritten using the relation given by Wehausen and Laitone (1960) to be equal to the imaginary part of the integral form of the Green function which can be calculated directly. The real part of the Green function has a singularity at $R/h = 0$ and takes a significantly large number of terms in the infinite sum to give a converged solution. However, for the domain $R/h > 0.5$ a six decimal accuracy can be achieved by using approximately $6h/R$ number of terms. To be consistent, the algorithm is developed to iterate until a 7 significant digit accuracy is achieved.

A number of comparisons are made with published data to establish the validity of the code. Figure. 1 shows for the case $Kh = 5$ the comparison between the series and integral Green function with that of Li (2001) and Monacella (1966).

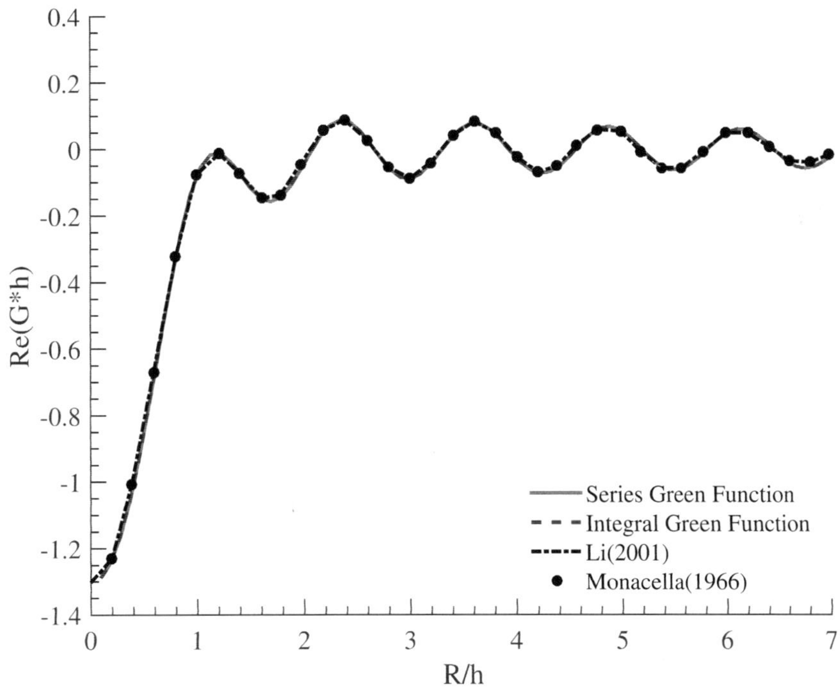

Figure. 1. The real part of Gh when $Kh = 5.0$, $kh = 5.000454$ and $0 < R/h < 7$

2.1.3 Forces and motions of the body

The forces and moments on the body are obtained by integrating the pressure over the submerged body surface. The pressure is obtained from the velocity potential by applying the Bernoulli's equation

$$P = \frac{1}{2}\rho U^2 - \rho\frac{\partial\Phi}{\partial t} - \frac{1}{2}\rho|\nabla\Phi|^2 - \rho g z \tag{2.10}$$

The hydrodynamic force can be obtained by:

$$F_{Hj} = -\int_S P n_j ds \qquad j = 1,2,\ldots,6 \tag{2.11}$$

The added mass and damping is obtained by integrating the radiation pressure over the hull as:

$$A^0_{jk} = -\frac{\rho}{\omega_e}\int_S Im(\phi_k)n_j ds \qquad B^0_{jk} = -\rho\int_S Re(\phi_k)n_j ds \tag{2.12}$$

followed by forward speed corrections as described in Guha and Falzarano (2015a). The wave excitation force due to incident wave also known as the Froude-Krylov force is obtained as:

$$F_I = i\omega_I\rho\int_S \phi_I n_j ds \tag{2.13}$$

and the force due to diffracted wave is:

$$\begin{aligned}
F_D &= \rho\int_S (i\omega_e n_j - U m_j)\phi_D ds \\
&= -\rho\int_S \phi^0_j \frac{\partial\phi_I}{\partial n} ds \qquad \text{for } j = 1,2,3,4 \\
&= -\rho\int_S \phi^0_j \frac{\partial\phi_I}{\partial n} ds + \frac{\rho U}{i\omega_e}\int_S \phi^0_3 \frac{\partial\phi_I}{\partial n} ds \qquad \text{for } j = 5 \\
&= -\rho\int_S \phi^0_j \frac{\partial\phi_I}{\partial n} ds - \frac{\rho U}{i\omega_e}\int_S \phi^0_2 \frac{\partial\phi_I}{\partial n} ds \qquad \text{for } j = 6
\end{aligned} \tag{2.14}$$

The calculated added mass, damping and wave excitation forces are used to solve the equation of motion to get the vessel response as:

$$\sum_{k=1}^{6}[-\omega_e^2(M_{jk}+A_{jk}) + i\omega_e B_{jk} + C_{jk}]\eta_k = F^I_j + F^D_j \qquad j = 1,2,\ldots,6 \tag{2.15}$$

where M_{jk} is the mass matrix, C_{jk} is the hydrostatic stiffness matrix and η_k the vessel response in k^{th} mode of motion.

2.2 Time domain simulation considering large amplitude motion

The equations of motion describing a floating body are in general nonlinear and time varying. The fully nonlinear equations of motion of a general body have been derived in a variety of references including Abkowitz (1969), Lewandowski (2004), Somayajula and Falzarano (2015a) etc. In the vectorized form the equations of motion are given by

$$m[\ddot{\boldsymbol{\xi}} + \dot{\boldsymbol{\omega}} \times (\boldsymbol{x_G} - \boldsymbol{\xi}) + \boldsymbol{\omega} \times [\boldsymbol{\omega} \times (\boldsymbol{x_G} - \boldsymbol{\xi})]] = \boldsymbol{F} \tag{2.16}$$

$$I\dot{\boldsymbol{\omega}}' + \boldsymbol{\omega}' \times I\boldsymbol{\omega}' = R[\boldsymbol{M} - \boldsymbol{x_G} \times \boldsymbol{F}] \tag{2.17}$$

where m is the mass of the vessel, I is the moment of inertia matrix, $\boldsymbol{\xi}$ is the 3×1 vector representing the translational displacement vector, $\boldsymbol{x}_G$ represents the position vector of center of gravity in the inertial frame of reference, $\boldsymbol{\omega}$ and $\boldsymbol{\omega}'$ represent the angular velocity in the inertial frame of reference and body fixed frame respectively. $\boldsymbol{F}$ and $\boldsymbol{M}$ denote the external force and moment applied to the ship system. The detailed derivation of the equations of motion can be found in Somayajula and Falzarano (2015b). However, if nonlinearities due to inertial coupling are neglected, then theses equations can be reduced to the form shown in (2.15) for the case of a single regular incident wave.

In order to solve for the nonlinear equations of motion for corresponding motions, the nonlinear force and moment vectors need to be evaluated. The external forces and moments can be combined into a single 6×1 vector F and can be expressed as the sum of forces due to incident Froude Krylov $\{F_I\}$, scattering $\{F_S\}$, radiation $\{F_{Rad}\}$, viscous $\{F_V\}$ and restoring forces $\{F_{Res}\}$ as shown in (2.18).

$$\{F\} = \begin{Bmatrix} \boldsymbol{F} \\ \boldsymbol{M} \end{Bmatrix} = \{F_I\} + \{F_S\} + \{F_{Rad}\} + \{F_V\} + \{F_{Res}\} \tag{2.18}$$

The scattering and radiation forces are assumed to be linear and obtained using the frequency domain formulation described in section 2.1. While the scattering forces in the frequency domain are transformed into time domain using a Fourier transform, the radiation forces are more involved. In the frequency domain the radiation forces are expressed in terms of the frequency dependent added mass and radiation damping and are given by (2.19).

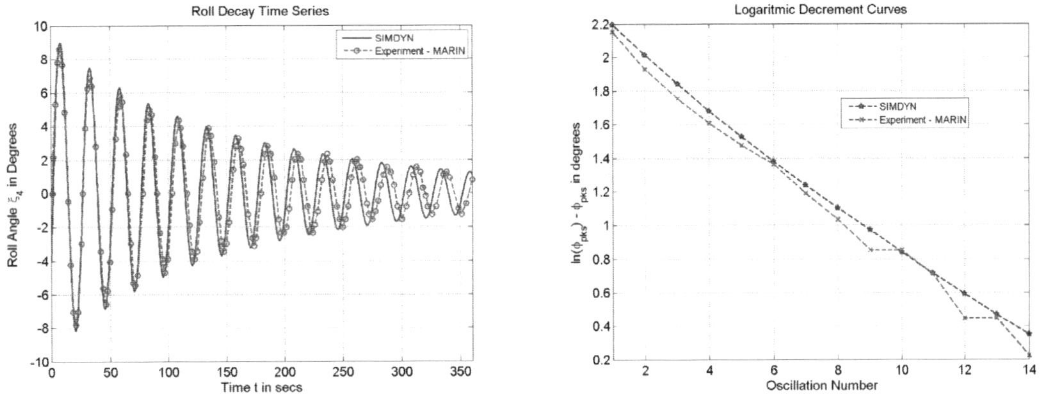

(a) Roll Decay Test (b) Roll Decrement Curve

Figure. 2. Comparison of roll decay test from numerical simulation with experiments

$$\{F_{Rad}\} = -[A(\omega)]\{\ddot{\xi}\} - [B(\omega)]\{\dot{\xi}\} \tag{2.19}$$

However, in the time domain they are represented as a convolution integral given by (2.20) where $[K(\tau)]$ represents the 6×6 impulse response function (IRF) matrix. The IRFs $[K(\tau)]$ are related to the frequency dependent radiation damping $[B(\omega)]$ and can be obtained using (2.21). For more details on the numerical implementation of IRF calculation and calculation of the convolution integral please refer to Somayajula (2015).

$$\{F_{Rad}\} = -[A(\infty)]\{\ddot{\xi}\} - [B(\infty)]\{\dot{\xi}\} - \int_{-\infty}^{t} [K(t-\tau)]\{\dot{\xi}(\tau)\} d\tau \tag{2.20}$$

$$[K(\tau)] = \frac{2}{\pi} \int_{0}^{\infty} [B(\omega) - B(\infty)] \cos(\omega\tau) d\omega \tag{2.21}$$

While the scattering and radiation forces are considered to be linear, the developed simulation tool allows for the consideration of both linear and nonlinear Froude-Krylov and hydrostatics depending on user's choice. Two approaches are implemented here for considering nonlinearities due to Froude Krylov and hydrostatics. The first approach involves using the standard stretching formulation proposed by Wheeler (1970) where the linear wave kinematics are stretched to the instantaneous incident wave elevation and the corresponding pressure due to the stretched kinematics is integrated around the hull to give the nonlinear force vector at each time step (Somayajula and Falzarano, 2015a,b). This approach is necessary for simulating phenomenon which violate the assumption of small amplitude of motion utilized by linear theory, for example parametric excitation described in section 4.

The second approach is to use a consistent second order potential theory, which result in the capture of sum and difference frequency forces arising from a bi-chromatic incident waves. The resulting second order forces and moments are usually an order of magnitude smaller than the linear theory forces and moments and play a significant role when the natural frequency of the vessel being analyzed is outside the wave frequency zone (which is a typical case for offshore structures). Details of calculating the second order drift forces are discussed in section 2.3. It is however, important to note that including nonlinear Froude-Krylov and hydrostatic forces due to Wheeler stretching should not be used in conjunction with second order drift forces as it results in double counting the non-linearities.

The viscous damping force is obtained by the standard approach suggested by Ikeda et al. (1978) which separates the effective linear damping into its components - skin friction damping, wave damping, eddy damping, lift damping and bilge keel damping. Each of these components except the wave damping (obtained exactly from frequency domain analysis) are calculated using empirical relations. The empirical expressions are well documented in Falzarano et al. (2015). Figure. 2 shows a comparison of the roll decay test obtained from numerical simulation using empirical model with the experiments performed at MARIN for the Pram hull form. The comparison demonstrates that the empirical predictions work fairly well for a general ship shaped structure.

Once all the components of the force vector are evaluated, the differential equations (2.16) and (2.17) governing the of motion are solved using a 4th order Runge-Kutta numerical integration scheme to obtain the nonlinear motions.

2.3 Calculation of the second order mean drift forces and the Quadratic Transfer Functions

The influence of second order mean drift forces are proven to be of critical importance in the prediction of stability criteria for floating bodies as well in determining the integrity of the mooring system design. Numerical prediction of the mean drift forces rely primarily upon two methods. First, the far field method proposed by Maruo (1960) which is based on diffracted and radiated wave energy and the momentum flux at infinity. Second is the near field method first proposed by Boese (1970) and later modified by Pinkster (1979) and Faltinsen et al. (1980). This method is more intuitive and gives forces in all 6DOF compared to the former which can only provide forces in 3DOF. In the near field method, a perturbation approach is applied to differentiate zeroth, first and second order quantities and the forces and moments are simply obtained by integrating the hydrodynamic pressure on the body surface.

2.3.1 Mean drift forces

The velocity potential ϕ, wave elevation ζ, relative wave elevation ζ_r, body motion amplitude η and the pressure p are perturbed using a small parameter ε of the order of the wave slope assuming small amplitude oscillation of the body about a mean position.

$$\begin{aligned} &\phi = \varepsilon\phi^{(1)} + \varepsilon^2\phi^{(2)} + \dots \qquad &&\zeta = \varepsilon\zeta^{(1)} + \varepsilon^2\zeta^{(2)} + \dots \qquad &&\zeta_r = \varepsilon\zeta_r^{(1)} + \varepsilon^2\zeta_r^{(2)} + \dots \\ &\eta = \varepsilon\eta^{(1)} + \varepsilon^2\eta^{(2)} + \dots \qquad &&p = p^{(0)} + \varepsilon p^{(1)} + \varepsilon^2 p^{(2)} + \dots \end{aligned} \tag{2.22}$$

Substituting above expressions into the Bernoulli's equation (2.10) gives the expression for zeroth, first and second order pressures. Similarly, the perturbation of the hydrodynamic force in equation (2.11) give:

$$\mathbf{F} = -\left(\int_{S_0} ds + \int_{wl} \zeta_r dl\right)\left(p^{(0)} + \varepsilon p^{(1)} + \varepsilon^2 p^{(2)}\right)\left(\mathbf{n}^{(0)} + \varepsilon(\theta^{(1)} \times \mathbf{n}^{(0)}) + \varepsilon^2 H\mathbf{n}^{(0)}\right) \tag{2.23}$$

Separating terms with ε^2 gives the second order force equation as:

$$\begin{aligned} \mathbf{F}^{(2)} = &-\int_{wl} \frac{1}{2}\rho g \left(\zeta_r^{(1)}\right)^2 \frac{\mathbf{n}^{(0)}}{\sqrt{1-n_3^2}} dl + \int_{S_0} \rho \left(\frac{\partial\phi^{(2)}}{\partial t} - U\frac{\partial\phi^{(2)}}{\partial x}\right)\mathbf{n}^{(0)} ds \\ &+ \int_{S_0} \frac{\rho}{2}\left\{\left(\frac{\partial\phi^{(1)}}{\partial x}\right)^2 + \left(\frac{\partial\phi^{(1)}}{\partial y}\right)^2 + \left(\frac{\partial\phi^{(1)}}{\partial z}\right)^2\right\}\mathbf{n}^{(0)} ds \\ &+ \int_{S_0} i\omega_e\rho \left\{(\eta_1 - \eta_6 y_B + \eta_5 z_B)\frac{\partial\phi^{(1)}}{\partial x}\right. \\ &\left. + (\eta_2 + \eta_6 x_B - \eta_4 z_B)\frac{\partial\phi^{(1)}}{\partial y} + (\eta_3 - \eta_5 x_B + \eta_4 y_B)\frac{\partial\phi^{(1)}}{\partial z}\right\}\mathbf{n}^{(0)} ds \\ &- \rho g A^{(0)}\left[\eta_4\eta_6 x_{B,f} + \eta_5\eta_6 y_{B,f} + \frac{1}{2}(\eta_4^2 + \eta_5^2)Z_0\right]\hat{k} \\ &- \omega_e^2\left\{-\eta_2\eta_6 m + \eta_4\eta_6 m z_g - \eta_6\eta_6 m x_g + \eta_3\eta_5 m + \eta_4\eta_5 y_g - \eta_5\eta_5 m x_g\right\}\hat{\imath} \\ &- \omega_e^2\left\{\eta_1\eta_6 m + \eta_5\eta_6 m z_g - \eta_6\eta_6 m y_g - \eta_3\eta_4 m - \eta_4\eta_4 m y_g + \eta_4\eta_5 m x_g\right\}\hat{\jmath} \\ &- \omega_e^2\left\{-\eta_1\eta_5 m - \eta_5\eta_5 m z_g + \eta_5\eta_6 m y_g + \eta_2\eta_4 m - \eta_4\eta_4 m z_g + \eta_4\eta_6 m x_g\right\}\hat{k} \end{aligned} \tag{2.24}$$

Taking a time average over one wave period cancels the term with the second order potential and the rest of the terms can be evaluated from the linear potential theory. Similarly the second order moment terms can be obtained, but excluded from here for brevity (See Guha (2016)). These forces are known as the mean drift forces.

2.3.2 Quadratic Transfer Functions

The second order force and moment obtained from perturbation theory accommodate interaction between two wave trains with different amplitudes and frequencies. The interaction between two regular waves (bi-chromatic waves) gives rise to forces and moments at difference and sum frequencies. The total second order force (given by (2.24)) and moment separated into components due to the sum and difference frequency effects as represented below.

$$\boldsymbol{F}^{(2)}(\omega_m, \omega_n) = \boldsymbol{F}^{(2d)}(\omega_m, \omega_n) + \boldsymbol{F}^{(2s)}(\omega_m, \omega_n) \tag{2.25}$$

(2.26)

where $\boldsymbol{F}^{(2s)}(\omega_m, \omega_n)$ and $\boldsymbol{F}^{(2d)}(\omega_m, \omega_n)$ are the sum and difference frequency components of the second order force respectively. If the two regular wave trains in a bi-chromatic wave have amplitudes A_m and A_n, a set of second order transfer functions can be defined as shown below. These transfer functions are also referred to as quadratic transfer functions (QTFs).

$$Q_i^{(2d)}(\omega_m, \omega_n) = \frac{F_i^{(2d)}(\omega_m, \omega_n)}{A_m A_n} \qquad Q_i^{(2s)}(\omega_m, \omega_n) = \frac{F_i^{(2s)}(\omega_m, \omega_n)}{A_m A_n} \qquad \text{for } i = 1,2,3 \tag{2.27}$$

$$Q_i^{(2d)}(\omega_m, \omega_n) = \frac{M_{i-3}^{(2d)}(\omega_m, \omega_n)}{A_m A_n} \qquad Q_i^{(2s)}(\omega_m, \omega_n) = \frac{M_{i-3}^{(2s)}(\omega_m, \omega_n)}{A_m A_n} \qquad \text{for } i = 4,5,6 \tag{2.28}$$

For most of the offshore structures (except TLPs) such as spars, semi-submersibles and FPSOs, the difference frequency effects are significantly important than the sum frequency effects. It is practically observed that the largest contribution to the second order forces usually comes from frequency pairs which are close by ($\omega_m \approx \omega_n$). For such cases, Newman suggested an approximation to calculate the difference frequency QTF by using just the mean drift coefficients. The original approximation suggested by Newman is given by:

$$Q^{(2d)}(\omega_m, \omega_n) = Q^{(2d)}(\omega_m, \omega_m) \tag{2.29}$$

However, other forms of Newman approximation which have become quite accepted include the arithmetic mean and geometric mean models shown in (2.30) and (2.31) respectively.

$$Q^{(2d)}(\omega_m, \omega_n) = \frac{1}{2}\left(Q^{(2d)}(\omega_m, \omega_m) + Q^{(2d)}(\omega_n, \omega_n)\right) \tag{2.30}$$

$$Q^{(2d)}(\omega_m, \omega_n) = \text{sign}(Q^{(2d)})\sqrt{Q^{(2d)}(\omega_m, \omega_m)Q^{(2d)}(\omega_n, \omega_n)} \tag{2.31}$$

The second order force in time domain is obtained by taking a double Fourier transform of the product of QTF with the pairs of harmonic amplitudes of the wave train and is shown in (2.32).

$$F_i^{(2d)}(t) = \text{Re}\left\{\sum_{m=1}^{N}\sum_{n=1}^{N} A_m A_n^* Q_i^{(2d)}(\omega_m, \omega_n) e^{i(\omega_m - \omega_n)t}\right\} \tag{2.32}$$

where A_m and A_n represent the amplitudes of the mth and n^{th} components of an irregular wave train given by (2.33).

$$\eta(t) = \text{Re}\left\{\sum_{m=1}^{N} A_m e^{i\omega_m t}\right\} \tag{2.33}$$

However, there are more efficient ways to implement the above formulation numerically where instead of taking a double Fourier transform, the calculation can be reduced to a single Fourier transform over a series sum (Duarte and Sarmento, 2014; Somayajula, 2015).

3 VALIDATIONS

3.1 Finite depth calculations in frequency domain

The new finite depth Green function developed is incorporated in the frequency domain tool MDLHydroD (Guha (2012); Guha and Falzarano (2013, 2015a)). This enabled the program to be able to predict hydrodynamic loads and corresponding motion for submerged or floating bodies in deep and finite water depth with or without forward speed. The unique capability of consideration of the hull emergence angle in prediction of added resistance (Somayajula et al. (2014); Guha and Falzarano (2015b)) also becomes available for the finite waterdepth condition. A number of validation studies have been performed comparing calculated results with experimental as well as other commercially available numerical tools and found to be in good agreement.

Here, a truncated cylinder of radius 1m and draft 0.5m is considered. The water depth of 1m corresponds to the depth to draft ratio(h/T) of 2 is chosen. The vertical center of gravity is set at $z = 0$ and the radius of gyrations in roll, pitch and yaw are considered to be 1m. The frequency domain results are obtained for wave heading 150 degrees and zero forward speed. Figure. 3 shows the panel model of the cylinder. The calculated results are compared with an industry standard program (Lee (2013)). Figure. 4 - 7 shows the comparison of the response amplitude operators for surge, heave, roll and pitch mode of motions and Figure. 8 shows the

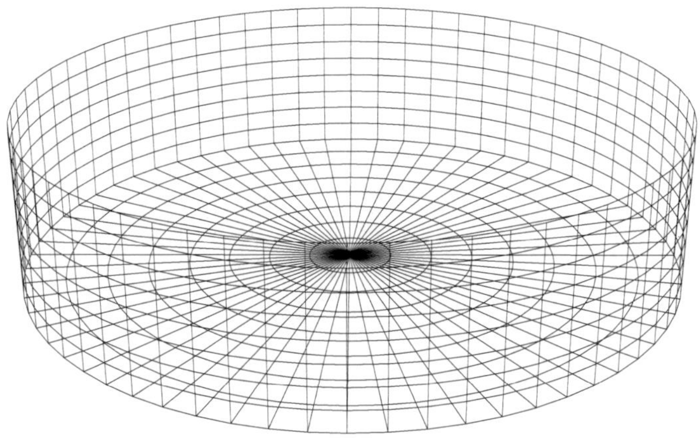

Figure. 3. Panel model of the cylinder

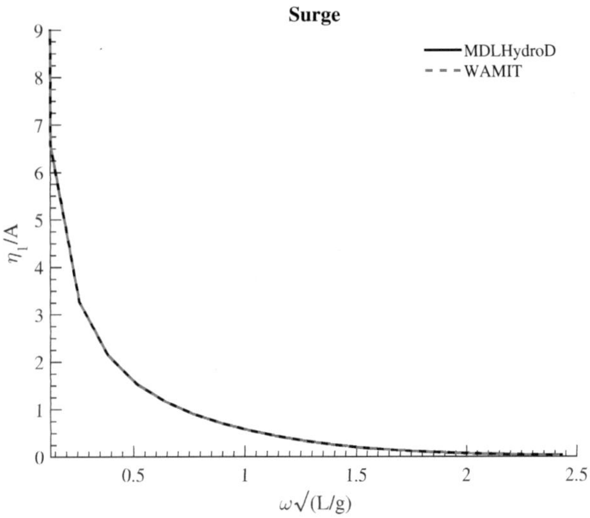

(a) Surge RAO amplitude

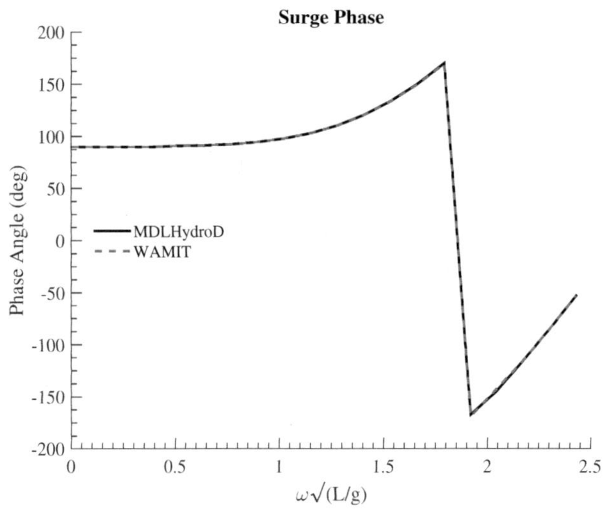

(b) Surge RAO phase

Figure. 4. Surge RAO comparison for cylinder at $Fn = 0$ and $h/T = 2$

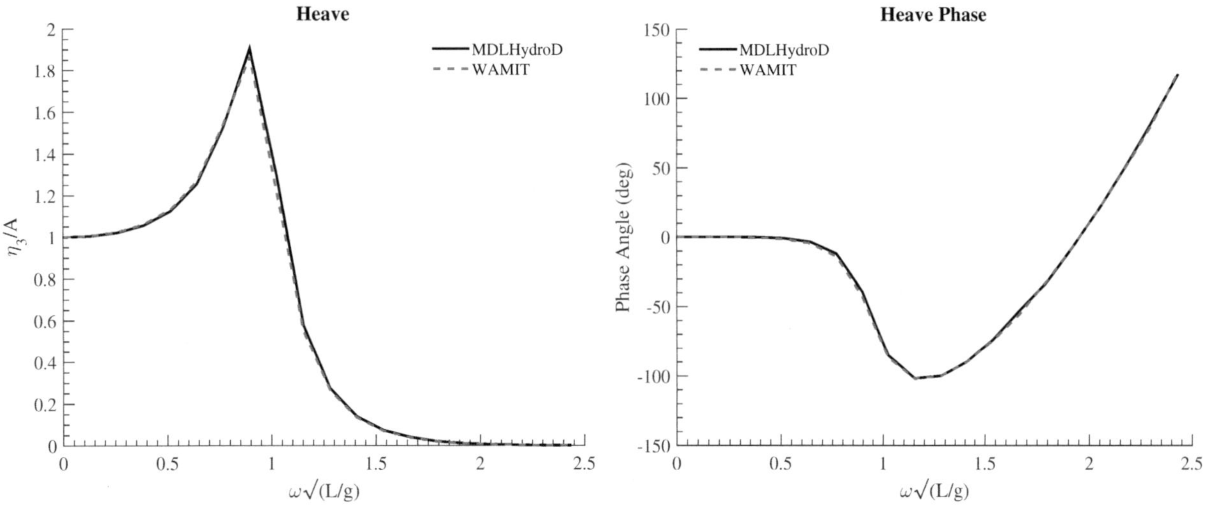

(a) Heave RAO amplitude (b) Heave RAO amplitude

Figure. 5. Heave RAO comparison for cylinder at $Fn = 0$ and $h/T = 2$

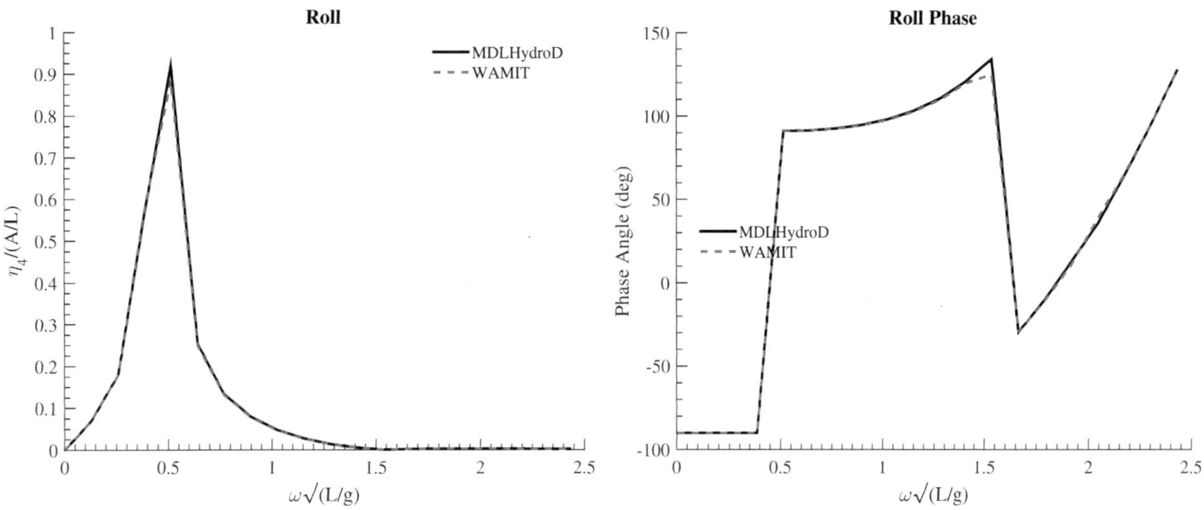

(a) Roll RAO amplitude (b) Roll RAO phase

Figure. 6. Roll RAO comparison for cylinder at $Fn = 0$ and $h/T = 2$

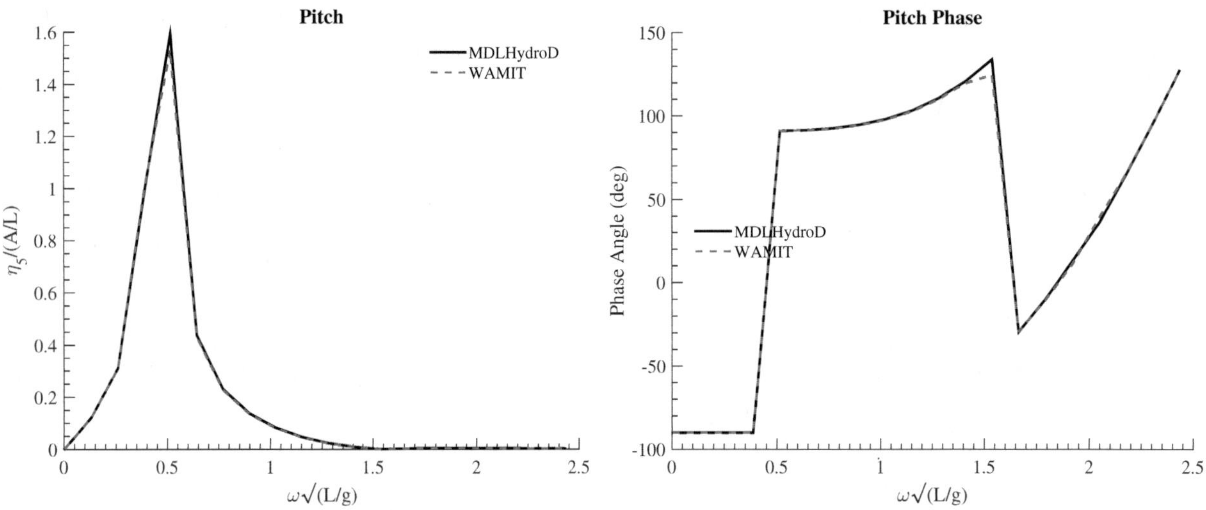

(a) Pitch RAO amplitude (b) Pitch RAO phase

Figure. 7. Pitch RAO comparison for cylinder at $Fn = 0$ and $h/T = 2$

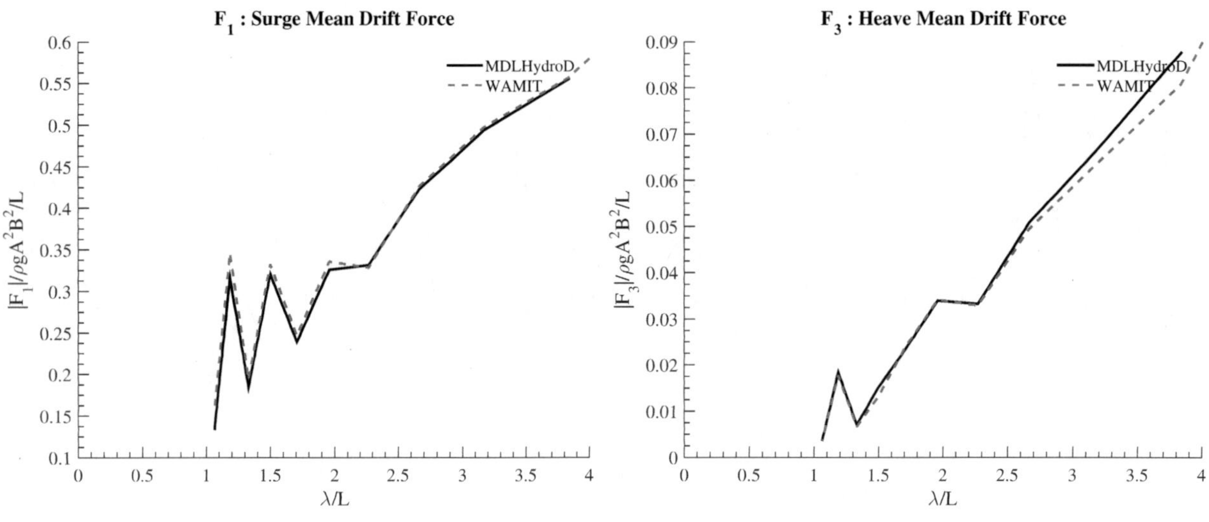

(a) Surge mean drift force (b) Heave mean drift force

Figure. 8. Comparison of mean drift forces for cylinder at $Fn = 0$ and $h/T = 2$

comparison of the mean drift force in surge and heave. In all of the cases, the results were found to be in excellent agreement with the industry standard code.

The finite depth forward speed motion prediction results are relatively scarce in the open literature. Here the added mass coefficients are compared in Figure. 9 with published results of Huijsmans, R.H.M. and Dallinga (1983) for the Series 60 hull with $C_b = 0.7$. The depth to draft ratio $h/T = 1.15$ is considered and forward speed corresponding to $Fn = 0.2$ is selected. The results were found to be satisfactory. Further validation results are presented in Guha (2016).

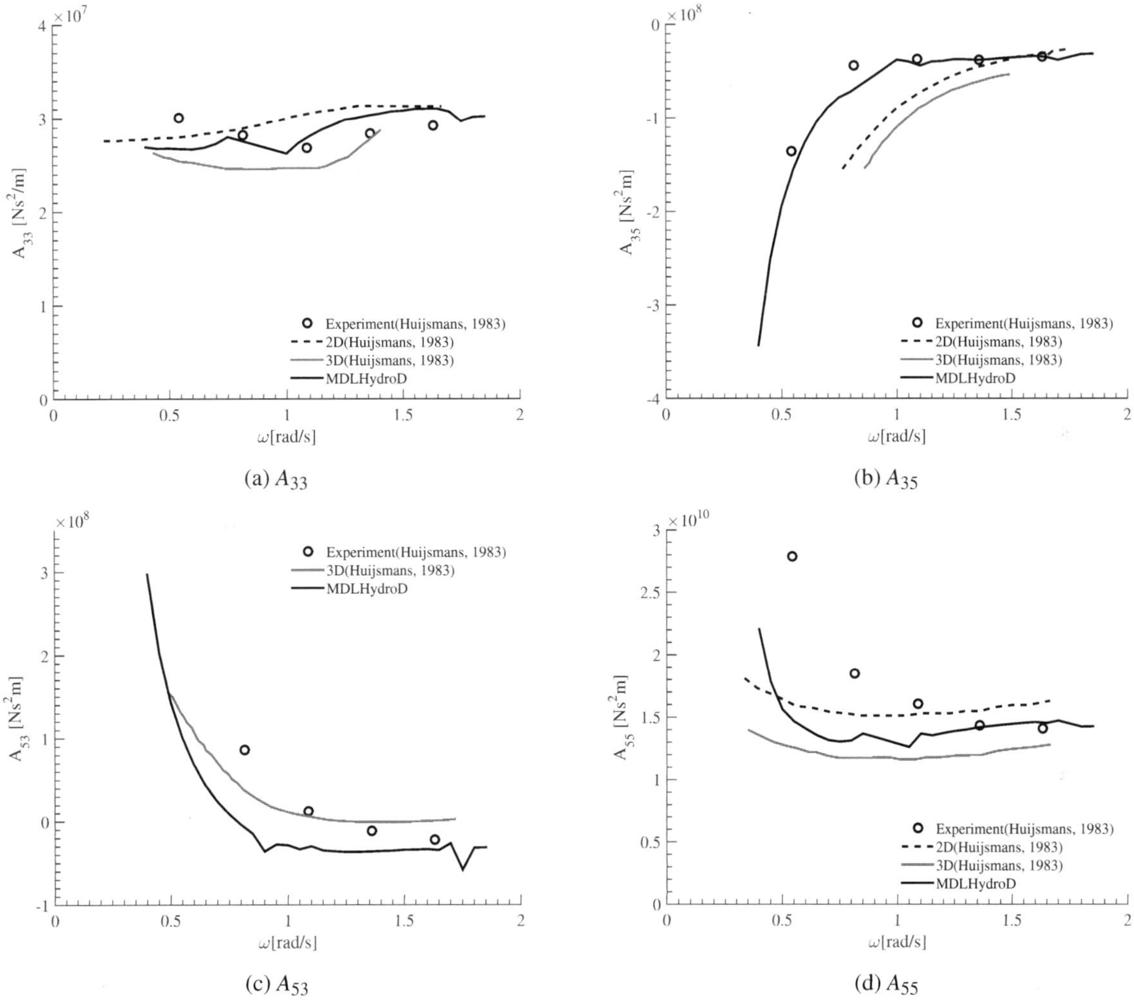

(a) A_{33} (b) A_{35} (c) A_{53} (d) A_{55}

Figure. 9. Comparison of added mass for Series 60 ($Cb = 0.7$) hull at $Fn = 0.2$ and $h/T = 1.15$

3.2 Time domain results

In addition to the frequency domain tool, an in-house time domain tool SIMDYN has also been developed to simulate the nonlinear equations of motion of a general floating body in time domain while including:

- Nonlinear Froude-Krylov forces
- Nonlinear hydrostatics
- Viscous forces and moments

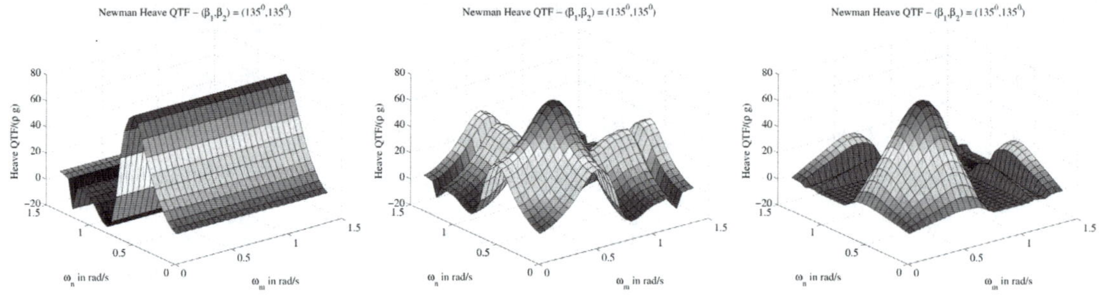

(a) Original (b) Arithmetic Mean (c) Geometric Mean

Figure. 11. Heave QTF using different Newman approximations

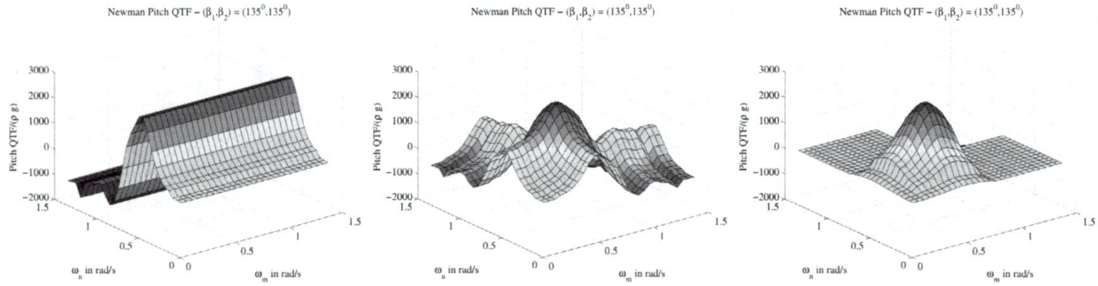

(a) Original (b) Arithmetic Mean (c) Geometric Mean

Figure. 12. Pitch QTF using different Newman approximations

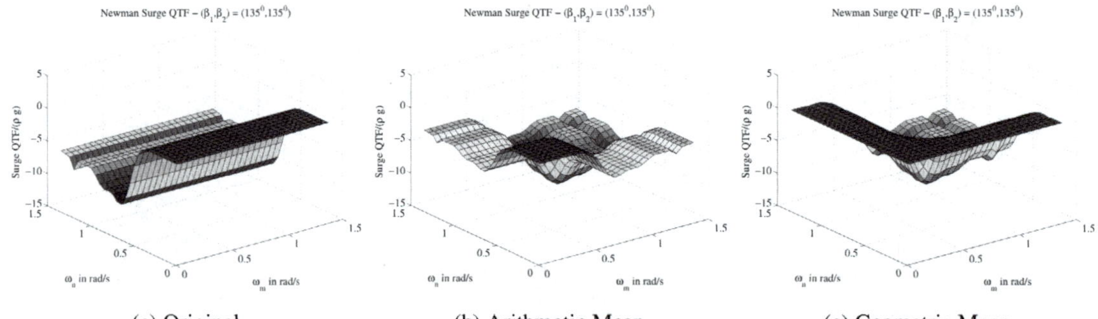

(a) Original (b) Arithmetic Mean (c) Geometric Mean

Figure. 10. Surge QTF using different Newman approximations

Similar to the frequency domain tool, the time domain simulation program has also been extensively validated against experimental data and other existing commercial codes. Figure. 10, Figure. 11 and Figure. 12 shows the surge, heave and pitch QTFs computed using different forms of the Newman approximation for the KCS container ship (hull form publicly available).

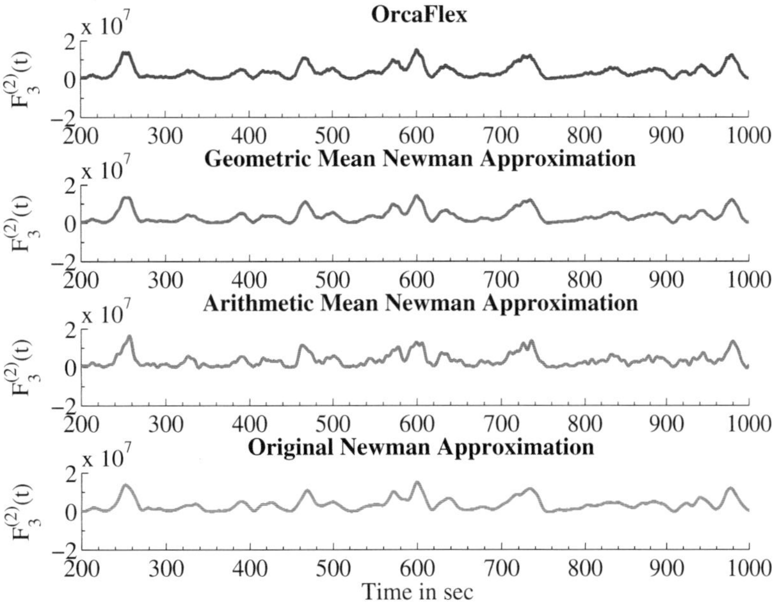

Figure. 14. Comparison of 2nd order heave force in time domain

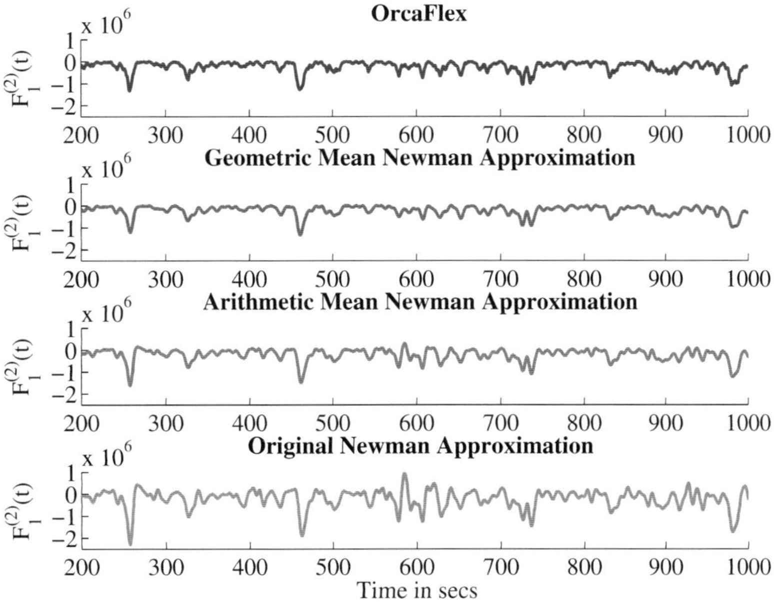

Figure. 13. Comparison of 2nd order surge force in time domain

The corresponding second order surge force, heave force and pitch moment in time domain have been compared against the results from the commercially available software OrcaFlex (Orcina, 2015) and are shown in Figure. 13, Figure. 14 and Figure. 15 respectively. It is important to note that OrcaFlex uses the geometric mean Newman approximation to calculate the second order drift forces and moments. It can be seen from the comparisons that the second order forces and moments obtained by geometric mean Newman approximation in the current study agree well with the the results obtained from OrcaFlex.

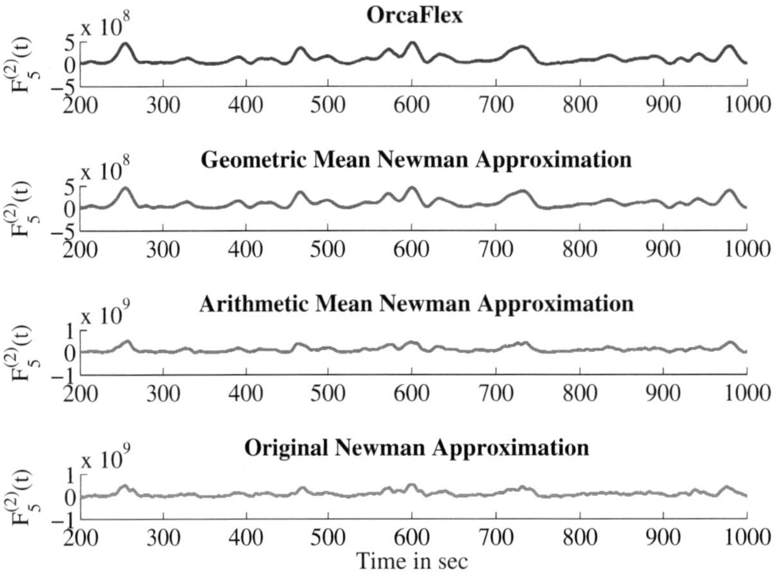

Figure. 15. Comparison of 2nd order pitch moment in time domain

4 PARAMETRIC EXCITATION IN SINGLE COLUMN FLOATER

While the previous section demonstrated the application of the frequency and time domain tool using a consistent second order formulation, this section demonstrates an application of the nonlinear time domain simulation while including the nonlinear Froude-Krylov and hydrostatics using a Wheeler stretched kinematics. It is reiterated that the Wheeler stretched approach although not order consistent is very robust in simulating instabilities experienced by ships and offshore platforms, especially involving parametric excitation which cannot be captured using the second order approach.

Parametric excitation is characterized by a large amplitude of motion even when the direct exciting force is close to zero. In general, it is caused due to a time varying parameter of the system such as stiffness or damping. It has recently been observed to be a serious concern for container ships in both regular (Moideen et al., 2012, 2013) and irregular waves (Moideen et al., 2014; Somayajula and Falzarano, 2014a; Somayajula et al., 2014). However, recent studies by Haslum et al. (1999); Haslum (2000) have shown that even certain offshore platforms such as classic spars and single column floaters are susceptible to parametric excitation.

In this section we demonstrate this effect for a single column floater (SCF) originally proposed by ABB (Huang et al., 2000). A rendered surface model of the SCF is shown in Figure. 16 and its particulars are listed in Table 1. The SCF includes a moonpool through both the hard tank and the soft tank and its details can be found in Chou et al. (2004). The moonpool has been considered while simulating the motions numerically. As the objective of this study is to demonstrate that the discussed simulation tool is able to capture the dynamics of instability, the truss elements connecting the two tanks (which provide quite a bit of damping) are not included in the numerical model. If additional external damping is included, the amplitude of the motion will be reduced. Also the presence of additional linear damping will require considerably higher level of excitation to excite the structure into an instability (damped Mathieu/Hill phenomenon). For a detailed drawing describing the dimensions please refer to Chou et al. (2004).

Table 1. Particulars of the Single Column Floater

Particulars	Values
Draft (m)	62.48
Displacement (tonnes)	38507.54
Freeboard (m)	12.19
KG (m)	20.88
Roll and Pitch Radius of Gyration (m)	45.47

Figure. 16. Single Column Floater

4.1 Critical frequency for parametric excitation

The parametric instability in spar type platforms is exhibited as a response amplification in the coupled heave and pitch modes. For a clearer understanding, lets consider a spar platform being excited by regular waves with a frequency ω different from the natural heave frequency ω_{n3}. Due to the low damping in heave mode of motion the heave response is composed of the superposition of the steady state motion at wave frequency ω and the transient motion at heave natural frequency ω_{n3}. The superposition of the transient response with the steady state motion leads to a heave response envelope oscillating with frequency $|\omega - \omega_{n3}|$. Haslum (2000) showed that if the frequency associated with the envelope of the heave motion coincides with the pitch natural frequency $\omega - \omega_{n3} = \omega_{n5}$, the system experiences as parametric excitation in pitch mode of motion.

$$\omega_{n5} = \omega - \omega_{n3} \tag{4.1}$$

The critical wave period at which this effect is seen can be expressed in terms of the heave and pitch natural periods and is given by:

$$T_{\text{critical}} = \frac{2\pi}{\omega} = \frac{2\pi}{\omega_{n3} + \omega_{n5}} = \frac{1}{\frac{1}{T_{n3}} + \frac{1}{T_{n5}}} \tag{4.2}$$

However, it is also clear that the difference frequency between the incident wave and Mathieu excited pitch motion coincides with the heave natural frequency which leads to a resonant amplification of the heave motion. The resonant heave motion leads to a larger amplitude of heave envelope process which feeds back into the Mathieu pitch motion and the cycle continues.

4.2 Simulation of instability for SCF

The time domain simulation tool developed in subsection 2.2 is applied to simulate the motions of the single column floater in both regular and irregular waves. Figure. 17 and Figure. 18 show the comparisons of heave and pitch responses from the nonlinear simulation with linear theory. Note that the motions are specified about an origin located at the center of gravity of the vessel. It can be seen that both the heave and the pitch motions show an exponential rise from the steady state linear solution which is also a charecteristic exhibited by parametric response.

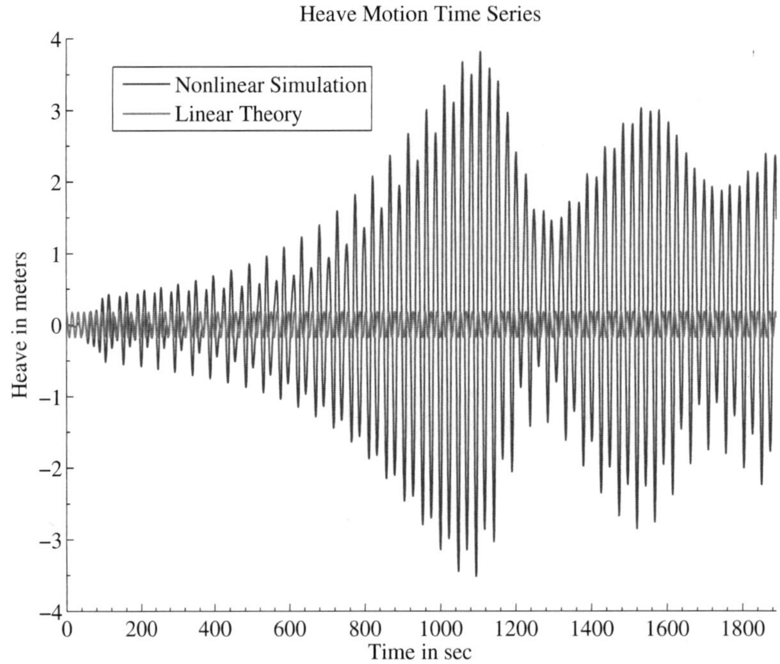

Figure. 17. SCF heave response in regular waves with $H = 8\ m$ and $T = 15.73\ s$

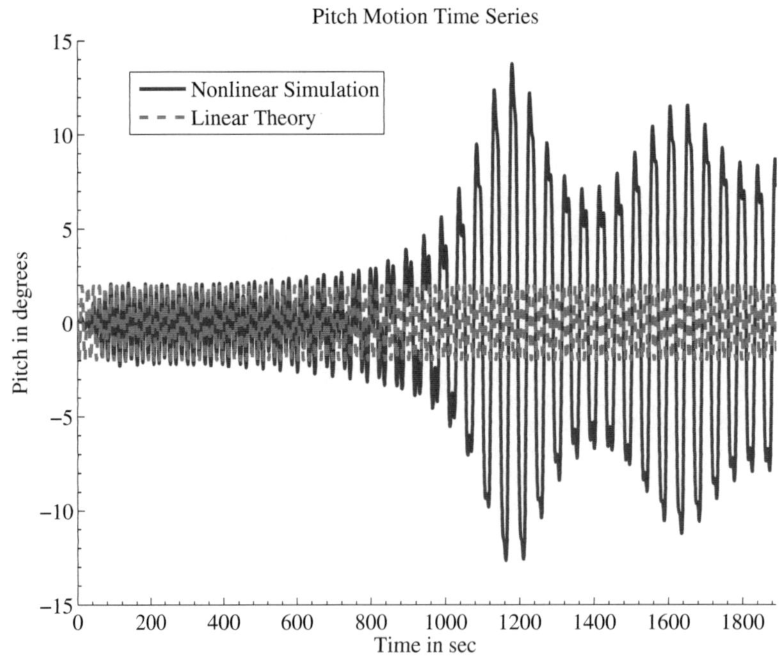

Figure. 18. SCF pitch response in regular waves with $H = 8\ m$ and $T = 15.73\ s$

5 CONCLUSIONS

A new efficient finite water depth Green function has been developed and incorporated in a 3D potential flow solver. The paper presents key expressions used in developing the numerical finite depth Green function and compares with published results. The motion prediction for a floating cylinder at shallow water depth is compared with results obtained from a commercial software. Furthermore, the hydrodynamic coefficients at forward speed in finite water depth is also presented for a ship shaped structure.

The paper also discusses the development of a nonlinear time domain simulation tool to simulate the nonlinear motions of a generalized floater. Two specific approaches to calculate the nonlinear force vector are discussed along with examples. It is shown that both the approaches of calculating the nonlinear force and moment vectors - stretching kinematics and second order analysis - are needed to allow for simulation of the various phenomena observed in marine and offshore structure analysis. The second order model has been validated against a commercially available software. The developed tool has also been specifically applied to the problem of Mathieu type instability in SCFs to demonstrate that the simulation program is able to capture the relevant dynamics of the system. While the two approaches are applicable to different types of problems, the option of having a combination of the two approaches allows for a strong hydrodynamics suite of programs capable of handling all marine and offshore analysis issues.

ACKNOWLEDGEMENTS

This work was supported in part by the American Bureau of Shipping and by the Office of Naval Research (ONR Grant N00014-13-1-0756). The authors would like to thank Dr. Richard Korpus for sharing his insight on ship performance and Dr. Robert Brizzolara for continuing his support.

REFERENCES

MA Abkowitz. *Stability and Motion Control of Ocean Vehicles.* MIT Press, Cambridge, MA, 1969.

Peter Boese. *Eine einfache Methode zur Berechnung der Widerstandserh{ö}hung eines Schiffes im Seegang*, volume 17. Inst. f{ü}r Schiffbau der Univ. Hamburg, 1970.

Frank Chou, John Chianis, and Xinyu Zhang. SELF INSTALLED SINGLE COLUMN FLOATER. In *23rd International Conference on Offshore Mechanics and Arctic Engineering*, Vancouver, 2004.

T Duarte and a J N a Sarmento. Effects of Second-Order Hydrodynamic Forces on Floating Offshore Wind Turbines. In *32nd ASME Wind Energy Symposium*, number January, National Harbor, Maryland, 2014. American Institute of Aeronautics and Astronautics.

OM Faltinsen, KJ Minsaas, N Liapis, and SO Skjørdal. Prediction of resistance and propulsion of a ship in a seaway. In *Proc. 13th Symposium on Naval Hydrodynamics*, 1980.

Jeffrey Falzarano, Abhilash Somayajula, and Robert Seah. An overview of the prediction methods for roll damping of ships. *Ocean Systems Engineering*, 5(2):55–76, 2015.

Amitava Guha. *Development of a Computer Program for Three Dimensional Frequency Domain Analysis of Zero Speed First Order Wave Body Interaction.* Master thesis, Texas A&M University, College Station, TX, 2012.

Amitava Guha. *Development and Application of a Potential Flow Computer Program: Determining First and Second Order Wave Forces at Zero and Forward Speed in Deep and Intermediate Water Depth.* Phd dissertation, Texas A&M University, College Station, TX, 2016.

Amitava Guha and Jeffrey Falzarano. Estimation of Hydrodynamic Forces and Motion of Ships with Steady Forward Speed. *Accepted in International Shipbuilding Progress*, pages 1–38, 2015a.

Amitava Guha and Jeffrey Falzarano. The effect of hull emergence angle on the near field formulation of added resistance. *Ocean Engineering*, 105(1):10–24, sep 2015b. ISSN 00298018. doi: 10.1016/j.oceaneng.2015.06.012.

Amitava Guha and Jeffrey Falzarano. Application of Multi Objective Genetic Algorithm in Ship Hull Optimization. *Ocean Systems Engineering*, 5(2):91–107, 2015c. doi: 10.12989/ose.2015.5.2.091.

Amitava Guha and Jeffrey M. Falzarano. Development of a Computer Program for Three Dimensional Analysis of Zero Speed First Order Wave Body Interaction in Frequency Domain. In *Proceedings of the ASME 2013 32nd International Conference on Ocean, Offshore and Arctic Engineering*, pages 1–9, Nantes, France, jun 2013. ASME. ISBN 978-0-7918-5539-3. doi: 10.1115/OMAE2013-11601.

Amitava Guha, Abhilash Somayajula, and Jeffrey M. Falzarano. Analysis of Causeway Ferry Dynamics for Safe Operation of Improved Navy Lighterage System. In *Proceedings of the 13th International Ship Stability Workshop*, pages 1–8, Brest, 2013.

H a Haslum, Umoe Olje, and O M Faltinsen. Alternative Shape of Spar Platforms for Use in Hostile Areas. *Proceedings of Offshore Technology Conference*, pages 1–12, 1999. ISSN 01603663.

HA Haslum. *Simplified methods applied to nonlinear motion of spar platforms*. Doctorate dissertation, Norwegian University of Science and Technology, 2000.

Herbjørn Haslum. Various applications of spar platforms. *70 Years Anniversary Seminar for Professors Odd M. Faltinsen and Torgeir Moan*, 2004.

J L Hess and A M Smith. Calculation of non-lifting potential flow about arbitrary three-dimensional bodies. *Journal of Ship Research*, 8(3):22–44, 1964.

Edward Huang, Shankar Bhat, Yong Luo, and Jun Zou. Evaluation of Dry Tree Platform Concepts. *Offshore Technology Conference*, pages 1–13, 2000. ISSN 01603663. doi: 10.4043/11899-MS.

R.P. Huijsmans, R.H.M. and Dallinga. Non-linear Ship Motions in Shallow Water. In *Ship and platform motion symposium*, 1983.

Yoshiho Ikeda, Yoji Himeno, and Nori Tanaka. A Prediction Method for Ship Roll Damping. Technical report, Department of Naval Architecture, University of Osaka Prefecture, 1978.

F John. On the motion of floating bodies I. *Communications on pure and applied mathematics*, 2(1):13–57, 1949.

F John. On the motion of floating bodies II. Simple harmonic motions. *Communications on pure and applied mathematics*, 3(1): 45–101, 1950.

C H Lee. WAMIT User Manual. Technical report, WAMIT, Inc., Chestnut Hill, MA, USA, 2013.

EM Lewandowski. *The dynamics of marine craft: maneuvering and seakeeping*. World Scientific, Singapore, 2004.

Lin Li. *Numerical seakeeping Predictions of Shallow Water Effect on Two Ship Interactions in Waves*. Phd dissertation, Dalhousie University, 2001.

C. M. Linton. Rapidly convergent representations for Green' functions for Laplace' equation. *Proceedings of the Royal Society A: Mathematical, Physical and Engineering Sciences*, 455(1985):1767–1797, 1999. ISSN 1364-5021. doi: 10.1098/rspa.1999.0379.

Yingyi Liu, Hidetsugu Iwashita, and Changhong Hu. A calculation method for finite depth free-surface green function. *International Journal of Naval Architecture and Ocean Engineering*, 7:375–389, 2015.

Hajime Maruo. Wave Resistance of a Ship in Regular Head Seas. *Bulletin of the Faculty of Engineering, Yokohama National University*, 9(March), 1960.

Hisham Moideen, Jeffrey M. Falzarano, and Abhilash Sharma Somayajula. Parametric roll of container ships in head waves. *Ocean Systems Engineering*, 2(4):239–255, dec 2012. ISSN 2093-6702. doi: 10.12989/ose.2012.2.4.239.

Hisham Moideen, Abhilash Somayajula, and Jeffrey M. Falzarano. Parametric Roll of High Speed Ships in Regular Waves. In *Proceedings of ASME 2013 32nd International Conferences on Ocean, Offshore and Arctic Engineering*, volume 5, page V005T06A095. ASME, jun 2013. ISBN 978-0-7918-5539-3. doi: 10.1115/OMAE2013-11602.

Hisham Moideen, Abhilash Somayajula, and Jeffrey M Falzarano. Application of Volterra Series Analysis for Parametric Rolling in Irregular Seas. *Journal of Ship Research*, 58(2):1–9, 2014. ISSN 15420604. doi: http://dx.doi.org/10.5957/JOSR.58.2.130047.

V. J. Monacella. The Disturbance Due to a Slender Ship Oscillating in Waves in a Fluid of Finite Depth. *Journal of Ship Research*, 10(4):242–252, 1966.

J.N. N. Newman. Numerical solutions of the water-wave dispersion relation. *Applied Ocean Research*, 12(1):14–18, jan 1990. ISSN 01411187. doi: 10.1016/S0141-1187(05)80013-6.

Orcina. OrcaFlex Manual v10.0a. Technical report, Orcina Ltd., Ulverston, Cumbria, UK, 2015.

M. K. Pidcock. The Calculation of Green's Functions in Three Dimensional Hydrodynamics Gravity Wave Problem. *International journal for numerical methods in fluids*, 5:891–909, 1985.

J.A. Pinkster. Mean And Low Frequency Wave Drifting Forces On Floating Structures. *Ocean engineering*, 6(1):593–615, 1979.

Abhilash Somayajula. *Ship Roll Dynamics in Irregular Seas*. Phd dissertation proposal, Texas A&M University, 2015.

Abhilash Somayajula and Jeffrey Falzarano. Large-amplitude time-domain simulation tool for marine and offshore motion prediction. *Marine Systems & Ocean Technology*, 10(1):1–17, 2015a. doi: 10.1007/s40868-015-0002-7.

Abhilash Somayajula and Jeffrey M. Falzarano. Non-linear Dynamics of Parametric Roll of Container Ship in Irregular Seas. In *Proceedings of ASME 2014 33rd International Conferences on Ocean, Offshore and Arctic Engineering*, pages 1–10, San Francisco, 2014a. ISBN 9780791845493. doi: 10.1115/OMAE2014-24186.

Abhilash Somayajula and Jeffrey M Falzarano. Non-Gaussian Analysis Methods for Planing Craft Motion. *International Journal of Ocean Systems Engineering*, 4(4):293–308, 2014b.

Abhilash Somayajula, Amitava Guha, Jeffrey Falzarano, Ho-hwan Chun, and Kwang Hyo Jung. Added resistance and parametric roll prediction as a design criteria for energy efficient ships. *Ocean Systems Engineering*, 4(2):117–136, jun 2014. ISSN 2093-6702. doi: 10.12989/ose.2014.4.2.117.

Abhilash Sharma Somayajula and Jeffrey M Falzarano. Validation of Volterra Series Approach for Modelling Parametric Rolling of Ships. In *34th International Conference on Ocean, Offshore and Arctic Engineering*, St. John's, Newfoundland, Canada, 2015b.

J V Wehausen and E V Laitone. Surface waves. *Encyclopedia of Physics*, 9:446–815, 1960.

J. D. Wheeler. Method for Calculating Forces Produced by Irregular Waves. *Journal of Petroleum Technology*, 249:359–367, 1970.